College of Engineering
Center for Environmental Research and Technology
University of California
Riverside, California

Hydrogen Fuel for Surface Transportation

Joseph M. Norbeck
James W. Heffel
Thomas D. Durbin
Bassam Tabbara
John M. Bowden
Michelle C. Montano

Published by:
Society of Automotive Engineers, Inc.
400 Commonwealth Drive
Warrendale, PA 15096-0001
U.S.A.
Phone: (412) 776-4841
Fax: (412) 776-5760

Library of Congress Cataloging-in-Publication Data

Hydrogen fuel for surface transportation / Joseph M. Norbeck...
[et al.].
p. cm.
Includes bibliographical references and index.
ISBN 1-56091-684-2
1. Hydrogen cars. 2. Hydrogen as fuel. 3. Fuel cells.
I. Norbeck, Joeseph M., 1943- .
TL229.H9H94 1996
629.25'38--dc20 96-38602
CIP

ISBN 1-56091-684-2

SAE Order No. R-160

Dedication

To Dr. Robert M. Zweig, whose longstanding commitment
to the advancement of hydrogen as a clean fuel
has been a source of inspiration to many

Contents

Preface

This book has been written as part of an ongoing program in hydrogen fuel for surface transportation at the College of Engineering—Center for Environmental Research and Technology (CE-CERT) at the University of California, Riverside. Faculty, students, and staff who are working with new technologies for hydrogen production and hydrogen-fueled vehicles have developed this book as a comprehensive overview, reference, and database on the subject. The book provides background information on the advantages and disadvantages of hydrogen as a fuel for surface transportation, describes the current state of technology of hydrogen-fueled vehicles, and discusses the future requirements of the so-called "hydrogen economy." We believe that the reprints and introductory material of each chapter will be invaluable to new and experienced researchers in the field.

The hydrogen program at CE-CERT is directed by Joseph M. Norbeck, CE-CERT's Director, and Project Manager James W. Heffel. The development of this book was supervised by Steven E. Belinski, who was also the primary editor. The following individuals were responsible for writing the individual chapters: Thomas D. Durbin contributed to all the chapters; Bassam Tabbara contributed to Chapters 3, 6, and 7; John M. Bowden contributed to Chapters 4 and 6; and Michelle C. Montano contributed to Chapter 5. Helen Ku assisted with the preparation of the final manuscript.

This book is based on information collected in a comprehensive literature survey which was performed as part of a larger project funded by the South Coast Air Quality Management District (SCAQMD). The opinions, findings, recommendations and/or conclusions do not necessarily represent the views of the SCAQMD. The SCAQMD has not approved or disapproved of this book's contents, nor has the SCAQMD passed judgement upon the accuracy or adequacy of the information presented herein.

Since the field of hydrogen for surface transportation is evolving so rapidly, information on current developments in the field of hydrogen-fueled vehicles and the hydrogen economy can be obtained directly from CE-CERT at (909) 781-5791.

Joseph M. Norbeck
James W. Heffel
Thomas D. Durbin
Bassam Tabbara
John M. Bowden
Michelle C. Montano

Riverside, California
April 1995

Chapter 1

Introduction

Over the past two decades there has been considerable effort in the United States to develop and introduce alternative transportation fuels to replace conventional fuels such as gasoline and diesel. Environmental issues, most notably air pollution, are among the principal driving forces behind this movement. Emissions from transportation sources are currently the dominant source of air pollution, representing 70% of carbon monoxide, 41% of nitrogen oxides (NO_x), and 38% of hydrocarbon emissions in the United States [1]. The transportation sector also accounts for about 30% of the man-made emissions of the greenhouse gas CO_2 in the United States [1] and about 25% of man-made CO_2 emissions globally [2]. While the effects of increasing concentrations of CO_2 and other greenhouse gases on global temperature and climate patterns are difficult to quantify, many scientists argue that ignoring the possibility of dramatic climate change is unwise (for a discussion of the uncertainties surrounding the greenhouse effect refer to [3-8]).

The push to expand the use of alternative fuels in the United States is also motivated by the need for energy security. The United States is the world's largest energy consumer and its second largest producer. Current trends indicate that both U.S. consumption and the percentage of oil that is supplied by imports will continue to increase. Estimates by the Energy Information Administration predict that the percentage of U.S. petroleum supplied by imports will increase from 39 percent in 1990 to 60 percent by the year 2010. Total imports of petroleum are thus projected to increase from 6.7 million barrels per day in 1990 to 12.8 million barrels in 2010 [9]. This is particularly problematic given that the politically unstable Middle East region holds a majority of the proven resources of crude oil and will likely have a stronghold on the oil market for the foreseeable future.

Over the years, these concerns and governmental regulations have stimulated research and development programs for alternative fuels and alternative fuel vehicles. To date, the bulk of this research has focused on carbon-based fuels such as reformulated gasoline, methanol, and natural gas. While vehicles utilizing such fuels have demonstrated the ability to meet the stringent new California standards (transitional low emission vehicle [TLEV], low emission vehicle [LEV], and ultra-low emission vehicle [ULEV]), carbon-based fuels will always release carbon dioxide and some amount of hydrocarbons as part of the combustion process. Additionally, carbon-based fuels such as reformulated gasoline and natural gas are produced from nonrenewable sources and thus will eventually be subject to the problem of limited availability.

One alternative to carbon-based fuels is hydrogen. For years, hydrogen advocates have promoted the use of hydrogen as a long-range solution to the problems associated with fossil fuel use. This idea is even found in Jules Verne's "The Mysterious Island," a science fiction novel written in the 1870s. In one passage he writes, "...water will one day be employed as fuel, that hydrogen and oxygen which constitute it, used singly or together, will furnish an inexhaustible source of heat and light, of an intensity of which coal is not capable.... I believe, then, that when the deposits of coal are exhausted, we shall heat and warm ourselves with water. Water will be the coal of the future."

Hydrogen has been used extensively since it was first isolated by Henry Cavendish in 1766, and it is currently used in a number of industrial applications. Although hydrogen is not abundant as a gas on earth, it can be synthesized from coal, oil, or natural gas, or obtained from water using electrolysis. As a fuel, hydrogen emits only minimal amounts of pollutants. When it is burned in an internal combustion engine, the primary combustion product is water with no CO_2. Although some NO_x emissions are formed when hydrogen is used in an internal combustion engine, hydrogen is still the cleanest fuel for internal combustion engine applications. Alternatively, hydrogen can be used essentially emission-free in a fuel cell.

Since the 1970s, modern researchers have investigated the potential of using hydrogen as a fuel for surface transportation. Today, these research efforts continue to gain momentum. Mercedes-Benz, BMW, and Mazda are all demonstrating hydrogen vehicles. The Big Three automakers in the United States (Ford, General Motors, and Chrysler) are working on the development of fuel cells for vehicular applications. There are also several programs in Europe to investigate the use of hydrogen as a fuel for vehicles, including the EUREKA fuel cell demonstration project, the Swiss-German cooperative "hydrogen powered applications using seasonal and weekly surplus of electricity" (HYPASSE) program, and the Euro Quebec Hydro Hydrogen Pilot Project (EQHHPP) to look at shipping hydrogen produced from off-peak hydropower in Canada to Europe for use as a fuel.

Although there has been rapid progress in the development of hydrogen-related technologies, a number of obstacles must still be overcome before hydrogen can be considered a viable and competitive fuel. The purpose of this book is to review the accomplishments of research in this area over the past two decades. While hydrogen can be used in a range of applications as a fuel, such as for aeronautical and space transportation, the focus here is on the use of hydrogen for surface transportation and, in particular, the use of hydrogen in vehicular applications.

This book is designed to serve as a reference for practicing engineers as well as those who are planning to study the potential of hydrogen as an alternative transportation fuel. Each of the chapters is composed of two main sections: the first section is an introduction to the specific topic of the chapter, while the second contains recent articles about research on the topic. In the reprinted articles, the reader will find a more detailed description of some of the topics covered in the chapter. The book also contains a comprehensive reference list arranged by subject, which is given in Appendix A.

The book begins by surveying, in Chapters 2 and 3, the development of power plants (specifically the internal combustion engine and the fuel cell) for hydrogen vehicles. The hydrogen fuel internal combustion engine is probably the most developed hydrogen power plant. An internal combustion engine operated on hydrogen has different characteristics than that operated on gasoline, however, due to hydrogen's unique properties. Hydrogen engines have a tendency to ignite prematurely and produce less power if liquid hydrogen or direct injection technologies are not used. Advanced fuel delivery techniques and engine modifications have been investigated by researchers in an effort to alleviate these problems. As a result of hydrogen's low volumetric energy density, onboard storage systems for hydrogen also experience some combination of weight and/or volume penalty, leading to reductions in performance and trunk or passenger space, as discussed in Chapter 2.

The use of fuel cells in vehicles is one of the most rapidly advancing areas of hydrogen transportation research. These electrochemical engines offer the potential for emission-free operation and have higher efficiencies than standard heat engines. Fuel cell technology is still not fully developed, however. The materials used to make fuel cells are expensive and include platinum and specialized polymers. There has been some important research leading to cost reductions in fuel cells, however, especially in the reduction of the amount of platinum used. Another important focus of fuel cell research is the need to increase the power density of the fuel cells themselves. These issues are discussed in Chapter 3.

An important issue over the longer term will be the development of an extensive hydrogen infrastructure including large-scale hydrogen production and distribution. These issues are addressed in Chapters 4 and 6. Chapter 4 reviews the technical aspects of hydrogen production, while Chapter 6 focuses on economic and practical issues that will have to be addressed to develop a hydrogen infrastructure. Hydrogen produced from fossil fuels is currently more costly than other fuels such as natural gas or gasoline, and producing hydrogen from renewable sources is even more expensive. Technologies for producing hydrogen from renewable sources are advancing rapidly, however, and could be available at competitive prices for some applications in the foreseeable future. Developing a large-scale distribution system is another major obstacle to achieving a so-called hydrogen economy. Thus, hydrogen will initially have to be introduced into smaller-scale markets, such as fleet operations.

Aside from these technical and economic issues, the issue of public perception is important. Currently, hydrogen has a poor reputation with regard to safety, due in part to the visibility of events such as the Hindenburg accident. Although hydrogen poses some unique dangers due to its properties, it does not appear that these dangers are necessarily any greater than those of using other fuels, given proper precautions. In fact, hydrogen's overall safety record to date is quite good. The transition from using hydrogen for more specialized applications to use on a widespread marketplace is a major one, however, and will require a thorough demonstration of hydrogen's safe use in a vehicle. The important aspects of hydrogen safety are reviewed in Chapter 5.

References

1. U.S. Department of Transportation, Federal Railroad Administration, Office of Policy, 1993.

2. Okken, P.A., "A Case for Alternative Transport Fuels," *Energy Policy*, Vol. 19, p. 400, 1991.

3. Schlesinger, M.E., "Greenhouse Policy," *Research and Exploration*, Vol. 9, p. 159, 1993.

4. Houghton, J.T., Jenkins, G.J., and Ephraums, J.J., *Climate Change 1992: The Supplementary Report to the IPCC Scientific Assessment*, Cambridge University Press, Cambridge, 1992.

5. Lindzen, R., "Absence of Scientific Basis," *Research and Exploration*, Vol. 9, p. 191, 1993.

6. Rosenzweig, C. and Hillel, D., "Agriculture in a Greenhouse World," *Research and Exploration*, Vol. 9, p. 208, 1993.

7. Schneider, S.H., "Degrees of Certainty," *Research and Exploration*, Vol. 9, p. 173, 1993.

8. Karl, T.R., "Missing Pieces of the Puzzle," *Research and Exploration*, Vol. 9, p. 234, 1993.

9. Energy Information Administration, *Annual Energy Outlook 1994*, Report No. DOE/EIA-0383(94), 1994.

Chapter 2

Hydrogen Engines and Vehicles: Characteristics and Development

2.1 Introduction

The earliest attempt at developing a hydrogen engine was reported in 1820 by Reverend W. Cecil. Cecil presented his work before the Cambridge Philosophical Society in a paper entitled "On the Application of Hydrogen Gas to Produce Moving Power in Machinery" [1]. The engine itself operated on the vacuum principle, where the power is produced by using atmospheric pressure to drive a piston back against a vacuum. The partial vacuum is created by burning a hydrogen/air mixture, allowing it to expand, and then cooling it. Although the engine reportedly ran satisfactorily, vacuum engines, in general, never became practical [2].

There are several references to hydrogen engine development in the 1850s including an English patent application in 1854 and a prototype engine built in 1856 by Benini [2]. During his work with combustion engines in the 1860s and 1870s, Otto also supposedly used a synthetic producer gas for fuel, which probably had a hydrogen content of over 50% [3]. It was reported that Otto did experiments with gasoline also, but found it dangerous to work with, prompting him to return to using gaseous fuels. The development of the carburetor, however, initiated a new era in which gasoline could be used both practically and safely. Interest in other fuels subsided after this development [2].

Interest in hydrogen intensified during the 1920s and 1930s in Europe [4, 5]. In 1923, John Burdon Sanderson Haldane, a Scottish scientist, predicted that England would at some point use wind power to form hydrogen through electrolysis which could then be used for transportation and other applications. He suggested that the hydrogen could be stored underground as a liquid in vacuum-jacketed storage vessels. Fifteen years later, I.I. Sikorski, the developer of the first helicopter, suggested that using hydrogen as fuel, an aircraft's range could be extended to the point where it could circumnavigate the globe without refueling.

During this same time period there was also significant use of hydrogen gas in airships [4, 5]. Both sides used balloons buoyed by hydrogen during World War I to see behind enemy lines, and a number of lighter-than-air ships were built, tested, and used after the war. The Graf Zeppelin was one of the more prominent of these airships, and it successfully made a number of trips between Europe and the United States during its nine-year operational life, which ended in 1937. The Hindenburg itself made ten successful transatlantic flights before its infamous end in 1937. There was also interest in the idea of using hydrogen in the engines used to power these airships. Since the airships became lighter as they traveled, due to fuel consumption, it was necessary to vent hydrogen anyway to maintain the proper level of buoyancy. If this vented hydrogen could be used in an engine, the amount of fuel required by the airships would be reduced. This scheme was adopted on some tests, but was never used extensively in routine operation.

There was also extensive work in the development of hydrogen as a fuel for surface transportation during this period. One of the key personalities responsible for this drive was the German engineer, Rudolf Erren, who worked in Germany and in England. Erren himself estimated that over 1000 vehicles were converted to hydrogen or gasoline/hydrogen operation in Germany and England as a result of his efforts. He promoted the use of hydrogen as a fuel for a variety of transportation purposes including submarines, trucks, buses, rail cars, and torpedoes [4]. Unfortunately, his records were lost during World War II. Although interest in hydrogen fuel waned immediately following the end of World War II, some research into hydrogen engines continued, most notably that of R. O. King in Canada [6, 7].

Hydrogen has since been used extensively in the space program since it has the best energy-to-weight ratio of any fuel. Liquid hydrogen is the fuel of choice for rocket engines, and has been utilized in the upper stages of launch vehicles on many space missions including the Apollo missions to the moon, Skylab, the Viking missions to Mars, and the Voyager mission to Saturn. The main engine of the Space Shuttle is also fueled by hydrogen. In the 1950s, hydrogen was also used in a Martin B-57 bomber to determine the potential of using hydrogen fuel to increase the operating altitude and range of jets. A number of other concept planes utilizing hydrogen were also developed over the years [4].

In the 1970s, there was a resurgence of research into the possibilities of hydrogen-fueled transportation due to concerns over the availability of petroleum fuels and environmental pollution. Programs were initiated in several countries including Japan, West Germany, and the United States. Today, research into the development of hydrogen transportation vehicles is as extensive as ever, with programs throughout the world including Europe, Australia, Canada, Japan, and the United States.

In this chapter, hydrogen internal combustion engine technology and on-board storage techniques are discussed. The hydrogen vehicles themselves are presented separately in Chapter 7, along with hydrogen fuel cell vehicles to allow for easy reference. The remainder of this chapter is organized as follows. In section 2, the pertinent combustion properties of hydrogen are discussed. In section 3, some of the special characteristics of hydrogen engines are discussed. Finally, in section 4, a discussion of on-board hydrogen storage is given.

2.2 General Properties of Hydrogen as a Fuel

The unique properties of hydrogen engines are a direct consequence of the properties of hydrogen itself. In this section, the properties of hydrogen as they relate to combustion and engine phenomena will be discussed. The properties of hydrogen as a fuel are listed in Table 2 in the reprint by Das at the end of this chapter along with those of gasoline and methane.

Limits of flammability

The limits of flammability are one of the most important properties of a fuel. These parameters are a measure of the range of fuel/air ratios over which an engine can operate. Hydrogen has a wide range of flammability in comparison with other fuels. One of the significant advantages of this is that hydrogen can run on a 'lean' mixture (see the limits of flammability based on equivalence ratio). This is a mixture in which the amount of fuel is less than the theoretical, stoichiometric, or chemically correct amount needed for combustion with a given amount of air, i.e., an air-rich mixture. Generally, fuel economy is greater and the combustion reaction is more complete when an engine is run on slightly lean mixtures. Additionally, the final combustion temperature can be lowered by using ultra-lean mixtures, reducing the amount of engine out nitrogen oxide emissions. Some compromise must be made, however, as lean operation can significantly reduce the power output of an engine due to a reduction in the volumetric heating value of the fuel-air mixture (see section 2.2).

Hydrogen engines can be run more effectively on excessively lean mixtures than gasoline engines. For gasoline, the lowest percentage of fuel per volume of air that can be used to obtain combustion is 1%. Given that this value is not significantly less than the stoichiometric value of 1.76%, the ability of gasoline engines to run on lean mixtures is limited. For hydrogen, on the other hand, as little as 4% hydrogen by volume with air produces a combustible mixture, compared to the 29.53% needed for stoichiometric operation. Thus, a hydrogen engine can be run on ultra-lean mixtures or mixture considerably leaner than those used in gasoline engines. In terms of equivalence ratio, or the ratio of the actual fuel-air ratio to the stoichiometric value, hydrogen is theoretically operational down to an equivalence ratio of 0.1, whereas gasoline is only operational down to an equivalence ratio of 0.7. In practice, however, hydrogen engines typically cannot operate at equivalence ratios below 0.2 [8]. The variability of the limits of flammability for hydrogen as a function of temperature are shown in Fig. 2.1.

Minimum Ignition Energy

The minimum energy required for ignition is another important property of a fuel. For hydrogen, the minimum energy required for ignition is about an order of magnitude less than that required for gasoline. This enables hydro-

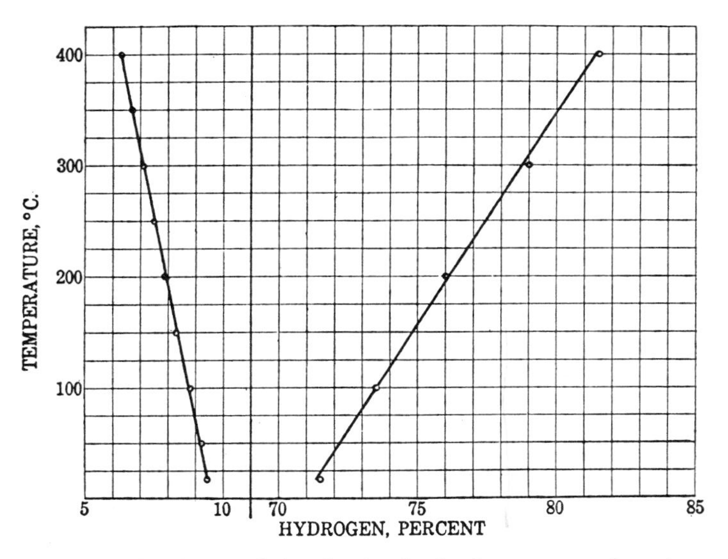

Figure 2.1. Flammability limits for hydrogen as a function of temperature. (Source: ref. [9]).

gen engines to run well on lean mixtures and ensures prompt ignition. Unfortunately, since very little energy is necessary to ignite a hydrogen combustion reaction, and almost any hydrogen/air mixture can be ignited due to the wide limits of flammability of hydrogen, hot gases and hot spots on the cylinder can serve as sources of ignition, creating problems of premature ignition and flashback. The minimum ignition energy, as a function of equivalence ratio, is shown in Fig. 2.2 for hydrogen-air and methane-air mixtures.

Quenching Gap or Distance

In the combustion chamber, the combustion flame is typically extinguished a certain distance (called the "quenching distance") from the cylinder wall due to heat losses. For hydrogen, the quenching distance is less than that of gasoline, so the flame comes closer to the wall before it is extinguished. Thus, it is more difficult to quench a hydrogen flame than a gasoline flame. The smaller quenching distance can also increase the tendency for backfire since the flame from a hydrogen/air mixture can more readily get past a nearly closed intake valve than the flame from a hydrocarbon/air mixture.

Self-ignition temperature

The self-ignition temperature is the temperature that a combustible mixture must reach before it will be ignited without an external source of energy. For hydrogen, the self-ignition temperature is relatively high. This has important implications when the charge is compressed. In

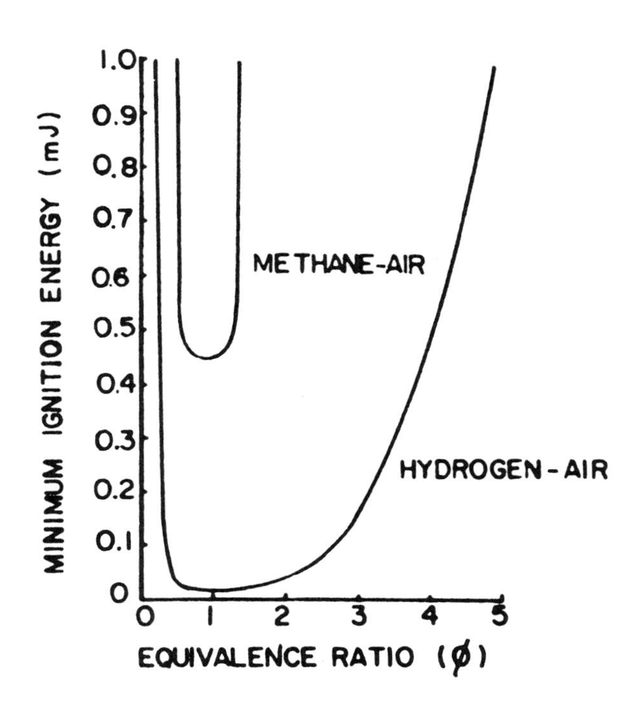

Figure 2.2. Minimum ignition energy as a function of equivalence ratio for hydrogen-air and methane-air mixtures. (Source: ref. [9]).

fact, the self-ignition temperature is an important factor in determining what compression ratios can be used by an engine, since the temperature rise during compression is related to the compression ratio via the following relation:

$$T2 = T1\left(\frac{V1}{V2}\right)^{\gamma-1}$$

where V1/V2 is the compression ratio, γ is the ratio of specific heats, and T1 and T2 are the initial and final temperatures, respectively. Thus, the compression ratio is limited by the temperature T2, in that this temperature cannot exceed the self-ignition temperature without causing premature ignition. The high self-ignition temperature of hydrogen allows larger compression ratios to be used in a hydrogen engine without increasing the final combustion temperature beyond the self-ignition temperature and causing premature ignition. The higher compression ratio is important because it is related to the thermal efficiency of the system as discussed later in this chapter. On the other hand, hydrogen is difficult to ignite in a compression ignition or diesel configuration, because the temperatures needed for this type of ignition are relatively high.

Flame Speed

As shown in Table 1 in the reprint by Das, the flame speed of hydrogen is nearly an order of magnitude higher than that of gasoline. This means that, for stoichiometric mixtures, hydrogen engines can more closely approach the thermodynamically ideal engine cycle. At leaner mixtures, however, the flame velocity decreases significantly. The flame front velocity of a hydrogen/air mixture as a function of equivalence ratio is shown in Fig. 1 in the reprint by Das at the end of this chapter.

Diffusivity

Hydrogen's diffusivity, or its ability to disperse in air, is considerably greater than gasoline's. The high diffusivity of hydrogen is advantageous for two main reasons. First, it facilitates the formation of a uniform mixture of fuel and air. Secondly, if a hydrogen leak does develop, the hydrogen will disperse rapidly. Thus, unsafe conditions can either be avoided or minimized.

Density

Hydrogen has an extremely low density. This creates two problems: (1.) a very large volume is necessary to store enough hydrogen to give a vehicle an adequate driving range, and (2.) the energy density of a hydrogen air charge, and hence the power output, is reduced as discussed below.

2.3 Special Characteristics of a Hydrogen Engine

Preignition problems

One of the primary problems that has been encountered in the development of operational hydrogen engines is premature ignition. Premature ignition occurs when the cylinder charge becomes ignited before the ignition by the spark plug and results in an inefficient, rough running engine. Backfire conditions can also develop if the premature ignition occurs near the fuel intake valve and the resultant flame travels back into the induction system. Premature ignition is a much greater problem in hydrogen-fueled engines than in gasoline-fueled engines, due to hydrogen's low minimum ignition energy and wide flammability range.

There have been a number of studies aimed at determining the cause of preignition. Some of the earliest investigations were carried out by King, et al [2, 6, 7]. Their results suggested that preignition could be caused by "hot spots" in the combustion chamber, such as on a spark plug or exhaust valve, or carbon deposits. Other research by Mishchenko has shown that backfire can occur when there is overlap between the opening of the intake and exhaust valves [10]. Although the intake and exhaust valves should theoretically open and close only at the bottom or top of the piston, there is typically an overlap period during which the valves are both open. Since the cylinder is filled with residual exhaust gases at this point, this researcher concluded that contact between the hydrogen-air mixture and residual exhaust gases was the cause of backfire.

Preignition is still an important issue with more modern hydrogen engines, although advances have been made over the last 20 years. Most modern fuel metering systems deliver the fuel and air separately to the combustion chamber. Thus, hydrogen can be added to the air mixture at a point where the conditions for preignition are less favorable. Techniques which thermally dilute the charge, such as water injection or residual exhaust gas recirculation (EGR), are also often added to hydrogen engines to help control premature ignition. Other researchers have argued that the most effective method of solving the problems of preignition and knock would be to redesign the combustion chamber and coolant systems to accommodate hydrogen's unique combustion properties [11]. Fuel delivery systems, thermal dilution techniques, and other designs for hydrogen engines are all discussed below.

Fuel Delivery Systems

Four general fuel delivery systems have been used in hydrogen engines. These are: carburetion, inlet manifold injection, inlet port injection, and direct cylinder injection. The first three techniques involve forming the fuel-air mixture during the intake stroke, either through the carburetor, in the intake manifold, or through an inlet port. By directly injecting fuel into the cylinder, fuel delivery can be controlled to take place after the closure of the intake valve. A brief review of some of the developments of different fuel delivery systems is given here. A more complete review is given in the reprint by Das at the end of this chapter.

The simplest method of delivering fuel to a hydrogen engine is via a carburetor. Although carburetion is no longer a viable technology for modern vehicles, there were several advantages to using a carburetor for early hydrogen engine developments. This system is similar to that used for carburetted gasoline engines, which made it easy to convert a standard gasoline engine to a dual-fueled gasoline/hydrogen or simply a hydrogen engine. This application does not require a sophisticated high-pressure injector either. The disadvantage to this technique is that engines which use carburetors are more susceptible to irregular combustion due to preignition and backfire problems. Additionally, the power output of an ideal hydrogen engine with a carburetor is about 15% lower than that of a comparable gasoline engine, as discussed below.

In early work on hydrogen engines, hydrogen/air charges were typically formed using a carburetor. Ricardo performed some of the first rigorous experiments in the 1920s using a single-cylinder engine with a carburetor [12]. He experienced severe preignition and flashback at equivalence ratios greater than unity, however. He concluded that hydrogen fuel was impractical for most uses. Captain R.O. King, et al also experienced preignition problems when they used carburetors with hydrogen engines [6, 7]. To eliminate these problems, they made several modifications to the engine, such as using "cold" spark plugs and an "aged" sodium-filled exhaust valve.

Troubled by the dual problems of lower power output and continual preignition problems, a number of researchers began exploring alternatives to carburetion. One solution to this problem is a technique called inlet manifold injection. This technique involves injecting fuel directly into the intake manifold rather than drawing fuel through the carburetor. Typically, the timing of the fuel injection is controlled so that the hydrogen is not injected into the manifold until after the beginning of the intake stroke, at a point where conditions are much less severe and the probability for premature ignition is reduced. The air, which is injected separately at the beginning of the intake stroke, dilutes the hot residual gases and cools off any hot spots. A similar technique is inlet port injection. With this system, fuel and air are injected from separate ports into the combustion chamber during the intake stroke and, as such, are not premixed in the intake manifold. Both of these methods can be used to reduce preignition problems, as has been demonstrated by a number of different researchers.

A number of techniques with a common theme of separating the fuel and air streams were developed in the 1970s and 1980s. Swain and Adt at the University of Miami carried out experiments using a method known as the Hydrogen Induction Technique where hydrogen was supplied separately from the air through a flow tube close to the intake valve [13]. Their reports demonstrated that inlet injection techniques could be used successfully to suppress preignition. Researchers at the University of Miami eventually became involved in a more comprehensive program, where they extensively investigated the performance of 19 different hydrogen engine configurations [14, 15]. Lynch, et al also investigated a technique called parallel induction [16]. The basic concept, shown in Fig. 2.3, is similar to inlet port injection in that fuel and air enter the combustion chamber during the intake stroke, but are not premixed in the intake manifold. The engines used in Lynch's tests also incorporated turbochargers with aftercooling to increase the power output of the engine.

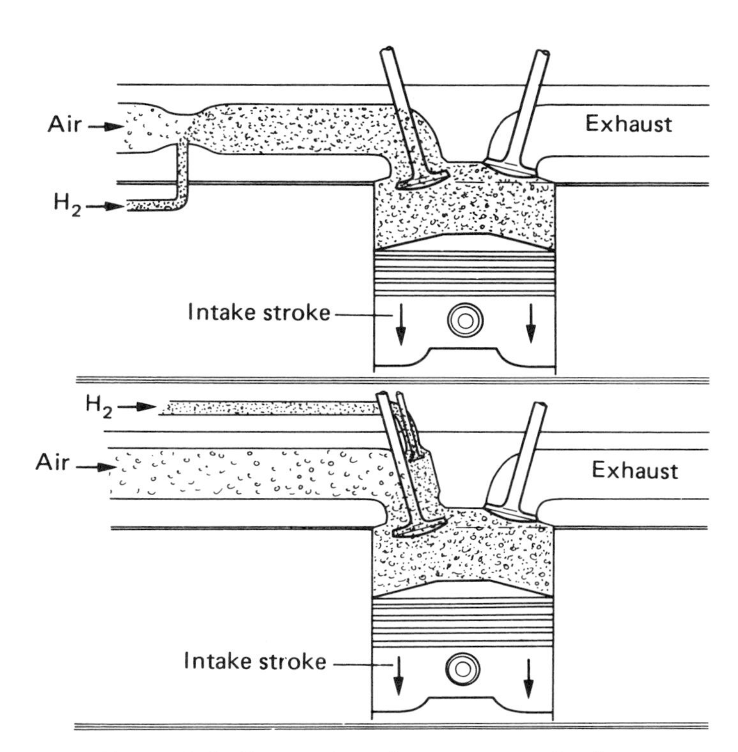

Figure 2.3. Comparison between premixed or "carbureted" hydrogen engines and port-injection designs like the parallel induction system. (Source: ref. [16]).

One advantage of using inlet injection, as opposed to more advanced direct injection, is its technical simplicity. This has been demonstrated in several studies. In the 1970s, MacCarley and Van Vorst extensively studied the prospects of using different fuel delivery systems to control backfire problems using a converted U.S. Postal Service Jeep [9, 17]. They found that the performance of the port injection system was more favorable than that of the direct injection system since it was not subject to any problems related to poor fuel/air mixing.

More recently, timed manifold injection has been advocated by researchers at the Indian Institute of Technology in New Delhi, India [18-20]. These researchers found that engines using a timed manifold induction system enjoyed some of the benefits of both compression ignition and spark ignition engines and operated essentially free of combustion problems. These engines had power outputs comparable to that of a spark ignition engine, while achieving diesel-like quality control and thermal efficiency. Studies were also done with low pressure, direct cylinder injection, but it was found that the engine was more susceptible to incomplete combustion when using this technique.

As hydrogen engines continued to progress, direct injection into the combustion cylinder stroke became more commonly used. Utilizing this technique, premature ignition during the intake stroke and backfire can be completely avoided if the fuel is injected after the intake valve is closed. The power output can also potentially be increased with this method to be 20% more than that of a gasoline engine and 42% more than that of a pre-mixed gaseous hydrogen mixture (see below).

It appears that Rudolf Erren used direct injection in some of his work. According to Kurt Weil, technical director of the Deutsche Erren Studiengesellschaft until 1938, "by injecting high pressure hydrogen in a special way into air or oxygen and igniting it even in diesel engines" Erren achieved a number of desirable combustion features [3, 21]. In the 1940s, Oehmichen conducted experiments on a single cylinder direct injection engine to determine its thermal efficiency and effective equivalence ratios of operation [3, 22, 23]. He achieved some successful results with the engine, but he did not use equivalence ratios greater than 0.67, where the probability of preignition increases greatly.

One of the most longstanding and important proponents of direct injection fuel delivery is a Japanese group led by S. Furuhama at the Musashi Institute of Technology in Tokyo. Since 1970, this group has developed a series of vehicles which have operated on hydrogen. Some earlier versions of the Musashi vehicles were based on a two-stroke engine fueled by high pressure hydrogen which was directly injected into the combustion chamber [24-26]. A platinum wire wrapped around a small piece of porcelain was used as a source for hot ignition. A special liquid hydrogen pump was designed to supply the high pressure hydrogen. In a more recent vehicle, the Musashi-8, a four-cylinder diesel engine with a spark ignition system was used [8]. More details about the hydrogen fueling system used in the Musashi-8 vehicle are given in a reprint by Furuhama at the end of this chapter.

The development of modern higher pressure injectors for more advanced fuel delivery systems has required considerable research, as problems associated with the high operating pressure and the precise control of the injection parameters necessary under these conditions must be overcome. The valve must travel considerably faster than a low pressure injector valve in order to obtain shorter injection times. The leak rate must also be controlled so that it is below the minimum amount needed to cause ignition before the main injection. This can be an especially difficult problem since the hydrogen molecule is small in size and has a high rate of diffusivity, enabling it to escape from even the smallest leaks. Finally, mechanical parts of the injector itself must be able to withstand severe thermal conditions and pressures as high as 10 MPa [19]. Many other groups have contributed to the development of hydrogen injectors to their present level of technology. Included among these are groups in the United States [22, 27], Canada [28-30], and New Zealand [31]. These efforts are discussed in greater detail in the cited references and in the reprint by Das.

Thermal Dilution to Eliminate Premature Ignition Problems

Preignition conditions can also be curbed using thermal dilution techniques such as exhaust gas recirculation (EGR) or water injection. As the name implies, an EGR system recycles a certain percentage of exhaust gases, diverting them back into the intake manifold. The introduction of recirculated exhaust gases helps to reduce the temperature of potential hot spots, reducing the possibility of preignition. Additionally, recirculating exhaust gases helps to reduce the peak combustion temperature, which reduces NO_x emissions. Typically 25% to 30% recirculation of exhaust gas is effective in eliminating backfire [10]. The power output of the engine is reduced when using EGR, however, as exhaust gases displace a portion of the available volume in the combustion cylinder, thus reducing the amount of air and fuel that can be drawn into the cylinder.

Another technique for reducing combustion problems is water injection, which acts as a thermal dilutent. The teams of Woolley and Anderson [32] and Koelsch and Clark [33] conducted early investigations on water injection [3]. Woolley and Anderson studied mechanical delivery systems and found that spraying water into the hydrogen stream prior to mixing with air produced better results than injecting water into the hydrogen/air mixture in the intake manifold. A potential problem with this type of system is that water can get mixed with the oil, so care must be taken to ensure that seals are leak-free. The University of Miami also found that improvements in cooling the cylinder head itself via extra water cooling and high conductivity materials to dissipate heat can also reduce backfire [14, 15].

Combustion Chamber and Coolant System Designs

Researchers at the University of Miami have suggested that redesigning the combustion chamber and coolant systems for hydrogen could provide a more effective way of controlling preignition and knock than redesigning the fuel delivery system [11]. They argued that a disk-shaped combustion chamber could be used to reduce turbulence in the chamber. The disk shape would help produce low radial and tangential velocity components and would not tend to amplify inlet swirl during compression. Since unburned hydrocarbons are not a concern in hydrogen engines, a large bore-to-stroke ratio could be used with this engine. To accommodate the range of flame speeds that would be found over the range of equivalence ratios to be used, two spark plugs would be needed. Also important would be the design of a coolant system that was capable of providing uniform flow rates to all locations needing cooling. Additional measures to decrease the probability of preignition could include the use of two small exhaust values as opposed to a single large valve, and the development of an effective scavenging system.

Diesel Configuration Engines

A considerable amount of research has been devoted to the development of hydrogen diesel or compression ignition engines. In fact, many of the researchers who have worked on the development of high pressure direct hydrogen injection have utilized diesel configurations [22, 24, 30]. Compression ignition engines can serve the requirements of applications which require a larger power output, such as heavy traction vehicles and stationary power supply units. Diesel engines also have the advantage of superior thermal efficiency and durability.

One problem with using hydrogen in a compression ignition engine is that it has a relatively high autoignition temperature. This makes it extremely difficult to ignite a hydrogen/air mixture without the aid of some additional ignition source. Most attempts to utilize hydrogen via pure compression ignition have failed for this reason. In early studies, Homan, et al were unable to obtain compression ignition even using compression ratios as high as 29 [34]. These researchers were able to achieve ignition with a glow plug, however. Welch and Wallace also had to install a glow plug to achieve ignition in investigations they performed with an air cooled single-cylinder four-stroke diesel engine [35].

In a more recent study, Wong attempted direct diesel injection in a single-cylinder engine which was modified to retain heat [36]. Ceramic parts were used to hold heat in the combustion region, leading to higher temperatures in the combustion chamber. Unfortunately, under conditions where lubricant combustion and cylinder hot spots were eliminated, the maximum compression temperature obtained was 800 K. Since this is below the hydrogen autoignition temperature, only sporadic compression ignition of hydrogen was achievable.

Ikegami, et al reported that compression ignition could be achieved by preliminarily adding fuel to the combustion chamber through either pilot injection or a small leak [37]. They proposed that hot residual gases from the previous cycle serve as an ignition source in this configuration. Once the correct preliminary fueling rate was determined, the test engine ran knock-free over a wide range of operating conditions. Ikegami, et al made their discovery, in part, when they realized that the successful compression ignition they had reported in a previous study was the result of a minute leak in the hydrogen fuel injector.

Quality Control

A distinct advantage of using hydrogen as a fuel, with its wide range of flammability, is that the fuel-to-air ratio or the "quality" of the charge mixture can easily be varied to meet different driving conditions or loads. This scheme of controlling the engine power output by changing the fuel-air ratio is called quality control. In contrast, for a gasoline engine with a carburetor, the fuel-to-air ratio is kept more or less constant throughout the driving range and the power output is controlled by varying the amount of charge which is inducted into the intake manifold. In other words, the "quantity" of the charge is controlled. The overall efficiency of an engine is reduced when the quantity of the mixture is controlled since the amount of air inducted into the system is limited. These losses are particularly significant at low speed or idling conditions, which is where a major portion of actual driving time is spent.

Thermal efficiency

The overall thermal efficiency of a hydrogen engine is typically greater than that of a gasoline engine. This can be understood from the theoretical expression which relates the thermal efficiency to the compression ratio and the ratio of the specific heats of the mixture as follows:

$$\eta = 1 - \left(\frac{1}{r}\right)^{\gamma-1}$$

where r is the compression ratio, γ is the ratio of specific heats (Cp/Cv), and η is the efficiency. This equation shows that the thermal efficiency can be improved by increasing either the compression ratio or the specific heat ratio. In hydrogen engines, larger compression ratios than those used in gasoline engines can be used since the self-ignition temperature of hydrogen is so high. Hydrogen engines can also operate more effectively under lean mixtures. This means that the temperature of the burnt gases can be lowered resulting in higher specific heat ratios. Thus, the combined effect of larger compression ratios and larger specific heat ratios increases the thermal efficiency for hydrogen fueled engines [18]. Note that this expression for efficiency is for an ideal engine; in an actual engine the efficiency is typically lower due to mechanical and heat losses.

Emissions

The most important advantage of hydrogen vehicles is that they emit fewer pollutants than comparable gasoline vehicles. For a hydrogen engine, the principle exhaust products are water and some nitrogen oxides (NO_x). Emissions of unburned hydrocarbons, carbon monoxide (CO), carbon dioxide (CO_2), oxides of sulfur (SO_x), and smoke from hydrogen vehicles are either not observed or are much lower than those from gasoline vehicles. If a hydrogen engine burns excess oil, hydrocarbon and CO emissions can become significant, but they are still less than the emissions from a gasoline engine on a relative scale [38]. Small amounts of hydrogen peroxide have also been observed in the exhaust of hydrogen engines that were operated very inefficiently. Such inefficiencies should not occur in a properly functioning engine, however [39]. Hydrogen itself can also be emitted from an engine, but this is not a problem since hydrogen is non-toxic and not involved in any smog producing reactions.

NO_x are the most significant emissions of concern from a hydrogen vehicle. Unfortunately, NO_x emissions can have an adverse affect on air quality through the formation of ozone or acid rain. The formation of nitrogen oxides in a combustion engine is a consequence of using ambient air, which contains about 80% nitrogen, in the combustion chamber. When the combustion temperature is high enough, nitrogen and oxygen from the air portion of the mixture will combine to form NO_x. The formation of NO_x in the combustion reaction is primarily a function of three variables: (1.) the reaction temperature; (2.) the reaction duration; and (3.) the availability of oxygen. An increase in any of these variables leads to an increase in NO_x emissions. While it is true that some NO_x will be formed in the combustion chamber of a hydrogen internal engine, the NO_x level can be brought down using several methods. Hydrogen engines can be run on very lean mixtures to reduce the combustion temperature, for example. NO_x emissions can also be lowered by cooling the combustion environment using techniques such as water injection, exhaust gas recirculation, or using cryogenic fuel, which acts as a heat sink.

A comparison of emissions from several hydrogen engines and vehicles and their gasoline counterparts is given in Table 2.1.

Power Output

Another important parameter in evaluating the performance of an engine is the power output. This is derived by converting the chemical energy of the fuel into mechanical energy. A comparison of the heat output of hydrogen and gasoline engines is shown in Fig. 2.4 for various conditions. A stoichiometric mixture of gasoline and air and gaseous hydrogen and air pre-mixed externally occupy ~2% and 30% of the cylinder volume, respectively. Under these conditions, the energy of the hydrogen mixture is only 85% that of the gasoline mixture, resulting in about a 15% reduction in power. The problem is compounded even further when the engine is operated on lean mixtures or experiences preignition.

The loss of power using an externally mixed hydrogen/air mixture can be conceptualized in terms of engine displacement. For example, a 2.3-liter gasoline engine has a power output comparable to a larger 2.7-liter hydrogen engine at stoichiometric conditions or a 6-liter hydrogen engine operating at an equivalence ratio of 0.4. Similarly, a 2.3-liter hydrogen engine operating at an equivalence ratio of 0.4 produces a power output equivalent to that of only a one-liter gasoline engine [41].

The power output of a hydrogen engine can be improved using more advanced fuel injection techniques or liquid

Table 2.1: Emissions of hydrogen and same model gasoline vehicles

Vehicle or engine	Fuel	Emission Control	Test Procedure	NO_x (in g/Km)	CO	HC	CO_2
UCLA Gremlin, 1972*	CH_2	PCV, EGR	EPA, (unspecified)	0.52	0.27	0.12	6.04
UCLA Gremlin, 1974	CH_2	PCV, EGR	-	0.02	0.09	0.06	-
1972 Volkswagon	-	No CC	-	0.16	0.15	0.04	-
Dodge D-50 pick-up*	CH_2	PCV only	FTP	1.19	0.44	0.26	8.66
Dodge D-50 pick-up †	CH_2	PCV only	FTP	0.32	0.06	0.02	0.83
Dodge "D" pick-ups ‡	gasoline	EGR, CC	EPA city cycle	1.80	2.73	0.28	370
MB 310 van	hydride	No CC	FTP	1.40	0.29	0.10	8
Musashi-2 ¤	LH_2	No CC	CVS full-cycle	2.50	0.18	0.05	-
Ford 4-cyl.	Hydride	No CC ¥	FTP	0.96	0.55	1.01	-
Ford 4-cyl.	gasoline	No CC ¥	FTP	1.42	21.9	1.41	-
BMW 745i, 6 cyl.	LH_2	No CC	ECE cycle I	0.35	-	-	-
BMW 745i, 6 cyl.	LH_2	No CC	ECE cycle I	0.27	-	-	-
BMW 745i, 6 cyl.	gasoline	No CC	ECE cycle I	2.50	-	-	-
MB 280 TE wagon**	H_2/gasoline	No CC	FTP	0.52	1.87	2.17	-
UCR Ford Ranger[a]	CH_2	No CC	FTP	0.23	0.0	0.0	1.88
Mazda Miata	Hydride	No CC	FTP	0.08	0.04	0.01 ††	
Emissions Standards (in g/Km for comparison):							
Federal Tier 0	gasoline	CC ‡‡	FTP	0.62	2.11	0.25 (THC)	
Federal Tier 1	gasoline	CC ‡‡	FTP	0.25	2.11	0.16 ††	
CA TLEV			FTP	0.25	2.11	0.08 ††	
CA LEV			FTP	0.12	2.11	0.05 ††	
CA ULEV			FTP	0.12	1.05	0.02 ††	

Notes:

All emissions are in gm/km

CC = catalytic converter: ECE=European test cycle; EGR=exhaust gas recirculation; FTP=Federal Test Procedure; MB=Mercedes Benz; PCV = postive crankcase ventilation; CH_2 = compressed gaseous hydrogen; LEV = Low Emission Vehicle; TLEV = Transition Low Emission Vehicle; ULEV = Ultra-Low Emission Vehicle; N/A=not applicable; '-' = not reported.

a initial baseline test results (unoptimized)

* Very high carbon emissions due to excess oil consumption

† Average of two emissions tests

‡ Zero-mileage emissions tests by EPA for 1987 Chrysler "D" series pick-up trucks. Most vehicles equipped with oxidation catalysts only; some had 3-way catalysts. Hydrogen vehicle was a 1979 model with at least 30,000 km on it.

¤ Engine was out of tune and backfired easily when the emission test was performed.
¥ Inference from ref. [38], considering relatively high exhaust emissions on gasoline.
¶ Data from ECE given in refs. as grams/test only. Peschka suggests that the ECE cycle is 1.013 km [40].
** Dual-fuel operation, with 100% H_2 @ idle, 30-60% H_2 @ part load, and 0% H_2 @ full load. Relatively high HC emissions due to use of gasoline enrichment at start-up, until engine and coolant were hot enough to release hydrogen from the metal hydride.
†† NMOG or NMHC emissions.
‡‡ Most vehicles have a 3-way catalytic converter, exhaust gas recirculation, and evaporative control canister, and more.
(Source: from ref. [38] and sources therein, [41] and [42])

hydrogen, as shown in Fig. 2.4. If liquid hydrogen is premixed rather than gaseous hydrogen, the amount of hydrogen that can be induced into the combustion cylinder can be increased by approximately one-third and the power output increased by about 37%. The output can be further increased by directly injecting hydrogen into the cylinder under high pressure. Using this technique, the maximum amount of both air and hydrogen are introduced into the combustion chamber resulting in a 20% increase in power compared to a gasoline engine using a carburetor and a 42% increase in power compared to a pre-mixed gaseous hydrogen mixture. It should be noted that the pressure provided by a hydride storage system would be insufficient for high pressure injection.

The gap between the power output of a gasoline engine and a hydrogen engine can also be reduced with engine modifications, such as a blower (a belt-driven supercharger) or a turbocharger (an exhaust-driven supercharger), or by using a larger displacement engine. The primary problem with a blower is that it requires energy to operate and hence reduces the efficiency of the engine. While turbochargers typically work well at higher speeds, their performance at lower engine speeds can be less than desirable. At lower engine speeds, the performance of a turbocharger is hampered because the density and temperature of the gas being pumped through it is too low to provide for adequate operation. This is an even greater problem when the engine is operated in lean mode since this lowers the exhaust gas temperature and density. The turbocharging performance can be improved by utilizing dual turbochargers, where a small turbocharger is used for low engine speeds and a second, larger turbo is used for higher engine speeds, or by using an electric motor to assist in spinning up the turbo. The primary disadvantages of larger displacement engines are that they are heavier and require more space in the engine compartment, and hence must be sized appropriately.

The actual performance of the vehicle depends not only on the power output of the vehicle's engine but also on the vehicle's weight. As discussed in section 2.4, the weight of a hydrogen vehicle will typically be greater than that of a gasoline vehicle due to the added weight of the on-board

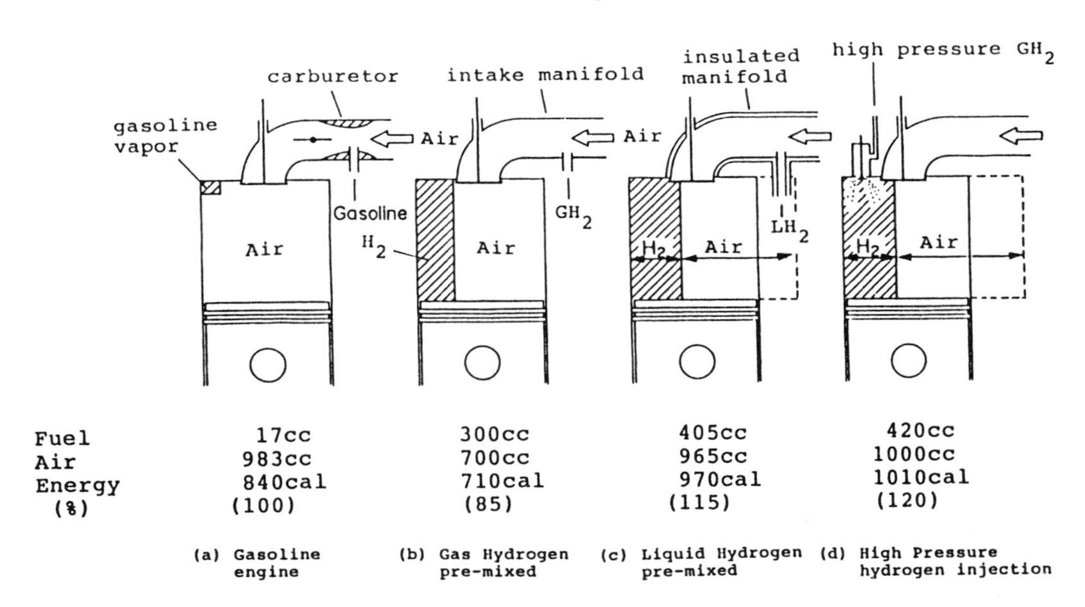

Fig. 2.4 Comparison of maximum calorific value (output) of a fuel-air mixture in a 1000 cc displacement volume engine. (Source: ref. [43])

storage system. This extra weight will adversely affect the performance of a hydrogen vehicle. A comparison of the power output of hydrogen prototype vehicles to their gasoline counterparts is given in Table 2.2.

Dual-fuel Engine Applications

A number of researchers have investigated the possibility of combining hydrogen with other fuels in dual-fuel applications. There are a number of advantages of using hydrogen in small controllable amounts with other fuels. Hydrogen can provide benefits during the combustion cycle. With its fast flame speed, hydrogen can reduce the burning time of the primary fuel mixture, providing for greater efficiency. Hydrogen is also advantageous when used in conjunction with fuels that don't have good cold start capability. Dual-fuel applications are reviewed briefly here and in the reprint by Das.

At the Jet Propulsion Laboratory (JPL) of the California Institute of Technology in the 1970s, researchers compared the characteristics of engines run on hydrogen, gasoline, and mixtures of the two fuels [44]. The hydrogen/gasoline mixtures were formed by a catalytic partial oxidation hydrogen gas generator. The results showed that hydrogen was the most efficient fuel for any NO_x level and that the engine could be operated on lower equivalence ratios as more hydrogen was supplemented for gasoline. Additionally, only hydrogen or hydrogen-supplemented fuel mixtures could meet the 0.4 g/mile NO_x level. Hydrogen/gasoline dual-fuel engines were also the subject of several studies done at the General Motors Research Laboratory [45, 46].

Mercedes-Benz has utilized gasoline/hydrogen dual-fuel engines in some of its prototype vehicles [47, 48]. Developed in cooperation with the University of Kaiserslautern, the engines for these vehicles were designed to utilize the positive combustibility characteristics of hydrogen while avoiding the pitfalls of pure hydrogen operation. For start-up and idle conditions, the engine operated purely on hydrogen. At partial load conditions, the engine was operated on a mixture of hydrogen and gasoline. The high excess air operation possible with hydrogen at low and part loads could be used to reduce the overall formation rate of NO_x, increase the thermal efficiency of the engine, and reduce fuel consumption. At full load, pure gasoline operation was used to avoid the power loss associated with the lower per volume calorific value of hydrogen.

Recent studies by Mathur, et al have investigated the use of hydrogen in a diesel/hydrogen dual-fuel compression ignition engine [49]. It was found that hydrogen could be substituted for diesel fuel up to the point where hydrogen constituted 38% of the total fuel mixture with no resulting loss in the system efficiency and only nominal loss in power output. In order to eliminate knocking, nitrogen, helium, and water were used as diluents. Use of water injection increased the optimal substitution of hydrogen to 66% with minimal losses in power output and thermal

Table 2.2 Performance Comparison of Hydrogen Vehicles to Gasoline Vehicles

Vehicle	Fuel Storage	Power (% rel. to gasoline)
1979 LNAL/DFVLR Buick	LH_2	Acceleration twice as long
Musashi-4, 3-cyl.	LH_2	+25% max. power
Musashi-5, 3 cyl., CI	LH_2	Up to 25% more power
Musashi-7, CI truck	LH_2	+11% max. power
4 cyl. Toyota lab engine	GH_2	Up to 13% more power
4-cyl. car	70 kg metal hydride	23% less power
4-cyl. BMW 520	LH_2, external mixing	11% less max. power
6-cyl. BMW 745i	LH_2, direct injection	6% less max. power

(Source: ref [38] and references therein)

efficiency. It was also found that NO_x, smoke, and exhaust emissions were reduced when water injection was used. Nitrogen was found to be the best diluent from an engine performance standpoint. Using nitrogen, the amount of hydrogen substitution could be increased to 48% with no loss in thermal efficiency and only nominal loses in power output. Other studies of dual diesel/hydrogen operation include those by Gopal, et al [50] and Weigang [51].

Recently, considerable attention has been devoted to studying the combustion characteristics of mixtures of natural gas (or methane) and hydrogen. As there is renewed interest in compressed natural gas as a replacement fuel for gasoline, there are prospects for such mixtures becoming a valuable alternative fuel in the future. Researchers at Hydrogen Consultants, Inc. (HCI) have been investigating hythane, a mixture comprised of 0-20% hydrogen with methane, since the late 1980s. The efforts at HCI have included a number of prototype vehicles converted to operate on natural gas and hydrogen mixtures [52]. Researchers at the Florida Solar Energy Center have also studied methane/hydrogen mixtures [53]. Their research has focused on a mixture they call HY-TEST, which contains between 20-50% hydrogen with methane. Similar to mixtures of hydrogen with other fuels, these mixtures allow for leaner combustion than pure methane [52, 54]. By increasing the range of lean operations of the fuel, such mixtures can be effectively used to reduce NO_X emissions [52, 53]. A reprint by Raman, et al which further discusses the possible benefits of methane/hydrogen is included at the end of this chapter. Recent studies of methane/hydrogen mixtures are also discussed in references [53-55].

Other researchers have investigated the possibility of a methanol/hydrogen dual-fuel spark ignition engine [56]. It was found that improved combustion characteristics, relative to pure methanol operation, could be obtained when methanol was supplemented with hydrogen. In particular, problems associated with the high latent heat of evaporation and the high ignition temperature of methanol, such as difficulty with cold starting, could be suppressed. The addition of hydrogen increased the flame velocity and shortened the ignition lag, thereby increasing the thermal efficiency. The thermal efficiency could also be increased with this mixture by increasing the compression ratio. One problem with this engine was aldehyde emissions.

IC Engines for Hybrid Vehicles

Lawrence Livermore, Sandia Livermore, and Los Alamos National Laboratories are currently involved in a joint project to develop an optimized hydrogen-fueled engine for series hybrid automobiles. Systems analysis, engine design, and kinetics modeling are being conducted by Lawrence Livermore. Performance and emission testing is being carried out by Sandia. Los Alamos is involved with computational fluid dynamics and combustion modeling. A first generation optimized hydrogen engine head has been developed and fabricated for use with a single-cylinder Onan engine. The head features a 14.8:1 compression ratio, dual ignition, water cooling, two valves, and an open quiescent combustion chamber to minimize heat transfer losses. Initial testing shows promise of achieving an indicated efficiency of 42 to 46% and emissions of less than 100 ppm NO_x [57].

2.4 On-Board Storage Systems

One of the greatest challenges in developing a hydrogen-fueled vehicle is designing a satisfactory on-board storage system for the fuel. The principal problem with storing hydrogen is that it has such a low density. Thus, most hydrogen storage systems are considerably bulkier and/or heavier than those used for conventional gasoline or diesel fuels. Hydrogen is typically stored on a vehicle as a liquid in cryogenic containers, a gas bound to certain metals (hydride), or as a high pressure compressed gas. A comparison of the characteristics of these three different systems, along with those of a reference 18.9 ℓ (five-gallon) gasoline tank, is given in Table 2.3. These and other techniques are discussed further in the following section.

Liquid Hydrogen

Liquid hydrogen (LH_2) storage has been the subject of considerable development in Japan, Germany, and the United States. The conditions necessary to store liquid hydrogen are fairly severe, however, requiring a temperature of 20 K at $2x10^5$ Pa. To maintain these conditions, liquid hydrogen is typically stored in a double-walled, super-insulating vessel. Hydrogen can be drawn from either the liquid or gas phase of these vessels and delivered to the engine. These vessels help to minimize the transfer of heat from the outside world, thus reducing the boil-off hydrogen. Evaporation rates with these vessels are typically on the order of 2% per day or less. It is also significant to note that an amount of energy equivalent to

Table 2.3 On-Board Hydrogen Storage Comparison

	Gasoline Reference	Liquid Hydrogen (20 K)	Hydride FeTi (1.2%)	Compressed Hydrogen (20.7-69.0 MPa)
Btu	629,500	629,500	629,500	629,500
Fuel wt (kg)	13.9	4.7	4.7	4.7
Tank wt (kg)	6.3	18.6	547.5	63.3-86
Total fuel system weight (kg)	20.4	23	552	67.9-90.5
Volume (ℓ)	18.9	177.9	189.3	408.8-227.2

(Source: ref. [58])

40% of the heating value of the hydrogen itself is lost during the energy intensive liquification process [59].

One of the advantages of LH_2, as opposed to other on-board hydrogen systems, is its weight. As shown in Table 2.3, an advanced LH_2 system weighs slightly more than a comparable gasoline system. Liquid hydrogen is considerably more bulky than gasoline, however. Complete LH_2 systems can range 6 to 10 times larger than a gasoline tank holding the equivalent amount of energy, reducing storage and/or passenger space [38, 57]. This ratio is slightly less on an equal distance basis, as hydrogen can be used more efficiently in an engine than gasoline.

Another advantage of using LH_2 is that it allows hydrogen to be delivered to the engine cold, i.e., -80°C [8]. This offers several performance advantages. The volumetric efficiency of the engine can be increased, for example, as cold hydrogen occupies less volume than an equivalent amount of warmer hydrogen. This, in essence, allows more of the fuel-air mixture to be inducted into the cylinder, producing a higher energy charge mixture. Secondly, the level of NO_x in the exhaust can be reduced since cryogenic hydrogen acts like a heat sink and reduces the temperatures in the combustion chamber.

More details on advances in liquid hydrogen technology are given in reprints at the end of the chapter by Furuhama and by Peschka and Escher. A more extensive discussion of LH_2 on-board storage is also given in reference [60].

Metal Hydrides

Metal hydride storage systems are based on the idea that gaseous hydrogen readily absorbs in metals, forming a weak chemical bond. Metal hydrides are typically in a granular or powder form and, thus, have a large surface area and large capacity for storage. To release gaseous hydrogen from the metal, the hydride is heated to a certain temperature. Exhaust heat from the engine, carried by cooling water or exhaust gases, is commonly used for this purpose. There are several advantages to using hydride systems. The metal hydride system is easier to deal with than other on-board storage systems, in that neither high pressures nor cryogenic temperatures are necessary for operation. Hydrogen obtained from a hydride storage system is insufficiently pressured for use in high pressure injection applications, however, a metal hydride system is also one of the safest alternatives for storing hydrogen.

The biggest disadvantage of hydride systems is that they have a low mass energy density and thus tend to be very heavy. In fact, hydride storage units, including the hydride and cooling system, can range from 120-485 kg (265-1070 lb), containing only 0.5 to 2% hydrogen by weight. The weight of hydride storage units is about 10 to 20 times greater than gasoline tanks for the same energy content. Besides affecting performance, the high storage weight limits the range of the car to 150-300 km (93-186 miles). Complete hydride systems can also be very large, ranging from 100 to 300 ℓ in volume, creating additional space problems [38].

The capacity of the hydride system can be reduced due to contamination. Povel, et al found that the capacity of a Ti/V/Mn hydride was reduced by 2% after only two filling cycles using H_2 gas of 99.995% purity [48]. This would lead to a 50% reduction in storage capacity after only 68 fillings. Oxygen and water contamination was the primary cause of the loss in storage capacity. Storage capacity lost due to water contamination was regained when the hydride was reactivated by heating, however. Other problems observed by these researchers include a tendency for the alloying material to expand when hydrogen is added.

Several criteria are used to evaluate hydride systems. The most important is that a system has a minimal mass with satisfactory operating characteristics. The hydride should have a high absorption capacity and a high density, utilize a minimal amount of heat to desorb or release the hydrogen, and have a low cost. In addition, a system must have adequate characteristics of pressure change in the temperature range of 293 to 473 K. No hydride system developed to date is outstanding on all accounts. The properties of the typical hydrides are given in Table 2.4. In general, these hydrides can be divided into two categories: low temperature hydrides, with hydrogen desorption temperatures up to 373 K, and high temperature hydrides.

Among the low temperature hydrides, ferrotitanium is one of the most promising candidates. It has a good absorption ability and a low cost [61]. When Mn is added to the ferrotitanium the lattice expands, leading to a stabilization of hydrogen phases and a lowering of the storage pressure. The maximum capacity of hydrogen for a $Fe_{0.8}TiMn_{0.2}$ hydride is about 1.9% of the total mass at 2 MPa and 273 to 293 K. Hydrogen can be released from these hydrides using only heat from exhaust gases. In fact, the amount of heat available from the exhaust gases exceeds the amount necessary to liberate hydrogen from a low temperature hydride by a factor of 2 or 3. To avoid overheating the hydride tank, it is necessary to release the surplus heat, which can be done by regulating the amount of exhaust gases passing through the hydride tank. One problem with these hydrides is that the bulk density of Fe-Ti is high, making these systems heavy.

Table 2.4 Hydride Characteristics

Characteristics	$Ti_2Ni\text{-}H_{2,5}$	$FeTi\text{-}H_2$	$VH\text{-}VH_2$	$LaNi_5\text{-}H_{6,7}$	$Mg_2Cu\text{-}H_3$	$Mg_2Ni\text{-}H_4$	Mg-H
Absorption ability of hydrogen in % of the alloy mass	1.61	1.87	1.92	1.55	2.67	3.71	8.25
Heat at desorption $KJkg^{-1}$, H_2	15.8	14.92	19.12	15.59	32.7	30.6	38.8
Equilibrium pressure at 293K (when charging) in MPa	0.55	0.29	0.21	0.12	not formed		
Charging	easy			very difficult		difficult	very difficult
Temperature desorption K							
at p=1.0 MPa	307.0	325	326.0	346	591.0	623.0	635.0
at p=0.15 MPa	270.0	280	288.0	294	522.0	540.0	569.0
Occurrence of the metal	often	very often	very seldom	seldom		often	
Safety	safe				highly inflammable	safe	highly inflammable
Hydride mass equivalent to the energy of 1 liter gasoline	16.75	14.45	14.0	17.4	14.65	7.28	3.2
Alloy mass necessary to accumulate 2.5 kg H_2	155.0	134	130.0	161.0		67.5	324.0
Mass of the combustion device for 2.5 kg H_2 (the mass of concentration with 0.4 of the mass of the hydride)	217.0	188	182.0	225.0		95.0	50.0

(Source: ref. [61])

Higher temperature hydrides cannot be run merely on exhaust gases and, thus, are typically used in combination with low temperature hydrides. The characteristics of such a combined system are given in Table 2.5. Since high temperature hydrides are typically lighter than the lower temperature hydrides, the weight of the overall system can be reduced by using combined systems. Higher temperature systems do have a tendency to break up, however. The optimal systems appear to be Mg-Mg_2Ni. The addition of Ni into this primarily Mg mixture is done in order to increase the speed of the reaction kinetics. The capacity of the hydride decreases as the Ni content increases, however, making it necessary to keep the Ni content as low as possible.

Hydride storage systems have played an important role in the Daimler-Benz hydrogen vehicle program since the 1970s. The vehicles developed by Daimler-Benz for the Berlin fleet test program from 1984 to 1988 were all equipped with metal hydride storage. The hydride container used by Daimler-Benz in this test program is shown in Fig. 2.5 [48]. The container consists of 19 high-grade steel tubes filled with an alloy of primarily titanium, vanadium, and manganese ($Ti_{0.98}$ $Zr_{0.02}$ $V_{0.43}$ $Fe_{0.09}$ $Cr_{0.05}$ $Mn_{1.5}$). A liquid media flows through the storage unit in a longitudinal direction between the tubes and the outer casing to heat the hydride and release hydrogen. The system was capable of storing a maximum of 1.5 kg (3.3 lbs.) of hydrogen and had a total weight of 142 kg (313 lb).

The Berlin fleet test program also served as a testing ground for the refueling capabilities of hydride systems. For this test fleet, the hydrogen refueling station was placed centrally at the Berlin gas works. At this site hydrogen was obtained from the city gas, which contained ~50% hydrogen by volume, using a pressure surging adsorber (see Chapter 4 for details). The hydrogen tanks were continuously checked to ensure that the hydrogen was >99.9995% pure. Hydrogen was administered into the hydride tank via a quick-fitting coupling under a pressure of 5 MPa [48]. The vehicle could be refueled to 80% capacity in under ten minutes. Refueling times of 10-20 minutes were recommended, however, for more complete refueling [62].

Compressed gas

Probably the most straightforward way to store hydrogen is as a compressed gas in a high-pressure vessel. These vessels are typically aluminum cylinders wrapped with fiberglass. Other designs being tested include aluminum cylinders wrapped with kevlar or graphite, plastic liners wrapped with graphite, and high-strength aluminum cylinders. Volume and mass are both issues with storing hydrogen as a compressed gas, however. As seen in Table 2.3, pressurized hydrogen systems at approximately 20 MPa weigh nearly three times more than what a comparable liquid hydrogen system does, and occupy more than twice the volume. This volume can be reduced by increasing the pressure of the compressed gas, potentially up to as high as 55 MPa (8000 psi) [63]. For such higher pressure vessels, cost and safety will be issues. More energy will also be required to compress the hydrogen gas to these pressures.

There are plausible scenarios for gaseous hydrogen use, however. Figure 2.6 shows a side view comparison of a standard gasoline vehicle and two hydrogen vehicles. A hydrogen vehicle with a 161 km (100 mile) range, comparable to the range of current electric vehicles, would require two hydrogen tanks weighting about 45 kg (100 lb) each. As shown, the vehicle will lose approximately half of its trunk space and weigh an additional 100 kg (222 lb) compared to the gasoline vehicle. A hydrogen vehicle with a 322 km (200 mile) range is also feasible. For a 322 km range, two more tanks would be necessary, which would essentially eliminate the trunk space of the vehicle. The development of more efficient fuel cells will enable the range of hydrogen vehicles to be expanded. The storage needed for a fuel cell vehicle with a range of

Table 2.5 The characteristics of the combined hydrides

Coefficients	FeTi, 100%	FeTi, 70% Mg_2Ni_2, 30%	FeTi, 60% MgNi, 40%	FeTi, 50% Mg_2Ni, 50%	FeTi, 70% Mg, 30%	FeTi, 60% Mg, 40%	FeTi, 50% Mg, 50%
Mass, kg	188	160	150	140	146	132	118
Mass decrease compared to FeTi, %	0	15	20	25.5	22.5	30	37

(Source: ref. [61])

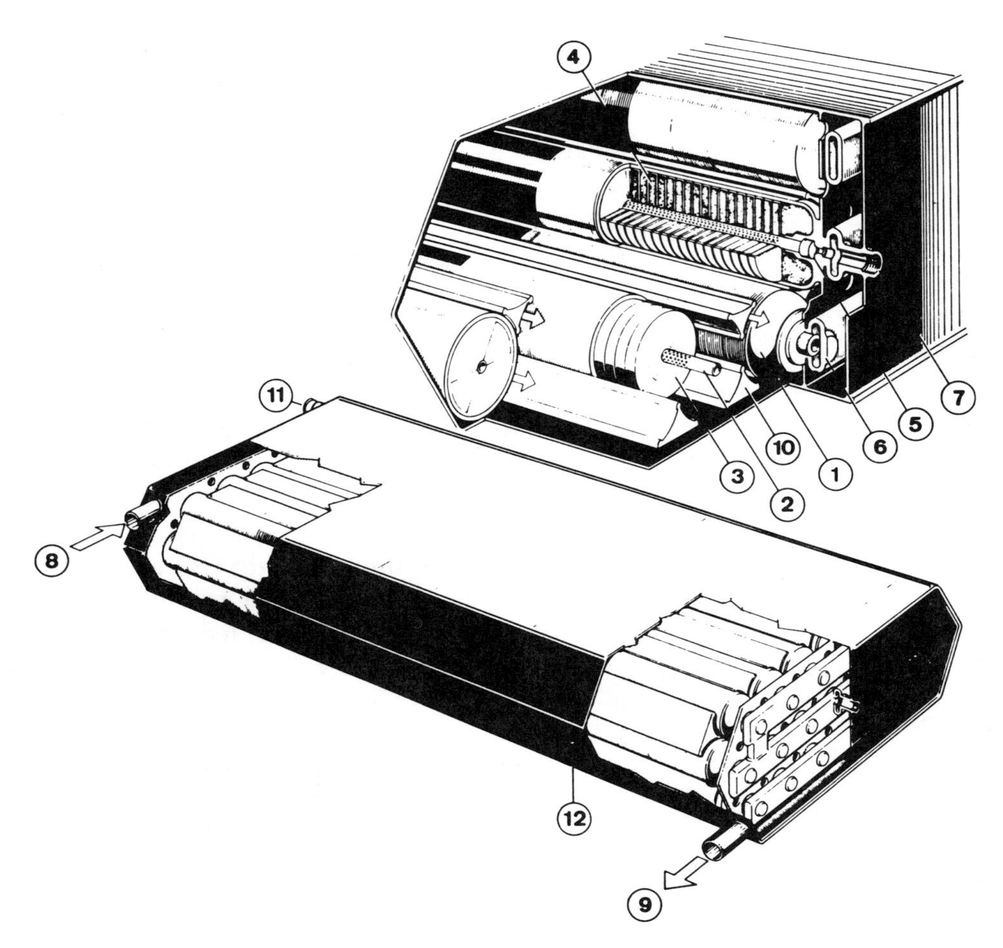

Fig. 2.5. Hydrogen storage unit by Mannesmann. (Source: ref. [48])

about 403 km (250 miles) is equivalent to that of an internal combustion engine hydrogen vehicle with a range of 161 miles.

Steam-Oxidation of Iron

H-Power Corporation has been developing a novel method of forming hydrogen on-board via the steam-oxidation of iron. Hydrogen is formed by reacting steam with sponge iron, the raw ingredient for steel-making furnaces, to form rust:

$$Fe + H_2O \Leftrightarrow FeO + H_2$$
$$3FeO + H_2O \Leftrightarrow Fe_3O_4 + H_2$$

After the tank of iron is completely rusted, it is exchanged at the refueling station with a new tank of iron. The rust can then be converted back to pure iron using methods already established on an industrial scale to produce sponge iron [64]. The steam and heat needed for the on-board reaction could potentially be provided by the exhaust of a fuel cell and waste heat. The weight and volume of a sponge iron system might be similar to that of compressed gaseous hydrogen at higher pressures (55 MPa or 8000 psi). There are still a number of issues that remain with this technology, however, including the fact that a catalyst is needed when the reaction is performed at practical temperatures, i.e., 80 to 200° C [65].

Activated Carbon Storage

Another technique of storing hydrogen is by adsorbing it onto a carbon surface [66, 67]. With this technique the amount of adsorption can be increased as the temperature is lowered, but cryogenic temperatures are not needed. Researchers at Syracuse University have developed a system based on supertreated carbon that can absorb a large number of carbon molecules at -123° C and 55 MPa [68].

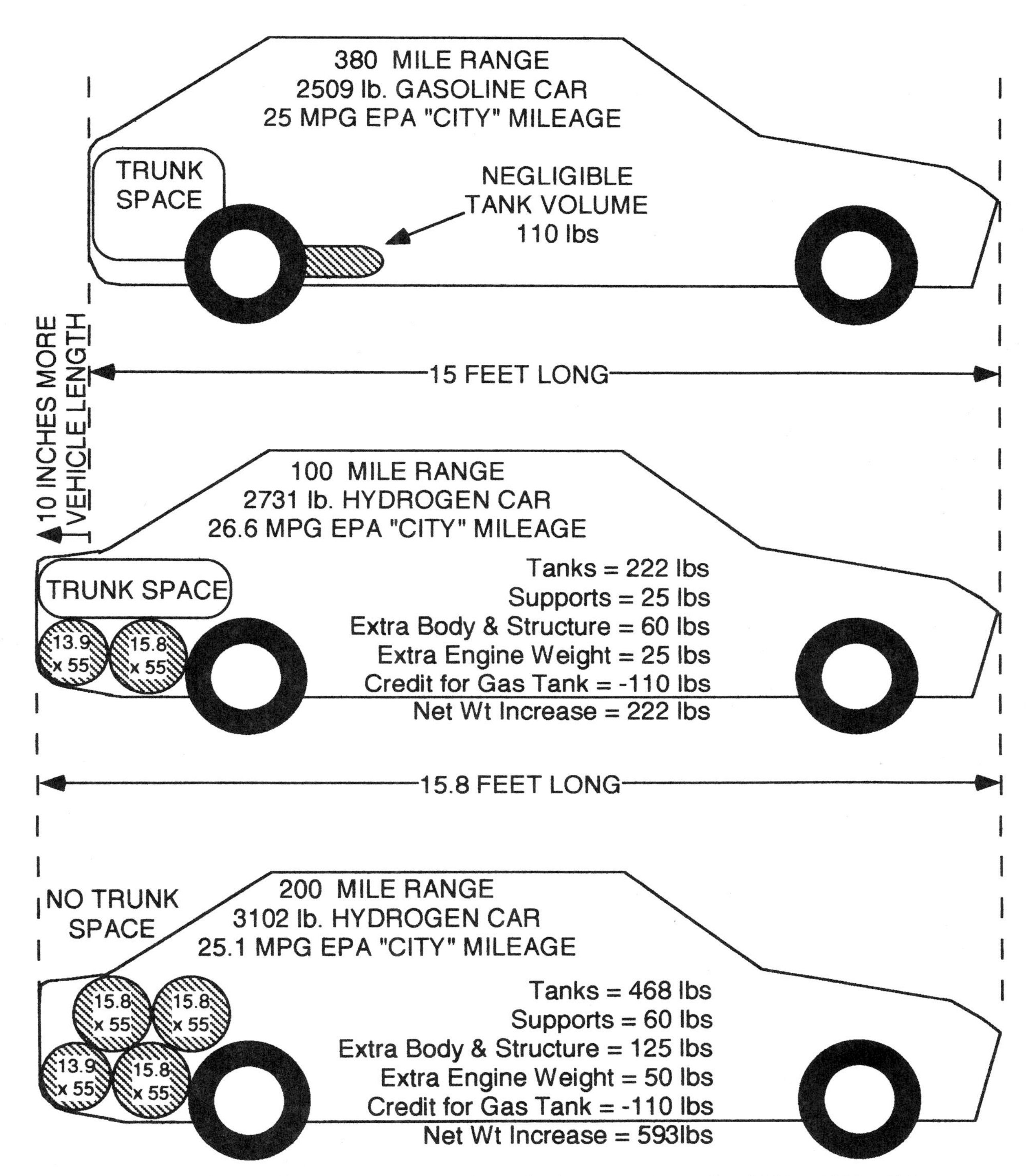

Figure 2.6 Comparison of a gasoline compact car with two hydrogen options; 1) a 161 km (100 mile) range compact that competes with EVs and 2) a 322 km (200 mile) range compact that loses its trunk to compete more closely with the gasoline vehicle. Engine sizes are increased in the hydrogen vehicles to give an equal acceleration. The sketches are to correct relative scale for 13.9" × 55" and 15.8" × 55" composite pressure vessels made by Brunswick Corp. (Source: Ref. [69])

Glass Microspheres

Hydrogen can also be stored under high pressure in glass spheres. The basis of this storage method is that hydrogen can be released by heating the microcapsules, which increases the rate of diffusion of hydrogen through the glass. At room temperature, the diffusivity of hydrogen through the glass capsules is sufficiently small that the hydrogen would not diffuse out. Microspheres have not been developed for commercial applications, however.

2.5 Conclusion

Research on hydrogen engines and vehicles has come a long way in the past several decades. Although some NO_x emissions are formed when hydrogen is used in an internal combustion engine, hydrogen is still the cleanest fuel for internal combustion engine applications. Hydrogen engines can also operate more efficiently than their gasoline counterparts. With the development of more advanced fuel delivery systems, combustion related problems such as premature ignition can be controlled. The primary technical problems for hydrogen vehicles are reduced power output, due to the lower volumetric energy density of a hydrogen/air mixture, and increased vehicle weight due to fuel storage. Engine modifications such as adding turbochargers, intercoolers, or blowers can increase the power output, but each has drawbacks. The power output can also be increased using high-pressure direct injection and/or liquid hydrogen. Both of these methods require more advanced systems, however, such as high-pressure fuel injectors and, in the case of LH_2, a high-pressure liquid hydrogen pump and insulated liquid hydrogen storage tank. Although there are several viable methods of storing hydrogen, each suffers from some combination of increased volume, cutting into storage or trunk space, and/or increased weight, leading to reduced vehicle performance.

References

1. Cecil, W., "On the application of hydrogen gas to produce a moving power in machinery; with a description of an engine which is moved by the pressure of the atmosphere, upon a vacuum caused by explosions of hydrogen and atmospheric air," *Trans. Cambridge Philos. Soc.*, Vol. 1, p. 217, 1822.

2. Van Vorst, W.D. and Woolley, R.L., "Hydrogen-fueled surface transportion," in *Hydrogen; Its Technology and Implications*, Edited by K. E. Cox and D. K. Williamson Jr., CRC Press, Boca Raton, FL, p. 1979.

3. Das, L.M., "Hydrogen Engines: A View of the Past and a Look into the Future," *Int. J. Hydrogen Energy*, Vol. 15, p. 425, 1990.

4. Hoffmann, P., *The Forever Fuel: The Story of Hydrogen*, Westview Press, Boulder, Colorado, 1981.

5. Williams, L.O., *Hydrogen Power: An Introduction to Hydrogen Energy and its Applications*, Pergamon Press, Oxford, U.K., 1980.

6. King, R.O. and Rand, M., "The oxidation, decomposition, ignition, and detonation of fuel vapours and gases. XXVII. The hydrogen engine.," *Canadian J. Technol.*, Vol. 33, p. 445, 1955.

7. King, R.O., Wallace, W.A. and Mahapatra, B., "The oxidation, ignition and detonation of fuel vapours and gases. V. The hydrogen engine and detonation of the end gas by the ignition effect of carbon nuclei formed by pyrolysis of lubricating oil vapor," Can. J. Res., Sect. F., Vol. 26, p. 264, 1948..

8. Furuhama, S., "Trend of Social Requirements and Technological Development of Hydrogen-Fueled Automobiles," *JSAE review*, Vol. 13, p. 4, 1992.

9. MacCarley, C.A., "Electronic Fuel Injection Techniques for Hydrogen Fueled I.C. Engines," M.S. Thesis in Engineering, University of California, Los Angeles, 1978.

10. Mishchenko, A.I., "The Ways of the Setting Up of Automobile-Type Hydrogen Engine Performance," in *Hydrogen Energy Progress V*, Edited by T. N. Veziroglu and J. B. Taylor, Pergamon Press, Elmsford, NY, p. 1529, 1984.

11. Swain, M.R., Swain, M.N. and Adt, R.R., "Considerations in the Design of an Expensive Hydrogen-Fueled Engine," SAE Technical Paper No. 881630, Society of Automotive Engineers, Warrendale, PA, 1988.

12. Ricardo, H.R., "Further Note on Fuel Research," *Proc. Inst. of Automob. Eng. of London*, Vol. 5, p. 327, 1924.

13. Swain, M.R. and Adt, R.R., "The Hydrogen-Air Fueled Automobile," in *7th Intersociety Energy Conversion Engineering Conference*, American Chemical Society, Washington, D.C., p. 1382, 1972.

14. Swain, M.R., Pappas, J.M., Adt, R.R. and Escher, W.J.D., "Hydrogen engine design data-base summary," in *18th Intersociety Energy Conversion Engineering Conference*, American Institute of Chemical Engineers, New York, NY, p. 536, 1983.

15. Swain, M.R., Adt, R.R. and Pappas, J.M., Experimental Hydrogen-fueled Automotive Engine Design Database Project, U.S. Department of Energy, Washington D.C., Report No. DOE/CS/31212-1, 1983.

16. Lynch, F.E., "Parallel induction: a simple fuel control method for hydrogen engines," *Int. J. Hydrogen Energy*, Vol. 8, p. 721, 1983.

17. MacCarley, C.A. and Van Vorst, W.D., "Electronic fuel injection techniques for hydrogen powered I.C. engines," *Int. J. Hydrogen Energy*, Vol. 5, p. 179, 1980.

18. Mathur, H.B. and Das, L.M., "Performance Characteristics of a Hydrogen Fueled S.I. Engine Using Timed Manifold Injection," *Int. J. Hydrogen Energy*, Vol. 16, p. 115, 1991.

19. Das, L.M., "Fuel Induction Techniques for a Hydrogen Operated Engine," *Int. J. Hydrogen Energy*, Vol. 15, p. 833, 1990.

20. Das, L.M., "Studies on timed manifold injection in hydrogen operated spark ignition engine: performance, combustion and exhaust emission characteristics," Ph.D. Thesis, Indian Institute of Technology, New Delhi, India, 1987.

21. Weil, K.H., "The hydrogen I.C. engine—its origin and future in the emerging-transportation-environment system," in *7th Intersociety Energy Conversion Engineering Conference*, American Chemical Society, Washington, D.C., p. 1355, 1972.

22. Homan, H.S., De Boer, P.C.T. and McLean, W.J., "The effect of fuel injection on NO_x emissions and undesirable combustion for hydrogen-fueled piston engines," *Int. J. Hydrogen Energy*, Vol. 8, p. 131, 1983.

23. Oehmichen, M., "Wasserstoff als Motortreib-mittel," in *Verein Deutsche Ingenieur, Deutsche Kraft-fahrtforschung*, Verlag GMBH, Berlin, 1942.

24. Furuhama, S. and Fukuma, T., "High Output Power Hydrogen Engine with High Pressure Fuel Injection, Hot Surface Ignition and Turbo-Charging," in *Hydrogen Energy Progress V*, Edited by T. N. Veziroglu and J. B. Taylor, Pergamon Press, Elmsford, NY, p. 1493, 1984.

25. Furuhama, S. and Kobayashi, Y., "Hydrogen Cars with LH_2-Tank, LH_2-Pump and Cold GH_2-Injection Two-Stroke Engine," SAE Technical Paper No. 820349, Warrendale, PA, 1982.

26. Furuhama, S., "Hydrogen Engine Systems for Land Vehicles," in *Hydrogen Energy Progress VII*, Edited by T. N. Veziroglu and A. N. Protsenko, Pergamon Press, Elmsford, NY, p. 1841, 1988.

27. Varde, K.S. and Frame, G.A., "Development of a High Pressure Hydrogen Injection for SI Engine and Results of Engine Behavior," in *Hydrogen Energy Progress V*, Edited by T. N. Veziroglu and J. B. Taylor, Pergamon Press, Elmsford, NY, p. 1505, 1984.

28. Krepec, T., Tebelis, T. and Kwok, C., "Fuel Control Systems for Hydrogen-Fueled Automotive Combustion Engines--A Prognosis," *Int. J. Hydrogen Energy*, Vol. 9, p. 109, 1984.

29. Krepec, T., Giannacopoulos, T. and Miele, D., "New Electronically Controlled Hydrogen-Gas Injector Development and Testing," in *Hydrogen Energy Progress VI*, Edited by T. N. Veziroglu, N. Getoff and P. Weinzierl, Pergamon Press, Elmsford, NY, p. 1087, 1986.

30. Krepec, T., Giannacopoulos, T. and Miele, D., "New electronically controlled hydrogen-gas injector development and testing," *Int. J. Hydrogen Energy*, Vol. 12, p. 855, 1987.

31. Glasson, N. and Green, R., "High Pressure Hydrogen Injection," in *Hydrogen Energy Progress IX*, Edited by T. N. Veziroglu and C. D.-J. Pottier, International

Association for Hydrogen Energy, Coral Gables, FL, p. 1285, 1992.

32. Woolley, R.L. and Anderson, V.R., "Hydrogen Engine NO_x Control by Water Induction," Billings Energy Corporation Publication No. 77001, 1977.

33. Koelsch, R.K. and Clark, S.J., "A comparison of hydrogen and propane fuelling of internal combustion engine," SAE Paper No. 790677, Society of Autmotive Engineers, Warrendale, PA, 1979.

34. Homan, H.S., Reynolds, R.K., De Boer, P.C.T. and McLean, W.J., "Hydrogen-fuelled diesel engine without timed ignition," *Int. J. Hydrogen Energy*, Vol. 4, p. 315, 1979.

35. Welch, A.B. and Wallace, J.S., Performance characteristics of a hydrogen-fuelled diesel engine with ignition assistance Final Report, NRCC, Report No. DSS Contract File No. 24 SU.31155-2-2664. Serial No. ISU 82-00340, 1986.

36. Wong, J.K.S., "Compression Ignition of Hydrogen in a Direct Injection Diesel Engine Modified to Operate as a Low-Heat-Rejection Engine," *Int. J. Hydrogen Energy*, Vol. 15, p. 507, 1990.

37. Ikegami, M., Miwa, K. and Shioji, M., "A study of hydrogen fuelled compression ignition engines," *Int. J. Hydrogen Energy*, Vol. 7, p. 341, 1982.

38. DeLuchi, M.A., "Hydrogen Vehicles: An Evaluation of Fuel Storage, Performance, Safety, Environmental Impacts, and Cost," *Int. J. Hydrogen Energy*, Vol. 14, p. 81, 1989.

39. Swain, M.R., Swain, M.N., Leisz, A. and Adt, R.R., "Hydrogen Peroxide Emissions from a Hydrogen Fueled Engine," *Int. J. Hydrogen Energy*, Vol. 15, p. 263, 1990.

40. Peschka, W., "Liquid Hydrogen as a Vehicular Fuel—A Challenge for Cryogenic Engineering," in *Hydrogen Energy Progress IV*, Edited by T. N. Veziroglu, W. D. Van Vorst and J. H. Kelley, Pergamon Press, Elmsford, NY, p. 1053, 1982.

41. Heffel, J.W. and Norbeck, J.M., "Preliminary Evaluation of UC Riverside's Hydrogen Truck," presented at the 6th Annual U.S. Hydrogen Meeting, Alexandria, VA, 1995.

42. Kludjian, V.Z., "Hydrogen for Vehicles—Mazda's Hydrogen Vehicle Development Program," presented at the National Hydrogen Association's 5th Annual U.S. Hydrogen Meeting, Washington, D.C., 1994.

43. Furuhama, S., "Hydrogen Engine Systems for Land Vehicles," *Int. J. Hydrogen Energy*, Vol. 14, p. 907, 1989.

44. Finegold, J.G., "Hydrogen: primary or supplementary fuel for automotive engines," *Int. J. Hydrogen Energy*, Vol. 3, p. 83, 1978.

45. Parks, F.B., "A Single-Cylinder Engine Study of Hydrogen-Rich Fuels," SAE Technical Paper No. 760099, Society of Automotive Engineers, Warrendale, PA, 1976.

46. Stebar, R.F. and Parks, F.B., Emission Control with Lean Operation using Hydrogen-supplemented Fuel, General Motors Research Publication No. 1537, 1974.

47. May, H. and Gwinner, D., "Possibilities of improving exhaust emissions and energy consumption in mixed hydrogen-gasoline operation," *Int. J. Hydrogen Energy*, Vol. 8, p. 121, 1983.

48. Povel, R., Topler, J., Withalm, G. and Halene, C., "Hydrogen drive in field testing," in *Hydrogen Energy Progress V*, Edited by T. N. Veziroglu and J. B. Taylor, Pergamon Press, Elmsford, NY, p. 1563, 1984.

49. Mathur, H.B., Das, L.M. and Patro, T.N., "Hydrogen Fuel Utilization in CI Engine Powered End Utility Systems," *Int. J. Hydrogen Energy*, Vol. 17, p. 369, 1992.

50. Gopal, G., Rao, P.S., Gopalakrishnan, K.V. and Murthy, B.S., "Use of hydrogen in dual-fuel engines," *Int. J. Hydrogen Energy*, Vol. 7, p. 267, 1982.

51. Weigang, W. and Lainfang, Z., "The Research on Internal Combustion Engine with the Mixed Fuel of

Diesel and Hydrogen," in *International Symposium on Hydrogen Systems*, Edited by T. N. Veziroglu, Z. Yajie and B. Deyou, China Academic Publishers, Beijing, China, p. 83, 1985.

52. Raman, V., Hansel, J., Fulton, J., Lynch, F. and Bruderly, D., "Hythane—An Ultraclean Transportation Fuel," in *Hydrogen Energy Progress X*, Edited by D. L. Block and T. N. Veziroglu, International Association for Hydrogen Energy, Coral Gables, FL, p. 1797, 1994.

53. Hoeskstra, R.L., Collier, K. and Mulligan, N., "Demonstration of Hydrogen Mixed Gas Vehicles," in *Hydrogen Energy Progress X*, Edited by D. L. Block and T. N. Veziroglu, International Association for Hydrogen Energy, Coral Gables, FL, p. 1781, 1994.

54. Swain, M.R., Yusuf, M.J., Dulger, Z. and Swain, M.N., "The Effects of Hydrogen Addition on Natural Gas Engine Operation," SAE Technical Paper No. 932775, Society of Automotive Engineers, Warrendale, PA, 1993.

55. Wallace, J.S. and Cattelan, A.I., "Hythane and CNG Fuelled Engine Exhaust Emission Comparison," in *Hydrogen Energy Progress X*, Edited by D. L. Block and T. N. Veziroglu, International Association for Hydrogen Energy, Coral Gables, FL, p. 1761, 1994.

56. Du, T.-S., Li, J.-D. and Lu, Y.-Q., "Experimantal study on spark engine burning methanol-hydrogen mixed fuel," in *Hydrogen Energy Progress VI*, Edited by T.N. Veziroglu, N. Getoff, P. Weinzierl, Pergamon Press, Elmsford, NY, p. 1073, 1986.

57. Smith, J.R., Aceves, S., and Van Blarigan, P., "Series hybrid vehicles and optimized hydrogen engine design," presented at the 1995 SAE Future Transportation Technology Conference & Exhibition, Costa Mesa, CA, August, 1995.

58. Kukkonen, C.A. and Shelef, M., "Hydrogen as an Alternative Automobile Fuel: 1993 Update," SAE Technical Paper No. 940766, Society of Automotive Engineers, Warrendale, PA, 1994.

59. Yamane, K., Hiruma, M., Watanabe, T., Kondo, T., Hikino, K., Hashimoto, T. and Furuhama, S., "Some performance of engine and cooling system on LH_2 refrigerator van Musashi-9," in *Hydrogen Energy Progress X*, Edited by D. L. Block and T. N. Veziroglu, International Association for Hydrogen Energy, Coral Gables, FL, p. 1825, 1994.

60. Peschka, W., *Liquid Hydrogen—Fuel of the Future*, Springer-Verlag, Wein, New York, 1992.

61. Petkov, T., Veziroglu, T.N. and Sheffield, J.W., "An Outlook of Hydrogen as an Automotive Fuel," *Int. J. Hydrogen Energy*, Vol. 14, p. 449, 1989.

62. Feucht, K., Hurich, W., Komoschinski, N. and Povel, R., "Hydrogen Drive for Road Vehicles—Results from the Fleet Test Run in Berlin," in *Hydrogen Energy Progress VI*, Edited by T. N. Veziroglu, N. Getoff, and P. Weinzierl, Pergamon Press, Elmsford, NY, p. 1079, 1986.

63. DeLuchi, M.A. and Ogden, J.M., "Solar-Hydrogen Fuel-Cell Vehicles," *Transportation Research, Part A, Policy and Practice*, Vol. 27A, p. 255, 1993.

64. Mayersohn, N.S., "The Outlook for Hydrogen," *Popular Science*, p. 67, October 1993.

65. Ogden, J. and Nitsch, J., "Solar Hydrogen," in *Renewable Energy: Sources for Fuels and Electricity*, Edited by T. B. Johansson, H. Kelly, A. K. N. Reddy, and R. H. Williams, Island Press, Washington, D.C., p. 925, 1993.

66. "New Method for Storing Hydrogen Fuel Developed," *New Technology Week*, p. 5, 1991.

67. Hynek, S., Fuller, W., Bentley, J. and McCullough, J., "Hydrogen Storage by Carbon Sorption," in *Hydrogen Energy Progress X*, Edited by D. L. Block and T. N. Veziroglu, International Association for Hydrogen Energy, Coral Gables, FL, p. 985, 1994.

68. Young, K.S., "Advanced Composites Storage Containment for Hydrogen," *Int. J. Hydrogen Energy*, Vol. 17, p. 505, 1992.

69. Courtesy of Lynch, F.E., Hydrogen Consultants, Inc., 1995.

Fuel Induction Techniques for a Hydrogen Operated Engine

L. M. Das

Indian Institute of Technology

Abstract—It is practically impossible to replace the internal combustion engines which have already become an indispensable and integral part of our present day life style, particularly in the transportation and agricultural sectors. Unfortunately, the survival of these engines has, of late, been threatened by the dual problems of the fuel crisis and environmental pollution. Therefore, to sustain the present growth rate of civilization, a non-depletable, clean fuel must be expeditiously sought. Hydrogen exactly caters to these specified needs. Hydrogen, even though "renewable" and "clean-burning" it does give rise to some undesirable combustion problems in an engine operation, such as backfire, pre-ignition, knocking and rapid rate of pressure rise. It has been experimentally evaluated that the fuel induction technique (FIT) does play a very dominant role in obtaining smooth engine operation. This paper discusses such various possible modes. Research work carried out by different investigators has been highlighted.

INTRODUCTION

A fuel has an infinite supply potential. It can be generated from water using any non-fossil energy source and upon combustion it produces water which goes back to the earth's water supply system from where it came. From an environmental standpoint, it is exceptionally clean.

The above-mentioned characteristics define a very desirable fuel and hydrogen does possess these characteristics. So situations arising out of the present-day energy crisis do not affect the hydrogen-fuel-system. As far as engine operation is concerned, a total hydrogen-fuelled engine will not emit unburnt hydrocarbons, CO, particulate matter, sulphur dioxide, smoke etc. From several practical considerations hydrogen is safer compared to conventional petroleum fuels. Being very light, leaking hydrogen rises up very rapidly through the air, thus creating an explosion possibility only to the space immediately above the leak. On the other hand, spilled gasoline creates safety-related problems which do persist for a long time. Because of low emissivity characteristics, radiation hazards from a hydrogen flame are of lesser consequence as compared to a gasoline flame.

While judging the suitability of hydrogen as an engine fuel, it will be desirable to compare its various physical and chemical properties with other conventional engine fuels as given in Tables 1–3. It is evident that petroleum fuels are liquid at room temperature whereas hydrogen remains a gas even at a much lower temperature (i.e. −253°C). The flammability limits, ranges of equivalence ratios over which the engine system is operable, auto-ignition temperature and minimum ignition energy are some of the properties which determine the suitability of the fuel for engine application. However, since some combustion characteristics of hydrogen fuel set it completely apart from other conventional fuels, unless these properties are appropriately exploited to an advantage for improved engine characteristics, they might give rise to various unwanted combustion problems.

UNDESIRABLE COMBUSTION PROBLEMS

Figure 1 [1, 2] shows the ranges of equivalence ratios suitable for hydrogen engine operation. A close look at the properties of the fuel brings in some very important points with respect to engine operation. Interestingly, most properties of hydrogen fuel if appropriately exploited to a point of advantage, could prove extremely desirable. On the other hand, the same property, if wrongly used, could be fatal.

The ignition energy required to ignite an air–fuel mixture depends very much on the air-fuel or equivalence ratio—hydrogen has an extremely low ignition energy compared to gasoline. This is a very crucial property. On one hand, the low minimum ignition energy enables the conventional ignition system to be effective with a very low energy spark whereas at the same time it makes the system susceptible to surface ignition. Surface ignition is a highly undesirable combustion phenomenon because it precipitates flashback, pre-ignition and rapid rates of pressure rise. Based on the lower flammability limit, hydrogen seems to be superior to gasoline, but a small leakage from a hydrogen operated system brings in the problem of safety. As far as the quenching distance is concerned, hydrogen combustion which can be initiated with a low energy spark, becomes difficult to quench. Because of the smaller quenching distance of hydrogen, a flame in a hydrogen–air mixture

Table 1. Physical and chemical properties of various fuels (values generally accepted from literature)

Properties	Gasoline	Hydrogen	Ammonia	Methanol	Ethanol
Molecular weight	91.4	2.02	17.03	32.04	46.07
Heat of combustion (net) MJ kg^{-1}	43.4	120.1	18.6	20.1	26.9
Stoichiometric mixture					
Mass air/Mass fuel	14.5	34.3	6.1	65	9.0
Maximum laminar					
flame speed m s^{-1}	0.37	2.91	0.010	0.52	—
Adiabatic flame temp., °C	2637	2756	2484	2576	2594
Octane number					
Research	91–100	130^{+}	130	110	106
Motor	82–94			87	89

escapes more readily past an even nearly closed intake valve than a hydrocarbon–air mixture.

The minimum ignition energy required for ignition (0.02 mJ) of a hydrogen–air mixture has often been responsible for the fresh charge being ignited and thereby causing a flame that propagates through the induction system giving rise to backfire. The simplest method to avoid backfire is to ensure the absence of combustible mixture in the intake manifold. A reduction of temperature level could also prove very effective. On the other hand, conditions leading to pre-ignition could be disposed of by preparing a late hydrogen–air mixture. These can be achieved by various methods such as (i) use of leaner mixtures, (ii) exhaust gas recirculation, (iii) intake air cooling (by liquid hydrogen or by water) and (iv) reduction of valve overlap.

Several investigators have adopted various means to combat the effect of these phenomena in a hydrogen operated engine. The mode of mixture preparation has been found to be quite important in determining the overall operational characteristics of a hydrogen engine.

FUEL INDUCTION TECHNIQUES (FIT)

The fuel induction techniques have been found to be playing a very dominant and sensitive role in determining the performance characteristics of an I.C. Engine. The 'FIT' for an S.I. engine can be classified into four categories such as Carburetion, Inlet Manifold Injection, Inlet Port Injection and Direct Cylinder Injection. These conventional methods of 'FIT' could also be applied to engine operation with a non-conventional alternative fuel, such as hydrogen. Of these methods; carburetion by the use of a gas carburettor has been the simplest and the oldest technique. In a gasoline-fuelled engine, the volume occupied by the fuel is about 1.7% of the mixture whereas a carburetted hydrogen engine, using gaseous hydrogen, results in a power output loss of 15%. Thus, apart from eliminating unwanted combustion symptoms, fuel induction techniques have also been quite effective in compensating for the power loss. Injection of hydrogen into the inlet manifold offers an alternative to the conventional load control method by throttling. This method uses the typical properties of hydrogen fuel (such as wide flammability limits) to a point of advantage. It also possesses the ability to initiate fuel delivery at a timing position sometime after the beginning of intake stroke. The system could be so designed that the intake manifold does not contain any combustible mixture thereby avoiding extreme situation leading to undesirable combustion phenomena. The arrangement for air being inducted prior to fuel delivery, has two very important roles to play. It provides a pre-cooling effect and thus renders inoperative the pre-ignition sources that could be present on the surface. Secondly, this also helps to quench or at least to dilute any hot residual

Table 2. Combustion properties of hydrogen, methane and gasoline (values generally accepted from literature)

Property	Hydrogen	Methane	Gasoline
Limits of flammability in air, vol%	4.0–75.0	5.3–15.0	1.0–7.6
Stoichiometric composition in air, vol%	29.53	9.48	1.76
Minimum energy for ignition in air, mJ	0.02	0.29	0.24
Autoignition temperature, K	858	813	501–744
Flame temperature in air, K	2318	2148	2470
Burning velocity in NTP air, cm s^{-1}	265–325	37–45	37–43
Quenching gap in NTP air, cm	0.064	0.203	0.2
Percentage of thermal energy radiated from			
flame to surrounding, %	17–25	23–33	30–42
Diffusivity in air, $cm^2 s^{-1}$	0.63	0.2	0.08
Normalised flame Emmisivity 2000 K, 1 atm	1.00	1.7	1.7
Limits of flammability (equivalence ratio)	0.1–7.1	0.53–1.7	0.7–3.8

Table 3. Lean flammability limits of various fuels (values generally accepted from literature)

Fuel	Chemical formula	Lean flammability limits Volume per cent	Equivalence ratio
Methane	CH_4	5.3	0.53
Propane	C_3H_8	2.2	0.54
Pentane	C_5H_{12}	1.5	0.58
Octane	C_8H_{18}	1.0	0.60
Benzene	C_6H_6	1.4	0.51
Methanol	CH_3OH	7.3	0.56
Hydrogen	H_2	4.0	0.10

combustion products that could be present in the compression space near TDC. In a carburetted engine system, the valve overlap between the exhaust and the intake stroke can bring the fuel–air charge into contact with the residual hot gases. However, if by any chance pre-ignition does take place during intake stroke, it will have much lesser consequence as compared to that occurring in a carburetted engine. Some investigators have also carried out research on intake port injection. In such a system both air and fuel enter the combustion chamber during the intake stroke, but are not pre-mixed in the intake manifold.

Direct cylinder injection of hydrogen into the combustion chamber does have all the benefits of the late injection as characterized by manifold injection. In addition, the system permits for fuel delivery after the closure of the intake valve and thus, intrinsically precludes the possibility of backfire. However, as described later, the injection system will have to cater to some stringent requirements in respect of the severe thermal environment which the injector is bound to encounter. Besides, all the mechanical parts which form part of the injection system must be able to withstand such a high pressure, say to the tune of about 100 atm. When considering a practical automobile, maintaining a high pressure such as about 100 bars, in a vehicle for onboard storage methods raises serious problems. However a detailed discussion on vehicular storage methods is beyond the scope of the paper.

ACHIEVEMENTS AND GAPS

Researchers throughout the world have been working persistently for decades and hence most of the benefits and problems of hydrogen engines have already been identified.

A definite conclusion which can be drawn from these research results is that the undesirable combustion phenomena have greatly impeded the practical achievement of a common hydrogen-fuelled autovehicle: and the mode of fuel induction from one method to other has very seriously influenced the situation.

In the earliest phase of hydrogen engine research Ricardo [3], King *et al.* [4–6] had all adopted the carburetted technique, primarily with a view to achieve

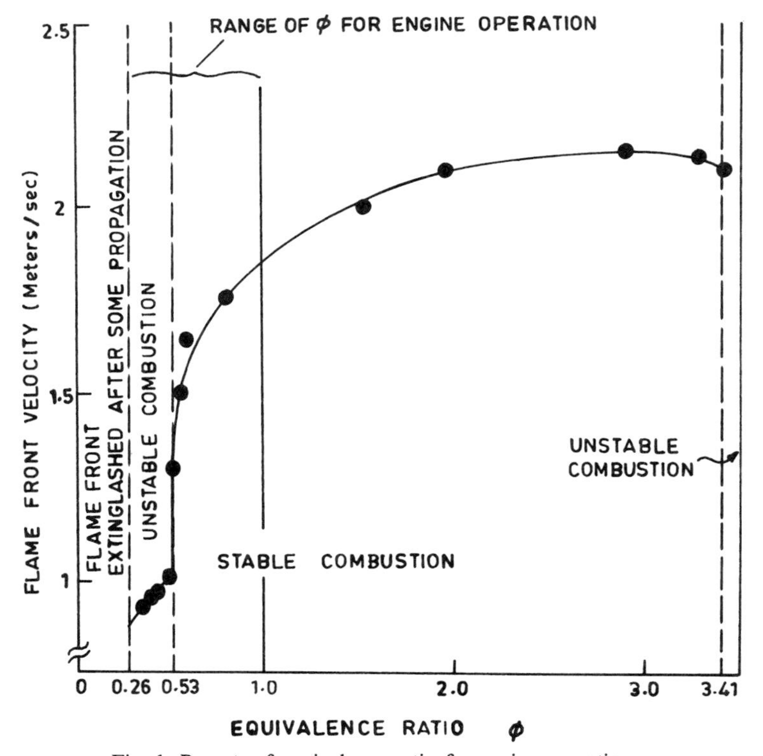

Fig. 1. Ranges of equivalence ratio for engine operation.

hydrogen-fuelled engine operation. Ricardo is reported to have encountered the problems of "popping back into the carburettor" and was unable to get rid of this problem even at the compression ratio as low as 3.8. Thus he concluded hydrogen to be impractical for most uses. King's work also centered around the carburetted engine and he also experienced severe backfire and pre-ignition problems. Special investigations were carried out to identify the causes of these phenomena. Efforts were made to ensure elimination of backfire and pre-ignition phenomena caused by free floating carbon particles, carbon deposits and cylinder hot spots. Conditions suspected to be promoting backfire were deliberately created inside the engine cylinder to arrive at definite conclusions. King used "cold spark plugs" and an aged sodium filled valve.

Hydrogen engine research did suffer a setback for a long period because of the availability of sufficient petroleum-based fuels. However, in the latter part of the 1970s when the dual problem of petroleum fuel depletion and environmental pollution assumed significance, hydrogen was again tried as an alternative fuel by many investigators for its infinite source potential and non-polluting characteristics. Because of the simplicity of engine configuration obtained only by the use of a gas carburettor and the requirement of low pressure for hydrogen induction, these investigators probably used hydrogen carburetion as the FIT.

As emphasized earlier, these carburetted versions of the engine systems, apart from developing low power outputs (as compared to the gasoline-fuelled engines) also exhibited severe operational combustion-related problems such as backfire, pre-ignition, combustion knock and rapid rate of pressure rise. It is the teething problem of backfire (which persisted in the carburetted hydrogen engines, and was extremely difficult to eliminate in most operating engine conditions) which prompted several other researchers to try out alternative modes of fuel induction. Swain and Adt [7, 8] tried out a method of "Hydrogen Induction Technique" (HIT) in which hydrogen was supplied through the passage on the intake valve. Their reports verify the effectiveness of HIT over conventional carburetion technique in overcoming backfire and pre-ignition problems.

Lynch [9] has suggested "Parallel Induction" which has proved successful in getting over the problems associated with backfire. Broadly speaking, this is a method similar to intake port injection. He has also reported another method of hydrogen induction technique through a copper tube placed inside the air intake port. A sleeve-type valve-seat mechanism built on the original intake valve is used to control the system. This method of delayed hydrogen admission proved quite effective in suppressing the undesirable combustion phenomena.

Bindon *et al.* [10] successfully tried out a novel technique of providing a quality-controlled mixture through a lean burn carburettor specifically developed for hydrogen operations. Timed port injection was also tried successfully by these researchers to eliminate the presence of the combustible mixture in the intake manifold in the proportions that could cause backfire. Varde and Frame [11, 12] developed a system of electronically controlled fuel injection in which injection was designed to take place close to the intake valve when the valve was open. This method was adopted chiefly to ensure the lack of hydrogen–air mixture entering the engine. Research results of these investigations show that it was thus possible to achieve a higher thermal engine efficiency as compared to the carburetted operation of either gasoline or even hydrogen. In some combustion related studies he also compared the variation of rate of pressure rise as well as the flame speed with respect to equivalence ratio as shown in Figs 2 and 3.

McCarley and Van Vorst [13] carried out extensive experiments on a hydrogen engine adopting both a port injection and direct injection system. The engine system configuration, designed to ensure quality control, did prove quite effective in ensuring a backfire free operation. It was further established that the fuel delivery in the injection system was not solely governed by the intake air flow. Hence it is always possible to optimally design a system based on various engine parameters and thus avoid conditions leading to backfire. As far as pollution aspects are concerned it has been found the $(NO)_x$ formation could be minimized by a precise control of equivalence ratio. Figure 4 shows the performance parameters of an electronic hydrogen injection system developed by McCarley and Van Vorst. In some experiments; they had incorporated an additional water induction system. Wooley *et al.* [14] had also taken recourse to water induction as an approach to reduce the frequency of backfire occurrence.

Some work has been reported in a practical hydrogen-fuelled automobile [15] using port injection. Watson *et al.* [16] have also been reported to have overcome the problem of backfire in a hydrogen engine with delayed port admission of hydrogen at relatively low pressures. At higher load conditions they avoided backfire with the use of water injection.

In I.I.T., Delhi [17] four different types of fuel induction techniques were tried as shown in Table 4. Depending upon the experimental condition it was found to be virtually impossible to get rid of flashback either in carburetion or in a continued manifold injection system. The other two methods such as timed manifold injection and low pressure direct cylinder injection were subjected to elaborate experimental investigation. Both these methods, by definition, should preclude the problems of flashback and pre-ignition either by supplying hydrogen gas directly into the cylinder after the closure of the intake valve (in LPDI) or by introducing hydrogen at an appropriate time in the manifold and at an appropriate location so that hydrogen is introduced after the potential hot spots are cooled again. Two different designs of injection systems were developed for carrying out the experiments. It was observed that, compared to LPDI, the TMI system required a less sophisticated design of the injector, as the former needed the injector to be capable of surviving in the severe thermal environment

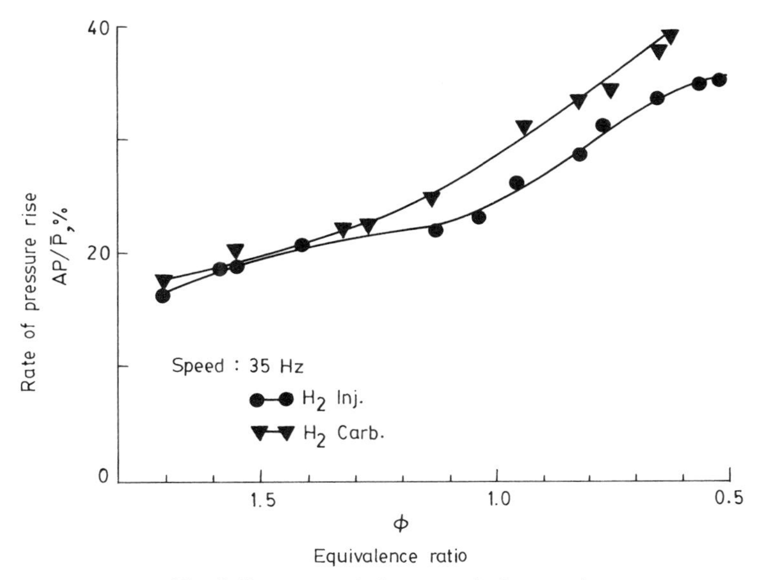

Fig. 2. Pressure variation vs equivalence ratio.

of the combustion chamber. Therefore leaking of the injector tip seemed to be a constant problem in most of the preliminary experiments which, of course, could ultimately be eliminated by proper choice of material subjected to heat treatment processes. On the other hand such a problem almost did not exist in a TMI system. In addition to this, LPDI seemed to exhibit problems of incomplete combustion, probably due to such a short time allowed for the mixing of hydrogen and air to take place. Such a situation did not arise in the case of TMI where mixing was proper and complete. Figure 5 shows the TMI configuration adopted for the experiment at IIT, Delhi. Timed manifold injection was observed to have possessed certain specific advantageous features with regard to other modes of fuelling techniques. In the entire range of experimental investigation hydrogen was supplied to the engine system in a gaseous phase thus leading to conditions of uniform and rapid mixing. Furthermore, the system did offer the option of adopting fuel injection being delayed to a point after the intake air has begun. Such a method does help in adequately cooling down the potential hot-spots which are quite

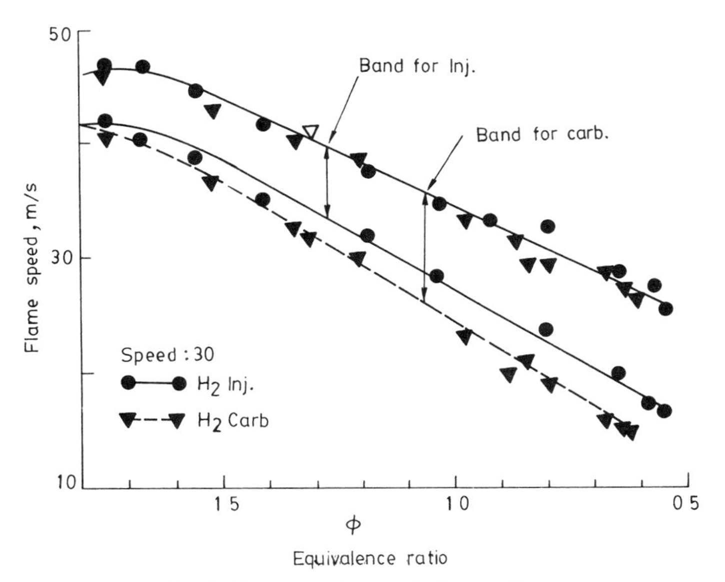

Fig. 3. Flame speed vs equivalence ratio.

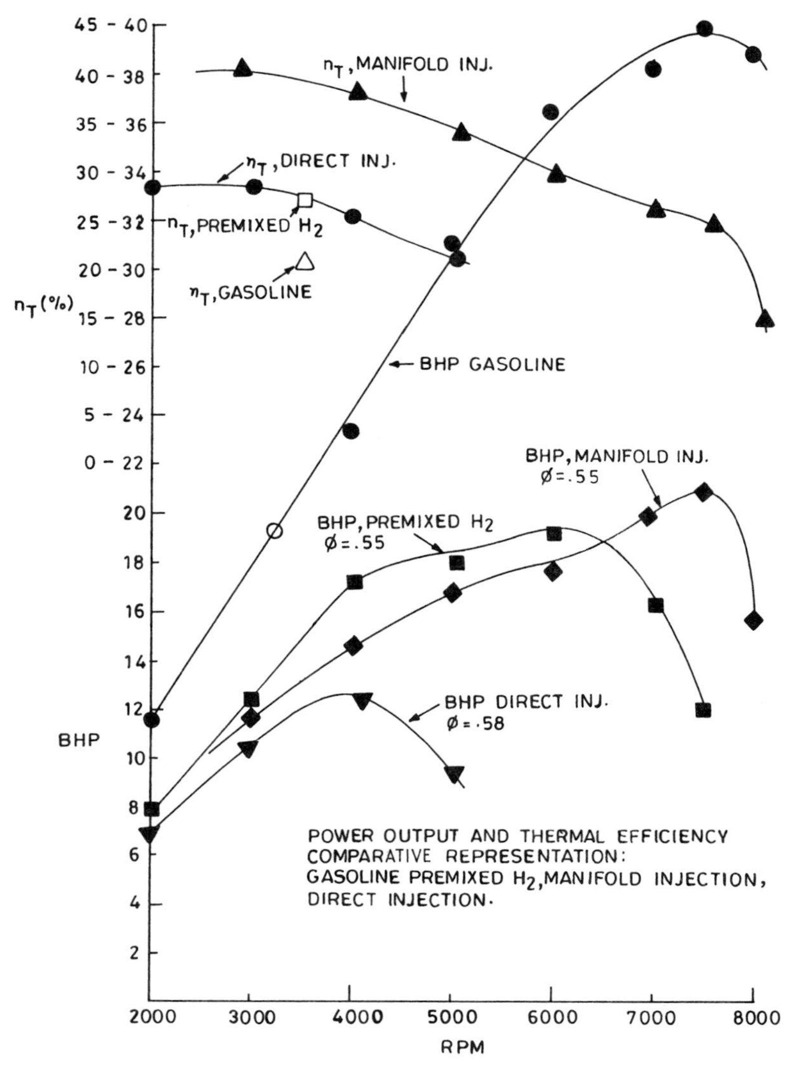

Fig. 4. Performance parameters of manifold and direct injection hydrogen engine.

often responsible for causing thermally-induced backfire. However, if by chance, inspite of all these preventive measures, backfire takes place, it would definitely cause much less damage than the one caused in a carburetted version of the engine.

A series of exhaustive experiments were conducted earlier on the same engine using carburetion as the fueling mode. A comparative evaluation of both carburetted and TMI configuration indicated that TMI version of the engine was able to achieve an increase of 4.2% in indicated thermal efficiency and almost a 20% rise in peak power output [17]. The experimental arrangement shown in Fig. 5 exhibited a unique operational feature. It permited the flexibility of adopting diesel-like quality governing and achieving the efficiency of a diesel engine while developing a specific output comparable to an S.I. engine.

The technique of direct cylinder injection has been tried as an effective step against the undesirable combustion phenomena since very early phase of hydrogen engine research by Erren [18]. A little later Oehmichen [19] carried out an extensive work in a hydrogen engine and was successful in circumventing problems of backfire and pre-ignition. The hydraulically operated hydrogen injection system developed by Varde and Frame [12] was also applied to a direct cylinder injection configuration with the injection scheduled to occur during the compression stroke. Besides exhibiting good performance characteristics the system is reported to have given lower levels of pollutants as compared to that of gasoline fuelled spark ignition engine.

Homan [20] carried out experiments on a hydrogen-fuelled engine using a LIRIAM (Late Injection, Rapid Ignition and Mixing) technique. A large number of

Table 4. Mixture formation methodologies investigated

S. No.	Mixture formation	Classification	Hydrogen flow timing	Supply pressure
1.	Continuous carburetion (CC)	Pre-IVC	Continuous flow	A little above atmospheric
2.	Continuous manifold injection (CMI)	Pre-IVC	Continuous flow	Slightly greater than atmospheric
3.	Timed manifold injection (TMI)	Pre-IVC	Hydrogen flow commences after the opening of the intake valve but completed prior to IVC.	1.4 to 5.5 kgf cm^{-2}
4.	Low pressure direct cylinder injection (LPDI)	Post-IVC	Hydrogen flow commences after the intake valve closure and is completed before significant compression pressure rise.	2 to 8 kgf cm^{-2}

operating parameters and their influence on performance, exhaust emission as well as combustion characteristics of the engine were thoroughly investigated. Operational characteristics with conditions of the least pollution and minimum undesirable combustion symptoms were experimentally evaluated. Figure 6 shows the injector developed by Homan. However, this work clearly describes the inherent problems that arose in the design and development of the injector.

Suzuki's work [21] was carried out with a low pressure, direct cylinder injection system. The system had utilized the advantage of early mixture preparation. However, it was conclusively realized through a series of experiments that hydrogen induction into the combustion chamber is a more effective step to avoid the backfire tendency, particularly at low speeds. Furuhama and his team of researchers have been carrying out persistent hydrogen engine research, with various engine configuration, for a very long time. They are reported to have experienced [22, 23–27] problems of high pressure rise rate and incomplete combustion. This is believed to have occurred due to heterogeneous mixture formation as the injection was scheduled to take place at the end of compression stroke (in the almost stagnant condition of engine cylinder). Due to the low density of hydrogen and limited time available for mixing at the end of compression stroke, direct injection has a definite disadvantage. This has also been the experience with many other researchers. As an alternative to try out the elimination of these effects, Furuhama *et al.* developed a system in which hydrogen was injected onto a hot surface near TDC to achieve a diesel-like combustion. In such a system combustion took place while hydrogen was being injected in a turbo-charged engine which showed good performance characteristics. They were successful also in reducing the effect of noise and vibration in these experiments. Furuhama and his team also experimentally evaluated that the volumetric efficiency of a hydrogen engine could be increased by 15% over that of a gasoline fuelled engine, with the liquid hydrogen supplied to the intake manifold. This figure indicates that power output and volumetric efficiency can be still further increased by injecting hydrogen directly into the cylinder.

Fig. 5. Cam-actuated timed manifold injection engine configuration.

Murray and Schoeppel [28–30] developed hydrogen injection techniques in small single-cylinder industrial engines. For higher power output a relatively high pressure of 66 atm was required. Their work also showed that knocking combustion was more prevalent when significant amounts of hydrogen entered the cylinder prior to ignition by spark. Theirs is the first reported work on $(NO)_x$ emissions from hydrogen engines.

Marotono and Dini [31] chose a two-stroke Piaggio engine of 200 c.c. to carry out hydrogen operation mainly with a view to achieve better performance characteristics. The three parameters such as quantity of injected hydrogen, total injection timing and injection timing before TDC were found to be critical in determining the performance characteristics of the engine. Their investigation showed that a good mixing obtained with suitable diffusers and injection advances were extremely important to achieve excellent engine performance characteristics.

COMPRESSION IGNITION ENGINE

Ikegami *et al.* [32] investigated hydrogen combustion in a conventional swirl chamber type diesel engine. It has been reported by these investigators that hydrogen-fuelled diesel combustion could be achieved to a limited extent because of the auto-ignition characteristics of the

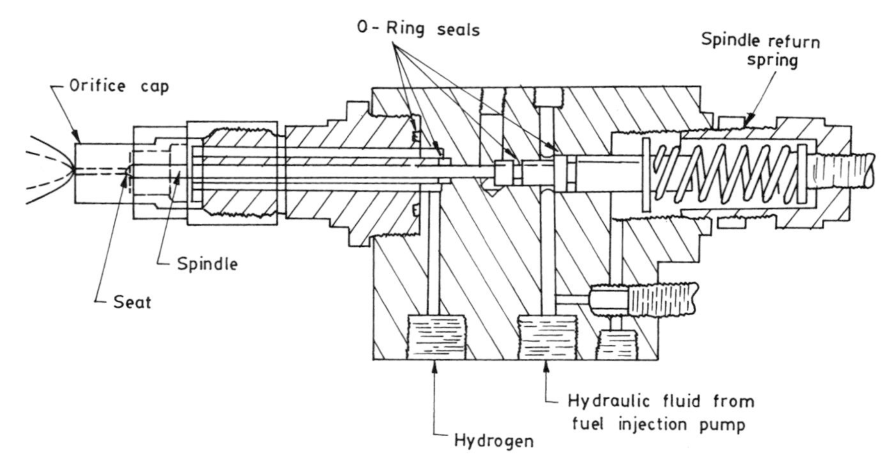

Fig. 6. The hydrogen injector.

fuel. An interesting observation made in this work was that once the swirl chamber was vitiated either by one small leakage or by a pilot injection, smooth combustion could be attained. A pilot injection ensures ignition and also reduces the ignition delay to some extent. A small leakage from the injector most often exhibited similar effects. Sometimes pronounced improved effects have been observed on the ignition. It has found that a definite amount of leakage, once established, permitted the engine to run without any symptoms of knocking over a fairly wide range of operation. However, an excessive introduction of the preliminary fuel may cause auto-ignition by itself thus giving rise to rough combustion. In this work, the conditions required to ensure smooth burning have been thoroughly studied by varying the amount and time of pre-injections and the quantity of fuel leakage.

In another attempt [25] a closed cycle engine system was simulated by supplying a 21% oxygen mixture to the test engine. The engine operation was observed to be extremely satisfactory without any ignition aid. The engine system operating with an oxygen–argon charge exhibited substantial gain in indicated thermal efficiency. Figure 7 shows the hydrogen gas injector.

Tebelis and Krepec [33] developed a gas injection system based on the principle of microprocessor control. The opening time of such a system was controlled by solenoid actuation. However, while designing a gas injection system, it must be borne in mind that a basic difference does exist in the injection characteristics of a conventional liquid fuel with that of a gaseous fuel. The bulk modulus of elasticity for liquids is relatively higher in comparison to that of gases. Hence it is not possible to utilize the conventional diesel injection system for a gaseous fuel application. Therefore, many investigators had used a hydraulically operated injection system. The injection system developed at I.I.T. Delhi [17] could either be hydraulically operated or cam-actuated. Tebelis and Krepec proposed three different configurations of an electronic injection system with different control flexibil-

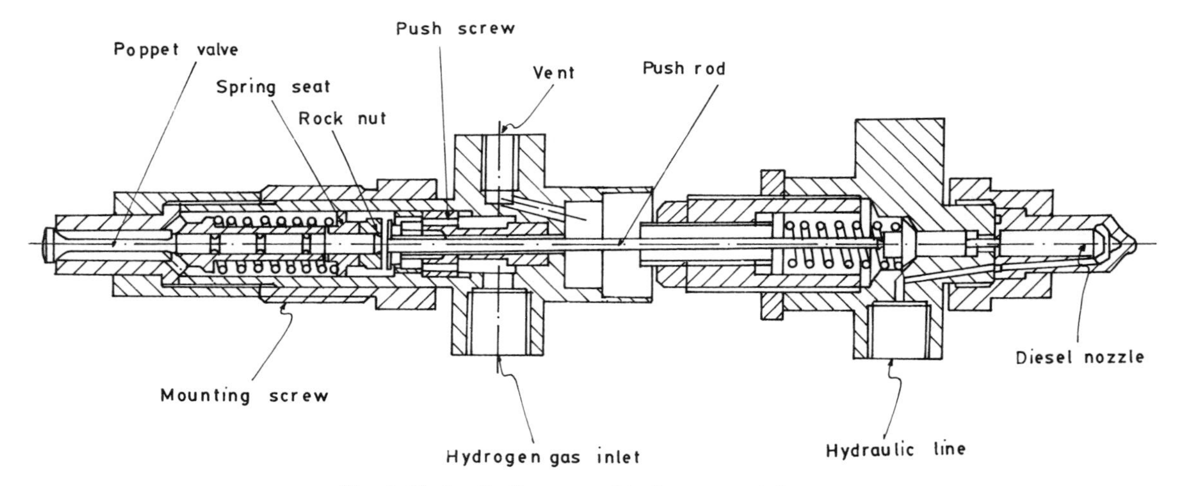

Fig. 7. Hydraulically actuated hydrogen gas injector.

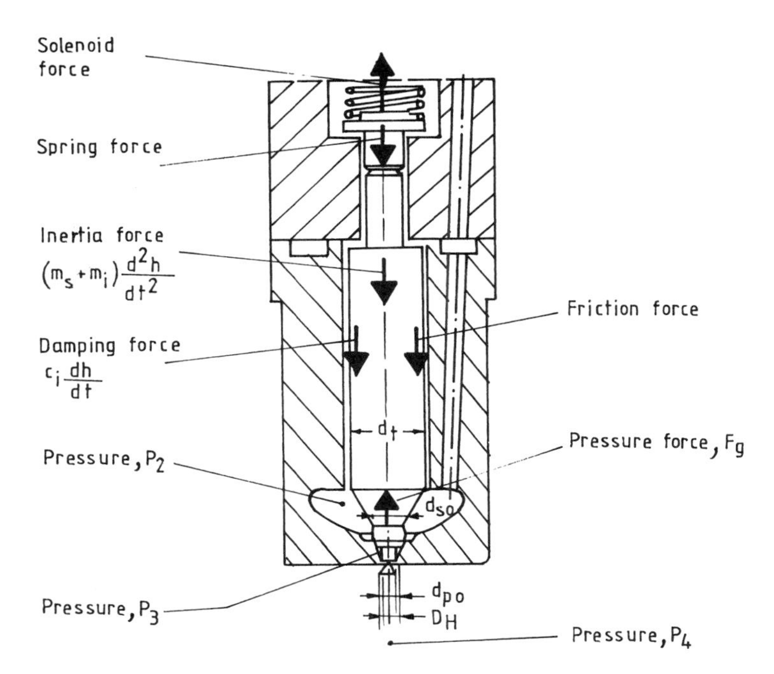

d_t = Needle core diameter
d_{po} = Pintle diameter
d_{so} = Needle seat diameter
D_H = Nozzle hole diameter

Fig. 8. Schematic of a pintle nozzle injector with forces acting on the needle.

ity. The pintle nozzle type of injector was used in all three cases. The three configurations of the injection system were injector only system (IOS), metering valve-injector system (MVIS) and control valve-injector system (CVIS). The schematic diagrams of these configurations are shown in Fig. 8. They have also developed a mathematical model to simulate the dynamic response of the system. The calculated needle movement very closely agreed with the values obtained practically through experiments. However, these investigators have prescribed further work in reducing the mass of moving parts, decreasing the volume and minimizing the fuel leakage.

CONCLUSIONS

The consistent research efforts and their outcome clearly show that a mixture formation method plays a decisive role in the practical emergence of a future hydrogen specific engine.

Future developments of such engines depend a lot also on the mode of storage and supply system. Using cryogenic hydrogen supplied from a liquid hydrogen tank has the prospects of increase in volumetric efficiency and thus the power output. It also reduces the specific fuel consumption as well as the level of $(NO)_x$ emissions. The limits of backfire are further lowered.

Late fuel injection, on the other hand, is a very promising fuel induction technique as it does preclude the possibility of backfire, the century-old problem which has been bothering the hydrogen researchers. This technique could also be adopted to both two-stroke as well as four stroke engines.

An appropriate TMI system designed specifically on the basis of hydrogen's combustion characteristics for a particular engine configuration ensures smooth engine operational characteristics without any undesirable combustion phenomena. However, all those characteristics have been evaluated in converted engines. So, an integrated fuel induction and storage method must be designed for a hydrogen-specific engine which can embrace the benefits of good performance, least exhaust emission and controlled combustion characteristics of an ideal engine system.

REFERENCES

1. J. Breton, Ann. Office Natl. Combustible Liquids, **11**, 487 Theses Faculte des Sciences, Univ. Nancy.
2. S. Wendlandt, *Physik chem.* **110**, 637 (1924).
3. H. R. Ricardo, Further note on fuel research. *Proc. Inst. Auto Engrs*, Vol. XVIII, Part 1, pp. 327–341 (1924).
4. R. O. King, S. V. Hays, A. B. Allan, R. W. P Anderson and E. J. Wacker, The hydrogen engine: combustion knock and related flame velocity. *Trans. Engng, Inst. Canada (EIC)*, Vol. 2, No. 4 (1958).
5. R. O. King and M. Rand, The Oxidation, decomposition, ignition and detonation of fuel vapours and gases—XXVII. The hydrogen engine. *Can. J. Technol.* **33** (1955).
6. R. O. King, W. A. Wallace and B. Mahapatra, The oxidation, ignition and detonation of fuel vapours and gases—V. The hydrogen engine and detonation of the gas by the igniting effect of carbon nuclei formed by pyrolysis of lubricating oil vapour. *Can. J. Technol.* **34**, 00 (1957).
7. M. R. Swain and R. R. Adt, The hydrogen-air fueled automobile. *Proc. 7th Intersociety Energy Conversion Engineering Conf.*, p 1382 (1972).

8. M. R. Swain, John M. Pappas, Robert R. Adt Jr and William J. D. Escher, Hydrogen-fuelled Automotive Engine Experimental Testing to provide an initial Design Data-Base. SAE 810350 (1981).
9. F. E. Lynch, Parallel induction: A simple fuel control method for hydrogen engines. *Int. J. Hydrogen Energy* **8**, 721 (1983).
10. J. Bindon, J. Hind, J. Simmons, P. Mahlknecht and C. Williams, The development of a lean-burning carburettor for a hydrogen-powered vehicle. *Int. J. Hydrogen Energy* **10**, 297 (1985).
11. K. S. Varde and G. A. Frame, A study of combustion and engine performance using electronic hydrogen fuel injection. Paper presented at the *World Hydrogen Energy Conf.*—IV, Pasadena, California (1982).
12. K. S. Varde and G. A. Frame, Development of a high pressure hydrogen injection for S.I. engines and results of engine behaviour. *5th World Hydrogen Energy Conf.*, Toronto, Canada (1984).
13. C. A. MacCarley and W. D. Van Vorst, Electronic fuel injection techniques for hydrogen-powered engines. *Int. J. Hydrogen Energy* **5**, 179 (1980).
14. R. L. Wolley and D. L. Henriksen, Water induction in hydrogen powered IC engines. *Hydrogen Economy Miami Energy* (THEME) Conference, Miami Beach, Florida, U.S.A. (1974).
15. R. Povel, J. Topler, G. Withalm and C. Halene, Hydrogen drive in field testing. *Proc. 5th World Hydrogen Energy Conf.*, Toronto, Canada (1984).
16. H. C. Watson, E. E. Milkins, W. R. B. Martin and J. Edsell, An Australian hydrogen car. *Proc. 5th Hydrogen Energy Conf.*, Toronto, Canada (1984).
17. L. M. Das, Studies on Timed Manifold Injection in Hydrogen operated Spark Ignition Engine: performance combustion and Exhaust Emission Characteristics. PhD Thesis, IIT, Delhi (1987).
18. R. A. Erren and W. H. Campbell, Hydrogen: a commercial fuel for internal combustion engines and other purposes. *J. Inst. Fuel* **6**, 277 (1933).
19. M. Oehmichen, Wasserstoff als Motortreibmittel. *Deutsche Kraftfahrt-forschung*, Heft 68. VDI-Verlag GmbH, Berlin (1942).
20. H. S. Homan, An Experimental Study of Reciprocating Internal Combustion Engines Operated on Hydrogen. PhD Thesis, Cornell University (1978).
21. K. Suzuki, Y. Uchiyama and J. Hama, Research of hydrogen fueled spark ignition engine. *Proc. 3rd World Hydrogen Energy Conf.*, Tokyo, Japan (1980).
22. S. Furuhama and Yoshiyuki Kobayashi, Development of a hot-surface-ignition hydrogen injection two-stroke engine, *Proc. 4th World Hydrogen Energy Conf.*, Vol 3, p. 1009, Pasadena, California (1982).
23. S. Furuhama and Fukuma Takao, High output power hydrogen engine with high pressure fuel injection, hot surface ignition and turbocharging. *Hydrogen Energy Progr.* V, Pergamon Press (1984).
24. Y. Kobayashi, S. Furuhama, M. Iida and Y. Enomoto, LH_2 Car with a two-stroke direct injection engine and LH_2 pump. *Hydrogen Energy Progress* (1980).
25. S. Furuhama, M. Hiruma and Y. Enomoto, Development of a liquid hydrogen car. *Int. J. Hydrogen Energy* **3**, 61 (1978).
26. S. Furuhama and H. Azuma, Hydrogen injection two-stroke spark ignition engine. *Proc. 2nd World Hydrogen Energy Conf.* (1978).
27. S. Furuhama, Y. Kobayashi and M. Iida, A LH_2 Engine Fuel System on Board Cold GH_2 Injection into Two-Stroke Engine with LH_2-pump. ASME Paper No. 81-HT-81 (1981).
28. R. G. Murray and Schoeppel, Emission and performance characteristics of an air-breathing hydrogen-fuelled internal combustion engine. *Proc. Sixth Inter-society Energy Conversion Engng. Conf.* (IECEC), Paper No. 719009 (1971).
29. R. G. Murray, R. J. Schoeppel and C. L. Gray, The hydrogen engine in perspective. *Proc. Seventh IECEC*, San Diego, California, Paper 729216 (1972).
30. R. J. Schoeppel, Design Criteria for Hydrogen Burning Engines. Final Report, Environmental Protection Agency Contract No. EHS 70–103 (1971).
31. L. Martorano and D. Dini, Hydrogen injection in two-stroke reciprocating gas engines. *Int. J. Hydrogen Energy* **8**, 935 (1983).
32. M. Ikegami, K. Miwa and M. Shioji, A Study of hydrogen fuelled compression ignition engines. *Int. J. Hydrogen Energy* **7**, 341 (1982).
33. T. Tebelis and T. Krepec, A concept of electronically controlled hydrogen-gas injector for high speed compression ignition engine, *Proc. Hydrogen from Renewable Energy*, Cape Canaveral, pp. 397–405 (1985).

Trend of Social Requirements and Technological Development of Hydrogen-Fueled Automobiles

Shoichi Furuhama
Musashi Institute of Technology

1. Introduction

Automobiles have become an indispensable means for transportation of both people and goods and will continue to increase their important roles in keeping with the expansion of the world economy. However, there are some issues which may force the world to reverse this trend, such as the potential difficulty in obtaining a stable oil supply and the increase of exhaust matter from oil combustion causing urban pollution and damaging the global ecology.

In order to cope with this situation there have been many studies, developments and trials of various types of automibile engines and their peripheral issues but among them only hydrogen-fueled vehicles seems to be a practical means from the current scientific and technological points of view. Although various types of systems have been studied for hydrogen-fueled engines for automobiles, they tend to converge on hydrogen injection systems.

2. Hydrogen as clean energy for automobile use

Table 1 shows a comparison of the amount of carbon dioxide in the exhaust gas of various types of fuel when they are completely burned off for the same amount of energy. Light oil, gasoline and methanol develop almost same amount of carbon dioxide. The amounts for coal and natural gas are 1.5 times and 0.8 times of that of gasoline respectively, and it can be safely concluded that there is no significant difference among these fuels, whilst hydrogen develops zero carbon dioxide. Therefore, it should be the best to use hydrogen obtained from water by solar energy or some other natural energy. Since there is no established major production process for fuel hydrogen, it is too early to assess its cost. At the moment the hydrogen available to Japanese researchers is very costly but it is reported that there is no great difference between costs of liquid hydrogen in the United States and gasoline in Japan respectively for the same amount of energy.

Fuel	Molecular formula	Calorific value	CO_2	Comparison with gasoline
Coal	C	8100	3.54	1.56
Diesel fuel	$C_{16}H_{34}$	10590	2.28	1.01
Gasoline	C_8H_{18}	10630	2.27	1.00
Methanol	CH_3OH	4770	2.26	0.99
Natural gas	CH_4	11930	1.80	0.79
Hydrogen	H_2	28700	0	0

Table 1 CO_2 [kg/(Q)] as a combustion product of various fuels
Q = heat value of 1 liter of gasoline

The next issue to study is a comparison of hydrogen-

Shoichi FURUHAMA
Prof. and Rector
Musashi Institute of Technology

CURRICULUM VITAE

Shoichi Furuhama has long been studying phenomena concerning lubrication for pistons and piston rings, and the results of his studies and the measuring techniques invented by him have been widely utilized, contributing greatly to internal combustion engineering and automobiles. He was also among the first to notice the potential of hydrogen as a non-polluting automobile fuel and began the study of hydrogen-burning engines. Clarifying the phenomena of hydrogen combustion, he proposed an engine in which pressurized hydrogen was directly injected into the cylinder and ignited with an ignition plug. In addition, he started the feasibility study of hydrogen automobiles when the first energy crisis occurred, and proved the feasibility through his original research. He was awarded the Scientific Contribution Prize on October 18, 1990 at the Okinawa Macaw Hotel during the 1990 JSAE Autumn Convention.

* Received 5th October, 1991

fueled cars and other clean energy cars. First of all, solar energy cars may be the best because they use solar energy directly. The total amount of solar energy delivered onto the whole earth's surface every day is huge. At the present time, about 5 billion people are consuming an average of 2 to 3kw of energy per/person constantly in the world. However, even if 10 billion people consume 10kw each, the total solar energy supply is well over several hundred times that. However, it is difficult to utilize solar energy efficiently because its energy density is extremely low and it is only around 0.8kw/m^2 even in the broad daylight of a fine day. An automobile with 6m^2 of solar receiving area and 20% light-to-electricity conversion efficiency has optimum motor power of merely 1HP or 0.7kw.

The next candidate should be electric automobiles, but their practical applications may be very limited because the batteries used in them are prohibitivery heavy, as indicated in Table 2. Even if their weight were reduced to one tenth of the current weight, they would still weigh 5 times as much as a tank of gasoline or 2 times as much as a tank of liquid hydrogen with the same amount of energy.

Fuel tank	Content		Tank weight (kg)	Total weight (kg)
	Volume (1)	Weight (kg)		
Gasoline	30	22	5	27
Methanol	62	49	8	57
Hydrogen				
MH		8.2	764	772
HP (15MPa)	670	8.2	755	763
LH_2	115	8.2	65	73
Battery(*)				1360

(*) Assuming that the energy density is 40Wh/kg and that the efficiency of the battery's conversion to power is 5 times higher than that of the gasoline engine.

Table 2 Comparison of fuel storage weights corresponding to 30 liters of gasoline

In the case of a hydrogen engine using liquid hydrogen, the total weight of the fuel itself and the fuel tank is 2.7 times and 1.3 times that of a gasoline engine and a methanol engine respectively and we can consider this to be within the practical range. The expected output power of a hydrogen engine with an external mixing system is only 85% of that of a gasoline engine of the same stroke volume and is rather low. The major reason for this low output power is that the intake air volume is limited to only 70% that of a gasoline engine because the air to fuel ratio is 34.2:1 and 2.3:1 for weight and volume respectively. On the other hand, with an internal mixing system which takes in air only whilst hydrogen is injected into the cylinder during the compression stroke, an output as high as 120% that of a gasoline engine can be expected. Therefore, an internal-mixing hydrogen engine has basic features superior to those of a gasoline engine and seems to be the most promising engine for clean cars.

3. Fuel tank for hydrogen

The weight of hydrogen is 0.37 times of that of gasoline for the same amount of energy. However, its volume is 3,000 times that of gasoline. In order to cope with the above fact, several different transportation methods as listed below have been developed.

(I) Metal Hydride (MH)

As indicated in Table 2 the total weight of the fuel and the tank is too heavy. Also, it is difficult to release the fuel at a pressure high enough for injection into a combustion chamber.

(II) High Pressure Tank (HP)

This is convenient for injection but when the fuel remaining in the tank drops below the injection pressure, it cannot be consumed. As in the case of MH, the weight of the system is too heavy. Studies on development of lightweight materials and an ultra-high-pressure storage method are being carried out.

(III) Liquid Hydrogen (LH_2)

This system is light in weight and a high-pressure pump can easily provide the pressure necessary for injection. However, at present, it is too costly to liquefy hydrogen and further improvement is imperative in this area as well as in low temperature technology with regard to heat insulating tanks and others.

(IV) Low-temperature HP tank

This system was designed by Krepec et al. of Canada[1] and is indicated in Fig.1. The fuel pressure in this system is at the same level as that of HP. The fuel temperature in this system is 100k while the LN_2 temperature is 77k, and heat insulation is much easier than for LH_2. The capacity of the low-temperatue tank is three times of that of a normal-temperature HP tank. In case of hydrogen injection at 10MPa pressure and 300k temperature from the tank of 20MPa and 300k interior state, half of the contents of the tank remain and cannot be used, while, the remaining fuel is reduced to 1/6 in case of 100k initial temperature in

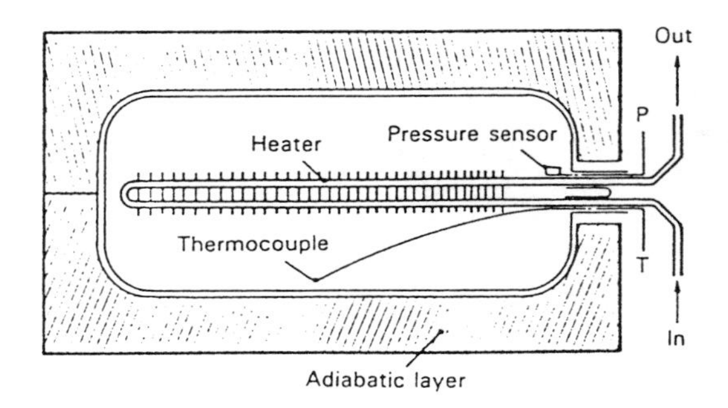

Fig.1 High-pressure and low-temperature H_2 tank with temperature control unit

the tank. However, this requires precise control of temperature and pressure matching, and still the weight is too heavy.

Considering the above, it can be safely assumed that the LH_2 system is the only one with potential for hydrogen-fueled transportation. In order to make the system practical it is necessary to develop a more compact fuel tank with around 1% a day of blow-off loss due to heat inflow during engine shutdowns. However, the hydrogen gas blow-off can be used of supplying it into the intake manifold when the engine is operating.

4. External mixing system and backfire

The lean limit of hydrogen for combustion is 0.135 (it is around 0.2 even in an actual engine), while it is 0.7 for gasoline, so that precise control of the air-to-fuel mixing ratio by means of a carburetor is not required and an external mixing system of very simple construction as indicated in Fig.2 is widely used. With this system the theoretical output power is as low as 85% of that of a gasoline engine and virtually backfire takes place at around 50% of the level of a gasoline engine and places a limit on the output power.

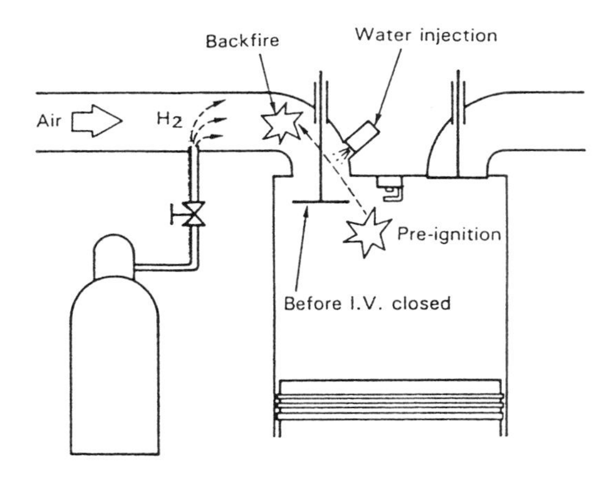

Fig.2 External mixing H_2 engine with backfire

On the basis of the experiments of the author of this article et al.[2], backfire can be defined as a phenomenon whereby pre-ignition advances at every cycle and finally it occurs before the intake valve is closed, so that the flame spreads to the hydrogen mixed with air in the intake manifold, through the gap of the valve, causing an explosion. The author et al. assumed that the ignition plug was the cause of the preignition noted above but King et al.[3] have put forward the theory that red-hot soot from oil is the cause. Backfire tends to take place at the time when an engine is started or accelerated but the actual causal factor has not yet confirmed.

Various countermeasures have been proposed in order to prevent it but none of them is sufficiently reliable. The Benz[4] claims that water injection has good effect but the amount of water required for the system is rather large and deterioration of performance and of the life of the engine can be anticipated with this method.

5. Internal mixing

The internal mixing system in which air only is taken in first and then hydrogen is injected into the cylinder, never has any backfire and a kind of supercharging effect provides 120% of the output power of a gasoline engine, while the external mixing system provides only 50%.

As to injection timing there will be two choices as described below:

(I) Low-pressure injection: Hydrogen is injected upon closing of the intake valve for rather a long period at a low pressure of around 1MPa.

(II) High-pressure injection: Fuel is injected when the piston is around TDC for a short period at a pressure higher than 8MPa, as in the case of a diesel engine.

The study of Furuhama et al.[5] conducted on two-stroke engines indicates, as shown in Fig.3, that the maximum output power available from the low-pressure injection system with normal-temperature hydrogen is around the same level as that of gasoline engines, but output power is restricted by pre-ignition. Injection of low-temperature hydrogen of 0—50°C could suppressed pre-ignition and provided the theoretical maximum power. However, further experiments revealed that the suppression effect was not so signifi-

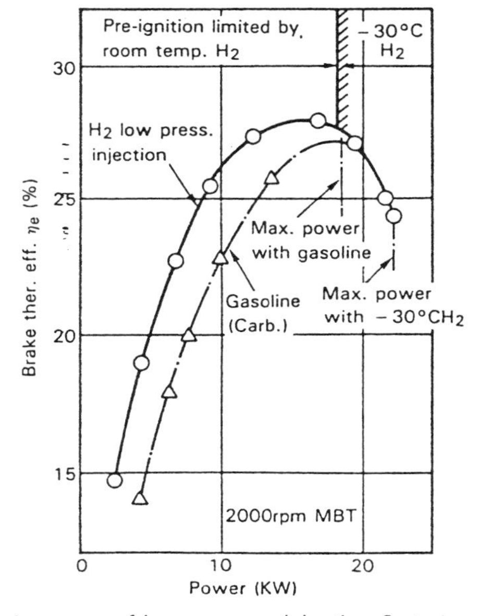

Fig.3 Performance of low-pressure injection 2-stroke engine and effect of cold H_2 injection (loss of fuel bypass eliminated)

cant for 4-stroke engines. Under partial load, the hydrogen engine can be operated with super-lean combustion and high thermal efficiency can be achieved.

Since the self-ignition point of hydrogen is as high as 580°C, compression alone will not cause any ignition in the high-pressure injection system even if the compression ratio is high. Therefore, ignition by a hot surface or a spark is needed. Also, since the system is using gas fuel, a special means is required for control of operation of the injection valve. With the high-pressure injection system there is no problem concerning startability at low temperature and the most ideal compression ratio can be selected for maximum performances, low maximam pressure and minimum NOx [6] production. Since the system cuases neither pre-ignition nor knocking, it can be applied to any size of engine, large or small.

6. Examples of the LH_2 pump high-pressure injection method

6.1. System

The study referred to in section 3 revealed that the LH_2 would only be a practical system when the tank weight was redeced. In order to obtain high output power, it is most logical to obtain the high pressure required for the injection by means of a liquid hydrogen pump, because liquid hydrogen is of much higher density than gaseous hydrogen and is incompressible so that even a small pump of low driving power can provide the pressure. Therefore, for hydrogen-fueled engine systems for automobiles a combination of a liquid hydrogen tank, a liquid hydrogen pump, high-pressure hydrogen gas injection equipment and spark ignition equipment is considered most ideal. Each of above components will be described in the following section.

6.2. Fuel system

Figure 4 shows an example of a fuel system mounted in a rear truck. The space between the stainless steel double walls of the liquid hydrogen tank is evacuated down to 10^{-7}Torr for adiabatic reasons and the outside of the inner wall is lined with paper to minimize heat radiation (this construction is called super-insulation). The tank capacity is 82 liters (equivalent to 21 liters of gasoline) and the tank weights 42kg.

A liquid hydrogen pump is inserted in the tank and this reciprocating pump is driven by a DC motor through a crank shaft mounted on the top of the tank. The motor rotation speed is computer controlled in order to match the amount of discharge rate of the pump to the hydrogen consumption of the engine by keeping the surge tank pressure constant at a predetermined level. Also, in order to eliminate any effect of the ambient temperature on the injection amount, the temperature in the tank is kept at the atmospheric level. The liquid hydrogen pump involves two inherent problems. One is that at temperatures below −253°C, all gases except helium become solid and even air changes its form to become granular. Therefore, the friction surfaces of the piston and cylinder must be made of a synthetic material having self-lubricating properties. The second problem is that synthetic material has a much larger coefficient of thermal expansion of above synthetic material than that of metal, while it is desirable to make the cylinder with metal in order to fasten it strongly to the outer part. The leakage prevention devices such as piston rings receive hydraulic pressure on their back surface and they generate high frictional heat so that the

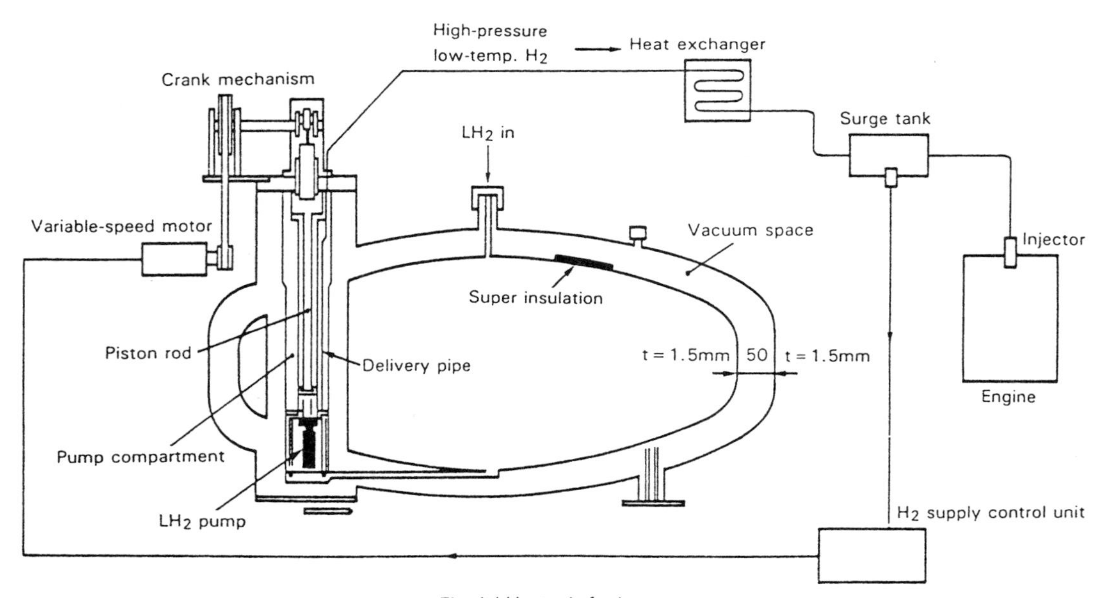

Fig.4 LH_2 tank fuel system

liquid hydrogen taken in is easily vaporized. Therefore, leakage of high-pressure liquid hydrogen must be prevented with a gap of 1.5–2 micrometers between the cylinder and the piston.

In order to cope with the severe conditions described above, the thermal shrinkage of the inner surface of the stainless steel cylinder must match that of the outer surface of the piston. Therefore, Furuhama et al.[7] have developed a new pump of unique construction in which a piece of invar alloy is incorporated in the synthetic material collar of the piston's outer part for correction of the shrinkage. The D×S of above pump is set at 15×15mm. The pump provides enough hydrogen for engine output power of 100HP or 75kw at 800rpm pump revolution rate. Peschka of the DLR (German Space and Aviation Laboratory)[8] recently developed a high-pressure liquid hydrogen pump which used almost the same principle as above.

A new high-pressure hydrogen injector recently developed by Furuhama et al. is indicated in Fig.5. High-pressure oil delivered from a high-pressure diesel injection pump, *P*, is fed to a hydraulic oil inlet (1) and then the hydraulic pressure (2) thrusts a piston (4) and opens a valve (6) to start injection. Since the high-pressure oil leaks through a small relief valve (3), valve (6) is automatically closed by the high-pressure hydrogen (5) when oil feed to (1) ceases. Therefore, the injection time period can be kept within the range of that of the diesel injection. Also, with this construction, the need for a strong coil spring is eliminated and operation of each injector can be easily controlled to the same movement.

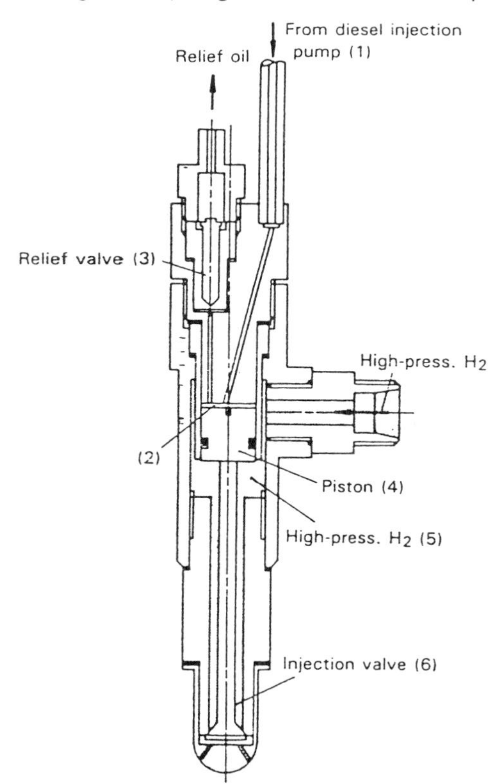

Fig.5 High-pressure H_2 injector without coil spring

6.3. Ignition and combustion

(1) Hot surface and spark ignition

When a hydrogen gas jet is ignited by directing it onto a high-temperature surface, ignition timing is determined by the injection timing and it is rather simple to control. A study of Schlieren high-speed pictures in a constant volume container[9] revealed that over 900°C surface temperature is required to obtain the ignition indicated in Fig.6. However, from the viewpoint of durability it is desirable to keep the surface temperature below 800°C. Fuel consumption for the electric power to maintain the required surface temperature is not negligible and also it is known that a large, heavy battery must be used for this purpose.

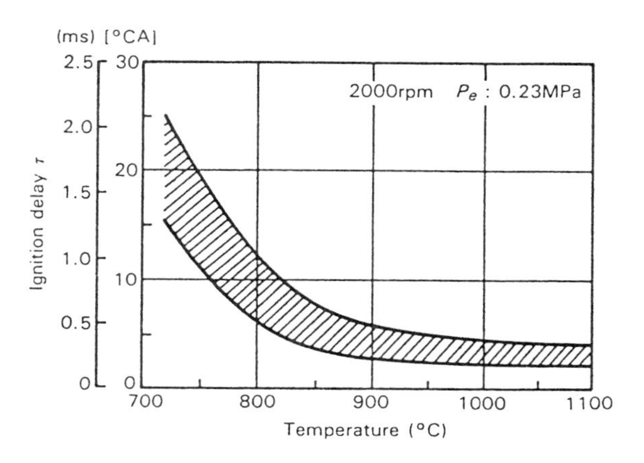

Fig.6 Effect of hot surface temperature on ignition delay τ.

On the other hand, the electric spark method requires precise matching of two factors, injection timing and ignition timing. However, all other potential problems noted above can be resolved in this method, and thus spark ignition seems to be the superior method at the moment.

(2) Ignition timing

Because the hydrogen jet density is much lower than that of air in the combustion chamber, the speed of the hydrogen jet is rather low. This fact is illustrated by the Schlieren pictures of Fig.7. Therefore, some tricks are needed in order to obtain a short ignition delay τ, which will give stable ignition and smooth pressure rise. As indicated in Fig.8, τ varies following the change of τ_0, which is the time period in which a jet travels from the injection hole to the spark gap, and it is necessary to make the distance between the above two points as short as 3mm.

The other issue to be dealt with is the spark timing for each jet. Figure 9(a) represents the shortest τ, where

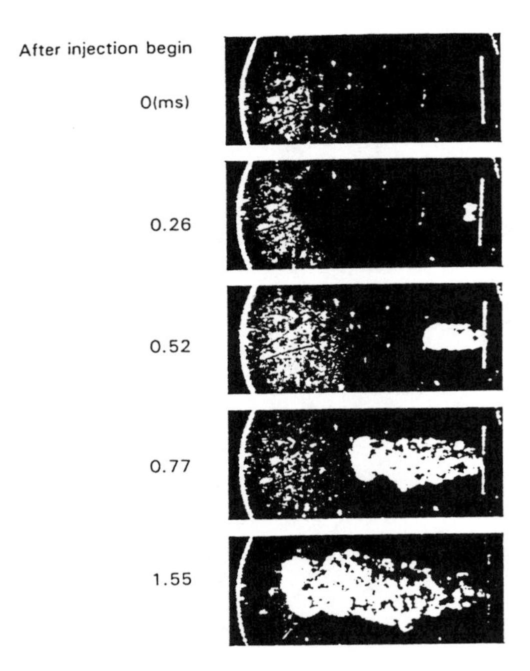

Fig.7 Schlieren photograph of H_2 jet

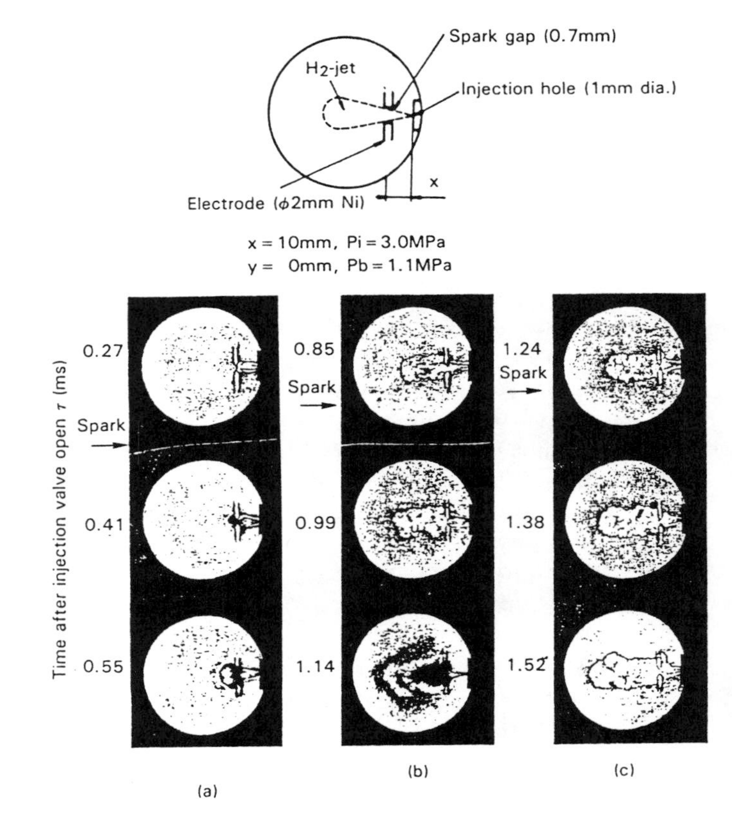

Fig.9 Ignitability of hydrogen jet

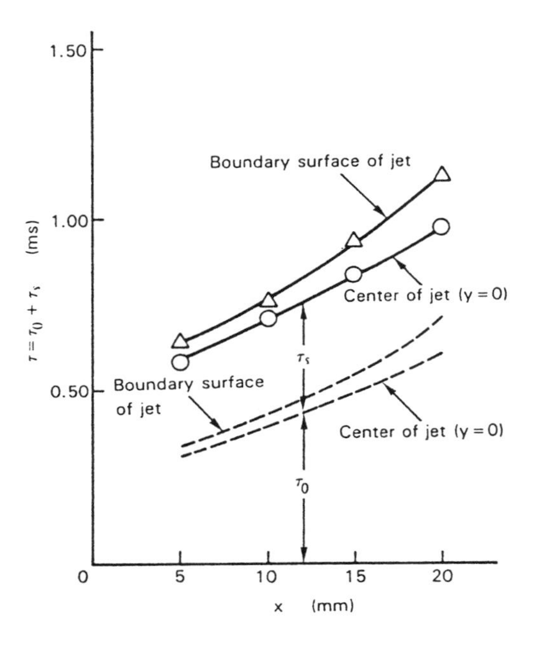

Fig.8 Ignition delay τ, arrival time to the gap, and time τ_s from arrival to ignition

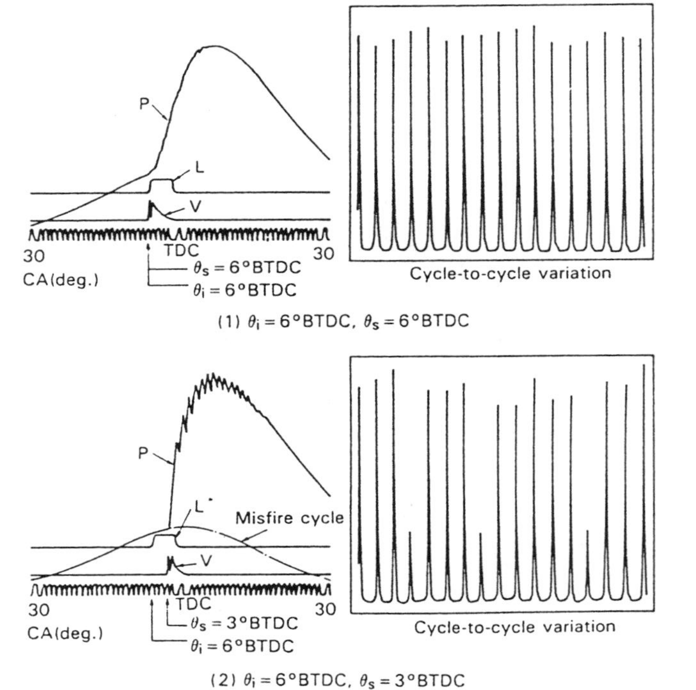

Fig.10 Indicator diagram showing effect of spark timing on the combustion (BMEP = 0.45MPa at 1000rpm)

the spark occurs at the moment when the tip of the jet arrives at the spark gap, while Fig.9(b) shows a slightly delayed spark with a slightly longer τ. When the spark is further delayed until the center of a jet arrives at the spark gap as indicated in Fig.9(c), misfire may occur because of overrich mixing ratio. Figure 9(a) represents the case where the timing of injection start, θ_i, and the timing of the spark, θ_s, exactly match each other, i.e. $\theta_s = \theta_i$. Figure 10 shows experiments related to above. Figure (1) indicates the case of $\theta_s = \theta_i$ where pressure increase is very smooth and also the cycle-to-cycle variation is rather small. In other words it represents an ideal result. On the other hand Fig.(2) represents the case of delay of θ_s by 3° CA, where $dp/d\theta$ is too big, with extreme pressure vibrations and occasional misfires. Based on above experiments it can be concluded that it is desirable to keep the ignition timing $\theta_s = \theta_i$ with a tolerance of +/-0.05ms.

(3) Flame spreading

Since hydrogen is a gaseous fuel and its ignitable mixing ratio range is extremely wide, people are apt to assume that with a high-pressure injection system a flame should spread rapidly and completely in a combustion chamber. However, the reality is opposite to the above assumption and the need to increase combustion efficiency is a major problem to be resolved for this injection method. Figure 11 shows Schlieren photographs of flame propagation for time t from the beginning of injection. The hydrogen is ignited at $t = 0.65$ms but the flame is not transferred to the adjacent jet until $t = 1.56$ms, and the time lag should be regarded as actual ignition delay which causes occurrence of pressure wave. As indicated in Fig.12[10], it is known that after increase of injection pressure P_i, mixing is accelerated and combustion efficiency is improved accordingly. Therefore, the optimum compression ratio corresponding to P_i, the shape of the combustion chamber and other factors should be specially considered for the high-pressure hydrogen injection system.

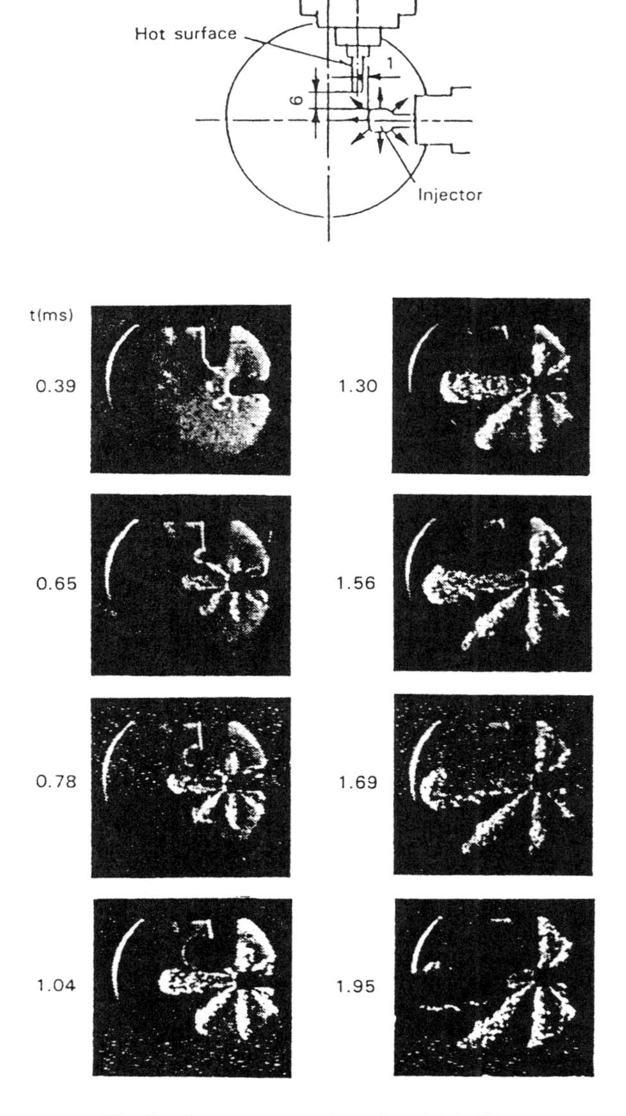

Fig.11 Flame propagation of multiple H_2 jets

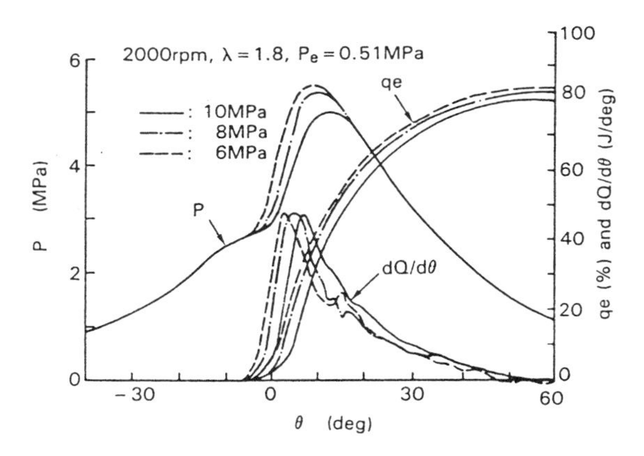

Fig.12 Effect of injection pressure on combustibility

(4) Partial premixing of the hydrogen fuel

If a small part of the fuel is supplied to the intake manifold, ignition delay is shortened greatly and combustion is improved as shown in Fig.13. However, excess air ratio for only premixing of the hydrogen λ_p under the range of 6—7 may cause abnormal combustion. This method provides the benefit of eliminating the loss of blow-off hydrogen from the liquid hydrogen tank during engine operation because it can be used as a part of the premixing fuel.

(5) Suppression of NO_x

Complete elimination of NO_x still has not been achieved to date. However, a combination of the methods listed below will certainly decrease it to a somewhat lower level.

(a) EGR: Figure 14 shows an example of experiments with a high-pressure injection engine, indicating that the NO_x reduction effect is very high even in the case of 15% EGR rate.

(b) Cooled hydrogen injection: Figure 15 shows the experimental results of low-pressure injection in a 2-stroke engine with hydrogen cooled to −80°C and it is seen that the NO_x output is reduced to one half that of normal-temperature injection. It can be assumed that the effect with high-pressure injection may be more significant. This method can be implemented only when liquid hydrogen is used.

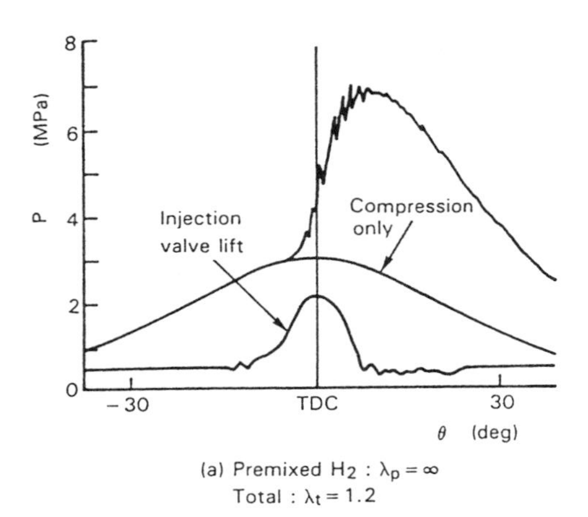

(a) Premixed H2 : $\lambda_p = \infty$
Total : $\lambda_t = 1.2$

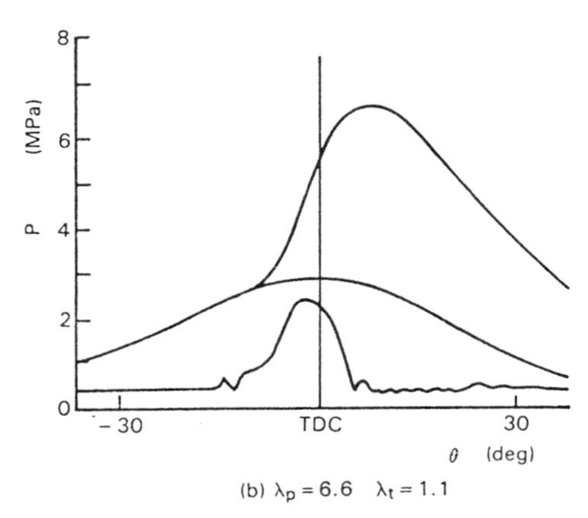

(b) $\lambda_p = 6.6$ $\lambda_t = 1.1$

Fig.13 Effect of premixing H_2 at 2000rpm, Pe = 0.74MPa

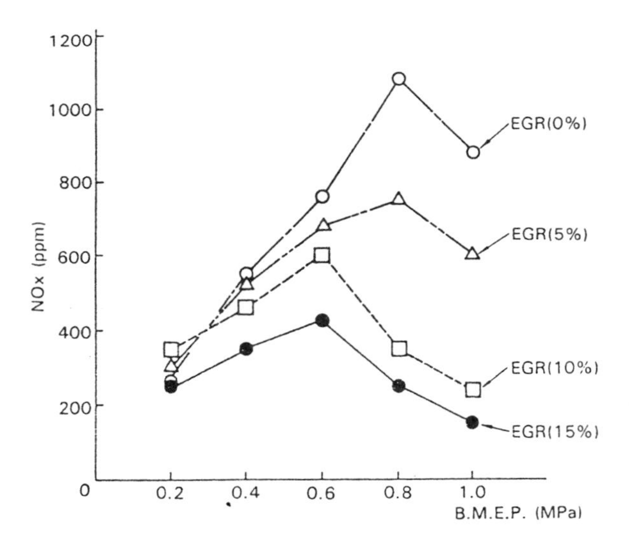

Fig.14 Effect of EGR on NOx reduction (1200rpm)

c) Decrease of NO_x emission also can be achieved to certain degree by changing the combustion mode by some means such as injection timing delay, minimizing compression ratio while maintaining the original engine performances, or partial fuel premixing.

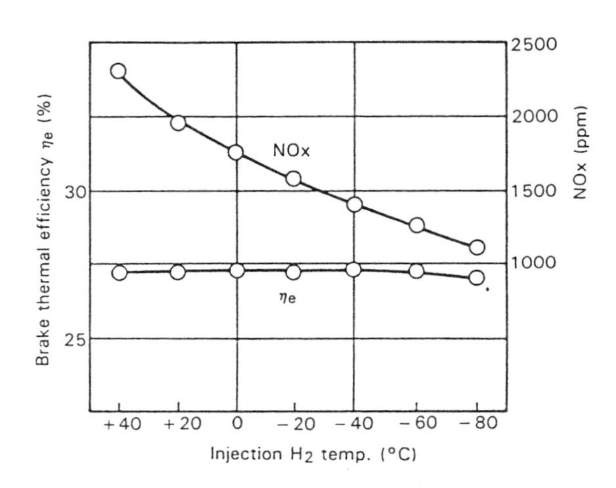

Fig.15 Effect of injection H_2 temperature on NOx emission (low-pressure injection 2-stroke engine)

7. Musashi-8

The Musashi Institute of Technology started its study of hydrogen engines in 1970 and it developed its 8th hydrogen-fueled automobile for the WHEC-8 held in 1990 in Hawaii. A 3-liter 4-cylinder diesel engine was modified for a high-pressure hydrogen injection spark ignition system, and it was installed in a Fairlady sportscar with the courtesy of Nissan Motor Co.,Ltd. Figure 16 shows the Musashi-8 cruising on a test course for experimental purposes, while Fig.17 shows all of its systems. The major points of the vehicle are summarized below.

(1) Liquid hydrogen tank: Capacity — 100 liters (equivalent to 26 liters of gasoline). Weight — 60kgs.
(2) The liquid hydrogen pump and the injection valves are the same as those described in section 6.2 but the P_i has been improved to 10MPa in this model from the 8MPa of the Musashi-7.

Fig.16 View of Musashi-8 running on Nissan Test Course

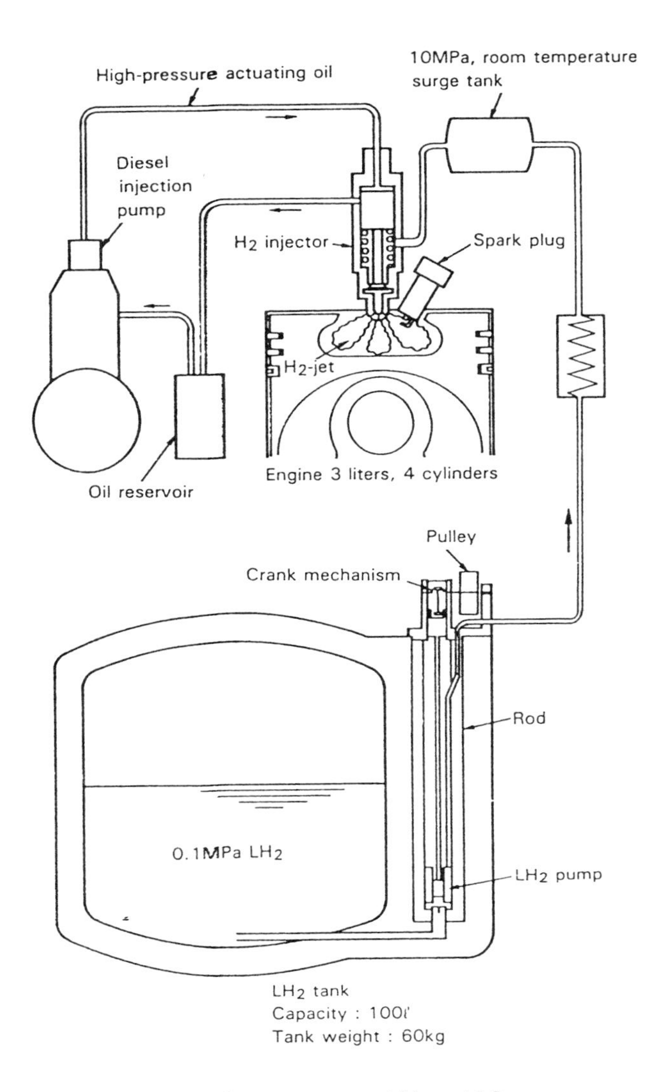

Fig.17 Power system of Musashi-8

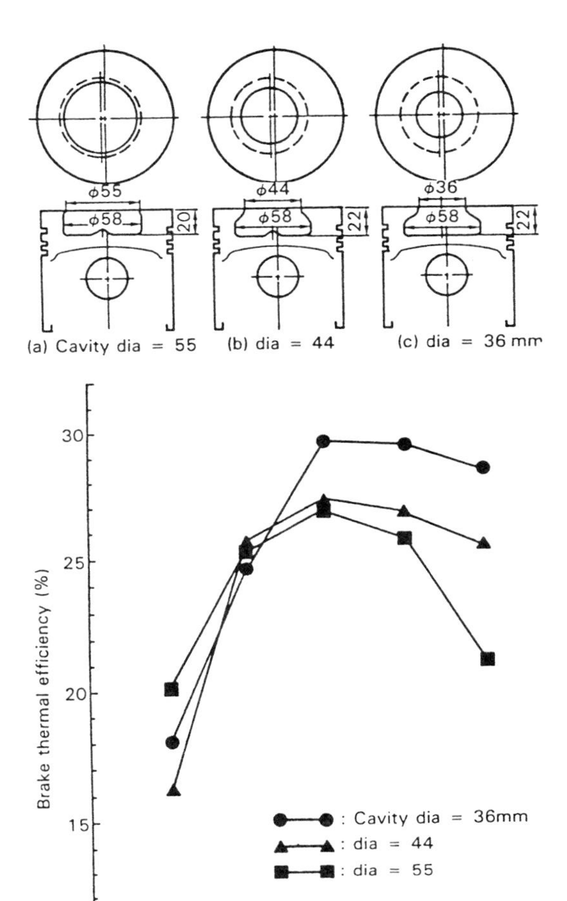

Fig.18 The effect of cavity diameter on thermal efficiency

(3) On the basis of the experimental results indicated in Fig.18, the piston cavity inlet diameter was reduced to 36mm from the 55mm of the original diesel engine.

(4) The optimum compression ratio was 12:1 for Musashi-7[6] but with the increase of P_i the optimum level changed to 13.5:1 and this figure was adopted for the new engine.

(5) Partially premixed fuel is supplied to the intake manifold.

(6) The control mode of the injection timing was modified from adjustment of number of revolutions which was used on the original diesel engine.

(7) Introduction of spark ignition: A new electronic control system was developed to make the timing $\theta_s = \theta_i$.

With the modifications indicated above, very stable cruising at a maximum speed of 130km/h was confirmed in an experiment open to the public.

8. Conclusion

In conclusion, the above discussion can be summarized as follows:

(1) Hydrogen seems to be the best fuel for future automobiles, except for some special applications.

(2) For the utilization of hydrogen, the system uses a liquid hydrogen tank and a liquid hydrogen pump. High-pressure hydrogen injection and spark ignition appear to be the most appropriate methods.

(3) Musashi-8 has successfully confirmed the above assumptions.

(4) As described above, the basic system of a hydrogen-fueled automobile has been established and it can be put into practical production in rather short period if the social need arises. The specific time period needed for the vehicle's introduction will be highly dependent on the strength of the social demand for it.

References

(1) T. Krepec, D. Miele and C. Lisio, "New Concept of Hydrogen Fuel Storage and Supply for Automotive Application", Proc. 7th WHEC, Moscow, P.1127 (Sept. 1988)

(2) Furuhama, S. and Yamane, K., "Combustion Characteristics of Hydrogen-Fueled Spark Ignition Engines", Trans. of JSAE, No.6, P.54 (Nov. 1973)

(3) R.O. King, W.A. Wallace and B. Mahapatra, "The Oxidation, Ignition and Detonation of Fuel Vapors and Gases, V. The H_2 Engine and the Nuclear Theory of Ignition." Can. J. Research, F.26, P.264 (1948)

(4) F. Wolpers, W. Gelse and G. Witholm, "Comparative Investigation of a Hydrogen Engine with External Mixture Formation which can either be Operated with Cryogenic Hydrogen or Non-cryogenic Hydrogen and Water Injection", Proc. 7th WHEC Moscow, P.2119 (Sept. 1988)

(5) Furuhama, S., "High-Performance Hydrogen-Fueled Engines", ASME and JSME Thermal Engineering Conf. Vol.4, P.99 (1983)

(6) Furuhama, S. and Fukuma, T., "Liquid-Hydrogen-Fueled Diesel Automobile with Liquid Hydrogen Pump", Advances in Cryogenic Eng., Vol.31, P.1047 (1986)

(7) Furuhama, S., Kobayashi, Y. and Iida, M., "On-Board LH_2 Engine Fuel System", Heat Transfer D. of ASME, 20th ASME-I. Mech.E, Milwaukee, (Aug. 1981)

(8) W. Peschka, "Hydrogen Combustion in Tomorrow's Energy Technology", Proc. 6th WHEC, Vienna, P.1019 (July 1986)

(9) Fukuma, T., Fujita, T., P. Pichainarong and Furuhama, S., SAE Paper 861579

(10) P. Pichainarong, Iwata, T. and Furuhama, S., "Study of Thermodynamic Analysis in Hydrogen Injection Engines", Proc. 8th WHEC, Hawaii, P.1275 (July 1990)

Hythane - An Ultraclean Transportation Fuel

V. Raman and J. Hansel
Air Products and Chemicals, Inc.
J. Fulton and F. Lynch
Hydrogen Consultants, Inc.
D. Bruderly
Bruderly Engineering Associates

Abstract

Hydrogen has strong effects on the combustion of natural gas in internal combustion engines. In the lean burn range, hydrogen enables NOx and THC reductions by allowing leaner operation at part load conditions. Hydrogen also makes performance improvements with very lean mixtures that can be sacrificed for additional NOx reduction. Steady state natural gas engine tests were unable to resolve significant changes on adding hydrogen during stoichiometric operation with a three-way catalyst but transient vehicle tests (FTPs) consistently show strong effects.

1. INTRODUCTION

The following is a summary of several projects involving mixtures of hydrogen with natural gas carried out since 1989 by the authors in cooperation with others, acknowledged at the end of this paper. The effects of hydrogen on the combustion of hydrocarbons has been studied in bomb experiments and engines since the dawn of combustion science (e.g., reference[1]). The ten-page limit of this paper makes it infeasible to acknowledge such a large data base adequately. The combustion literature (e.g., ref.[2]) provides summaries of significant contributions toward understanding how hydrogen affects combustion of other fuels, including methane, and partial combustion products, like carbon monoxide. Several relevant engine studies during the last 20 years are discussed in reference[3]. Recent publications in this area include references[4,5,6].

[1]Burstall, A. F., "Experiments on the Behaviour (sic) of Various Fuels in a High Speed Internal Combustion Engine", Institution of Automobile Engineers, Vol. 22, (1927). This work included methane, hydrogen and mixtures of the two in "town gas".

[2] Lewis, B. and von Elbe, *Combustion, Flames and Explosions of Gases*, 2nd ed. Academic Press, New York (1961).

[3] Lynch, F.E. and Egan, G.J., "Near Term Introduction of Clean Hydrogen Vehicles via H_2-CNG Blends", Proc. 4th Canadian Hydrogen Workshop, Toronto (1989)

[4] Meyer, R. C. and Hedrick, J.C., "Advanced Gas Prime Mover Concepts", Final Report SwRI-1613, for Gas Research Inst., Cont. No. 5086-233-1442 (1990)

[5]Swain, M. R. et al., "The Effects of Hydrogen Addition on Natural Gas Engine Operation", SAE Paper No. 932775 (1993)

[6]Cattelan, A., et al, "Hythane and CNG Fueled Engine Exhaust Emission Comparison", Proc. 10th World Hydrogen Energy Conf., Miami (1994)

The literature and the information provided below support the conclusion that hydrogen has strong effects on the combustion of natural gas. The combination of hydrogen with natural gas is known as Hythane ®[7]. Most recent work in this area has focused on dilute concentrations of hydrogen, generally less than 10% by energy content[8]. However, the concept spans the entire transition from natural gas to pure hydrogen in a renewable energy future.

At low levels of penetration by hydrogen into the transportation energy marketplace, it is possible to derive "leverage" from the special properties of hydrogen. For example, if 5% of the buses in a fleet were fueled by pure hydrogen and those buses produced negligible emissions, the reduction in fleet emissions would be 5%. If the same quantity of hydrogen were spread over the whole bus fleet in a way that reduced emissions by 10%, the benefit derived from the hydrogen would be doubled-- i.e., a leverage factor of 2. The following data show that significant leverage is possible by using hydrogen as an additive to natural gas.

There are two distinct modes of operation of internal combustion engines where hydrogen is useful; 1) lean burn and 2) stoichiometric combustion with three-way catalysis. Lean burn engines operate near the limit of excess air where combustion temperatures and nitrogen oxides (NOx) emissions are relatively low. Lean burn NOx control is limited by increasing total hydrocarbon (THC) emissions and losses in fuel economy that signal the onset of erratic combustion near the lean limit. Hydrogen extends the lean limit of natural gas, thereby enabling lower NOx emissions without excessive THC. Hydrogen also improves fuel economy and engine performance near the lean limit, thus providing other forms of leverage.

Stoichiometric combustion requires precise control of the air/fuel ratio near the chemically correct proportions. This mode of combustion depends upon three-way catalysis to complete the combustion process. Residual oxygen and NOx react with CO and THC on the catalyst. The three-way catalyst (TWC) has spawned a revolution in low emission gasoline engines, largely within the last decade. While this emission control strategy is equally applicable with natural gas or Hythane, the highly precise gaseous fuel control systems that are necessary to achieve optimal emissions control are in their infancy.

The post-catalyst emissions levels observed in steady state operation with natural gas or Hythane and a TWC are so low that the effects of the hydrogen additive, if any, are below test resolution. However, under the transient operating conditions of the Federal light duty vehicle test procedure (FTP), significant differences between natural gas and Hythane emissions have been consistently observed by four EPA certified high and low altitude laboratories in four different vehicles.

[7] Hythane is a registered trademark of Hydrogen Consultants, Inc.

[8] Hydrogen content may be specified by volume, mass or energy content. Energy content is the preferred basis because it directly indicates the proportion of natural gas displaced by hydrogen. It may also be used directly in cost analyses, e.g.; 5% H_2 x H_2 \$/ MJ + 95% NG x NG \$/MJ = Hythane \$/MJ.

2. LEAN BURN

All of the data presented in this section are from a 1992-93 study by Hydrogen Consultants, Inc. (HCI) and Colorado State University (CSU) for the Department of Energy and the National Renewable Energy Laboratory (DOE/NREL). The test engine is a 5.7 liter GM V-8 with an Impco gas mixer "carburetor" and a Kenics static mixing tube with internal vanes that eliminate cylinder-to-cylinder mixture variations. Spark timing was set to the minimum advance for best torque (1% below peak torque) unless otherwise noted. All tests were conducted at 2500 rpm. Light load data were taken at 35 kPa of manifold air pressure. Wide open throttle data, at CSU's altitude, were taken at 80 kPa. Englehard contributed the palladium-based catalyst used in the study. It was aged through 18,000 miles of on-road use and dynamometer testing with natural gas, Hythane and gasoline. APCI donated the hydrogen and methane for the study and collaborated technically in the project. Colorado Department of Health contributed technical advise and assistance with the emissions instrumentation.

Figure 2-1 shows steady state brake-specific NOx vs. THC with varying amounts of hydrogen in methane at light load. As the mixture is leaned from ϕ[9] = 0.9 to $\phi = 0.7$, NOx falls rapidly at the expense of increased THC. In reducing NOx emissions to 1 gram/kW-h with methane, the THC emissions rise to about 15 grams/kW-h. As hydrogen content increases, the THC emissions fall. At the 1 gram/kW-h NOx level, the THC emissions are cut in half somewhere between 15 and 30 volume % hydrogen. The corresponding hydrogen energy content is between 5 and 10% so a 50% reduction in THC emissions represents a leverage factor between 10 and 5 respectively, e.g., 50% reduction ÷ 5% H_2 = 10.

Not apparent in Figure 2-1 are increases in engine performance that result from adding hydrogen. Figure 2-2 compares BMEP between methane and Hythane in the lean burn range. The advantage of Hythane below $\phi = 0.7$ results largely from a thermal efficiency increase. An additional advantage of the hydrogen additive is increased mixture energy density. Although hydrogen contains less energy per unit volume than methane, it produces more energy per unit of oxygen consumed in the combustion process. With $\phi > 0.7$, the former effect is dominant and methane/air mixtures contain more energy. With $\phi < 0.7$, the latter effect takes over and hydrogen/air mixtures contain more energy. The changes in BMEP that should result from 30% hydrogen by volume are indicated by the "theoretical" line in Figure 2-2.

Figure 2-3 shows what happens to NOx when the BMEP advantage of Hythane is sacrificed by retarding the spark advance until methane and Hythane produce equal BMEP. The same effect is illustrated in Figure 2-4, along with the corresponding BMEP[10] data. Figure 2-4 shows that a good oxidation catalyst

[9] ϕ = is the fuel/air equivalence ratio, relative to chemically correct or "stoichiometric" proportions. $\phi < 1$ is said to be "lean" combustion, $\phi > 1$ is said to be "rich" combustion.

[10] BMEP is "brake mean effective pressure", a measure of engine torque that is normalized to eliminate the effect of engine size. A small single-cylinder engine and a large V-8 can be equitably compared in terms of BMEP but not in terms of torque.

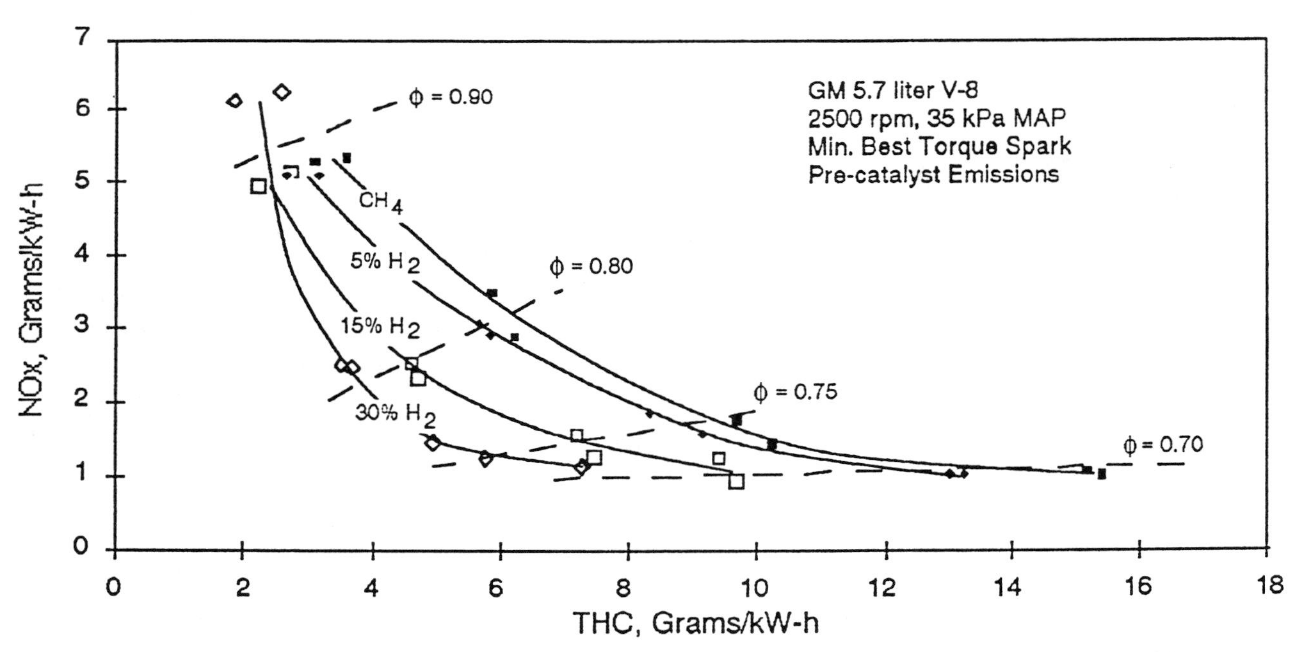

Figure 2-1. At any given NOx level, THC emissions are significantly less with Hythane. The dashed lines indicate the equivalence ratio corresponding to each set of points.

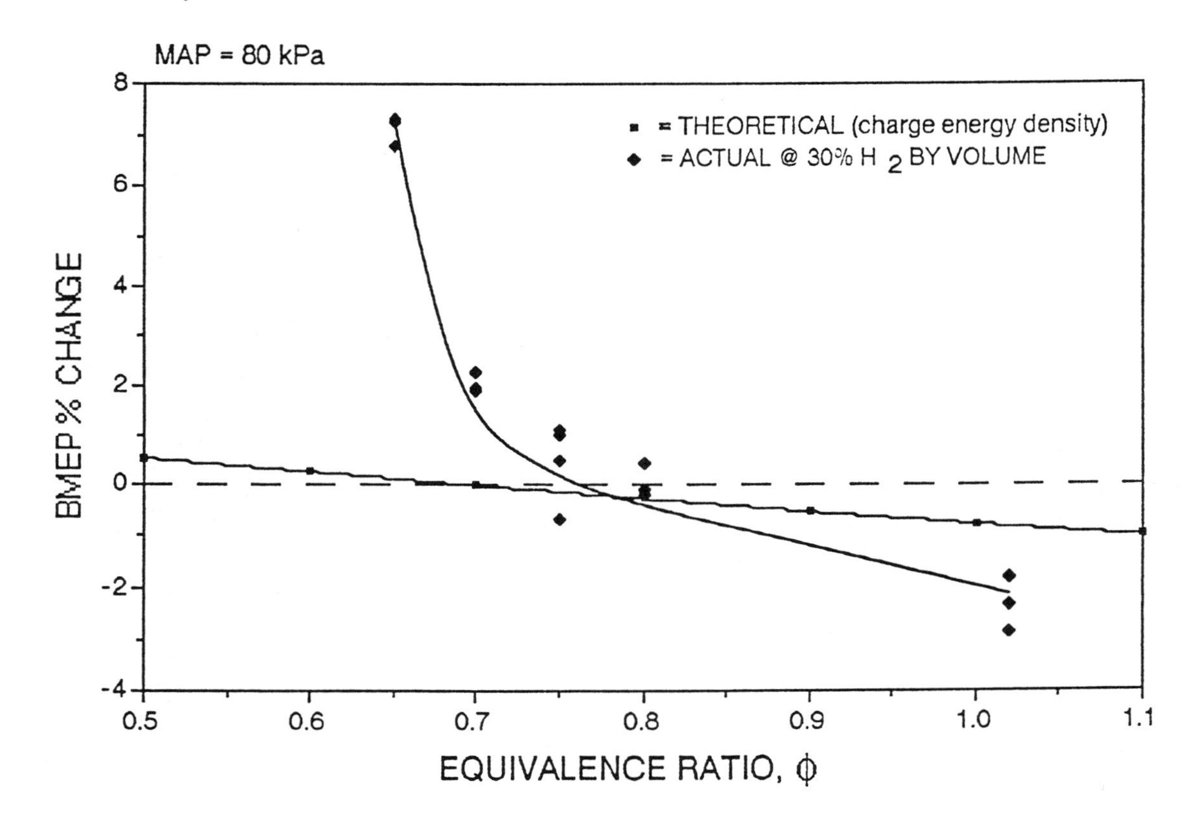

Figure 2-2. The actual improvement in BMEP with 30% H_2 by volume (10% by energy) far exceeds the small increase in charge energy density.

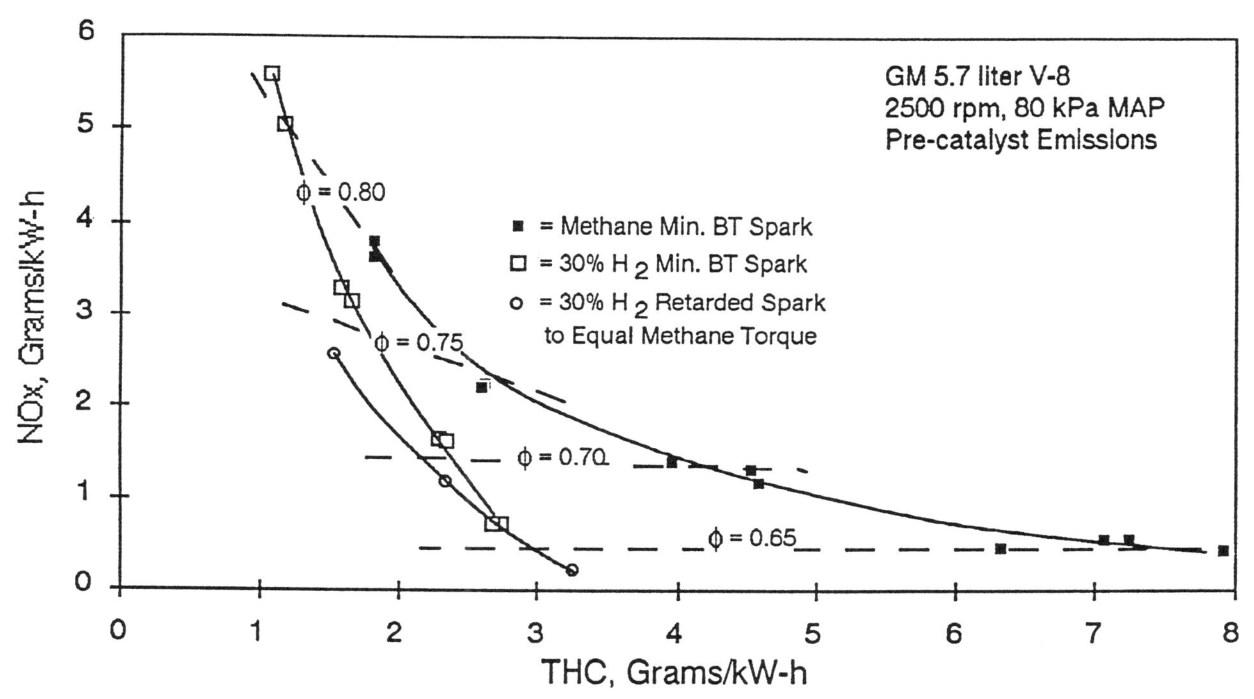

Figure 2-3. At wide open throttle (MAP = 80 kPa), the thermal efficiency advantage of Hythane may be sacrificed to reduce NOx.

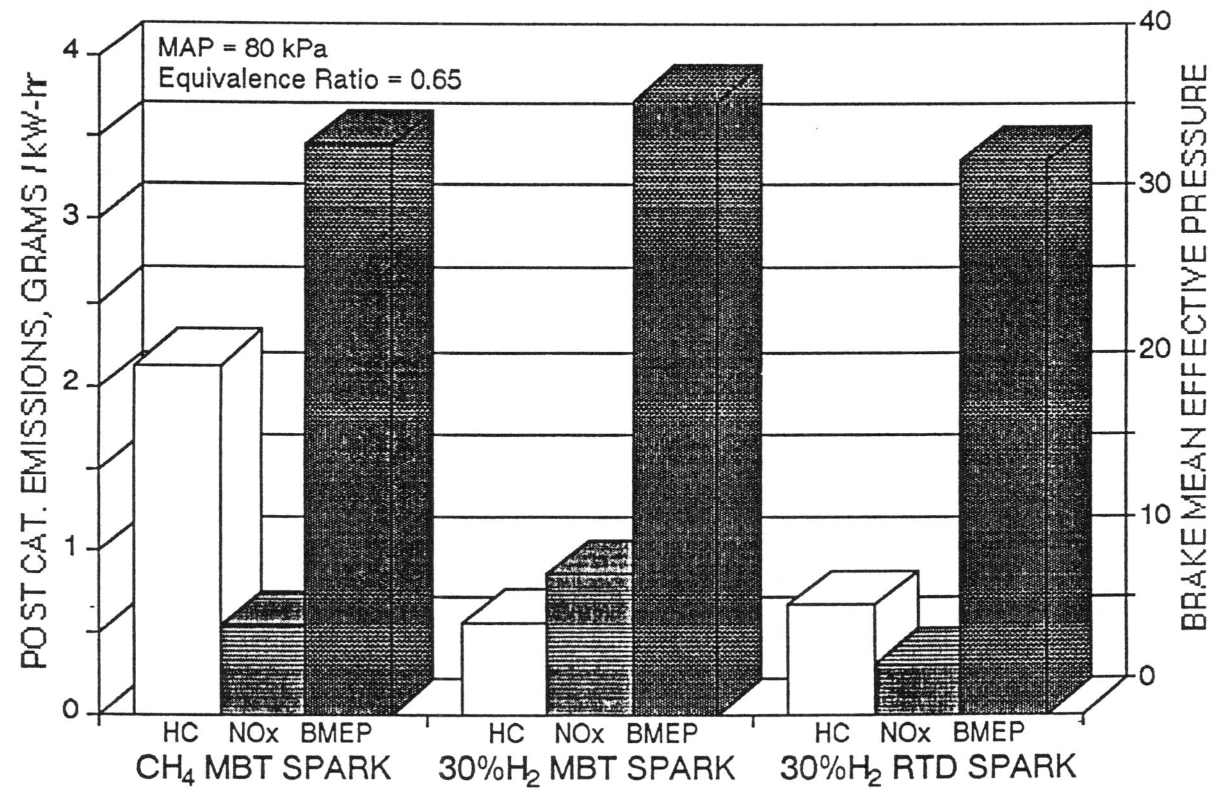

Figure 2-4. Post-catalyst emissions show that sacrificing the torque advantage of Hythane nets further emissions reductions relative to methane.

decreases THC emissions significantly, relative to the pre-catalyst data of Figure 2-3. It is nonetheless apparent that the hydrogen additive enables improvements in both NOx and HC emissions in the lean burn range.

Pre-catalyst CO emissions follow a trend comparable to THC emissions, showing significant hydrogen leverage. Post-catalyst CO emissions, with or without hydrogen, were below the limits of the instrumentation in the lean-burn portion of the DOE/NREL study.

3. STOICHIOMETRIC WITH THREE-WAY CATALYSIS

This section presents results from several Hythane projects. The DOE/NREL study, discussed above, included steady state tests with near-stoichiometric air/fuel mixtures. There have also been four separate studies of Hythane in vehicles via, steady state dyno tests, the Federal Test Procedure (FTP) and Hot 505 tests (a subset of the FTP). The project participants and funding sources are listed in Table 3-I below. All but the 1989 American Lung Association project have utilized closed loop exhaust oxygen feedback control, based on Impco's FCP1 analog controller with interface electronics from HCI.

Going into the DOE/NREL study, the HCI/CSU team was aware of the fact that zirconia-based exhaust oxygen sensors were miscalibrated for natural gas and Hythane. The usual control voltage for gasoline engines with zirconia oxygen sensors is 450 mV. Steady state tests at HCI and "Hot 505" tests at Colorado Department of Health during the Denver Hythane Project had shown that increasing the control voltage with natural gas or Hythane out to the ~800 mV limit of the sensor brought NOx downward significantly with minor effects on CO and HC. Development work with CSU's Natural Gas Challenge truck confirmed this. The usual 450 mV control voltage is too lean for optimal use of a three-way (HC, CO, NOx) catalyst with natural gas.

The DOE/NREL study employed a wide range zirconia-based oxygen sensor to determine equivalence ratio. The instrument is called an AFRecorder. It was found that this instrument was also miscalibrated for methane or Hythane by 1.5%. The peak CO_2 emissions, with or without hydrogen, occurred with an AFRecorder indication of ϕ = 1.015. CO emissions also showed a sharp increase at an AFR indication of ϕ = 1.015. Slightly past this, at an AFRecorder reading of ϕ = 1.020, the NOx emissions hit a minimum value. Taken together, the CO_2, CO and NOx data confirm that the AFRecorder is reading 1.5% richer than true stoichiometry (true ϕ = 1.000 when AFRecorder says 1.015) and that the true ϕ for NOx minimization is ϕ = 1.005 (1.020 minus the 0.015 error).

Figure 3-1 below shows the sharp post-catalyst NOx vs. CO tradeoff that occurs as the mixture passes through stoichiometric. A 1.5% error in ϕ is extremely significant for emissions control purposes. For example, at a true ϕ = 0.985, CO is below test sensitivity and NOx is 9 g/kW-hr. At a true ϕ = 1.015, NOx is below test sensitivity and CO approaches 18 g/kW-hr. At the "sweet spot" (true ϕ = 1.005), both

Table 3-I. Hythane Vehicle Test Projects to Date.

<u>HCI/American Lung:</u>	(1989)
Vehicle:	1979 Dodge D-50 Pickup, 2.6 liter 4-Cylinder, Turbo Intercooled, HCI Parallel Induction, Open Loop
Funding:	HCI Internal R&D
Participants:	American Lung Assn. (Riverside, CA), Tren Fuels, CSU Nat'l Ctr. Vehicle Emissions Control and Safety, HCI
<u>HCI Hythane Prototype:</u>	(1990 - present)
Vehicle:	1991 Chevrolet S-10 Pickup, 2.5 liter 4-Cylinder Impco CA125 mixer, FCP1 Controller, HCI Interface
Funding:	HCI Internal R&D
Participants:	California Air Resources Board, City and County of Denver, Colorado Dept. of Health, Impco, Englehard, CSU Nat'l Ctr. Vehicle Emissions Control and Safety, Tren Fuels, HCI
<u>Denver Hythane Project:</u>	(1991 - 1993)
Vehicles:	Three 1991 Chevrolet Cheyenne Pickups, 5.7 liter V-8 Impco CA300 mixer, FCP1 Controller, HCI Interface
Funding:	Urban Consortium (DOE), Air Products & Chemicals, Public Service Co. of Colorado
Participants:	City and County of Denver (lead), Air Products & Chemicals, Colorado Dept. of Health, Public Service Co. of Colorado, HCI
<u>PEO Hythane Project:</u>	(1992 - 1993)
Vehicles:	1992 Chevrolet Van, 4.3 liter V-6 Impco CA300 mixer, FCP1 Controller, HCI Interface
Funding:	Pennsylvania Energy Office, Cost Sharing; National Fuel Gas (Erie, PA) Air Products & Chemicals, Bruderly Engineering, HCI
Participants:	Bruderly Engineering (lead), Air Products & Chemicals National Fuel Gas, U.S. EPA (Ann Arbor), HCI

NOx an CO are within test-to-test scatter of the instrumentation.

Any differences between methane and Hythane in Figure 3-1 are below test sensitivity. This does not mean that there are no differences. For example, diluted sample NOx data past $\phi = 1.005$ are typically 1 ppm ± 2-3 ppm. Some NOx readings are slightly negative. Therefore, differences of the magnitude observed in the transient tests discussed below are "within scatter".

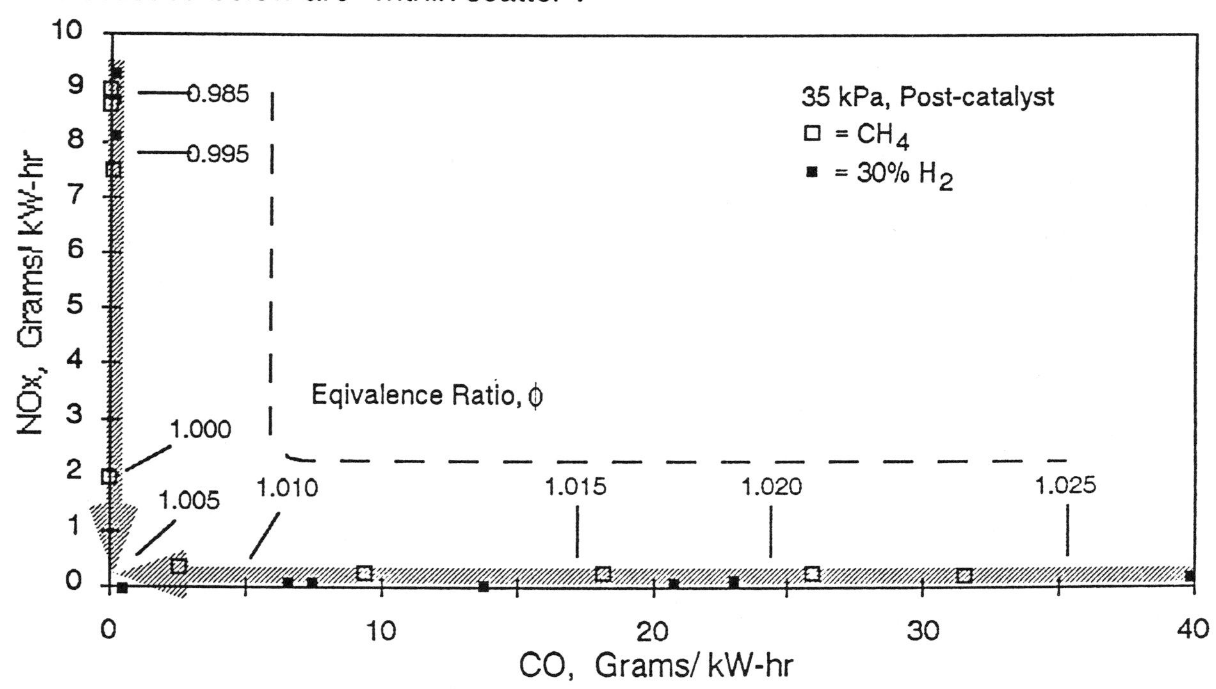

Figure 3-1. NOx vs. CO characteristic of the three-way catalyst. The slightly lower NOx shown for Hythane is within the noise level of the instruments.

Recent FTP transient vehicle tests, conducted by EPA (Ann Arbor) on the PEO Hythane van, are typical of the results obtained earlier at Colorado Department of Health, California Air Resources Board and the National Center for Vehicle Emissions Control and Safety. At any fixed setting within the range of a zirconia oxygen sensor, Hythane produces lower CO emissions at the expense of an increase in NOx. The unofficial[11] test data are shown in Figure 3-2. It is noteworthy that the nonmethane hydrocarbon emissions are virtually zero.

A set of Hot 505 tests was run during the Denver Hythane Project with varying oxygen sensor control voltage. The purpose of the tests was to learn whether sacrificing the CO advantage of Hythane would result in a NOx advantage. The results are shown in Figure 3-3. By shifting the calibration of the control system from 440 mV with natural gas to 762 mV with HY5[12] it was possible to get a simultaneous reduction

[11] The EPA report is not yet available. The data in Figure 3-2 are from notes taken by F. Lynch of HCI during the tests.

[12] "HY5", a nickname contributed by Dr. Steven Foute of City and County of Denver, means 5% hydrogen by energy content (15% by volume) in natural gas.

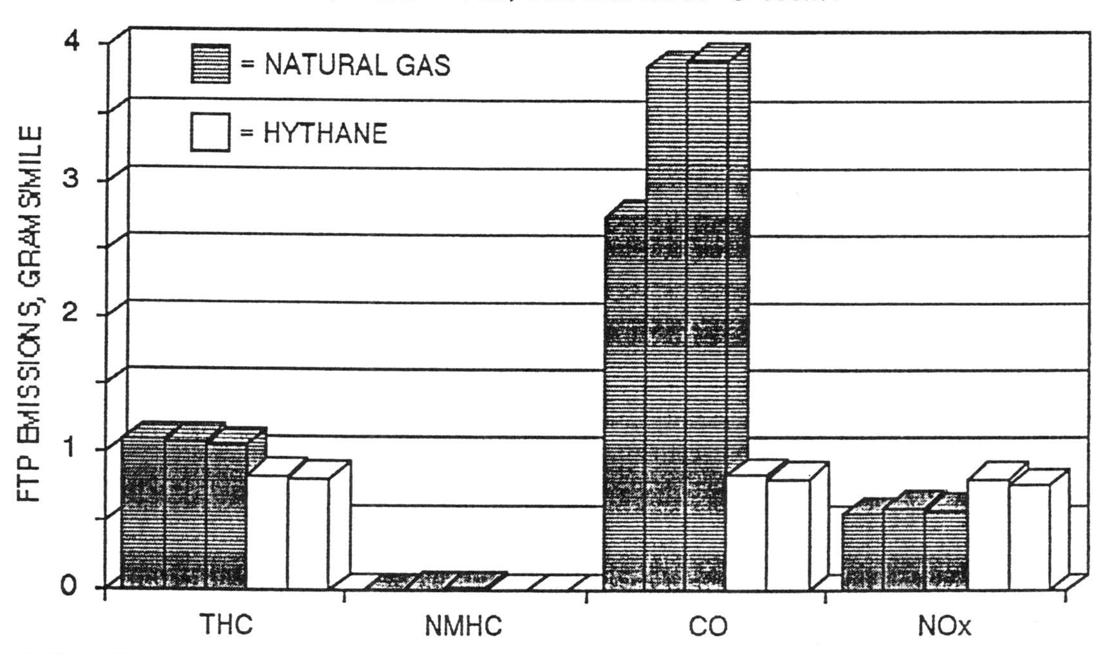

Figure 3-2. Preliminary FTP emissions test results from the PEO Hythane Project.

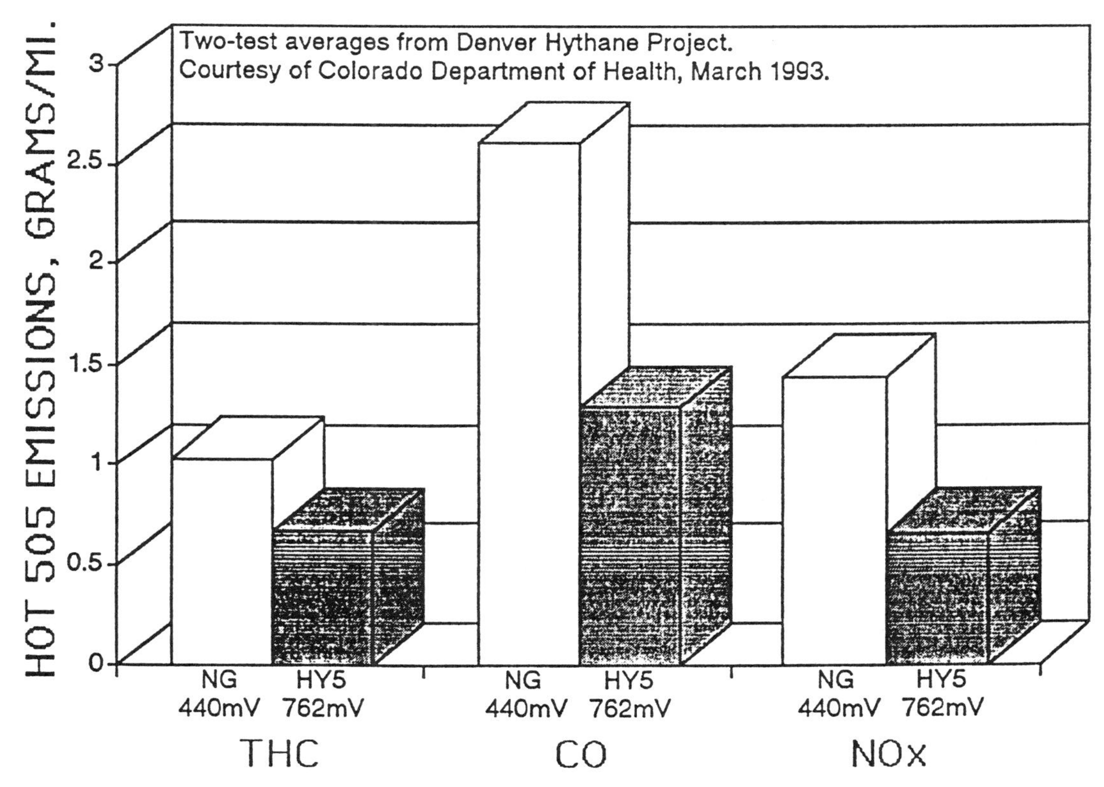

Figure 3-3. Hot 505 emissions with natural gas and HY5 at different oxygen sensor control voltages.

in both CO and NOx. Additional tests with natural gas at 762 mV showed NOx comparable to HY5 but much higher CO.

Based on the results shown in Figure 3-3, it appears possible to get a leverage factor of about 10 on NOx and CO, i.e., 5% hydrogen by energy content makes 50% reduction in emissions. The THC leverage is about 3. The reason for this leverage is not apparent from the steady state tests reported above. The present hypothesis is that hydrogen improves combustion stability with the standard gasoline ignition and exhaust recycle schedules used in all transient tests to date.

4.0 CONCLUSIONS

Hydrogen is a useful additive for natural gas that enables leaner operation under part load conditions and improves BMEP at wide open throttle near the lean limit. Both of these properties of Hythane are useful for NOx reduction without exceeding the normal THC emissions or fuel consumption observed with natural gas.

The advantages of Hythane for operation with near stoichiometric mixtures and a three-way catalyst are not completely understood. Strong effects, consistently observed in transient tests of four vehicles by four different laboratories, were not apparent in steady state engine tests. Further work is needed to understand and fully exploit the effects observed in transient tests.

5.0 ACKNOWLEDGEMENTS

The authors gratefully acknowledge the contribution of the following individuals and their organizations in conducting the projects and obtaining the data reported in this paper:

Colorado State University;
Dr. Bryan Willson, Brad Boender, Jamie Schneider, Jason Yost,

Colorado Department of Health;
Ron Ragazzi

City and County of Denver;
Dr. Steven Foute, Deborah Kielian, Terry Henry

National Fuel Gas;
John Groth, Ronald Gray

U. S. EPA;
Tony Barth, Bob Moss

Germany's Contribution to the Demonstrated Technical Feasibility of the Liquid-Hydrogen Fueled Passenger Automobile (931812)

Walter Peschka
DLR, Stuttgart
William J. D. Escher
NASA Headquarters

ABSTRACT

The German aerospace research establishment, the DLR, in Stuttgart, is conducting an extended program in advanced terrestrial energy/ environment subjects, including transportation. These technology development and demonstration activities generally pivot off in-hand aerospace accomplishments and know-how. A major project presently being concluded, after nearly two decades of work by DLR staff and their research and industrial colleagues in Europe and the United States, has addressed the use of liquid hydrogen (a staple aerospace fuel) for automotive vehicle applications.

This work recognizes that, in the strategic view, the non-fossil (fossil in the nearer term) production of hydrogen from water is mankind's leading prospect for achieving energy independence from the world's depleting fossil-energy resources. At the same time, hydrogen is conclusively the most environmentally benign chemical fuel possible. For example, its carbon dioxide effluent is zero.

The paper describes the very considerable contribution of Germany in advancing the technical status of cryogenic liquid hydrogen-powered demonstration vehicles and their supporting fuel-supply infrastructure. Complete technical feasibility has been documented for three critical subsystem areas upon which the ultimate practicability of the *Hydrogen Car* centers: 1) Liquid hydrogen self-serve fueling stations, 2) Vehicle onboard liquid hydrogen storage containment and processing equipment, and 3) Automotive engines which operate competently on hydrogen using a number of distinctly different hydrogen/air mixture formation techniques.

INTRODUCTION

Rationale for Using Hydrogen as a Chemical Fuel

Hydrogen represents the only practical, technically feasible *carbon-free* chemical fuel. Apart from any secondary combustion products from such sources as engine lubricants, nitrogen oxides are the only pollutant of importance in airbreathing combustion applications, such as automotive engines. And as demonstrated in several hydrogen-fueled research and demonstration automotive engines, NOx levels can be significantly suppressed below that readily achievable with hydrocarbon fuels.

Significantly, *no* oxides of carbon are produced, neither carbon monoxide as a presently regulated pollutant, nor carbon dioxide, the leading greenhouse gas associated with global-warming trends.

Along with its unique environmentally benign nature, hydrogen fuel can be readily produced from non-fossil primary energy resources, such as solar and nuclear sources. These sources can be harnessed to carry out water-splitting reactions (e.g., water electrolysis) wherein only water feedstock is required, and hydrogen and oxygen gas is produced. It is observed, however, that present economics lead generally to fossil-fuel based hydrogen production. A leading example is the methane steam-reforming process utilizing natural gas and water (as steam) to produce hydrogen, with a carbon dioxide effluent. Water electrolysis, a technologically mature and process efficient (ca. 70 - 85 % conversion efficiency range) method is employed today in many "niche applications". However, with its production costs tied mainly to the local price of electricity, electrolytic hydrogen is substantially more expensive than fossil-fuel produced hydrogen.

Today, outside of its prevalent use as a superior space propulsion fuel in cryogenic hydrogen/oxygen applications such as the U.S. Space Shuttle, hydrogen finds essentially no commercial/industrial fuel use. Today's economics, and the long-existing hydrocarbon fuel supply and utilization infrastructure maintain this status quo.

About Hydrogen's Cost Today

With regard to cost, it remains to be seen that there is no market for liquid hydrogen in Europe at this time. This is clearly shown in a comparison of costs for liquid hydrogen between the USA and Europe, exemplified by the Federal Republic of Germany. The numbers for the USA correspond with costs of 0.8 US dollars per gallon as obtained from large scale liquefaction plants (30 t per day) for space projects. The costs in the Federal Republic of Germany are currently higher but will decrease considerably due to European space projects (Ariane).

Despite frequently expressed opinions, the energy from hydrogen cannot be obtained at a better price than that of any other energy carrier, be it petroleum, natural gas or anything else. Non-fossil produced hydrogen, say, prospective nuclear-electrolytic hydrogen, is estimated to be substantially more expensive than fossil-produced hydrogen, e.g., through steam-reforming of natural gas. But in the long run, there will be price increases associated with diminishing high-quality fossil energy supplies, or those that occur for other reasons (e.g., cartel actions). Comparatively, then, non-fossil hydrogen prices may then become competitive.

Compared to other carriers of secondary energy, hydrogen, because it is essentially environmentally benign in its use, will only be economically acceptable when fuel-use *environmental costs* have to be included in future economical considerations. If the resultant environmental costs are taken into consideration, for example with fossil energy carriers, <u>hydrogen would currently be competitive</u>. Admittedly, the quantification of environmental damage is still a socially and, hence, politically sensitive problem and therefore difficult to solve. Thus, in many cases it is currently difficult to effectively address this question in a competently responsive way.

Innovative hydrogen production from fossil energy carriers is an interesting alternative with respect to environmental pollution because hydrogen as a fuel, with the exception of NOx, does not cause environmentally detrimental vehicle exhaust emissions. The pollutants resulting from the production can be eliminated centrally with increased economy and efficiency. The catalytic decomposition (cracking) of methane is one method for the production of hydrogen from fossil raw materials which may well become more important in the future.

Taking a Long-Range Strategic View of the Energy Future

But if a long-range strategic view of the world energy/environment situation is taken, it can be argued that there will be a long-term transition ahead in which hydrogen fuel, along with electricity, will play an increasing role in energy supplies, initially complementing traditional hydrocarbon fuels and ultimately supplanting them. Not so many decades from now, the Earth's remaining high-quality hydrocarbons, e.g., oil and natural gas, will be considered too valuable as chemical feedstocks and precursor materials to be simply burned. Concomitantly, if this vision is to be realized, the heavy environmental-degradation burden of a fossil-energy economy will be finally lifted once and for all.

The leading spokesman professional organization advocating hydrogen energy is the International Association for Hydrogen Energy (IAHE). On the side of governments, the International Energy Agency (IEA) has supported hydrogen-energy research and demonstration activities since the time of the oil-embargo induced crisis of the mid 1970's (The IAHE was established in 1975.) Somewhat in the vein of trade associations, the U.S. has the National Hydrogen Association (NHA), which actively promotes the hydrogen-energy cause from its headquarters in Washington, D.C. Equivalent groups are active in numerous other countries.

Future Vision: A Hydrogen Energy System

In pursuit of this prospect, energy-system researchers and planners worldwide have long studied aspects of the overall Hydrogen Energy System represented in the simplified pictorial diagram of Figure 1. This "systems view" reflects the entire cycle of hydrogen-energy processing: 1) production, 2) delivery and 3) use. As pictured here there are many technical alternatives available, and markets to be potentially served over this entire energy system.

Transportation,the subject of this paper, constituted as it is with a large set of "mobility systems", usually taken as individual vehicles and vehicle fleets, represents a very exciting and technically challenging one of the four hydrogen-energy user sectors called out in the system diagram; it is brought out in further detail in Figure 2.

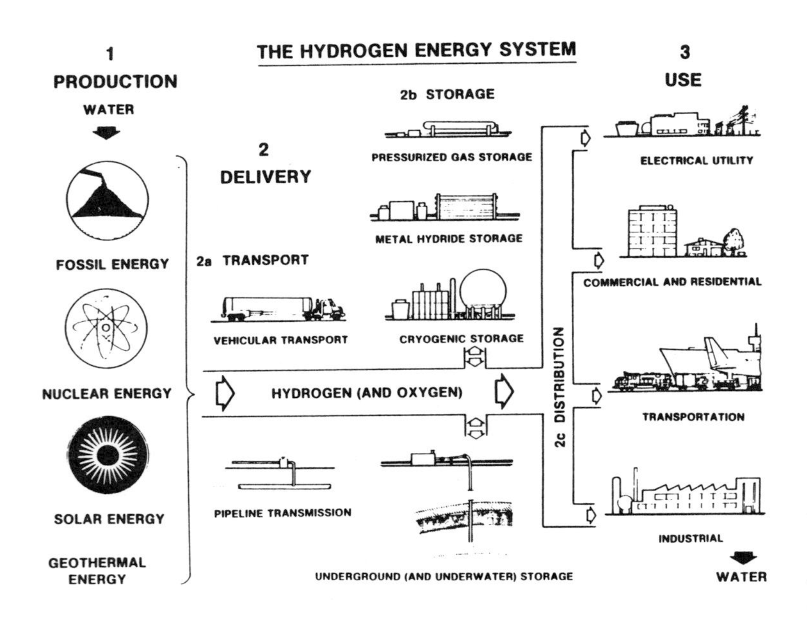

Figure 1 General System Schematic for a Projected Future Hydrogen Energy System

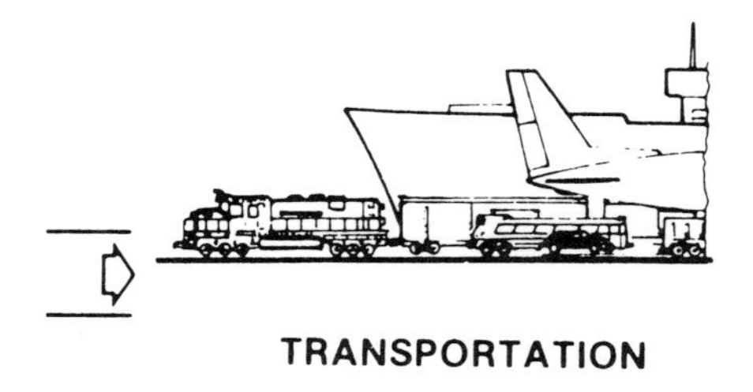

Figure 2 Transportation Using Sector of the Hydrogen Energy System

Mobility applications, the central subject which the SAE involves itself as a professional society, generally require storable, transportable chemical fuels. As discussed earlier, today these are mainly liquid hydrocarbons, prominently: gasoline, diesel and aircraft-turbine fuels. Although hydrogen, like natural gas, is conventionally thought of as a *gaseous* material,quite like natural gas (as liquefied natural gas or LNG), hydrogen can be cryogenically processed into its liquid state, liquid hydrogen or LH_2.

Space Program Needs Created Today's *Cryohydrogen* Availability

Characteristically, liquid hydrogen is the designated form used in the space transportation propulsion field. It is normally combusted in rocket engines with cryogenic liquid oxygen, producing and extremely energetic steam-based exhaust, the nearly invisible plume emitted supersonically from such space engines as the Pratt & Whitney RL-10 and the Rocketdyne SSME (Space Shuttle Main Engine).

In fact, the impetus to develop today's cryogenic liquid hydrogen to the present state-of-the-art level (e.g., routine rail- and over-the-road semitrailer transport of LH2 by the industrial gas community) was the basic need of the space propulsion community for the liquid cryogenic form of hydrogen. The large-scale use of hydrogen in America's Apollo moon program greatly expanded the industrial gas industry support base, as well as creating large-scale government facilities for storing and handling LH2.

The "story" of how liquid hydrogen was rapidly developed from the status of a "laboratory curiosity" to a routinely produced tonnage industrial chemical, as stimulated mainly by space propulsion needs for a higher energy fuel (than, say, kerosene or RP-1 rocket fuel) is authoritatively related in John Sloop's NASA-sponsored book, "Liquid Hydrogen as a Propulsion Fuel 1945 - 1959" /0/.* This work stands as a key reference in the lead author's book on cryogenic hydrogen as a future fuel /1/.

Although, as stated earlier, the fuel-use of hydrogen (as LH2) is today limited to space transportation applications in the U.S. and elsewhere (Europe's *Ariane* satellite launcher series, Russia's Shuttle-equivalent, Energia system), broad use of hydrogen fuel in the transportation sector has been under assessment for many decades.

Rather extensive studies of aircraft, rail, road-vehicles, ship and pipeline transportation systems have been conducted, with hardware demonstration projects carried out in most of these application areas. Cryogenic liquid hydrogen, mandatory for aircraft usage, is generally noted by these transportation studies to be the necessary, or at least the preferred, form of hydrogen for use as a mobility fuel. (Exception: hydrogen pipeline compressors, which can use the high-pressure gaseous form being transported.)

* Numbers within slash marks, thus, (/ /) refer to individual references cited which are listed at the end of the paper

Most prominent as a vehicle type addressed in this set of studies of hydrogen-fueled transportation systems, and, especially, in the numerous research and demonstration projects conducted to date worldwide, is the passenger automobile -- the *hydrogen car*. This brings us to the specific subject of the paper.

Automotive Applications of Cryogenic Liquid Hydrogen Fuel

Specifically, it is *automotive applications* of cryogenic liquid hydrogen that is addressed here, namely that extensive effort mounted in Germany by its national aerospace research establishment, the Deutsche Forschungsanstalt fuer Luft- und Raumfahrt e.V. (literally translated: the German research establishment for air- and space-flight). Formerly, the DFVLR, the designating acronym is now, DLR.

In the work to be summarized here, the DLR was joined by a number of German industry partners in the research and demonstration efforts carried out since about 1975, namely the automotive firm of BMW, and the industrial gas equipment firms, Messer Griesheim and Cryomec (Basel).

In view of this paper's dominant focus on the automotive use of cryogenic liquid hydrogen, let it be briefly stated at the beginning that rather extensive attention has also been given to *non-cryogenic* hydrogen. This work, including driving demonstrations, embraces metal-hydride storage, high-pressure gas containment, chemical carriers of hydrogen (e.g., methyl cyclohexane-to-toluene processing), and other innovative means.

However, it is the authors' basic thesis, predicated largely on what the space programs (taken internationally) have accomplished in demonstrating the technical practicability and safety of liquid hydrogen systems and processing, that it will, in fact, be the cryogenic form of hydrogen which will mainly be used to service the future transportation sector. It is this predispositional orientation which guided the work carried out in Germany over the past two decades now to be reported.

Given that the German work to be described concentrated on the automobile application (for a number of practical reasons: equipment size, hardware availability, strong popular interest, etc.), nevertheless, as a final authors' comment in this introduction section, let it be stated that initial applications of liquid hydrogen fuel in the transportation sector is, in our judgment, *not* likely to be in the private automobile sector in general.

Rather, applications to larger fleet-type vehicles: trucks, buses, perhaps commercial aircraft, is expected to come first. This will be the case for reasons of operating economics and the relative ease of establishing viable and safe fueling-infrastructures, and better-controlled equipment operations and maintenance means. It is with this thought that the reported DLR research and demonstration work included candidate onboard cryogenic hydrogen pump evaluations involving truck/bus class, rather than typical automotive-application hydrogen flow-rates.

The Three Critical Hydrogen Automobile-Use Issues

Perhaps the essence of what is held up in the paper as "Germany's Contribution to the Demonstrated Technical Feasibility . . . ", is that the three key technical areas of concern most often questioned by both automotive engineers and the general public, when considering prospects for the "hydrogen car", each have been carried to what the authors believe to be a successful conclusion, by the DLR and those collaborating industrial firms involved in this work.

These technical areas (issues) are:

- Hydrogen fueling means and operation
- Onboard hydrogen storage means
- Engines operating competently on hydrogen

As often heard -- for each issue -- are these typical comments:

Hydrogen fueling - "Too complicated and dangerous; it would take a trained white-lab-coated expert to fuel a hydrogen vehicle."

Onboard storage- "Impractiicably large, heavy, complicated and expensive fuel containment hardware would be needed." "Hydrogen's basic physical properties will limit vehicle driving range unacceptably." "Too dangerous in the case of accidents."

Hydrogen engines - "Hydrogen's combustion characteristics inevitably lead to induction-system backfiring, rough combustion, loss of power and high NOx levels."

. . . and so on. The reader is invited to add his own thoughts on the subject at this point! And then read on.

Hydrogen Vehicle Fueling Means

Figure 3 shows a DLR prepared liquid hydrogen fueled BMW automobile being refueled at one of several experimental fueling facilities which have been successfully demonstrated. Fully automatic filling stations which can be safely operated, even by technically nontrained people, have already been developed, tested and demonstrated /1, 8-10/. Operational error is prevented by a microprocessor controlled electronics device.(Figs. 3 & 4). Basically the experience acquired with LH2 is also applicable for the use of LNG.

The automatic refueling system must meet the following requirements:

1. Operational error must not lead to a dangerous situation. In particular, the release of liquid hydrogen into the air must be avoided.

2. The refueling process must stop automatically upon failure of the electronics or the hydrogen carrying lines.

3. Refueling must stop automatically when the vehicle tank is full.

4. It must be possible to stop the refueling process at any time by external command, i.e., partial refueling must be possible.

5. The interior of lines carrying hydrogen and valves must be protected from air and moisture.

6. The combination of starting the engine, followed by the departure of the vehicle, must be prevented automatically as long as there is any physical connection between the vehicle and filling station.

7. Connections between the filling station and the vehicle should not be able to be interrupted during the refueling process.

A total of three filling stations were built and tested within the scope of these projects, the first two in somewhat simplified versions. One was used for refueling test vehicles in a joint project between Los Alamos National Laboratory and the DLR). The third filling station (Fig. 3), which exhibited all of the previously described features in laboratory tests, has yet to reach the full demonstration stage because lack of funding. It has, however, served as a basis for further development by the cryoindustry.

With systems working according to this principle, a total refueling time (line cooldown included) for a cold 140 liter fuel tank of between 4 and 4.5 minutes is state of the art,which agrees well with the required amount of time for refueling with conventional fuel /1, 3/. Fifteen minutes are required for cases where the tank is at ambient temperature requiring extensive cooldown.

Vehicular Onboard Containment of Hydrogen

General Transport and Storage Technical Status

The storage and safe handling of large amounts of liquid hydrogen is state of the art, particularly in the United States due to developments in space technology as supported by the industrial gas industry, which has in the meanwhile expanded cryogenic liquid hydrogen service widely throughout the commercial sector, e.g., for controlled atmospheres required for semiconductor manufacturing.

Now common industrial practice, the transportation of hydrogen in its liquid state is more economical with any substantial level of product demand and is not nearly as burdensome as its alternatives. This becomes immediately apparent when it is considered that, with transportation via pressurized gas cylinders (or, projectively, similarly heavy hydride storage containers), the weight of the hydrogen is only about 1.5% of that of the storage container. For the transportation of 70 kg of liquid hydrogen for instance, which requires about a 1,000 Liter container weighing from 200 to 300 kg, about 5 tons of "containerization" is required with the use of gas bottles or iron-titanium hydride.

With this in mind, it makes much more sense to transport liquid hydrogen to customers who require substantial amounts of pure gaseous hydrogen, converting it into the required high- or low-pressure gas at the customer site, using high-pressure liquid-to-gas converters in a manner similar to that long in use use with liquid nitrogen or liquid oxygen servicing systems.

Automotive Onboard Storage Containers

Development of liquid hydrogen tanks started at the DLR in 1974 and has been carried out since 1982 by the cryoindustry /1/. Tanks, suitable for motor vehicles, with a capacity of up to approximately 150 Liter, are now available for use in test vehicles. This corresponds to a gasoline equivalence of a 33 liter tank. A representative vehicle cryohydrogen container and its adjunct controls and "plumbing are reflected in Figures 5, 6 and 7.

Such vacuum-jacketed, multilayer insulated tanks made of aluminum alloy and fitted with vapor cooled shields, weighing 50-60 kg, have an evaporation rate of less than 1.5% per day when using glass fiber or carbon fiber composites for the inner vessel suspension. They supply either cryogenic gaseous or liquid hydrogen using available automatically controlled instrumentation, valves, safety valves, tube systems, and heat exchanger, suitable for motor vehicles, or hydrogen at ambient temperature, up to a pressure of approximately 0.5 MPa, using a compact heat exchange warmed by the engine coolant circuit.

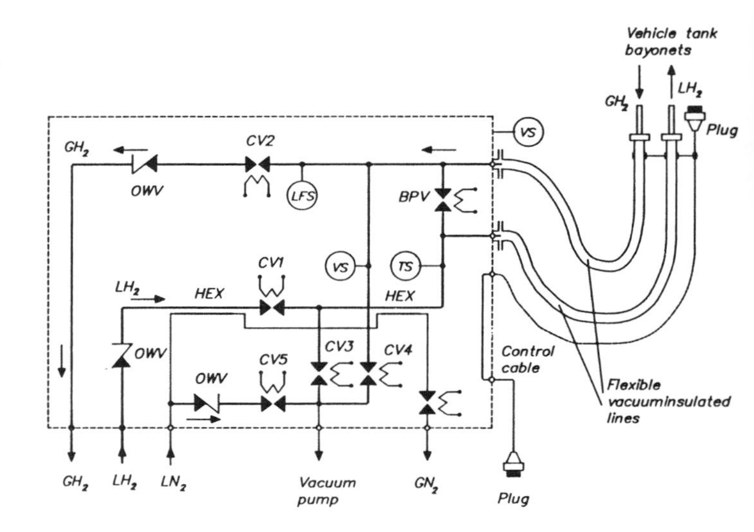

Figure 4 Control and Flow Schematic Diagram of Automatic Filling Station (shown in Fig. 3)

Figure 3 Automatic Self-serve Liquid Hydrogen Filling Station with BMW 745i Demonstrator

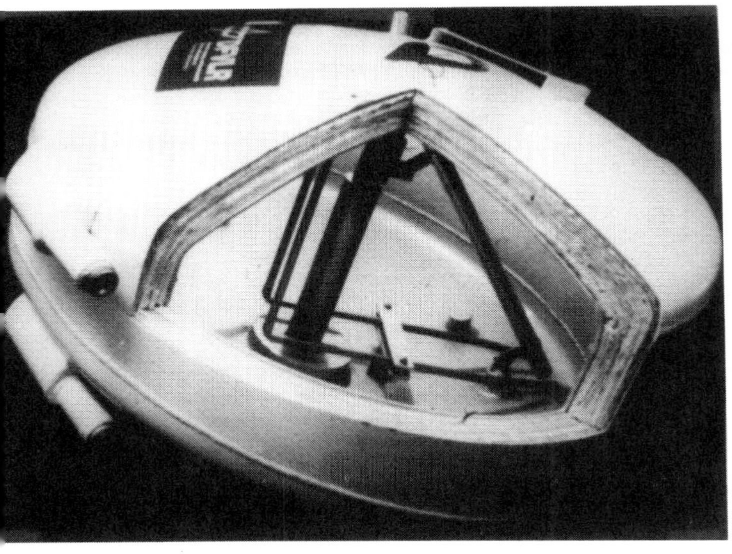

Figure 5 Vehicle Liquid Hydrogen Fuel Container (sectioned to show dewar construction)

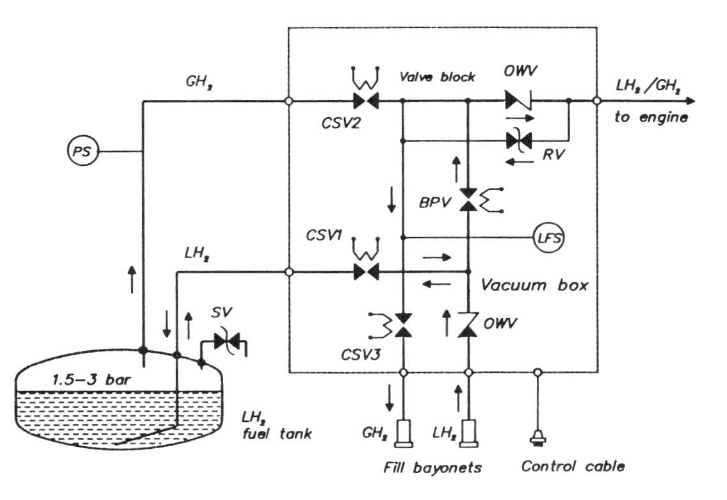

Figure 6 Control and Flow Schematic of Fuel Container to Engine Interface

The external design of these flat tanks has either a full rotational symmetry, i.e., cylindrical with convex ends /1/, where the cylindrical part can be practically nonexistent /7/, or, for an improved installational fit, they may be tapered, again using rotationally symmetric convex ends (Fig. 5). Tanks of this design are installed inside the trunk at the rear of the vehicle, which can only be tolerated for test vehicles in view of the lost trunk space.

The first development steps were begun at the DLR for tank forms similar in shape to flat prisms, which are more compatible with fully serviceable conventional vehicles /10,1/. This type of container could be comformally fitted just aft of the vehicle rear seat and ahead of the trunk, thereby not intruding on trunk space. This location is also advantageous from the container protection standpoint in case of severe collision damage, and thus maximizes vehicle safety.

It should be noted that the basic dewar-type vacuum-jacketed multilayer insulation construction of such containers (see Figure 5) make them quite robust and capable of absorbing high-g impacts

Figure 7 Container Vacuum Box with Solenoid, Check and Relief Valves (see Fig. 6)

without loss of structural integrity. This is considered a large safety plus-feature. (see later section on hydrogen vehicle safety.)

Cryogenic Hydrogen Onboard Fuel Processing

Depending on the type of mixture formation employed by the engine (both external and internal mixture formation are discussed in the next section), the fuel conditioning system has to provide preconditioned hydrogen (pressure, temperature). For example, either gaseous hydrogen at ambient temperature and a low pressure of 0.3-0.5 MPa or, cryogenic hydrogen between 40 and 100 K at approximately 0.3-0.5 MPa suffices for external mixture formation type engines.

Conversely, cryogenic hydrogen between 20-40 K under high pressure may be required for internal mixture formation (direct cylinder injection). This requires one or more small cryogenic hydrogen pumps as will be described.

Typically, for Otto-cycle engines and two stroke engines with injection during the compression stroke, hydrogen supply pressures between 1 and 3 MPa is usually necessary, whereas for "late start" near top-dead-center injection (approximately 5° b.t.d.c.), a pressure level of considerably more than 10 MPa typically has to be provided.

Cryogenic Fuel-conditioning System: High-pressure Cryopumps

(Note: This subject is also covered in greater detail in the major section of the paper, that describing the extensive DLR engine work embracing high-pressure injection)

While, with external mixture formation the pressure of about 0.3 - 0.4 MPa, which is necessary for in-manifold injection, is either already available in the fuel tank or can be produced by a small piston or diaphragm pump that is integrated into the tank, more sophisticated cryogenic liquid pumping equipment is required for internal mixture formation applications, particularly those with late- start injection. They can, however, be designed for high operational redundancy /40, 21, 22/.

With "hybrid" (external plus internal) mixture formation, a unique approach investigated by the DLR, a certain gaseous fraction, typically produced by LH2 high pressure pumps, can be then used for the external mixture formation phase. The production of pressures as high as 20 MPa can either be obtained directly from the input pressure level of the fuel tank or that which is produced by an LH2-pump that is directly driven by the engine, or via a "cold compressor" /46./The latter would be supplied with supercritical hydrogen at about 1.5 MPa and 35 K, via a boost pump integrated into the fuel tank.

Later Figure 22 shows a compact cryogenic hydrogen pump developed by Cryomec AG on a test stand at the DLR. It was developed through the miniaturization of a larger, proven and tested unit. It provides a pressure of up to 25 MPa at a discharge rate of 6 l/min (at 450 strokes/min). In addition, since LH_2 as boiling liquid, tends to form vapor during the intake phase of the pump, thereby reducing the volumetric efficiency of the pump, and also impeding well-controlled pressure build-up during the delivery stroke, measures need to be taken in order to guarantee proper liquid filling during the intake phase.

In Japan's Musashi Institute of Technology related work, as referenced here, besides a relatively small stroke rate (300-500 strokes per minute), their system consisted of a pressurized tank, intake valves with the lowest possible pressure loss, as well as a pump design with a stationary piston and moving cylinder, equipped with a prechamber. During the intake phase the cylinder moves into the liquid through, due to inertial effects, substantially enhancing the filling of the prechamber and pump cylinder /23/.

DLR developments led to LH_2 reciprocating pumps with gas film bearing pistons /29, 35, 22/, as well as pistons /10 / with two sleeves made of PTFE graphite that compensate for radial expansion. The unit was physically separated from the LH_2 tank. Developments at Cryomec-AG Basel lead to small compact piston pumps for liquid hydrogen that could be installed primarily in vehicles like trucks and buses. They led to a successful DLR test program with LH_2 up to a pressure of about 240 bar at a LH_2 flow rate up to about 6 to 8 liters per minute.

The cryogenic hydrogen pump development aspects of the DLR effort, as more directly associated with engine mixture formation needs, is discussed in further detail in the following section of the paper.

Hydrogen Automotive Engine Developments: Mixture Formation is the Key

During engine operation the unique characteristics of hydrogen mentioned previously carry strong design considerations with regard to combustion as well as mixture formation alternatives. The very wide volume-dependent ignition range of H_2-air mixtures of from about 5% to 75% (fuel/air equivalence ratio = 0.07 to 9) enables quite lean operation with greatly reduced NOx emissions. As is generally known, engine power-level control by varying equivalence ratio ("quality control") is feasible, with a considerable reduction in air-side throttle and breathing losses. In this variable-leaneness operating regard, hydrogen differs considerably from other gaseous fuels such as natural gas or propane, as well as gasoline.

<u>Mixture Formation with Ambient-Temperature Hydrogen</u>

> (<u>Note</u>: the special advantages of *cryohydrogen* usage, in contrast to using ambient-temperature hydrogen, are covered in the next major subsection)

As noted above, substantial reduction of NOx emissions can be achieved with lean mixture concepts without using catalysts for exhaust after-treatment. Lean mixture concepts can be achieved much more effectively with hydrogen than with conventional fuel, due to the significantly wider ignition range of hydrogen/air mixtures).

Supercharging is a candidate measure to compensate for the loss of specific power output which is related to the lean mixture concepts. (Another approach is to increase engine displacement.) However, due to lower exhaust gas temperature with lean hydrogen operation (approximately 650° C) under partial load, there is usually not enough energy available from the exhaust gas for meeting turbocharger turbine demands. This "turbocharger hole" can be effectively bridged with an additional engine mechanically-driven centrifugal compressor (for example, the MKL unit supplied by the firm, Zahnradfabrik Friedrichshafen AG).

<u>External Mixture Formation in General</u>: In contrast to conventional fuels, the hydrogen fraction in a stoichiometric mixture at ambient temperature (1 kg H_2 to 34.3 kg air) corresponds to about 30% of the mixture volume. Concerning the mixture volumetric heat value (2890 J/L) this results in a corresponding power loss at the engine compared to conventional dense-liquid fuel (3900 J/L) under otherwise similar conditions.

External mixture formation with ambient gaseous hydrogen, which for LH_2-storage requires a heat exchanger for evaporation and warming up the hydrogen, is currently a standard procedure for the operation of LH_2-engines in general. To achieve satisfactory engine operation several additional measures are necessary to prevent uncontrolled preignition and backfiring into the intake manifold. Departing from conventional continuous central injection, these measures include primarily continuous individual intake manifold injection directly at the cylinder intake port and sequential intake manifold injection via timed mechanically, hydraulically or electromagnetically actuated injectors, again, at the individual intake valve ports (such is referred to as "decentralized injection".).

When opening these injectors at the intake stroke (about 40° a.t.d.c.) the problems of the uncontrolled preignition at hot spots in the combustion space, or because of hot residual gas can be reduced considerably. In this manner, the cooling effect of the intake air is used, which can be improved by additional supercharging of the engine under partial load. Nevertheless, under full load, especially at low engine r.p.m., additional cooling measures, for example water injection into the intake manifold, are necessary in order to guarantee satisfactory engine operation under all conditions /4, 11, 34/.

Without adequate cooling measures and at a brake mean effective pressure above about 6.5-7 bar H_2/air mixtures tend toward uncontrolled preignition at equivalence ratios > 0.7, directly with the hot residual gas as well as at hot spots in the combustion chamber where catalytic effects play a considerable role. Backfiring of the combustion into the intake manifold results either in unsatisfactory engine operation or even complete engine failure. Through appropriate cooling measures, for example, through modification of the cooling water routes in order to achieve increased exhaust-side cooling of the combustion chamber, sodium- or lithium-filled exhaust valves, as well as water injection into the combustion chamber via intake air, about 70% of the power of conventional liquid fuel operation can be attained, using external mixture formation of ambient temperature H_2.

In this case, however, he torque flow is especially unsatisfactory with low engine speeds. Water injection, which probably can only be used in experimental engines, can perhaps be replaced by equivalent amounts of exhaust gas recirculation (EGR) due to its resulting suppression of uncontrolled preignition. Although EGR has the advantage of improved mixture formation, it also results in an additional power and torque loss as a result of an increase in charge temperature and decrease in the oxygen fraction of the mixture.

<u>Centralized Mixture Formation:</u> Although centralized mixture formation results in a very good mixture homogeneity, with hydrogen, it characteristically results in intolerable operating behavior because the entire intake manifold is affected during backfiring.

<u>Decentralized Mixture Formation:</u> With decentralized individual cylinder fuel injection directly at the intake valve port, with an appropriately large intake manifold volume, only the part of the corresponding intake manifold that is directly attached to the cylinder intake port contains a combustible mixture, so that the charge that remains in the intake manifold of the remaining cylinders does not backfire. However, it can be difficult to transfer the cylinder charge to the intake manifold, which in fuel precharging must have an appropriate length. With a tuned intake, tube charging is adequate, but with ram-pipe supercharging it is not. (Here each cylinder has its own intake pipe of a specific length, which is connected to an air distributor).

The design of the intake manifold is more critical with hydrogen than with conventional fuel but in this manner the required increase in the torque flow at low engine speeds (in a regular automobile engine about 2000 rpm) can be achieved somewhat easier /16, 17/.

As a consequence of low energy content in the exhaust gas due to larger partial load efficiency and lower volumetric heat value, exhaust gas turbocharging is less suitable with hydrogen operation than with conventional fuel despite an unthrottled air supply. Although the recognized turbocharger deficiency can be diminished through the reduction in the flow orifice of the turbines case, this results in increased choking of the exhaust gas and in additional problems with uncontrolled preignition with the hot residual gas.

Any problems with the hot residual gas could basically be reduced through an increase of the compression ratio (to the range of 10:1 to 11:1) /14/. This approach, however, is incompatible with supercharging. The general experience with Otto-cycle carburetor engines, which have been directly converted to hydrogen operation, has been that the original intake manifold often had to be significantly modified, or completely redesigned, because the resonance frequencies of the intake manifold were drastically changed by the H2-air mixture.

This can be the case with continuous injection, sequential injection with hydrogen precharging, and also with continuous decentralized intake manifold injection in which the respective load condition of the engine has to be taken into consideration. This is because the density of the H2-air mixture is lower than the air suction and thus the sound velocity in the mixture increases considerably (Fig. 8). This may result in significant consequences especially with mixtures that are insufficiently homogenized having zones with large hydrogen fractions (locally above stoichiometric, fuel-rich conditions). Tests to determine if variable intake manifold geometry /15/ could result in improvements would be very informative.

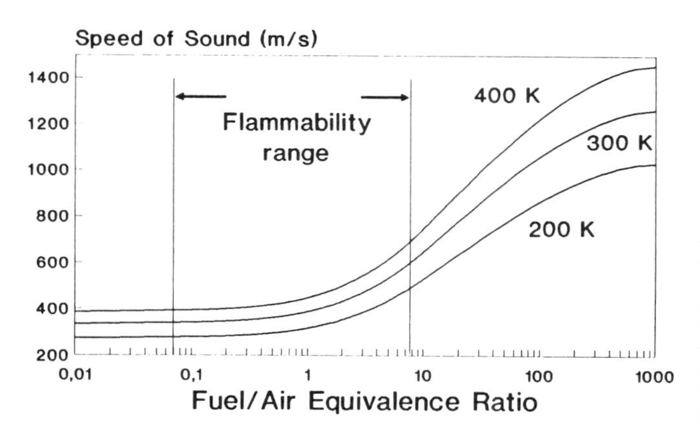

Figure 8 Sound Velocity of Hydrogen/Air Mixtures for a Range of Temperatures

<u>Sequential Single Cylinder Injection:</u> Sequential single cylinder injection into the intake port with an open intake valve does not have any impact on mixture physical qualities because the hydrogen is added directly to the air flow at the cylinder intake thus avoiding mixture formations in the intake manifold to the greatest possible extent. Constant mixture characteristics, a vitally important prerequisite for good homogeneity of the cylinder charge, are present when the air mass flow rate during the intake phase remains constant, the hydrogen injection takes place independent of rotation speed within a fixed range of crankshaft revolution angles, and fuel injection is controlled by a throttle valve or timing valve. An intake air mass flow rate that varies periodically during the intake phase, as in tuned intake tube charging (maximum) or ram pipe supercharging (resonance dip) /12, 13, 16/ can result in local excessive fuel concentrations in the mixture that can lead to uncontrolled preignition.

For steady hydrogen operation with external mixture formation, sequential single cylinder injection /14, 17-19/ provides the best possible continuity of power and torque. The use of digital engine management can also improve intermittent operation (Fig. 9). Considering that external mixture formation using ambient-temperature hydrogen represents an acceptably simple process, a further technically complicated concept, although demonstrated, may not be needed , and it is not clear that further, substantial improvements can be achieved for two-valve engines.

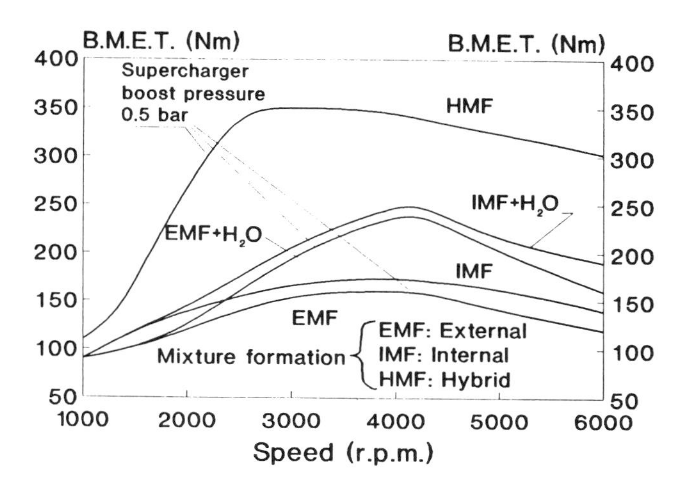

Figure 9 Brake Mean Effective Torque of BMW M30 Engine with External and Internal Hydrogen/Air Mixture Formation, with and without Water Injection (using ambient temperature hydrogen)

<u>Internal Mixture Formation at Ambient Temperature:</u> Internal mixture formation prevents intake air displacement by the bulk of the gaseous fuel. When choosing the injection start at about the beginning of the compression stroke there is still enough time left for injection and mixture formation in which no great demands are made on the injection pressure (1.0 - 1.5 MPa) /18, 19/. However, uncontrolled preignition during the compression stroke cannot be eliminated in this way. This can only be achieved with late start injection (about 5^0 b.t.d.c.). For injection, mixing, ignition and combustion at optimum power ,there are only about 5 ms available with high speed engines at 5000 rpm /18, 20/.

Compared to conventional liquid fuels, with hydrogen at ambient temperature, at each stroke, considerably larger fuel volumes have to be injected in about the same short period of time (maximum of as little as 2-3 ms). This requires correspondingly large injection pressures in the range of 15-20 MPa. Further problems occur with injectors that have to be actuated via a linear drive. This is because, when opening the injector by means of pressure from pulsed fuel admission, there is no satisfactory fuel mass flow rate control due to the compressibility of the gaseous hydrogen being processed.

The volume of ambient-temperature fuel that is injected at each power stroke requires controlling orifice areas, which, despite the high injection pressure, are far larger than comparable-sized Diesel injectors. This inevitably leads to considerably larger moving masses and the associated driving-power and control problems. Although the fuel injection pressure should be chosen as high as practically achievable, in order to reduce the opening diameter as previously discussed, the LIRIAM process (late injection rapid ignition and mixing), which was described in /7, 20/, requires, as well, increased turbulence in the combustion chamber, after an extremely short mixing time, considering the large volume of fuel, with the combustion air.

Producing hydrogen pressures on board a vehicle between 15 MPa to 20 MPa and more, can only be achieved with acceptable expenditure in construction size and drive power with high pressure pumps for liquid hydrogen /21, 22/. Such units have been successfully demonstrated by Furuhama in test vehicles with two-stroke and four-stroke engines /23-25, 30-33/. Therefore, the use of internal mixture formation with late start injection in vehicles can only be practicably realized with liquid hydrogen.

Mixture Formation with Cryogenic Hydrogen

The advantage of liquid hydrogen with regard to its onboard storage feasibility have been long recognized, and have been widely demonstrated /1, 28, 29/ (See later Table 1). With respect to mixture formation, it is the fundamental cryogenic properties of the fuel that are important /35, 21, 1/. Dependent upon the specific mixture formation process, there are different ways to capitalize on the use of cryogenic hydrogen (as opposed to ambient-temperature hydrogen). It centers on the fact that in automotive-sized engines hydrogen is mainly used as cold gas (boiling point 20.4 K) and then, above all can be used supercritically close to hydrogen's critical temperature and pressure (33 K, 1.28 MPa) for mixture formation. This avoids two-phase flow difficulties, eliminating the liquid vaporization step in the mixture formation process.

As a result of the low mass flow rate and correspondingly small vaporization enthalpy (440 J/g) it is practically impossible to inject *liquid* hydrogen, per se, into the combustion chamber of small-displacement automotive engines. However, this possibility cannot be excluded for large steady-operation industrial engines or ship engines, without further investigation.

External Mixture Formation: Initial test results with cryogenic hydrogen external mixture formation /14, 36/ showed that, without reverting to water injection, basically the same low exhaust emission as with water injected ambient-temperature fuel, can be achieved. Furthermore, due to charge cooling by the cryogenic hydrogen, engine power output which corresponds to normal gasoline operation can be achieved in dynamometer testing. In addition to cryogenic mixture formation via continuous individual intake port injection, timed cryogenic port injection via mechanically actuated injectors as well as electromagnetically actuated cryogenic injectors led to comparable results /37/.

Electromagnetically actuated cryogenic injectors show the advantage of power control through control of the opening period (modified gasoline injection units, e.g., L-jetronic or Motronic) but such production devices still require substantial development work to achieve equipment reliability and longevity in hydrogen service. Mixture of the intake air with cryogenic hydrogen and injection of liquid hydrogen directly into the intake air results in a temperature drop in the fuel/air mixture and thus, due to mixture density, in supercharging effect for the engine. As has been demonstrated by Furuhama /38/ for the first time, as compared to conventional liquid fuel operation, power losses with hydrogen can be substantially compensated for.

According to Figure 10, for a stoichiometric fuel/air mixture ratio, a maximum temperature drop of about 80°C can be achieved for a homogeneous mixture that corresponds to an increase in density and the charge volumetric heat value of about 36%. For this condition, the cryogenic or liquid hydrogen serving as a heat sink is approximately equivalent to the heat sink provided by water injection. The drop in the mixture temperature results in a cooling effect in the combustion chamber with a reduction in the tendency of the mixture toward uncontrolled preignition.

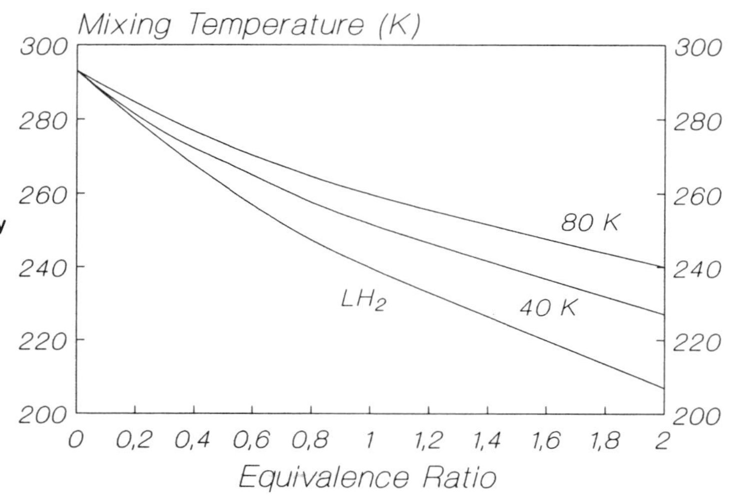

Figure 10 Mixture Cooling Effect of Using Cryogenic Hydrogen (at various temperatures) as a Function of Fuel/Air Equivalence Ratio

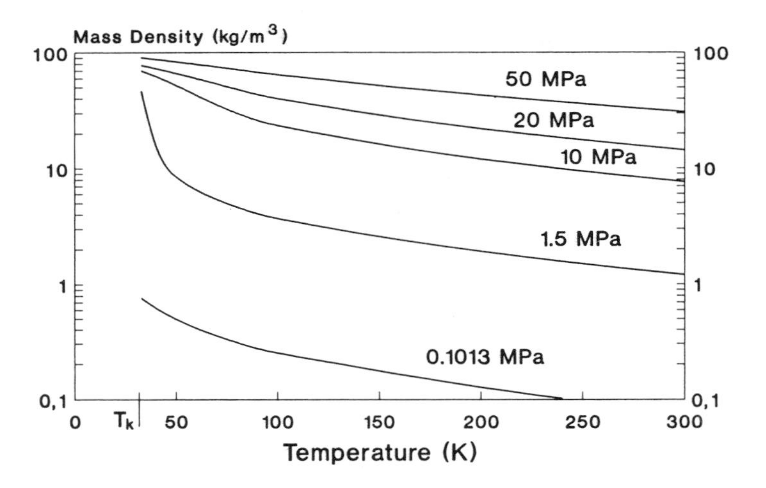

Figure 11 Cryogenic Hydrogen Mass Density as a Function of Temperature and Pressure

With a nonhomogeneous mixture in front of the intake valve the local charge temperature drops even more in any fuel-rich stratified mixture zones. This could, however, induce ignition delay, which during the initial phase of the mixture homogenization in the combustion chamber, contributes to greater reduction in the tendency toward uncontrolled preignition, to lower flame temperature and thus to a reduction of NOx emissions.
Because the density of the cryogenic hydrogen (see Fig. 11) is considerably greater than ambient hydrogen, the fuel volume that must be injected is considerably reduced.

Thus, with fuel precharging there is a smaller mismatch of the resonance frequencies for an intake manifold designed solely for air. Experience with decentralized continuous and decentralized sequential external cryogenic mixture formation with a 2-liter, four-cylinder test engine showed that under normal conditions and operating at about near-stoichiometric, the equivalent levels of specific torque and power achieved with gasoline operation, can be attained under steady operation (Fig. 12 and 13). But, under partial load,and above all, in the lower engine speed range,there were considerable dips in the torque.

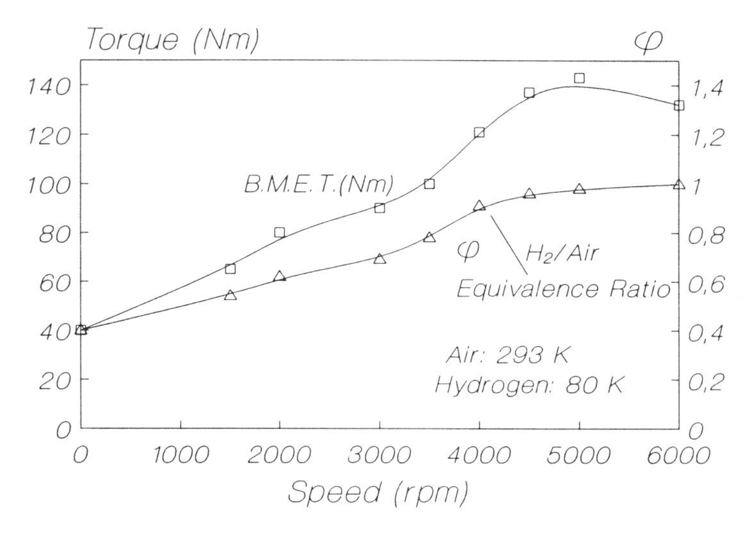

Figure 12 Torque and Fuel/Air Equivalence Ratio as a Function of Engine Speed for Decentralized Cryogenic Hydrogen Fuel Injection

Figure 13 2-Liter Displacement, 4-Cylinder Test Engine with Decentralized Cryogenic Hydrogen Injection (see Fig. 12)

Anticipated problems with water-vapor condensation and possible ice formation in the intake air did not occur in the test engine described in /39/. Evidently, the cryogenic hydrogen/air mixture did not stay in the inlet parts of the intake manifold for the period of time that is necessary for this problem to be experienced.

Experience to date shows that cryogenic external mixture formation may primarily be most advantageous in engine operation with sequential injection, but further technical complexities such as that of mechanical supercharging or improved tuned-intake tube charging or ram pipe charging may not be avoidable.

Cryogenic Internal Mixture Formation: Cryogenic internal mixture formation with injection start at the beginning of the compression stroke does not offer any significant advantages over operating with ambient-temperature hydrogen. Primarily, uncontrolled preignition and thus, loss of sustainable torque under partial load cannot be eliminated or circumvented in this manner (see Fig. 9).

According to current experience, cryogenic internal mixture formation with late start injection, via mechanically or hydraulically actuated injectors for cryogenic hydrogen, is the most promising method to develop hydrogen-fueled piston engines with driving qualities which, with respect to driver expectations, closely correspond to conventionally fueled Otto-cycle and Diesel engines, while being innately superior to them with regard to low polluting exhaust emissions /14, 40,1, 30, 31/.
Developments by Furuhama demonstrate that it is possible to apply this same approach to two-stroke gas engines with electronic ignition /1/ as well as to Diesel engines with glow plugs as an ignition primer /24, 33/.

These results show that the residual gases at the charge exchange of the two stroke engines correspond to the exhaust gas recirculation with four stroke engines. Together with charge cooling by cryogenic hydrogen, there is a considerable reduction in nitric oxide emission. However, it turns out that at an injection pressure of about 8 MPa with injection initiation directly after the closing of the air inlet by the piston, the efficiency and power of the test engines decreased considerably below a mixture temperature of 150 K.

This is assessed to be due to the insufficient homogeneity of the hydrogen/air mixture and the resultant lower,and spatially inconsistent speed of the flame-front propagation. This, in turn, results in both a reduction in the peak pressure, with less than optimal temporal position of the maximum pressure, as well as cyclic variations in power out put. Something similar can be observed in four cylinder engines. Remedial action can take the form of intensified mixing, increased combustion chamber turbulence and appropriate injector design changes.

According to results to date and in contrast to often expressed opinion, considerable problems still exist in achieving a suitable distribution of hydrogen mixture homogeneity in general, and in particular with late start injection, which above all requires that a strong focus on the development of suitable cryogenic injectors for hydrogen.

In contrast to conventional neat-liquid diesel injectors, mechanizing successful cryogenic hydrogen injection requires that a cold gas with a substantial volume has to be injected into the high-pressure combustion space, and has to be effectively distributed. Regarding jet formation and gas mixing with air in the combustion space, totally different conditions have to be fulfilled than the case with dense-liquid diesel fuel. This result from the basic gaseous (however cold) nature of the injected medium, the large density difference between fuel and air, and the considerably higher flame-propagation speed, as well as completely different turbulence characteristics, all of which influence mixing and combustion.

Evidently, hydrogen engines with internal mixture formation and late start injection share several problems with conventional diesel engines. Increased nitric oxide emissions and insufficient homogeneity of the fuel/air mixture require require lean combustion,whereas basically stoichiometric combustion with hydrogen appears possible.This requires additional measures, such as injection into an already existing flame front, be taken to both reduce nitric oxide emission and improve efficiency.

Due to the short-time total injection periods available in high-speed engines (about 5ms at 5000 rpm), maximum injection periods of 3 ms must be achieved with a variation in the amount of fuel per stroke of the order of 6:1 to 10:1 (for power-level control. Minimal injection periods of about 0.3 ms to 0.5 ms have to be attained. Clearly, the development of these types of individually controlled injectors, designed for a gaseous medium, in general presents a basic developmental challenge in the development of H2-engines using internal mixture formation /18, 37, 41, 42, 39, 60/

These very short time periods require an injection pressure between about 15 MPa to about 20 MPa. On one hand this is to provide a greater initial density to the H2-flow that leaves the injector at sonic speed and thus a larger impulse. On the other, this is to keep the volume of the gas that has to be injected as small as possible in order to be able to utilize small injector orifices and thus allow for a small moving mass. Figure 14 and Figure 15 show the critical mass flow density as well as sound velocity of cryogenic hydrogen as a function of pressure /43/.

A result of the considerable deviation from the ideal gas state is the characteristic increase in the velocity of sound, despite the drop in temperature, when approaching the critical point of hydrogen (33 K, 1.2 MPa). Here, the advantages of using cryogenic hydrogen under high pressure are obvious. Figure 16 shows the principle and Figure 17 shows one of the injectors for cryogenic hydrogen designed for 15 MPa. A forerunner of these injectors, though only designed for an injection pressure of 1.5 MPa and early injection start, was originally developed for a mutual project with BMW for its six-cylinder M30 engine, for which BMW has provided to the DLR a correspondingly modified cylinder head.

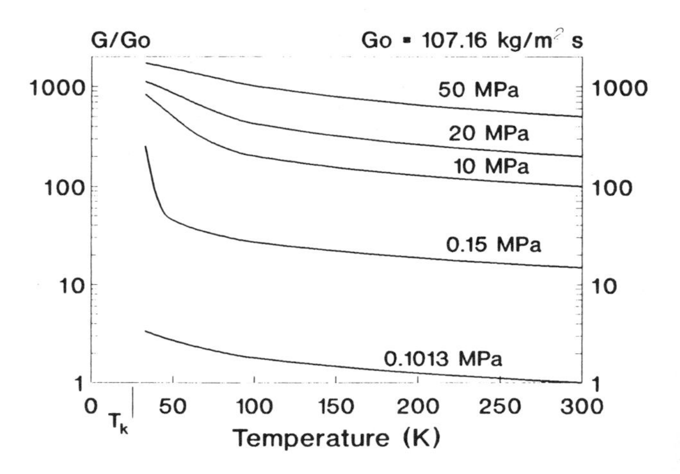

Figure 14 Normalized Hydrogen Critical Mass-flow Density as a Function of Fuel Temperature and Pressure

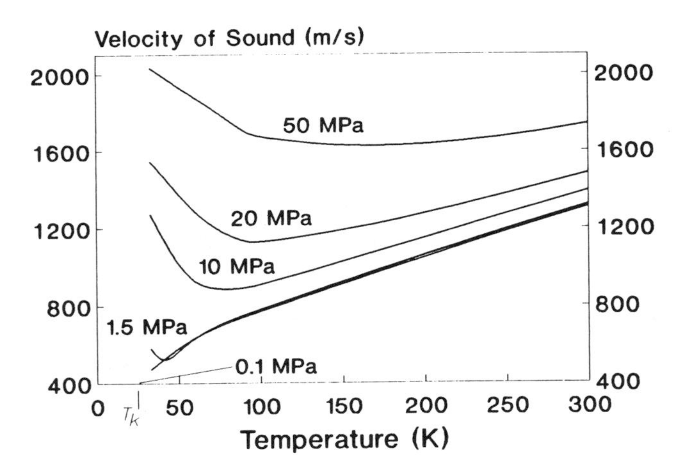

Figure 15 Sound Velocity in Hydrogen as a Function of Temperature and Pressure

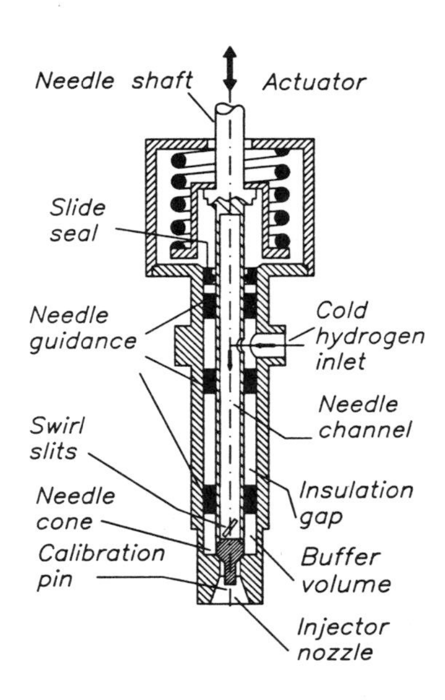

Figure 16 Cross-sectional View of Cryogenic Hydrogen Fuel Injector Designed for 15 Mpa Supply Pressure

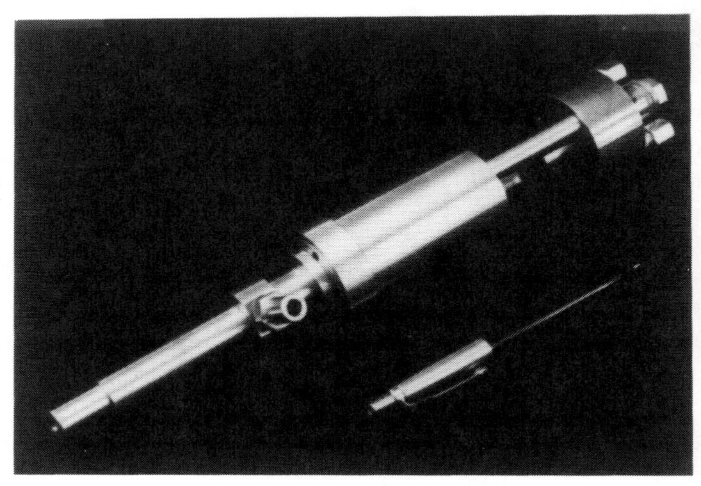

Figure 17 Photograph of Fuel Injector (Ref. Fig. 16)

Cryogenic Hydrogen Injector: The 15 MPa injector shown in Figure 16 consists of an injector shaft with injector nozzle and hollow injector needle that is guided in sleeves made of PTFE bronze.The hollow injector needle which conducts the hydrogen, is closed by means of spring tension against the cylinder pressure and is opened via a hydraulic linear actuator which admits pressure pulses. The pressure pulses, with changing pulse width, can be supplied via a diesel fuel injector pump or via an electrohydraulic device as is currently done with test vehicles. Thus, injection pressures for liquid hydrogen in the range from 10-15 MPa and above can be controlled safely /1, 23, 29-31, 35, 40, 41/.

The injector needle is pushed into the seat of the injector nozzle by a spring that is guided through a spring retainer made of titanium, and closes against the cylinder pressure. The cryogenic hydrogen flows in the upper part of the injector shaft into the hollow needle and, via three diagonally arranged swirl slits, into the buffer space. From there it flows into the nozzle via the opened needle seat and calibration pin, and then into the combustion chamber.

The injector was thermally designed so that the heat transfer to the cryogenic hydrogen would be as small as possible and, without high-vacuum class insulation, that is very difficult to provide in this application. Adequate thermal insulation of the hydrogen flowing into the hollow needle is, however, attained by exposing the injected hydrogen solely to that gas at rest in the insulation gap around the outside of the needle. Under suppressed convection, gaseous hydrogen has a heat conductivity near 10^{-3} W/cm.K. (This is compared to stainless steel with about 6.10^{-2} W/cm.K, aluminum 3 W/cm.K, PTFE 3.10^{-3} W/cm.K and air 2.10^{-4} W/cm.K.)

The maximum measured heat flow to the cryogenic hydrogen was about 35 W. For the heat flow coming from the combustion chamber wall (400°C) via the nozzle and the needle seat (1 mm width) about 2.5 W were obtained. This small value is obtained because a considerable portion of the heat flow in the bypass takes place through the outer part of the nozzle and the cylindrical injector shaft on the cylinder head (120°C). With an H_2 mass flow of about 0.8 g/s per cylinder (full load), the hydrogen warms up about 10 K, where the decrease in the mass flow has a more significant impact under partial load.

Linear Drive: An adequate prototype was developed for a hydraulic linear drive to operate this injector. A diagram of the hydraulic element with lapped piston is show in Figure 18. Hydraulic liquid under high pulsed pressure, supplied by a synchronously arranged Diesel injection pump, flows to the bottom side of the piston via inlet holes and lifts the piston and the injector needle, which is connected to it by the shaft, against the spring tension. The maximum stroke (up to about 1 mm) is constrained by a piston impact plate. When the bottom edge of the piston clears the drain hole, pressure no longer increases and the piston comes to rest, while the hydraulic fluid, which is regulated by the pressure pump, escapes via the drain hole. By appropriately matching the bottom side of the piston, oscillations of the piston during the opening phase can be avoided.

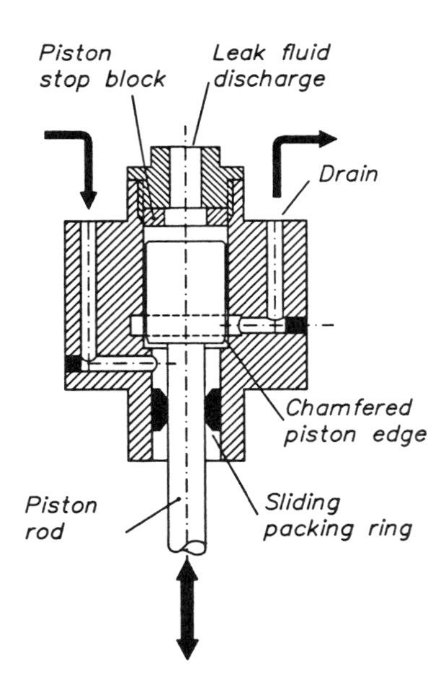

Figure 18 Hydraulically-actuated Linear Drive Unit for Operating Fuel Injector (Ref. Fig. 16)

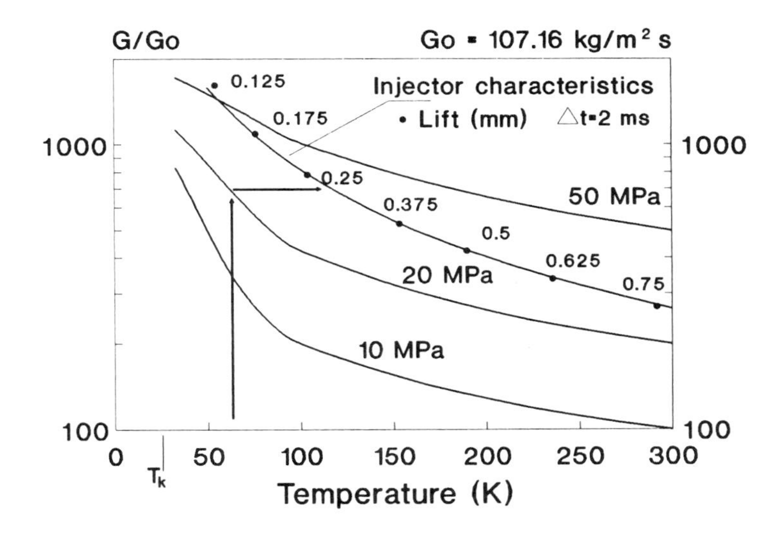

Figure 19 Normalized Steady Fuel Mass-flow Rate for Various Injector Lift Distances as a Function of Injector Inlet Pressure and Temperature

Figure 19 shows the steady mass flow rate as a function of the stroke for ambient hydrogen. Maximum mass flow rate is obtained at about a 0.8 mm stroke. With operation of the injector a minimum opening period of 1 ms was achieved by using the pressure pump. 0.4 ms of this are for the opening and closing process (needle movement).

As can be seen in Figure 14, the mass flow rate increases considerably with cryogenic hydrogen. The maximum values that are necessary for the BMW M30 test engine can be attained with an inlet pressure of about 20 MPa with a needle stroke of 0.25 mm and an opening period of about 3 ms (Fig. 19). Due to the shorter stroke, there was also a shorter opening and closing process, each of about 0.15 ms so that the shortest reproducible opening period that could be attained was about 0.5 ms. A comparison shows that with hydrogen at 80 K and 20 MPa pressure, the volume that had to be injected was at most (at stoichiometric) about 0.57 cm^3 compared to about 0.1 cm^3 with equivalent gasoline and Diesel fuel amounts. Cryogenic mixture formation under high pressure leads to fuel amounts to be injected per stroke of almost the same order.

Injector Actuation: Actuation via hydraulic pressure pulses has its lower limit with opening periods of about 0.5 ms with conventional in-line pumps. Here, the elasticity of the pressure lines as well as transient time effects also could have a negative effect on the control of pulse shape. Another "desirable" reduction in the pulse duration seems to be attainable in parallel with the development of appropriate hydraulic systems like distribution pumps, or pump elements, perhaps, that are directly actuated via the camshaft of the engine, as well as by selecting other linear drives.

Selecting extremely fast electromagnetic linear drives in an appropriate configuration is one way that has already been tried using ambient-temperature hydrogen /44, 45/. A disadvantage of this approach is the typical large reactive power requirements of this type of drive during the opening phase that can range between about 1-2 kW per injector. In this case appropriate power driving electronics with energy recovery features is required.

Piezohydraulic Drive: A potentially interesting process for a fast linear drive could be derived from electrostriction of piezoceramic materials like, for example, barium titanate, etc. Here, cutoff frequencies of up to about 100 kHz can be realized, in principle. However, the very small stroke of less than 0.1 mm, which is about the maximum presently technically feasible, has to be increased to about 0.5 mm. Apart from purely mechanical designs, hydraulic transmissions would be primarily suitable here.

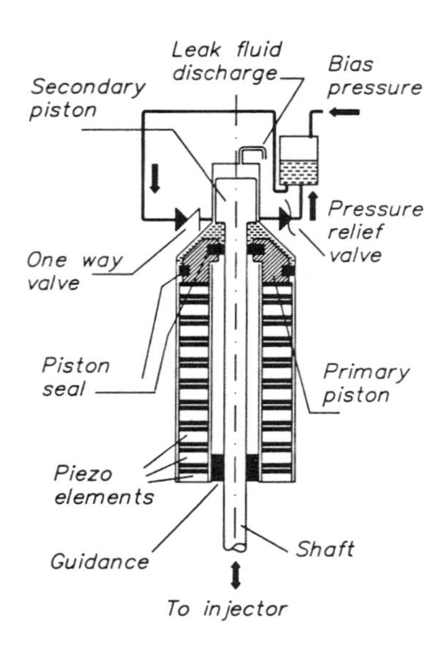

Figure 20 Cross-sectional View of Piezohydraulic Drive

Figure 20 is a diagram of a linear piezohydraulic drive on a test stand where a primary piston is moved by the displacement caused by the piezoelements. This primary piston produces a stroke of an appropriate size by means of a corresponding smaller secondary piston. The hydraulic system can be very compact and can be designed with cutoff frequencies far above the highest pulse frequencies used for injector activation. Static pressure admission of the hydraulic system from the pressure of the H2-fuel system reduces elastic effects of containment walls and compressibility of hydraulic liquid. Experience to date is, however, lacking concerning long-term behavior of those piezoceramic materials that may be suitable for this innovative type of linear drive.

Flow Geometry: As opposed to the Diesel engine and Otto engine with direct liquid-fuel injection, in which the evaporation of the fuel plays an important role, with supercritical cryogenic hydrogen an adequate volume of a cold gas has to be injected into the combustion chamber and very rapidly distributed. Here too, the formation of the physical fuel stream that is to be injected into the combustion chamber for effective mixing with the compressed air is ultimately important in the combustion process.

In addition to providing for adequate turbulence in the combustion chamber, a decisive factor for a quick mixing and homogenization process is to have as large a jet impulse as possible. This dictates that the mass density of the fuel at the stream boundaries be as close as possible to the density of the air in the combustion chamber in order to achieve optimal impulse exchange during mixing. An increase in the stream-surface area can be attained by placing from 8-10 individual radial or fan-shaped hydrogen streams passing over the combustion chamber. In this case, with one of the individual streams passing over the glow plug or electrodes of the spark plug to assure rapid ignition /23, 24, 31/.

However, from a practical standpoint, the fluid density within the injected hydrogen jet can only partially be brought to the air density in the combustion chamber. At the same pressure and temperature the air density is about 14 times higher than hydrogen. With liquids this corresponds to the density ratio of water to mercury. By using cryogenic hydrogen under high pressure the conditions near the stream exit improve considerably. Here, however, the stream rapidly expands with increased length under pressure reduction, temperature increase and expansion of the average mixture path. Appropriate adjustments of the injector design and combustion chamber geometry can provide more optimal conditions for effecting the mixing and combustion process.

Looking ahead, effort should be made to design a combustion chamber, which for the most part, is inside the cylinder head and symmetrically situated with respect to the cylinder's axis. It should have pressure zones near the periphery of the piston-face surface and include a hull in the piston floor. In this manner with two valve engines, the hydrogen injector as well as the spark plug can be located in the cylinder head with a tolerably small inclination of their axes to the cylinder axis. Also, with appropriate flow geometry, as in direct injection Diesel and methanol engines, the best conditions for rapid fuel/air mixing are attained, which with hydrogen leads to even shorter mixture times since no evaporation process is necessary.

Though intended, it was not possible to fully implement this design in the DLR BMW test engine shown in Figure 21. Engine and cylinder head were originally designed under the auspices of a joint collaborative project with BMW for internal mixture formation with early injection start. Here, the modification of the cylinder head mainly permitted the placement of the injector with a fuel stream that passes over the glow plug toward the exhaust valve. The injection of six streams in the direction of the cylinder inlet flow, which is determined by the relative location of the inlet and exhaust valve, was accomplished by means of an injector nozzle flap together with its associated bore holes for the flow exit.

Figure 21 BMW 6-cylinder Test Engine Arranged for Hybrid Mixture Formation (both low-pressure gaseous and high-pressure cryogenic hydrogen injectors)

Considering the fundamental importance of cryogenic hydrogen onboard high-pressure pumping, it was of interest to explore applicable developments within the cryogenic equipment manufacturing and cryogenic industrial gas industries (the "cryoindustry" community). Cryomec Basel has developed such machinery for liquid hydrogen service, however at a substantially larger scale than fitted automotive applications of interest. In a cooperative effort with the DLR, Cryomec engineered a single-cylinder "miniaturized" variant of their line of multicylinder cryopumps which was tested successfully at the DLR. This unit is shown in Figure 22 on the test stand. Still, it is sized more for potential commercial vehicle service (trucks, buses) providing higher flow rates than used in smaller automobile class applications.

Hybrid Mixture Formation: In this dual fuel admission approach, a combination of external and internal mixture formation is applied. Via external mixture formation, the cylinder is filled with a lean mixture (equivalence ratio < 0.6) that does not demonstrate any tendencies toward uncontrolled preignition or any substantial NOx formation. This is accomplished by means of a sequential intake manifold injection of hydrogen below 0.4 - 0.5 MPa. The mixture is ignited at about 40° b.t.d.c. Then, using internal mixture formation, high-pressure cryogenic hydrogen is injected into the flame front at about 5° b.t.d.c. under 20 MPa. Due to the conditions of the flame turbulence and the resulting flame front, problems regarding rapid mixture and ignition of the late injected fuel can now be disregarded. In the interest of necessitating only a correspondingly moderate combustion pressure rise thus (avoiding rough combustion,or knocking), the trend of the fuel injection over time and thus the combustion function, controls the cylinder pressure.

Figure 21 shows the low pressure solenoid injectors mounted on the newly developed intake manifold as well as cryogenic high pressure injectors that are mounted directly on the cylinder head. The high pressure injectors are hydraulically actuated by the diesel injection pump acting as an actuation- pressure pump. It is located on the right side below the engine. In the foreground of the right upper part of the figure can be seen the vacuum insulated cryogenic hydrogen line. From the limited work accomplished to date, it appears quite possible that hybrid mixture formation for hydrogen engines enables time-averaged near-stoichiometric (high power output) combustion, with very low NOx emissions and high operating efficiencies.

The hybrid mixture formation approach is further discussed in the later section, "Advanced Concepts for Liquid Hydrogen-fueled Vehicles."

Figure 22 *Cryomec-AG (Basel)* Compact Cryogenic Hydrogen Piston-type Pump (sized for truck/bus-class service)

Automotive Test Vehicles Using Liquid Hydrogen Fuel

As listed in Table 1, several liquid hydrogen test vehicles have been developed and tested since 1971 in the USA, Japan and the Federal Republic of Germany /1/. Vehicles were initially fitted with laboratory or commercial containers for liquid hydrogen whereas, in most cases, external mixture formation with ambient gaseous hydrogen was used for the engine operation.

TABLE 1

Automotive Test Vehicles Using Liquid Hydrogen Fuel

1. 1971:	Perris Smogless Automotive Association, /51/
2. 1973:	Billings Energy Cooperation /5/
3. 1973/74:	Los Alamos National Laboratory (LANL) /52/
4. 1973/74:	University of California, Los Angeles (UCLA) /53/
5. 1975:	Musashi Institute of Technology, Tokyo /54/
6. 1978/79:	DLR Stuttgart and University Stuttgart FKFS /55/
7. 1979:	Los Alamos National Lab./DLR joint project /8/
8. 1980:	Musashi Institute of Technology (Musashi V)/56/
9. 1981:	DLR Stuttgart /57/
10. 1982:	Musashi Institute of Technology (Musashi V)/58/
11. 1984:	DLR Stuttgart/BMW, joint project /7/
12. 1984:	Musashi Institute of Technology (Musashi VI) /59/
13. 1986:	BMW/DLR, joint project /39/
14. 1986:	Musashi Institute of Technology (Musashi VII) /30-32/
15. 1988:	BMW, new edition of the 1984-vehicle (27)

According to the authors' knowledge, vehicle (1), a Ford F250 pickup truck with a 6-cylinder engine, was the first automotive vehicle ever driven with onboard liquid hydrogen storage. A unique non-airbreathing engine conversion was demonstrated which ran on hydrogen/oxygen reactants. This precludes NOx emissions altogether. Liquid hydrogen and liquid oxygen were stored in Union Carbide-Linde commercial storage containers. Evidently, there are no other publications of test programs and corresponding test results from the Perris group since its pioneering work in 1971.

Up until 1988, Vehicles (9) and (11) were the only two operational liquid hydrogen test vehicles in Europe. A BMW test vehicle (long version of the new 750i) started undergoing tests in 1988. It has electronically controlled continuous individual intake port injection and a centrifugal supercharging compressor operated from the engine by a high-speed gear developed by the firm, Zahnradfabrik Friedrichshafen. It has a cylindrical 100 Liter liquid hydrogen storage tank supplied by Messer Griesheim GmbH.

The complete cryogenic fuel conditioning system, including a heat exchanger for warming liquid hydrogen up to ambient temperature, was designed and manufactured in accordance with the successfully proven LH2 fuel conditioning system of the DLR test vehicle (11) also by Messer Griesheim GmbH.

Further, BMW has developed a vehicle of this type without the charging compressor. Engine power data has also not been released. Vehicle (13), the BMW-DLR Joint Project Test Vehicle, a BMW 745i, was exhibited at the 6th World Hydrogen Conference in Vienna in 1988. It was the first LH2 test vehicle in Europe with direct cylinder injection of cryogenic hydrogen. In this process, cold hydrogen (40-100 K) is injected directly into the combustion chamber by hydraulically actuated cryogenic injectors at a maximum pressure of 2 MPa and electrically ignited. (Cryogenic internal mixture formation with early-start injection at approximately 150^{0} b.t.d.c.). The development of the cryogenic injectors, as well as the LH2 intermediate-pressure pump, installed outside the LH2 tank and driven via a hydraulic motor, were the essential DLR contributions to this vehicle. Due to technical difficulties associated with the project this promising system could not be tested.

Safety Aspects of Liquid Hydrogen-fueled Vehicles

No fully credible statements can currently be made about effects of accidents on the LH2 fuel system and specific damage that can arise. The implied database simply does not exist. On one hand, there are parallels to liquid propane gas (LPG) and liquid methane (LNG), but on the other hand these fuels cannot be directly compared to liquid hydrogen because, for example, their behavior when openly spilled and dissipating in the air is completely different.

Furthermore, from NASA experience /47/ in dealing with liquid hydrogen it can be inferred that, in accidents to date with LH2 tank vehicles, the chain of events of accidents with other flammable liquids, for example gasoline, liquid propane gas or ethylene, instead of LH2, in most cases would have had more dramatic consequences. In the case of commercial aircraft applications, liquid hydrogen is estimated to be substantially *safer* than either conventional Jet-A (kerosene) or liquid methane (approximated by LNG) in the instance of a survivable crash accident /1/.

In order to obtain a reliable answer to the questions posed here,clearly it will be necessary to gain considerable further experience, and to carry out corresponding demonstration vehicle driving test programs.

From carrying out spill experiments with both small and large amounts of liquid hydrogen as well as from results of tests carried out with liquid hydrogen in aeronautics, statements can be made which support the opinion that liquid hydrogen would not be worse in this regard than gasoline or liquid propane gas when weighing its advantages and disadvantages /48/. This statement also corresponds to the actual driving experience to date involving a very limited number of accidents experienced with LH2-test vehicles. A more detailed accounting of this experience will be found in /1/.

According to the extensive amount of literature concerning technical safety questions in dealing with hydrogen, which has been available for some time, the safe handling of hydrogen is currently state-of-the-art fact in industry and commerce practice. In this regard hydrogen does not differ extensively from other liquid and gaseous energy carriers. All are potentially hazardous, as is any fuel. Hydrogen-air reactions deflagrate in the open and therefore result in a small increase in pressure,generally resulting in a relatively low degree of destruction.

In enclosed or partially enclosed areas, any fuel-gas/air mixture can result in a detonation, which -- if it occurs -- can cause high-level destruction. In this process, characteristic features of hydrogen are its high flame speed in air, wide range of flammability and detonability. However, it is generally observed that the high destructive pressures in detonations, only occur in enclosed areas. However, these facts do not hinder the author in /49/ from certifying that hydrogen-air mixtures in enclosed areas demonstrate a typical deflagration behavior, that is, almost no tendency toward detonation.

In contrast to gasoline and methane in the vapor phase, gaseous hydrogen tends to disperse very quickly in the air. This is a result of hydrogen's strong buoyancy due to its low density (about 1/12 of the density of air) and the associated free turbulent convection which occurs naturally. This dissipative action is not caused by the diffusion processes which take place at a much slower pace as alleged in some articles concerning hydrogen safety (see for example /50/).

The positive experience with the widespread use of town gas in earlier times, which contains a high percentage of hydrogen, as well as -- more recently -- with liquid hydrogen, now a staple fuel of the space programs, internationally. The safety record of liquid hydrogen associated industrial practitioners is impressively good as well. Also, as observed above, on a much more limited basis, those automotive vehicles used to demonstrate the applicability of liquid hydrogen in transportation to date have not revealed any innate or insurmountable technical safety problems /1, 47, 48/. (See Table 1.)

Advanced Concept Liquid Hydrogen-fueled Vehicles

What can now be proposed as "the way to go" in any further development of the liquid hydrogen-fueled automotive vehicle?

Based on experience to date with test vehicles and test engines, the importance of the following concepts becomes evident when taking advantage of the cryogenic properties of liquid hydrogen in order to improve engine operation while reducing NOx emissions:

- Cryogenic external mixture formation.
- Cryogenic internal mixture formation and electronic ignition as well as early- and late-start injection.
- Combination of external with internal mixture formation (hybrid system).

<u>Hybrid Mixture Formation</u> -The third technical approach, previously introduced, has been embraced in the general layout and flow schematic diagram of Figure 23, a projected, "advanced concept" vehicle power system. Aspects of this concept are further discussed below for consideration by today's advanced automotive engineering community.

Cryogenic internal mixture formation with late start injection corresponds to conventional Diesel operation and, in view of its controlled combustion characteristics, offers the largest potential for full-power, low pollution, or perhaps, essentially *pollution-free* combustion. Whereas, with cryogenic external mixture formation, the pressure level in the fuel tank is sufficient for engine operation, fuel pressurization with reciprocating cryopumps is necessary for effecting hydrogen internal mixture formation, as previously indicated.

With early-start injection, this can be effectively achieved with an electrically or hydraulically driven one-stage liquid hydrogen pump installed in the immediate vicinity of, but separated from, the fuel tank. In order to avoid two phase flow during the intake stroke of the pump an adequate pressure level in the fuel tank (approximately 0.1-0.2 MPa) and a low stroke rate (300-400/min) are necessary in addition to supercooling of the cryogenic liquid to avoid vapor formation at the pump inlet.

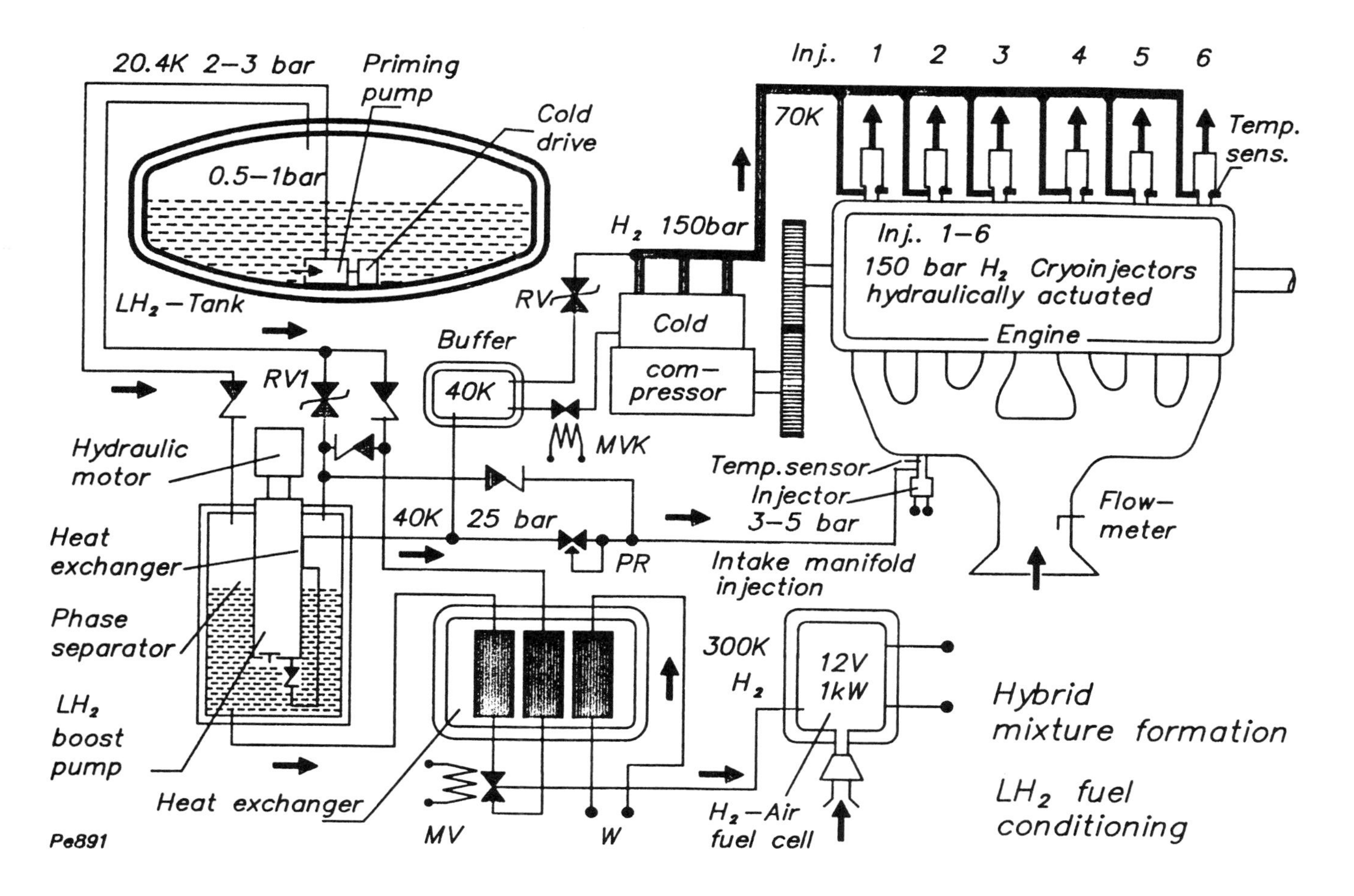

Figure 23 Overall Vehicle Control and Flow Schematic Diagram for an Envisioned Hybrid Mixture Formation Type Cryogenic Hydrogen Automotive Power System

The fuel control for the engine is effectively implemented via variation of the opening period of the cryogenic fuel injectors. Application of digital-control overall engine management also enables dual-fuel operation, i.e., the use of either hydrogen or gasoline as the fuel, without a large amount of transition effort. Furthermore, it should be noted that such dual fuel operation is considered a practical necessity in the routine conduct of over-the-road hydrogen vehicle driving tests. Each DLR test vehicles for instance (Table 1), permits gasoline operation as well as hydrogen operation using digital engine management to facilitate the switchover.

With the external mixture formation facet of the hybrid fueling approach readily handled by virtue of inherent tank pressure availability, let us consider the technical approach for effecting internal mixture formation and late start injection (see Figure 23):

Contrary to current experience, pressurization up to the necessary injection pressure of over 10 MPa is achieved in a minimum of two pumping stages (Fig. 23). The electrically or hydraulically driven LH_2 pump is maintained next to the LH_2 tank for the purpose of supercritical fuel pressurization up to approximately 1.5-2.5 MPa and warming of hydrogen to the critical temperature (33 K) , thus avoiding two-phase conditions /46/.

At this point, the fluid cryogenic hydrogen reaches the engine compartment via the cryogenic feed line where it is brought to the necessary injection pressure via a multicylinder, high pressure cold compressor which is directly driven by the engine. The design of this cold compressor is not critical with regard to flow orifices. Furthermore, most of the compression work (1-2 kW) is efficiently performed directly by the vehicle's engine.

This system approach demonstrates high redundancy despite a considerably sophisticated design. For example, upon failure of either the cold compressor or intermediate pressure pump, engine operation is then still possible with reduced power output. Upon simultaneous failure of the intermediate and high-pressure pump, emergency operation can still be carried out via external mixture formation by means of the LH_2 tank pressure. The same procedure can be applied to the start-up cooldown phase transient, to be experienced with warmed up pumps after longer periods of vehicle non-operation.

Early Start Injection May be Preferred for the Automobile - Cryogenic late-start internal mixture formation is seen to be especially applicable for engines with a larger piston displacement than normally found in passenger vehicles, such as used in commercial vehicles such as trucks and buses. In fact, due to its considerable complexity as well as the increased need for space on the cylinder head, this approach may not be optimal for light-duty passenger automobiles.

If so, electronically controlled cryogenic internal mixture formation with early *start* injection, possibly in combination with timed external mixture formation, would seem to be especially attractive for use in engines with medium piston displacement and medium power output.This avoids the need for the very high-pressure processing equipment otherwise needed (as reflected in Figure 23). Here it is tacitly assumed that the internal combustion engine will still be the optimal solution for automotive vehicle power for a long time. (Hydrogen is also applicable to other engine types, e.g., gas turbines.)

Onboard Cryohydrogen Storage Container Variants - While in commercial vehicles with centralized infrastructure (refueling) and nearly 24-hour-operation, relatively simple LH2-tanks with an acceptably high boiloff rate (5-8% per day) can definitely be used; stricter requirements have to be fulfilled for use in smaller passenger vehicles.

Based on current experience with hydrogen storage onboard vehicles, a tank with nearly prismatic shape, installed behind the back seat, is an attractive solution with regard to safety aspects, which leaves about 60% of the trunk space available in the vehicle. Efforts should be made to achieve a maximum boiloff rate of about 1-1.5% per day for passenger vehicles. As reported here, this level of performance has been achieved in the several cylindrical containers developed and demonstrated. But even this small amount of boil-off hydrogen is not wasted.

The vented hydrogen can be used to produce electricity via a small (yet to be developed) fuel cell battery, one suitable for automotive auxiliary power application. With it, the starter battery can also be recharged when the vehicle is not in operation. It is possible that the currently conventional engine-driven electric alternator does not have to be used at all. This would also definitely be attractive from an energy conversion efficiency standpoint because the overall fuel related efficiency of the electrical power generation is very low in automotive vehicles, especially in urban traffic. [Fuel cell efficiencies of > 60% are available vs. engine-alternator values estimated to be < 20%]

Another possibility which would avoid the emission of vaporized hydrogen into the environment from a long-stationary vehicle is through the use of a small supplementary metal hydride storage unit. The absorbed hydrogen would be later released, while driving, by heat conducted from the exhaust gas of the engine. It would be fed to the engine in addition to the fuel deriving from the LH2-tank. In extending this concept, such a supplementary hydrogen storage means could serve as an emergency reserve permitting limited-range driving with an empty LH2-tank.

The advanced concepts for LH2-vehicles described here, along with the basic accomplishments covered in the paper, can be seen to offer a sound basis for the further development of currently operational test vehicles, and for the field demonstration of vehicle fleets of vehicles which use liquid hydrogen fuel. For such limited application, only a minimal LH2 supply infrastructure would be needed.

The *perceived* impracticability of using existing gasoline station type operations for liquid hydrogen service seems to be, next to the question of economical hydrogen production and delivery through the infrastructure, one of the main arguments raised against instigating a near/medium-term, general introduction of hydrogen in private automotive traffic operations, despite the fact that hydrogen is obviously far less harmful to the environment and has long-term advantages regarding availability (see "Introduction").

But the DLR work described here provides more than a convincing argument that this hydrogen-servicing perception, along with others generally held, may now be seriously questioned. Technical feasibility for such developments is now in hand.

Summary

Terrestrial vehicles will continue to use internal combustion engines to a greater or lesser degree, whereas battery-electric drives will continue to be reserved for special applications. Liquid hydrogen is a candidate alternative fuel offering unique strategic advantages with regard to the environment and assured long-term energy supply. It is shown to be favorably compatible with future automotive mobility needs.

Efforts conducted by the DLR over the past two decades, focusing on cryogenic liquid hydrogen automotive technology development and demonstration, have directly and positively responded to the three areas of technical concern, which, with regard to this approach, are often evidenced by today's transportation community:

1) Hydrogen fueling means and operation,

2) Onboard hydrogen storage and processing means,

3) Internal combustion engine design and operation.

Technical feasibility for each of these areas, in the realistic hardware context of liquid-hydrogen fueled passenger automobile development and operations, has now been convincingly demonstrated.

Cryogenic liquid hydrogen is clearly the preferred form of this fuel from the standpoint of both general transportation-energy transport and storage requirements. More specifically it is found to be highly advantageous for both vehicle refueling operations and onboard storage. With regard to propulsion means, hydrogen/air mixture formation in internal-combustion engines can also be significantly improved by using the cryogenic form, rather than ambient-temperature gas.

In addition to external mixture formation that leads to increased power per piston displacement levels and reduced NOx emissions as a result of charge cooling with cryogenic hydrogen, considerable improvements become available with *internal* mixture formation calling for injection of H2 at medium to high pressure levels. Practicably, needed levels of fuel pressurization can only be accomplished through LH2 processing with onboard cryopumps.

The beneficial cryogenic characteristics of hydrogen such as its higher injectant density, and its considerable charge-cooling effect enhance the fuel injection and mixing process and, thereby, the basic combustion process. Hybrid mixture formation, providing a possible optimal integration of external *and* internal mixture formation with hydrogen, appears very attractive with respect to power and torque production, all at the highest efficiency level.

About the Authors

Prof. Dr.Ing.habil, Walter Peschka, a career technical staff member of the Deutsche Forschungsanstalt fuer Luft- und Raumfahrt (DLR), Stuttgart, Germany, is the lead author of this paper. Over the past two decades he has focused his professional activities on strategic issues of energy and the environment, specializing on aspects of hydrogen energy applications. His book, *Liquid Hydrogen, Fuel of the Future*, originally published in German and now available in an English version /1/ is considered a landmark publication on the subject addressed in this paper. The extensive liquid hydrogen-fueled automotive technology development and demonstration accomplishments of the DLR, reviewed in the paper, were both originally instigated and personally led by Dr. Peschka, throughout the course of the work.

Mr. William J.D. Escher, associate author, is currently a program manager in the Transportation Division, Office of Advanced Concepts and Technology in the Headquarters organization of the National Aeronautics and Space Administration, Washington, D.C., U.S.A. He has specialized in aerospace propulsion and power technology development, and its potential application to terrestrial systems. Earlier, he was involved as an independent researcher in broad aspects of the projected Hydrogen Energy System, and is the founding secretary of the International Association for Hydrogen Energy. As a special adviser to the U.S. Department of Energy in the 1970's through the mid-1980's, he actively supported hydrogen facets of DOE's Alternative Fuels Utilization Program. He is a member of the AIAA and *SAE International.*

References

/0/ Sloop, J.L., Liquid Hydrogen as a Propulsion Fuel, 1945-1959, National Aeronautics and Space Administration Publication SP-4404, 1978

/1/ Peschka, W.: Liquid Hydrogen, Fuel of the Future. 303 pp, Springer-Wien, New York, 1992

/2/ Pohlenz, J.B.; Stine, L.O.: New Process Promises Low Cost Hydrogen. Oil and Gas J., 60, 82-85, (1962) see also: Pohlenz, J.B.; Stine, L.O.: New Process Makes Hydrogen from Fuel Gases, Hydrocarbon Processing and Petrol. Ref., 41, 191-194, (1962)

/3/ Peschka, W.: The Use of Hydrogen for Vehicles, Wasserstoff als Kraftstoff für bodengebun dene Fahrzeuge. Paper in English. VDI Rep. Nr. 1020, 279-300, 1992

/4/ Povel, R.; Töpler, J.; Withalm,G.; Halene, C.: Hydrogen Drive in Field Testing. Hydrogen Energy Progress V, Vol. 4. London/New York: Pergamon Press 1984, 1563-1578.

/5/ Billings, R.E.: Hydrogen Storage for Automobiles Using Cryogenics and Metal Hydrides. Proc. The Hydrogen Economy (Miami) Energy (THEME) Conf., 1974

/6/ Stewart, W.F.: Hydrogen as a Vehicular Fuel. In: Williamson, K.D., Jr., Edeskuty, F.J. (eds.): Recent Developments in Hydrogen Technology, Vol.2, pp.69-146, Cleveland Ohio CRC Press (1986).

/7/ Peschka, W.: Liquid Hydrogen for Automotive Vehicles - Status and Development in Germany. Cryogenic Processes and Equipment. New York: ASME 1984, 97-104, Libr. of Congr. Cat. Card. No. 84-72999

/8/ Stewart, W.F.: Operating Experience with a Liquid Hydrogen Fueled Buick and Refueling System. Proc. 4th WHEC, Vol. 3, (1982), 1071-1093. Int.J. Hydrogen Energy 9, 1984, 525-538

/9/ Peschka, W.: The Status of Handling and Storage Techniques for Liquid Hydrogen in Motor Vehicles. Int.J.HydrogenEnergy, Vol.12, No.11, 753-764, 1987

/10/ Peschka, W.: Hydrogen Combustion in Tomorrow's Energy Technology Int.J.Hydrogen Energy, Vol.12, No.7; 481-500, 1987

/11/ Withalm, G., Gelse, W.: The Mercedes Benz Hydrogen Engine for Application in a Fleet Vehicle. In: Proc., 6th World Hydrogen Energy Conf., Vol.3, pp. 1185-1198 Oxford: Pergamon Press (1986).

/12/ Fiala, E., Willumeit, H.P.: Schwingungen in Gaswechselleitungen von Kolbenmaschinen, MTZ 28, p.144-151, (1967)

/13/ Seifert, H.: Die charakteristischen Merkmale der Schwingrohr- und Resonanzaufladung. SAEpaper 82032 (1982)

/14/ Binder, K., Withalm, G.: Mixture Formation and Combustion in Interaction with the Hydrogen Storage Technology. In: Proc. 3th World Hydrogen Energy Conf., Tokyo, Vol. 2, pp.1103-1117 Oxford : Pergamon Press 1980.

/15/ Lenz, H.P.; Duelli, H.: Neue stufenlos längenvariable Sauganlage für optimalen Drehmomentverlauf eines Einspritzmotors. ATZ 89, 6, (1987)

/16/ Lenz, H.P.: Mixture Formation in Spark-Ignition Engines. Springer Verlag Wien New York, SAE Warrendale, PA, 400 p, 1992

/17 May, H., Gwinner, D.: Möglichkeiten der Verbesserung von Abgasemissionen und Energieverbrauch bei Wasserstoff-Benzin-Mischbetrieb, MTZ, 4, (1981)

/18/ McCarley, C. A., Van Vorst, W.D.: Electronic Fuel Injection Techniques for Hydrogen Powered I.C. Engines. Int. J. Hydrogen Energy 5, 179-204 (1980).

/19/ Adt, R. R., Jr., Swain, M. R., Pappas, J. M.: Hydrogen Engine Performance Project. U.S. Dept. of Energy (DOE), Second Annual Report, Contr.No. EC-77C03-1212, 1980. See further DOE/CS/31212-1, Washington DC., (1983)

/20/ Homan, H. S., De Boer, P. T. C., McLean, W. J.: The Effect of Fuel Injection on NOx-Emissions and Undesirable Combustion for Hydrogen-Fueled Piston Engines. Int. J. Hydrogen Energy 8/ No. 2 (1983).

/21/ Peschka, W.: Liquid Hydrogen Reciprocating Pumps for Automotive Application. In: Adv. Cryog. Eng., Vol. 35 B, pp. 1783 - 1790, New York: Plenum Press (1990)

/22/ Peschka, W.: Liquid Hydrogen Pumps for Automotive Application. Int.J.Hydrogen Energy, 15, pp. 817-825, (1990)

/23/ Furuhama, S., Fukuma, T., Kashima, T.: Liquid Hydrogen Fuel Supply System for Hot Surface Ignition Turbocharged Engine. In: Cryogenic Processes and Equipment, ASME, pp. 105 - 114 (1984).

/24/ Furuhama, S., Fukuma, T.: Liquid Hydrogen Fueled Diesel Automobile with Liquid Hydrogen Pumps. In: Adv. Cryog. Eng., Vol. 31, pp. 1047 - 1056 New York: Plenum Press 1986.

/25/ Furuhama, S.: Hydrogen Engine Systems for Land Vehicles, Int. J. Hydrogen Energy. 14/ No. 12, 907 - 914 (1989).

/26/Stewart, W. F.: Operating Fueled Buick and Refueling System, Int J. Hydrogen Energy, 9/ No. 6, 525 - 538 (1984).

/27/ Regar, N., Strobl, W., Heuser R.: Pkw-Antriebe mit Elektro- und Wasserstoffspeicher (Stand der Entwicklungen und Perspektiven). In: Proc. 2. Aachener Kolloqium, Fahrzeug- und Motorentechnik, pp. 463-483 (1989).

/28/ Peschka, W.: Liquid Hydrogen for Automotive Vehicles-Status and Development in Germany. In: Cryogenic Processes and Equipment, ASME, pp. 97 - 104 (1984).

/29/ Peschka, W.: The Status of Handling and Storage Techniques for Liquid Hydrogen in Motor Vehicles, Int. J. Hydrogen Energy, 12, 753-764, (1987).

/30/ Takiguchi, M., Furuhama ,S.: Combustion Improvement of Liquid Hydrogen Engine for Medium Duty Trucks. SAE-Techn. paper 870535 (1987)

/31/ Furuhama, S., Fukuma, T.: High Output Power Hydrogen Engine with High Pressure Fuel Injection, Hot Surface Ignition and Turbocharging. Int.J.Hydr.En. 11, 399-407, (1986)

/32/ Furuhama, S.: Hydrogen Engine Systems for Land Vehicles, Int. J. Hydrogen Energy. 14/ No. 12, 907 - 914 (1989).

/33 /Takiguchi, M., Furuhama S.: Combustion Improvement of Liquid Hydrogen Engine for Medium Duty Trucks. SAE- Techn. paper 870535, (1987)

/34/ Withalm, G., Gelse, W.: The Mercedes Benz Hydrogen Engine for Application in a Fleet Vehicle. In: Proc., 6th World Hydrogen Energy Conf., Vol.3, pp. 1185-1198 Oxford: Pergamon Press (1986).

/35/ Peschka, W.: Liquid Hydrogen-Cryofuel in Ground Transportation. In: Adv. Cryog. Eng., Vol. 31, pp. 1035 - 1046 New York Plenum Press 1986.

/36/ Peschka, W.: Liquid Hydrogen as a Vehicular Fuel - A Challenge for Cryogenic Engineering. Int. J. Hydrogen Energy 9, 515-525 (1984)

/37 /McCarley, C. A.: Development of a High Speed Injection Valve for Electronic Hydrogen Fuel Injection. In: Proc., 3rd World Hydrogen Energy Conf., Vol. 2, pp. 1119-1134 New York: Pergamon Press 1980.

/38/ Furuhama, S., Hiruma, M., Enemoto, Y.: Development of a Liquid Hydrogen Car. Int. J. Hydrogen Energy 3, 61-81 (1978).

/39/ Peschka, W., Nieratschker, W.: Experience and Special Aspects on Mixture Formation of an Otto-Engine Converted for Hydrogen Operation. Int.J.Hydrogen Energy, 11/ No. 10, 653 - 660 (1986).

/40/ Furuhama, S., Fukuma, T.: High Output Power Hydrogen Engine with High Pressure Fuel Injection, Hot Surface Ignition and Turbocharging Int.J.Hydrogen Energy, Vol.11, No.6, 399-407, 1986

/41/ Krepec, T., Giannacopoulos, T., Miele, D.: New Electronically Controlled Hydrogen-Gas Injection Development and Testing. Int. J. Hydrogen Energy, 12/ No. 12 (1987).

/42/ Varde, K. S., Frame, G. M.: A Study of Combustion and Engine Performance Using Electronic Hydrogen Fuel Injection, Int. J. Hydrogen Energy, 9/ No. 4, 327 - 332 (1984).

/43/ Mc. Carty, R. D., Hord, J., Roder, H. M.: Selected Properties of Hydrogen. NBS Monograph 168 (1981).

/44/Seilly, A.H.: Colenoid Actuators- A New Concept in Extremely Fast Acting Solenoids. SAE-paper 810 462, (1981)

/45/ Krepec, T., Giannacopoulos, T., Miele, D.: New Electronically Controlled Hydrogen-Gas Injector Development and Testing. Int. J. Hydrogen Energy, 12/ No. 12, 855 - 862 (1987).

/46/ Peschka, W.: Cryogenic Fuel Technology and Elements of Automotive Propulsion Systems. Adv. in Cryogenic Engineering 37, Plenum Press, New York, 1992

/47/ Ordin, P.M.: Review of Hydrogen Accidents and Incidents in NASA Operations NASA-Tm-X71565, 1974

/48/ Hord, J.: Is Hydrogen Safe? NBS-Technical Note 690, 1976

/49/ Winter, C.J.: Mit Wasserstoff-Antrieb ins nächste Jahrhundert. Luft-und Raumfahrt, 4, pp.26-31, 1989.

/50/ Fischer, M., Eichert, H.: Sicherheitsaspekte des Wasserstoffs. DFVLR-Nachrichten, 34, pp.33-37, 1981.

/51/ Underwood, O.L.; Dieges, P.B.: Hydrogen and Oxygen Combustion for Pollution Free Operation of Existing Standard Automotive Engines. Proc. 6th IECEC, SAE-Paper 7/9046, 1971.

/52/ Stewart, W.F.; Edeskuty, F.J.; Williamson, K.D.; Lutgen, H.N.: Operating Experience with a Liquid Hydrogen Fueled Vehicle. Los Alamos Scientific Lab. LA-VR-74-1637, erschienen auch in: Adv. Cryog.Eng. 20, 1974, 82-89

/53/ Finegold, J.G.; Van Vorst, W.D. et al.: The UCLA Hydrogen Car: Design, Construction and Performance. Trans. SAE 730507, 1973

/54/ Furuhama, S.; Hiruma, M.; Enemoto, K.: Development of a Liquid Hydrogen Car. Int. J. Hydrogen Energy 3, 1978, 61-81

/55/ Peschka, W.; Carpetis, C.: Cryogenic Hydrogen Storage and Refueling for Automobiles. Int. J. Hydrogen Energy 5 (1982), 619-626

/56/ Furuhama, S.; Kobayashi, Y.: A Liquid Hydrogen Car with a Two Stroke Direct Injection Engine and LH2-Pump. Int. J. Hydrogen Energy 7, 1982, 809-820

/57/ Peschka, W.: Liquid Hydrogen for Automotive Vehicles - Experimental Results. ASME-Paper No. 81-HT-83 (1981)

/58/ Furuhama, S.; Kobayashi, Y.: Development of Hot Surface Ignition Hydrogen Injection TwoStroke Engine. Proc. 4th WHEC, Vol. 3, Adv. in Hydrogen Energy, 1009-1020, New York: Pergamon Press 1982

/59/ Furuhama, S.; Fukuma, T.; Kashima, T.: Liquid Hydrogen Fuel Supply System to Hot Surface Ignition Turbocharged Engine. Cryogenic Processes and Equipment - 1984, ASME, 1984, 105114.

/60/ Varde, K. S., Frame, G. A.: Development of a High-Pressure Hydrogen Injection for SI Engine and Results of Engine Behaviour. Int. J. Hydrogen Energy, 10/ No. 11, 743 - 748 (1985).

Chapter 3
Fuel Cells

3.1 Introduction

Historical Background

Sir William R. Grove, a trained lawyer and knighted judge, is generally regarded as the discoverer of the fuel cell. Grove constructed the world's first fuel cell in 1839, using hydrogen for fuel, oxygen as the oxidizing agent, and dilute sulfuric acid as the electrolyte. After this experiment, scientists of the 19th century scrambled to develop fuel cells that used different fuels and electrolytes. Mond and Langer made important contributions to the development of hydrogen fuel cells by introducing various designs that would ensure as much contact as possible between the gases and the electrolyte [1]. Another scientist, A. C. Becquerel, worked to develop a carbon-burning cell that employed a fused salt electrolyte. The cell featured a carbon and a platinum electrode immersed in a fused nitrate.

Near the turn of the century, the immense popularity of electrical energy and the low efficiencies of power generation (about 2.6% efficiency for coal plants at the time) resulted in another push for the development of an efficient source of electricity. It was in this setting that Dr. W. W. Jacques developed a fuel cell for power generation that used carbon as a fuel. The cell reportedly had an efficiency of about 80%. Jacques failed to give reliable information on cell operation, overvoltages, and duration, however.

The first major fuel cell development project in the 20th century was launched by Francis T. Bacon, an engineer associated with Cambridge University in England and a descendant of the famous seventeenth century philosopher and scientist by the same name. Bacon focused his research on the H_2/O_2 alkaline fuel cell using nickel electrodes because he thought that the expensive platinum catalyst systems used in earlier research would prevent mass commercialization of the fuel cell. Bacon argued that nickel, at about 205° C, was sufficiently active to eliminate the need for any other catalyst. He was also the first to introduce porous gas diffusion electrodes to increase the contact between the gases and the electrolyte. Bacon's research culminated in 1959 with the demonstration of a remarkable 5 kW system capable of simultaneously powering a welding machine, a circular saw, and a two-ton capacity fork-lift truck [2].

The Bacon fuel cell system served as the foundation for systems used in the Gemini and Apollo space flights. NASA's interest in space applications pushed the development of fuel cells even further. General Electric (GE) and United Technologies Corporation (UTC), under contracts from NASA, soon became the leaders in fuel cell technology.

It wasn't until the energy crisis of 1973 and the Arab oil embargo, however, that these companies started to seriously consider terrestrial applications for fuel cells. Because of the energy crisis, the Energy Research and Development Administration (ERDA), now the U.S. Department of Energy (DOE), and the Electric Power Research Institute (EPRI) stressed the need for conservation of petroleum-derived fuels and for increasing the utilization of coal, natural gas, and renewable energy resources [3]. At that point, a push for fuel cell powered vehicles started that continues to this day.

Potential Benefits of Fuel Cells

Fuel cells are conceptually simple, and their efficiency, modularity, environmental characteristics, and siting flexibility permit their use in a variety of applications [4]. Applications of fuel cells range from large-scale power plants to vehicular applications to 1.5 V batteries for small gadgets. The operation of a fuel cell is also completely reversible, i.e., a vehicle could be fueled with hydrogen overnight by operating the cell in reverse as an electrolyzer (refer to Chapter 4). All one needs is a source of electric power and a source of water.

The principal advantage of fuel cells is that they offer the potential to substantially lower, if not eliminate, conventional pollutants. A fuel cell using hydrogen directly as a fuel would produce only water during operation. If the hydrogen was supplied via the reformation of methanol, the only products aside from water would be CO_2, evaporative emissions from the methanol fuel tank, and pollutants from the reformer itself. These pollutants would be only a fraction of the emissions from a more conventional internal combustion engine operating on gasoline, however [5]. Since the operational temperatures of fuel cells are considerably less that those of conventional combustion engines, the formation of nitrogen oxides is also eliminated.

Fuel cells can also achieve higher efficiencies than conventional combustion engines. The efficiency of a heat engine is limited by thermodynamic principles to about 35% as purported by Sadi Carnot's *Réflexions Sur La Puissance Motrice du Feu* (Reflections on the Moving Power of Fire). The fuel cell, on the other hand, obtains its energy through a direct and reversible chemical process and is, therefore, not subject to the Carnot cycle limitation (see discussion in Section 3.3). This allows fuel cells to operate at an efficiency two to three times greater than that of a conventional heat engine.

The difficulties of developing an adequate fuel cell system for vehicles should not be underestimated, however. While considerable progress has been made in the development of fuel cell technologies over the past decade, and especially in the last five years, many barriers remain, including cost, weight, and bulkiness. Some prototype vehicles have demonstrated that fuel cell vehicles can have low emissions and a reasonable range, but development, design, and operational problems will have to be solved before fuel cells can be mass-produced. Still, in view of the potential of fuel cells to operate at high efficiencies with no emissions, it is likely that development work and the scope of applications for fuel cells will continue to grow into the next century.

3.2 What is a Fuel Cell?

A fuel cell is a galvanic cell (i.e., a battery) which converts chemical energy into electrical energy by means of an electrochemical process. As with a battery, a fuel cell is a device in which electron transfer is forced to take place through an external pathway rather than directly between reactants. A fuel cell operates by taking in fuel and an oxidant (usually pure oxygen or oxygen in air) at separate electrodes and converting the excess chemical energy into direct current (DC) electricity. The fuel source is typically either hydrogen or a hydrogen carrier such as natural gas or methanol. Fuel cells produce minimal pollution when fueled with a hydrogen carrier and zero pollution when operated on hydrogen. In the case of a H_2/O_2 fuel cell, water and heat are the only by-products.

Unlike conventional batteries, a fuel cell does not utilize the material that it is composed of as a fuel source. A fuel cell takes a fuel from the *outside* and converts it into electrical energy. Thus, fuel cells do not suffer from some of the limitations of conventional batteries in that they do not have to be discarded, as in the case of non-rechargeable batteries, or recharged, as in the case of rechargeable batteries.

Design and Operation of a Generic Fuel Cell

A fuel cell consists of two porous electrodes where the energy conversion process takes place, and a solid or liquid electrolyte which conducts ions to form a closed circuit. A generalized hydrogen fuel cell is shown in Fig. 3.1.

The fuel molecules, in this case hydrogen, are introduced on the anode side of the fuel cell. At the anode, which is sometimes called the fuel electrode, the hydrogen mole-

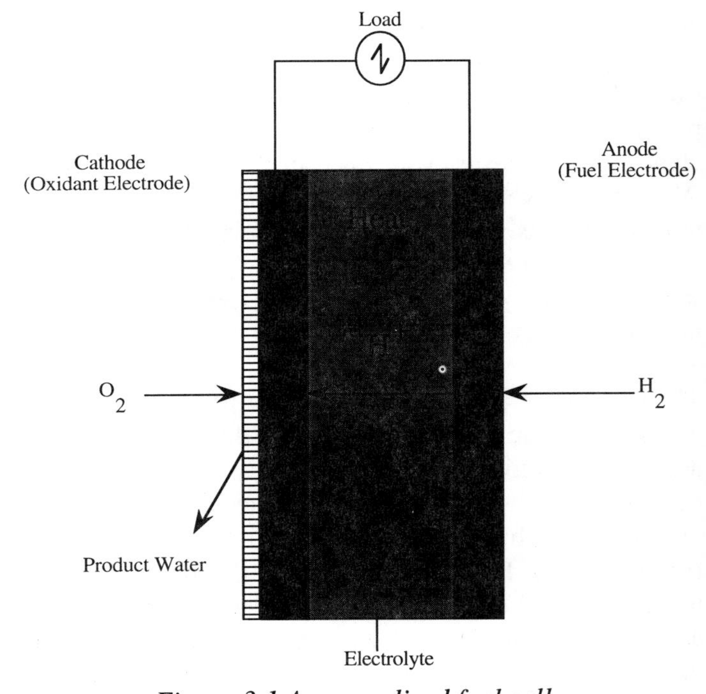

Figure 3.1 A generalized fuel cell.

cules give up their electrons, resulting in the formation of positively charged hydrogen ions as follows:

$$H_2 \Rightarrow 2H^+ + 2e^-$$

These electrons are subsequently drawn to the cathode by the oxidizing agent, oxygen in this case, resulting in the formation of negatively charged oxygen ions. As the electrons move through the external circuit connecting the cathode and anode, electrical energy is created which can be used as a source of power.

Such a process would be slowed considerably if there were no mechanism to redistribute the ions formed at the electrodes. To allow the migration of ions from one electrode to the other, an electrolyte is introduced between the two electrodes. The electrolyte provides a pathway for hydrogen ions (protons) to pass from the anode to the cathode (oxygen) side of the cell. There, hydrogen ions combine catalytically with oxygen, which has diffused to the membrane-catalyst interface through the porous cathode, and electrons from the adjacent cell to form water, according to the following reaction:

$$4H^+ + O_2 + 4e^- \Rightarrow 2H_2O$$

Thus, the overall reaction in the fuel cell is simply described as follows:

$$2H_2 + O_2 \Rightarrow 2H_2O$$

3.3 Theoretical Background for Fuel Cells

Efficiency

One of the most important advantages of fuel cells is their ability to operate at higher efficiencies than heat engines. This is a consequence of the fact that a fuel cell directly converts chemical energy to electrical energy, while a Carnot engine converts chemical energy to mechanical energy using heat as an intermediary. In the case of a standard heat engine, the efficiency is determined by the amount of useful work, W, done by a given quantity of input heat, Q_{in}. This useful work, W, is a function of the amount of input heat, Q_{in}, minus the quantity of heat that is rejected to the outside surroundings, Q_{out}. Thus, the efficiency can be described as follows:

$$\eta = W/Q_{in} = (Q_{in} - Q_{out})/Q_{in}$$

Since the heat of the system can be given in terms of the entropy, this equation can be rewritten in terms of T_{in} and T_{out} for the system as follows:

$$\eta = (T_{in} - T_{out})/T_{in}$$

While the existence of a perfect heat engine is not prohibited by this relationship, if the condition $T_{out}=0$ can be met, in practical situations some heat is always rejected. In fact, the limitations of the Carnot cycle can be particularly significant in practical systems. While combustion temperatures, T_{in}, can be as high as 2100K or more, actual operating temperatures are generally much lower. The cold reservoir for a heat engine is the ambient temperature, or about 300K [6].

The theoretical limitations of the heat engine are avoided when a fuel cell is used because a fuel cell converts chemical energy directly into electrical energy. For the fuel cell, the amount of work performed by the 'engine' is simply related to the change in the enthalpy of the system, ΔH, minus the amount of heat exchanged with the surroundings, which is TΔS. Thus, the usable work for the system is the change in the Gibbs free energy for the system:

$$W = -\Delta H - Q = -(\Delta H - T\Delta S) = -\Delta G$$

The efficiency is then the amount of usable work that is produced by a given change in enthalpy.

$$\eta = \Delta G/\Delta H$$

The efficiencies derived from this relationship are typically higher than those for the heat engine, primarily due to the fact that the amount of heat exchanged with the surroundings is relatively small compared with the change in enthalpy. For some chemical reactions this efficiency actually exceeds 100%. This does not imply that the fundamental laws of thermodynamics are violated, however, but rather that the system absorbs heat from the surroundings and converts it into electrical power [6].

Operational Characteristics of a Fuel Cell

Of course, there are a number of loss mechanisms, such as energy required by auxiliary equipment, power conditioning (dc-to-ac conversion) and fuel processing, that bring the total efficiency of a process or system down below its ideal efficiency. The fuel cell must also operate under real conditions with practical current densities and reasonable reaction rates. As larger currents are drawn, and the deviation from ideal thermodynamically reversible conditions becomes greater, both the efficiency of the system and the output voltage drop.

The relationship between current density, output voltage, and efficiency for a real system is given in Fig. 3.2. The curve of current vs. voltage can be characterized by three

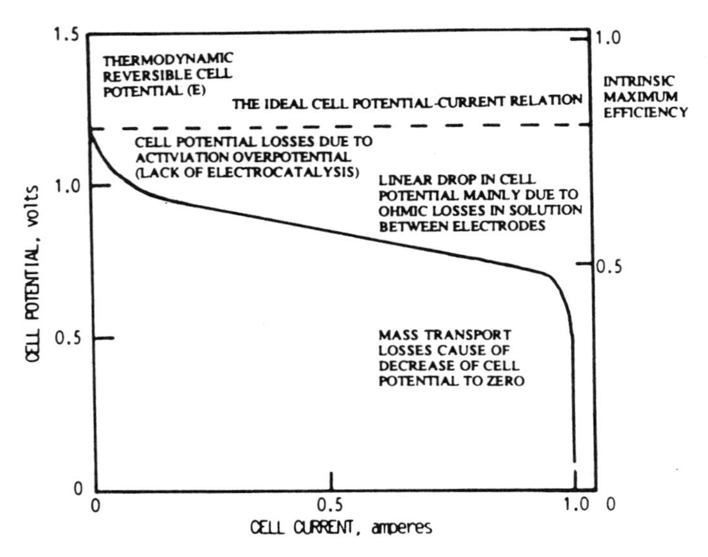

Figure 3.2 Typical plot of cell potential vs. current for fuel cells, illustrating regions of control by various types of overpotentials. (Source: ref. [7])

different regimes which are characteristic of three different sources of overpotential (a term used to describe the deviation of the potential from its ideal value). At low current densities, the predominant mechanism for cell potential loss is activation overpotential. Activation overpotential is the reduction in the potential that results from the need to bias the electrodes to compensate for the lack of electrocatalytic activity [6].

In the wide middle region of the graph, the output potential drops linearly as the current is increased. This is due to ohmic losses that occur within the electrolyte and the electrodes of the cell. At higher current densities, the voltage rapidly drops off as the current is increased due to mass transport losses. Mass transport losses result when the current in the cell is increased and the reactants become consumed at greater rates than they can be supplied, while the products accumulate at greater rates than they can be removed. At a certain point, the concentration of the reactants at the electrode surface, along with the external cell voltage, drops to zero.

In order to obtain an adequate power density (which is given by the multiplication of the current density and the voltage), fuel cells are generally operated in the ohmic overpotential region. In this region the actual efficiency, given by the ratio of the operating voltage to the ideal theoretical voltage, is less than the theoretical value, as seen in Fig. 3.2. While the ideal efficiency for a H_2-O_2 fuel cell is 83% if the product is liquid water and 94% if the product is water vapor, operating fuel cells achieve presently efficiencies that are closer to 40-45% [8].

3.4 Types of Fuel Cells

Various types of fuel cells have been developed since Sir William Grove invented the fuel cell in 1839. Some of the important features of various fuel cell types are given in Table 3.1.

In general, fuel cells are separated into two categories: those that operate at low temperatures and those that operate at higher temperatures. Fuel cells with an alkaline or solid polymer electrolyte are generally classified as low temperature fuel cells, while Phosphoric Acid fuel cells (PAFCs), Solid Oxide Fuel Cells (SOFCs), and Molten Carbonate Fuel Cells (MCFCs) are generally considered high-temperature fuel cells. The two most common fuel cells in use today are the Alkaline Fuel Cell, which has been used extensively by the U.S. space program, and the Phosphoric Acid Fuel Cell. The Alkaline fuel cell is a rel-

Table 3.1 Fuel Cell Comparison

Fuel Cell Type	Operating Temperature (°C)[a]	Power Density (kW/liter)[b]	Power Density (kW/kg)[b]	CO_2 Tolerant[a]	CO Tolerance[a]
Solid Oxide	1000	1-4	1-8	yes	good
Molten Carbonate	600	-	-	yes	good
Phosphoric Acid	150-205	0.16	0.12	yes	fair
Alkaline	65-220	0.1-1.5	0.1-1.5	no	poor
Solid Polymer	25-120	0.1-1.5	0.1-1.5	yes	poor

(Source: [a] from ref. [9] [b] from ref. [10])

atively specialized fuel cell system which requires special treatment of the fuel and air supplies to remove carbon dioxide. The Solid Polymer Fuel Cell is currently receiving considerable attention, as it is one of the most promising fuel cell technologies for vehicular applications. At this point, the SOFC and the MCFC are primarily being developed for stationary utility applications, such as power plants.

3.5 The Principles of Low Temperature Alkaline Fuel Cells

Alkaline fuel cells are one of the most developed fuel cell technologies. An important advantage of Alkaline fuel cells is that even when using cheaper non-noble metals such as nickel as electrocatalysts, they exhibit good performance. Other advantages of alkaline fuel cells include their high efficiency, relative ease of operation, low weight and volume, and reliable performance. Unfortunately, the electrolyte of an alkaline fuel cell is so CO_2 intolerant that vehicle systems would probably have to be supplied with either pure oxygen or air scrubbed of CO_2 [10]. Even the small levels of CO_2 in the air (about 350 ppm) are sufficient to carbonate the electrolyte and form solid deposits in the porous electrode. The extra cost and space requirement of storing pure oxygen or removing CO_2 from the air is one of the most important issues with regards to using alkaline fuel cells in vehicles.

The electrolyte for an alkaline fuel cell can either be mobile or an immobile electrolyte based in a porous matrix such as asbestos. Mobile alkaline electrolyte fuel cells use an electrolyte that continuously circulates between the electrodes. The principle of operation of this fuel cell is shown in Fig. 3.3. The products of the H_2-O_2 reaction (water and heat) dilute and heat the liquid electrolyte, but are removed from the cell as the electrolyte circulates [11]. In an immobile electrolyte system, the electrolyte does not continuously circulate, and hence water and heat must be removed by other means. Water is removed from the cell by hydrogen gas circulation, where water evaporates into the hydrogen stream supplied at the electrode and is subsequently condensed. The waste heat is removed via circulating coolant [11]. The principle of operation of an alkaline matrix fuel cell is shown in Fig. 3.4. Alkaline matrix fuel cells developed by United Technologies Corporation/International have been used for space shuttle flights since the 1970s [3]. More information on the hydrogen/air alkaline matrix fuel cell is given in the reprint by Staschewski at the end of this chapter.

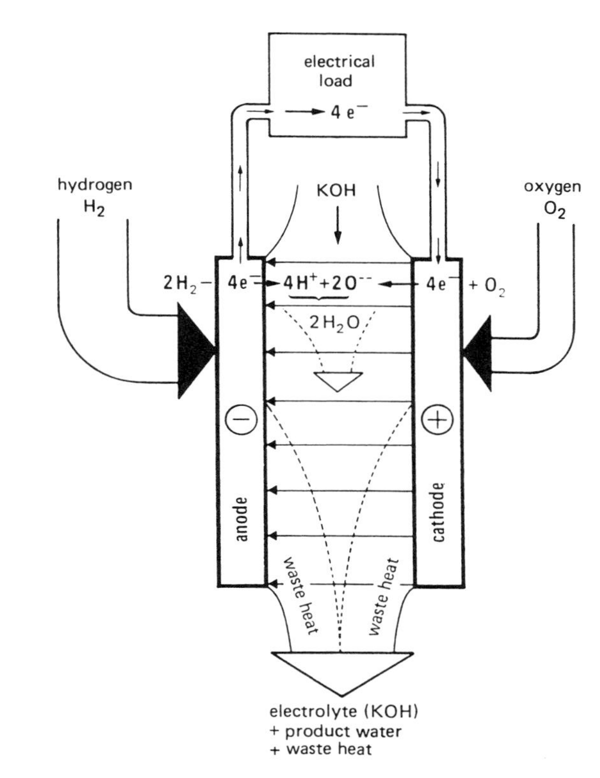

Figure 3.3. Principle of the alkaline fuel cell with mobile electrolyte. (Source: ref. [11])

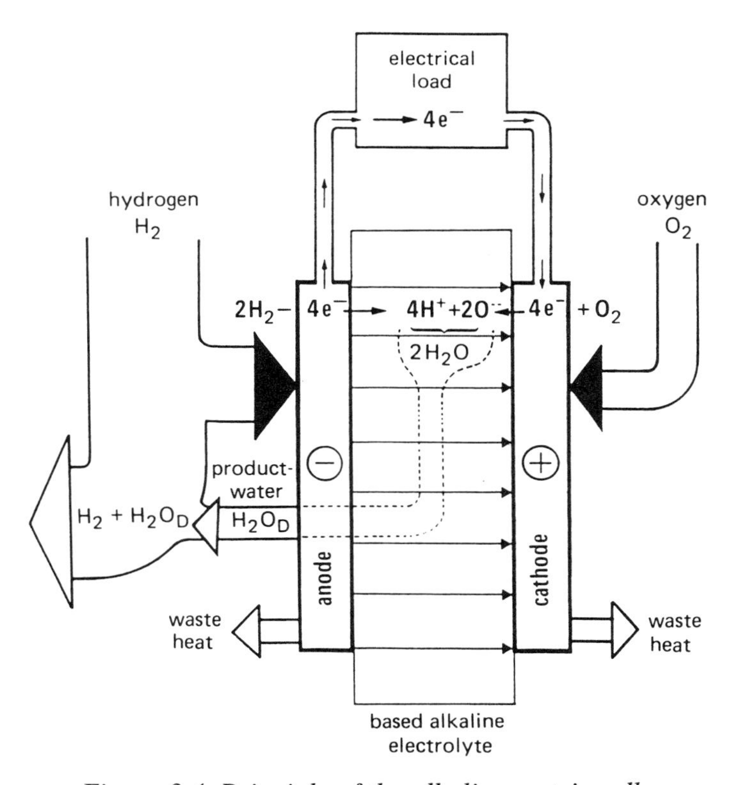

Figure 3.4. Principle of the alkaline matrix cell. (Source: ref. [11])

3.6 The Solid Polymer Fuel Cell (SPFC) or Proton Exchange Membrane (PEM) Fuel Cell

Historical Background

The Solid Polymer Fuel Cell (SPFC) is probably the most attractive for vehicle applications. It was first developed by General Electric (GE) for NASA in the 1960s. The perceived advantages of the SPFC for space applications were its high energy density compared to batteries, the absence of corrosive liquid electrolytes, the relative simplicity of the stack design, and the ruggedness of the system [12]. This technology initially suffered from a limited operating lifetime. By 1964, however, the improvements GE had made on the membrane electrolyte gave the cell an approximate lifetime of 500 hours. This was satisfactory to NASA and the cells were used in seven Gemini missions.

In the mid-1960s, GE adapted DuPont's Nafion for use in the SPFC. This Teflon-like material exhibited a substantially improved operating lifetime, in excess of 57,000 hours. But NASA experts, by this time, had selected the alkaline fuel cell for use in the Apollo program because they believed that the SPFC was intrinsically resistive and that alkaline fuel cells would be better for the new program. This, in effect, put the SPFC on the shelf for the next 20 years. With the exception of some work at Los Alamos National Laboratory, the SPFC lay dormant until 1984 [12].

The 1980s witnessed a renaissance in the development of the Solid Polymer Fuel Cell (SPFC). This was due in part to the fact that, in 1983, the Canadian Department of National Defense determined that the SPFC might satisfy some of the growing military needs for compact power supplies if it could be engineered for terrestrial applications and manufactured at a lower cost. This pushed Canadian companies such as Ballard Technologies to further develop SPFCs. Since then, there has been a considerable interest in developing the SPFC technology at various other locations including Los Alamos National Laboratory, Texas A&M University, H-Power, Hamilton Standard, International Fuel Cells Corporation (IFC) in the U.S., and at Siemens in Germany.

Description of SPFC Operation

The SPFC consists of two porous electrodes which are bonded to either side of a thin sheet of an ion-conducting polymer, the solid polymer electrolyte (SPE). The electrolyte used here is a proton exchange material (which is discussed in more detail below). The backs of the electrodes are contacted by plates which contain channels through which hydrogen and oxygen are supplied. The electrical contact to the electrodes is also made through these plates [12]. The water by-product is produced at the cathode side for an SPFC and is removed from the cell by an oxygen loop or by static water management. Waste heat, just as in the alkaline matrix cell, is removed by coolant circulation [11]. A picture diagram showing SPFC operation is given in Fig. 3.5, and more detailed diagrams of SPFCs are given in Figs. 3.6 and 3.7.

Electrolyte

The polymer electrolyte of a SPFC is a plastic-like membrane which ranges in thickness from 50 to 175 μm. These membranes are composed of perfluorosulfonic acids, which are Teflon-like fluorocarbon polymers with side chains ending in sulfonic acid groups. Such membranes (commonly referred to as the PEM [Proton

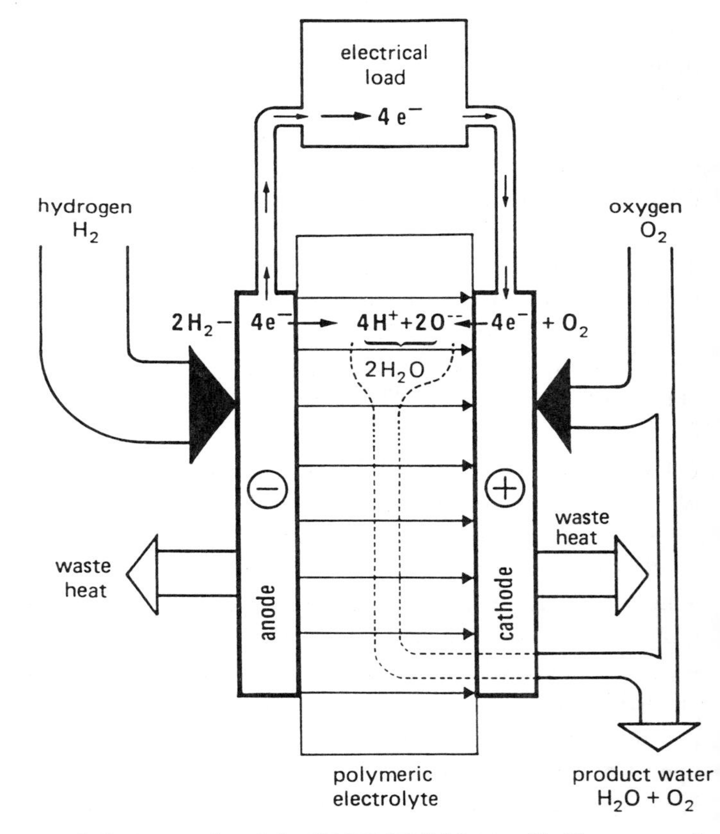

Fig. 3.5. Principle of the SPFC/PEM fuel cell. (Source: ref. [11])

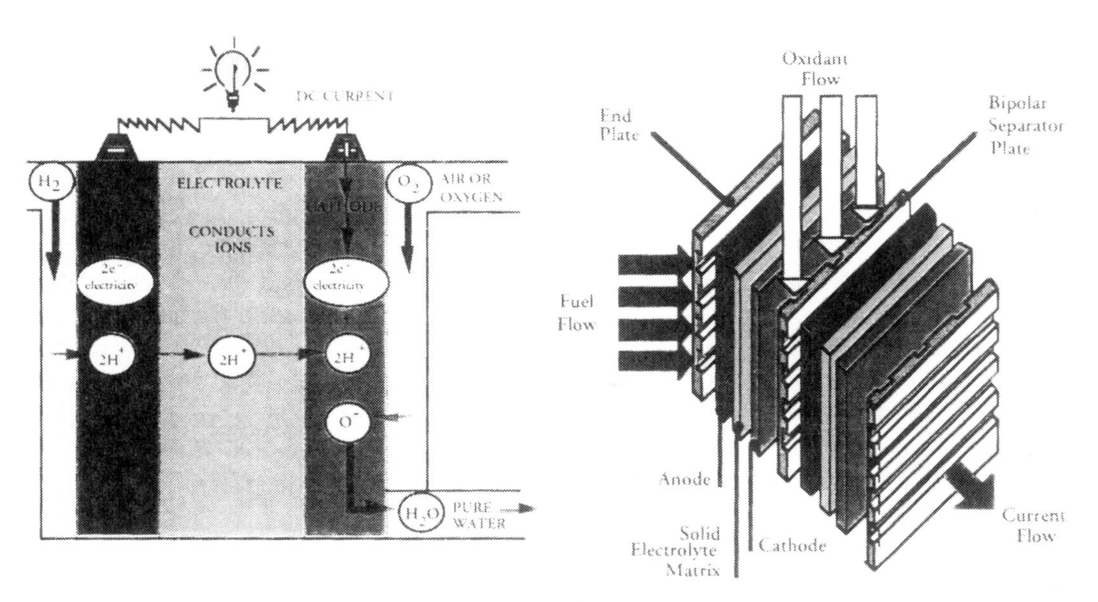

Fig. 3.6. Energy Partners' FuelCell™ (Courtesy: Energy Partners)

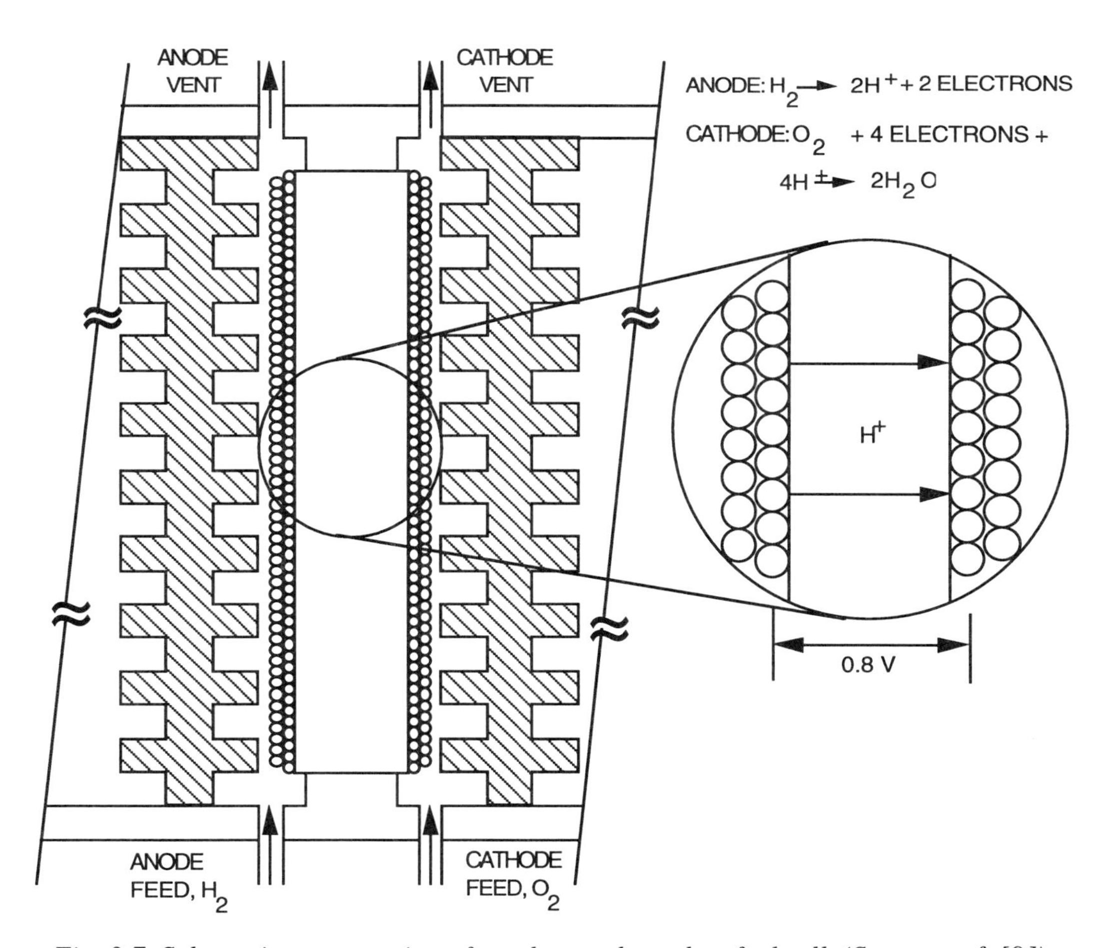

Fig. 3.7. Schematic cross section of a polymer electrolyte fuel cell. (Source: ref. [9])

Exchange Membrane] or the IEM [Ion Exchange Membrane]) are electrical insulators, but excellent conductors of hydrogen ions.

A number of different types of membranes have been used in fuel cells over the years, including membranes manufactured by Dow and DuPont. Until recently, Nafion 117 (a trademark of E.I. du Pont de Nemours and Company) was the most extensively used membrane in SPFCs. In 1987, however, Ballard Technologies Corporation demonstrated that SPFC performance could be improved significantly using a new membrane developed by Dow. The Dow membrane is a sulfonated fluorocarbon polymer, like Nafion, that can be used to generate larger current and power densities in fuel cells [12, 13].

Studies have also shown that the Dow membranes have a lower specific resistivity than Nafion. This can be attributed to the fact that there are a greater number of sulfonic acid groups for a given weight in the Dow polymer membrane than in the Nafion polymer membrane. This can be seen by comparing the chemical composition of the Dow membrane [7]:

$$CF_2{=}CFOCF_2CF_2SO_3H$$

with that of the Nafion membrane:

$$CF_2{=}CFOCF_2CFOCF_2CF_2SO_3H$$

It is important to note that preserving the correct water content in the membrane is necessary to maintain proper protonic conductivity and thus avoid resistive losses in the cell. Imbalance between the production and evaporation rates of water can result in either flooding of the electrodes or membrane dehydration, both of which severely limit performance. In the case of dehydration, the adherence of the membrane to the fuel cell electrode can be adversely affected [14].

Electrodes

The electrodes themselves are porous and allow for effective gas diffusion. They typically have three layers: a catalyzed or active layer; a hydrophobic diffusion layer with high porosity to minimize diffusion-related problems; and a backing such as a nickel screen or carbon cloth which also acts as a current collector. Electrodes must be sufficiently conductive to pass the generated current to the adjacent cell.

The catalyst layer is added to the electrode to increase its effectiveness. The catalyst must have good contact with both the electrical and protonic conductors while maintaining access to the gas. Until recently, it was believed that the high costs and high levels of platinum needed in the catalyst layer would preclude the use of SPFC in commercially viable vehicles. Recent work at the Los Alamos National Laboratory and at Texas A&M University has demonstrated, however, that electrodes with low loadings of platinum can provide performance similar to that of electrodes with much higher levels of platinum [7, 9]. The approach used by researchers at the Los Alamos National Laboratory involves interpenetrating platinum into a carbon support, as illustrated in Fig. 3.8. In Fig. 3.8(a) the catalyst particles are ineffective because protons from the membrane cannot migrate across the carbon surface. By impregnating the pores of the electrode with a thin layer of protonic conductor, as displayed in Fig. 3.8(b), the supported catalyst becomes highly effective. Using this technique, performance comparable to that of electrodes with a high platinum catalyst loading can be achieved with approximately 1/20th the amount of platinum [9].

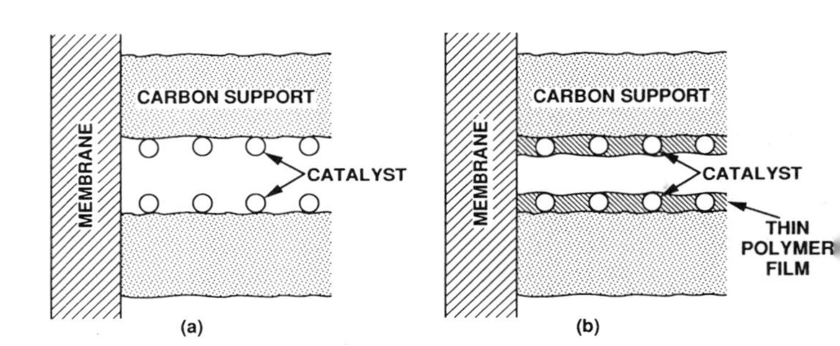

Figure 3.8 Schematic representation of the technique developed at Los Alamos for use of supported platinum catalysts in polymer electrolyte fuel cells. (Source: ref. [9])

Other work on electrode development has been done by CNR Institute for Transformation and Storage of Energy in Italy. The focus of this work has been the development of an SPFC technology based on a proprietary double-layer electrode formed by a catalyst layer overlapped to the diffusion layer [15]. This group has done some studies on how the "loading" of an electrode with both Nafion and a hydrophobic agent, such as polyfluoroethylenepropylene, can influence cell performance. Loading of an appropriate amount of electrolyte or Nafion into the pores of the catalyst, for example, can improve the contact between the catalyst particles and the electrolyte. It was found that loading the diffusion layer with polyfluoroethylenepropylene, the hydrophobic agent, influences the mass transport properties of the electrodes and the cell activity. Nafion loading of the electrode, on the other hand, primarily influenced the ionic resistance of the cell. A maximum in anode and cathode activities was obtained using an appropriate amount of Nafion loading, in this case 0.9 mg/cm^2. At a Nafion loading lower than this level, Nafion distributed into the pores of the catalyst layer decreasing the pore volume value. Above this level, the Nafion formed a surface film on the electrode. The effects of varying the Nafion loading on the anodic and cathodic voltage and the ionic resistance are shown in Figs. 3.9 and 3.10, respectively.

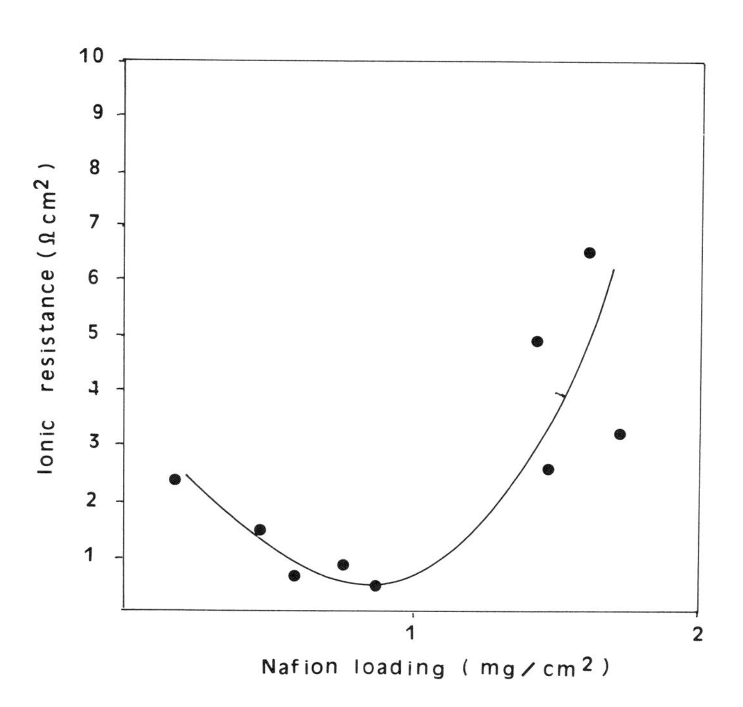

Figure 3.10 Effect of Nafion loading on ionic resistance. Test air flow at 50° C, 100% r. h. conditions. (Source: ref. [15])

3.7 High Temperature Fuel Cells

Introduction

High temperature fuel cells actively being developed for commercial applications include: Phosphoric Acid Fuel Cells (PAFCs), Molten Carbonate Fuel Cells (MCFCs), and the Solid Oxide Fuel Cells (SOFCs). These cells can tolerate CO_2 and, thus, can operate on fuel mixtures produced from coal gasification, reformed alcohols (see Section 3.8 on reformers), or natural gas.

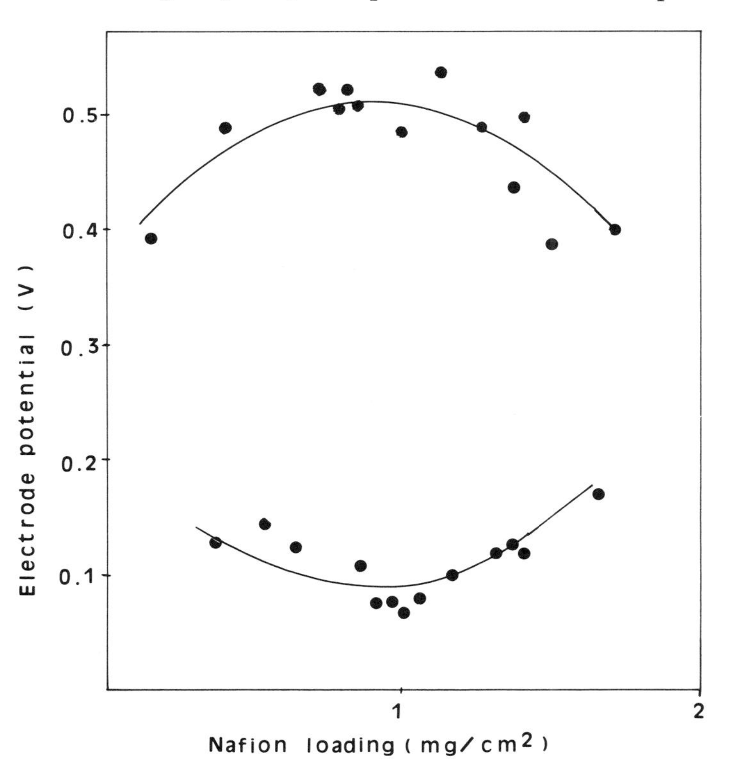

Figure 3.9 Effect of Nafion loading on anodic and cathodic voltage at 200 mA/cm2 for H2/O2 fuel cells at 50°C, 1 atm. (Source: ref. [5])

The SOFC and MCFC have both been developed mainly for stationary utility applications such as power generation, as opposed to vehicular applications. Such technologies are ready for commercial-scale demonstrations, and offer an environmentally friendly way of producing electricity at competitive costs [16]. PAFCs have been demonstrated for both stationary source and vehicular applications.

Phosphoric Acid Fuel Cell (PAFC)

The main reason for the development of the PAFC in the late 1960s and 1970s in the United States was the capabilities of its electrolyte. In particular, phosphoric acid is the only commonly available acid capable of functioning in a fuel cell at temperatures much higher than the boiling point of water [17]. Phosphoric acid is stable and has extremely low volatility, even at an operating temperature of 200° C. For vehicular applications, PAFCs have several drawbacks. They are large and heavy and thus not well suited for light-duty vehicles, although they may be satisfactory in heavy-duty applications [10]. PAFCs also have to be either warmed up before they are operated or be continuously maintained at an elevated temperature. PAFCs are also somewhat sensitive to CO contamination, thus levels of CO in reformed mixtures must be reduced to about 1-2% to obtain effective operation. It should be noted, however, that PAFCs have a much greater tolerance for CO than SPFCs.

At present, the PAFCs of International Fuel Cells (IFC), formerly known as United Technologies Corporation (UTC), are the most advanced technologically in terms of performance and rate of decay [17]. Earlier PAFCs, developed by UTC prior to 1978, were composed of a sandwiched configuration, as shown in Fig. 3.11. Here, the ribs in the cross-flow configuration are rotated through 90° for clarity. The phosphoric acid electrolyte, maintained at about 200° C, was contained in a porous hydrophilic matrix located between the catalyst layers of the anode and cathode. The electrolyte reservoir capacity of this "conventional" design used multiple carbon paper layers and was very limited. It involved a large number of carbon-to-carbon contacts that resulted in a high ohmic resistance.

This design was later supplanted by a "ribbed substrate" design where ribs are formed on the gas side of the electrode backing, as shown in Fig. 3.12. A catalyst layer and a half-thickness matrix is applied to the non-ribbed (flat) side of the porous substrate to form each electrode. The ribbed substrates typically have a thickness of about 1.8 mm. Impermeable flat graphite sheets, about 1 mm thick, act as bipolar gas separator plates between the anode and the cathode of adjoining cells. Cooling plates are placed approximately one every five to eight cells. About 1.5 kg

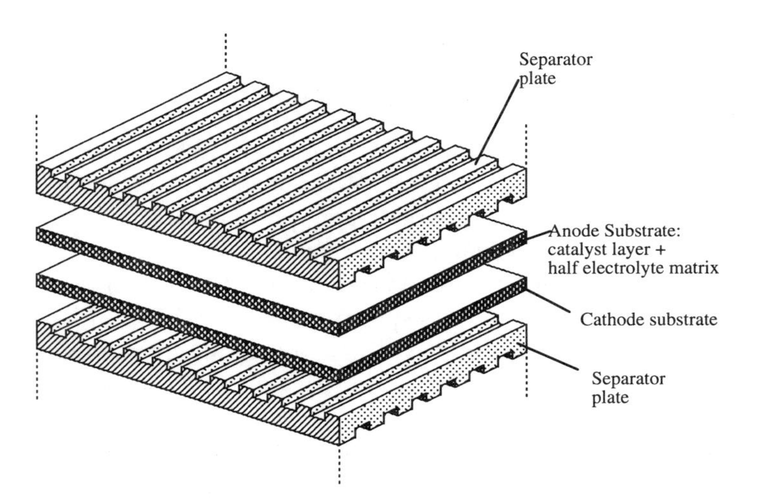

Figure 3.11 Sandwiched "conventional" PAFC configuration developed by UTC (ca 1978) (Source: ref. [17])

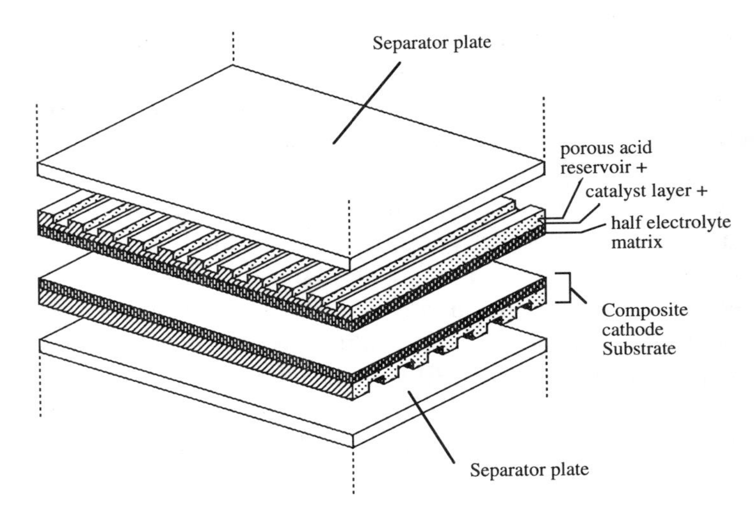

Figure 3.12 United technologies' PAFC ribbed substrate cell configuration (Source: ref. [17])

of electrolyte per m^2 can be stored in this cell, and it is estimated that this is sufficient for more than 40,000 hours of operation at 205° - 210° C and 8.3 x 10^5 Pa (8.2 atm) pressure [17].

Molten Carbonate Fuel Cell (MCFC)

In contrast to the PAFC, an external fuel processor is not necessary when using a MCFC because the MCFC operates at a temperature of approximately 1200° F (650° C). Because the MCFC has the capability of internally reforming hydrocarbons, the MCFC can achieve a high efficiency with extremely low emissions [18]. MCFCs are a promising power generation alternative because they can operate on natural gas or gas produced from the gasification of coal—the cleanest way of using this readily available fossil fuel.

The operation principle of the MCFC is shown in Fig. 3.13. At the cathode, O_2 and carbon dioxide (CO_2) react with available electrons to form carbonate ions (CO_3^{2-}), which migrate through the molten carbonate electrolyte towards the anode. There the carbonate ions react with hydrogen and carbon monoxide, internally reformed from water and the methane fuel. Electrons released in the anode reaction: $H_2 + CO_3^{2-} \Rightarrow CO_2 + H_2O + 2e^-$ produce a flow of electricity. CO_2 and water constitute the only emissions of the plant.

The main problems associated with the MCFC are the corrosion and sintering of its construction materials, which can result in the electrodes becoming dehydrated or

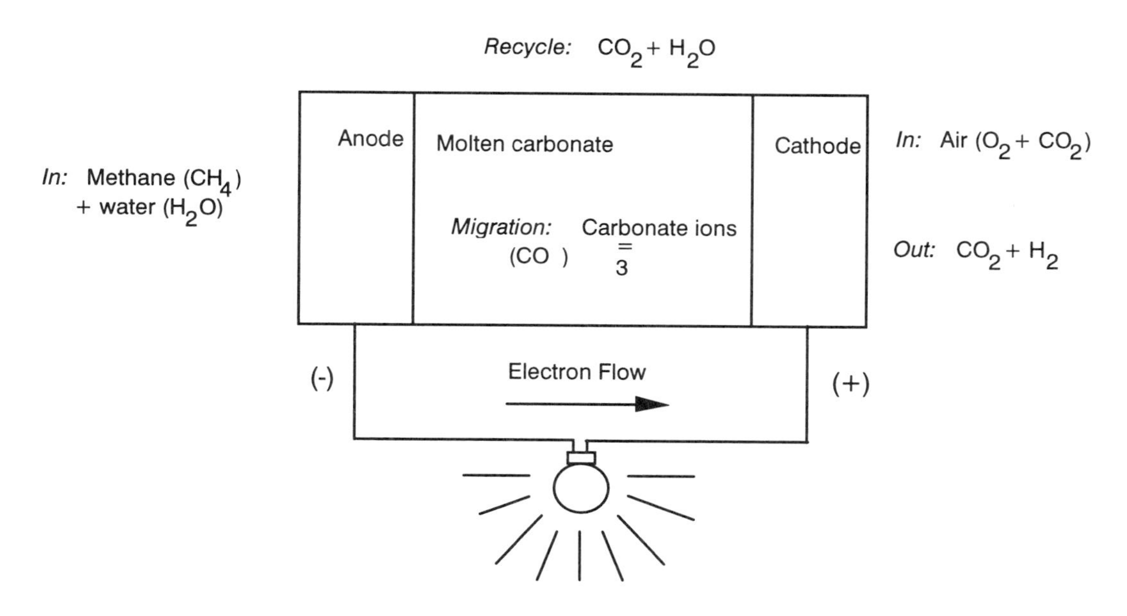

Figure 3.13. Operation principles of the molten carbonate fuel cell (MCFC) (Source: ref. [17])

flooded. Platinum group metals are one of the few materials with a proven resistance to molten carbonate electrolytes [19]. For more details on the MCFCs, refer to part one of the two-part reprint by Minh, included in this chapter.

Solid Oxide Fuel Cell (SOFC)

SOFCs use a solid oxide material, usually zirconia doped with other rare earth element oxides like yttria, as an electrolyte. This electrolyte, when heated to approximately 1000°C, conducts oxide ions at a sufficient rate such that it acts as a solid state electrolyte. No noble metal catalyst is needed. Westinghouse Electric Company is the main developer of this type of fuel cell, and is operating several pilot plants with a tubular configuration in Japan and the United States. Other SOFC configurations being investigated include a monolithic design, and a planar configuration [20, 21].

The operation principle of the SOFC is shown in Fig. 3.14 where:

The Cathode reaction is: $1/2\ O_2 + 2e^- \Rightarrow O^{2-}$

The Anode reaction is: $H_2 + O^{2-} \Rightarrow H_2O + 2e^-$

The Cell reaction is: $H_2 + 1/2\ O_2 \Rightarrow H_2O$

Since a solid oxide fuel cell operates by transport of oxide ions rather than fuel-derived ions, it can, in principle, be used to oxidize any gaseous fuel. Another advantage of SOFCs is that cell components can be made in a variety of configurations that may not be feasible in cells that employ liquid electrolytes. The SOFC also has a good tolerance for fuels with high levels of impurities, which makes this cell very attractive for operation on heavier fuels such as coal gas [17]. All these features can make the SOFC competitive for industrial applications. For more details on the high temperature SOFC refer to part two of the two part reprint by Minh, in this chapter.

3.8 Developing Fuel Cells for Vehicular Applications

The design of an appropriate fuel cell system could provide many benefits in transportation applications. Hydrogen Fuel Cell Electric Vehicles (FCEVs) could combine some of the best features of Battery-Powered Electric

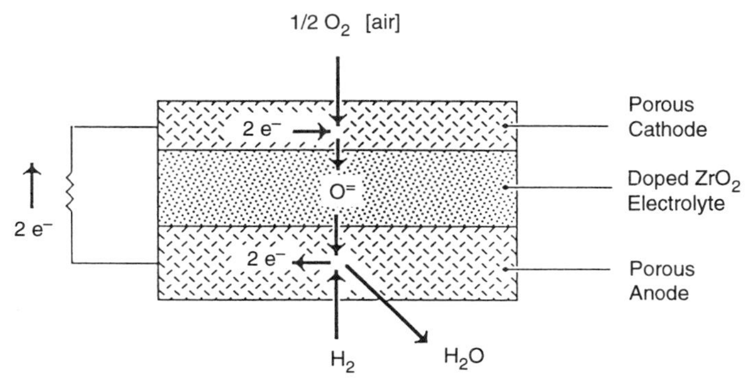

Figure 3.14 Operation principles of a solid oxide fuel cell (SOFC) (Source: ref. [17])

Vehicles (BPEVs) (zero emissions, quiet operation, and long-life) with the long range and fast refueling time of ICEVs [10]. Of the fuel cell technologies currently available, SPFCs appear to be the best suited for vehicular applications. The polymer electrolyte provides room temperature startup, elimination of many corrosion problems, and the potential for low resistance losses [22]. It also has a solid electrolyte which facilitates the sealing and safety of the fuel cell stack, thus preventing leakage and minimizing the corrosion of the cell components. SPFCs can tolerate CO_2 in the fuel stream making the use of reformed hydrocarbon fuels possible, but they are fairly intolerant of CO.

Parameters for a Solid Polymer Fuel Cell Vehicle

One of the most important requirements for a fuel cell system is that it provides sufficient power and current densities. At current densities on the order of 1 to 4 A/cm^2, concentration overpotential can become a significant problem with SPFCs [9]. If the reactant gases are unable to reach catalyst sites at sufficient rates, the reaction will not be sustained and the voltage will rapidly drop off to zero. Los Alamos National Laboratory (LANL) has been working on this problem for a number of years. From this research, they have developed an SPFC with low catalyst loading electrodes that is capable of delivering a power density of 0.9 W/cm^2 at a current density of 2 A/cm^2 using pressurized hydrogen and air.

Such cells can be stacked in series to produce more power. A set of design parameters for a stack of 133 such cells with an active area of 500 cm^2 each is shown in Table 3.2. Such a stack would provide a continuous power output of 20 kW (26.7 hp) (200 A at 100 V). At a higher current density of 1.8 A/cm^2, the power output could be increased to 60 kW (80 hp) [9]. This stack can meet the power requirements for a compact car with performance comparable to that of a 4-cylinder internal combustion engine (ICE). With regards to power output, it is important to note that in an electric vehicle, a battery can be used to supplement the power of a fuel cell during periods of peak operation, such as during acceleration. Hence, a fuel cell with a power rating that is less than that of a gasoline engine can still provide performance comparable to that of the gasoline engine.

Cost of a FCEV

One of the most critical challenges for fuel cell research is cost reduction, in particular, with regards to proton-conducting membranes and platinum-catalyzed electrodes. Until recently, it was thought that the platinum requirements for SPFCs, typically on the order of 4 mg/cm^2 [23], would be so great that fuel cells would be virtually impossible to market [5]. At such loadings, an automobile would require up to $10,000 worth of platinum alone [9]. Research laboratory advances at Los Alamos National Laboratory have successfully reduced the platinum requirement by a factor of 20, however.

With the relatively high performance of these low load electrocatalysts, the cost of platinum used in fuel cells can be reduced substantially. Perspectives on platinum costs for fuel cell vehicles in the literature vary. Researchers at Los Alamos National Laboratory estimate that using their

Table 3.2. Fuel cell parameters for a passenger vehicle

Net continuous power	20 kW		
Net peak power	60 kW		
Operating points			
Peak	0.50 V	1.8 A/cm^2	0.9 W/ cm^2
Continuous	0.75 V	0.4 A/cm^2	0.3 W/cm^2
Stack size			
Active area	500 cm^2		Diameter 25 cm
Cross section	1000 cm^2		Diameter 35 cm
Cell thickness	0.5 cm		
No. cells	133		
Stack length	66 cm		Total 75 cm
Stack voltage (nom.)	100 V		
Stack volume	0.072 m^3		
Stack density	1.0 g/cm^3		
Stack weight	75 Kg		Total 100 Kg

(Source: ref. [9])

thin film technology, an 80 kW peak power (20 kW nominal power) passenger car would require less than $500 worth of platinum (see reprint by Gottesfeld, et al). Srinivasan, et al, put the costs of noble metal catalysts at about $30-40 per kilowatt [8]. DeLuchi and Ogden estimate that the amount of platinum that will be needed for a vehicle using a 25 kW fuel cell will eventually only be about twice that needed in today's catalytic converters [10]. Allowing for a 50% increase in the price of platinum to account for the increased demand for platinum per vehicle, they predict that this would result in costs that would be less than $100 at the manufacturing level. Some other authors give substantially higher numbers for the amount of platinum that will be needed, however, Kukkonen and Shelef estimate that a 40 kW fuel cell would need at least 80 grams of platinum or about 30-60 times more than used in automobile catalytic converters [24].

The costs of polymer membranes is probably an even more important issue now. Currently, DuPont's Nafion and Dow's state-of-the-art membranes cost about $60-$200 per kilowatt [8]. These costs will probably have to be reduced by about a factor of 10 to 20 in order to produce affordable fuel cells (at approximately $200 per kilowatt). Such reductions could be possible if the membranes were to be marketed on a wider scale. The membrane electrolyte used in a SPFC-powered electric vehicle is manufactured in a fashion similar to that of many other common polymer membranes. If other polymer products such as Teflon and other petrochemical products can be used as a basis, the production costs of membrane electrodes could follow a classical learning curve, with costs declining by about 20% each time production doubles. If the membrane were marketed, and demand for it were to increase, some studies have indicated that membrane costs per car could drop from several thousand dollars per car to about $300 to $400 per car [5, 25].

Estimating the costs of actual vehicles requires the consideration of a number of other factors including vehicle lifetime, fuel costs, power plant efficiency, and any equipment that is either added to or left off the vehicle. Although the cost of prototype fuel cell electric vehicles is presently high, these costs should drop if FCEVs ever reach the production stage and establish markets. DeLuchi and Ogden have projected that if fuel cells can be successfully developed, their costs could be competitive with those of gasoline ICEVs on a life cycle basis. Prospects of longer life and lower costs for the electric drive train and improved fuel economy all lower the costs for FCEVs in these scenarios. A summary of the base case cost results obtained by these researchers is given in Table 3.3, for ICEVs, FCEVs, and BPEVs [10].

3.9 Systems Engineering Aspects

A complete fuel cell system consists of a number of different components, as shown in Fig. 3.15, including: the fuel cell stack itself; storage for fuel (either hydrogen or methanol); an air compressor, to pressurize the gases used in the fuel cell; a cooling system; a water management system, to keep the fuel cell saturated and to remove product water; and an electric motor and controller. If the vehicle stores methanol, instead of hydrogen, a reformer is also needed to convert the methanol into hydrogen for use in the fuel cell. A high-powered battery will probably also be needed to supply power during times of peak operation. While the general features of all FCEVs are similar, the design of a particular system depends on the specific requirements for the system such as [3]:

Table 3.3 Summary of cost results of FCEVs

	Gasoline ICEV 640-km range	FCEV 400-km range	FCEV 250-km range	BPEV 400-km range
Fuel retail price ($/gallon-equivalent)	1.18	2.97	3.04	2.57
Full retail price of vehicle ($)	17,302	25,446	23,183	28,247
Maintenance cost ($/year)	516	434	434	388
Life-cycle cost (cents/km)	21.45	21.33	20.94	22.96
Break-even gasoline price ($/gallon)	N/A	1.43	1.27	2.11

(Source: ref. [10])

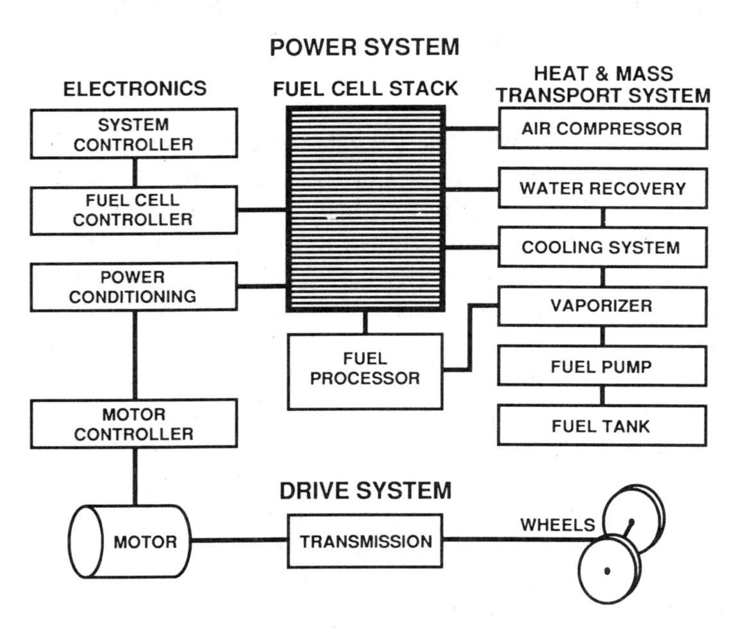

Figure 3.15. Block diagram of a fuel cell electric vehicle propulsion system showing the various subsystems that need to be integrated. (Adapted from ref. [9])

- The type of fuel cell and fuel processor is determined by the primary fuel (hydrogen, natural gas, coal, alcohol, petroleum).
- The method for removing by-products such as heat and water is determined, in part, by the operating temperature.
- The power rating and the need for power conditioning or a bottoming cycle (the latter two terms will be discussed below) are dependent on the application.
- Start-up time and peak load capability are particularly important considerations for transportation applications.

Electrical Storage System

The electrical storage system in a FCEV consists of a battery bank that can be charged by the fuel cell under low-load conditions, or with the energy that would otherwise be lost in braking using a "regenerative braking" system. The required electrical storage system for a FCEV would be larger than the battery used in conventional vehicles, but smaller than the batteries needed for Battery Powered Electric Vehicles (BPEVs) [5] (refer to reprints by Kaufman in this chapter and Billings, et al, and Prater in Chapter 7 for more details on FCEVs and their drive systems).

Reformers

Hydrogen can be stored on-board a vehicle in the form of hydrocarbons and then processed to a hydrogen-rich gas by a reformer. The advantage of this method of operation is that the problems of storing pure hydrogen on-board the vehicle are avoided (see Section 2.4). Methanol is, by far, the best hydrocarbon to use in the reforming process because it can be catalytically converted (on a Cu-Zn catalyst) to hydrogen and water at only 200° C to 300° C.

One of the problems with methanol reformers is that they have relatively long warm-up times and cannot change their power output rapidly enough to meet changing needs during driving. Los Alamos National Laboratory has developed a reformer that can provide better performance in these areas [9]. The design is shown in Fig. 3.16. This design uses an internal fan to recirculate the reformate through the catalyst bed. This keeps the catalyst bed at its optimal temperature. Heat can also be injected to increase the reforming rate.

Another problem with reforming systems is that the hydrocarbon-to-hydrogen reaction is often incomplete, leaving carbon monoxide in the fuel stream. Unfortunately, both PAFCs and SPFCs are sensitive to CO contamination. The maximum permissible carbon monoxide content for a fuel gas in current low temperature cells is low for PAFCs, about 3%, and even lower for SPFCs, around 100 ppmv. These small concentrations of CO can

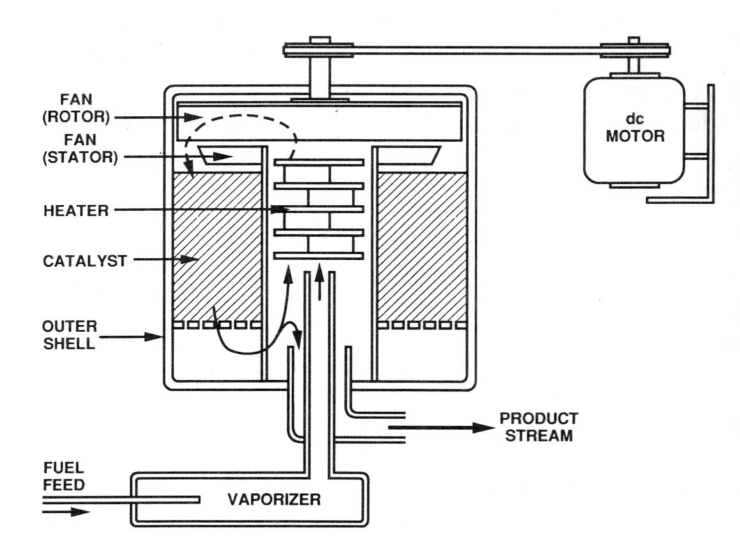

Figure 3.16 Schematic cross section of a recirculating-methanol-water reformer developed at Los Alamos. (Source: ref. [9])

poison the cell, and decrease cell performance. This problem and some of its solutions are discussed in detail in the reprint "Hydrogen from Methanol: Fuel Cells in Mobile Systems" by Ganser and Höhlein.

Bottoming Cycles

Bottoming cycles are used in large high-temperature fuel cell systems. Their function is to increase the efficiency of a power plant by transforming the waste heat generated from the fuel cell stack into useful electric energy. Gas and steam turbines are both used for bottoming cycles.

Power Conditioning

A power conditioner is needed for fuel cell systems used in utility applications. Power conditioners are used to maintain a constant alternating current output despite the variation in the direct current cell stack voltage as a function of the load.

Water Removal

Water is a by-product in any H_2/O_2 fuel cell system, and its removal is essential to prevent the electrodes from flooding or gas flow from becoming blocked. There are two types of water removal systems: *active* and *passive.* Active systems use the gas fuel stream to carry water vapor away from the electrodes, whereas passive systems use diffusion or natural convection to remove water. Active water removal is preferred for large fuel cell systems because it is capable of handling very large mass flows.

Cooling

Controlling the temperature at which the cell operates is important since temperature affects the hydration of the cell materials and the performance of the cells. Temperatures outside of the desired range can lead to a considerable loss in performance. As with water removal systems, there are two types of cooling systems: *active* and *passive.* In passive cooling, the system's excess heat is removed by conducting it away from the electrodes and radiating it into the surroundings. Active cooling systems usually use either gas or liquid as a cooling fluid.

Air Compressor

For practical systems, fuel cells must utilize air as the oxidant. While it is simplest to use air at atmospheric pressure in a fuel cell, the performance of a fuel cell can be increased by increasing the air pressure. Thus, the cost and size of a fuel cell needed to achieve a certain power level decrease if the operating pressure can be increased. However, the integration of an air compressor into a fuel cell system increases the complexity of the system and the compressor itself adds to the weight of the system.

3.10 Recent Developments with Fuel Cells

Recently, leading developers and utility companies in the U.S., Japan, and Europe have been actively involved in trying to introduce fuel cells into the commercial market.

Manufacturing and Systems Development

Japan has one of the most dynamic fuel cell programs. The Ministry of International Trade and Industry in Japan has set a target of 2250 MW of fuel cell capacity by the year 2000 [26]. Japan currently has the largest fuel cell operating capacity of any nation [27]. Japan also boasts the largest single operational fuel cell stack, generating 11 MW of power [26]. This phosphoric acid fuel cell demonstration plant is operated by the Tokyo Electric Power Company and is located in Ichihara City on Tokyo Bay. The plant supplies power to approximately 4000 homes. The plant was constructed through a joint venture between International Fuel Cells (IFC) and Toshiba Corporation.

Fuji Electric is one of the leading Japanese firms in the area of Phosphoric Acid Fuel Cell (PAFC) development. Fuji Electric has committed considerable resources towards commercialization of PAFCs in the areas of on-site cogeneration (50-500 kW), dispersed power plants for utilities (MW class), and vehicles (< 100 kW) [28]. As of August 1993, Fuji Electric had installed 5.5 MW of PAFC power plants and had orders for 1.9 MW more [29]. The company has constructed a factory with an annual manufacturing capacity of up to 15 MW. The largest unit being developed by Fuji Electric is a 5 MW plant that was expected to be ready for field delivery in 1995.

International Fuel Cells Corporation (IFC) and its subsidiary, the ONSI Corporation, have to date been the largest producers and installers of fuel cells in the world [29]. These activities include the 11 MW facility in Japan and a number of commercial 200 kW PAFC power plants that have been installed in the United States and elsewhere. As of August 1994, 69 such plants had been ordered [27]. The company plans to offer a new version of their 200 kW power plant in 1995.

Several MCFC demonstration programs are being actively pursued in the United States [30, 31]. The Energy Research Corporation (ERC) of Danbury, CT is building a

2 MW MCFC plant in Santa Clara, California. This $46 million project is receiving funding from five California utilities, the Electric Power Research Institute, the DOE, and the National Rural Electric Cooperative. The Demonstration is expected to last 1995 to 1998. ERC has already demonstrated an MCFC stack at the Pacific Gas and Electric Company in San Ramon. M-C Power Corporation of Burr Ridge, IL is also building a 250 kW MCFC scheduled to be operational by 1995 in the San Diego Gas & Electric Company's service area. The unit was designed by M-C Power with Bechtel Corporation and Stewart & Stevenson. Another M-C Power 250 kW MCFC plant will be operated at the Science and Technology Center of Unocal in Brea, CA. Several Japanese companies are also developing MCFCs including Ishika-Wajima-Harima Heavy Industries (IHI), Hitachi and Mitsubishi Electric Corporation (MELCO) [31].

Westinghouse Electric Corporation is one of the leading developers of SOFCs [32, 33]. The Westinghouse SOFC design is based on a tubular cell configuration. Westinghouse has two 25 kW units operating at the Rokka Island Test Center for Advanced Energy Systems near Kobe, Japan. One of these units has just completed its demonstration after over 7000 hours of operation [27]. In the U.S., Westinghouse has a 20 kW plant being operated by Southern California Edison and is planning a second 100 kW plant. Other companies testing planar designs for electrodes include Ceramatec, Technology Management, Inc., and Ztec of the United States and Japanese firms Fuji, Mitsubishi, and Murata. Allied-Signal Corporation is developing a monolithic cross flow design which could eventually prove to be more efficient, but it is proving difficult to fabricate.

At Argonne National Laboratory (ANL) new ion-conducting materials are being studied and synthesized for use in moderate-temperature SOFCs [34]. Researchers have discovered that empty spaces or 'tunnels' in the crystal structure of an electrolyte can facilitate the passage of oxide ions. Fuel cells are being built and tested which incorporate some of the new materials identified at ANL. This work has been ongoing since 1988.

There are also some significant ongoing development projects in Europe. In Germany, the Bavarian state utility and a consortium of industries have set up the Solar Wasserstoff Bayern AG research center. This center receives approximately $6.25 million annually for fuel cell research [27]. Ansaldo and IFC are operating a 1 MW power plant for the Azienda Energetica Municiple in Milan, Italy. A number of ONSI's 200 kW units are also in place throughout Europe [27, 29]. In the Netherlands, Energieonderzoek Centrum Nederland (ECN) is planning to operate two 250 kW demonstration units in 1995. One will operate on natural gas while the other will use a coal-derived synthesis gas [31]. Both Ballard Power Systems of Canada and Mitsubishi have also announced programs to develop SPFCs for stationary applications [23]. Information on additional demonstration projects can be found in refs. 27 and 29.

Transport Applications

This section briefly discusses ongoing projects which relate to the transportation sector. Specific hydrogen fuel cell vehicles are reviewed separately in Chapter 7.

H-Power Corporation is the prime contractor for a DOE-sponsored program to develop three prototype 25-passenger buses [5, 35, 36]. The first bus was unveiled in April of 1994 and is being tested. The second bus is expected to be fully integrated by the end of 1994 and a third will be integrated in the first quarter of 1995. The first bus has met the Transit Industry performance standards for acceleration and has a range of 190 miles. The buses are powered by PAFCs in conjunction with Ni-Cd batteries for additional power. The methanol-fueled, liquid-cooled PAFC subsystem has a net rated power of 47.5 kW and an overall fuel cell subsystem efficiency of about 42%. The fuel cell was developed by the Fuji Electric Co. Further details on this project are given in the reprint by Kaufman at the end of this chapter.

General Motors (GM) is currently involved in a multi-phase DOE-sponsored project to demonstrate the feasibility of PEM fuel cells and methanol reformers for transportation applications. This DOE program teams GM with members of the scientific community and leaders in the fuel cell industry to assist in transferring technology from the research arena to the industrial arena. Participants in this program include Allison Engine Co., Los Alamos National Laboratory, Dow Chemical Co., Ballard Power Systems, and DuPont Co. Phase one of the project culminated in the integration and testing of a complete 10 kW PEM fuel cell system [5, 37]. In phase two of this project, a 60 kW brassboard system (30 kW fuel cell and 30 kW battery) is expected to be developed and tested. Further details on this project are given in the reprint by Creveling at the end of this chapter.

Ford Motor Co. and Chrysler Corporation have just recently been awarded a total of $29 million from the DOE in contracts for fuel cell developments [38]. The Ford contract for $13.8 million will focus on reducing the cost, size, and weight of fuel cells. Ford will also investigate hydrogen production, distribution, storage, and dispensing. Chrysler was awarded $15 million which it will spend doing cooperative work with the Allied-Signal Corporation.

Los Alamos National Laboratory (LANL) is researching PEM fuel cells with the goal of improving the performance and reducing the cost of such cells to the point where their widespread use in transportation applications is feasible. Currently, LANL is looking at a number of issues including: (1.) optimizing the platinum/carbon electrode structure, (2.) optimizing the backing and flow-field, (3.) characterizing conductivity and water transport in polymer membranes, (4.) evaluating long-term endurance and materials stability, (5.) developing and validating an integrated model for the fuel cell membrane, electrodes, and backing, and (6.) identifying and solving the performance problems of methanol-powered fuel cells [39, 40]. Further details on these projects are given in the reprint by Gottesfeld, et al at the end of this chapter.

The Center for Electrochemical Systems and Hydrogen Research, Texas Engineering Experiment Station at Texas A&M University has been involved in PEM fuel cells projects since early 1988 [41]. Among the major issues addressed with these studies include the following: (1.) microelectrode techniques for the determination of electrode kinetic, mass transport, and ohmic parameters to permit PEM fuel cell performance analysis; (2.) means of increasing effective platinum utilization; (3.) evaluation of alternative membranes for PEM fuel cells; (4.) optimization of electrode and Membrane-Electrode-Assembly (MEA) structures and of PEM fuel cell operating conditions; (5.) theoretical and experimental analysis of PEM fuel cell thermal management and water management; and (6.) engineering design, development, and testing of multicell PEM fuel cell stacks. In the six years of study at Texas A&M, researchers have been able to reduce platinum loadings by a factor of 100, while achieving power densities three times those obtained during the mid-1980s. Further details on these projects are given in the reprint by Anand, et al at the end of this chapter.

There are a number of other demonstration projects of note. Ballard Power Systems of Canada has built and is testing a transit bus powered entirely by fuel cells and is currently in the process of constructing a second fuel cell bus [23]. Energy Partners of the United States has demonstrated a battery/fuel cell powered automobile called the Green Car [42]. Mercedes and Mazda have demonstrated, respectively, a fuel cell powered van and a fuel cell powered golf cart [23]. Fuel cell buses are also being developed as part of the EUREKA and Euro-Quebec Hydro Hydrogen Pilot Project (EQHHPP) programs [43, 44, 45]. The Belgian firm Electrochemical Energy Conversion (ELENCO) has been working on H_2/air fuel cell power plants (5 to 60 kW) for small stations and for city buses [46], and Siemens is developing a hydrogen/oxygen system for powering submarines [23]. Most of these programs are discussed in greater detail in Chapter 7, which profiles hydrogen vehicles.

3.11 Conclusion

Fuel cells offer a technology which can dramatically reduce air pollutant emissions for both stationary and mobile applications. Since fuel cells directly convert chemical energy into electrical energy, they can also attain higher efficiencies than standard heat engines. Cost is still a major issue with regard to marketing fuel cells, as they are composed of expensive materials. Recent breakthroughs have dramatically reduced the amount of platinum that fuel cells will require, however. Improving the energy density of fuel cells will be another important challenge for fuel cell researchers over the next decade. Given the potential of fuel cells for high efficiencies and zero emissions, fuel cell vehicles could meet transportation needs into the twenty-first century and beyond. Technical and economic obstacles still remain, however, that must be solved before such vehicles can be introduced into a wider marketplace.

References

1. Liebhafsky, H.A. and Cairns, E.J., *Fuel Cells and Fuel Batteries*, John Wiley & Sons, 1968.

2. Appleby, A.J. and Foulkes, F.R., *Fuel Cell Handbook*, Van Nostrand Reinhold, New York, NY 1989.

3. Srinivasan, S., "Fuel Cells for Extraterrestrial and Terrestrial Applications," *Journal of the Electrochemical Society*, Vol. 136, p. 41, 1989.

4. Trimble, K. and Woods, R., "Fuel Cell Applications and Market Opportunities," *Journal of Power Sources*, Vol. 29, p. 1990.

5. Kelly, H. and Williams, R.H., *Fuel Cells and the Future of the US Automobile*, Report for the Office of Technology Assessment, U.S. Congress, 1992.

6. Kartha, S. and Grimes, P., "Fuel Cells: Energy Conversion for the Next Century," *Physics Today*, Vol. 47, No. 11, p. 54, 1994.

7. Srinivasan, S., Manko, D.J., Koch, H. and et al, "Recent Advances in Solid Polymer Electrolyte Fuel Cell Technology with Low Platinum Loading Electrodes," *Journal of Power Sources*, Vol. 29, p. 367, 1990.

8. Srinivasan, S., Dave, B.B., Murugesmoorthi, K.A., Parthasarathy, A. and Appleby, A.J., "Overview of Fuel Cell Technology," in *Fuel Cell Systems*, Edited by L. J. M. J. Blomen and M. N. Mugerwa, Plenum Press, New York, NY, p. 37, 1993.

9. Lemons, R.A., "Fuel Cells for Transportation," *Journal of Power Sources*, Vol. 29, p. 251, 1990.

10. DeLuchi, M.A. and Ogden, J.M., "Solar-Hydrogen Fuel-Cell Vehicles," *Transportation Research, Part A, Policy and Practice*, Vol. 27A, p. 255, 1993.

11. Strasser, K., "The Design of Alkaline Fuel Cells," *Journal of Power Sources*, Vol. 29, p. 149, 1990.

12. Prater, K., "The Renaissance of the Solid Polymer Fuel Cell," *Journal of Power Sources*, Vol. 29, p. 239, 1990.

13. Verbrugge, M.W. and Hill, R.F., "Analysis of Promising Perfluorosulfonic Acid Membranes for Fuel-Cell Electrolytes," *Journal of Electrochemical Society*, Vol. 137, p. 3770, 1990.

14. Bernardi, D.M., "Water-Balance Calculations for Solid-Polymer Electrolyte Fuel Cells," *Journal of Electrochemical Society*, Vol. 137, p. 3344, 1990.

15. Staiti, P., Poltarzewski, Z., Alderucci, V., Maggio, G. and Giordano, N., "Solid Polymer Electrolyte Fuel Cell (SPEFC) Research and Development at the Institute CNR-TAE of Messina," in *Hydrogen Energy Progress IX*, Edited by T. N. Veziroglu and C. D.-J. Pottier, International Association for Hydrogen Energy, Coral Gables, FL, p. 1425, 1992.

16. Serfass, J.A., "The Future of Fuel Cells for Power Production," *Journal of Power Sources*, Vol. 29, p. 119, 1990.

17. Appleby, A.J., Richter, G.J., Selman, J.R. and Winsel, A., "Current Technology of PAFC, MCFC and SOFC Systems: Status of Present Fuel Cell Power Plants," in *Electrochemical Hydrogen Technologies*, Edited by H. Wendt, Elsevier Science Publishing Company Inc., New York, p. 425, 1990.

18. Douglas, J., "Fuel Cells for Urban Power," *EPRI Journal*, September, p. 5, 1991.

19. Cameron, D.S., "World Developments of Fuel Cells," *Int. J. Hydrogen Energy*, Vol. 15, p. 669, 1990.

20. Riley, B., "Solid Oxide Fuel Cells—The Next Stage," *Journal of Power Sources*, Vol. 29, p. 223, 1990.

21. Yoshida, T., Hoshina, T., Mukaizawa, I. and et al, "Properties of Partially Stabilized Zirconia Fuel Cell," *Journal of the Electrochemical Society*, Vol. 136, p. 2604, 1989.

22. Springer, T.E., Zawodzinki, T.A. and Gottesfeld, S., "Polymer Electrolyte Fuel Cell Model," *Journal of Electrochemical Society*, Vol. 138, p. 2334, 1991.

23. Prater, K.B., "Polymer Electrolyte Fuel Cells: A Review of Recent Developments," *Journal of Power Sources*, Vol. 51, p. 129, 1994.

24. Kukkonen, C.A. and Shelef, M., "Hydrogen as an Alternative Automobile Fuel: 1993 Update," SAE Technical Paper No. 940766, Society of Automotive Engineers, Warrendale, PA, 1994.

25. DeLuchi, M., "Hydrogen Fuel-Cell Vehicles," Institute of Transportation Studies University of California, Davis, Report No. UCD-ITS-RR-92-14, 1992.

26. Newman, A., "Fuel Cells Come of Age," *Environ. Sci. Technol.*, Vol. 26, p. 2085, 1992.

27. Hirschenhofer, J.H., "Fuel Cell Status, 1994," *IEEE AES Systems Magazine*, p. 10, November 1994.

28. Anahara, R., "Phosphoric Acid Fuel Cells (PAFCs) for Commercialization," *Int. J. Hydrogen Energy*, Vol. 17, p. 375, 1992.

29. Hirschenhofer, J.H., "Fuel Cell Technology Status," *IEEE AES Systems Magazine*, p. 21, November 1993.

30. "Fuel Cell Demonstration Planned," *EPRI Journal*, p. 35, September 1993.

31. Gillis, E., "Molten Carbonate Fuel Cell Technology," *EPRI Journal*, p. 34, April/May 1994.

32. Douglas, J., "Solid Futures in Fuel Cells," *EPRI Journal*, p. 6, March 1994.

33. Lipkin, R., "Firing Up Fuel Cells," *Science News*, Vol. 144, p. 314, 1993.

34. Goldstein, R., "Solid Oxide Fuel Cell Development," *EPRI Journal*, p. 32, October/November 1992.

35. Kaufman, A., "Phosphoric Acid Fuel Cell Bus Development," in the proceedings of *The Annual Automotive Technology Development Contractors' Coordination Meeting*, Edited by Society of Automotive Engineers, Warrendale, PA, p. 517, 1992.

36. "Alternative Fuels—The Fuel Cells are Coming," *H_2 Digest*, September/October 1994.

37. Creveling, H.F., "Proton Exchange Membrane (PEM) Fuel Cell System R&D for Transportation Applications," in proceedings of *The Automotive Technology Development Contractors' Coordinating Meeting*, Society of Automotive Engineers, Warrendale, PA, p. 485, 1992.

38. "Chrysler, Ford win fuel cell contracts," *Automotive News*, p. 6, July 18, 1994.

39. Gottesfield, S., Wilson, M.S., Zawodzinski, T. and Lemons, R.A., "Core Technology R&D for PEM Fuel Cells," in proceedings of *The Automotive Technology Development Contractors' Coordinating Meeting*, Society of Automotive Engineers, Warrendale, PA, p. 511, 1992.

40. Springer, T., Wilson, M., Zawodzinski, T., Derouin, C., Valerio, J. and Gottesfeld, S., PEM and Direct Methanol Fuel Cell R&D, in proceedings of *The Annual Automotive Technology Development Contractors' Coordination Meeting*, Society of Automotive Engineers, Warrendale, PA, 1994.

41. Anand, N.K., Appleby, A.J., Dhar, H.P., Ferreiera, A.C. and et al, "Recent Progress in Proton Exchange Membrane Fuel Cells at Texas A&M University," in *Hydrogen Energy Progress X*, International Hydrogen Energy Association, Coral Gables, FL, p. 1669, 1994.

42. Nadal, M. and Barbir, F., "Development of a Hydride Fuel Cell/Battery Powered Electric Vehicle," in *Hydrogen Energy Progress X*, International Association for Hydrogen Energy, Coral Gables, FL, p. 1427, 1994.

43. Vandenborre, H. and Sierens, R., "Greenbus: A Hydrogen Fuelled City Bus," in *Hydrogen Energy Progress X*, International Association for Hydrogen Energy, Coral Gables, FL, p. 1959, 1994.

44. DeGeeter, E., Van den Broeck, H., Bout, P., Groote Woortmann, M., Cornu, J.P., Peski, V., Dufour, A., and Marcenaro, B., "Eureka Fuel Cell Bus Demonstration Project," in *Hydrogen Energy Progress X*, International Association for Hydrogen Energy, Coral Gables, FL, p. 1457, 1994.

45. Marcenaro, B., EQHHPP FC BUS: Status of the Project and the Presentation of the First Experimental Results, in *Hydrogen Energy Progress X*, International Association for Hydrogen Energy, Coral Gables, FL, p. 1447, 1994.

46. Kordesch, K., Oliveira, J.C.T., Kalal, P. and et al, "Fuel Cell R&D—Toward a hydrogen economy," *Electric Vehicle Developments*, Vol. 8, p. 25, 1989.

Hydrogen - Air Fuel Cells of the Alkaline Matrix Type: Manufacture and Impregnation of Electrodes

D. Staschewski
Karlsruhe Nuclear Research Center (KfK)
Institute for Neutron Physics and Reactor Technics

Abstract—Lightweight fuel cells containing immobilized electrolyte proved to be decisive for the power systems of space vehicles, and are also candidates for terrestrial applications such as vehicular traction. Thus the KfK development of matrix fuel cells has been continued with particular respect to electrode performance. Electrodes with various collectors, PTFE bonding and platinum concentrations of the Pt/C catalyst have been manufactured and tested in terms of modified system parameters. Although this work has been confined to the laboratory scale, cell format and gas–water management are appropriate for extending the experimental set-up to a technical cell stack. Parallel to matrix cell fabrication a demonstration of alternative vehicular propulsion has been carried out in practice by driving an electric truck equipped with commercial alkaline fuel cell aggregates and fuelled by hydrogen and pure oxygen gas.

INTRODUCTION

The alkaline matrix concept of fuel cells (FC) has a long tradition, but at present is only utilized in the power plants of space vehicles. In Europe, this special line of FC technology has recently been boosted by the HERMES project. Although efficient FC stacks have been constructed on an industrial scale by European enterprises, no practicable solution was at hand for the ESA space shuttle. The German companies of Varta and Siemens for instance, are now challenged by HERMES developed FC aggregates using Raney-nickel and silver as catalysts for the partial electrochemical reactions of hydrogen and (pure) oxygen. These systems work with pumped KOH electrolyte. But owing to problems with gas bubbles in a flowing medium under microgravity and to weight savings and other economies, an immobilized KOH system was required for ESA. According to a selection limited to European FC types in 1989, the Siemens technology adapted to the matrix concept appeared to be most promising for HERMES [1].

Indeed, the alkaline matrix applied in NASA fuel cells played a very important role in the American space missions. During the last few years, however, an unexpected breakthrough occurred in FC research: the competing proton exchange membrane made from polyperfluorosulfonic acid (Nafion, SPE, now PEM and DEM) gained superiority [2]. There are reports on FC tests with current densities up to 5 A cm^{-2} [3], and expectations concerning the power-specific weight amount to 1 kg kW^{-1}. In future, acidic cells of the PEM type might push away even advanced alkaline systems. If the PEM cell is qualified for extraterrestrial surface missions [4], it should also be good for vehicular transportation on Earth, offering the advantage of optimal air utilization.

The FC activities of the Karlsruhe Nuclear Research Center at first arose from considerations about an efficient substitution of gasoline and diesel oil by hydrogen electrolytically produced at nuclear power stations [5]. Today solar radiation may be a more popular source of primary energy, but that does not affect the application sector where our activities have been settled. In order to investigate alternative vehicular traction based on H_2 fuel cells, a Volkswagen electrovan was equipped with Siemens FC stacks and gas cylinders and, provided with a licence plate, driven on public roads. Although rearrangement and operation of the hydrogen truck are not included in the author's tasks, some data and experiences of the operational team are noted in this paper.

On account of high costs and considerable power-specific stack weight FC traction power could only be installed at an inadequately low level, i.e. scheduled 3 × 7 kW, but in practice reduced to 17.5 kW because the open-circuit voltage had to be adapted to the chopper controller. The nominal weight of one compact 7-kW unit came to 85 kg. For driving pure oxygen was needed instead of air, which should preferably be used in cars to diminish operational expenses and save gas containers.

Obviously a more convenient solution has to be found for fuel cell traction, and therefore a second line of experimental work was established at the institute, which concentrated upon fabricating alkaline matrix cells fuelled by hydrogen and air. Methods and results of this activity have been presented at all WHE Conferences since Toronto 1984 [6–10]. The leading idea of matrix cell design was to apply lightweight plastics as spacers and to create joint gas spaces or plena for two different electrodes. Consequently edge current collection, i.e. the "monopolar" FC principle was favored. As for the choice of catalysts, the "salt catalyst" of Prototech (U.S.A.) containing 10% Pt on carbon served as a basic material. Owing to a minimum of man-power in the laboratory, time-consuming preparations and tests of home-made catalysts were not desirable. Nevertheless, the Prototech material was upgraded by platinum black precipitated on the water-suspended powder after reduction of Pt chloride with formic acid. Previous system studies mainly dealt with bilateral water removal at full matrix cell operation. In the last phase of investigation special attention was paid to the influence of collector thickness and conductive cell structures on electrode performance.

STRUCTURE AND FABRICATION OF ELECTRODES

As power-generating components of fuel cells, the electrodes have to meet many partly divergent requirements:

- mechanical stability under operational conditions;
- electronic conductivity, electric contact surface;
- capacity of electrolyte absorption, hydrophilic micropores;
- transport paths for ions, low electrolytic resistivity;
- porosity for gas diffusion, hydrophobic macropores;
- specific electrocatalysis;
- chemical stability against corrosion or oxidation.

Mostly the reactive layer consists of a coherent and conductive cavity structure backing or housing the catalyst and other dispersed ingredients. In our case the catalytic mixture of ingredients is directly filled into collectors made of gold-plated nickel screens. Fuel cells have been manufactured with single or double screens, the latter ones were metallically linked as a compound lattice by galvanization in a Ni−salt bath under mechanical pressure [8, 9]. Such combined dual collectors intensify the backing effect toward a cage-like inclusion, thus hindering catalyst movements or losses during cell fabrication and later cell operation with inevitable thermal cycles. Their exact planeness, strength and large contacting surface offer further advantages compared to single screen collectors of equal thickness which, however, have a lower weight, though also galvanized to cover the smooth wires with a rough nickel deposit. The collector thickness of tested electrodes ranged from 0.2 to 0.6 mm.

The ingredients of the powdery material are Pt/C catalyst, inert PTFE as binder and hydrophobic agent, and ammonium bicarbonate as pore-generating filler. Pulverization of the catalyst was carried out by high-speed stirring or by flotation of stirred powder in flowing air. Fine-grained Teflon, produced by drying an aqueous suspension of Hostaflon (Hoechst/Germany), is first dispersed within a definite batch of bicarbonate by means of the high-speed stirring apparatus. Then both ingredients are blended with the catalyst in the same way. Admixtures of liquid toluene transform the powder into a workable dough or lubricant which can be manually pressed into the collector screen. In doing so the frame zone of the collector has to be covered with a protective mask.

Heating at 110°C removes the solvent and filler bicarbonate completely, providing a formation of macropores in the catalytical layer which is proportional to the quantity of evaporated matter. These pores partly collapse when more material is squeezed upon the layer. Hence the filling procedure repeated with modified PTFE blending on both sides of the electrode results in establishing a gradient of Teflon concentration in the active material. The only difference between cathode and anode manufacturing consists in the PTFE graduation. According to our standard preparation, the cathodic catalyst contains batchwise 5%, 20% and 40% PTFE, while the anode is furnished with batchwise 0%, 10% and 20% PTFE, and thus is less hydrophobic. In the case of single-structured electrodes, compacting and smoothing of the compound layer are achieved by rolling or calendering the charged screen.

To improve the catalyst bonding by Teflon sintering, the electrodes are heated up to 340°C under argon. Afterwards the peripheral slit system surrounding the catalytical area of $20 \times 20\ cm^2$ is punched. Because now the connection between the internal area and edge zone is formed by 36 wire bridges of only 4−5 mm width, the conductive structure has to be reinforced. The remedy applied is soldering the bridges and fringes with tin which, however, increases the weight. In order to seal and isolate the tinned parts just as the reactive zone, the slits and all the voids left in the screen(s) must be carefully seal-glued. The next fabrication steps, necessary to get a functioning matrix cell but not elaborated in this special paper (compare [10]), include electrode extensions which possess ERP properties, i.e. an electrolyte storing capacity and large surface for evaporation of reaction water. It should be mentioned that all the structures and components within the fuel cell greatly influence the performance of the electrodes.

KOH IMPREGNATION OF ELECTRODES AND MATRIX

The electrolyte filling procedure takes place in a closed facility under vacuum of a rotating pump (Fig. 1). In this filling station the containers, valves and pipes are made of stainless steel with the exception of a sight tube and two glass bulbs. Provided with appropriate rubber gaskets, the single or multiple fuel cell is inserted and screwed between two bulky flanges. The lower flange B, rigidly linked with a 6 l container which stores 12 M KOH, was furnished with internal channels and ducts leading to the hydrogen slits of the cell. The upper flange A has ducts and openings corresponding to the system of air slits, and is connected with the vacuum conduits by means of flexible metallic tubes.

The electrolyte let down and passing the cell via the hydrogen plenum can be observed in the glass tube. After closure of the adjacent valve, a moderate pressure is applied upon the caustic liquid with nitrogen gas. This measure effects a quicker soaking and penetration of the reactive layers. When an excess of electrolyte drops into

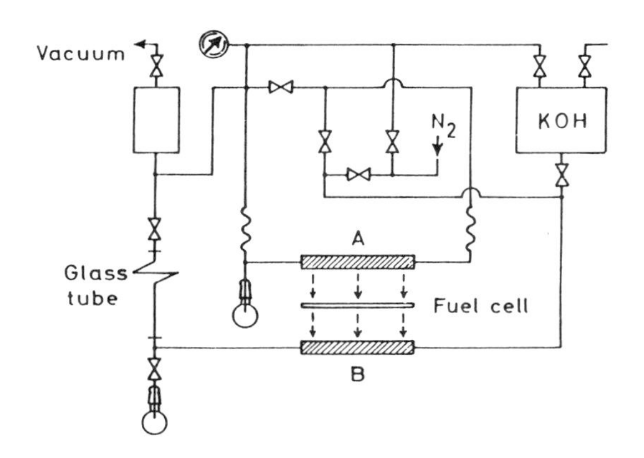

Fig. 1. KOH filling station for matrix fuel cells. A and B: stainless steel flanges with ducts corresponding to the standard slit system.

the upper glass bulb, the procedure can be finished by flushing the cell with nitrogen.

In the case of matrix cells operated with air, a thorough method for electrolyte renewal is crucial. Since only a small quantity of alkali can be stored in the central layers, the danger of KOH losses caused by leakages into the gas plena, often due to insufficient evaporation of reaction water, and by carbon dioxide absorption at the air-contacted side, is considerable. However, the KOH filling station ensures also regeneration of a fuel cell. In the course of cell testing, multiple replacements of diluted electrolyte have been successfully carried out. Frequently repeated filling of a cell is only limited by platinum extraction. After operational activation the catalyst tends to form colloid solutions, particularly with washing water.

MODIFIED TERMINAL BLOCKS FOR CELL TESTING

To run an alkaline matrix cell, specific peripheral conditions have to be established by means of terminal accessories. The start-up of cell operation first requires temperature conditioning which is achieved by heating a water circuit. If excess heat is produced during cell operation, this water flow must be brought to an external cooling system. Pumped air, depleted in CO_2, is preheated before entering the cell. Hydrogen gas circulates in an internal loop for removal of anodic water vapor. All these measures are conducted with terminal blocks which mainly consist of structured polysulfone plates and nickel sheets including mounted pumps, valves and connecting pieces. The two-block assembly has been always regarded as a precursor of a future multicell aggregate and is subject to continual improvements.

The conditioning block, schematically depicted in Fig. 2 (left side), contains a catalytical reaction zone D where a controlled combustion of H_2 takes place in the effluent air. Hydrogen is fed through a special duct into a supported space between two thin screens charged with PTFE-bonded Pt, Pd and Ni powder, stretched in a dual air chamber. The air, exhausted and traversing the final chambers, exchanges heat with the recirculated water. Regulation of the hydrogen input makes it possible to adjust the environmental temperature scheduled for cell testing. First warm-up experiments resulted in 30-min periods with a temperature rising from 25 to 80°C. After a test pause of some weeks,

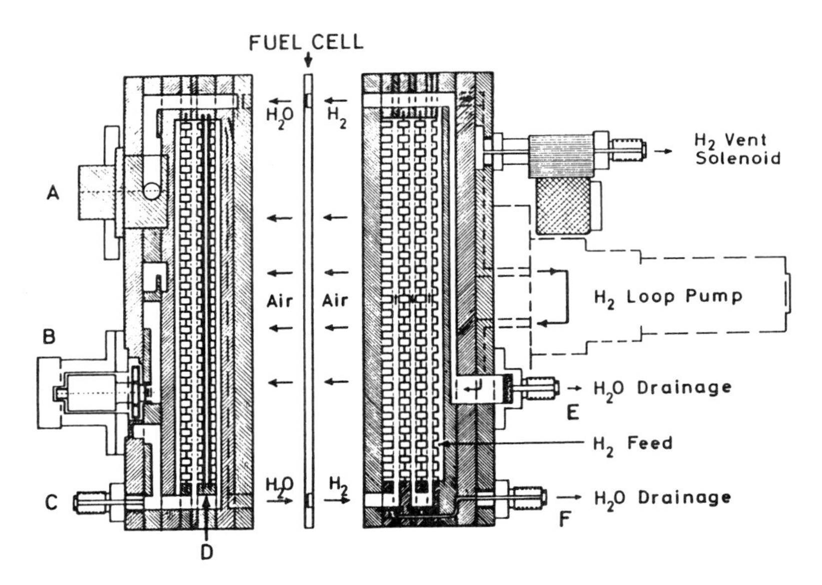

Fig. 2. Fuel cell test blocks as precursors of a multicell aggregate. Left side: the conditioning block, two-way valve for external or internal water conduct (A), water pump (B), exit for water pressure surveillance (C), chamber for catalytical combustion of hydrogen in flowing air (D). Right side: the heat exchange block of gases, drainage of the hydrogen loop pump (E), main loop drainage (F).

double the time was needed. The catalyzed combustion started with an enormous delay, apparently owing to tightening corrosion effects in the catalytic screens.

Direct hydrogen burning, without flames of course, appears to be obligatory as the heat source for cold starting in the case of the mobile autonomous FC stacks here envisaged. In the laboratory, however, electric heating is more convenient. For this reason other conditioning blocks have been equipped with heater foils. These commercially available resistors are tightly glued to enveloping nickel sheets by means of adhesive films. Using such heater elements we start fuel cell operation after 15 min of conditioning.

The heat exchange block for gases (Fig. 2, right side) carries both the hydrogen loop pump and the vibrational air pump. With reference to the drawing plane of Fig. 2 the air moves vertically through its chambers, while hydrogen traverses parallel to the plane. Drainage filters made of asbestos paper and determined to clear off the water condensed in the hydrogen loop have been sunk into the bottom parts below the gas slits. An advanced model of the loop block houses a plane water cooler between two H_2 chambers, thus allowing an enhanced condensation of loop water even in the case of experiments with immobile pure oxygen instead of flowing air. In the wake of discussions concerning matrix cells for HERMES, we introduced cell testing with oxygen gas. It proved to be a useful expansion of the test modes, supplying more specific information about the matrix system. Additionally, runs with oxygen or air overpressure relative to hydrogen feed pressure, i.e. 100–200 mbar over atmosphere, were conducted by choking the effluent oxidant. Since the test facility comprises a big water-cooled condenser preceding the throttling exit valve, the pressure of the flowing oxidants remained stable even during longer periods of operation.

PARAMETER STUDIES OF SINGLE CELL PERFORMANCE

Fuel cell performance results from a combined action of diverse effects dependent on structural and operational parameters. Because of temporal changes of the cell efficiency we have to distinguish between an initial phase with maximum activity and the long-term operation. In practice the latter is the only interesting phase. But it makes no sense to conduct extensive long-term runs if the initial cell power does not suffice. Thus only a few monocell tests have been prolonged up to 24 days of uninterrupted operation, yielding about 6 l of reaction water. The main work concentrated on short-time testing, mostly one week per cell. In this mode of probing it is not absolutely necessary to stabilize the electrolyte by means of additional porous cell components like reservoir plates (ERP) and PTFE membranes. Power decline in that phase can be clearly related to dilution and stripping of the electrolyte. Before this happens, there is time enough for studying the primary cell efficiency.

The catalytic effect of platinum on carbon, though intensified by higher Pt concentrations, did not dominate the performance measured with many of our cells. There was strong evidence that the decisive influence proceeded from the electronic conductivity inside and outside the electrodes. The catalytic carbon powder, entirely suitable as a baking and bonding material, was suspected of having a considerable specific resistance, and admixtures of platinum black should improve the current transfer within the baked porous catalyst. A compensation of the variant effects is demonstrated in the following comparison which relates cells to identically structured electrodes containing dual 0.4-mm screens and delivering practically the same maximum power of 30 W:

- MZ10—50% Pt/C, electrodes only nickel-plated, PTFE spacers;
- MZ12, 13—25% Pt/C, electrodes gold-plated, PTFE spacers;
- MZ20—10% Pt/C, electrodes gold-plated, gold-plated 0.6 mm nickel screens as spacers.

In this example the gradually diminished platinum content does not affect the power output on account of improvements in the conductive cell structure. That verifies the enormous importance of current collection in the extremely low-ohmic system. To get technically interesting current densities, the development of edgewise collecting cells has to be directed toward more bulky metallic elements as indicated with MZ11 in Table 1. This design however would not be in compliance with the lightweight concept.

As regards the effect of gas diffusion at the air electrode, the utilization of air oxygen was found to be excellent. Offering over-stoichiometric oxygen we obtained saturation plateaus of the measured power (Fig. 3). In addition, switching from air to pure oxygen resulted in a relative

Table 1. Properties of edgewise collecting electrodes

Fuel cell	Nickel screen Structure	Nickel screen Thickness (mm)	Total weight* (g)	Platinum content Pt/C	Platinum content mg Pt cm^{-2}	Maximum power (W) with air	Maximum power (W) with pure oxygen
MZ18	Single†	0.2	46	25%	1.83	19	21
MZ12	Double	0.4	93	25%	1.56	28	—
MZ20	Double	0.4	(170)‡	10%	0.61	32	42
MZ11	Double	0.6	118	25%	2.70	41	—

*Tinned but not glued screens.
†Nickel-plated and compacted by rolling.
‡Inclusive contacting gold-plated Ni spacers, 0.65 mm thick.

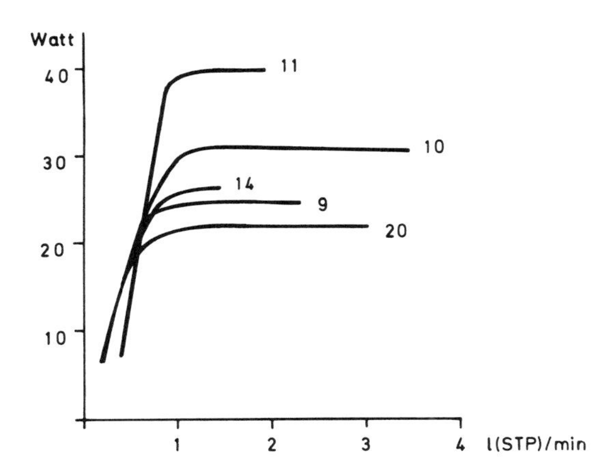

Fig. 3. Monocell power dependence on air flow rate. Saturation levels signed with cell numbers.

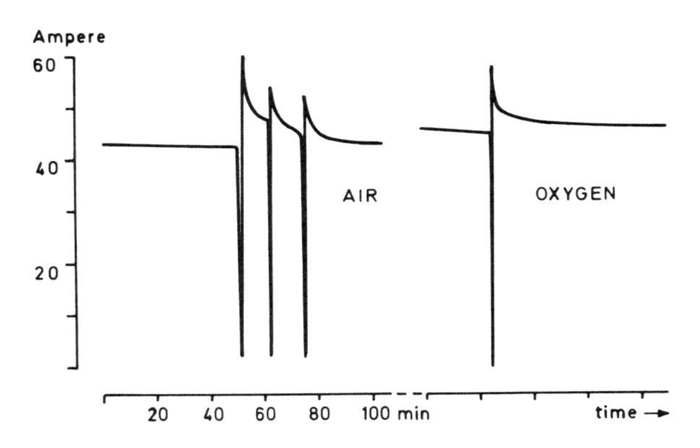

Fig. 5. Break effects with air and pure oxygen at maximum load of monocells containing 10% Pt on carbon. Air: short-time flow interruption causing a voltage decline from 0.5 to 0.01 V. Oxygen: short-time load cut increasing the voltage from 0.5 to 1.01 V (open circuit).

power increase of only 10% with 0.2 mm electrodes, and of 30% with 0.4 mm electrodes (Fig. 4). A special response of the system is revealed by oxygen-related break effects. At full load an interruption of the air flow or disconnection of the load resistor for some seconds led to a temporary power boost. This activation may be partly influenced by electrolyte shifting, but is mainly due to cathodic peroxide formation and catalyst inhibition cancelled either by starvation (missing oxygen) or self-decomposition (missing current) of the intermediates (Fig. 5). The inhibition–relief effect, also responsible for the hysteresis of voltage–current curves, makes it plain that a more efficient peroxide-suppressing catalyst would greatly contribute to enhanced cell performances.

PRESSURE STRATEGY OF MATRIX CELL OPERATION

Short-time cell tests with a reversed pressure balance of hydrogen relative to oxygen or air did not yield any significant advantage of the one or other operation mode. Nevertheless a permanent stability of matrix cell performance can only be achieved by electrolyte-retaining structures extending the abundantly wetted electrode into the gas plenum, i.e. by an electrolyte reservoir plate. In alkaline fuel cells the reaction water is produced at the anodic hydrogen electrode. The cathode principally loses water by the water-consuming turnover of oxygen to OH^- ions and the transport of these ions which carry solvation water when migrating to the anode. If there is a hydrogen–oxygen pressure gradient, excessive liquid is additionally moved to the low-pressure side of the matrix. In our experiments maximum gas pressures are usually limited to 200 mbar in excess relative to atmosphere. In order to save energy in a FC system, considered as autonomous in future, the pressure of pumped air should be as low as possible. Realizing this requirement we must first study the case of "hydrogen pressure > air pressure" leading to three options:

(i) cathodic ERP/unprotected anode;
(ii) cathodic ERP/anodic ERP;
(iii) cathodic PTFE membrane/anodic ERP.

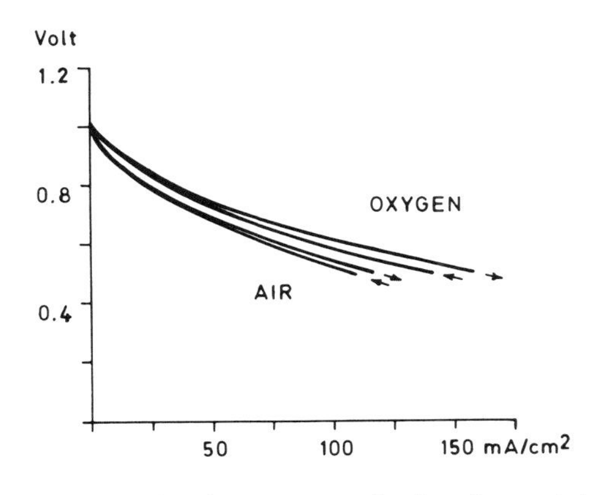

Fig. 4. Terminal voltage–current density characteristics of monocell MZ20 at 80°C after 24-h operation.

Option (i) may be the optimal strategy for cell operation with small gas pressures and moderate current densities, if the ERP surface is large enough to ensure a nearly exclusive evaporation of product water in the flowing air. High power, however, demands both electrodes equipped with ERPs and a hydrogen loop of considerable capacity. The third option, tested with two monocells, requires a stronger hydrophobic cathode with macropores resistant to flooding, because here the most concentrated electrolyte banks up under pressure toward the membrane barrier. On the anodic side the diluted electrolyte is forced to enter the ERP in favor of water removal with circulating hydrogen. This option is only possible with monopolar, i.e. edgewise, collecting fuel cells.

The reversed case of "hydrogen pressure < air pressure", demanding only an anodic ERP, is burdened with the handicap of higher air compression. But here a ventilation of electrolyte-free pores is favored at the cathode, thus providing intensified oxygen absorption and water evaporation. Cathodic water supply takes place by osmosis or rediffusion due to the KOH concentration gradient and in our system by induced electrolyte vibrations resulting from the swinging air pump. Drying-out of the cathode and matrix at extreme electric load can be met by pressure regulation or temperature control in the condensing hydrogen loop. This operation mode, most appropriate for matrix cells using pure oxygen, has been thoroughly studied in view of space applications [11].

THE ELECTRIC HYDROGEN TRUCK OF THE INSTITUTE

The KfK electrovan powered by fuel cells proceeded from a VW platform truck which had been equipped with a traction battery of 180 Ah and a DC chopper controller. The original lead–acid battery was replaced with a smaller one and three Siemens FC aggregates. The DC motor maintained had been designed for a rated power of 17 kW at 144 V and for peak power of 32 kW (3 min). Extensive installations necessary to accommodate the new system included storage cylinders with compressed gases, the water pump and air blowers of the aggregate-cooling circuit, a power contactor, a lot of pressure gages, fittings, safety valves, protecting containers, etc. Hydrogen as fuel was stored in two steel cylinders containing in total 1.5 kg H_2 at 200 bar. Oxygen and nitrogen for supporting pressure cushions were held each in one cylinder. The weight data of Table 2 refer to the official licensing of August 1988 and represent the initial state of arrangement which, far from being optimal, was subject to perpetual changes.

The experimental hydrogen truck has been operated according to the FC-battery hybrid concept using fuel cells as the main suppliers of traction energy under normal or moderate driving conditions and the lead–acid battery for intermittently high power levels. Since the Siemens units did not comprise any heating accessory and could be only warmed up by fuel cell operation, cold starting of the vehicle led to a rise of the electrolyte, implicating a possible flooding of the electrolyte reservoir tank. To get a sufficient performance of the evaporator stacks, the system's working temperature of 80°C had to be established. This mostly was achieved by employing an external load resistor. Well conditioned, the electrovan proved to be a reliable vehicle. Owing to the extremely unfavorable ratio of power to weight, vehicular acceleration and gradability barely complied with standard requirements. It should be mentioned that only 5% of the rated power was needed for ventilators, water pump and other accessories. The top speed of steady 70 km h^{-1} could be attained on plane highways with 8% power share of the lead–acid battery, resulting in a driving range of 120 km. Urban driving however reduced the range to approximately 60 km. Until February 1990 the total mileage amounted to 1940 km, corresponding to 220 h of fuel cell operation.

Fig. 6. Uncovered KfK hydrogen truck exhibiting fuel cell units, radiators, gas cylinders and accessory equipment.

At present, a reorganization of the gas storing system is under way. Substitution of the pressure cylinders by insulated Dewar vessels for liquid hydrogen and oxygen is expected to save weight, to facilitate fuel and oxidant handling, and to prolong the driving range.

CONCLUSION AND OUTLOOK

Experimental studies of the main effects responsible for matrix cell efficiency elucidated the crucial role of the current-collecting structures. As edgewise collection suffers greatly from transverse ohmic resistances of the electrodes, conductive ERP and spacer structures are envisaged

Table 2. Preliminary weight balance of the KfK electrovan

Fuel cell aggregates, 175 single cells, 17.5 kW rated power	230 kg
Traction battery, Pb-sulfuric acid, 90 Ah	282 kg
Two cylinders for hydrogen, one cylinder for oxygen, each 10 m^3 gas (STP), and one cylinder for nitrogen, 2 m^3 gas (STP)	222 kg
Additional vehicle roof, removable	100 kg
Accessory equipment for control, surveillance and safety	621 kg
Vehicular curb weight	2880 kg
Driver and payload	195 kg
Total permissible weight	3075 kg

to improve the cell performance. This leads, in practice, to a bipolar cell design with heavy metallic layers. As regards the most promising development of PEM fuel cells, a transition from alkaline matrix to proton exchange membrane is desirable especially with respect to an ideal utilization of air in FC-powered vehicles, and is now realized in our laboratory.

The KfK vehicular experimentation program once more demonstrated the technical feasibility of fuel cell traction based on hydrogen as a fuel, but also revealed a striking disproportion between FC aggregate weight and accessory weight which to a considerable extent is enforced by official safety instructions. Compared to gasoline cars of conventional design, the performance of the hydrogen truck is not at all impressive. But the experience gained with that clumsy electrovan may help to conceive the principles of alternative vehicular traction in practice and to realize a better technical compromise in the future.

Acknowledgements—The author would like to thank K. Schretzmann as head of the vehicle program for his approval of presenting data of the hydrogen truck in this publication. Also acknowledged is the excellent help provided by S. Gaukel, E. Kurz, K.-D. Schorb and E. Wachter.

REFERENCES

1. H. Gehrke, *Proc. European Space Power Conference*, Madrid, Spain, Vol. 1, pp. 203–209 (2–6 October 1989).
2. S. Srinivasan, E. A. Ticianelli, C. R. Derouin and S. Gottesfeld, *Proc. Fuel Cell Seminar*, Long Beach, CA, pp. 324–327 (23–26 October 1988).
3. D. Watkins, K. Dircks and D. Epp, *Proc. Fuel Cell Seminar*, Long Beach, CA, pp. 350–355 (23–26 October 1988).
4. J. Huff, J. Hedstrom and N. Vanderborgh, *Proc. European Space Power Conference*, Madrid, Spain, Vol. 1, pp. 217–219 (2–6 October 1989).
5. M. Dalle Donne, S. Dorner, G. Kessler and K. Schretzmann, *Int. J. Hydrogen Energy* **8**, 949–960 (1983).
6. D. Staschewski, *Proc. 5th World Hydrogen Energy Conf.*, Toronto, Canada, Vol. 4, pp. 1677–1684 (15–20 July 1984).
7. D. Staschewski, *Int. J. Hydrogen Energy* **11**, 279–283 (1986).
8. D. Staschewski, *Proc. 6th World Hydrogen Energy Conf.*, Vienna, Austria, Vol. 3, pp. 1266–1272 (20–24 July 1986).
9. D. Staschewski, *Int. J. Hydrogen Energy* **13**, 633–638 (1988).
10. D. Staschewski, *Proc. 7th World Hydrogen Energy Conf.*, Moscow, C.I.S., Vol. 1, pp. 357–366 (25–29 September 1988).
11. K. Kikuchi, T. Ozeki and Y. Yoshida, *Space Solar Power Rev.* **5**, 179–188 (1985).

High-Temperature Fuel Cells: Part I

How the Molten Carbonate Cell Works, and the Materials That Make It Possible

Nguyen Quang Minh

High-temperature fuel cells operate at temperatures greater than 600 °C, producing high-quality by-product heat suitable for use in cogeneration schemes or bottoming cycles. Because of their high operating temperature, however, high-temperature fuel cells present several significant technical challenges. Both the advantages and the technical challenges are summarized in Table 1. The special attraction of a high-temperature fuel cell is that common fuels such as methane can be reformed directly within the fuel cell (internal reforming). Thus, the need for fuel-processing subsystems is essentially eliminated, and a system based on high-temperature fuel cells can be simple and inexpensive.

Two types of high-temperature fuel cells, molten carbonate fuel cells (MCFCs) and solid oxide fuel cells (SOFCs), are under development for a variety of applications. Potential commercial markets include electric utility, cogeneration, and on-site applications. Steady progress is being made toward commercialization of MCFCs. SOFCs, considered an advanced concept, are at an early stage of development. Table 2 lists the operational characteristics and unique features of these cells. This paper describes the operating principles, the basic components and designs of the fuel cell stack, and the technological status of the MCFC. The SOFC will be discussed next month in Part 2.

Molten carbonate fuel cells

Carbonates are the only molten electrolytes that do not change in the presence of the electrochemical combustion of hydrogen and carbon monoxide, thus providing the basis for the MCFC. The MCFC consists of two porous electrodes separated by a molten carbonate electrolyte, which serves to conduct carbonate ions from the cathode (the oxidant electrode) to the anode (the fuel electrode). The fuel for the MCFC is a mixture of H_2 and CO, and the oxidant is a mixture of O_2 and CO_2. The overall reactions at the MCFC anode and cathode are as follows:

Anode $H_2 + CO_3^{2-} = H_2O + CO_2 + 2e^-$

$CO + CO_3^{2-} = 2CO_2 + 2e^-$

Cathode $\frac{1}{2}O_2 + CO_2 + 2e^- = CO_3^{2-}$

Figure 1 shows the reactions in an MCFC (*2*).

The state-of-the-art MCFC consists of an Ni anode, a Li-doped NiO cathode, and a lithium aluminate ($LiAlO_2$) matrix (or electrolyte support) filled with lithium and potassium carbonates as the electrolyte (referred to as the electrolyte structure or the tile) (Figure 2). MCFCs are not operated as single units; rather they are stacked one on top of the other to build up voltage. In a stack of cells, separator plates (also called bipolar plates), current collectors, and cell-to-cell seals are required (Figure 3) (*2*). MCFC stacks are operated at a temperature of about 650 °C and at a pressure of 1 to 10 atm.

The most interesting component of the MCFC is the electrolyte structure. The structure is composed of a molten mixture of lithium and potassium carbonates contained in a lithium aluminate ($LiAlO_2$) support. The molten carbonate mixture is retained in the ceramic matrix by capillary forces. At the fuel cell operating temperature, the electrolyte structure is a thick paste, and the paste provides gas seals (called the wet seal) at the edges of the cell. The MCFC electrolyte support is a matrix of fine ceramic particles (an aggregate of fine particles without interparticle bonding) that supports and contains

An intriguing technology

A fuel cell is an electrochemical device that converts the chemical energy of a fuel directly to electricity (and heat) without combustion as an intermediate step. The principle of fuel cell operation was first reported by Sir William Grove in 1839 *(1)*. Last year marked the 150th anniversary of his discovery.

Fuel cells offer several potential advantages over other electricity-generating systems: substantially higher conversion efficiency of fuel energy to electricity, modular construction, high efficiency at part load, minimal siting restriction, potential for cogeneration (simultaneous production of heat and electrical power), and much lower production of pollutants (including acid rain precursors).

The first fuel cell employed dilute sulfuric acid as the electrolyte and operated at room temperature. Since that time, fuel cell technology has grown from small laboratory-scale devices to large demonstration units, and has expanded into nonaqueous and solid electrolytes and high-temperature and -pressure operation. (See also Lindström's CHEMTECH series: 1988, August, p. 490; September, p. 553; November, p. 686; 1989, January, p. 44; February, p. 122.)

***Table 1.* High-temperature fuel cells—Advantages and technical challenges**

Advantages
Fast reaction kinetics
High efficiency
Internal reforming
High-temperature by-product heat
Technical challenges
Development of suitable materials
Corrosion and endurance
Fabrication

Table 2. **Features of high-temperature fuel cells**

	Molten carbonate fuel cell	Solid oxide fuel cell
Electrolyte	Li_2CO_3-K_2CO_3	Yttria-stabilized ZrO_2
(Electrolyte support)	($LiAlO_2$)	
Anode	Ni	Ni/Y_2O_3-ZrO_2 cermet
Cathode	Li-doped NiO	Sr-doped $LaMnO_3$
Bipolar plate/interconnect	310 SS clad with Ni	Doped $LaCrO_3$
Fuel (electroactive)	H_2, CO	H_2, CO
Oxidant (electroactive)	$CO_2 + O_2$	O_2
Operating temperature	650 °C	1000 °C
Operating pressure	1–10 atm	1 atm
Contaminant tolerance	< ppm sulfur	Not yet defined
Stack configuration	Planar	Tubular, monolithic, or planar

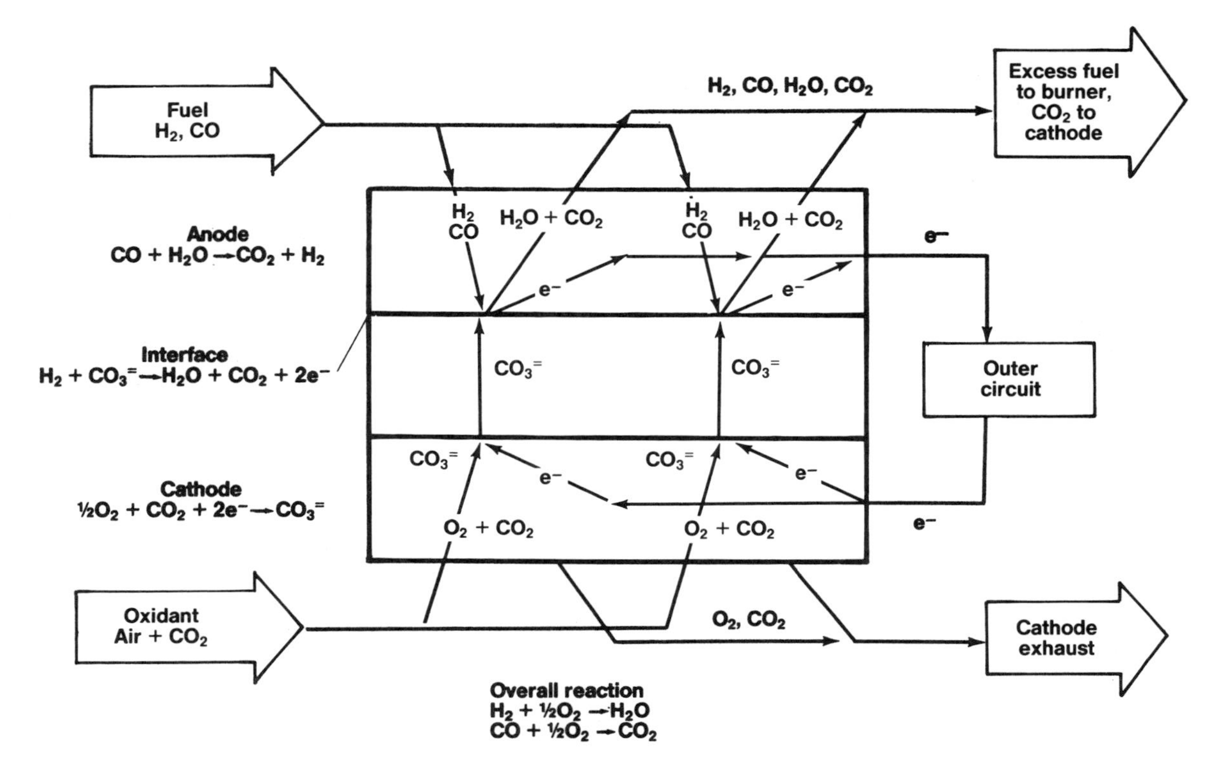

***Figure 1.* Schematic diagram of molten carbonate fuel cell reactions.** (2)

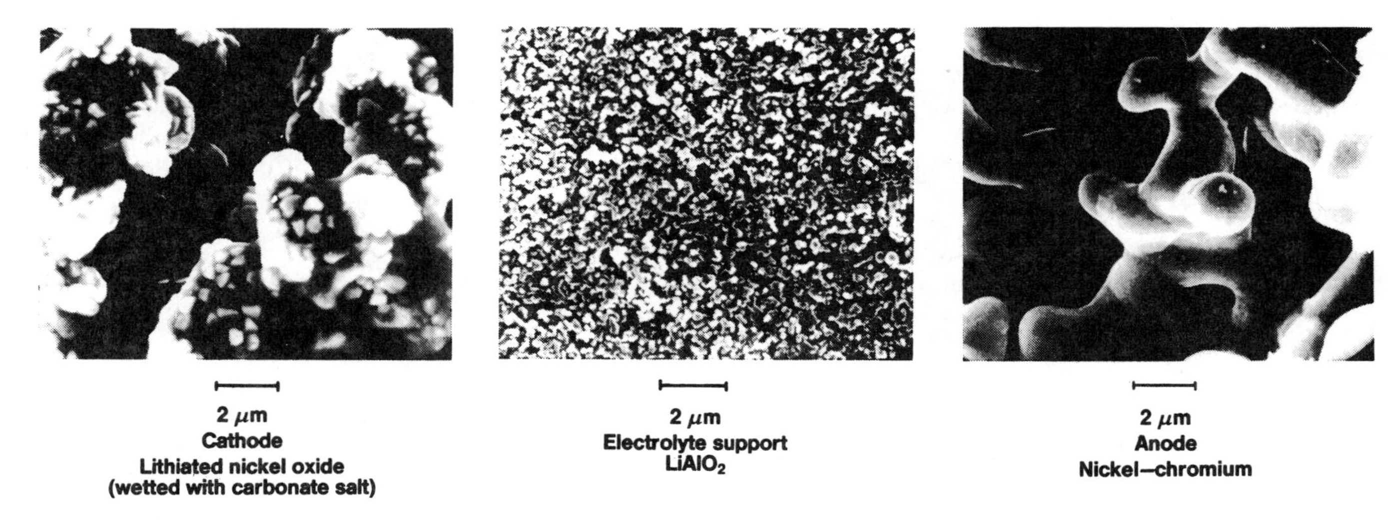

***Figure 2.* Micrographs of molten carbonate fuel cell electrodes and electrolyte support.** (2)

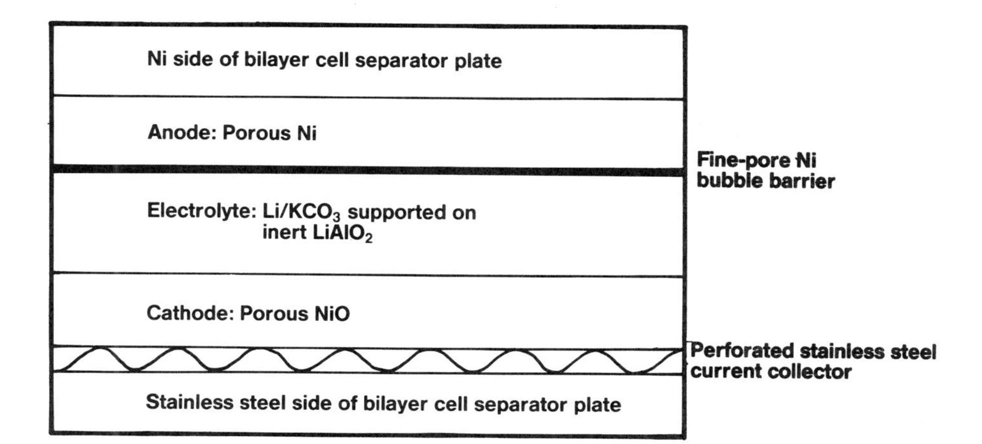

Figure 3. **Molten carbonate fuel cell stack components.** (2)

the molten carbonate mixture. The material for this support must be chemically inert against molten carbonates, adjacent cell components, and reactant gases, and must be morphologically stable. The carbonate retention capability of the support is a function of the morphology of the $LiAlO_2$ particles. Because the electrolyte structure is in contact with the porous anode and cathode, the $LiAlO_2$ matrix must have a narrow pore size distribution with a small mean pore size, typically 0.1 to 0.5 micron (3), to control distribution of the carbonate in the various cell components.

The strength and stiffness of the electrolyte structure depends on the relative amounts of carbonate and lithium aluminate. At low carbonate contents, the structure is rigid (essentially particle-to-particle contact); at high carbonate contents, the structure is fluid (a slurry); and in a narrow intermediate range, the structure is plastic (*4*). The composition of the plastic region is dependent on the particle size distribution and moves to higher carbonate contents for smaller particles. At present, the electrolyte structure typically consists of about 40 wt% $LiAlO_2$ and 60 wt% carbonate. MCFC electrolyte supports have almost always been made from finely divided submicron $LiAlO_2$ particles. The size, shape, and distribution of the particles control carbonate retention, mechanical properties, and effective ionic conductivity of the electrolyte structure. The most effective shape for the $LiAlO_2$ particles appears to be long rods or fibers of submicron diameter. Three allotropic forms of $LiAlO_2$ (alpha, beta, and gamma) have been reported; the gamma form is the most stable in the MCFC environment (*5*). $LiAlO_2$ has a low solubility in molten carbonates; therefore particle growth and phase transformation occur very slowly if at all. As changes in pore size affect carbonate electrolyte distribution, particle growth would affect the electrolyte. However, this has not been observed in MCFC testing to date.

The molten carbonate electrolyte is an alkaline carbonate mixture and is a liquid at the cell operating temperature (650 °C). The basic properties that must be considered in the selection of a carbonate electrolyte are conductivity, gas solubility, surface tension, vapor pressure, and corrosivity (*6*). At present, the preferred electrolyte is a binary lithium–potassium carbonate consisting of 62 mol% Li_2CO_3 and 38 mol% K_2CO_3. This electrolyte composition was chosen based on actual test results and may not yet be optimum.

The majority of the carbonate is contained in the electrolyte support. However, the carbonate is also distributed in the anode and cathode so as to allow the conducting of ions to or from the reaction site. The carbonate content in each component is primarily controlled by its relative pore size distribution, surface tension, wetting angles of the carbonate with the electrode material, and pressure difference between the anode and cathode compartments. These factors can be tailored and controlled to give appropriate carbonate distribution within the fuel cell.

The material currently used for MCFC cathodes is NiO. Nickel oxide is a p-type semiconductor; when it is used as the MCFC cathode, the material is doped in situ with lithium from the Li_2CO_3 in the molten electrolyte. Lithium doping (lithiation) significantly decreases the resistivity of NiO. With an equilibrium lithium dopant content in NiO of about 2 at. % at 650 °C, the resistivity of the doped material is about 0.2 ohm cm (*7*).

The NiO cathode is fabricated from a porous nickel plaque, which is oxidized and lithiated in situ. The porosity of the plaque before oxidation is about 70–80%. The nickel plaque has a mean pore size of 6–10 microns. After oxidation and lithiation, the NiO cathode appears to consist of agglomerates of nickel oxide particles, with the electrolyte filling the space between individual particles (micropores). The agglomerates are separated by gas-filled spaces (macropores), which correspond to the pores of the original metal plaque. Because lithiated NiO is completely wetted with the carbonate electrolyte, the agglomerates are covered with a thin film of electrolyte. Gases diffuse through the film to react at the electrode, and therefore it is important to avoid flooding the entire electrode. The MCFC NiO cathode is sensitive to carbonate electrolyte flooding and to the degree of filling, and operates best within 15–30% filling of total pore volume (*8*).

Cathode polarization is affected by cathode thickness, as ohmic losses in both the liquid (electrolyte) and solid (electrode) phases increase as the cathode thickness increases. Diffusional losses in the gas phase also increase with cathode thickness. On the other hand, activation and liquid phase diffusion losses decrease as the cathode thickness increases. The optimum cathode thickness is attained when the total losses are at a minimum, which for NiO cathodes is at a thickness of 0.8 mm (*9*).

Our current MCFC anode is 1-mm-thick nickel with 60–70% porosity and a 5-micron median pore size (*10*). The anode is generally fabricated by sintering nickel powder. Pure nickel tends to sinter at fuel cell operating conditions, resulting in loss of surface area and pore growth. Additives have been used to control this sintering. For example, addition of a few percent chromium oxide to nickel has been shown to effectively prevent anode sintering (*11*) by the formation of submicron $LiCrO_2$ on the nickel surface. The anode structure is also susceptible to deformation (creep) under compressive load during normal stack operation, i.e., shrinkage of anode thickness that occurs when cells are under load. Anode creep results in decreased porosity, increased contact resistance, and leaks. Oxide dispersion (e.g., Al_2O_3) will strengthen the nickel anode sufficiently to avoid this creep (*12*).

The polarization of the nickel anode is small compared to that of the cathode, probably owing to the rapid kinetics of fuel oxidation. As a result, the nickel anode is relatively insensitive to the degree of filling by the electrolyte. This insensitivity is fortunate; the anode can be used as an electrolyte reservoir. In this way, one can compensate for the long-term loss of electrolyte from corrosion and volatilization.

Steam reforming of methane and light hydrocarbons is possible at the MCFC anode because nickel is a catalyst for the reforming reactions. The performance of MCFC anodes is very sensitive to sulfur. The anode cannot tolerate more than 1–5 ppm of sulfur in the fuel gas without suffering a significant performance loss (*13*).

Stacking the cells

The MCFC stack design is based exclusively on a planar geometrical configuration. The MCFC stack is an assembly of repeating unit cells in electrical series; each unit cell has a total thickness of about 5 mm. Typical thickness of the cell components are: anode 0.8–1.5 mm, electrolyte 0.5–1.8 mm, cathode 0.4–1.5 mm, and bipolar plate 0.3 mm. In addition, MCFC stacks have bubble barriers and current collectors.

The bipolar plate provides electrical contact between adjacent anode and cathode in an MCFC stack. The material for the plate must be resistant to molten carbonate attack and compatible with other cell components. At present, stainless steel 310 is employed as a base metal for the separator plate (*14*) owing, in part, to its relatively low cost. In MCFC stacks, SS 310 plates are coated with a thin film of molten carbonate and are in contact with the cathode and oxidant on one side and with the anode and fuel gas on the other side. The 310 stainless steel has satisfactory corrosion resistance in the cathode environment, where a relatively stable and protective oxide scale forms in the presence of molten carbonate. However, this material can be severely corroded in the reducing environment of the anode and requires an expensive nickel-clad stainless steel.

The MCFC bubble barrier is a thin membrane positioned between the anode and the electrolyte to provide support and serve as a deterrent to gas crossover if cracks develop in the electrolyte structure. Nickel is generally the preferred material for the bubble barrier. The bubble barrier must contain pores that are filled with molten carbonate to provide continuous ionic transport. In order to be fully impregnated with carbonate electrolyte, bubble barrier pores must be finer than the anode pores.

Current collectors are generally made of stainless steel (cathode) or nickel plated steel (anode). The cathode current collector is a corrugated or perforated plate that also provides oxidant passages and cathode support. The anode current collector is a corrugated plate that is also used to provide fuel gas passages. In certain stack designs, the anode is ribbed to provide gas passages, and in this case, the bipolar plate serves as a current collector.

In an MCFC stack, cells can be supplied with reactant gases in crossflow, coflow, or counterflow configurations. Gas manifolding can be either external (Figure 4) or internal (Figure 5) (*14*).

What we still need to learn

Commercial MCFC stacks are expected to range in cell area from 4 to16 ft^2. The number of cells per stack is expected to range from 75 to 700, giving a total stack height of 10 to 200 inches. An MCFC-based electrical generating system will contain one or more of these stacks, each of which will have an output of 5 kW to 1 MW. Although there are as yet no stacks in operation of the size envisioned for commercialization, subscale stacks (1 ft^2, 20 cells) and short stacks (8 ft^2, 20 cells) have been tested for thousands of hours and have exhibited excellent performance (e.g., 0.70 V at 200 mA/cm^2 with a simulated coal gasifier fuel and 57% fuel utilization) (*15*).

For commercial applications, an MCFC stack must meet certain performance, lifetime, and cost criteria. Several things have been shown to cause excessive performance decay and shorten the desired lifetime of the fuel cell (*16*).

NiO cathode dissolution. Nickel oxide has a small degree of solubility in the carbonate electrolyte in the fuel cell cathode environment (about 10–15 ppm). However,

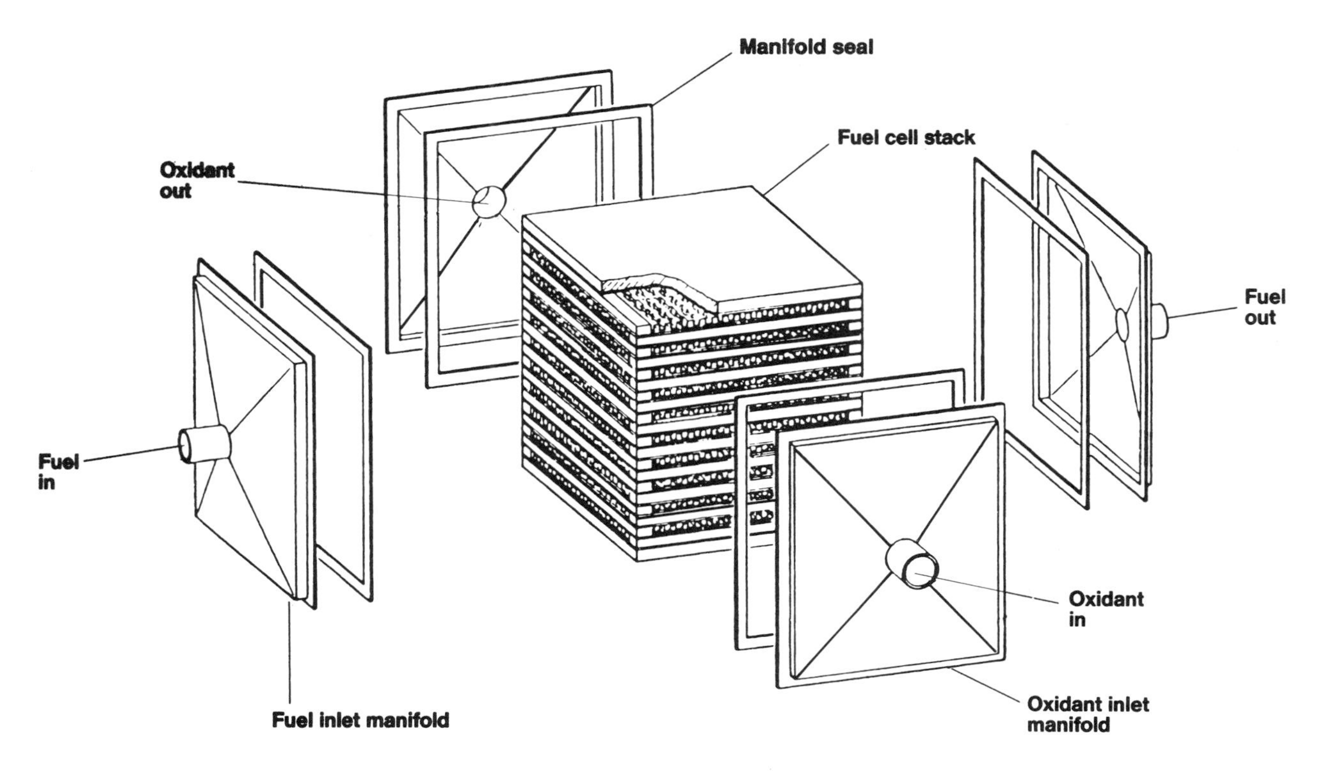

Figure 4. **External manifold molten carbonate fuel cell stack.** (14)

the dissolved nickel ions diffuse, under a concentration gradient, from the cathode toward the anode. At some location between the electrodes and under the influence of reducing conditions caused by the anode gas, the dissolved nickel precipitates as nickel metal. The precipitation of nickel creates a sink for the nickel ions, which facilitates further NiO dissolution. Thus, the dissolution of NiO can be a major life-limiting factor for the MCFC (*17*). The present industrial efforts attempt to prolong NiO life by employing low partial pressure of CO_2 and high-Li_2CO_3 electrolyte (to reduce nickel solubility). Additives (e.g., MgO) to the carbonate electrolyte to lower NiO dissolution have been tried as well. Two materials, $LiFeO_2$ and Li_2MnO_3, have been studied extensively (*18*) as potential replacement MCFC cathode materials.

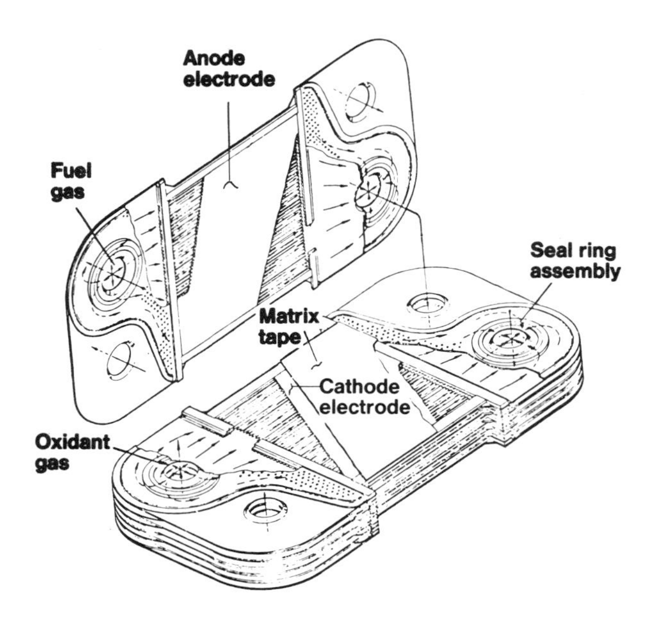

Figure 5. **Internal manifold molten carbonate fuel cell stack.** (14)

Bipolar plate corrosion. In MCFCs, the bipolar plates are coated with a thin film of molten carbonate electrolyte and are in contact with the cathode and the oxidant gas on one side and with the anode and fuel gas on the other side. They are therefore susceptible to hot corrosion attack (*19*), which causes decline in cell performance and can lead to cell failure by allowing direct reaction of oxidant and fuel. No satisfactory single material has been identified for MCFC bipolar plates, but high nickel alloys (e.g., INCO 825) and stainless steel (e.g., SS 310) have been found to have sufficient corrosion resistance in the cathode environment to be used as long as they are nickel-cladded on the anode side for corrosion protection.

Electrolyte loss. Long-term stability of MCFC cell and stack performance depends on limiting electrolyte loss from corrosion and volatilization. Corrosion loss is largely limited to the first 2000 h of operation. Loss by volatilization is a slow but continuing process. These two losses are considered to be manageable (excess electrolyte is stored or added) at least up to 10,000 h (*20*). Another problem is electrolyte leakage through the wet seal. This happens through a complicated process of charge and carbonate ion transfers between anode and cathode, with the net result being leaked electrolyte along the manifold gaskets to the top of the stack. Several engineering approaches to minimize electrolyte migration are being developed.

Electrode deformation. In MCFCs, the anode and cathode structures should be dimensionally stable. Any compacting of the structures not only decreases the active surface area but may cause loss of contact and high resistances between components. The porous nickel anode sinters at the cell operating temperature and is compressed as a result of the cell holding force. Nickel anodes have been stabilized by dispersing small amounts of certain oxides such as chromia and alumina in nickel. Recently, with the advent of dimensionally stable anodes, the compaction of the NiO cathode has been observed (*21*) at levels that are unacceptable for stacks used in commercial applications. Efforts to improve cathode compaction resistance emphasize heat treating and oxidation.

Sulfur contamination. Performance loss in an MCFC through the presence of sulfur in the fuel and oxidant gases occurs primarily at the anode. The sensitivity of the nickel anode to hydrogen sulfide has been demonstrated in many studies and cell tests. The presence of sulfur dioxide in the oxidant does not affect cathode polarization, but affects anode performance because sulfur dioxide in the oxidant will react to produce sulfate in the electrolyte and then be transported to the anode. In a cell, a concentration gradient of sulfate ions between the cathode and anode is established as hydrogen in the anode reduces sulfate to hydrogen sulfide. There is a general agreement on the negative impact of hydrogen sulfide on fuel cell performance, but the exact mechanism responsible for the loss is not clear. The tolerance limit for hydrogen sulfide for an MCFC with clean fuel is not firmly established, but it is likely to be on the order of less than 5 ppm. With 100% circulation of anode exhaust to the cathode, the tolerance limit will be on the order of 10^{-5} ppm H_2S in the fuel. Therefore, some sulfur removal from the recycled stream will be required. A cost benefit may be derived by replacing the nickel anode with a more sulfur-tolerant material.

More to come

High-temperature fuel cells will be an integral element of future means of generating electricity. Molten carbonate fuel cell (MCFC) technology has made steady progress and is moving toward large and tall stack demonstration. Solid oxide fuel cell (SOFC) technology is still at an early stage of development. However, SOFC research has received much attention recently, reflecting widening interest. Next month we will describe this promising technology.

References

(1) Grove, W. R. *Philosophical Magazine* **1839,** *14,* 127.
(2) Morgantown Energy Technology Center. *Fuel Cells—Technology Status Report;* U. S. Department of Energy: Report DOE/METC-87/0257, 1986.
(3) Maru, H. C.; Paetsch, L.; Pigeaud, A. In *Proceedings of the Symposium on Molten Carbonate Fuel Cell Technology;* Selman, J. R.; Claar, T. D., Eds.; The Electrochemical Society: Pennington, N.J., 1984; p. 20.
(4) Pierce, R. D. In *Fuel Cells: Technology Status and Applications;* Institute of Gas Technology: Chicago, 1982; p. 67.
(5) Kinoshita, K. In *Proceedings of the DOE/EPRI Workshop on Molten Carbonate Fuel Cells;* Electric Power Research Institute Report: EPRI WS-78-135, 1979, p. 4.
(6) Selman, J. R.; Maru, H. C. In *Advances in Molten Salt Chemistry;* Vol. 4; Mamantov, G.; Braunstein, J., Eds.; Plenum: New York, 1981; p. 159.
(7) Pierce, R. D.; Smith, J. L.; Poeppel, R. B. In *Proceedings of the Symposium on Molten Carbonate Fuel Cell Technology;* Selman, J. R.; Claar, T. D., Eds.; The Electrochemical Society: Pennington, N.J., 1984; p. 147.
(8) Selman, J. R.; Marianowski, L.G. In *Molten Salt Technology;* Lovering, D. G., Ed.; Plenum: New York, 1982; p. 323.
(9) Bregoli, L. J.; Kunz, H. R. *J. Electrochem. Soc.* **1982,** *129,* 2711.
(10) Appleby, A. J.; Foulkes, F. R. *Fuel Cell Handbook;* Van Nostrand Reinhold: New York, 1989.
(11) Marianowski, L. G.; Donado, R. A.; Maru, H. C. U. S. Patent 4 247 604; January 27, 1981.
(12) Erickson, D. S.; Ong, E. T.; Donado, R. *1986 Fuel Cell Seminar;* Courtesy Associates: Washington, D.C., 1986; p. 168.
(13) Smith, S. W.; Kunz, H. R.; Vogel, W. M.; Szymanski, S. J. In *Proceedings of the Symposium on Molten Carbonate Fuel Cell Technology;* Selman, J. R.; Claar, T. D., Eds.; The Electrochemical Society: Pennington, N.J., 1984; p. 246.
(14) General Electric. *Development of Molten Carbonate Fuel Cell Power Plant;* U. S. Department of Energy: Report DOE/ET/17019-20, 1985.
(15) Reiser, C. A.; Johnson, W. H. *1988 Fuel Cell Seminar;* Courtesy Associates: Washington, D.C., 1988; p. 384.
(16) Minh, N. Q. In *Proceedings of the Joint International Symposium on Molten Salts;* Mamantov, G.; Blander, M.; Hussey, C.; Mamantov, C.; Saboungi, M-L.; Wilkes, J., Eds.; The Electrochemical Society: Pennington, N.J., 1987; p. 677.
(17) Minh, N. Q. *J. Power Sources* **1988,** *24,* 1.
(18) Smith, J. L.; Kucera, G. H.; Minh, N. Q. *1986 Fuel Cell Seminar;* Courtesy Associates: Washington, D.C., 1986; p. 163.
(19) Singh, P. In *Proceedings of the Symposium on Corrosion in Batteries and Fuel Cells and Corrosion in Solar Energy Systems;* Johnson, C. J.; Pohlman, S. L., Eds.; The Electrochemical Society: Pennington, N.J., 1983; p. 124.
(20) *Assessment of Research Needs for Advanced Fuel Cells;* U. S. Department of Energy: Report DOE/ER/30060-T1, 1985; p. 171.
(21) International Fuel Cells. *Development of Molten Carbonate Fuel Cell Power Plant Technology, Quarterly Technical Progress Report No. 22, January–March 1985;* U.S. Department of Energy: Report DOE/ET/15440-1944, 1985.

Nguyen Q. Minh works at Allied-Signal Aerospace Company, AiResearch Los Angeles Division (2525 W. 190th Street, Torrance, CA 90504-6099; 213-512-3515). Before joining Allied-Signal, he was at Argonne National Laboratory. His research interests include fuel cells and other electrochemical energy systems, high-temperature electrochemistry, molten salts, ceramic processing, and chemical metallurgy. He holds several patents and has authored or coauthored two book chapters and more than 50 technical papers.

High-Temperature Fuel Cells: Part II

The Solid Oxide Cell

Nguyen Quang Minh

The high-temperature fuel cell promises high energy conversion efficiency without the need for external fuel processing. The molten carbonate fuel cell (MCFC), based on a Li_2CO_3–K_2CO_3 electrolyte, was described last month (CHEMTECH, January, p. 32). Here we will discuss the materials and design for the solid oxide fuel cell, a promising technology that is still in the early development stage.

The solid oxide fuel cell (SOFC) is an all-solid-state fuel cell consisting of two porous ceramic electrodes separated by a dense oxide ion-conducting ceramic electrolyte. In the SOFC, fuel (hydrogen or carbon monoxide) and oxidant (oxygen) produce a dc current by combining electrochemically across the solid oxide electrolyte. Oxygen fed to the cathode accepts electrons from the external circuit to form oxide ions. The oxide ions are conducted through the electrolyte to the anode. At the anode, oxide ions combine with hydrogen (or CO) in the fuel to form water (or CO_2), liberating electrons. Electrons flow from the anode through the external circuit to the cathode. The overall reactions at the SOFC electrodes are as follows:

Anode $H_2 + O^= \rightarrow H_2O + 2e^-$

$CO + O^= \rightarrow CO_2 + 2e^-$

Cathode $\frac{1}{2}O_2 + 2e^- \rightarrow O^=$

Figure 1 shows the reactions in the SOFC.

The materials most commonly used for SOFC components are stabilized zirconia for electrolytes, nickel/zirconia cermet for anodes, and doped lanthanum manganite for cathodes (Table 1). Examples of anode, electrolyte, and cathode microstructures are given in Figure 2 (*1*). In the SOFC stack, doped $LaCrO_3$ is used as interconnects to connect cells in electrical series. Present SOFCs are operated at about 1000 °C and ambient atmosphere.

Stack components

Electrolyte. The electrolyte in an SOFC must have proper conductivity, phase and conductivity stability in oxidizing and reducing environments, and chemical compatibility with other cell components at both operating and processing temperatures and atmospheres. The conductivity of the electrolyte has to be ionic over the range of oxygen pressures (1 to 10^{-18} atm) expected in operating fuel cells. The electrolyte must be dense to prevent mixing of the fuel and oxidant gases and, in addition, have thermal expansion that matches other cell components to avoid delamination and cracking of the ceramic components during thermal cycling.

Stabilized zirconia possesses adequate oxygen ion conductivity in oxidizing and reducing atmospheres and has been used successfully as the electrolyte in SOFCs. Zirconia, in its pure form, is not a good electrolyte, primarily because its ionic conductivity is too low. Doping with lower valent metal oxides such as yttria stabilizes the high-temperature cubic fluorite structure of zirconia and increases its oxygen vacancy concentration (*2*). This enhances ionic conductivity and leads to an extended oxygen partial pressure range of ionic conduction, making stabilized zirconia suitable for use as an electrolyte in SOFCs. The ionic conductivity of yttria-stabilized zirconia (8 mol% yttria) is about 0.09 $ohm^{-1} \cdot cm^{-1}$ at 1000 °C. Yttria-stabilized zirconia has a thermal expansion coefficient of about 10.5×10^{-6} cm/cm from room temperature to 1000 °C (*3*). Long-term behavior of stabilized zirconia under fuel-cell operating conditions has been extensively tested and the required properties of the material have been shown not to be affected after long-term testing at 1000 °C.

Cathode. Owing to the high operating temperature of the SOFC, only noble metals and electronic conducting oxides can be used as cathode materials. Noble metals such as platinum, palladium, or silver are unsuitable because of prohibitive cost and insufficient long-term stability owing to vaporization. Many oxides are also unsuitable because of their thermal expansion mismatch with the electrolyte and their lack of conductivity. At present, Sr-doped $LaMnO_3$ is most commonly used as SOFC cathode material.

Lanthanum manganite is known to be a p-type perovskite oxide and shows reversible oxidation–reduction behavior (*4*). The material can have oxygen excess or deficiency depending on specific conditions. $LaMnO_3$ begins to lose oxygen and experiences substantial decrease in conductivity at oxygen partial pressures $< 10^{-12}$ atm. In very reducing conditions at 1000 °C, $LaMnO_3$ decomposes into La_2O_3 and MnO. Thus, in fabrication of the SOFC, it is desirable to process $LaMnO_3$ in oxidizing atmospheres.

The conductivity of $LaMnO_3$ can be enhanced by doping with a lower valence ion such as strontium. The Sr-doped $LaMnO_3$ (10 mol% Sr) material currently used for SOFCs has a conductivity of 130 $ohm^{-1} \cdot cm^{-1}$ at 1000 °C in air (*4*). The thermal expansion of this material is about 12

***Table 1.* Characteristics of the solid oxide fuel cell**

Electrolyte	Yttria-stabilized ZrO_2
Anode	Ni/Y_2O_3–ZrO_2 cermet
Cathode	Sr-doped $LaMnO_3$
Bipolar plate/ interconnect	Doped $LaCrO_3$
Fuel (electroactive)	H_2, CO
Oxidant (electroactive)	O_2
Operating temperature	1000 °C
Operating pressure	1 atm
Contaminant tolerance	Not yet defined
Stack configuration	Tubular, monolithic, or planar

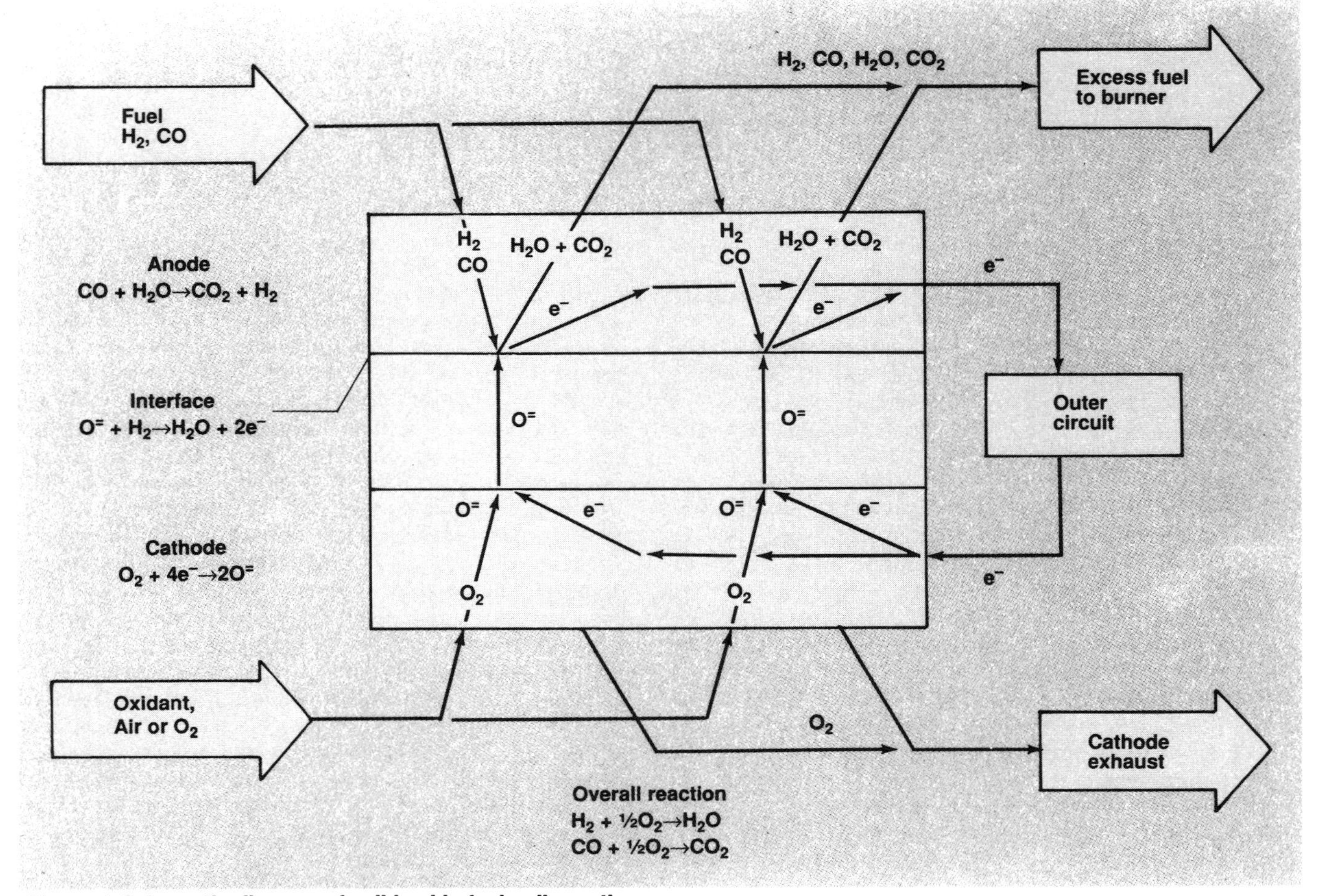

***Figure 1.* Schematic diagram of solid oxide fuel cell reactions**

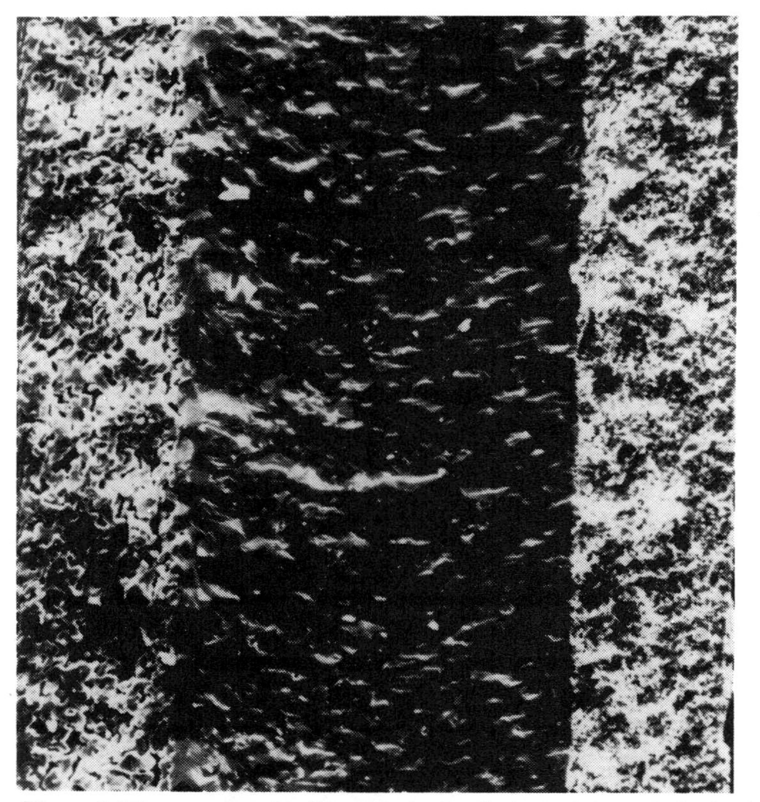

***Figure 2.* Micrographs of solid oxide fuel cell electrolyte and electrodes**

For more information

A fuel cell is an electrochemical device that converts the chemical energy of a fuel directly to electricity (and heat) without combustion as an intermediate step. The principle of fuel cell operation was first reported by Sir William Grove in 1839. Two years ago, we marked the 150th anniversary of his discovery.

The first part of this article (CHEMTECH, January 1991, p. 32), discussed the general characteristics of high-temperature fuel cells and described in detail the molten carbonate fuel cell (MCFC). For more information on fuel cells, see also CHEMTECH's five-part Lindström series (1988, August, p. 490; September, p. 553; November, p. 686; 1989; January, p. 44; February, p. 122.)

$\times 10^{-6}$ cm/cm from room temperature to 1000 °C (5). Manganese is known to be a mobile species at high temperatures and can easily diffuse into the electrolyte and other cell components, changing the electrical characteristics or the structure of the cathode or other components. Fabrication temperature is generally limited to below 1400 °C to minimize this migration, and there is no report on any significant manganese effects for cells operated up to 10,000 hours.

Anode. Because of the reducing atmosphere of the fuel gas, metals can be used as electrode materials for SOFC anodes. Nickel and cobalt particularly have been found to be the most suitable, and at present, nickel is commonly used. The thermal expansion coefficient of nickel is considerably larger than that of the zirconia electrolyte; a significant thermal expansion mismatch between the anode and other components can cause delamination and cracking in the fuel cell. Also, nickel sinters at the cell operating temperature, leading to a decrease in active surface area and loss of porosity. These problems have been circumvented by incorporating zirconia particles in the anode to develop a skeleton of zirconia on which the metal is coated. The zirconia in the SOFC anode supports the nickel metal particles, inhibits coarsening of the metallic particles at the fuel cell operating temperature, and provides an anode thermal expansion coefficient acceptably close to those of the other cell components. A compromise between conductivity and thermal expansion is required when the nickel loading in the anode is determined. About 30 vol% Ni is needed to maintain the required level of conductivity (300 $ohm^{-1} \cdot cm^{-1}$ (*6*) while minimizing the degree of thermal expansion mismatch.

Nickel/zirconia cermet has been shown to possess sufficient catalytic activity to reform hydrocarbons or natural gases internally. The presence of sulfur-containing species in the fuel can have deleterious effects on performance of the anode, causing unacceptable loss of cell voltage. The sulfur tolerance of SOFC anodes at the cell operating temperature is not well known, although SOFCs are known to be more sulfur-tolerant than MCFCs.

Interconnect. The SOFC interconnect must be stable in both oxidizing and reducing environments and must have as high an electronic conductivity as possible. The interconnect must also have low porosity to prevent gas cross-leaking between the electrodes. These stringent requirements eliminate all but a few metals and oxide systems from consideration. At present, doped $LaCrO_3$ has been commonly used as the interconnect, as it possesses low reactivity with other components, good electrical conductivity, and compositional stability. Various substitutions in the lanthanum chromite improve the thermal expansion match and conductivity under SOFC operating conditions.

Previous studies on $LaCrO_3$ have shown that it has intrinsic p-type conductivity because of formation of cation vacancies. The conductivity of the material can be enhanced by substituting a divalent ion on the lanthanum and chromium sites. For example, $LaCrO_3$ doped with 10 mol% Sr on the lanthanum site has a conductivity of about 25 $ohm^{-1} \cdot cm^{-1}$, and $LaCrO_3$ doped with 10 mol% Mg on the chromium site has a conductivity of about 10 $ohm^{-1} \cdot cm^{-1}$ at 1000 °C in air (*7*, *8*), compared with a conductivity of about 0.5 $ohm^{-1} \cdot cm^{-1}$ for undoped $LaCrO_3$. The conductivity of doped $LaCrO_3$ is reduced about 10-fold when the oxygen pressure is reduced from atmospheric pressure to the fuel condition. However, $LaCrO_3$ exposed to fuel on one side and to air on the other side has high enough conductivity to be used as the interconnect material in SOFCs. The thermal expansion coefficient of doped $LaCrO_3$ is dependent on the type and amount of dopant present and can be tailored to match those of zirconia. For example, the thermal expansion of strontium-doped materials is about 10–11 $\times 10^{-6}$ cm/cm from room temperature to 1000 °C, matching that of the zirconium electrolyte (*5*).

$LaCrO_3$, like all chromium compounds, does not densify at temperatures below 1700 °C and oxygen activities above 10^{-9} atm (*9*). The reason for the low sinterability is that the predominant mass transport during firing is an evaporation–condensation mechanism, which leads to coarsening of the original particles without densification. Thus, for cofiring the $LaCrO_3$ interconnect with other cell components, it is necessary to develop methods to sinter the material below 1400 °C in a relatively oxidizing atmosphere like air.

Stacking the cells

In the SOFC, where components are all solid, cells and stacks can be designed and fabricated into configurations that may not be feasible with cells employing liquid electrolytes. Thus, SOFC stack designs can be different from the flat plate type commonly used in other fuel cell systems. For other fuel cells, the components can be fabricated separately and then assembled to form a cell. The process conditions for the fabrication of each component can be selected independently. This cannot be done for the SOFC, because if the components are built up one by one, the conditions for each successive layer must be tailored to avoid altering the properties of the preceding layer. If all the components are formed together in the green (unfired) state, then all the components must be sintered under the same conditions.

At present, four common configurations have been proposed and fabricated: the seal-less tubular design, the segmented cell-in-series design, the monolithic design, and the flat plate (planar) design. The designs differ in the extent of dissipative losses within the cells, in the manner of sealing between fuel and oxidant channels, and in

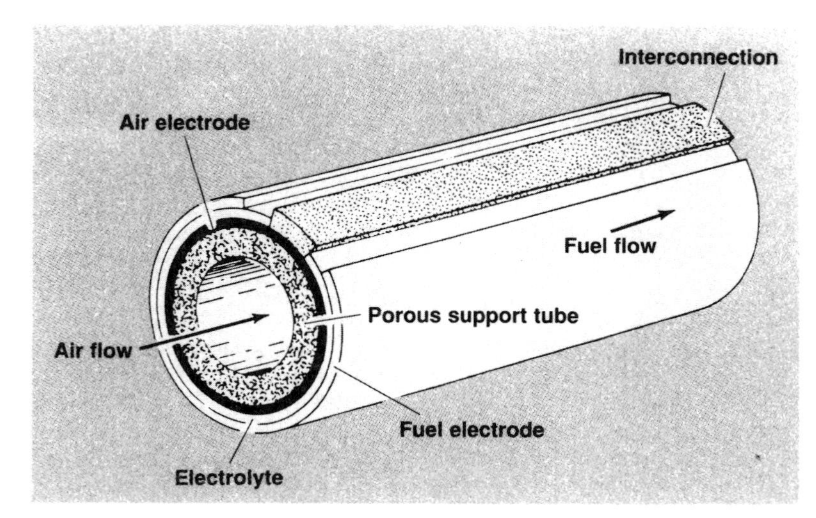

***Figure 3.* Seal-less tubular solid oxide fuel cell**

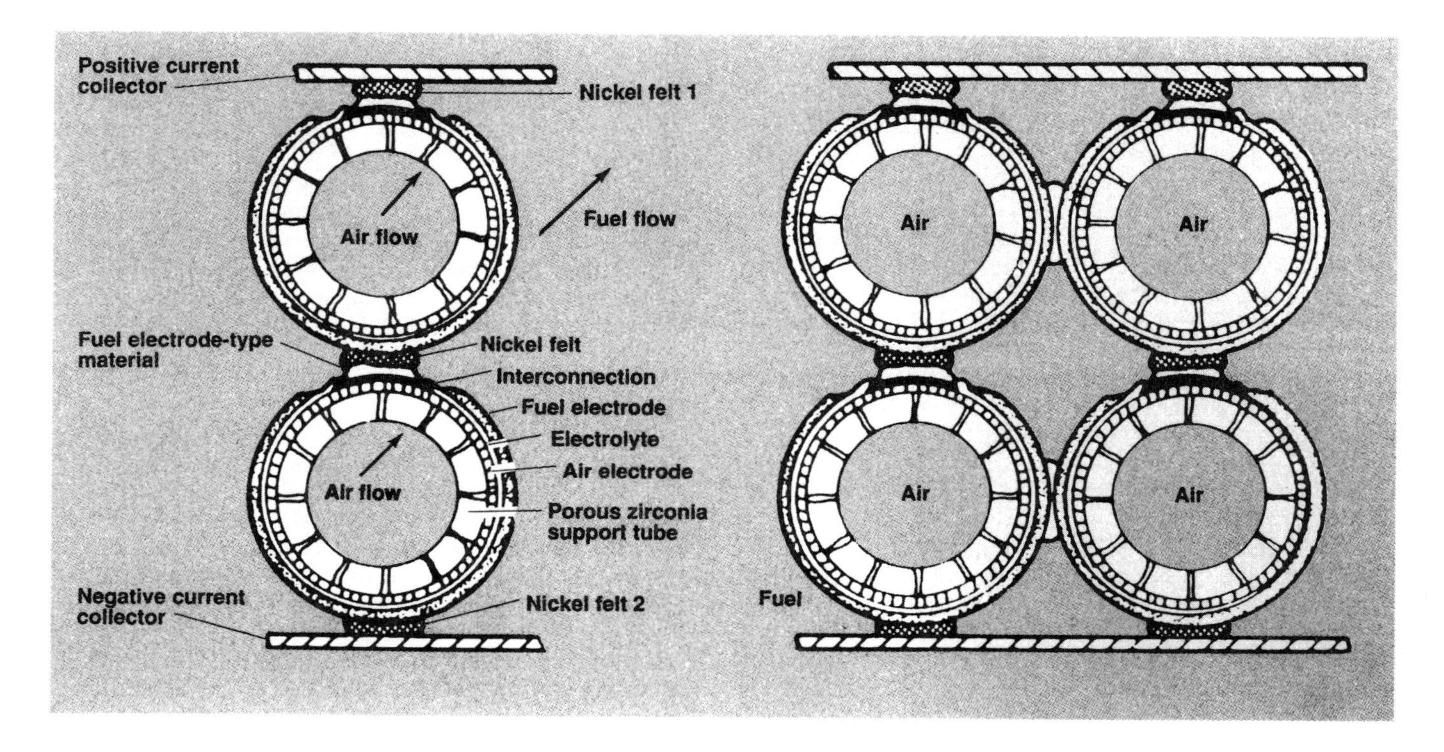

***Figure 4.* Electrical connections in seal-less tubular solid oxide fuel cells.** Nickel felt is pressed nickel metal fiber used in this SOFC design for electrical connection. Nickel felt 1 is the contact to the positive air electrode via the interconnection and the fuel electrode type material over the interconnection is the positive contact surface of a single cell. Nickel felt 2 is the contact to the negative fuel electrode.

making cell-to-cell electrical connections in a stack of cells. The ease of fabrication and assembly varies among the designs, but the differences are hard to quantify at this stage of development.

Seal-less tubular design. This design is the most advanced among the several SOFC concepts proposed (Figure 3). In this design, the cell is configured on a porous (30% porosity) calcia-stabilized zirconia support tube, which has an i.d. of 12–13 mm, a 1- to 1.5-mm wall thickness, and a 38- to 50-cm length. The tube is overlaid with a 0.5–1.0 mm porous cathode layer. A 50-μm gas-tight electrolyte layer covers the cathode except in a strip along the entire active cell length. This strip of exposed cathode is covered with a 30-μm gas-tight interconnect layer. The 100-μm anode covers the entire electrolyte surface. The support tube is closed at one end. Oxidant is introduced to the cell through an injection tube near the closed end, where it traverses and exits from the open annulus between the support tube and the oxidant injector tube. One distinct feature of this tubular design is that it has no seals, so the problems with gas-tight seals for ceramics at high temperatures are eliminated. Stacks are formed by combining cell units in bundles (Figure 4) (*10*).

The principal dissipative losses in seal-less tubular SOFCs are cell internal resistance and gas diffusion limitations. Significant internal resistance losses arise from

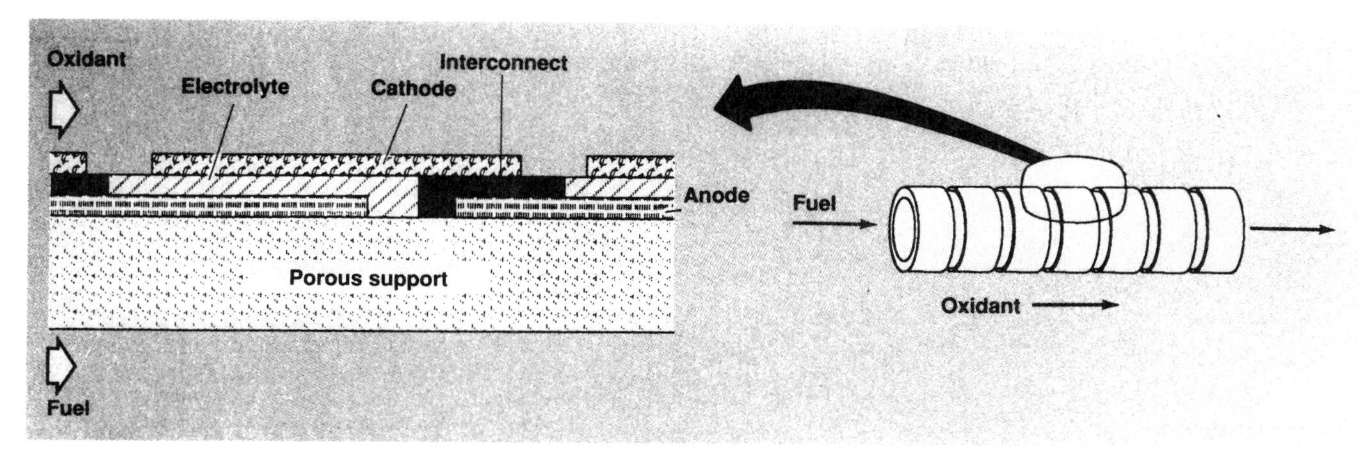

Figure 5. **Segmented cell-in-series design of solid oxide fuel cell (banded structure on tubular support)**

the relatively long path for current through the tubular cell, especially the long current path in the plane of the cathode. The thick support tube restricts the amount of oxygen that can be transported to the cathode/electrolyte interface, resulting in a loss in cell performance. Gas diffusion through the support tube can become the rate-limiting step and set the upper or limiting current for the cell.

Seal-less tubular SOFCs are fabricated by a manufacturing process involving several fabrication techniques. The support tube is formed by extrusion (*11*) because it offers potential manufacturing cost savings. The cathode and anode are applied on the support tube by a slurry coating technique (*12*) and the electrolyte and interconnect are fabricated by the electrochemical vapor deposition (EVD) process (*13*). Extrusion and slurry coating techniques are well-established, but we have needed to do considerable development work to scale up the EVD process for this purpose.

Segmented cell-in-series design. This design consists of individual cells connected in electrical and gas flow series. The cells are either arranged as a thin banded structure on a porous support tube (Figure 5) (*14*) or fitted one into the other to form a tubular self-supporting structure (bell-and-spigot configuration) (Figure 6) (*15*). The interconnect provides sealing (and electrical contact) between the anode of one cell and the cathode of the next cell. In this design, the fuel flows from one cell to the next in the tubular stack of cells. In the configuration where the support tube (made either of zirconia or alumina) is used, cells can be made very thin, with component thicknesses on the order of 40 μm. In the bell-and-spigot configuration, individual cells form into short cylinders of about 1.5 cm in diameter. The cells are about 0.3 mm thick to provide structural support.

Four or five cells in this kind of fuel gas flow series can generate about 10% more power than a single cell of the same total active surface area. The benefit of adding still more cells diminishes rapidly after five or six cells. The cell length in this design is kept relatively short in order to minimize the current path. In spite of the short cell length,

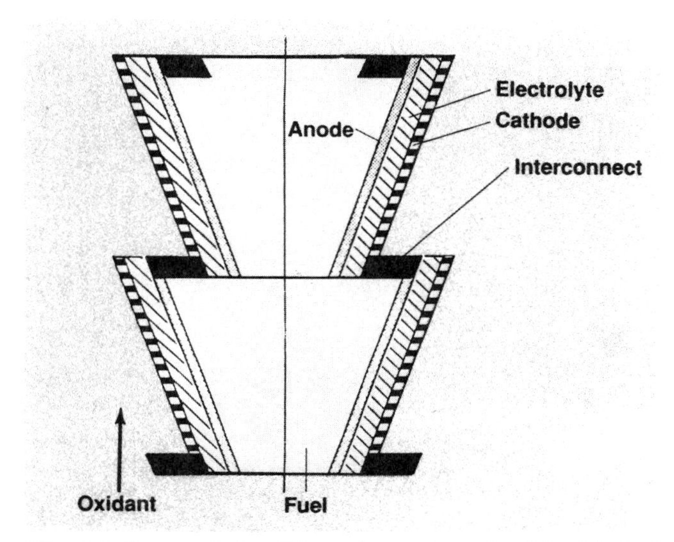

Figure 6. **Segmented cell-in-series design of solid oxide fuel cell (bell-and-spigot)**

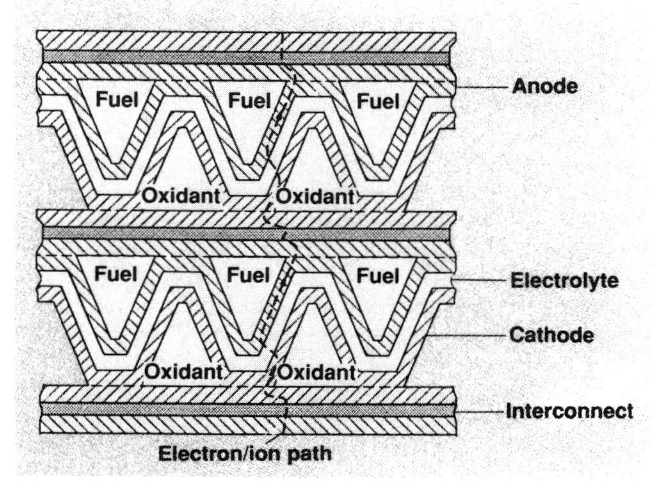

Figure 7. **Coflow monolithic solid oxide fuel cell.** Cell-to-cell distance is 1 to 2 mm.

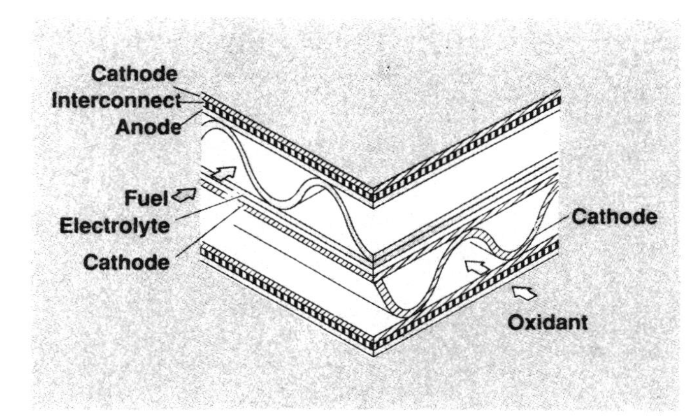

Figure 8. **Crossflow monolithic solid oxide fuel cell**

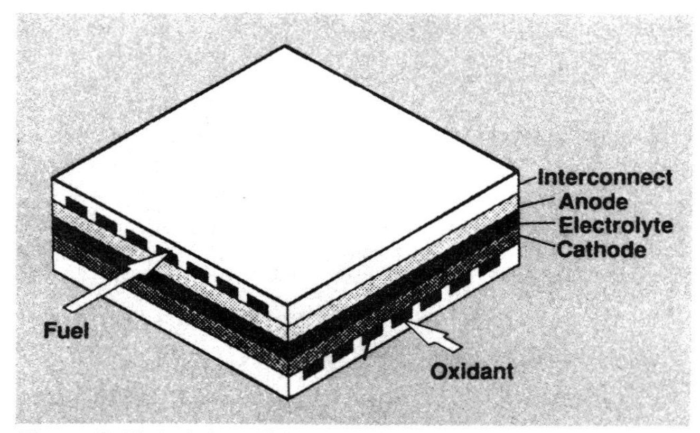

Figure 9. **Flat plate solid oxide fuel cell**

the major dissipative losses in this design are cell internal resistances. The use of a thick self-supporting electrolyte in the bell-and-spigot configuration can significantly increase the losses. On the other hand, a self-supporting electrolyte reduces the losses arising from gas transport through an inert support tube.

To manufacture segmented cells on a support tube, the electrolyte and interconnect are fabricated onto the tube by vapor deposition or plasma spraying methods (*16*). Self-supporting electrolytes of the bell-and-spigot are fabricated by axial pressing (*17*). Diffusion welding is used to make the interconnection seal between electrolyte cylinders. The electrodes are applied by spray coating.

Monolithic design. The monolithic design employs thin ceramic components of the SOFC to form a compact solid-state structure. In the coflow version of the monolithic SOFC (MSOFC) (*18*), the fuel cell consists of a honeycomb-like array of adjacent fuel and oxidant channels that resembles corrugated paperboard (Figure 7). The walls of the channels are formed from multilayer composites of the cell components. The coflow monolithic fuel cell is made of two types of laminated structures, each composed of three ceramics: anode/electrolyte/cathode and anode/interconnect/cathode. The anode/electrolyte/cathode composite is appropriately corrugated and is stacked alternately between flat anode/interconnnect/cathode composites. The typical thickness of cell components is 25 to 100 μm and the distance from cell to cell is 1 to 2 mm. In the crossflow version of the MSOFC (*19*), fuel and oxidant channels are formed from corrugated anode and cathode layers, respectively, and fuel and oxidant flows are at right angles to each other (Figure 8). The anode/electrolyte/cathode and anode/interconnect/cathode composites are flat and stacked alternately between corrugated anode and cathode layers.

Small cells, self-supporting corrugated structures, and thin components are the key features of the MSOFC. The monolithic design maximizes the active surface area of the fuel cell. For example, a monolithic SOFC with channels about 1 mm in height has a ratio of active surface area to volume of about 10 cm^2/cm^3 (*20*), about seven times the ratio for tubular SOFCs of 1 cm in diameter. As a result, the MSOFC can produce the same power as other fuel cells many times as large. The small size of individual cells reduces voltage losses that are due to internal electrical resistance. The lower internal resistance of the monolithic fuel cell improves cell efficiency.

The development of suitable materials and fabrication processes is critical to the success of the monolithic design. The proposed fabrication process for the MSOFC involves forming the fuel cell as a green body and cofiring the cell at elevated temperatures to form a sintered structure. Tape casting (*21*) and tape calendering (*22*) are being developed for making the thin ceramic layers and composites required for the monolithic fuel cell. Because the fuel cell is made by cofiring, an important aspect in the fabrication of the MSOFC is to develop methodologies for matching thermal expansion and firing shrinkage of the four cell components under the same firing conditions. Any significant mismatch in the thermal expansion and shrinkage can cause stress in the fired bodies and result in cracking.

Flat plate design. This design, common in other types of fuel cells, consists of anode, electrolyte, cathode, and interconnect configured as thin, flat plates (Figure 9). Individual cell thickness (mainly electrolyte thickness) is typically thicker than 250 microns in order to be self-supporting. The interconnect, having ribs on both sides, forms gas flow channels and serves as a bipolar gas separator contacting the anode and cathode of adjoining cells.

The flat plate design offers improved power density relative to the tubular and segmented cell-in-series designs but requires high-temperature gas seals at the edges of the plates. Compressive seals have been proposed; however, the unforgiving nature of a compressive seal can lead to a nonuniform stress distribution on the ceramic and cracking of the layers. Further, seals may limit the height of a cell stack. There is a higher probability for mismatches in tolerances (creating unacceptable stress levels) in taller stacks. Fabrication and assembly appear to be simpler for the flat plate design as compared with the other designs. The electrolyte and interconnect layers are made by tape casting and CVD (*23*, *24*). The electrodes are applied by the slurry method, by screen printing, or by plasma spraying. Fuel cell stacks are formed by stacking up layers much like other fuel cell technologies.

Our progress and promise

SOFCs of various designs have been fabricated and tested on a laboratory scale, and the results obtained so far have demonstrated the feasibility of the technology. Tested cells have exhibited excellent and stable performance for up to 10,000 hours. For example, typical performance of seal-less tubular SOFCs is 0.65 V at 250 mA/cm^2 on synthetic reformate gas (67% H_2, 22% CO, 11% H_2O) with 85% fuel utilization (*25*). The monolithic design has achieved a high current density of 2.2 A/cm^2 (*26*). The largest SOFC device tested to date is a 5-kW stack of 325 seal-less tubular cells (arranged in six strings of 54 cells). The stack provided approximately 80 A at 64 V. The critical issues facing SOFC technologies are the development of suitable materials and fabrication processes. Much research and development in these areas is required before the SOFC becomes practical.

High-temperature fuel cells allow the clean and efficient use of fossil fuels and provide a number of important operational and economical advantages. Molten carbonate fuel cell technology has made steady progress and is moving toward large and tall stack demonstration. Solid oxide fuel cell technology is still at an early stage of development, but has received much attention recently. High-temperature fuel cells will be an integral element in the future as we look to generate electricity from a variety of fuels.

Nguyen Q. Minh works at Allied-Signal Aerospace Company, AiResearch Los Angeles Division (2525 W. 190th Street, Torrance, CA 90504-6099; 213-512-3515). Before joining Allied-Signal, he was at Argonne National Laboratory. His research interests include fuel cells and other electrochemical energy systems, high-temperature electrochemistry, molten salts, ceramic processing, and chemical metallurgy. He holds several patents and has authored or coauthored two book chapters and more than 50 technical papers.

References

(1) McPheeters, C. C.; Dees, D. W.; Doris, S. E.; Picciolo, J. J. *1988 Fuel Cell Seminar;* Courtesy Associates: Washington, D.C., 1988; p. 29.
(2) Subbarao, E. C.; Maiti, H. S. *Solid State Ionics* **1984,** *11,* 317.
(3) Stevens, R. *An Introduction to Zirconia;* Magnesium Elektron Ltd.: London, 1983.
(4) Anderson, H. U.; Kuo, J. H.; Sparlin, D. M. In *Proceedings of the First International Symposium on Solid Oxide Fuel Cells;* Singhal, S. C., Ed.; The Electrochemical Society: Pennington, N.J., 1989; p. 111.
(5) Srilomsak, S.; Schilling, D. P.; Anderson, H. U. In *Proceedings of the First International Symposium on Solid Oxide Fuel Cells;* Singhal, S. C., Ed.; The Electrochemical Society: Pennington, N.J., 1989; p. 129.
(6) Dees, D. W.; Claar, T. D.; Easler, T. E.; Fee, D. C.; Mrazek, F. C. *J. Electrochem. Soc.* **1987,** *134,* 2142.
(7) Koc, R.; Anderson, H. U.; Howard, S. A. In *Proceedings of the First International Symposium on Solid Oxide Fuel Cells;* Singhal, S. C., Ed.; The Electrochemical Society: Pennington, N.J., 1989; p. 220.
(8) Singhal, S. C.; Ruka, R. J.; Sinharoy, S. *Interconnection Materials Development for Solid Oxide Fuel Cells;* U.S. Department of Energy Report, DOE/MC/21184-1985, 1985.
(9) Group, L.; Anderson, H. U. *J. Am. Ceram. Soc.* **1976,** *59,* 449.
(10) Westinghouse Electric Corporation. *Solid Oxide Fuel Cell Power Generation System: The Status of The Cell Technology—A Topical Report;* U.S. Department of Energy Report, DOE/ET/17089-15, 1984.
(11) Rossing, B. R. In *Proceedings of the Conference on High-Temperature Solid Oxide Electrolytes;* Brookhaven National Laboratory Report BNL 51728, 1983; p. 45.
(12) Bratton, R. J.; Reichner, P.; Montgomery, L. W. *1986 Fuel Cell Seminar;* Courtesy Associates: Washington, D.C., 1986; p. 80.
(13) Isenberg, A. O. In *Proceedings of the Symposium on Electrode Materials and Processes for Energy Conversion and Storage;* McIntyre, J. D. E.; Srinivasan, S.; Will, F. G., Eds.; The Electrochemical Society: Pennington, N.J., 1977; p. 572.
(14) Feduska, W.; Isenberg, A. O. *J. Power Sources* **1983,** *10,* 89.
(15) Rohr, F. J. In *Solid Electrolytes;* Hagenmuller, P.; Van Gool, W., Eds.; Academic: New York, 1976; p. 431.
(16) Ohno, Y.; Nagata, S.; Sato, H. In *Proceedings of the 15th Intersociety Energy Conversion Engineering Conference;* American Institute of Aeronautics and Astronautics: New York, 1980; p. 881.
(17) Dönitz, W.; Schmidberger, R. *Int. J. Hydrogen Energy* **1982,** *7,* 321.
(18) Ackerman, J. P.; Young, J. E. U.S. Patent 4,476,198; October 9, 1984.
(19) Poeppel, R. B.; Dusek, J. T. U.S. Patent 4,476,196; October 9, 1984.
(20) Fee, D. C.; Steunenberg, R. K.; Claar, T. D.; Poeppel, R. B.; Ackerman, J. P. *1983 Fuel Cell Seminar;* Courtesy Associates: Washington, D.C., 1983; p. 74.
(21) McPheeters, C. C.; Fee, D. C.; Poeppel, R. B.; Claar, T. D.; Busch, D. E.; Flandermeyer, B. K.; Easler, T. E.; Dusek, J. T.; Picciolo, J. J. *1986 Fuel Cell Seminar;* Courtesy Associates: Washington, D.C., 1986; p. 44.
(22) Minh, N. Q.; Horne, C. R.; Liu, F.; Staszak, P. R.; Stillwagon, T. L.; Van Ackeren, J. J. In *Proceedings of the First International Symposium on Solid Oxide Fuel Cells;* Singhal, S. C., Ed.; The Electrochemical Society: Pennington, N.J., 1989; p. 307.
(23) Lessing, P. A.; Tai, L. W.; Klemm, K. A. In *Proceedings of the First International Symposium on Solid Oxide Fuel Cells;* Singhal, S. C., Ed.; The Electrochemical Society: Pennington, N.J., 1989; p. 337.
(24) Milliken, C.; Khankar, A. In *Proceedings of the First International Symposium on Solid Oxide Fuel Cells;* Singhal, S. C. Ed.; The Electrochemical Society: Pennington, N.J., 1989; p. 361.
(25) Morgantown Energy Technology Center. *Fuel Cells—Technology Status Report;* U.S. Department of Energy: Report DOE/METC-89/0266, 1988.
(26) Fee, D. C.; Blackburn, P. E.; Busch, D. E.; Claar, T. D.; Dees, D. W.; Dusek, J.; Easler, T. E.; Ellingson, W. A.; Flandermeyer, B. K.; Fousek, R. J.; Heiberger, J. J.; Kraft, T. E.; Majumdar, S.; McPheeters, C. C.; Mrazek, F. C.; Picciolo, J. J.; Poeppel, R. B.; Zwick, S. A. In *Proceedings of the 21st Intersociety Energy Conversion Engineering Conference;* Vol. 3; American Chemical Society: Washington, D.C., 1986; p. 1634.

Hydrogen From Methanol: Fuel Cells in Mobile Systems

B. Ganser and B. Höhlein

Institute of Energy Process Engineering (IEV)

Abstract

The progress currently emerging in the optimization of low-temperature fuel cells with respect to efficiency, emission behaviour and lifetime has set a vigorous development in motion aimed at applying these fuel cells in motor vehicles. The hydrogen supply for a low-temperature fuel cell obtained by reforming a methanol/water mixture into a synthesis gas rich in hydrogen requires a precise adjustment of the reforming unit to the fuel cell under the special boundary conditions of a mobile system. On the basis of initial results from extensive theoretical and experimental studies on methanol reforming, various optimization criteria and possible approaches will be discussed.

1. Environmental Problems Due to Transport

Although in the Federal Republic of Germany road transport only causes about 20 % of the total carbon dioxide emissions, it accounts for 50 % and more of the emissions of nitrogen oxides, carbon monoxide and hydrocarbons subject to a legal limit (SCHNEIDER 1991). Added to this are particles, benzene and other emissions as well as secondary pollutants and emissions in the form of photooxidants (ozone, PAN, aldehydes) as a consequence of road traffic and low efficiencies in conventional gasoline- and diesel-driven vehicles, thus focusing efforts towards emission reduction on transport as an energy consumption sector.

2. Proposals for Novel Transport Systems

The problem to be solved can be outlined by the requirement of system improvements with a view to emission reduction, improved emission quality and improved efficiencies for the energy consumption sector of road traffic. With respect to specific emissions and specific fuel consumption, the guidelines laid down in the USA have already been adopted by Western European car manufacturers (HÖHLEIN 1991). Various approaches (Fig. 1) are being discussed; some prototypes for various new development lines are already being tested on the road.

Of the drive units in the figure, only the conventional internal combustion engine has achieved a stage of development permitting commercial utilization. Not only the emission balance, which depends decisively on the fuel used, is decisive for an ecological evaluation of the internal combustion engine but rather the efficiency with which energy is converted, that is to say an assessment of the entire energy convertion chain.

The difficulties in developing efficient batteries with long-term stability providing acceptable operating values mean that in spite of intensive efforts only systems with lead accumulators are available on the market at the moment for electric vehicles. These circumstances lead to the development of hybrid concepts combining the mature technology of the internal combustion engine with small electric drive units so that the user can rely upon having a functioning drive unit in such a vehicle in all situations. In view of these difficulties and the tightening up of exhaust gas regulations in Europe and the United States leading to future emission standards (Ultra Low Emission Vehicles standard) which cannot be fulfilled by vehicles with internal combustion engines, the urgency becomes apparent for the development of new efficient and low-emission drive systems for motor vehicles. Progress currently apparent in optimizing low-temperature fuel cells with respect to efficiency, emission behaviour and lifetime has set a vigorous development in motion aiming at an application of these cells in motor vehicles. Figure 2 shows the possible incorporation of low-temperature fuel cells in such a drive concept.

Mobile systems with fuel cells as energy convertion units must demonstrate a higher efficiency than conventional vehicle propulsion systems and the energy carriers used must meet the criteria laid down for energy density (Wh/(kg system) and Wh/(l system)), safety, handling and environmental compatibility of the entire energy convertion chain. In the same way as conventional mobile systems they must ensure rapid and simple availability, low susceptibility to failure and easy maintenance. The improvement potential of such a novel system is to be found in the option for improved overall efficiency with at the same time a minimization of emissions.

3. Methanol as an energy carrier for fuel cells

Based on the present level of fuel cell development, hydrogen would appear to be the most appropriate energy carrier and at the same time a suitable feed gas for a low-temperature fuel cell. However, in 1978 NASA already discovered in connection with the provision of hydrogen for onboard systems that methanol or a methanol/hydrogen mixture combined with reforming to hydrogen provides advantages with respect to storage and availability in comparison to hydrogen (BRABBS 1978):

- With respect to the criteria for mobile systems (tank weight, tank size, energy required for storage) methanol is better suited than hydrogen but nevertheless not as appropriate as gasoline/diesel (Figure 3).

- At temperatures of 200-300°Celsius methanol can be catalytically converted with water into hydrogen and carbon dioxide.

- Methanol can be reformed to provide hydrogen in mobile systems by using the waste heat and/or anode waste gas from a fuel cell.

- Logistic problems are more easily solved for methanol in comparison to hydrogen.

- Methanol can be produced from natural gas or hydrogen (via hydropower, nuclear energy or solar energy) and carbon dioxide (out of waste gases from technical processes).

Using methanol as the energy carrier and hydrogen as the feed gas for a fuel cell, a complete mobile system must comprise a reformer and a fuel cell with corresponding peripheral units (energy storage system) (Figure 2). An appropriate fuel cell would be a cell with a phosphoric acid electrolyte (PAFC) or with a polymer membrane as the electrolyte (PEMFC); the latter being finally preferable for mobile systems due to the immobile electrolyte at an operating temperature below 100° Celsius (MENZER et al. 1991).

Proceeding from the goal of supplying a fuel cell with hydrogen from a methanol/water vapour reforming system, tasks arise which, apart from a systems analysis study of the entire system, refer to the development of materials and, in particular catalysts within the fuel cells, the development of process engineering for fuel supply to the fuel cell - methanol reforming - and the direct interaction beween feed gas and fuel cell. Methanol reforming for hydrogen production - hydrogen as the feed gas for fuel cells - will be primarily discussed in the following, as studied at the Institute of Energy Process Engineering at the Research Centre Jülich.

4. Experimental and Theoretical Consideration of Methanol Reforming

Methanol reforming with or without water vapour is a process commercially available from several engineering companies and used on an industrial scale to produce pure hydrogen and synthesis gases (CO, CO_2, H_2, CH_4) - for example as town gas for peak load demand (RESTIN et al. 1985).

4.1 Thermodynamics

The reaction system for the production of gases containing hydrogen is based on the following stoichiometric relations:

Thermal splitting: $CH_3OH \longleftrightarrow CO_2 + 2H_2$ +91.3 kJ/mol

Steam reforming: $CH_3OH + H_2O \longleftrightarrow CO_2 + H_2$ +49.7 kJ/mol

One of the two independent stoichiometric equations specified above for the endothermic convertion of methanol into a synthesis gas can also be replaced by the equation for the homogeneous water gas reaction. Methane may occur as a by-product.

Homogeneous water gas reaction (shift) $CO + H_2O \longleftrightarrow CO_2 + H_2$ -41,6 kJ/mol

Methanation $CO + 3H_2 \longleftrightarrow CH_4 + H_2O$ -205.21 kJ/mol

The use of synthesis gases as feed gases for fuel cells must comply with the particular boundary conditions of the fuel cell under consideration, especially concerning the gas quality. The criteria for using a methanol reformer combined with a fuel cell are shown in Figure 4. Additional information is still required concerning the optimization of methanol reforming in conjunction with a fuel cell.

4.2 Simultaneous Equilibria

It can be shown for the first two stoichiometric equations mentioned above that apart from pressure and temperature in particular the molar methanol/water ratio plays a significant role in the simultaneous equilibrium. Thermal splitting without the addition of water vapour leads to a synthesis gas rich in CO (CO>30%) which is not suitable as a fuel gas for a low-temperature fuel cell since the PAFC requires CO contents in the feed gas < 3 vol.% and PEMFC CO contents < 100 ppmv at the present state of the art. Reforming of a methanol/water mixture permits a reduction of the CO fraction in the feed gas in favour of carbon dioxide and hydrogen. Figures 5 and 6 show the calculated equilibrium composition of the product gas under the given boundary conditions. At a molar ratio of $n_{H2O}/n_{Ch3OH} = 1.3$, the equilibrium value of the CO content in the dry product gas is about 1 vol.% at 200°C and 5 bar.

Extensive experiments at IEV on the reforming of methanol/water mixtures (n_{H2O}/n_{MeOH}=1.3) with commercial copper-zinc or copper-chromium catalysts have indicated that the selectivity of the catalyst for the homogeneous water gas shift reaction depends to a great extent on the reaction temperature.

It becomes clear from the relationship between the carbon monoxide content of the dry product gas and the reaction temperature shown in Fig. 7 that only subsequent gas treatment can ensure the CO content of <100 ppm required for a PEM fuel cell. The possible options for subsequent gas treatment in order to reduce the CO content by establishing separate catalyst and/or temperature zones are as follows:

1. Conversion of CO with H_2O by a high catalyst selectivity for the homogeneous water gas shift reaction

The calculation of the equilibrium values (see Figures 5 and 6) for the shift reaction shows that either the reaction temperature has to be extremely low or the water/methanol ratio very high in order to achieve the desired concentrations. A very low reaction temperature requires a catalyst active and selective at these temperatures which is not available at present, whereas a high water/methanol ratio does not appear meaningful with respect to process engineering and energy.

2. Conversion of CO with H_2 by a high catalyst selectivity for the methanation reaction

The equilibrium calculations for the methanation of a synthesis gas with 3 % carbon monoxide carried out at IEV result in a reduction of the CO content to almost 0 (<10 ppmv) at a temperature of 200°C. If the methane resulting in the fuel cell can be treated as an inert gas then methanation represents a very efficient procedure to reduce the carbon monoxide content. A further advantage of subsequent methanation would be that this catalyst does convert unconverted methanol from the reforming reaction into carbon dioxide, hydrogen or methane and the probability of poisoning of the fuel cell anode by methanol can therefore be considerably reduced.

3. Conversion of CO with $\frac{1}{2}O_2$ by selective oxidation of the adsorbed carbon monoxide

The selective oxidation of CO to CO_2 takes place in the presence of a precious metal catalyst at whose surface carbon monoxide is present in an adsorbed form. Selective oxidation is brought about by the introduction of oxygen (air) and the contents of carbon monoxide achievable are very small (LEMONS 1990). The simultaneous presence of noble metal, hydrogen, carbon monoxide and oxygen means that extremely strict safety measures have to be taken in order to provide adequate precautions against an explosion of the gas mixture.

On the basis of the considerations described above, a reforming reactor equipped with a subsequent catalyst bed for methanation is being set up at IEV in order to demonstrate the desired reduction of carbon monoxide in actual operation.

4.3 Criteria for Reformer Design

The quantity of hydrogen required to supply the overall system shown in Figure 2 must be provided from the most compact and lightest reformer possible, which is why great significance is attached to catalyst activity. Figure 8 shows the relationship between the quantity of hydrogen which can be generated per litre of catalyst and hour as a function of the experimental temperature. The catalysts tested in our experiments are the copper-zinc or copper-chromium catalysts already mentioned used for commercial hydrogen production from methanol in large-scale plants. Depending on the experimental temperature chosen, differences of up to 50 % result between the individual catalysts although these become smaller with increasing temperature. The adjustment of a constant catalyst load for all experiments facilitates the allocation of the methanol convertion rate as a function of experimental temperature shown in Figure 9.

On the basis of the experimental results presented it becomes apparent that the required high methanol conversion (>95 %) and the associated quantity of hydrogen which can be generated per litre catalyst is only achieved at temperatures above 260°C which is quite unacceptable with respect to the CO content of the product gas as a feed gas for low-temperature fuel cells. It is thus not possible to observe the following requirements in parallel:

- low CO content
- high methanol conversion
- low reaction temperature.

In order to illustrate this problem area, Figure 10 shows the qualitative relationship beween the reaction temperature, methanol conversion and CO content of the dry product gas by using our experimental results and equilibrium values. As already discussed, a separate subsequent gas treatment is indispensable for the combination of a methanol reformer with a PEM fuel cell so that in this case a higher reforming temperature can be selected since a catalytic burner represents a suitable heat source for this temperature level.

In our own laboratory up to 140 mol of hydrogen was produced per hour and litre of catalyst volume using industrial catalysts in the integral reactor, irrespective of ageing effects and residence times. This corresponds to an electric power output of about 3.7 kW for 80 % hydrogen convertion in the fuel cell operated at an electrical efficiency of 50 %.

5. Outlook

The approaches and initial results of significance for the application of methanol reforming for fuel cell feed gas production have been discussed in this paper. Assuming that the aim of our work is the application of passenger cars with electric engines and a top power of 20 kW (equivalent to 5 l of gasoline per 100 km) then based on our own data on methanol reforming this leads to a catalyst volume of 5 to 7 litres (without considering catalyst ageing) and a tank volume for the methanol/water mixture which is about three times as large as a comparable gasoline or diesel tank irrespective of the range.

These rough values and the prospect of a clearly improved emission balance in comparison to conventional systems - including electric vehicles with battery storage based on the current electricity mix in Germany - and of improved overall efficiency are adequate reasons for further study of the process engineering of methanol reforming in conjunction with a separate subsequent gas treatment and the integration of a catalytic burner and of optimizing combined operation with fuel cells in mobile systems.

REFERENCES

SCHNEIDER, H. (1991) Auto- und Umwelt Perspektiven für das Jahr 2000 ATZ Automobiltechnische Zeitung 93(1991)6, 354-362

BRABBS, T. A. (1978) Catalytic Decomposition of Methanol for Onboard Hydrogen Generation, NASA Technical Paper 1247, Lewis Research Center Cleveland, Ohio, USA 1978

HÖHLEIN, B. (1991) Neue Energieträger für den Verkehr, Monographien des Forschungszentrums Jülich, Vol. 5, Jülich, 1991

LEMONS, R. A. (1990) Fuel Cells For Transportation, Journal of Power Sources 29 (1990)251-264

MENZER, R., GANSER, B., HÖHLEIN, B. Niedertemperatur-Brennstoffzelle mit Methanol-Reformierung zur Brenngaserzeugung, Forschungszentrum Jülich, Internal Report KFA-IEV-IB-10-90, Jülich, 1990

RESTIN, K., JURKAT, P.,A., HILLER, H. (1985) Peaking Gas from Methanol, Proceedings: 16th World Gas Conference, Munich, 1985

HÄUSSINGER, P., LOHMÜLLER, R., WATSON, A. M. (1989) Hydrogen in Ullmann's Encyclopedia of Industrial Chemistry, VDH Verlagsgesellschaft GmbH, Weinheim, 1989

DAIMLER BENZ Wasserstoff - ein alternativer Treibstoff, Daimler Benz AG, Public Relations Dept., Stuttgart

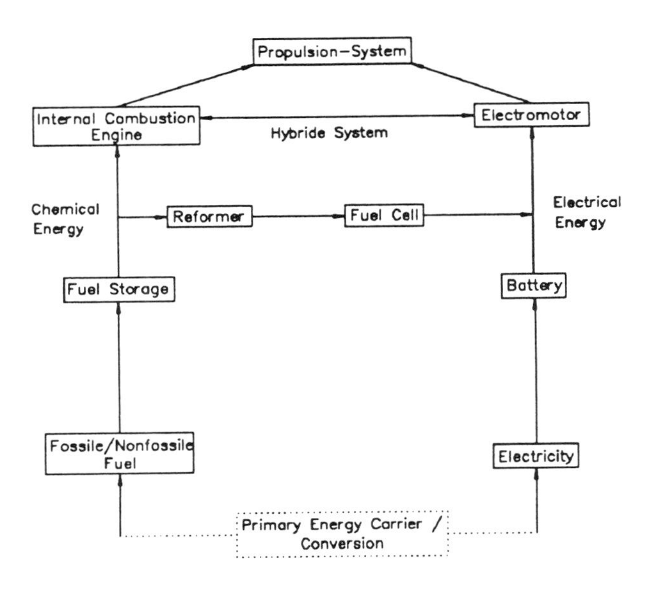

Fig. 1: Fundamental energy conversion for transportation applications

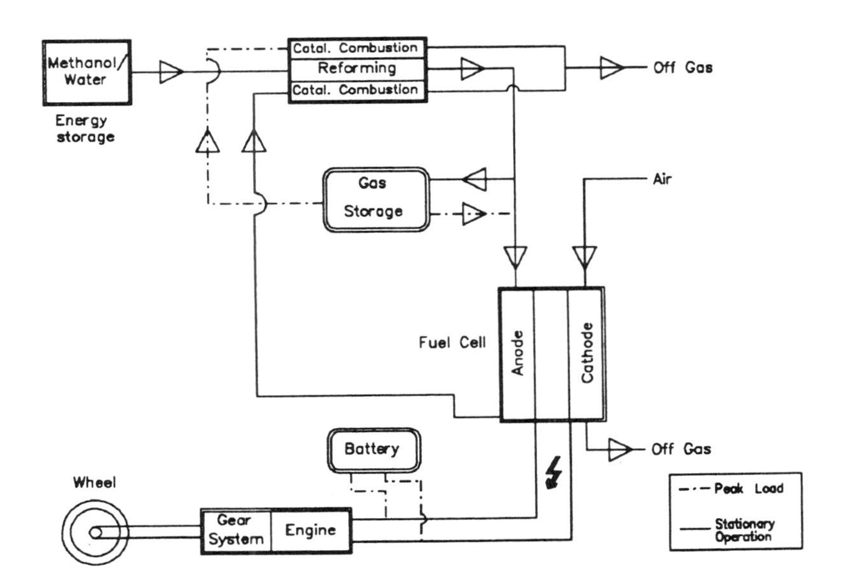

Fig. 2: Fuel cell system design

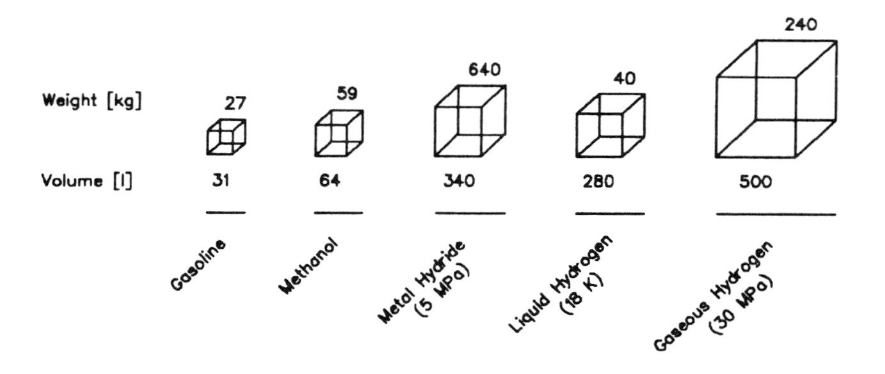

Fig. 3: Comparison of energy storage systems corresponding to the energy content (LHV) of 301 of gasoline; (HÄUSSINGER 1989; DAIMLER BENZ)

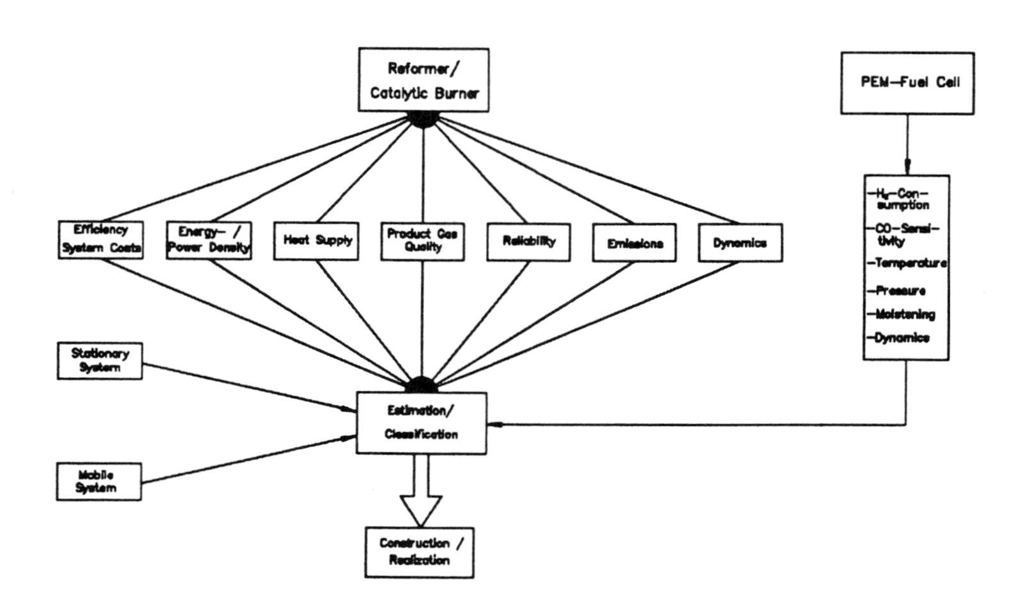

Fig. 4: Criteria for a Reformer Concept

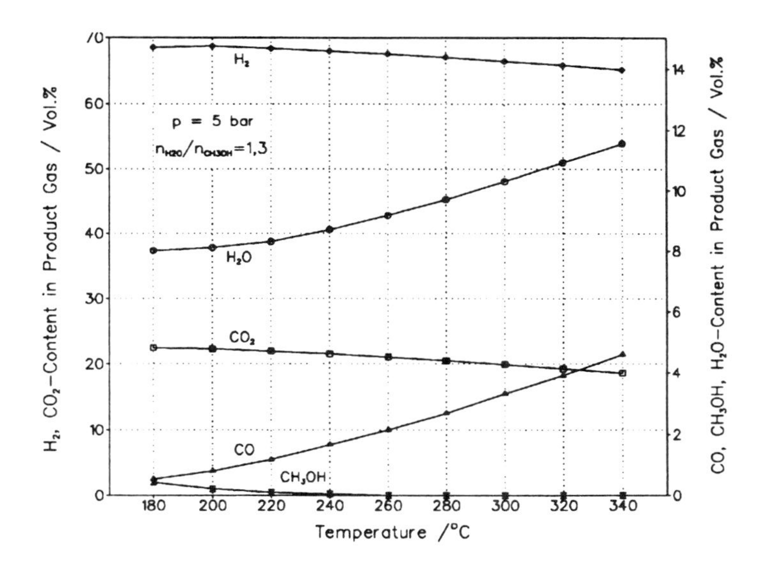

Fig. 5: Thermodynamic equilibrium as a function of temperature

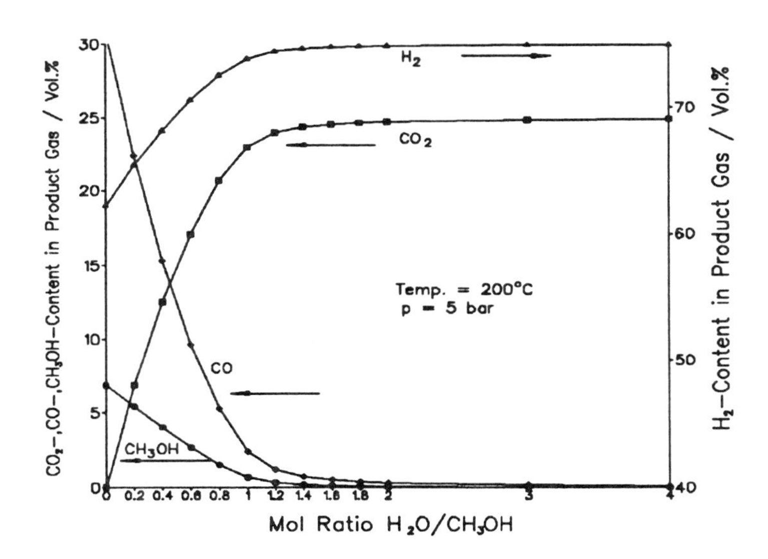

Fig. 6: Thermodynamic equilibrium as a function of water/methanol ratio

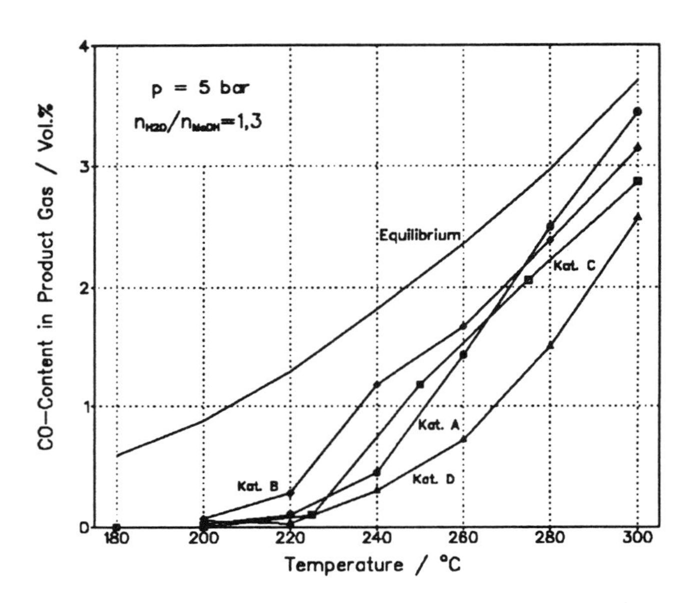

Fig. 7: Experimental data on CO-content in dry product gas as a function of temperature (at constant catalyst load)

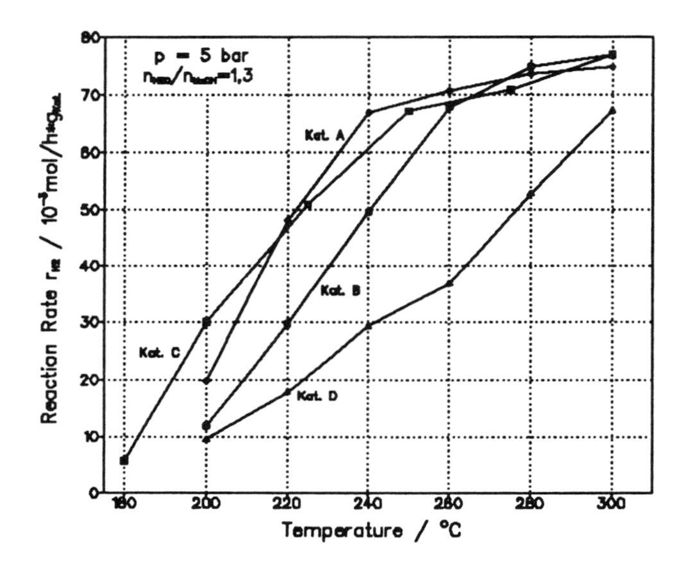

Fig. 8: Experimental data on reaction rate as a function of temperature (at constant catalyst load)

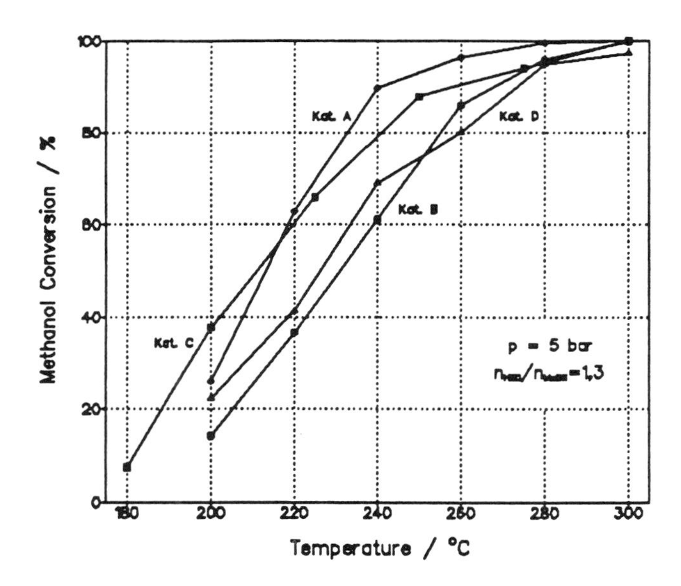

Fig. 9: Experimental data on methanol conversion as a function of temperature (at constant catalyst load)

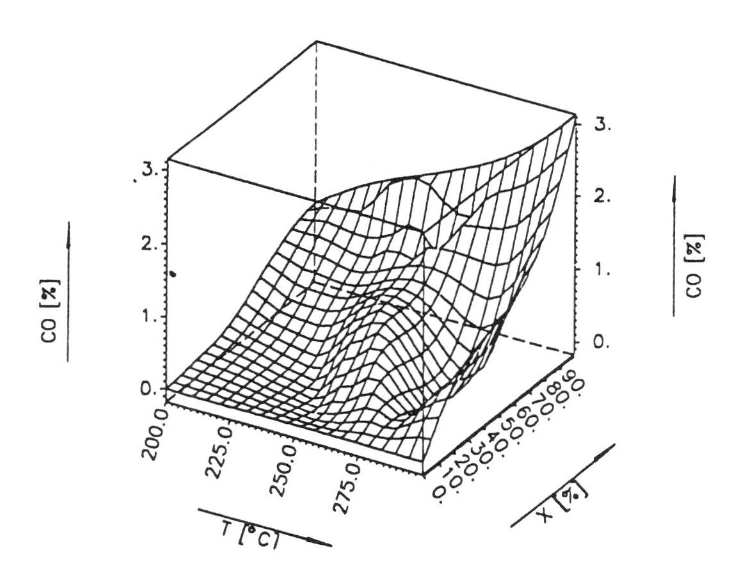

Fig. 10: 3D graph for the interrelation between temperature, co content and methanol conversion (at constant catalyst load)

Phosphoric Acid Fuel Cell Bus Development

Arthur Kaufman
H Power Corp.

ABSTRACT

This paper describes recent developments, current status, and key upcoming activities in a project that will provide three test-bed transit buses powered by a phosphoric acid fuel cell (PAFC)/battery electric system. The current Phase II project will utilize liquid-cooled PAFC technology that was successfully demonstrated at half-scale (25kW) in a brassboard system during Phase I. Fabrication of the first test-bed bus (TBB-1) is underway. The primary activities for the first half of 1993 will involve subsystem fabrication. System integration will take place during the third quarter, and rollout of TBB-1 is scheduled for October 1993.

Development of phosphoric acid fuel cell (PAFC) powered transit buses is currently being carried out under a U.S. Department of Energy contract.* This Phase II effort will yield three test-bed transit buses incorporating a PAFC/battery hybrid electric power source. The PAFC subsystem will be based on technology implemented for Phase I in the form of a 25kW brassboard unit.

The methanol-fueled, liquid-cooled PAFC system, which supplies all of the primary energy for test-bed bus (TBB) operation, produces no particulates and extremely low emissions of nitrogen oxides and other pollutants. The designated mode of operation allows the PAFC, which will be the highest-cost segment of the bus' propulsion system, to be rated at little more than the average power draw; a battery provides the incremental power required for acceleration and hillclimbing. Also, the overall fuel efficiency can be enhanced by utilizing the battery to harness portions of the energy available from braking and deceleration.

The TBBs to be fabricated in the Phase II effort will employ existing fuel cell, battery, and powertrain technologies. The primary objective of Phase II is to integrate these technologies and demonstrate the viability of the resulting propulsion system in a small, high-quality transit bus. The 29-ft TBB has been designed to meet all DOT performance requirements. It must be capable of operating on all urban routes - including the Georgetown University Transportation Society (GUTS) Arlington Loop, the Los Angeles RTD Line 16 route, and the DOT Transit Coach Duty Cycle. The TBB design configuration is shown in Table 1.

H Power Corp., in addition to its project management functions, is responsible for integration and testing of the TBB propulsion system. Subcontractors include: Bus Manufacturing U.S.A., Inc. (bus design and fabrication); Booz • Allen & Hamilton, Inc. (bus system integration); Fuji Electric Co., Ltd. (fuel cell hardware); Soleq Corporation (powertrain and controls); and Transportation Manufacturing Corporation (transit bus industry guidance and 40-ft bus conceptual design).

The PAFC subsystem design is to be completed before the end of 1992. This subsystem will supply approximately 55kW gross power at a nominal 115V (175 cells at 0.66 V/cell; 240 mA/cm^2; 2000 cm^2 active area) and 190°C. The overall design efficiency is 38% (based on methanol LHV). An internally-fired catalytic reformer will convert a premixed methanol/water stream (vaporized using fuel cell waste heat) to a hydrogen-rich fuel cell anode gas (at about 260°C); approximately 80% of the hydrogen will be consumed in the fuel cell, with the balance combusted to sustain the endothermic reforming reaction. The bulk of the fuel cell waste heat (at about 140-160°C) can be utilized to provide cabin heating. The fuel cell subsystem operation is depicted schematically in Figure 1, and its specifications are presented in Table 2.

A step-up d.c.-d.c. converter is being designed to match the fuel cell output voltage to the battery voltage (generally in the 200-280V range). The efficiency expected from the converter is 97%. The features of the converter are summarized in Table 3.

Two types of commercially-available batteries are being evaluated for use in the Phase II project: (i) lead-acid diesel-cranking and (ii) nickel-cadmium. Rigorous life-testing is being conducted in multiple test stations at Argonne National Laboratory to aid in the battery selection process. The test programs realistically

* Project sponsored by the U.S. Department of Energy (DOE), the U.S. Department of Transportation (DOT), and California's South Coast Air Quality Management District (SCAQMD) under DOE Contract No. DE-AC02-91CH10447.

Table 1

TBB CONFIGURATION

Length:	29 ft. (8.84 m)
Width:	96 in. (2.44 m)
Height:	121 in. (3.06 m)
Curb Weight:	22,678 lbs. (10,296 kg)
Seating Capacity:	27 passengers + driver
Standees:	11 passengers
Physically Impaired:	2 wheelchair locations* and wheelchair lift
Wheelbase:	195 in. (4.95 m)
Floor Height:	32 in. (0.81 m)
Ground Clearance:	6 in. (0.15m) min.
Useful Life:	12 yrs with fuel cell replacement every 10,000 hrs.
TBB Oper. Temp. Range:	-10°F to 115°F (-23°C to 46°C)
Depot Storage Temp. Range:	32.0°F to 122°F (0°C to 50°C)
Angle of Approach:	10°
Angle of Departure	10°
Range:	150 miles (241.35 km)

* 21 conventional seats accessible when 2 wheelchairs in use

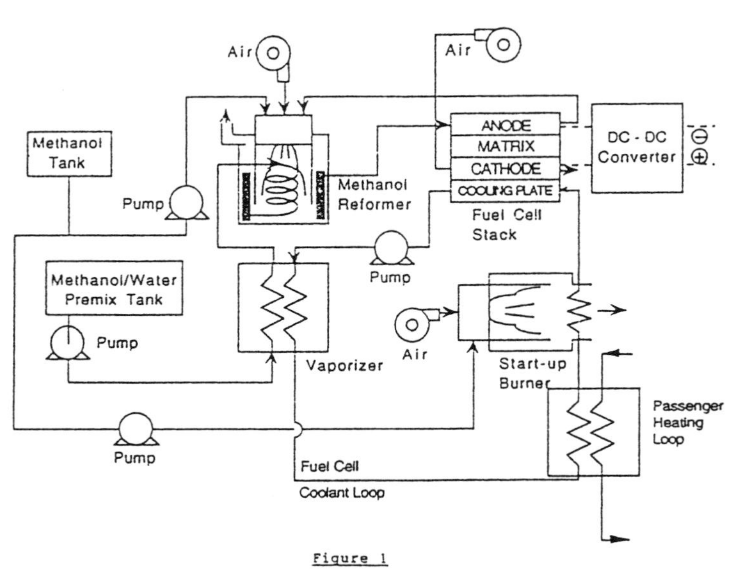

Figure 1

Phosphoric Acid Fuel Cell Subsystem

Table 2

FUEL CELL SUBSYSTEM CONFIGURATION

Type:	PAFC (liquid cooled)
Size:	1300mm W x 1150mm D x 2100mm H (6.89 ft x 4.26 ft x 4.85 ft)
Weight:	1780 kg (3921 lb)
Gross Power Output:	55.2 kW (74 hp)
Net Power Output:	50.0 kW (67 hp) (excluding d.c.-d.c. converter losses)
Output Voltage:	115 VDC
Methanol:	Commercial grade methanol with > 99.6% purity (no sulfur).
Water:	Deionized Water: $\kappa = 1.0\ \mu S/cm$, Cl < 1 ppm
Methanol Consumption at 50 kW power:	21.9 kg/hr (48.2 lb/hr)
Water Consumption at 50 kW power:	18.5 kg/hr (40.7 lb/hr)
Start-Up Time*:	30 min.
Transient Response Rate:	8 kW/min. power up, 8 kW/min. power down.

*The fuel cell start-up time is the time required for the fuel cell to go from off at ambient temperature to the operating temperature at which power output becomes available.

Table 3

DC-DC CONVERTER

NOMINAL RATING:	60 kW
MAXIMUM INPUT CURRENT:	480 A (at nominal 115 V)
MAXIMUM INPUT VOLTAGE:	180 V
NOMINAL OUTPUT VOLTAGE:	216 V
FREQUENCY:	≈ 5 kHz (equivalent)
EFFICIENCY:	95% min. (est. 97%)
SIZE:	18.9 in. x 23.6 in. x 56.3 in. high
WEIGHT:	165 lb

simulate conditions that will be encountered in a particularly severe duty cycle (GUTS), with varying degrees of regenerative braking. A decision on the battery for the first test-bed bus (TBB-1) will be made before the end of 1992.

A modified version of the General Electric CD-407 motor has been selected to provide the TBB traction power. A customized armature will allow the motor to operate at up to 3800 rpm, thereby avoiding the need for a multi-speed transmission; the axle is designed such that the TBB will reach 55 mph (an infrequent event in typical transit bus usage) at 3500 rpm. The design features for the traction motor are summarized in Table 4.

Table 4

Traction Motor Design Specifications

- System Voltage: 216 VDC
- Peak Input Power: 120kW
- Rated Running Efficiency: Approx. 91%
- Base Speed: Approx. 1000 RPM
- Maximum Torque: Approx. 800 ft-lb
- Forced Air Cooling: 800 CFM, 3-Phase AC
- Output Shaft: 30 Degree Involute

The TBB body/frame design has been approved, and the ordering of parts for fabrication has begun. The TBB will accommodate 27 seated passengers and 11 standees. Provisions have been made for two wheelchairs (displacing some conventional seats when used) and a wheelchair lift to meet Americans with Disabilities Act requirements.

A key activity for the near term is the design and development of the system controller. System control strategies and logic diagrams have been prepared; control algorithms are in an advanced stage of development. The PAFC brassboard unit from Phase I is being used to verify system performance and to confirm the validity of candidate control strategies. A comparison of brassboard power source experimental and simulation data is shown in Figure 2.

Brassboard Data and Simulation: Power

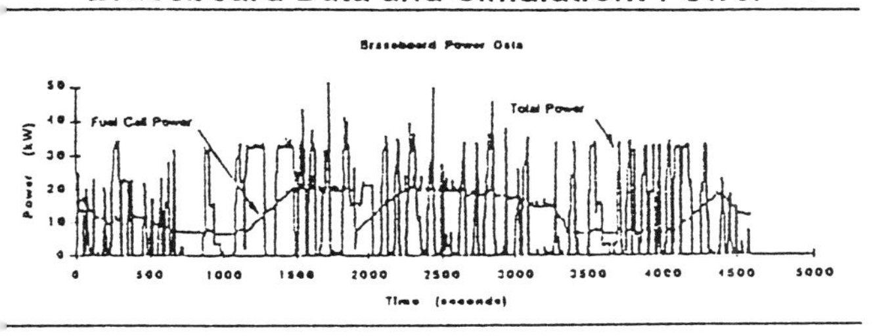

Figure 2

The primary activities for the first half of 1993 will involve subsystem fabrication. System integration will be carried out during the summer (location of subsystems aboard bus is shown in Figure 3); and rollout of TBB-1 is expected in October 1993.

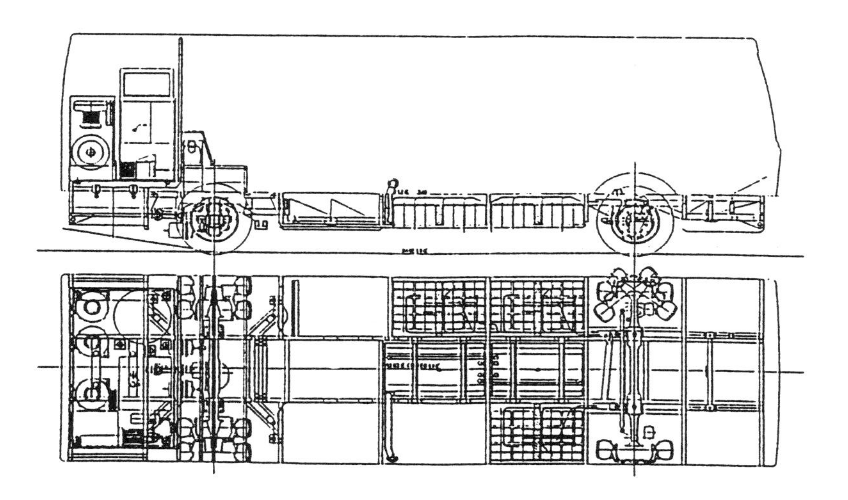

Figure 3

Location of Subsystems within Bus

Proton Exchange Membrane (PEM) Fuel Cell System R&D for Transportation Applications

H. F. Creveling
General Motors Corp.

ABSTRACT

Increasing environmental and strategic fuel concerns, and a potentially changing fuel base, continue to dictate the need for advanced vehicular power plants that exhibit improved fuel efficiency, reduced emissions, and alternate fuel capability. The electric vehicle has been suggested as one means of decoupling petroleum from the transportation sector and dramatically reducing vehicular environmental emissions. Some form of a hybrid power system could be beneficial in extending the overall range and load carrying capacity of today's battery-powered electric vehicles. Among the candidate power plants for the engine/charging component within the hybrid system, the indirect methanol proton-exchange-membrane (PEM) fuel cell is potentially the most efficient, as the device is unrestricted by heat-engine (Carnot cycle) limitations. Further, the fuel cell system directly produces electric power, thus eliminating the need for a generating system. The fuel cell system presents several technical challenges which require systematic research and development effort before it is ready for commercial vehicular applications.

Recent developments have created the potential of achieving breakthroughs in PEM technology. Although system technology is immature and material/component costs require improvement, the technology to build and successfully operate a fuel cell system on the component level can be demonstrated. Continuing work at Los Alamos National Laboratories, the Dow Chemical Co., Ballard Power Systems, and General Motors (GM) now indicates that material and component costs have the potential to become competitive with those of the internal combustion engines that may be required to meet extremely stringent future emission and fuel economy standards.

GM envisions a multiphase program leading to proof-of-concept testing of a fuel cell/hybrid vehicle. The initial program phase currently under contract emphasizes the development of technologies critical to the success of this propulsion system concept. This phase is designed to produce a methanol-fueled, 10-kW power source that demonstrates system feasibility for transportation applications. Key elements of this effort are tasks addressing the fuel cell, a fuel processor, electronic controls, a gas pressurizing system, and versatile management of water and heat. System requirements include low thermal inertia, a proper battery (to help with start-up and transients), and other ancillaries. Contract effort to date has addressed engineering development and testing of the fuel processor, performance testing of the Ballard fuel cell stacks, design of the prototype system, and development of digital controls for the system. Conceptual design studies were done to evaluate potential application of fuel cell/battery hybrid power systems in passenger cars, vans, and an urban bus. Component research and development has been directed toward membrane improvements, use of low catalyst loadings, and the design of advanced fuel cells. The work is being done by General Motors Corporation, under contract to the U.S. Department of Energy (DOE), Assistant Secretary for Conservation and Renewable Energy, Office of Transportation Technologies (Contract No. DE-AC02-90CH10435).

INTRODUCTION

General Motors continually evaluates promising vehicular power plant technologies. Those with outstanding potential to effect substantial payoffs in fuel economy, emissions reduction, and/or alternate fuel usage are of special interest. One such system is based on the PEM fuel cell.

Fuel cells are electrochemical engines, not batteries, which oxidize a fuel through electrochemical reactions, versus the thermochemical reactions found in the internal combustion engines of today's automobiles. In so doing, these devices transform the chemical energy of the fuel directly to electrical energy, without the heat-engine's Carnot cycle limitations on attainable efficiency in the production of work. Fuel cell engines are typically configured as stacks of individual fuel cells, and have been designed and demonstrated in five basic technology types. Applications have included alkaline technology in space vehicle power; phosphoric acid technologies in transportation; and phosphoric acid, molten carbonate, and solid oxide fuel cells in stationary power.

The PEM fuel cell offers outstanding potential for further transportation applications. The advantages of this emerging PEM technology are found in its low operating pressure and temperature; in the potential for demonstrating viable material and manufacturing costs; in its ability to accommodate start-up and transient operation typical of vehicular applications; and in the safety advantages of its basic operation and design. PEM fuel cells have demonstrated thousands of hours of reliable service in aerospace applications with hydrogen as a fuel. More recently, improved technologies necessary for transportation applications have been demonstrated.

FUTURE TRANSPORTATION POWER PLANTS

Future power plants for transportation applications will be required to demonstrate substantial advantages over current engines. Principal areas of improvement will be increased energy efficiency, reduced emissions, and the capability to operate on alternate fuels.

The fuel cell power plant is an electrochemical engine, and is thus unrestricted by the heat-engine limitations of the Second Law of Thermodynamics. System integration must be provided in such a way as to accomplish the thermal and mass management required to maximize the efficiency of the electrochemical engine system, including fuel processor and ancillary equipment. When this is done, the electrochemical engine is a very efficient candidate for vehicular applications.

Furthermore, this system offers near-zero regulated emissions and reduction in greenhouse emissions (CO_2) by virtue of its improved energy efficiency. The electrochemical engine system can be configured to operate on hydrogen directly or on a processed-fuel stream. In this program, methanol is the fuel of choice, and steam reforming is selected as the fuel processing technology, because of its relatively high yield of available hydrogen.

Electrochemical engine technology is immature, particularly when the requirements of transportation applications are considered. Transportation applications require substantial improvements in system power density, rapid transient response, including cold start-up, and major reductions in cost through improvements in design, material use, etc.

ELECTROCHEMICAL ENGINE SYSTEM

A simplified schematic of a methanol-fueled PEM fuel cell power plant is presented in Figure 1. The fuel cell power plant consists of three basic sections:

- fuel processor
- power section (fuel cell)
- power conditioner section (inverter)

The fuel processor uses steam to catalytically transform methanol to a hydrogen-rich gaseous fuel.

The power section is a series-stack of fuel cells which converts the processed fuel plus air directly into electricity. The power conditioner section converts the power output for use in driving traction motors, and/or battery charging, typically in a fuel cell/battery hybrid vehicle. Note in Figure 1 that the fuel cell's by-products are only heat and water.

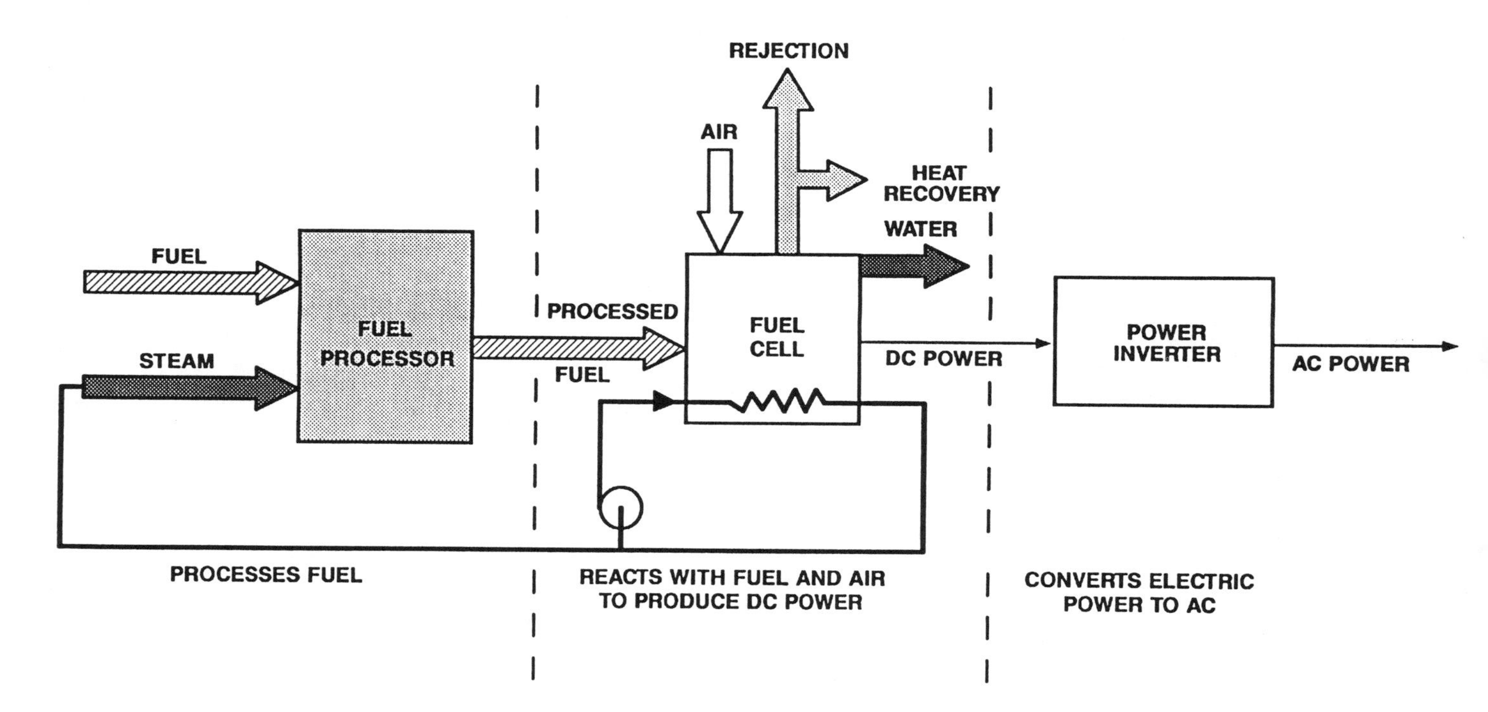

Figure 1. Electrochemical engine system.

As noted earlier the fuel cell power plant may be an excellent candidate for hybrid vehicular application. Fuel cell stacks are capable of operating at voltages and current fluxes that yield power density levels comparable to, and efficiencies nearly double those of, current vehicular engines. The efficiency of a PEM fuel cell system increases at part-power operation and can produce a substantial improvement in driving cycle efficiency; since the automobile is typically a part-load device, it stands to benefit very greatly from the fuel cell's high efficiency at low power.

HYBRID CONCEPTUAL VEHICLE

Vehicle and power system studies have shown that the PEM fuel cell system is a candidate for a hybrid propulsion package, utilizing both the fuel cell system and a battery pack of suitable power density. Such a propulsion package could be configured as shown in Figure 2, with the battery pack accommodating system start-up and vehicle transient-operation requirements and the whole system providing major improvements in both fuel economy and emissions. Clearly, any design and application will be evaluated primarily in terms of its ability to satisfy the requirements of cost, performance, and reliability.

Vehicle and mission analysis, coupled with performance calculations for the PEM electrochemical engine system, indicate that a PEM fuel cell powered hybrid passenger vehicle can demonstrate an energy efficiency nearly double that of its current counterpart over the same driving cycle. By virtue of this sharp improvement in energy efficiency, a 40% reduction in CO_2 emissions is predicted, along with a reduction in regulated emissions to less than 10% of those produced by the cleanest current power plants. Ideally, regulated emissions from the PEM electrochemical engine can be reduced to nearly zero, even while using a processed fuel such as methanol in an "indirect methanol" system such as that shown schematically in Figure 1.

PROGRAM LOGIC AND SCOPE

GM has conceptualized a program to research, develop, and demonstrate a fuel cell propulsion system in a passenger vehicle through a multiphase, eight-year effort, summarized in Figure 3, where the current DOE/GM contracted program is shown in the shaded portion of the figure. Overall program scope develops an advanced methanol/air PEM fuel cell power plant, addressing all goals of a competitive transportation application, and culminates in an actual fuel cell/hybrid vehicle demonstration.

The elements of the current (Phase 1) program activity are the essential first steps in accomplishing these goals. Current program work not only involves component research and development to accomplish new levels of fuel cell and fuel processor performance, but also addresses the whole subject of controls and system ancillaries required to accomplish the current critical technical challenge: integration and demonstration of a power system capable of rapid start-up and transient response. The initial 10-kW system will use commercially available PEM fuel cell stacks. Other components, including an advanced high-performance methanol fuel processor, will be based on the results of currently active research and development efforts and will be the key factors in the prototype system's transient response characteristics.

The Phase 1 program work plan includes a comprehensive assessment and modeling of vehicle mission requirements and the consequent hybrid vehicle power packages meeting these requirements. Passenger cars, vans, and urban buses have been considered separately. A "Reference Powertrain Design" will be produced for the selected vehicle. Also, technical and economic fuel issues are being assessed. Furthermore, GM/Allison Gas Turbine is constructing a transportation-oriented fuel cell facility during the course of this program.

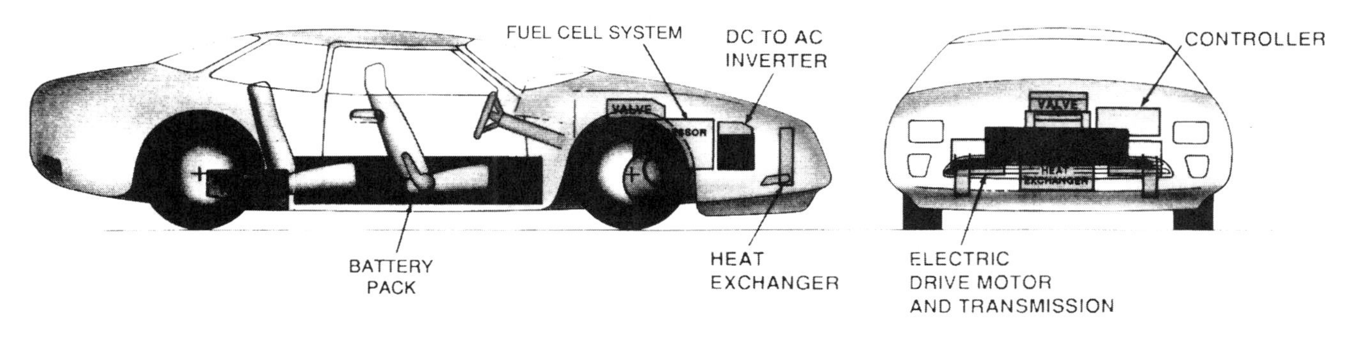

- POTENTIAL, VS. CURRENT POWER PLANTS
  - ENERGY EFFICIENCY UP 90%
  - REGULATED EMISSIONS DOWN 90%
  - CO_2 EMISSIONS DOWN 40+%

* CRITERIA: COST, PERFORMANCE, RELIABILITY

Figure 2. Hybrid conceptual vehicle.

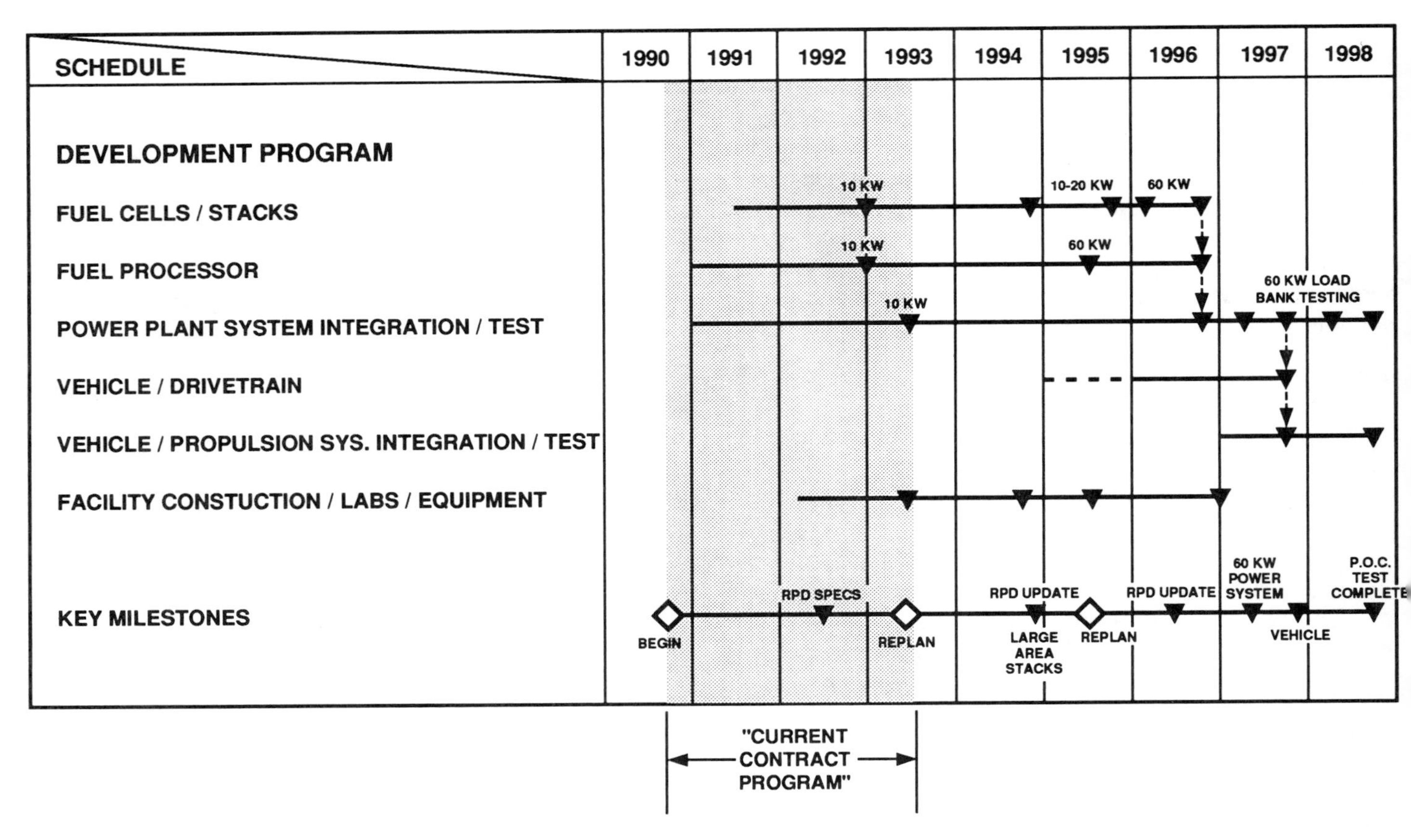

Figure 3. Program preliminary milestone schedule.

PROGRAM TEAM ORGANIZATION

DOE/GM PROGRAM TEAM - Work on the DOE/GM program Research and Development of a PEM Fuel Cell System for Transportation Applications (Contract DE-AC02-90CH10435) is being done by the program team shown in Figure 4. GM's Allison Gas Turbine Division is prime contractor to the DOE through the DOE's Office of Transportation Technologies. GM is addressing a spectrum of basic technologies and engineering issues associated with producing competitive transportation applications of the PEM fuel cell; several GM staffs and divisions are active in GM's total integration of the program effort. Subcontractors to GM are additional required elements of the program team. Each subcontractor was chosen for its expertise and commitment to meeting the program's objectives. Particularly notable as a subcontractor to GM on the program team is the Los Alamos National Laboratory (LANL). This collaboration is accomplished through the GM-LANL Joint Fuel Cell Development Center, located at LANL and staffed by employees of GM's Allison Gas Turbine Division and LANL.

Major participants in the program team are briefly described in the following paragraphs.

General Motors Corporation - General Motors' Allison Gas Turbine Division is in the program lead role, based on its capabilities in systems integration and in the multiple design disciplines required in producing advanced power systems. Additional General Motors support is provided by the Physical Chemistry and Vehicle System Research departments of the General Motors Research and Development Center and by the AC-Rochester Division.

- At General Motors Research basic research in fuel cell and fuel processing technology are program activities in the Physical Chemistry Department. Vehicles systems configuration, modeling, mission analyses, etc. are areas of activity in the Vehicle Systems Research Department.
- GM's AC-Rochester Division has a strong technical capability in the managing and delivery of fuels on-board GM's motor vehicles, and in the research, design, development, and manufacture of such components as pumps, sensors, injectors, and catalytic conversion devices. AC-Rochester plays a major role in the design and development of methanol reformer components and other power plant subsystems.

The Los Alamos National Laboratory - Los Alamos National Laboratory, operated by the University of California for the U.S. Department of Energy, has been funded since 1977 to conduct programs that explore fuel cells for transportation applications. Initially, LANL established the technical feasibility of this concept through simulation of the performance of a variety of vehicle classes (passenger cars, vans, buses) using state-of-the-art fuel cell parameters. Subsequently, LANL con-

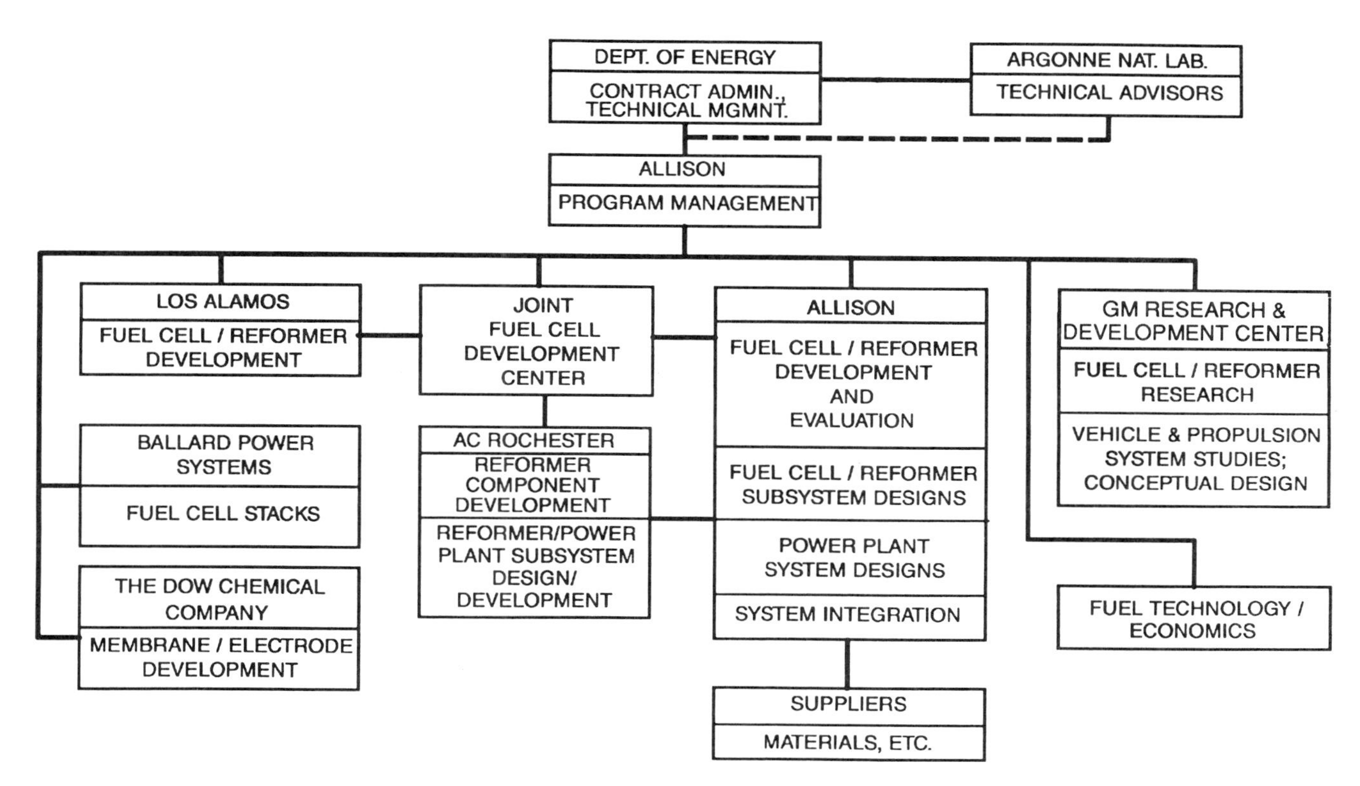

Figure 4. Program team functional organization.

tracted with several fuel cell manufacturers to establish projected performance for a series of fuel cell technologies with the emphasis on passenger car application. As a result of this study, LANL identified the PEM fuel cell technology as the most promising for electrically propelled passenger transport. In this concept, system power was derived using liquid methanol, which was processed on-board to a hydrogen-rich gas.

Another area explored by LANL was upgrading the fuel processor product gas quality. Because of the low operating temperature, the PEM fuel cell is sensitive to carbon monoxide (CO), which contaminates the catalyst surface and degrades fuel cell performance. LANL, under a previous contract to General Motors Corporation, explored new approaches for CO removal. One of these, preferential oxidation, converts CO to inert CO_2, even in the presence of large quantities of hydrogen. This new design involved concepts of heat transfer and flow control required for successful CO removal.

The Dow Chemical Company - The Dow Chemical Company currently operates numerous processes which utilize membranes for either separation purposes or ion exchange and/or transport. As a result, the company has extensive experience in all areas of not only membrane research and development but in the actual synthesis of the polymeric materials. The large research and process research groups that have been set up give Dow the capability to develop advanced concepts and materials such as the membranes proposed in this effort. Dow has developed significant related membrane technology for chlor-alkali processes.

Ballard Power Systems - Ballard is a developer and manufacturer of PEM fuel cells. Ballard's PEM fuel cell hardware and fabrication techniques have produced a fourfold improvement in power increase and size reduction. The Ballard PEM fuel cell is one-tenth the size of an equivalent phosphoric acid fuel cell.

SUMMARY OF ACCOMPLISHMENTS

The status of program activity is summarized for each of the major program technical tasks. Each of these tasks complements the activities in other program work areas and contributes directly to achieving the program's overall goals.

SYSTEM CONCEPTUAL DESIGN STUDY -

- The electrochemical engine system model of LANL has been further developed and refined as a result of being used to determine and optimize system configurations for transportation applications of the indirect methanol-fueled PEM technology. System efficiency calculations for part load operation are a basis for vehicle propulsion control strategy.
- A hybrid vehicle simulation model, proprietary to GM and identified collectively as VSIM, has been used in modeling a variety of vehicles, calculating vehicle energy usage (fuel economy, distribution of losses), vehicle performance characteristics (gradeabiity, acceleration, and range), plus emissions. Figure 5 illustrates the inputs and outputs of the VSIM model. A series hybrid vehicle con-

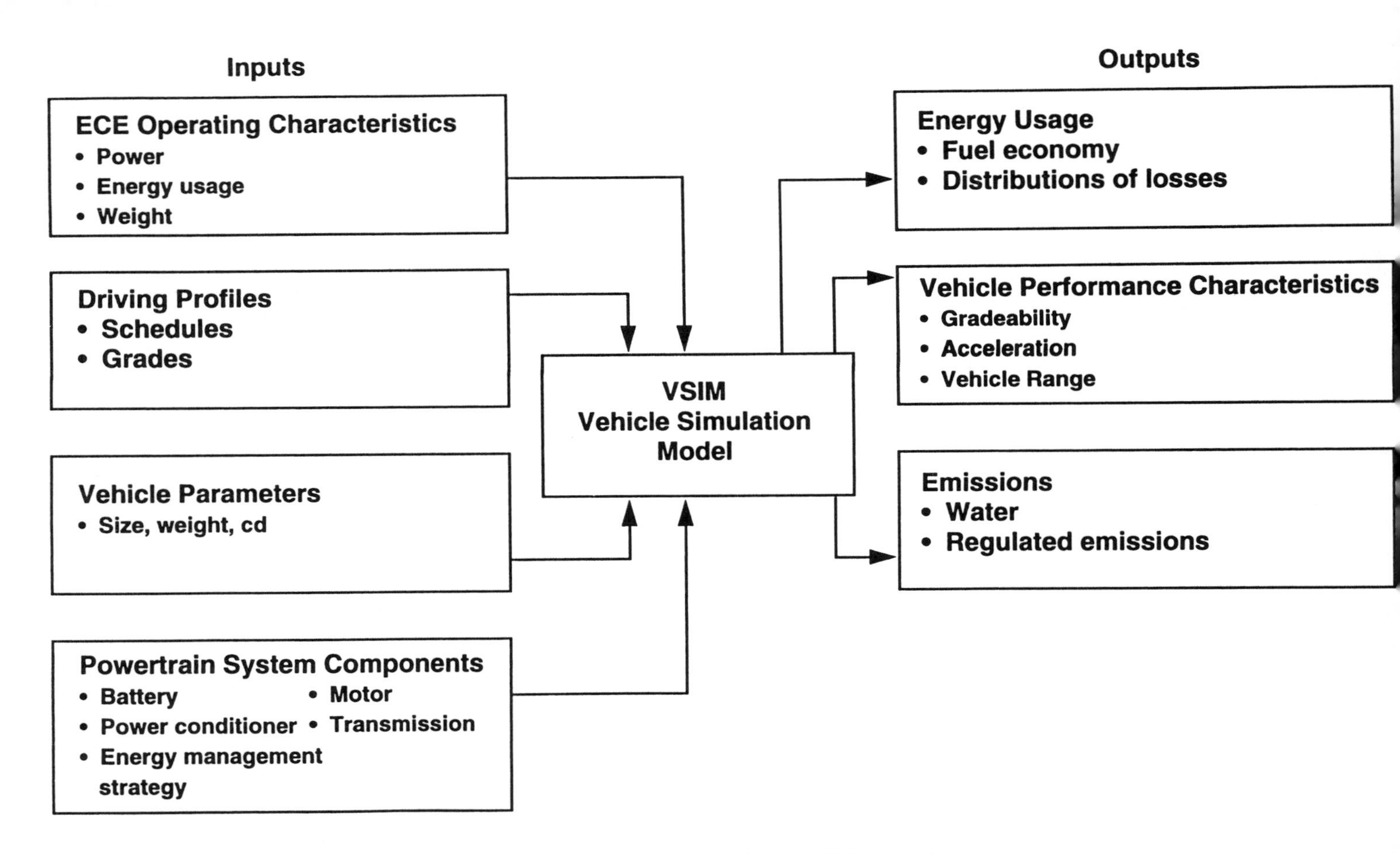

Figure 5. Hybrid vehicle propulsion model.

figuration with regenerative braking was chosen. The fuel cell was sized to meet long-term vehicle gradeability requirements; the battery was sized for maximum performance (matching conventionally powered vehicle 0 to 60 mph) and separately for similar performance (meeting the relatively modest acceleration requirements of the Federal Urban Driving Schedule during warmup).
With vehicle mission requirements defined and a vehicle evaluation (scoring) procedure established, four passenger vehicle types in each of the two performance classes were evaluated and the minivan was recommended as the vehicle choice for further parametric tradeoff analysis and for the Reference Powertrain Design.

- The infrastructure and commercialization study has thoroughly reviewed fuel and vehicle scenarios and requirements, with the goal of identifying the competitive requirements of implementing PEM fuel cell technology in the transportation sector and of assessing the corresponding infrastructure requirements and options, especially for fuel choice and supply. In addition to the vehicles studied in this program's vehicle simulations, the railroad locomotive has emerged as a candidate for fuel cell power as a means of dramatically reducing regulated emissions.

COMPONENT RESEARCH AND DEVELOPMENT - System component research and development has addressed a wide variety of issues, while focusing on problem solutions and needed design improvements for both the fuel cell and fuel processor.

Accomplishments in fuel cell research and development include:

- advanced membrane and electrode assemblies
  - low platinum loadings demonstrated
  - superior membrane materials developed
- membrane and electrode evaluation
  - new test facility for single cell measurements
  - excellent single cell performance

Fuel processor research and development results include:

- heterogeneous catalyst development for CO control
  - monolithic catalyst system tested
  - control strategies identified
- fuel metering hardware for water and methanol feeds
  - new fuel injector hardware developed and tested

Figure 6 shows monolithic catalyst samples typical of those developed by the AC Rochester Division of GM and utilized in achieving improved performance of fuel processor components.

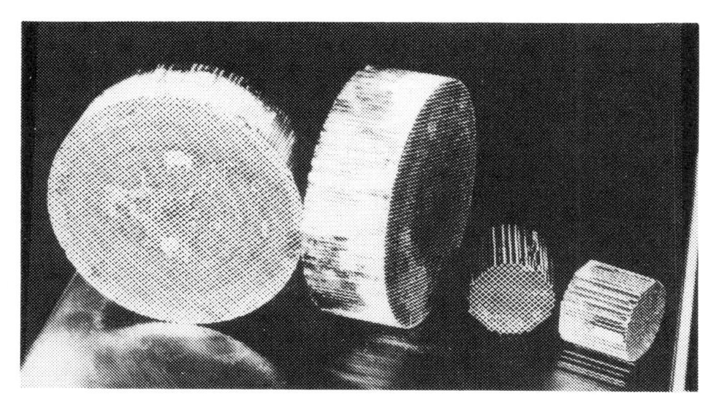

Figure 6. Monolith-support catalyst modules.

FUEL CELL STACK DEVELOPMENT - Fuel cell stack development has focused on testing and evaluation of 5-kW Ballard power generation stacks for Phase I of this contract program. Figure 7 illustrates this commercially available stack. Stack development work at the Joint Development Center has been fully supported by Ballard and has resulted in

- operation of Ballard stacks optimized for hydrogen and air
- testing of Ballard stacks with simulated reformate fuel

These stacks are advanced designs, using Dow membranes and addressing the particular requirements of reformate fuel.

FUEL PROCESSOR DEVELOPMENT - Fuel processor development work has focused on design and demonstration of reformer/vaporizer, shift reactor, and preferential oxidation components for the methanol processor required to demonstrate feasibility for transportation applications of the power system. Work during the current year has

- developed fuel/water metering system
- demonstrated rapid transient response
- demonstrated potential for rapid start-up
- developed monolith catalysts for reformer, shifter, and large preferential oxidation
- demonstrated viability of compact fuel processor

Figure 7. Ballard power generation stack.

Figure 8 shows the vaporizer and reformer modules of the Mark II fuel processor for the 10-kW electrochemical engine system.

SYSTEM ANCILLARIES, SENSORS, AND CONTROLS - This segment of program activity is necessarily broad in its overall scope, but has focused on the following major elements of the current program requirements

- control hardware and software has been developed in conjunction with reformer and fuel cell integration and testing
- a unique combustor design is in progress for the next generation vaporizer/reformer
- an oil-free scroll compressor has been selected for future system development

10-KW POWER SOURCE INTEGRATION - The 10-kW system integration and evaluation activity to this point has produced an "interim" fuel stack design conceived by Ballard in response to the requirements of reformate fuel, and has also resulted in near-complete development of the Mark II fuel processor subsystem, using electrical heating elements as an interim heat source.

SYSTEM COMPONENT TEST ACTIVITY

Extensive system component testing has been performed at the Joint Development Center. The fuel processor technology has been developed through separate design and test of the processor's major components. As noted earlier, steam fuel processing is selected here because of the relatively high yield of hydrogen. The fuel processor's ability to respond to rapid transients in system load determines the response time for the whole electrochemical engine system. The fuel processor consists of three principal components: the vaporizer and reformer unit; the shift reactor; and the preferential oxidation (PROX) unit. Of these, the vaporizer/reformer determines the fuel processor's ability to follow rapid transients in electrochemical engine system load. For

Figure 8. Fuel processor component hardware.

this reason, vaporizer/reformer design and the accompanying strategies for methanol and water injection and control have received major emphasis in this effort. Test activity on the major elements of the power system is summarized in Figure 9, and has resulted in the accomplishments summarized earlier.

PROJECTED TECHNOLOGY

Concurrent with the Phase 1 program activity described here, technology projections for fuel cell stacks, fuel processor, and other system components in the 1994-95 time frame have been used as the bases for the conceptual design of an integrated, methanol-fueled PEM fuel cell power plant, configured specifically for the "engine" compartment of a concept hybrid passenger vehicle and shown in Figure 10. Dimensions of this system are 19 in. height, 32 in. width, 31 in. length; the system design is capable of a continuous maximum power output of 40 kW, with intermittent peak power output of 60 kW. The fuel processor must be sized to accommodate the intermittent peak system output; the sizing requirement is partially responsible for the processor being the larger of the two major, cylindrical modules in the conceptual design shown.

PROGRAM SUMMARY

The Phase 1 DOE/GM program work under the referenced contract is to result in a feasibility demonstration of a prototype 10-kW fuel cell power plant capable of steady-state and transient operations using methanol fuel. The critical technical challenge at this point is integration of a complete PEM fuel cell power system including controls to manage start-up and transient operation representative of vehicle requirements. Other elements of the program effort address vehicle and mission analysis and modeling; fuel specification, infrastructure, and the economic issues and requirements associated with commercialization of transportation applications; and advanced technologies necessary to reduce projected fuel cell system costs, improve performance and ensure sufficient reliability to meet the competitive requirements of mass-market transportation applications.

Each element of the program activity is a building block of preparation for the work remaining in projected follow-on program activities that will require critical choices among competing technologies, materials and designs, plus scale-up and packaging of a power system, and tests in proof-of-concept vehicles.

• FUEL PROCESSOR	
- VAPORIZER AND REFORMER	1200 HRS
- SHIFT REACTORS	640 HRS
- PROX REACTORS	
• FULL SCALE	400 HRS
• TEST BENCH	600 HRS
- ASSEMBLED UNITS	75 HRS
• FUEL CELLS	
- BALLARD STACKS	
• OPERATING ON HYDROGEN & AIR	145 HRS
• OPERATING ON SIMULATED REFORMATE & AIR	102 HRS
- REFERENCE CELLS	
• OPERATING ON HYDROGEN & AIR	100 HRS
• OPERATING ON SIMULATED REFORMATE & AIR	100 HRS

Figure 9. Test activity.

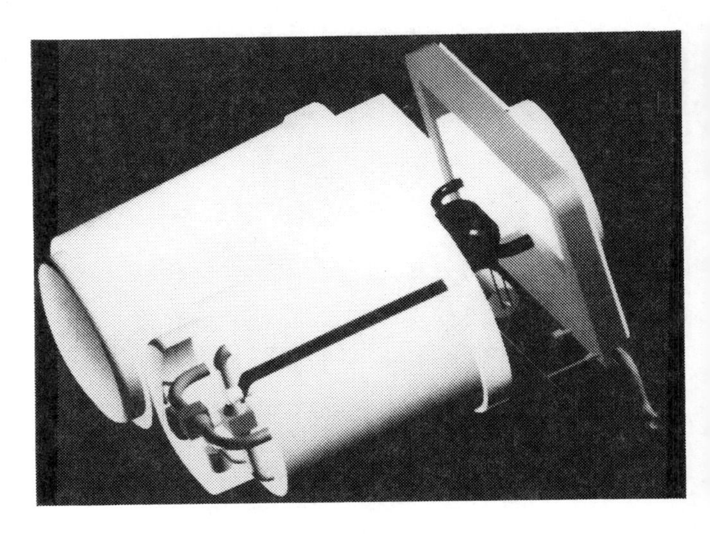

Figure 10. Conceptual power system for vehicle application.

Core Technology R&D for PEM Fuel Cells

Shimshon Gottesfeld, Mahlon S. Wilson,
Tom Zawodzinski and Ross A. Lemons
Los Alamos National Lab.

ABSTRACT

The objective of this program is to conduct the basic and applied research necessary to bring polymer electrolyte membrane (PEM) fuel cell technology to the performance and cost levels required for widespread use in transportation The specific goals are to reduce the intrinsic costs, to increase the power density, to optimize the system for operation on reformed organic fuels and air, and to achieve stable, efficient long-term operation.*

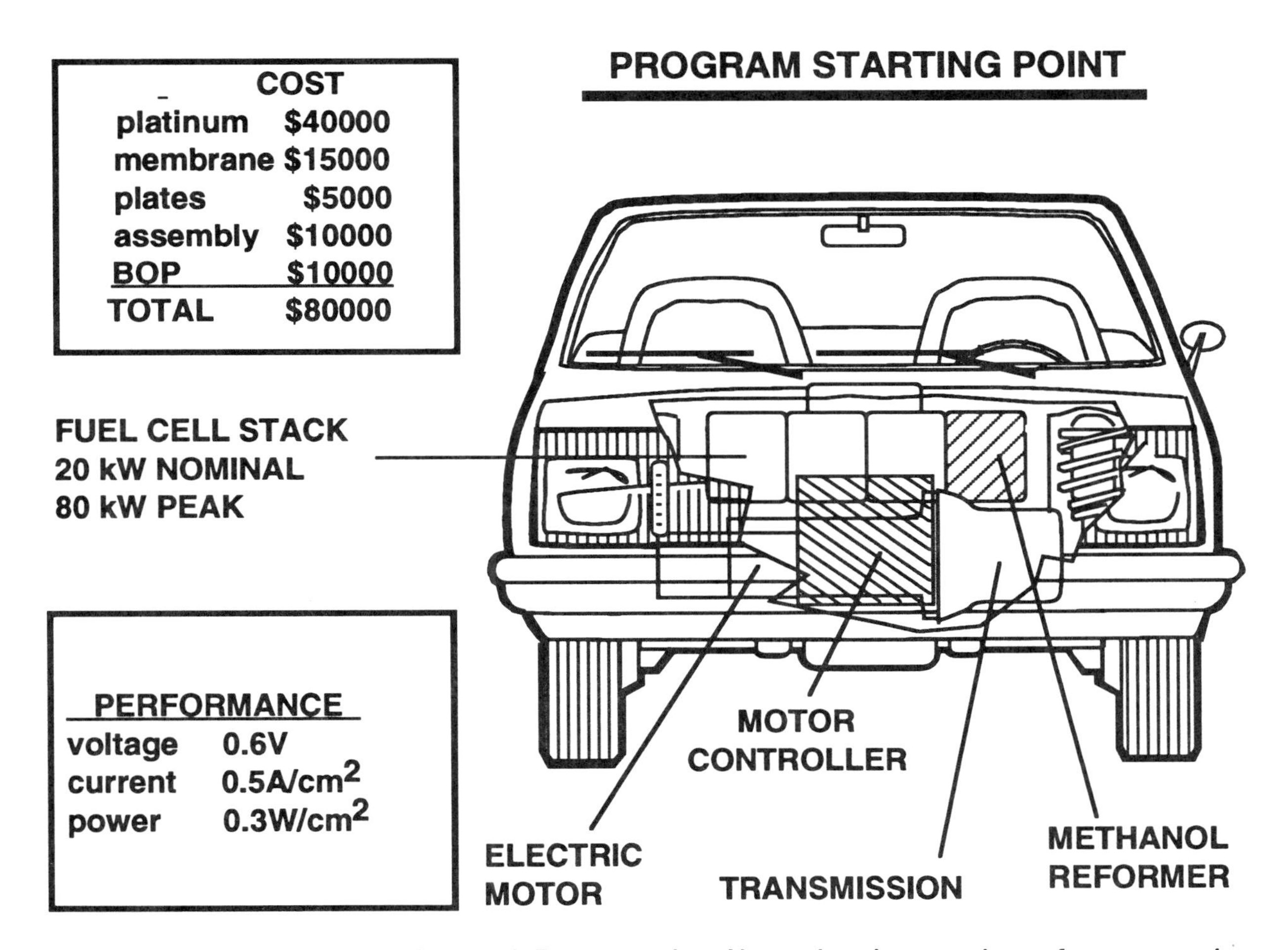

Fig. 1. The focus of the Core Research Program at Los Alamos is to improve the performance and reduce the cost of PEM fuel cells to the point where widespread use in transportation is feasible.

* Research sponsored by the U.S. DOE, Conservation and Renewable Energy, Office of Transportation Technology.

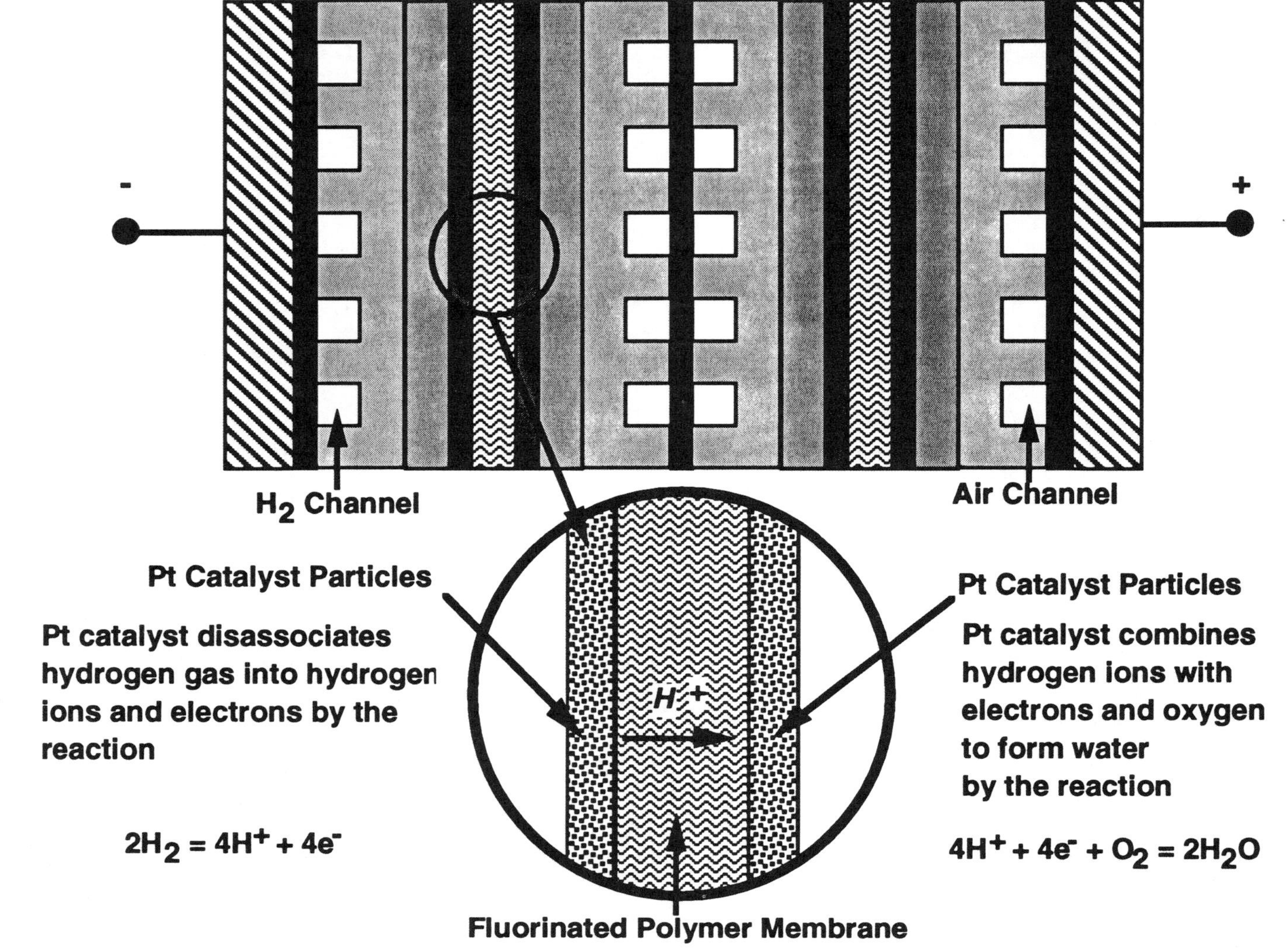

Fig. 2. The Polymer Electrolyte Membrane (PEM) fuel cell relies on a plastic-like fluorinated polymer membrane to conduct hydrogen ions from the anode to the cathode of the fuel cell. The chemical reactions on both sides of the cell are catalyzed by platinum or platinum alloys.

Four of the major areas of work in this program during the past year are described briefly in this paper. These areas are (1) optimizing the platinum/carbon electrode structure, (2) improving the performance on low-pressure air, (3) characterizing conductivity and water transport in polymer electrolyte membranes, and (4) evaluating long-term endurance and materials stability.

The cross section of a PEM fuel cell is shown schematically in Fig. 2. At the heart of the fuel cell is a fluorinated polymer membrane sandwiched between two electrodes. These electrodes are catalyzed with platinum to facilitate the electrochemical reactions.

The optimization of these Pt/C electrodes for PEM fuel cells has been a focus of effort at Los Alamos since 1984. In the original aerospace designs for the PEM fuel cell, platinum black was hot pressed directly into the membrane, as shown at the left of Fig. 3. This design required approximately 2 mg Pt/cm^2 for moderate performance.

Over the past 2 years, we have developed a thin-film electrode technology that uses carbon-supported platinum in a 4- to 5-µm-thick layer applied directly to the membrane as shown at the right of Fig. 3.

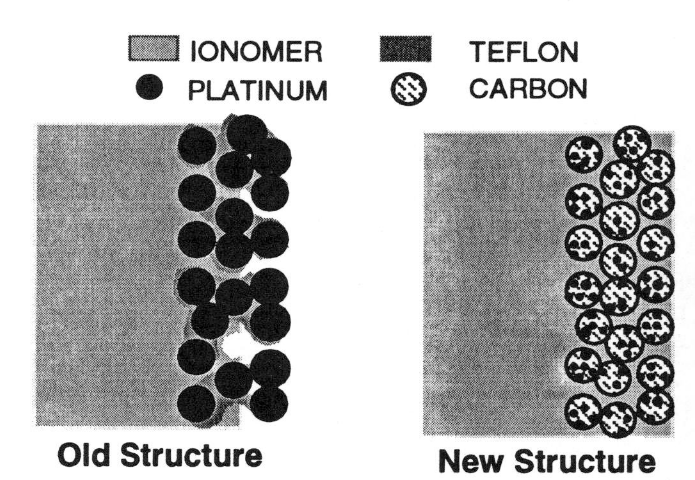

Fig. 3. The original design for the catalyzed electrodes for the PEM fuel cell used platinum black, which is very expensive. By using nanometer sized particles of platinum supported on carbon instead, the amount of platinum required can be reduced by more than a factor of 10.

As shown in Fig. 4, this electrode structure produces power densities near 1 W/cm^2 on H_2 and air, with a platinum loading of only 0.12 mg/cm^2. By comparison, the original platinum-black technology delivered a maximum power of less than 0.3 W/cm^2 on H_2 and air with more than 10 times as much platinum.

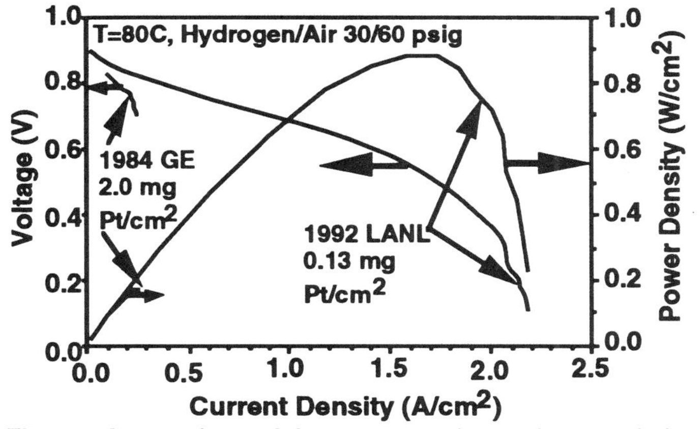

Fig. 4. Comparison of the current-voltage characteristic and the power performance of a Los Alamos, thin-film electrode with that of the early aerospace design that required more than 10-times as much platinum.

As shown in Fig. 5, the cost of platinum using the old technology would have been prohibitive for use in a fuel cell big enough to power a passenger car. This new thin-film technology using supported catalysts greatly reduces the cost. For an 80 kW peak-power passenger car, the thin-film technology would require less than $500 for platinum.

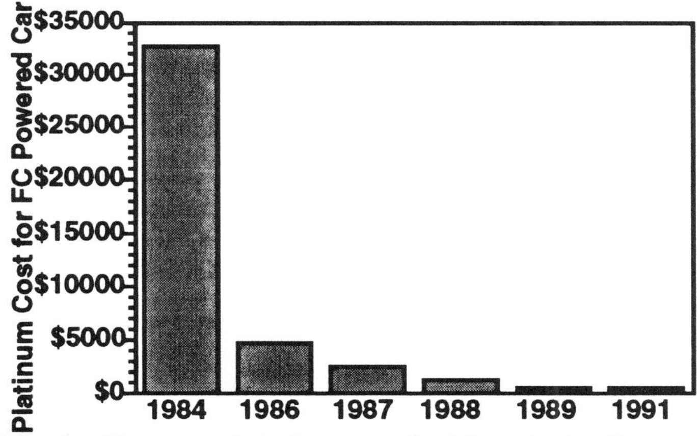

Fig. 5. The cost of platinum required for a fuel cell powered car has been reduced by almost a factor of 100 since 1984.

Although further improvements in platinum utilization would be valuable, the focus now is on improving the transport properties and on the low-cost manufacturability of the electrode structure. In the coming year, work will focus on improving the oxygen diffusivity and proton conductivity of the catalyst layer. Methods to apply thin, catalyst layers of highly dispersed platinum to ionomeric membranes of various types are being evaluated for performance, cost, stability, and reproducibility. The final goal of this effort is to produce membrane/electrode assemblies, with platinum loading smaller than 0.25 g/kW, that can be manufactured at a low cost in large areas and that exhibit stable performance over 2000 to 3000 hours of drive-cycle operation on methanol reformate and air.

In developing thin-film electrodes, we have made significant improvements in their performance on low-pressure air. When applied to PEM fuel cells the gas-diffusion electrode technology, originally developed for the phosphoric acid fuel cell, showed severe degradation in performance below about 2-atm air pressure. In contrast, the thin-film electrodes show a graceful decline in performance as the air pressure is lowered. As shown in Fig. 6. current densities greater than 800 mA/cm^2, at a cell voltage of 0.5 V, have been achieved using ambient pressure hydrogen and air.

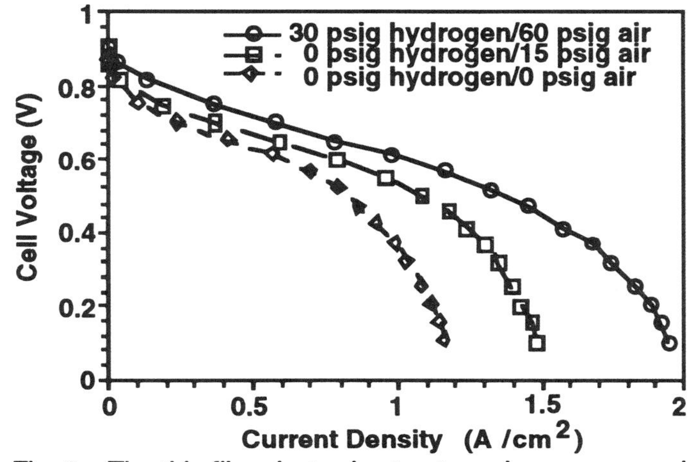

Fig. 6. The thin film electrode structures have very good performance on hydrogen and air. This performance declines gracefully as the air pressure is reduced.

Our future work will focus on the gas-diffusion backing and the flow field that have important effects on low-pressure air performance, and long-term stability, and that scale up to large cells. Low-pressure air performance is degraded by an inert nitrogen blanket within the gas diffusion backing that generates a diffusion barrier for oxygen access. We will attempt to develop alternative, low-profile structures that will minimize the thickness of this barrier to improve the low-pressure performance.

The conductivity of hydrogen ions and the transport of water are closely coupled in polymer electrolyte membranes in this type of fuel cell. It is very important to understand this relationship and how it is affected by the operating conditions of the fuel cell. Accordingly, we have measured water sorption, water diffusion, and water drag coefficients for a number of membranes under experimental conditions that mimic the conditions in an operating fuel cell as closely as possible. Substantial data have been accumulated for Nafion®117[1] , Membrane C[2], and the DOW experimental membrane[3] at both 30°C and 80°C.

---

[1] Nafion® is a product of E. I. Dupont and Nemours Co. U.S.A.

[2] Membrane C is a product of Chlorine Engineers, Japan.

[3] The DOW membrane is a product of Dow Chemical, U.S.A.

All of the transport properties of the membrane are linked to its water content. As shown in Fig. 7, this water content is controlled by the chemical activity of water to which the membrane is exposed. When exposed to gas containing water vapor, the membrane can take up anywhere from 2 to 14 water molecules for each of the ionic groups within the polymer, depending on the water concentration in the gas.

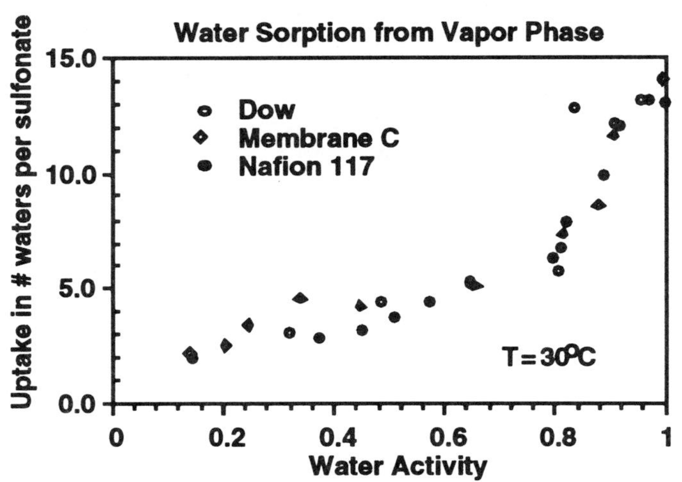

Fig. 7. The water content of polymer electrolyte membranes is strongly dependent on the water concentration in the gas to which the membrane is exposed.

This water content directly affects the transport of water as well as the hydrogen ion conductivity of the membrane. For example, Fig. 8 shows the dependence of the diffusion coefficient of water on the water content. Over the typical operating range of water content, each of the membranes tested has a comparable water diffusion coefficient, in the range of 1×10^{-6} to 6×10^{-6}-cm^2/s.

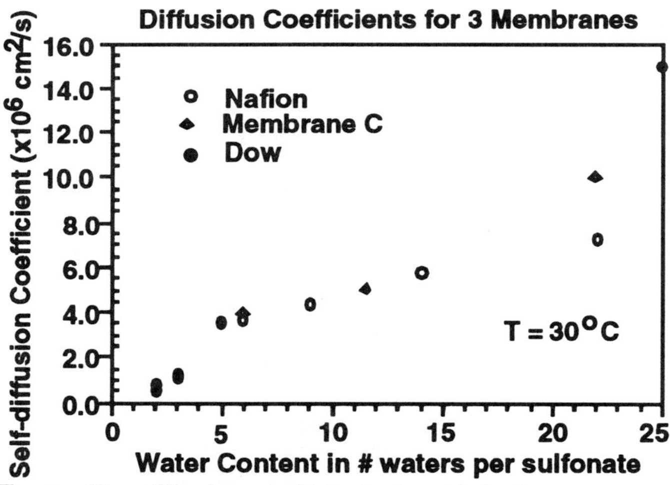

Fig. 8. The diffusion coefficient of water in the membrane is an important parameter because diffusion allows water that has been dragged through the membrane with the hydrogen ions to move back toward the anode, maintaining the ionic conductivity.

Of particular interest is the top, right-most point in Fig. 8. This point is for a DOW membrane exposed to liquid water rather than water vapor. When exposed to liquid water, this membrane can take up to 25 water molecules per sulfonate group. All of these membranes take up substantially more water when in contact with liquid water than they do when in contact with saturated water vapor. With this higher water content, the membrane has a higher water diffusion coefficient, facilitating water that has been dragged through the membrane with the hydrogen ions to move back toward the anode, maintaining the ionic conductivity.

The hydrogen ion conductivity is also a strong function of water content as shown in Fig. 9. Both the DOW and the Membrane C have higher conductivities than Nafion® 117 in this water content range. In contact with liquid water, the conductivity of all of these membranes increases to at least 90°C. In contrast, in contact with saturated water vapor, the membranes tend to lose water as the temperature is increased, resulting in a peak conductivity at approximately 80°C.

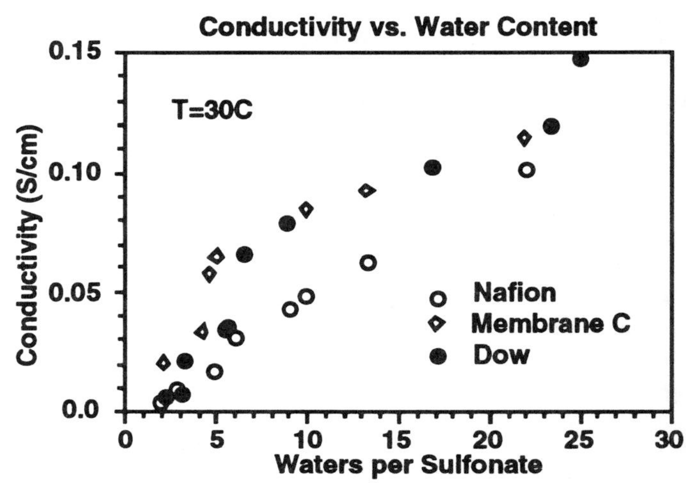

Fig. 9. The hydrogen ion conductivity of the membrane is a strong function of water content. At current densities above 0.5 A/cm^2, the ohmic loss from the ionic resistance is a large fraction of the loss in the fuel cell.

These data have been incorporated into a detailed microscopic model of water transport that can predict net water transport, water profiles, and membrane resistance as a function of operating conditions. During the coming year, we plan to extend these studies to other membranes and to a wider range of operating conditions. The membrane model will be integrated with the electrode and backing models to provide a predictive capability for the full membrane-electrode assembly.

During the past year we performed several single cell tests over 2000 hours on cells with very low platinum loadings. These tests were conducted with hydrogen and air, at a voltage output of 0.5 V. As shown in Figs. 10 and 11, in the longest test (over 4000 hours), using a Nafion® 117 membrane with thin-film electrodes and a platinum loading of 0.13 mg/cm^2, the current density dropped about 20% in the first 100 hours, then gradually decayed an additional 5% over the remaining 3000 hours. Understanding and eliminating this loss will be the central target of our work over the next year.

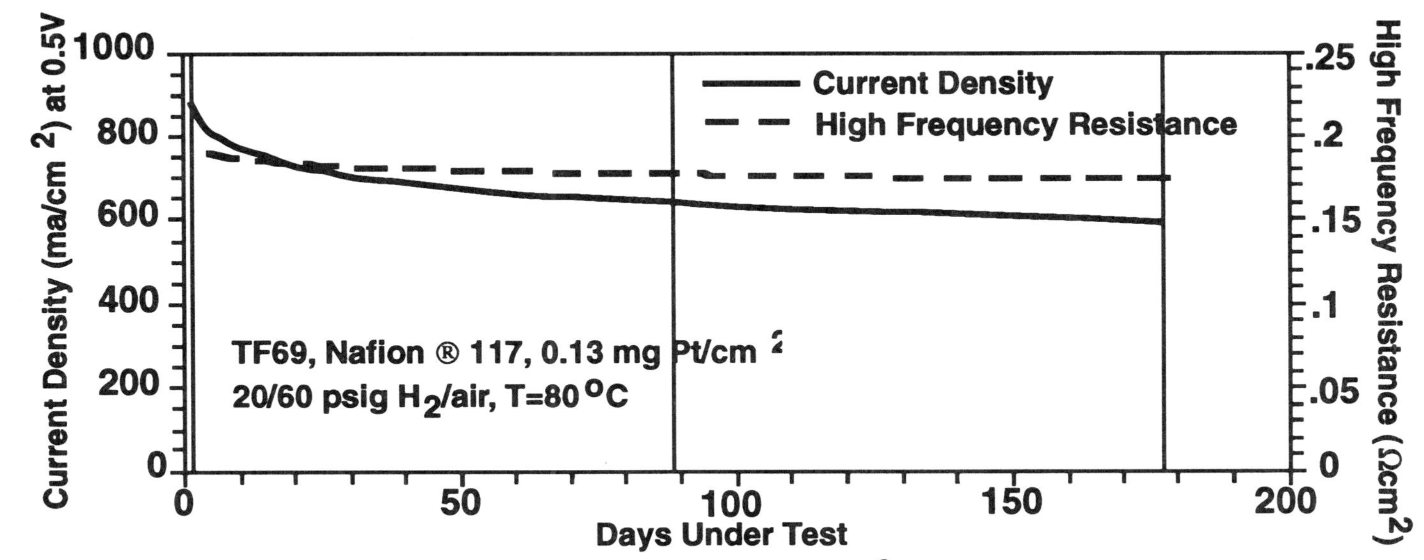

Fig. 10 PEM single fuel cells with catalyst loading of only 0.13mg Pt/cm^2 show only modest loss after continuous operation over 4000 hours. This is comparable to the operating time of an automobile over its useful life.

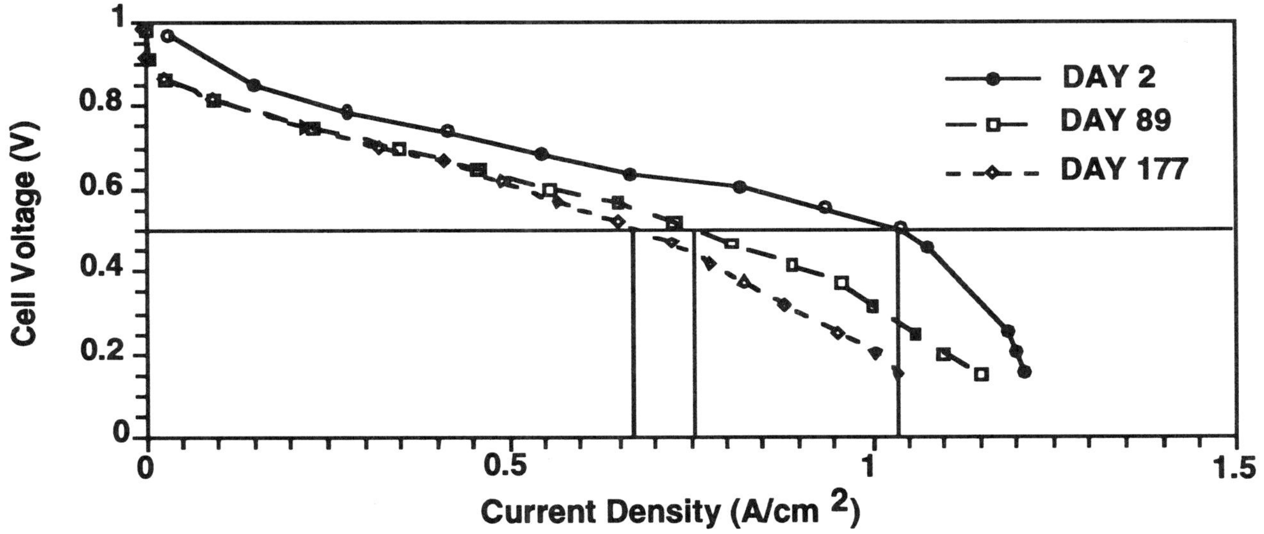

Fig. 11 Comparison of the current-voltage characteristic of the fuel cell at different stages of its life provides insights into the degradation mechanisms.

In summary, the Polymer Electrolyte Fuel Cell - Core Research Program, has made significant progress in characterizing and solving the key problems that obstruct the development of PEM fuel cells for widespread use in transportation. This work is being done in close consultation with U.S. industry to ensure that the results are made available for rapid incorporation into cost-effective fuel cells.

Recent Progress in Proton Exchange Membrane Fuel Cells at Texas A&M University

N. K. Anand,† A. J. Appleby,* H. P. Dhar,‡ A. C. Ferreira, * J. Kim,* S. Mukerjee,* A. Nandi,* A. Parthasarathy,* Y. W. Rho,* S. Somasandaran,† S. Srinivasan,* O. A. Velev,* and M. Wakizoe*

**Center for Electrochemical Systems and Hydrogen Research*
Texas Engineering Experiment Station (TEES)
†Department of Mechanical Engineering
Texas A&M University
‡BCS Technologies, Inc.

Abstract

Since 1988, work has been conducted at Texas A&M University to increase understanding of polarization losses, improve performance at atmospheric pressure with simultaneous reduction in platinum catalyst loading with an increase in its utilization. In six years, platinum loadings have been reduced by a factor of 100, at the same time tripling power density compared with mid-1980s results. A total platinum loading of 0.5 g/kW (cathode loading 0.05 mg/cm^2) now gives a nominal 0.74 V at 0.3 A/cm^2 (0.6 V at 0.8 A/cm^2, 0.48 V at 1.1 A/cm^2) on *non-humidified* air and hydrogen reactants at 1 atm absolute pressure (atma). Pressurized filter-press stacks are at a disadvantage because they require a heavy cooling system and stack components. An air-cooled atmospheric pressure stack can be designed with a nominal specific power of 1 kg/kW, using a 270 g/kW membrane-electrode assembly (MEA). How these performance improvements have been achieved is described.

1. INTRODUCTION

The SPE® System

The solid polymer electrolyte (SPE®) system was developed by the General Electric Company (GE) in the early 1970s as an evolutionary advance on the NASA Gemini project fuel cell of 1965. Its advantage over other acid fuel cells was the lack of a requirement for electrolyte isolation and management. An associated advantage was the fact that pure (indeed potable) water was the only liquid product.

During the late 1960s and 1970s, GE made two technological advances compared with the Gemini system. The first was the incorporation of Teflon® in the electrode structure, particularly in the cathode, to serve as a water-rejecting surface to prevent flooding by liquid product water at high current density. The second was the use of a single layer of perfluorinated polyethylene-propylene-methylethyl ether copolymer with terminal sulfonate groups, given the trade-name Nafion®. Both Teflon® and Nafion® were products of Du Pont de Nemours and Company, Wilmington, DE. Nafion® was discovered in 1962, examined for possible application in diaphragm cells in the chlor-alkali industry in 1962, first exploited in this application in 1964, and was first examined for use in fuel cells by GE at about the same time. The perfluorinated polymer was much more stable and more ionically conductive than the polystyrene-divinylbenzene copolymer which was previously used, and allow a current density about two orders of magnitude higher under comparable conditions with pure platinum black electrodes with platinum loadings of several mg/cm^2.

Penner Report Recommendations

The Department of Energy Advanced Fuel Cell Working Group Report (Penner Committee Report), published in January 1986, showed that much remained to be done to make the "solid polymer electrolyte" aqueous acid fuel cell (as it was then commonly called) into a practical system for widespread use [1]. Two major development problems of the SPE® system which impeded in such applications were identified. The first was the cost of the Nafion® membrane, then about $\$50/m^2$, or about \$750/kW at a then-typical atmospheric pressure hydrogen-air performance of 0.7 kW/m^2 at 0.7 V. The second was the inability of forming a three-dimensional three-phase-boundary interface between the solid Nafion® electrolyte film and an adequate amount of even pure platinum black catalyst. A continuous film of the latter required a minimum of about 3 mg/cm^2, at only a fraction of it was utilized in the three-phase boundary by the method then used for electrode-electrolyte film bonding, namely heating under pressure [1]. Based on atmospheric-pressure hydrogen-air performance in 1985, the cost of this platinum catalyst was greater than \$1000/kW. In consequence, the GE solid polymer electrolyte system (SPE®) was only suitable for use in specialized applications, usually requiring pressurized hydrogen and oxygen. Major emphasis was on its development as a high-current-density electrolyzer for life support systems.

Progress to 1988

In August 1984, work stopped on SPE® systems at GE's Wilmington, MA facility. The system had already been licensed to Siemens in Germany, and the technology for the United States, along with the SPE® trademark, was acquired by the Hamilton Standard Division of United Technologies Corporation. A program to improve the performance and practicality of the technology commenced at Los Alamos National Laboratory (LANL) in the early 1980s [2], the ultimate application being a fuel cell for an electric vehicle. A number of advances have been made there for the development of a system operating on methanol reformate. Pressurized operation on air is envisaged to increase the system power density. In 1986 Ballard Power Systems published data on a new experimental membrane from the Dow Chemical Company with chemistry similar to that of Nafion®, but of lower equivalent weight and with a thickness of only 75 μm dry, compared with 175 μm for Nafion® 117. With 4 mg/cm^2 platinum black electrodes, it gave a performance of 4 A/cm^2 at 0.5 V on pure hydrogen and oxygen at 5 atma pressure, over twice the current density of the best GE data reported under similar conditions [3]. This implied a more efficient interface, as well as a lower resistance, than that of GE cells.

GE SPE® cells were built on the membrane itself. What has become known as the Membrane-Electrode-Assembly (MEA) was built up by applying catalyst to the supporting membrane, together with appropriate conductive backing. This was required because in the GE application, the system was designed to operate under differential pressure of up to several atmospheres, so that the anode and cathode compartments required effective isolation. This approach has continued to be favored in recent work at LANL. For transportation applications this is not necessarily desirable, and an alternative approach was suggested as early as 1982 [4]. This proposed the use of an electrolyte gel with sufficient insolubility to remain in place, but which relied on mechnically-stable electrodes for mechanical integrity. This was labelled the "Nafion-soup" approach [1, 4]. This concept permitted the eventual use of less expensive polymer electrolytes. As a result of this and the considerations addressed in Ref. 1, it was suggested that electrodes might be impregnated with polymer by a wide range of means, including with solubilized electrolyte, with in-situ polymerization, or with polymer incorporated during electrode manufacture, to produce electrodes which could be associated with the same polymer, with other polymers, or with other acid electrolytes to form a fuel cell [5]. A similar, though much more specific approach was later used at LANL [6]. Such approaches to improving the internal electrolyte-catalysts-gas interface have permitted the use of carbon-supported platinum black electrodes with low platinum loading, still maintaining excellent performance.

2. PROGRESS AT TEXAS A&M UNIVERSITY/TEES, 1988-94

Task Summary

The Center for Electrochemical Systems and Hydrogen Research (CESHR) at TEES/Texas A&M University has had a series of proton exchange membrane fuel cell (PEMFC)* projects since early 1988. Major tasks in this work have been (i) to attain high power densities, high energy efficiencies, and long lifetimes in single cells with low platinum loading electrodes; (ii) to determine the most effective means of thermal and water management; and (iii) to develop multicell stacks with acceptable performance for both space and terrestrial applications, the latter with emphasis on road transportation. The tasks undertaken have been:

(i) Microelectrode techniques for the determination of electrode kinetic, mass transport, and ohmic parameters to permit PEMFC performance analysis;
(ii) Means of increasing effective platinum utilization;
(iii) Evaluation of alternative membranes for PEMFCs,
(iv) Optimization of electrode and MEA structures and of PEMFC operating conditions;
(v) Theoretical and experimental analysis of PEMFC thermal management and water management;
(vi) Engineering design, development and testing of multicell PEMFC stacks.

Microelectrode Determinations of Physical and Electrochemical Parameters

A novel micro-electrode technique has been developed which has allowed the determination of dioxygen reduction kinetic parameters at the platinum catalyst/PEM interface, mass transport parameters for dioxygen in the PEM; and PEM conductivity [7-10]. Experiments were conducted over a range of temperatures and pressure in experimental equipment shown schematically in Figure 1 [8-10]. Both pseudo-steady state (<10 mV/s) and transient electrochemical technique, including AC impedance spectroscopy were used in data acquisition. The electrode kinetic parameters for oxygen reduction at the platinum microelectrode-Nafion® interface were found to be similar to those occurring on platinum crystallites in PEMFCs (Table 1). The diffusion coefficient (D) for dioxygen in the membrane is less than that in aqueous electrolytes, but the solubility (C) of the gas is considerably higher in the membrane. Overall, the net result is that the product DC for dioxygen is about 10 times higher in the membrane than a dilute aqueous electrolyte. The specific conductivity of the Nafion® membrane is about the same as to of 0.1 M H_2SO_4. The parameters obtained are currently being used in PEMFC modeling analysis. This task was supported by the Texas Higher Education Coordinating Board Energy Research and Applications program (ERAP).

Increasing Effective PEMFC Platinum Utilization

Physical and electrochemical methods for localizing the platinum catalyst near the front surface of the electrode have been investigated as an alternative to sputter-deposition. The physical method consisted of brushing a thin layer of unsupported platinum black, suspended in Nafion solution on the active surface of a commercially-available electrode (E-TEK, Inc.), containing 0.4 mg Pt/cm^2.

---

* The abbreviated descriptive PEM is used to avoid conflict with the SPE® trademark, which is the property of Hamilton Standard Divison of United Technologies Corporation.

The amount of Pt thus deposited, was the same as previously used for sputter-deposition (0.05 mg/cm^2). In the second method, the same amount of Pt was electrodeposited onto E-TEK electrodes from a chloroplatinic acid bath. The performances of these electrodes were then evaluated in single cells. The electrode with the electro-deposited layer showed almost the same level of performance as the electrode with the sputtered layer [11-13].

A systematic study of the effect of platinum loading on PEMFC performance has been conducted. The results reveal that excellent performance may be obtained with high platinum utilization on electrodes with with only about 0.05 mg/cm^2 of carbon-supported catalyst (20 wt % platinum). Results are shown in Figure 2. These electrodes were prepared at CESHR using a rolling method [11, 12, 14].

In an effort to further improve performance at constant noble metal loading, platinum alloy electrocatalysts are being examined. Results in Figure 3 show enhanced performance with carbon-supported Pt-Cr, Pt-Mn, Pt-Ni, Pt-Fe, and Pt-Co electrocatalysts [15]. During 1993, graduate student S. Mukerjee collaborated with J. McBreen at the National Synchrotron Light Source, Brookhaven National Laboratory, to determine the effects of electronic and geometric factors on electrocatalysis using X-ray absorption spectroscopy, XANES (x-ray absorption near edge structure) and EXAFS (extended x-ray absorption fine structure). The combination of *in situ* XANES and EXAFS allows insight into the mechanism of dioxygen reduction electrocatalysis on Pt-Cr, Pt-Co, Pt-Ni, Pt-Fe and Pt-Mn). The XAS studies were made in both the transmission and fluorescent modes. Measurements were made at the Pt L_2 and L_3 edges and at the K edge of the alloying component. The *in situ* measurements were made at various potentials in the range from 0 to 1.0 V/RHE. The XANES results an emptying of d-states for all of the above electrocatalysts with the exception of Pt-Mn. Comparison of XANES and EXAFS results in two different potential regions (0.54 V and 0.84 V vs RHE) reveal different oxygen chemisorption behavior betwen Pt and the alloys at potentials greater than 0.8 V vs RHE. Correlation of the electrocatalytic activity for dioxygen reduction with d-band vacancies and Pt-Pt bond lengths show volcano-type relationships (c.f., Ref. 16).

Different portions of this task have been supported by Advanced Research Projects Agency (ARPA), ERAP-TEES and NASA.

Evaluation of Alternative PEM Electrolytes

The development of alternative PEM membranes with improved properties for fuel cell use is an important requirement. Desired physico-chemical characteristics include improved ionic conductivity, increased water retention, and higher chemical and electrochemical stability. Ionic conductivity may be improved by reducing the equivalent weight of the polymer, by increasing water retention, i.e., reducing water vapor pressure, and by the simple means of using thinner PEM layers. However, if thinner layers are used, increased cross-over of reactants may occur. The latter is associated with both the presence of pinholes, and with the diffusivity-solubility characteristics of the reactants at a given operating pressure. Hence, high electrochemical activity will be associated with higher cross-over. In effect, thickness and other membrane characteristics will be a particular compromise for any given PEMFC operating conditions.

A detailed evaluation of the Du Pont Nafion® family of membranes, which includes the 105 series (equivalent weight, 1000, thickness 5 mils or 125 μm), the Dow Chemical Company membrane and the Asahi Chemical Industry Company Ltd. Aciplex®-S membrane has been conducted. Results (Figure 4) show that:

(i) Dow and Aciplex®-S membranes have similar performance to ca. 2.0 A/cm^2 on H_2/O_2 and H_2/air;

(ii) At higher current densities, some mass transport limitations occur with Aciplex®-S;

(iii) Mass transport problems occur at lower current densities with Nafion® 117 [17].

The permeation method used for measuring hydrogen diffusion coefficients and solubilities in thin metallic membranes was adapted to determine mass transport parameters for the fuel cell reactants in Nafion® and in Aciplex®-S membranes. Transient and steady state currents were recorded as functions of the time for hydrogen oxidation and oxygen reduction for reactants diffusing through the PEM under differential pressure conditions. Analysis yields the D and C values (Table 2). D are higher for Nafion® than for Aciplex®-S membranes, while C vales are corresponding lower. However, the value of DC is higher for the Aciplex®-S than for Nafion®. In the 1980s, it was hoped that membranes might be developed which showed water-vapor-pressure suppression for operation well above 100°C. However, work at CESHR has shown that sulfonate, and bis perfluorosulfone imide groups behave similarly [18], as do phosphonate groups [19]. The high-temperature performance of phosphoric acid appears to be unique [18].

Portions of this work were supported by ARPA, NASA, and the Asahi Chemical Industry Co., Ltd. The investigators thank Du Pont de Nemours and Company, the Dow Chemical Company and the Asahi Chemical Industry Co. Ltd for furnishing the membranes used.

Optimization of Electrode-MEA Structures and PEMFC Operating Conditions

Accomplishments may be summarized as follows:

(i) Previous work at LANL showed that impregnation of the active layers of electrodes by a solution of Nafion® enhances the area of the three-dimensional reaction zone, substantially increasing supported or unsupported platinum catalyst utilization and allowing higher energy efficiencies and power densities to be attained [20].

(ii) hot-pressing electrodes to the PEM at a temperature slightly above the glass-transition temperature minimizes contact resistances and hence reduces ohmic loss;

(iii) humidification of hydrogen at about 10°C higher than the cell temperature results in higher PEMFC performance;

(iv) localization of platinum near the electrode front surface by using 20 or 40 wt % platinum-on-carbon catalyst, compared with typically 10 wt % in the PAFC [16], gives a thinner active layer. This provides more available catalytic sites per gram of catalyst and hence reduces activation, mass transport and ohmic overpotentials. Deposition of thin layers of platinum on the front surface of these electrodes further enhances performance [21].

(v) operation at elevated temperatures and pressures is essential to attain the highest possible performance in PEMFCs with low Pt loading electrodes, for example, 2.0 A/cm^2 at 0.6 V with pure hydrogen/oxygen and 2.0 A/cm^2 at 0.5 V with hydrogen/air.

Work is in progress to improve structural optimization of MEAs using Aciplex®-S membranes. Hot-pressing at higher temperatures about 20°C above the glass transition temperature further improves PEMFC performance. Another study is evaluating the effect of Teflon® content in the diffusion and active layers on the performances of PEMFCs. Results indicate that an electrode with 40 wt % of Teflon® in the diffusion layer and 30 wt % in the catalyst layer has a better performance than electrodes with lower Teflon® content. This task is supported by NASA, Asahi Chemical Industry Co. Ltd., and Mazda R&D of North America.

Thermal Management, Water Management and Mass Transport

Thermal Management: Thermal management of PEMFCs operating at high power densities is a major challenge [22]. At 2.0 A/cm^2, the electrical power density is 1.2 W/cm^2 while the enthalpy generation rate is 1.8 W/cm^2. The number of grams of coolant required per electrical Wh are given by $860[(V_T/V) - 1]/H$, where V_T is the thermoneutral potential (1.48 V for liquid water, 1.254 V for gaseous water), V is the operating potential, and H is the effective heat capacity of the coolant in calories/g. For convective cooling, H is equal to ΔTC_p, where ΔT is the temperature difference between the PEMFC and the sink and C_p is the specific heat of the coolant at constant pressure. For evaporative cooling, which is the subject of International Fuel Cells patents [23], H is the latent heat of vaporization. The formula shows that it is advantageous to reject gaseous water rather then liquid water from the system.

Process air cooling, separate air cooling, liquid cooling, and evaporative cooling were evaluated. For a PEMFC rejecting gaseous water at V = 0.7 V with ΔT = 20°C, 34 g of water or 97 g of air per Wh (i.e., 51 stoichs) will be required, compared with only 1.8 g of water (2.45 stoichs) if evaporative cooling is used. The air flow rates show that process air cooling is out of the question. Separate air cooling is satisfactory for PEMFCs operating at low current density (<300 mA/cm^2). For transportation applications, both liquid cooling and evaporative cooling require a convective radiator with a comparable cross-section to that for separate air-cooling, since total-loss evaporative cooling is impractical. Evaporative cooling may be carried out under vacuum (i.e., the system becomes a heat-pipe), or via process air or oxygen flow. The latter may be advantageous if process oxidant flow through evaporative cooling plates inserted in a multicell stack for humidification and heat removal, allowing simultaneous thermal and water management. These use two compartments separated by a porous ceramic or a hydrophilic membrane (e.g., Nafion®), and water is passed through one side and process oxidant through the other. Humidity will depend on gas flow rate and is effective for heat removal. The humidified gas then enters the cathode chamber for the electrochemical reaction where evaporation of product water causes further cooling. The method simultaneously humidifies the cathodic reactant and removes process heat.Liquid (water) cooling is used by Ballard Power Systems, Inc., the leading PEMFC developer, with recycle of cooling water circulation to evaporator plates combined with the stack where the incoming air is humidified. Liquid is effective at current densities as high as 2.0 A/cm^2.

Evaporative cooling was analyzed to determine the flow rate of oxygen or air to remove heat generated in the cell as a function of operating temperature at a current density of 1.0 A/cm^2. Different humidification levels were assumed. Figure 5 shows that this method is more favorable for air than for oxygen, because flow rates can be minimized. Higher operating temperatures (i..e., higher partial pressure at a given flow rate) reduce flow requirements. The required air or oxygen flow rate will increase with current density (see Table 3). An experiment was used to verify the results of the modeling study. Figure 6 shows a schematic view of the single cell used in this work. The theoretical analysis was confirmed by experiment up to 300 mA/cm^2. Simultaneous thermal and water management using evaporative cooling is more effective with air than with oxygen for the same reason as that for direct evaporative cooling.

Liquid cooling or evaporative cooling are more efficient than separate air cooling and are essential for compact pressurized stacks operating at high current density. However,.the requirement for radiator equivalent to the cross-sectional area of an air-cooled stack with separate air cooling and no radiator makes the latter system attractive for an optimized atmospheric pressure system.

Water Management: There is confusion in the literature concerning water transport from anode to cathode in the PEMFC. It is argued that protons drag water molecules through the electrolyte. If this is so, then proton conduction in PEM material is different from that in water, where very little or no net water transport occurs during the quantum mechanical jumping of protons between water molecules [24]. Fuel cells were operated over a range of current densities, temperatures and pressures with humidified hydrogen and dry cathodic reactant and water exiting the anode and cathode chambers was condensed. The amount of water exiting the cathode compartment corresponded to that expected for product water according to Faraday's laws, while water exiting the anode corresponded to that in the humidified gas. This experiment thus demonstrated no net transport of water from anode to cathode. The thermal and water management investigations were largely supported by ERAP-THECB, and resulted in two M.S. theses.

Mass Transport Diagnostics using O_2/He-Ar-N_2 Gas Mixtures: Mass transport overpotentials must be minimized when air is the cathodic reactant. A performance comparison between pure oxygen and air showed an increased slope in the pseudo-linear region of the E-i plot, which decreased with increased operating pressure. A departure from E-i plot linearity with a rapid fall in E with increasing current density when air is the cathodic reactant was noted at about 0.6 A/cm^2 at

1 atma and at about 1.2 A/cm^2 at 5 atma. In contrast, the E-i plot was linear up to several A/cm^2 with pure O_2 as the cathodic reactant. These effects were investigated using O_2/He, O_2/Ar and O_2/N_2 gas mixtures [25]. Results show that PEMFC performance with 20% O_2/80% Ar or N_2 is identical to that with air (Figure 7), however 20% O_2/80% He gives a longer pseudo-linear polarization region with a slightly lower slope. At 1 atma, O_2 concentrations over 40% show linear E-i plots with decreased slopes up to at least 1.0 A/cm^2 (Figure. 8). Thus, oxygen enrichment of air exceeding 40% O_2 using gas separation will decrease mass transport overpotential and increase power output.

We conducted a modeling analysis to interpret these results in terms of the effects of structure and composition of the diffusion layer of the electrode substrate and of the active catalyst layer on mass transport. A one-dimensional analysis showed that while mass transport problems in the film surrounding the electrocatalyst particles are minimal, the increase in linear region slope in the E-i plot results from decreased dioxygen mass transport in the catalyst layer. PEMFC performance may be improved by increasing the thickness of catalyst layer, giving a higher electrochemically active surface area, but a limit is imposed by mass transport and ohmic resistance. Operation at increased temperature will compensate performance loss at lower dioxygen partial pressures at 1 atma, but this becomes less at higher total pressures (e.g. 5 atma). Finally, operation at higher pressures significantly reduces mass transport overpotential because of increased dioxygen solubility. This work was supported by NASA-Lewis Research Center.

Another mass transport analysis determined PEMFC performance at different dioxygen or air flow rates. Results show that even at high current densities, there is little performance dependence on pure oxygen flow rate, but with air, there is a significant flow rate dependence at current densities over 300 mA/cm^2. If air is humidified to 70°C, the dioxygen partial pressure at 1 atma is only about 0.12 atm. At 50% oxygen utilization, this is reduced by a factor of 2 and product water will condense. To satisfy dioxygen requirements under these conditions at higher current densities, extremely high flow rates of air are necessary. Operation of stacks in this manner is impractical.

Operating Conditions in Practical Stacks: Since no water transport takes place across the stack, it should be possible to use product water for humidification. Using a technique developed at BCS Technologies, Inc [26]. In an extension of the ideas in Refs. 4 and 5, thin films of PEM (about 60 μm at a total weight of 6 mg/cm^2) may be deposited on electrodes to form a MEA [26]. The MEA can be operated dead-headed or a at low flow-rate on dry hydrogen, and on non-humidified oxidant at utilizations to give product water at partial pressures corresponding to those over a humid PEM. For example, 50% oxygen utilization in dry air at ambient pressure corresponds to operation at approximately 65°C. Some results are shown in Figure 9. The advantages of this mode of operation are obvious, particularly for transportation applications.

Engineering Design, Development and Testing of Multicell PEMFC Stacks

This task follows logically from studies to improve single-cell performance and to analyze thermal and water management. Scale-up from 5 cm^2 to 50 cm^2 resulted in no change in performance or lifetime. Scale-up to 150 cm^2 was accomplished in a 1 kW multicell stack in collaboration with Humboldt State University, California. Two 10 kW, 500 cm^2 stacks have been designed for Mechanical Technologies, Inc., Latham, NY (partly sponsored by the NASA Center for Space Power at TEES), and for the Korea Gas Corporation. Construction of these stacks is in progress. Lightweight stacks weighing about 1 kg/kW (bare stack weight) are planned. These will use 0.05-0.1 mg/cm^2 cathodes, and will operate at a nominal 0.7 V, 450 mA/cm^2 at ambient pressure. Operation under these conditions reduces stack weight compared with that of pressurized stacks, which require heavy filter-press components or a pressure vessel, together with liquid cooling. They also have an efficiency penalty of about 0.12 V/cell (at 1 A/cm^2, 0.6 V for compression work. We believe that the total system mass and volume of a hydrogen fuel cell will be reduced by operation at ambient pressure, which may also result in higher mean efficiency.

We should note that present production of the leading PEM material in the form of complex multilayer membranes for the chlor-alkali industry is in the tens of thousands of m^2 per year, i.e., the equivalent of 100,000 nominal kW, or 5,000 compact vehicles per year. Using the chemical industry's rule-of-thumb that a hundred-fold increase in production corresponds to a ten-fold decrease in cost, thinner layers of the electrolyte material may cost about 5% of today's $250/kW ($800/m^2, <$7,000/kg) based on the above nominal performance, i.e., $12.50/kW at a production rate of about one million compact cars per year, requiring under 10 metric tons of platinum, less than 10% of present production. This corresponds to $110 per vehicle.

REFERENCES

1. A. J. Appleby and E. B. Yeager, *Energy* **11**, 137 (1986).
2. R. J. Lawrence, Interim Report, LANL-29, *New Membrane-Catalyst Concept for Solid Polymer Electrolyte Systems,* No. 9-X53-D6272-1, University of California, Los Alamos National Laboratory, Los Alamos, NM (1984).
3. A. J. Appleby and F. R. Foulkes, *Fuel Cell Handbook,* Van Nostrand Reinhold, New York, 1987, p. 295.
4. A. J. Appleby, Proc. Workshop on Renewable Fuels and Alternative Power Sources for Transportation, Boulder, CO, June 1982, Solar Energy Research Institute, Golden, CO, p. 55; H. L. Chum and S. Srinivasan, ibid. Executive Summary, p. v.
5. A. J. Appleby, U.S. Patent No. 4,610,938 (September 1986).
6. I. D. Raistrick , U. S. Patent No. 4,876,115 (October 1989).
7. A. Parthasarathy, C. R. Martin, and S. Srinivasan, *J. Electrochem. Soc.* **138**, 916 (1991).
8. A. Parthsarathy, S. Srinivasan, A. J. Appleby, and C. R. Martin, *J. Electroanal.Chem.* **339**, 101 (1992).
9. A. Parthasarathy, S. Srinivasan, A. J. Appleby, and C. R. Martin, *J. Electrochem. Soc.* **139**, 2530 (1992).
10. A. Parthasarathy, S. Srinivasan, A. J. Appleby, and C. R. Martin, *J . Electrochem. Soc.***139**, 2856 (1992).
11. A. C. Ferreira, S. Srinivasan, and A. J. Appleby, *Ext. Abs., 181st Meeting Electrochem. Soc., St Louis, MO, May 1992,* The Electrochemical Society, Pennington, NJ (1992), p. 11.
12. A. C. Ferreira, S. Srinivasan, and A. J. Appleby, *Ext. Abs., 182nd Meeting Electrochem. Soc., Toronto, Ont, October 1992,* The Electrochemical Society, Pennington, NJ (1992), p. 163.
13. S. Srinivasan, A. C. Ferriera, A. Parthasarathy, O. A. Velev, and A. J. Appleby, *Methods to Attain High Energy Efficiencies and Long Lifetimes in Polymer Acid Fuel Cells,* Annual Report, Grant No. NAG 3-1125, NASA-Lewis Research Center, Cleveland OH, June 1992.
14. A. C. Ferreira and S. Srinivasan, *Ext. Abs., 185th Meeting Electrochem. Soc., San Francisco, CA, May 1994,* The Electrochemical Society, Pennington, NJ (1994), in press.
15. S. Mukerjee and S. Srinivasan, *J Electroanal. Chem.* **357**, 201 (1993).
16. A. J. Appleby, *Energy* **11**, 13 (1986).
17. M. Wakizoe, O. A. Velev, S. Srinivasan, and A. J. Appleby, *Ext. Abs., 182nd Meeting Electrochem. Soc., Toronto, Ont, October 1992,* The Electrochemical Society, Pennington, NJ (1992), p. 156.
18. A. J. Appleby, O. A. Velev, J-G. LeHelloco, A. Parthasarthy, S. Srinvasan, D. D. DesMarteau, M. S. Gillette, and J. K. Ghosh, *J. Electrochem. Soc.* **140**, 109-111 (1993).
19. D. A. Scarpiello, D. J. Burton, R. Guneratne, K. J. Espinosa, and D. L. Maricle, *Ext. Abs. 1990 Fuel Cell Seminar, Phoenix, AZ,* National Fuel Cell Coordinating Group, Washington, D.C., 1990, p. 480.
20. E. A. Ticianelli, J. G. Beery, and S. Srinivasan, *J.Appl.Electrochem.* **21**, 597 (1991).

21. E. A. Ticianelli, C. A. Derouin, and S. Srinivasan, *J. Electroanal. Chem.* **251**, 275 (1988).
22. H. Koch, A. Nandi, N. K. Anand, O. A. Velev, D. H. Swan, S. Srinivasan, and A. J. Appleby, *Ext. Abs., 179th Meeting Electrochem. Soc., Seattle, WA, October 1990,* The Electrochemical Society, Pennington, NJ (1992), p. 176.
23. R. A. Sanderson, U.S. Patent No. 3,498,844 (March 3, 1970); J. K. Stedman, U.S. Patent No. 3,761,316 (September 25, 1973); J. F. McElroy, U. S. Patent No. 4,795,683 (January 3, 1989).
24. B. E. Conway, J. O'M. Conway, and H. Linton, *J. Chem. Phys.* **24**, 834 (1956).
25. Y. W. Rho, O. A. Velev, S. Srinivasan, and A. J. Appleby, *Ext. Abs., 182nd Meeting Electrochem. Soc., Toronto, Ont, October 1992,* The Electrochemical Society, Pennington, NJ (1992), p. 732.
26. H. P. Dhar, United States Patent No. 5,242,764 (September 1993).

Table 1. Oxygen Reduction Kinetic Parameters at the Fuel Cell Electrode-PEM Interface

	Tafel slope, –b (mV/decade)	Exchange Current Density, j_o ($10^7 A/cm^2$)
Unsupported Pt black (10 mg/cm^2 on carbon cloth) (Electrochem. Inc.)/125µm Dow, 50°C, 1 atma	58	3.09
20 wt % Pt/C, Nafion®-impregnated, 0.4 mg/cm^2 (ETEK, Inc.)/125µm, 50°C, 1 atma	64	1.09
100µm Pt microelectrode/175µm Nafion®, 25°C, 1 atma Pseudo-steady-state voltammetry	63 118	0.0205 7.8
100µm Pt microelectrode/175µm Nafion®, 25°C, 1 atma Electrochemical impedance spectroscopy	65 133	0.409 5.9

Table 2. DC and ΔE Values for the Aciplex®-S and Nafion®-117 Proton Exchange Membranes

	Aciplex® -S				Nafion® - 117			
	D x 10^6 cm^2/S	C_o mM	DC_o x 10^6 mM-cm^2/S	$\Delta E^{\#}$	D x 10^6 cm^2/S	C_o mM	DC_o x 10^6 mM-cm^2/S	$\Delta E^{\#}$
H_2, 70°C	1.33	78.3	109.14	4.68	3.91	12.13	47.43	5.72
O_2, 70°C	3.11	6.68	20.77	3.41	8.54	1.81	15.46	3.39

Table 3. Mass Flow Rate of Ambient Air Required for Evaporative Cooling

Current Density (mA/cm^2)	Cell Potential (V)	Heat (W)	Air Stoichiometric Mass Required
200	0.702	3.488	1.42
400	0.604	8.384	2.34
600	0.513	14.527	2.71
800	0.426	21.881	3.00
1000	0.340	30.430	3.40

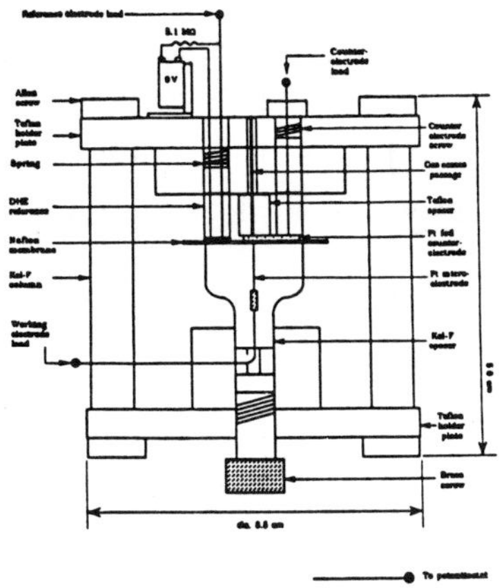

Fig. 1. Schematic of a micro-electrode set up for evaluation of mass transport and electrode kinetic parameters in a PEM environment

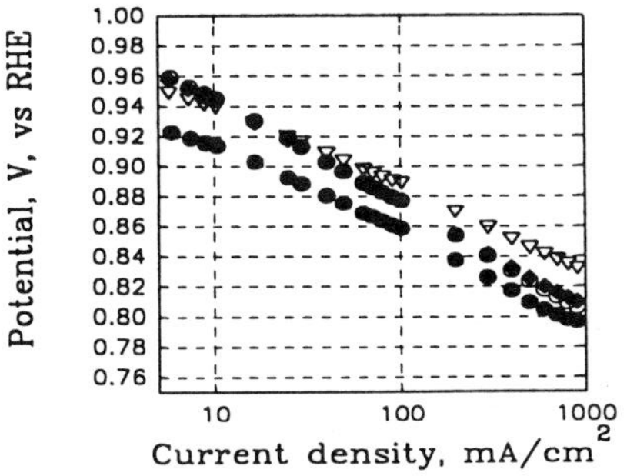

Fig. 3. *iR* corrected Tafel plots for oxygen reduction in proton exchange membrane fuel cells. Pt/C (●), Pt/Cr (∇), Pt/Co(◆) and Pt/Ni (O)

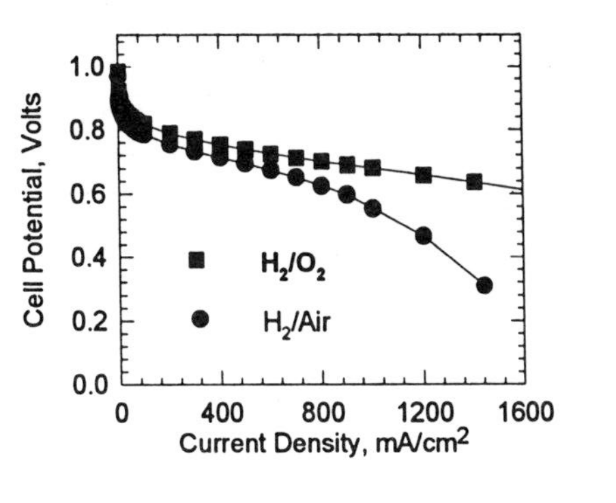

Fig.2.Cell Potential vs current density plots for single cell with 0.05 mgPt/cm^2 electrode, at 70°C and atmospheric pressure. Dow membrane. H_2/O_2, H_2/Air

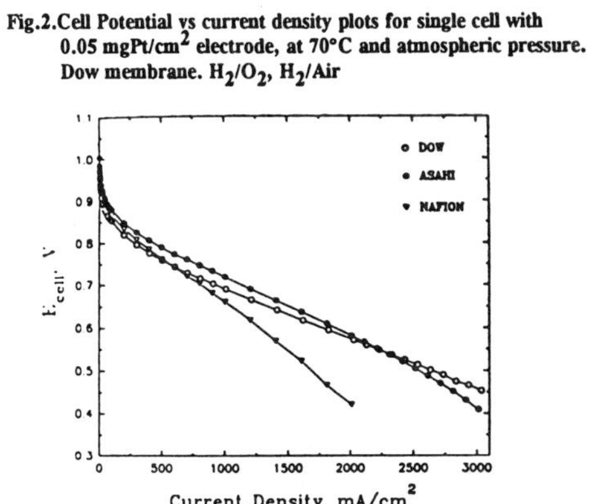

Fig. 4. Comparison of Dow Nafion and Aciplex-S membrane in a PEM fuel cell environment

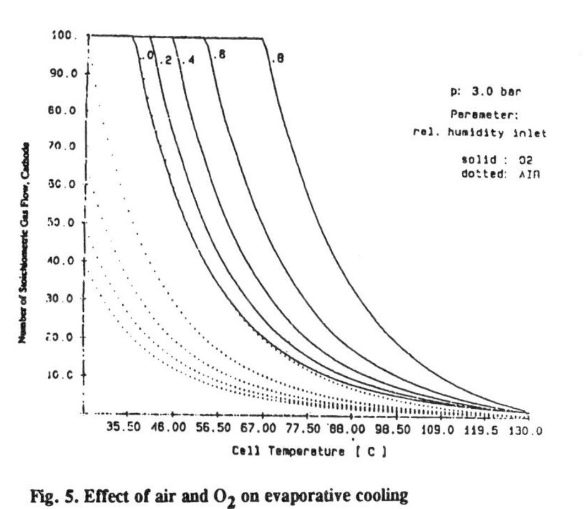

Fig. 5. Effect of air and O_2 on evaporative cooling

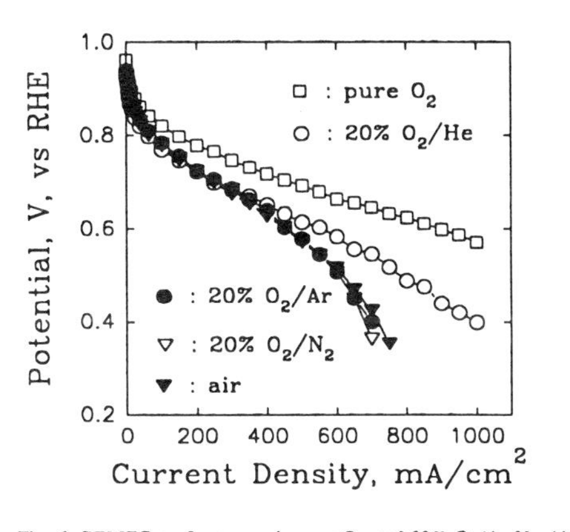

Fig. 6. PEMFC performance in pure O_2 and 20% O_2/Ar-N_2-Air-He

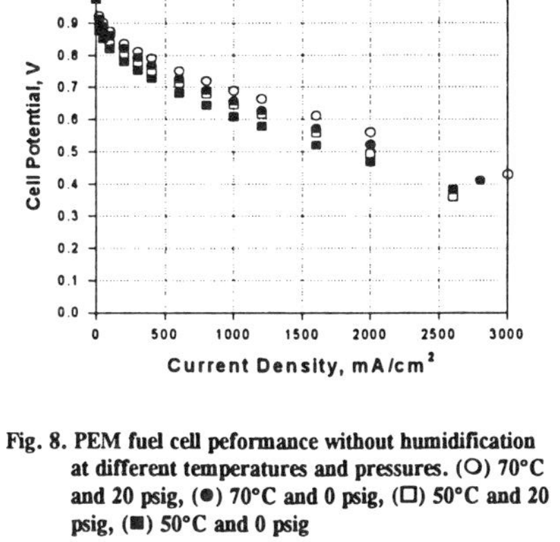

Fig. 7. Effect of percent O_2 in N_2 on PEM fuel cell performance

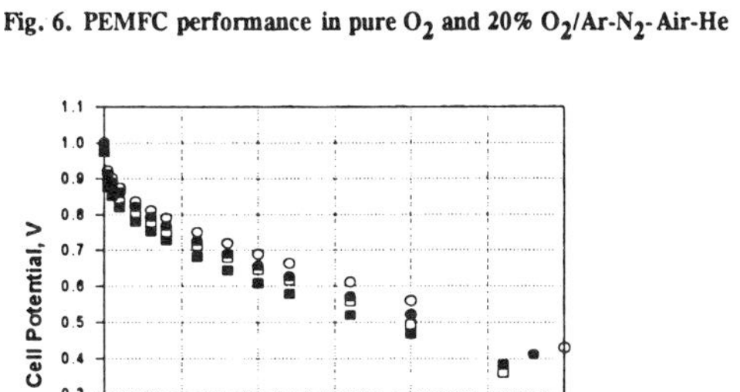

Fig. 8. PEM fuel cell peformance without humidification at different temperatures and pressures. (○) 70°C and 20 psig, (●) 70°C and 0 psig, (□) 50°C and 20 psig, (■) 50°C and 0 psig

Chapter 4

Hydrogen Production

4.1 Introduction

Hydrogen is currently produced for a number of industrial applications. It is an important component of the processes used to produce methanol and ammonia and the process used to refine petroleum. It is also used in various other manufacturing processes such as metallurgical processing and is the fuel of choice for rocket engines. Still, the total production of manufactured hydrogen represents only about 72×10^{12} BTUs of energy in comparison with the 18×10^{15} BTUs of natural gas and 18×10^{15} BTUs of petroleum that are used in the U.S. each year [1]. Industrial uses of hydrogen are given in Table 4.1.

While hydrogen is abundant on earth, it is only found in small quantities as a gas. Thus, the only way to meet society's need for hydrogen is to produce it by processing hydrogen rich materials. Hydrogen can be produced via a number of methods and from various different materials. Hydrogen production from fossil fuels is currently the most economical method of producing hydrogen and will likely remain so in the near future. Fossil fuels are limited in quantity, however, and pollutants are emitted when they are used to produce hydrogen.

As environmental issues become more important in society, the push to develop an economically feasible method of producing hydrogen from renewable sources continues. Water electrolysis is the ideal method for producing hydrogen, provided that electricity can be supplied from clean, renewable sources. Unfortunately, from an economic standpoint, hydrogen production from renewable energy sources is still not feasible on a large scale. In this chapter, hydrogen production from fossil fuels and alternatives to fossil fuel production are reviewed. Alternative hydrogen production techniques to be discussed include water electrolysis, thermochemical water decomposition, photoconversions, and production from biomass. The focus of this chapter will be on the technical details of the production processes themselves. The economic aspects of hydrogen production will be discussed separately in Chapter 6.

4.2 Hydrogen Production from Fossil Fuels

Introduction

Hydrogen can be produced from a number of different fossil fuel sources including natural gas, oil, and coal. Each fossil fuel requires specific techniques to release the hydrogen from the hydrocarbon and water molecules, with some techniques being more economical than others. Hydrogen production from fossil fuels is discussed in this subsection and in the reprint by Steinberg and Cheng.

Table 4.1 - Hydrogen Manufactured in the U.S., 1977

Uses	BTU ($\times 10^{12}$)	Percentage
Oil Refining	36	50
Ammonia production	27	37
Methanol production	6	9
Chemical manufacturing, rocket fuel, metallurgical processing, electronics, etc.	3	4
Total	72	100

(Source: ref. [1])

Steam Reformation

Steam reformation is, at present, the most efficient, economical, and widely used technique for the production of hydrogen. The basic process involves the conversion of a hydrocarbon and steam to hydrogen and carbon oxides. The modern process consists of four steps:

i. feedstock purification (principally to remove sulfur)
ii. steam reforming of hydrocarbons to form hydrogen and carbon oxides
iii. shift conversion of carbon monoxide to carbon dioxide
iv. purification (removal of CO_2, CO, and hydrocarbons)

The critical steps of the process are steps (ii) and (iii) which are described by the following chemical reactions:

ii. $C_nH_m + nH_2O \Rightarrow nCO + (n + m/2)H_2$
iii. $CO + H_2O \Leftrightarrow CO_2 + H_2$

Where n = 1 and m = 4 if the feedstock is methane, and n = 1 and m = 2.2 if it is naphtha. Only light hydrocarbons that are capable of being completely vaporized without leaving a carbon residue are used, since such residues can foul the catalysts used in the steam reforming process. Hydrocarbon feedstocks used in fossil fuel production include methane, naphtha, and No. 2 fuel oil [2].

Natural gas, primarily composed of methane, is the preferred feedstock for hydrogen production. Natural gas is easy to handle, has the highest hydrogen-to-carbon ratio, at 4:1 (CH_4), generally has low levels of sulfur contamination, and is presently inexpensive. Natural gas is also more reactive with water than other hydrocarbons (such as residual oil), and thus lower temperatures can be used when hydrogen is produced from natural gas.

The steam reformation reaction (ii) is highly endothermic, requiring temperatures ranging from 760-925° C, and pressures of approximately 2 MPa. To improve the efficiency of this reaction, catalysts are used. The catalyst is usually composed of NiO mixed with Al_2O_3, CaO, MgO, or other oxides. Catalysts are prepared through conventional techniques, either by precipitation-cement bonding, coprecipitation-sintering, or preformed ceramic carrier-impregnation methods. The desired properties of the catalysts are: high reducibility, high activity, high stability, resistance to carbon deposition, and good mechanical strength. In the natural gas reforming reaction, the catalyst has to be active and stable. For naphtha steam reforming, the catalyst must possess high resistance to carbon deposition as well as high activity and stability [3].

Since the steam reformation reaction is reversible, i.e., hydrogen can convert back to methane, an appropriate steam-to-carbon ratio and high temperatures are required to favor the production of reformed gas (CO and H_2). To ensure a minimum concentration of CH_4 in the reformed gas, the process typically employs a steam-to-carbon ratio of 3:5 at a process temperature of approximately 815° C and pressures up to 3.5 MPa. This allows for the conversion of more than 80% of the hydrocarbons to oxides of carbon at the outlet of the reformer [4]. The steam reformation process is energy intensive and accounts for a significant part of the total energy usage [5].

After the reformer process, the gas mixture is fed through a heat recovery step and reacted with excess steam as in (iii). This reaction is exothermic and is carried out at temperatures ranging from 200-400° C. This step reduces the CO content to levels of approximately 0.2-0.4% by volume.

Purification

The cold gas leaving the heat recovery step contains H_2, CO_2, any remaining CO, and other impurities that must be removed to obtain the desired purity. There are several processes which can be used to remove CO_2, including wet scrubbing processes, which use chemicals such as monoethanolamine, sulfinol, and hot potassium carbonate. The remaining CO and CO_2 can be removed by a methanation reaction, resulting in a final H_2 purity of approximately 97 to 98%.

Pressure swing adsorption (PSA) is another method used to remove impurities from the steam reformer off-gas. PSA is used to produce ultrapure (99.99+ mol%) hydrogen from crude hydrogen sources containing from 40-95 mol% hydrogen [6] by passing the raw gas through a series of activated carbon beds. Impurities such as CO_2, CO, H_2O, N_2, CH_4, C_2H_4, and other hydrocarbon impurities are selectively adsorbed into the beds leaving only the hydrogen product gas whose recovery ranges from 65-90%. The by-product gas is then used as an additional feed source for supplying heat to the reformer. These beds are regenerated by adiabatic depressurization at ambient temperature.

Partial Oxidation

A process which is similar to steam reformation is the production of hydrogen through partial oxidation (POX). In the partial oxidation process, steam, oxygen, and hydrocarbons are converted to hydrogen and oxides of carbon. POX does require additional facilities, however, such as an air separation plant (which separates oxygen from the nitrogen in the air) to provide oxygen, as well as larger shift and separation equipment for the conversion and removal of CO_2 [4].

The basic POX process involves five steps, four of which are similar to the steps in the steam reformation process. The only additional step occurs between steps (i) and (ii) of the steam reformation process. In the extra step, hydrocarbons are oxidized to provide sufficient energy to drive the process and release an additional amount to hydrogen. Pure oxygen must be used in this step, as nitrogen in the air will contaminate the hydrogen end product. The oxidation reaction proceeds as follows:

$$C_nH_m + n/2O_2 \Rightarrow nCO + m/2H_2$$

Because higher operating temperatures are used in noncatalytic POX (1150-1315° C) as opposed to steam reformation of natural gas (815° C), the equilibrium conditions for POX occur at much higher pressures. The synthesized gas thus exits at a high pressure (6.0 MPa or 880 psig), reducing or possibly eliminating the need for compression. The shift reaction and gas purification processes of POX are similar to those in steam reforming.

Feedstocks for POX range from methane to coal. Catalytic processes, which occur at temperatures of about 590° C, use feedstocks ranging from methane to naphtha. Noncatalytic POX, which occurs at temperatures of about 1150-1315° C, uses feedstocks such as methane, heavy oil, and coal. At present, oxidation coupling of methane is not economically feasible on an industrial scale [7]. Thus, heavy hydrocarbons, which do not react well with the catalysts necessary in steam reformation, are currently used as a feedstock for POX.

The drawbacks of using heavy hydrocarbons are lower hydrogen-to-carbon ratios, increased difficulties in transportation compared to methane, and increased impurities in the fuel. Feedstock prices of heavy hydrocarbons are lower than those of lighter hydrocarbons, however, and their availability will extend further into the future. Although the POX reactor itself costs less than the steam reformer, the overall costs of POX are greater than those of steam reformation. This can be attributed to the cost of the oxygen plant and the extra costs of removing sulfur and other impurities from the more contaminated feedstocks (heavy oil and coal) of POX.

Coal Gasification

The chemistry involved in the POX of heavy oils is similar to that of the POX of coal, which is commonly referred to as 'coal gasification.' More sophisticated and expensive plants are necessary to process coal since it is found in many different compositions. Coal used in coal gasification can be divided up into three broad categories: anthracite, bituminous, and lignite. Since the physical properties, amounts of sulfur contamination, and both quantity and composition of ash produced vary greatly in coal, it is difficult to use coal in a POX process. Coal, therefore, is the most expensive hydrocarbon to use for hydrogen production.

There are two types of coal gasification processes, those which are carried out at atmospheric pressure, such as the Koppers-Totzek process, and those which are carried out at higher pressures, such as the Texaco process. The Koppers-Totzek process operates at near ambient pressures, avoiding the difficulties of processing coal at higher pressures. After the raw gas is cooled in the waste recovery step and mixed with water to remove ash, however, it must be compressed. Hydrogen compression requires a great deal of energy, and therefore reduces the efficiency of the process. The Texaco process operates at a pressure near 5.5 MPa, avoiding the necessity for compression, and resulting in a more cost-effective process.

Steam-Iron Process

Before inexpensive natural gas made steam reformation a more attractive option, the steam-iron process was commonly used to produce industrial hydrogen. In the steam-iron process, steam is reacted with hot iron to produce hydrogen gas at process temperatures ranging from 815-870° C. The first step of the process involves coal gasification using steam and regular air. This reaction produces a gas mixture containing mostly CO and H_2. This resulting gas mixture is then introduced into a second chamber where it is used to produce FeO through the following reactions:

$$Fe_3O_4 + H_2 \Rightarrow 3FeO + H_2O$$
$$Fe_3O_4 + CO \Rightarrow 3FeO + CO_2$$

$$FeO + H_2 \Rightarrow Fe + H_2O$$
$$FeO + CO \Rightarrow Fe + CO_2$$

The excess CO and H_2 not used to reduce the iron oxide can be used to produce electrical energy, as this gas represents about 54% of the input coal heating value [4]. By producing electrical power from the gas that is not used, the overall hydrogen production cost can substantially be reduced.

The third step involves the transfer of the hot iron, which has just been reduced, to a steam-iron reactor, where the iron is oxidized in the presence of steam to produce Fe_3O_4 and a hydrogen-rich gas. This reaction requires temperatures of about 815-870° C and proceeds accordingly:

$$Fe + H_2O \Rightarrow FeO + H_2$$
$$3FeO + H_2O \Rightarrow Fe_3O_4 + H_2$$

The gas mixture is then purified by condensing out the steam and running the remaining gas through a methanation reaction, resulting in approximately 97% pure hydrogen gas. The Fe_3O_4 is then recycled back to the second step.

Even though a large amount of electrical power is produced, in addition to the hydrogen in the steam-iron process, production costs are only slightly less than those of the Texaco process. This is due to high plant costs and the fact that extra resources are needed to handle the extra iron oxide.

4.3 Water Electrolysis

Introduction

Electrolysis is a well-developed technology whose commercial use dates back to the 1890s [8]. Pure water and electrical energy are the only inputs necessary to produce hydrogen electrolytically. Under ideal circumstances nonpolluting renewable resources, such as solar energy or wind and hydro-power, would be used to generate the electric energy needed (the use of renewable sources to produce electricity is discussed in greater detail in Chapter 6). Unfortunately, such techniques are currently not economically feasible for large-scale hydrogen production. The use of electrical energy generated from fossil fuels is also expensive, leading to costs three to five times greater than hydrogen production directly from fossil fuels, and the generation of electrical energy from fossil fuels releases pollutants into the air.

Basic Concepts and Components

In simple terms, electrolysis is a process by which electricity is used to decompose water into its components—gaseous hydrogen and oxygen. The electrolysis process (see Fig. 4.1) uses two electrodes, one positive and one negative, which are immersed in a current-conducting electrolyte that may consist of a mixture of water with a salt, acid, or base. The most common electrolytes used are alkaline electrolytes, such as potassium hydroxide (KOH). These electrolytes are excellent conductors of electricity and present the fewest stability or corrosion-related problems. Salts are not used as they are unstable in electrochemical processes and acids are rarely used because they tend to be highly corrosive.

Using KOH as an example, the basic reaction at the negatively charged cathode proceeds as follows:

(1) $K^+ + e^- \Rightarrow K$
(2) $K + H_2O \Rightarrow K^+ + H + OH^-$
(3) $H + H \Rightarrow H_2$

The reaction is initiated when a positively charged potassium ion is reduced (1) and subsequently reacts with water to form a hydrogen atom and a hydroxyl ion (2). The highly reactive hydrogen atom then bonds to the metal of the cathode. This hydrogen atom can combine with another bound hydrogen atom to form a hydrogen molecule, which leaves the electrode as a gas (3).

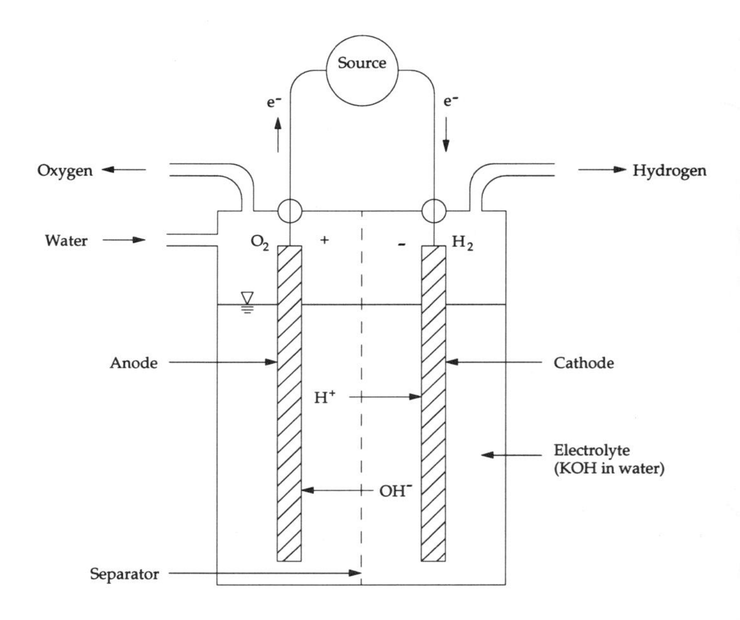

Figure 4.1. Typical electrolysis cell. (Source: ref. [9])

At the positive anode, a reaction similar to that at the cathode occurs, producing oxygen:

(1) $OH^- \Rightarrow OH + e^-$
(2) $2OH \Rightarrow H_2O + O$
(3) $O + O \Rightarrow O_2$

where oxygen is released in a manner similar to hydrogen.

In order to increase the rate of production of hydrogen gas, catalysts are applied to the surfaces of electrodes. Catalytic coatings have properties which allow for quick recombination of hydrogen at the electrode surfaces. Without a catalyst present, atomic hydrogen would build up on the electrode, reducing the current flow and slowing the production of hydrogen gas.

The effectiveness of catalytic materials is improved by increasing the ratio between the real and the apparent surface area of an electrode or by combining different catalytic components [10]. The surface area of the electrode can be increased significantly during manufacture by applying a special coating to the surface before it is immersed in the caustic solution. This coating typically contains two different substances. One of the substances is chosen for its ability to leach out into the caustic while the other substance, the catalyst, remains. As part of the coating leaches out into the caustic, the surface of the electrode becomes pitted, increasing its surface area. This results in a catalyst with a higher production rate and good chemical resistance to the caustic electrolyte.

Recombination rates and stability are important considerations when choosing materials for the cathode and anode. Nickel, coated with small quantities of platinum metals, is the most commonly used material for cathodes. These nickel-based cathodes have good hydrogen recombination rates, similar to those of pure platinum materials, but are much less expensive. For the anode, nickel and copper both show good activity rates for oxygen recombination. When coated with oxides of metals such as manganese, tungsten, and ruthenium, these materials perform well as anodes [9].

To prevent the intermixing of hydrogen and oxygen, a diaphragm, or gas separator, is used. It is designed to allow the free passage of ions while acting as a barrier to prevent the mixing of hydrogen and oxygen gas. The diaphragm is commonly asbestos-based, and is, therefore, only effective at relatively low temperatures (since asbestos diaphragms tend to break apart in a caustic environment at temperatures greater than 80° C).

Basic Design

At present, there are two primary types of industrial electrolyzers: the tank electrolyzer and the filter-press electrolyzer (Fig. 4.2). The tank type electrolyzer consists of alternating unipolar electrodes hung in cells containing electrolyte. These cells are separated by diaphragms to prevent the mixing of gases. The main advantage of using a tank type electrolyzer is that repairs can be made on individual tanks. Since each tank is a separate cell connected to the others in parallel, one cell can be shut down and repaired without affecting the others. This type of electrolyzer has a relatively large surface area, however, and requires more space than the filter-press design. Additionally, this type of electrolyzer cannot be operated at high temperatures, due to large heat losses which are a consequence of the large external tank surface area.

The filter-press electrolyzer is made up of alternate layers of bipolar electrodes and diaphragms. The cells are connected in 'series,' and thus are spatially more compact and can be operated at higher pressures (producing smaller bubbles) compared with the unipolar electrodes of the tank type electrolyzer. If one cell malfunctions, however, all of the cells must be shut down to commence repairs.

Theoretical Principles

One method of increasing the efficiency of electrolysis is by increasing the operational temperature of the process. This is the basis of some of the advanced electrolysis techniques that are discussed below. A brief review of thermodynamics of electrolysis is given here to help explain the possible benefits of performing electrolysis at elevated temperatures.

The electrolysis reaction, in its most fundamental form, is simply the splitting of water into its individual components, hydrogen and oxygen. The energy change for this reaction can be expressed through the following thermodynamic equation:

$$\Delta H = \Delta G + T\Delta S$$

where ΔH is the enthalpy change of the total energy involved, T is the absolute temperature, and ΔS is the entropy change. Here, the Gibb's free energy change, ΔG, is the minimum amount of work needed to drive the reaction (related to the electrical energy demand) while $T\Delta S$ is the amount of thermal energy required for the reaction

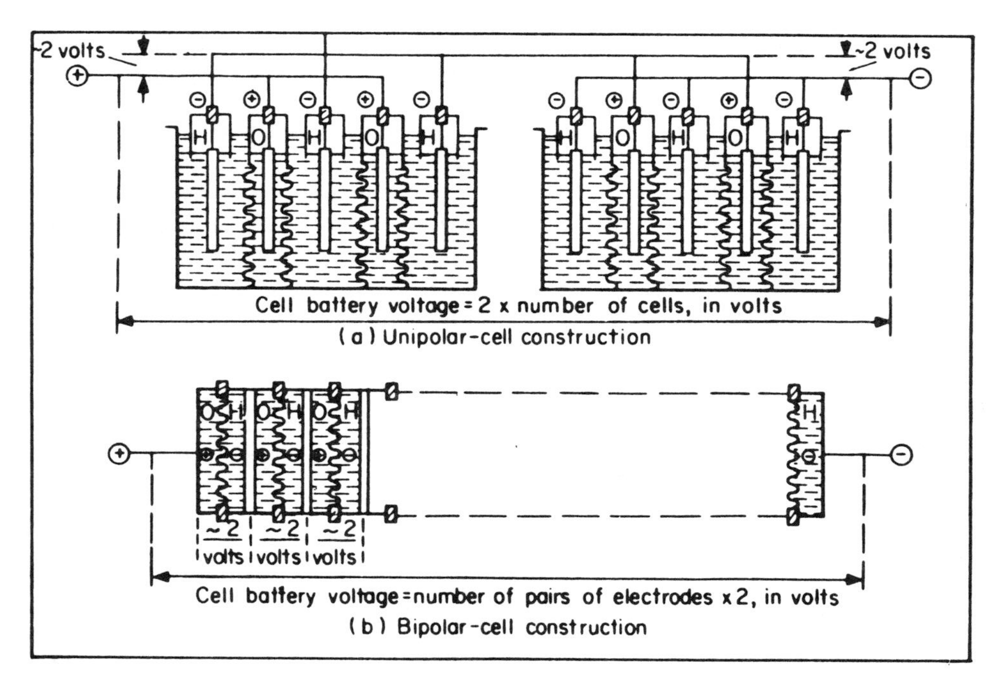

Figure 4.2. Schematic diagram showing unipolar (tank type) and bipolar (filter press type) cells (Source: ref. [11])

when a minimum amount of electrical energy is supplied [12]. The significance of the relationship between enthalpy, Gibb's free energy, and entropy is illustrated in Fig. 4.3, where these quantities for the water electrolysis reaction are plotted as a function of the reaction temperature. This figure clearly demonstrates that ΔG, or the amount of electrical energy needed for electrolysis, decreases as the reaction temperature increases. In other words, the amount of electrical energy needed to perform electrolysis is reduced at higher operational temperatures.

Figure 4.4, which shows a plot of the temperature-voltage relationship for water electrolysis, demonstrates another example of the interplay between thermal and electrical demands. The plot of voltage vs. temperature can be divided into three distinct regions. At a given temperature, hydrogen can not be produced below a certain voltage. At 25° C, the minimum theoretical electrical energy necessary for the splitting of water is 1.23 V. Under these conditions the reaction is endothermic, and heat must be supplied from the external surroundings for reaction to occur. Since the addition of heat to the process offsets the need for electrical energy, the electrical efficiency is highest at a voltage just above this threshold voltage. Under these conditions, however, very little current flows per unit area and the production rate of hydrogen is very low. In practical situations, the voltage must be increased above the theoretical reversible voltage to increase the current per unit area at an electrode, hence the ideal efficiency is never attained under real process conditions. If the voltage of the cell is increased to 1.47 V at 25° C, the process is isothermal, requiring no input heat, and operates at 100% electrical efficiency. At voltages higher than 1.47 V, heat is released into the surroundings as water is decomposed [9].

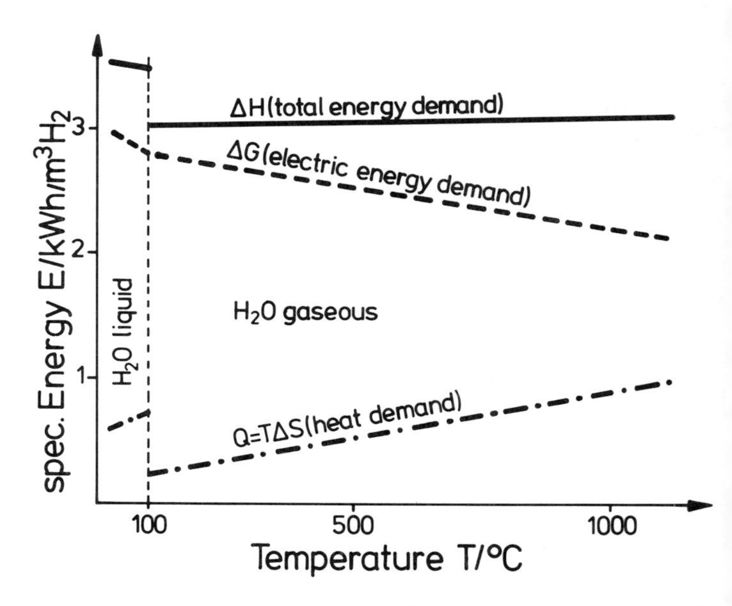

Figure 4.3. Thermodynamics of water splitting. (Source: ref. [13])

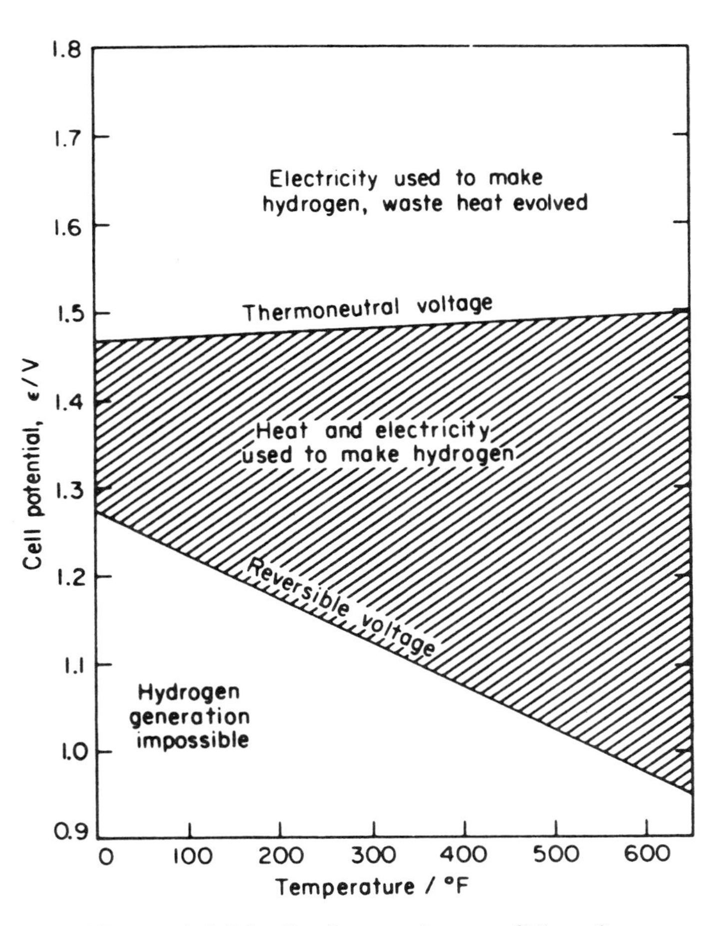

Figure 4.4. Idealized operating conditions from water electrolysis. (Source: ref.[11])

The significance of the tradeoffs between electrical and thermal energy is related to the fact that electrical energy is expensive. In fact, the largest cost in electrolytic production of hydrogen is the cost of electricity. At the average grid electricity price, electrolyzer electricity costs constitute more than 75% of the cost of liquid hydrogen produced with present technologies [14]. Thus, the costs of electrolytic production could conceivably be dramatically reduced by using heat to supplement the process.

Current Designs

As technology advances, the efficiency of the water electrolysis process continues to improve. These improvements are primarily the result of designs which allow operation at higher current densities, temperatures, and pressures. Currently, there are three promising advanced technologies under investigation: advanced alkaline water electrolysis (AWE), solid polymer electrolysis (SPE), and high temperature electrolysis (HTE). These technologies are briefly reviewed below and are discussed in greater detail in the reprint by Dutta at the end of this chapter.

Advanced Water Electrolysis (AWE)

Advanced Water Electrolysis (AWE) is similar to normal alkaline water electrolysis, thus it has the potential for application in today's industrial production facilities. The key to AWE is operating the cells at temperatures greater than 100° C, thus improving voltage efficiency and decreasing the amount of electrical energy required to produce hydrogen. The alkaline environment is very harsh, however, and gas separators, sealants, and other materials used in AWE must be resistant to high temperature alkaline corrosion over long periods of time. Inorganic membranes are one type of separator currently being used for AWE. AWE which uses such membranes is referred to as inorganic membrane electrolysis (IME). These membranes are composed of a thin, i.e., 0.3-0.4 mm, low resistance polyantimonic acid membrane, capable of surviving in harsh caustic environments. More information about AWE can be found in references [14-19].

Solid Polymer Electrolysis (SPE)

In SPE, a solid rather than a liquid electrolyte is used. The compactness, simplicity of design, operation, and maintenance, as well as the absence of a corrosive liquid electrolyte, have made SPE technology a viable alternative for large-scale commercial electrolysis [20]. One polymer currently being used is Nafion®, which was developed by the DuPont Corporation (refer to Chapter 3 for further details). Nafion® is a Teflon®-like polymer with sulfonic acid groups attached to a fully fluorinated polymer backbone. When saturated with water, this polymer becomes acidic and capable of conducting ions. Thus, it can be used in the same manner as a liquid electrolyte. Some advantages of Nafion® are its strength as a stable solid, its capability of being formed into very thin sheets of uniform thickness, and its resistance to gas penetration. These properties allow Nafion® to be used in unusually thin electrolysis cells, approximately 0.5 cm thick, without a diaphragm. Because Nafion® is a solid, it also performs well under high pressures (up to 4.0 MPa or 590 psi) and temperatures (80-150° C). There are some disadvantages that must be overcome before SPE becomes commercially feasible, however. Nafion® itself is very expensive and its use requires a high platinum catalyst loading and novel fabrication methods for current collectors and electrode-membrane assemblies. Further information about SPE is found in refs. [14, 21, 22].

High Temperature Electrolysis (HTE)

The advantages of HTE include its potential for greater electrical efficiency and the advantage of a solid electrolyte, which is noncorrosive and does not experience liquid and flow distribution problems. The high efficiency of HTE results from using high temperatures which, as previously discussed, decreases the need for electrical energy and minimizes the overvoltage. Because very high temperatures are needed, however, HTE requires a high temperature heat source as well as costly materials and fabrication techniques. HTE is carried out at close to 1000° C using vaporized water and a solid oxygen ion conducting electrolyte. Yttria stabilized Zirconia (Y_2O_3 + ZrO_2, abbreviated YSZ) is commonly used as the electrolyte [13]. Further advances are needed, however, before HTE will become practical for commercial uses.

4.4 Thermochemical Water Decomposition

Introduction

An alternative to splitting water with electricity is to use heat alone to thermochemically decompose water. By using heat as the primary source of energy, the resulting total theoretical efficiency of the process is approximately 50% [9]. Additionally, the high costs of electricity needed for electrolytic water decomposition can be avoided. In this section, the process of thermochemical water decomposition will be reviewed.

Basic Concepts

Thermochemical water decomposition is based on the idea of using heat to split water (Fig. 4.5). A water molecule can be split directly using high temperatures (>2,500° C). However, stable construction materials capable of withstanding such an extreme environment and a heat source capable of continuously delivering heat at such high temperatures are currently unavailable. In order to lower the temperature required to decompose water directly chemical reagents are used, often in multi-step chemical reactions.

Thermochemical water decomposition generally involves at least three steps: oxygen production, hydrogen production, and materials regeneration. By using three steps as opposed to one, the temperature for water decomposition can be reduced to approximately 700° C, a temperature readily attainable in industrial applications. An example of a general cycle is described below:

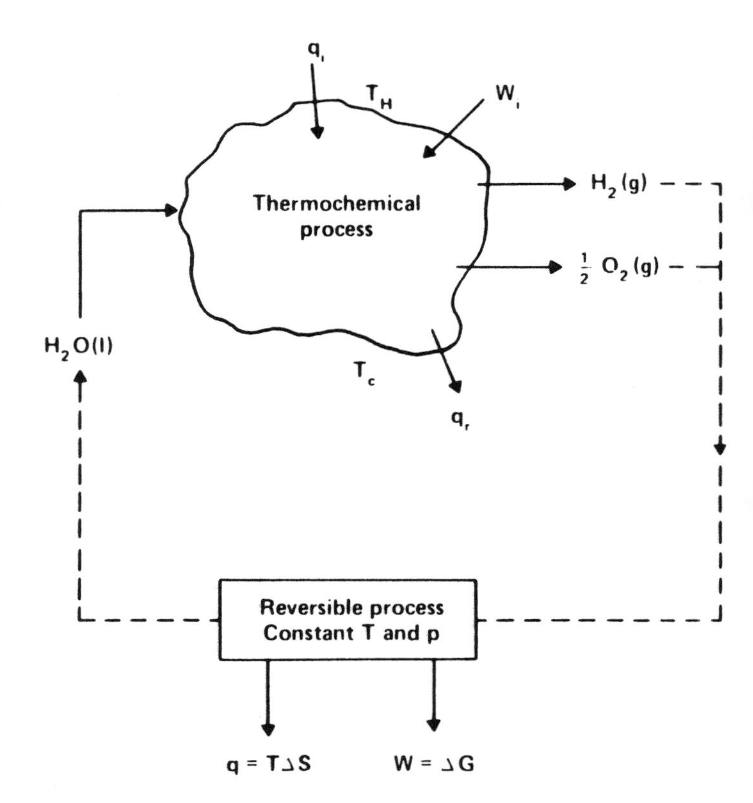

Figure 4.5. General thermochemical water decomposition process. (Source: ref. [23])

1) $AB + H_2O + Heat \Rightarrow AH_2 + BO$

2) $AH_2 + Heat \Rightarrow A + H_2$

3) $2BO + Heat \Rightarrow 2B + O_2$

4) $A + B + Heat \Rightarrow AB$

The reaction series is essentially an oxidation-reduction reaction and, thus, requires at least one element which can readily change its oxidation state. Elements such as iron, oxygen, and chlorine are, therefore, used in many cycles.

Criteria in Choosing a Chemical Reaction Series

A number of criteria must be met when selecting a multistep thermochemical process. The most important criterion is that the changes in Gibb's free energy of individual reactions approach zero within the temperature gradient considered. If one step in the reaction process has a large negative free energy, products formed could be too stable to be recycled. A cycle with such a reaction is not efficient [11]. Also, each reaction in the series must deliver the highest possible yield of product in the shortest amount of time. The product yield is primarily reduced by the fact that most reactions in thermochemical cycles do not go to completion.

The number of individual steps in the process should be as low as possible, since the overall yield of a reaction series is the mathematical product of each individual reaction. For example, a series of four steps, each with yields of 94%, would result in an overall yield of $(0.94)^4$ or ~78.1%. Increasing the number of individual steps also increases the amount of work required to separate and recycle unconverted reactants.

Each reaction must proceed at both a fast rate and at a rate which is comparable with the other reactions so that no shortage or buildup of products occurs. The most desirable series of reactions is one that can be performed continuously, with each reaction in the series progressing at an optimum rate. The control of the reaction rates is complicated by the requirements of specific temperatures, pressures, and flow rates. In addition, internal heat recovery must be considered. In other words, the excess energy from the exothermic parts of a process must be saved and supplied to the endothermic parts to provide maximum efficiency.

The separation of reaction products is a crucial part of any thermochemical reaction series, being both energy and cost intensive. Separation requirements must therefore be carefully considered when creating a cycle. The reaction cycle cannot result in any chemical by-products, as it would require a separate process sequence using unnecessary energy to remove and reintroduce such material.

Intermediate products should be relatively easy to handle. Cyclic processes involving the transfer of gases and liquids are the best, as gases and liquids can be quickly and cost-effectively transported within the plant. The intermediate products should not be highly corrosive, such as halogen acids, since such materials are difficult to handle. Additionally, components A and B in the reaction sequence described above must be available in reasonable quantities at a reasonable cost and should not be themselves hazardous or result in products that are hazardous to the environment.

Current Status

The world's first functioning thermochemical water-splitting apparatus, the Mark 13, began operation in the spring of 1978. Since then, several other bench-scale thermochemical water decomposers have been designed and tested. None have been successful enough to constitute the construction of an industrial scale plant. Thermochemical processes are still in the early stages of development, requiring more investment of time and money before an economically acceptable and environmentally safe process is found. Some thermochemical processes currently under study are listed below. A more detailed evaluation of one of these processes, UT-3 process, is given in the reprint by Aochi, et al at the end of this chapter.

Mark-10/Ispra [24].

1) $2H_2O + SO_2 + I_2 + 4NH_3 \Rightarrow 2NH_4I + (NH_4)_2SO_4$ (325K)

2) $2NH_4I \Rightarrow 2NH_3 + H_2 + I_2$ (900K)

3) $(NH_4)_2SO_4 + Na_2SO_4 \Rightarrow Na_2S_2O_7 + H_2O + 2NH_3$ (675K)

4) $Na_2S_2O_7 \Rightarrow SO_3 + Na_2SO_4$ (825K)

5) $SO_3 \Rightarrow SO_2 + 1/2O_2$ (1140K)

Mg-S-I Process/National Chemistry Laboratory for Industry [25].

1) $I_2(c) + SO_2(aq\ or\ g) + 2H_2O(l) \Rightarrow H_2SO_4(aq) + 2HI(aq)$ (343K)

2) $2MgO(c) + H_2SO_4(aq) + 2HI(aq) \Rightarrow MgSO_{4(aq)} + MgI_2(aq)$(343K)

3) $MgI_2(aq) \Rightarrow MgO(c) + 2HI(g) + nH_2O(g)$ (673K)

4) $MgSO_4(c) \Rightarrow MgO(c) + SO_2(g) + 1/2O_2(g)$ (1268K)

5) $2HI(g) \Rightarrow H_2(g) + I_2(g)$ (1268K)

UT-3 process/University of Tokyo [26].

1) $CaBr_{2(s)} + H_2O(g) \Rightarrow CaO(s) + 2HBr(g)$ (973-1023K)

2) $CaO(s) + Br_2(g) \Rightarrow CaBr_2(s) + 1/2O_2(g)$ (773-873K)

3) $Fe_3O_4(s) + 8HBr(g) \Rightarrow 3FeBr_2(s) + 4H_2O(g) + Br_2(g)$ (473-573K)

4) $3FeBr_2(s) + 4H_2O(g) \Rightarrow Fe_3O_4(s) + 6HBr(g) + H_2(g)$ (823-873K)

4.5 Photo Conversion

Introduction

Photovoltaic arrays, liquid semiconductor systems, and biological organisms have all been investigated as methods of collecting or absorbing solar energy. Once absorbed, this energy can be used to break apart water molecules into hydrogen and oxygen atoms. In this section, the application of photochemical and photobiological techniques to the production of hydrogen are discussed.

Photochemical Processes

In photochemical processes, semiconductors or in-solution metal complexes are the prime agents used to collect sunlight. More information on photoelectrochemical methods of hydrogen production can be found in the reprint by Getoff at the end of this chapter.

Semiconductor Systems

Semiconductors doped with small amounts of impurities are the most commonly used materials to capture sunlight. The typical configuration is formed when two doped semiconductors (one n-type semiconductor and the other a p-type semiconductor) are brought together to form a p-n junction. At the junction, the charges in the n- and p-type materials rearrange in a manner that forms a permanent electric field across the junction. When photons with an energy greater than the 'band gap' of the material strike the junction, an electron is freed to move around in the material (a corresponding vacancy on the parent atom, known as a hole, is also formed). The permanent electric field formed at the p-n junction forces the electron and hole in opposite directions. When connected to an external load, this electric current can be used to perform work. (A more detailed description of the physics behind this process can be found in ref. [27]).

A situation similar to that described for the p-n junction can occur when a photocathode and a photoanode are immersed in an aqueous electrolyte. The photocathode is typically a p-type material which has an excess of positive holes, while the photoanode is composed of an n-type material which has an excess of electrons. When these electrodes are placed in different chambers separated by a semipermeable membrane, a system known as a liquid-junction transducer is formed.

The reaction at the liquid semiconductor junction is initiated by the absorption of a photon of energy larger than the band gap, which creates an electron-hole pair. The positive hole subsequently reacts with water to create positive hydrogen ions and gaseous oxygen:

$$2h^+ + H_2O \Rightarrow 1/2O_2 + 2H^+$$

The hydrogen ions formed in this reaction penetrate the diaphragm and proceed towards the photocathode where they recombine with the electrons which are transported through the load circuit. Hydrogen gas is then formed when hydrogen atoms on the photocathode recombine. The photocathodic process can be expressed as follows:

$$2H^+ + 2e^- \Rightarrow H_2$$

Thus, the overall process results in water photoelectrolysis [28].

Both single and double photoelectrode systems have been developed for photoelectrochemical hydrogen production. Fujishima and Honda developed a system using a single TiO_2 photoanode with a platinum metal cathode. Kainthla, et al developed a two photoelectrode system which operates using p-InP(Pt) as the photocathode and Mn-oxide protected by n-GaAs as the photoanode, all in a cell containing an alkaline solution as the electrolyte [29]. This system is shown in Fig. 4.6. The maximum conversion efficiency for this system is 8.2%, under short circuit conditions when light is converted solely to chemical energy. For semiconductor-liquid interfaces in general, energy conversion efficiencies of around 12% have been reported [30].

In-Solution Metal Complexes

An alternative to semiconductor hydrogen production is the use of in-solution metal complexes such as photochemical catalysts. Fine suspensions of metals (e.g., TiO_2 of grain size 0.06 mm, 3-15 mg ml^{-1}) in alkaline or acidic environments can produce hydrogen and oxygen upon illumination. The suspended metal complexes absorb the photon energy creating an electric charge separation that drives the water-splitting reaction. Currently, this approach is less advanced than semiconductor methods

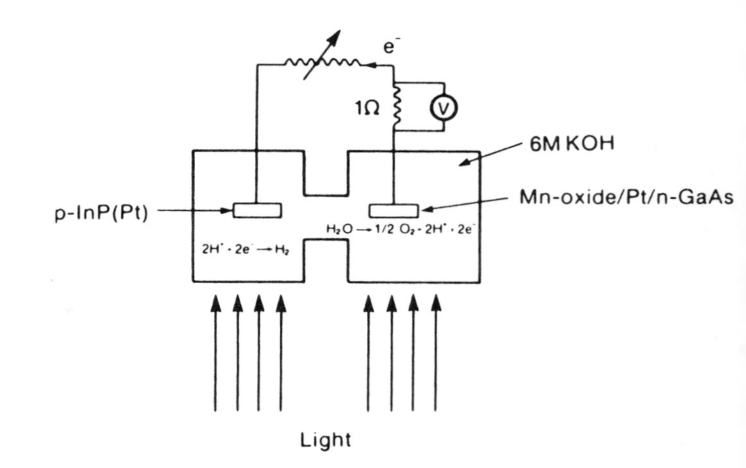

Figure 4.6. Two electrode self-driven photoelectro chemical cell for water electrolysis. (Source: ref. [29])

but offers the possibility of avoiding photocorrosion problems.

Photobiological Processes

In photobiological processes, a biological system is used to produce hydrogen from light. Photobiological organisms currently under study include cyanobacteria (blue-green algae), photosynthetic bacteria, and eukaryotic (green) algae. Hydrogen production by cyanobacteria is discussed in further detail in the reprint by Smith, et al at the end of this chapter.

Cyanobacteria

Cyanobacteria have two important hydrogen-metabolizing enzymes: nitrogenase and membrane-bound uptake hydrogenase. The relationship between the two enzymes is illustrated in Fig. 1 in the reprint by Smith, et al where the nitrogenase produces hydrogen molecules and the hydrogenase consumes them. To produce hydrogen, cyanobacteria must be kept in an anaerobic environment, otherwise they will consume their own hydrogen. Inert gases, such as argon, or partial vacuums are used to maintain an oxygen-free environment [31]. Even in such environments, however, oxygen, which is produced from water along with hydrogen, can still affect the efficiency of the reaction.

In some cells, almost all of the hydrogen that is evolved by the cyanobacteria is immediately reutilized by the hydrogenase enzymes [32]. Chemicals or genetic processes can be used to limit the effect of the hydrogenase. Carbon monoxide together with acetylene or low nickel ion concentrations can be used during the growth of specific cyanobacteria [33]. Hydrogenase activities of *A. cylindrica*, for example, require nickel ions as a necessary cofactor to function [34]. The hydrogenase for *A. cylindrica* are thus inhibited when given low concentrations of nickel ions, and the overall production of hydrogen increases. Genetics could also be used to produce hydrogenase-deficient strains of cyanobacteria. These genetically engineered cyanobacteria would be expected to produce hydrogen spontaneously [33].

Cyanobacteria are ideal organisms for producing hydrogen due to their ability to perform photosynthetic CO_2-fixation using water as an electron donor and their capability to thrive in a simple inorganic environment. Unfortunately, efficiency for the conversion of solar energy to hydrogen by cyanobacteria is only on the order of 0.1% [32].

Photosynthetic Bacteria

Photosynthetic bacteria can produce hydrogen at high rates from organic acids, alcohols, sugars, and a few sulfur compounds for longer periods than cyanobacteria [35]. Various electron donors from industrial and agricultural waste could be used to create an economically viable process involving photosynthetic bacteria. Non-sulfur photosynthetic bacteria, for example, produce hydrogen at the expense of simple organic acids or sugars. Their effectiveness depends on factors such as the availability of substrates, light irradiance, and temperature [36]. Materials such as whey, starches, sugar refinery waste, and distillery waste can be used as electron donors.

The photosynthetic bacteria are usually grown under anaerobic conditions and immobilized to maximize hydrogen production. Advantages of using photosynthetic bacteria over other biological systems include the ability to grow cells which produce hydrogen from organic compounds as well as substrate conversion efficiencies from organic acids to hydrogen in the range of 20-100% [37].

Eukaryotic Algae

In 1942, Gaffron and Rubin first demonstrated hydrogen photoevolution by unicellular algae. Eukaryotic algae differ from cyanobacteria or any other type of bacteria (prokaryotic cells) in that eukaryotes have a true nucleus enclosed by a membranous nuclear envelope [38]. Normally, eukaryotic aerobic photosynthesis uses carbon dioxide (CO_2) as the terminal electron acceptor, resulting in a CO_2 fixation compound as the energy-rich photoproduct [39]. Certain classes of eukaryotic algae, however, can evolve molecular hydrogen under appropriate physiological conditions. For instance, when placed in an anaerobic atmosphere, certain green algae are capable of synthesizing the enzyme hydrogenase [39]. Therefore, when both CO_2 and O_2 are removed from the environment, hydrogen evolution becomes the main pathway for the movement of electrons. Here, hydrogen ions serve as the electron acceptors, instead of CO_2, and water is reduced to molecular hydrogen and oxygen. Unfortunately, the light saturated rate of oxygen evolution is significantly reduced compared to when CO_2 is the terminal electron acceptor, and thus the rate of hydrogen production is very low.

4.6 Hydrogen Production from Biomass

Introduction

Biomass is defined as all plant and animal material, not including fossilized fuels, which can be converted into energy. Biomass is organic in nature and derived from a variety of sources including residues, wastes, and crops grown specifically for energy use. Biological methods and thermochemical conversions are the main processes used in extracting hydrogen from biomass. This section discusses different sources of biomass and techniques for extracting hydrogen from biomass. The use of pine and willow to produce hydrogen is discussed in the reprint by Timpe, et al at the end of this chapter.

Sources of Biomass

Sources of biomass for hydrogen production are diverse and abundant and include plant, animal, and human wastes and residues as follows.

Plants

Plants used for energy production include agricultural crop residues, forest residues, standing vegetation, and energy crops. Agricultural crop residues are the portion of the crop which is ordinarily left in the field after harvest or at the processing site (i.e., wheat and rice straws, corn stalks, husks, and sugarcane bagasse). Forest residues include logging residues and trees removed for reasons other than for use as lumber. Standing vegetation is wild and woody vegetation and trees, while energy crops are grown specifically for use as an energy source.

The purpose of using energy crops is to provide a renewable source of energy while using underutilized land to grow crops. In considering possible energy crops, many details must be examined, including: obtaining energy from the crop that exceeds the energy necessary to plant and process; crop-to-fuel processing that does not result in wastes harmful to the environment; and production of crops that do not cause deterioration to the land through nutrient depletion, damage to soil structure, or erosion of soil. Additionally, energy crops should not require fertilizers, pesticides, or weed killers made from fossil fuels. Energy crops which have been proposed include sugarcane, sweet sorghum, cassava, soybean, sunflower, euphorbia, trees, woody plants, napier grass, and many others [40].

As the population of the world increases, competition for land use will increase, making aquatic plants more attractive as an alternative to land grown energy crops. Aquatic plants considered as potential biomass feedstocks include water hyacinths, duckweed, and *Hydrilla* [40].

Animals

Animal wastes used as an energy source are primarily in the form of manure. The most successful manure collection systems occur where livestock is confined and collection occurs daily.

Human Wastes

Human wastes include municipal solid wastes (MSW), wastewater, and industrial wastewater. MSW includes all non-hazardous wastes that are initially solid. The organic fraction of MSW (which is the part that can be most effectively used for energy) is extracted from the general waste stream through size reduction and separation processes. Industrial wastes that can be used to produce hydrogen are generated by food processing plants (i.e., canneries, breweries, sugarmills, distilleries, etc.) and some chemical industries (organic plants and coal carbonization facilities). In wastewater (or sewage), the organic fraction is separated and dried by conventional wastewater treatment procedures. The organic fraction can then be converted to energy through various methods.

Methods of Hydrogen Recovery from Biomass

There are many methods for converting biomass to hydrogen. Modern technologies can convert biomass to a variety of different products including methane, methanol, ethanol, oils, charcoal, or hydrogen directly. Hydrogen can be extracted from all of these hydrocarbons via either steam reformation or coal gasification. Anaerobic digestion and thermochemical conversion techniques, however, will be the focus of this section.

Anaerobic Digestion

Anaerobic digestion is a versatile bioconversion process that is used to produce methane gas and other useful residues from a variety of homogeneous and heterogeneous carbonaceous materials [41]. The multistage process, in which the waste from one set of organisms is the food for another set, takes place in the absence of air (oxygen) and in the presence of a population of anaerobic bacteria [42].

Anaerobic digestion consists of three basic processes. In the first stage, cellulose and hemicellulose are broken down by enzymes to form soluble organic compounds. The soluble organics are then converted to hydrogen, carbon dioxide, formate, and acetate by acid-producing bacteria. Finally, methane is produced by methane-producing bacteria from the hydrogen, carbon dioxide, formate, and acetate.

Stirred-tank reactors are currently the mainstay of most digestion systems, but significant advances are being made to develop other designs for methane production [43]. Such designs include plug-flow, two-phase, and fixed-film digesters. An example of an anaerobic digester used to convert crops to methane is shown in Fig. 4.7. At ambient temperatures, gas production is very slow. Therefore, the reactor temperature must be increased (to 37° C for crop biomass degradation) through the use of thermophilic microorganisms, microorganism accumulation, or through direct thermal heating [44]. Wastes commonly used as food for the bacteria include sewage, municipal solid waste, manure, and energy crops.

Anaerobic digestion is the preferred process for gasification of high-moisture feeds, as the total energy produced from anaerobic digestion is usually greater than that produced from a thermal conversion process [45, 46]. Anaerobic digestion processes may have very low to negative net energy returns, however, depending on digester design and operation conditions (plant size, climatic conditions, and other factors) [41]. The resultant gas has a medium fuel value, with a concentration of methane slightly higher than its concentration of CO_2. Hydrogen can be obtained from the gas through steam reformation, but this would require too much energy to be cost effective. Thus, improved methods of hydrogen extraction from anaerobic digestion must be developed before anaerobic digestion becomes a valuable source of hydrogen production.

Thermochemical Conversions

The gasification of biomass is defined as the thermal decomposition of organic matter with the help of an auxiliary gas such as air, oxygen, or hydrogen to yield primarily 'gaseous' products. A general flow scheme for a biomass-to-hydrogen plant is given in Fig. 4.8. After the feedstock is introduced into the reactor, the feedstock reacts with steam and air (oxygen)—at a temperature dictated by the feedstock characteristics—to rapidly yield a gas mixture of hydrogen, carbon monoxide, carbon dioxide, water, and methane along with hydrogen sulfide and other *trace* impurities (as biomass is low in sulfur) [47]. The remaining gasification process is similar to the coal gasification process described in Section 4.1. A major disadvantage of direct biomass gasification, however, is the co-production of tars, phenolics, and acids [48]. The gas composition and other information on an atmospheric pilot wood gasifier developed by Le Creusot show hydrogen gas to be a major component of the gas composition. The feedstocks usable for biomass gasification include dried sewage sludge (dewatered, conditioned, and processed), wood, municipal solid waste, agricultural waste, and energy crops.

Similar to other methods of hydrogen production, catalysts are used to increase production efficiencies of biomass gasification. The use of catalysts in biomass gasification can result in large improvements in the process efficiency by reforming the product gas and cracking the tars formed. The advantages of catalytic steam gasification are that no oxygen or air is required, little or no tar is formed, and both the yields and efficiencies are greater than those obtained by conventional gasification [49]. Catalysts can be mixed with biomass or added in a second reactor. Materials used as catalysts include nickel and other elements or compounds, such as silica-alumina.

Gasification of biomass is more profitable if the process is done under higher pressures. The main advantage of performing gasification under pressure (approximately 3 MPa for methanol production) is that the equipment is less expensive and smaller in size. When methanol is being produced as a method of storing hydrogen, the use of higher gasification pressures will result in less energy consumed per unit volume of methanol synthesis.

The growth of sugar cane and sorghum on the 66.4 million acres of land that was out of production in the U.S. between 1981 and 1988 could have supplied enough energy to generate 34 percent of the non-nuclear power that was generated in 1986 [52]. Thus, biomass could potentially supply large amounts of hydrogen. The technology to effectively convert crops to hydrogen must first be developed. A number of biomass low-Joule gasification units are now available in the size ranges of 50-500 kg/hr biomass [53]. However, a significant amount of research and development is required if biomass gasification is to become a cost-effective pathway for hydrogen production. Separation and purification via selective membranes or catalytic processes must first be perfected, as well as more efficient gas-cleanup concepts for the removal of tars and oils [54].

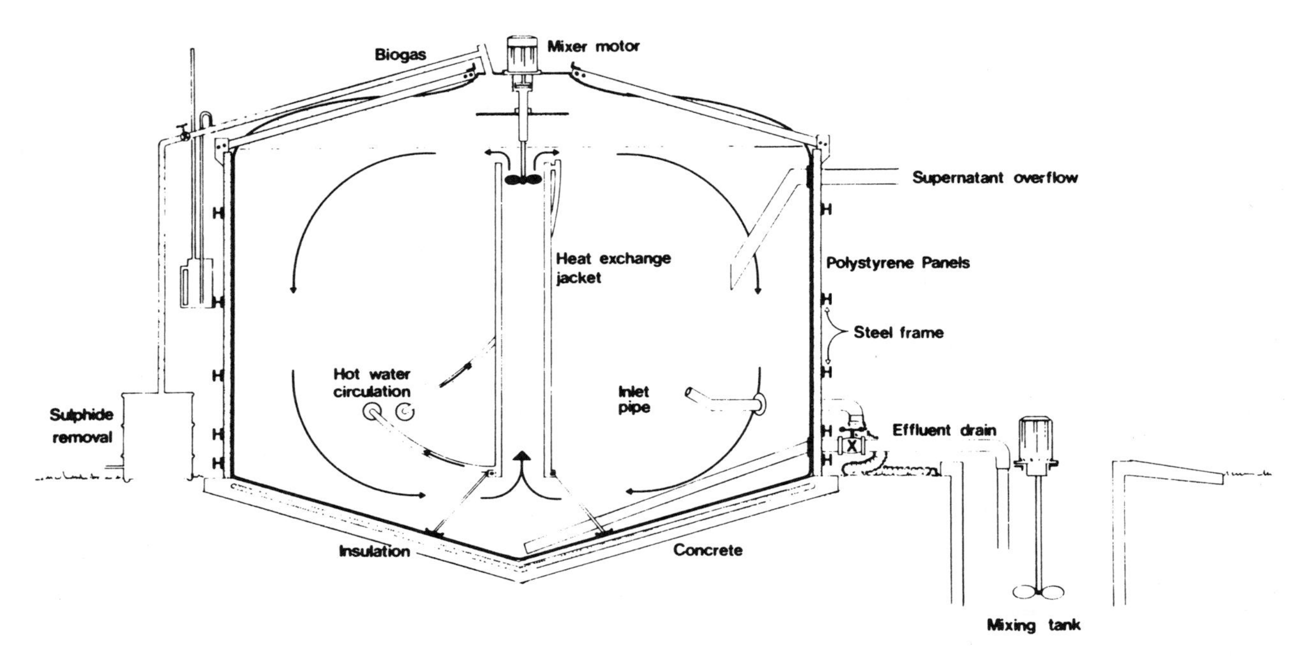

Figure 4.7. Cross-sectional view of the anaerobic digester used on the Invermay 'energy farm' to convert crops to biogas. (Sources: ref. [44])

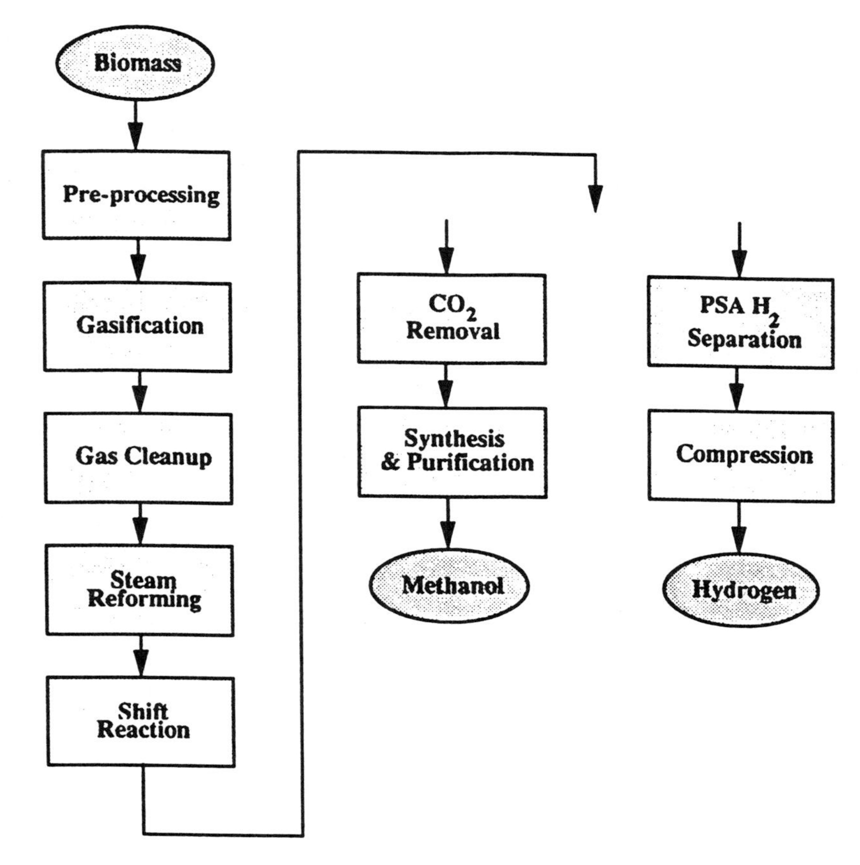

Figure 4.8. Block diagram of process configurations for the production of methanol or hydrogen from biomass. (Source: ref. [50])

4.7 Conclusion

Hydrogen can be produced by a variety of methods, including water electrolysis, thermochemical water decomposition, fossil fuels, photoconversions, and/or biomass. While the production of hydrogen from fossil fuels is currently the most economical method of producing hydrogen in large quantities, the fact that fossil fuels are non-renewable and emit pollutants when they burn makes them undesirable as a future source of hydrogen. For the long term, hydrogen will have to be produced from clean, renewable sources to make full use of hydrogen's advantages as a fuel. Water electrolysis or biomass gasification are good potential candidate processes. Biomass has the potential of becoming an important source of hydrogen in the future. Potential U.S. biomass supplies for energy purposes have been estimated to be the energy equivalent of 7 million barrels of oil per day—40 percent from biomass waste resources and 60 percent from biomass energy crops grown on 80 million acres of excess agricultural lands [55]. The land and water requirements for such biomass production would be substantial, however. In the future, water electrolysis may be the ultimate source for hydrogen production. Hydrogen produced from electricity is currently expensive, however, and even more so if it is derived from renewable sources. Thus, technical and economic advances are still needed before large-scale production of hydrogen from renewable sources is possible.

References

1. Kukkonen, C.A. and Shelef, M., "Hydrogen as an Alternative Automobile Fuel: 1993 Update," SAE Technical Paper No. 940766, Society of Automotive Engineers, Warrendale, PA, 1994.

2. Minet, R.G. and Desai, K., "Cost-effective Methods for Hydrogen Production," *Int. J. Hydrogen Energy*, Vol. 8, p. 285, 1983.

3. Hao Shu-ren Mao, P.-S. and Wang, J.-S., "The Performance of Naphtha Steam Reforming Catalyst Z409/Z405 and its Applications in Production of Ammonia and Hydrogen," in *Hydrogen Energy Progress IX*, Edited by T. N. Veziroglu and C. Derive-J. Pottier, International Association for Hydrogen Energy, Coral Gables, FL, p. 23, 1992.

4. Steinberg, M. and Cheng, H.C., "Modern and Prospective Technologies for Hydrogen Production from Fossil Fuels," *Int. J. Hydrogen Energy*, Vol. 14, p. 797, 1989.

5. Cromarty, B., "Modern Aspects of Steam Reforming for Hydrogen Plants," in *Hydrogen Energy Progress IX*, International Association for Hydrogen Energy, Coral Gables, FL, 1992.

6. Sircar, S., "Production of Hydrogen and Ammonia Synthesis Gas by Pressure Swing Adsorption," *Separation Science and Technology*, Vol. 25, p. 1087, 1990.

7. Petit, C., Libs, S., Roger, A.C. et al, "Hydrogen Production by Catalytic Oxidation of Methane," in *Hydrogen Energy Progress IX*, Edited by T. N. Veziroglu and C. Derive-J. Pottier, International Association for Hydrogen Energy, Coral Gables, FL, p. 53, 1992.

8. Hoffmann, P., *The Forever Fuel: The Story of Hydrogen*, Westview Press, Boulder, CO, 1981.

9. Williams, L.O., *Hydrogen Power An Introduction to Hydrogen Energy and its Applications*, Pergamon Press, Oxford, U.K., 1980.

10. Huot, J.Y., Trudeau, M.L. and Schulz, R., "Low Hydrogen Overpotential Nanocrystalline Ni-Mo Cathodes for Alkaline Water Electrolysis," *Journal of the Electrochemical Society*, Vol. 138, p. 1316, 1991.

11. McAuliffe, C.A., *Hydrogen and Energy*, Gulf Book Division, Houston, TX, 1980.

12. Dönitz, W. and Erdle, E., "High Temperature Electrolysis of Water Vapour—Status of Development and Perspectives for Application," in *Hydrogen Energy Progress V*, Edited by T. N. Veziroglu and J. B. Taylor, Pergamon Press, Elmsford, NY, p. 767, 1984.

13. Dönitz, W., Erdle, E., and Streicher, R., "High Temperature Electrochemical Technology for Hydrogen Production and Power Generation," in *Electrochemical Hydrogen Technologies*, Edited by H. Wendt, Elsevier Science Publishing Company Inc., New York, NY, p. 213, 1990.

14. Dutta, S., "Technology Assessment of Advanced Electrolytic Hydrogen Production," *Int. J. Hydrogen Energy*, Vol. 15, p. 379, 1990.

15. Dutta, S., Block, D.L. and Port, R.L., "Economic Assessment of Advanced Electrolytic Hydrogen Production," *Int. J. Hydrogen Energy*, Vol. 15, p. 387, 1990.

16. Vandenborre, H., Leysen, R., Tollenboom, J.P. and Baetslé, L. H., "On the Inorganic-Membrane-Electrolyte (IME) Water Electrolysis," in *Hydrogen Energy System*, Edited by T. N. Veziroglu and W. Seifritz, Pergamon Press, Elmsford, NY, p. 2379, 1978.

17. Vandenborre, H., Leysen, R., Nackaerts, H., et al, "Advanced Alkaline Water Electrolysis Using Inorganic-Membrane-Electrolyte (I.M.E.) Technology," in *Hydrogen Energy Progress V*, Edited by T. N. Veziroglu and J. B. Taylor, Pergamon Press, Elmsford, NY, p. 703, 1984.

18. Divisek, J., Malinowski, P., Mergel, J. and Schmitz, H., "Improved Construction of an Electrolytic Cell for Advanced Alkaline Water Electrolysis," in *Hydrogen Energy Progress V*, Edited by T. N. Veziroglu and J. B. Taylor, Pergamon Press, Elmsford, NY, p. 655, 1984.

19. Divisek, J., Malinowski, P., Mergel, J. and Schmitz, H., "Advanced Techniques for Alkaline Water Electrolysis," in *Hydrogen Energy Progress VI*, Edited by T. N. Veziroglu, N. Getoff and P. Weinzierl, Pergamon Press, Elmsford, NY, p. 258, 1986.

20. Schmittinger, P., "Hydrogen by Chlor-Alkali Electrolysis," in *Electrochemical Hydrogen Technologies*, Edited by H. Wendt, Elsevier Science Publishing Company Inc., New York, NY, p. 261, 1990.

21. Nuttall, L.J. and Russell, J.H., "Solid Polymer Electrolyte Water Electrolysis Development Status," in *Hydrogen Energy System*, Edited by T. N. Veziroglu and W. Seifritz, Pergamon Press, Elmsford, NY, p. 391, 1978.

22. Andolfatto, F., Durand, R., Michas, A., et al, "Solid Polymer Electrolyte Water Electrolysis: Electrocatalysis and Long Term Stability," in *Hydrogen Energy Progress IX*, Edited by T. N. Veziroglu and C. Derive-J. Pottier, International Association for Hydrogen Energy, Coral Gables, FL, p. 429, 1992.

23. Funk, J.E., "Thermochemical Water Decomposition: Current Status," in *Recent Developments in Hydrogen Technology*, Edited by K. D. J. Williamson and F. J. Edeskuty, CRC Press, Inc., Boca Raton, FL p. 1, 1986.

24. Rosen, M.A., "Thermodynamic Analysis of Hydrogen Production by Thermochemical Water Decomposition Using the ISPRA MARK-10 Cycle," in *Hydrogen Energy Progress VIII*, Edited by T. N. Veziroglu and P. K. Takahashi, Pergamon Press, Elmsford, NY, p. 701, 1990.

25. Mizuta, S. and Kumagai, T., "Progress Report on the Mg-S-I Thermochemical Water-Splitting Cycle—Continuous Flow Demonstration," in *Hydrogen Energy Progress VI*, Edited by T. N. Veziroglu, N. Getoff and P. Weinzierl, Pergamon Press, Elmsford, NY, p. 696, 1986.

26. Aihara, M., Sakurai, M. and Yoshida, K., "Reaction Improvement in the UT-3 Thermochemical Hydrogen Production Process," in *Hydrogen Energy Progress VIII*, Edited by T. N. Veziroglu and P. K. Takahashi, Pergamon Press, Elmsford, NY, p. 493, 1990.

27. Hsieh, J.S., *Solar Energy Engineering*, Prentice-Hall, Inc., Englewood Cliffs, NJ, 1986.

28. Ohta, T., "Photochemical and Photoelectrochemical Hydrogen Production from Water," in *Hydrogen Energy Progress VI*, Edited by T. N. Veziroglu, N. Getoff and P. Weinzierl, Pergamon Press, Elmsford, NY, p. 484, 1986.

29. Bockris, J.O. and Kainthla, R.C., "The Conversion of Light and Water to Hydrogen and Electric Power," in *Hydrogen Energy Progress VI*, Edited by T. N. Veziroglu, N. Getoff and P. Weinzierl, Pergamon Press, Elmsford, NY, p. 449, 1986.

30. Heller, A. et al, "Output Stability of n-CdSe/Na2S-S-NaOH/C Solar Cell," *J. Electrochem. Soc.*, Vol. 125, p. 1156, 1978.

31. Markov, S.A., Rao, K.K. and Hall, D.O., "A Hollow Fibre Photobioreactor for Continuous Production of Hydrogen by Immobilized Cyanobacteria Under Partial Vacuum," in *Hydrogen Energy Progress IX*, Edited by T. N. Veziroglu and C. D.-J. Pottier, International Association for Hydrogen Energy, Coral Gables, FL, p. 641, 1992.

32. Bothe, H. and Kentemich, T., "Potentialities of H_2 Production by Cyanobacteria for Solar Energy Conversion Programs," in *Hydrogen Energy Progress VIII*, Edited by T. N. Veziroglu and P. K. Takahashi, Pergamon Press, Elmsford, NY, p. 729, 1990.

33. Smith, G.D., Ewart, G.D. and Tucker, W., "Hydrogen Production by Cyanobacteria," in *Hydrogen Energy Progress VIII*, Edited by T. N. Veziroglu and P. K. Takahashi, Pergamon Press, Elmsford, NY, p. 735, 1990.

34. Daday, A., Mackerras, A.H. and Smith, G.D., *J. Gen. Microbiol.*, Vol. 131, p. 231, 1985.

35. Sasikala, K., Ramana, C.V. and Raghuveer Rao, P., "Photoproduction of Hydrogen from the Waste Water of Distillery by *Rhodobacter Sphaeroides* O. U. 001," *Int. J. Hydrogen Energy*, Vol. 17, p. 23, 1992.

36. Singh, S.P., Srivastava, S.C. and Pandey, K.D., "Hydrogen Production by Rhodoseudomonas at the Expense of Vegetable Starch, Sugarcane Juice and Whey," in *Hydrogen Energy Progress IX*, Edited by T. N. Veziroglu, C. Derive and J. Pottier, International Association for Hydrogen Energy, Coral Gables, FL, p. 615, 1992.

37. Sasikala, K., Ramana, C.V., Rao, P.R. and Subrahmanyam, M., "Effect of Gas Phase on the Photoproduction of Hydrogen and Substrate Conversion Efficiency in the Photosynthetic Bacterium *Rhodobactersphaeroides* O.U. 001," *Int. J. Hydrogen Energy*, Vol. 15, p. 795, 1990.

38. Campbell, N.A., *Biology*, The Benjamin/Cummings Publishing Company, Inc., Redwood City, 1990.

39. Greenbaum, E., "Hydrogen Production by Photosynthetic Water Splitting," in *Hydrogen Energy Progress VIII*, Edited by T. N. Veziroglu and P. K. Takahashi, Pergamon Press, Elmsford, NY, p. 743, 1990.

40. Ward, R.F., "Potential of Biomass," in *Fuel Gas Systems*, Edited by D. L. Wise, CRC Press, Inc., Boca Raton, FL, p. 1, 1983.

41. Ghosh, S., "Net Energy Production in Anaerobic Digestion," in *Energy from Biomass and Wastes V*, Institute of Gas Technology, Lake Buena Vista, FL, 1981.

42. Martin, J.H., "Operation of a Commercial Farm-Scale Plug-Flow Manure Digester Plant," in *Energy from Biomass and Wastes V*, Institute of Gas Technology, Lake Buena Vista, FL, 1981.

43. Klass, D.L., "Energy from Biomass and Wastes: 1980 Update," in *Energy from Biomass and Wastes V*, Institute of Gas Technology, Lake Buena Vista, FL, 1981.

44. Stewart, D.J., "Methane from Crop-Grown Biomass," in *Fuel Gas Systems*, Edited by D. L. Wise, CRC Press, Inc., Boca Raton, FL, p. 85, 1983.

45. Ghosh, S. and Klass, D.L., "Conversion of Urban Refuse to Substitute Natural Gas by the BIOGAS® Process," in proceedings from the Fourth Mineral Waste Utilization Symposium, Chicago, IL, 1974.

46. Ghosh, S. and Klass, D.L., "SNG from Refuse and Sewage Sludge by the BIOGAS® Process," in proceedings of the Symposium on Clean Fuels from Biomass, Sewage, Urban Refuse, and Agricultural Wastes, Orlando, FL, 1976.

47. Goyal, A. and Rehmat, A., "Fuel Evaluation for a Fluidized-Bed Gasification Process (U-Gas)," in *Clean Energy from Waste & Coal*, Edited by R. M. Khan, American Chemical Society, Washington, D.C., p. 58, 1991.

48. Edye, L.A., Richards, G.N. and Zheng, G., "Transition Metals as Catalysts for Pyrolysis and Gasification of Biomass," in *Clean Energy from Waste & Coal*, Edited by R. M. Khan, American Chemical Society, Washington, D.C., p. 90, 1991.

49. Mitchell, D.H. et al, "Methane/Methanol by Catalytic Gasification of Biomass," *CEP*, p. 53, 1980.

50. Ogden, J.M., "Renewable Hydrogen Transportation Fuels," presented at Solar Electric Vehicles '92, Boston, MA, 1992.

51. Chrysostome, G., Lemasle, J.M. and Ascab, G., Syngas Production from Wood: Gasification Unit of Clamecy, Elsevier Applied Science, Brussels, Belgium, 1986.

52. Hamrick, J.T., "Biomass-Fueled Gas Turbines," in *Clean Energy from Waste & Coal*, Edited by R. M. Khan, American Chemical Society, Washington, D.C., p. 78, 1991.

53. Beenackers, A.A.C.M. and Van Swaaij, W.P.M., "Gasification, Synthesis Gas Production and Direct Liquefaction of Biomass," in *Energy from Biomass I*, Elsevier Applied Science, Brussels, Belgium, 1986.

54 "Hydrogen Program Plan," Report by the U.S. Department of Energy, Office of Conservation and Renewable Energy, Report No. DOE/CH10093-147.

55. Johansson, T.B., Kelly, H., Reddy, A.K.N. and Williams, R.H., "Renewable Fuels and Electricity for a Growing World Economy: Defining and Achieving the Potential," in *Renewable Energy: Sources for Fuels and Electricity*, Edited by T. B. Johansson, H. Kelly, A. K. N. Reddy and R. H. Williams, p. 1, 1993.

Modern and Prospective Technologies for Hydrogen Production from Fossil Fuels

Meyer Steinberg and Hsing C. Cheng
Process Sciences Division, Brookhaven National Laboratory

Abstract

A study is presented assessing the technology and economics of hydrogen production by conventional and advanced processes. Six conventional processes are assessed including (1) steam reforming of natural gas, (2) partial oxidation of residual oil, (3) gasification of coal by Texaco process, (4) gasification of coal by Koppers-Totzek process, (5) steam-iron process and (6) water electrolysis. The advanced processes include (1) high temperature electrolysis of steam, (2) coal gasification and electrochemical shift, (3) integrated coal gasification and high temperature electrolysis, (4) thermal cracking of natural gas and (5) the HYDROCARB thermal conversion of coal. By-product hydrogen, thermochemical water splitting and high energy radiation, plasma and solar photovoltaic-water electrolysis production of H_2 are discussed.

It is concluded that steam reforming of methane is the most economic near-term process among the conventional processes. Processes based on conventional partial oxidation and coal gasification are two to three times more expensive than steam reforming of natural gas. New gas separation processes, such as pressure swing adsorption, improve the economics of these conventional processes. Integration of hydrogen production with other end-use processes has an influence on the overall economics of the system.

The advanced high temperature electrochemical systems suffer from high electrical energy and capital cost requirements. The thermochemical and high energy water splitting techniques are inherently lower in efficiency and more costly than the thermal conversion processes. The thermal cracking of methane is potentially the lowest cost process for hydrogen production. This is followed closely by the HYDROCARB coal cracking process. To reach full potential, the thermal cracking processes depend on taking credit for the clean carbon fuel by-product. As the cost of oil and gas inevitably increases in the next several decades, emphasis will be placed on processes making use of the world's reserve of coal.

1. INTRODUCTION

A study is presented herein assessing the technology and economics of hydrogen production by various means from fossil fuels. Included in the study are the well developed technologies that have been practiced on an industrial scale and the more advanced prospective processes. The technologies and process diagrams are described and estimates of the capital and the operating costs are discussed. A number of studies of hydrogen production costs have been made previously. This study attempts to bring together these estimates and to correlate the present (mid-1987) costs on a uniform economic basis.

The well developed conventional processes presented, include (1) Catalytic Steam Reforming of Natural Gas, (2) Partial Oxidation (POX) of Heavy Oil, (3) Coal Gasifications, (4) Steam-Iron Process, and (5) Electrolysis of water. Hydrogen as a by-product from other conventional sources is mentioned. Newer gas separation methods such as pressure swing adsorption for separation of hydrogen from process gas are also discussed. The unconventional advanced systems presented include (1) High Temperature Electrolysis (HTE) of Steam, (2) Electroconductive Membrane Process for Production of Hydrogen from Gasified Coal, (3) Combined Coal Gasification with High Temperature Electrolysis (CG-HTE), (4) Thermal Cracking of Natural Gas and (5) Thermal Cracking of Coal (HYDROCARB). Other advanced processes including thermochemical water-splitting high temperature plasma, photolytic and radiation systems are also briefly discussed.

The processes presented were all assessed at a hydrogen production capacity of 100×10^6 SCF/D. This capacity is large enough to take advantage of economy-of-scale and small enough to still be of use for synthetic fuel manufacture. The technologies and capital and operating cost estimates are compared including by-product credit values and some conclusions are reached.

2. HYDROGEN PRODUCTION BY CONVENTIONAL PROCESSES

Hydrogen by Catalytic Steam Reforming of Natural Gas

For several decades, steam reforming of hydrocarbons has been the most efficient, economical and widely used process for production of hydrogen and hydrogen/carbon-monoxide mixtures. The process basically involves a catalytic conversion of the hydrocarbon and steam to hydrogen and carbon oxides. Since the process works only with light hydrocarbons which can be vaporized completely without carbon formation, the feedstocks used range from methane, to naphtha to No. 2 fuel oil [1]. Fifty per cent of the hydrogen produced comes from the water (steam) when methane is used and 64.5% when naphtha is used.

A simplified basic flow diagram of the conventional steam reforming process is shown in Figure 1a. The process consists of three main steps: (1) synthesis gas generation, (2) water-gas shift and (3) gas purification. To protect the catalysts in the hydrogen plant, the hydrocarbons have to be desulphurized before being fed to the reformer. The desulphurized feedstock is then mixed with process steam and reacted over a nickel based catalyst contained inside of a system of high alloy steel tubes. The following reactions take place in the reformer:

$$C_n H_m + nH_2O \longrightarrow nCO + (n + m/2)H_2$$

$$CO + H_2O \rightleftharpoons CO_2 + H_2$$

$$CO + 3H_2 \rightleftharpoons CH_4 + H_2O$$

where n = 1 and m = 4, if the hydrocarbon feedstock is methane, and n = 1 and m = 2.2, if it is naphtha. The reforming reaction is strongly endothermic, with energy supplied by radiant combustion of fuel gas or oil. The metallurgy of the tubes usually limits the reaction temperature to 760-925°C (1400-1700°F). At the same time, the CO shift and methanation reactions quickly reach equilibrium at all points in the catalyst bed. The equilibrium composition of the reformed gas is favored by high steam to carbon ratio, low pressure and high temperature. To ensure a minimum concentration of CH_4 in the product gas, the process generally employs a steam to carbon ratio of 3 to 5 at a process temperature of around 815°C (1500°F) and pressures up to 3.5 Mpa (508 psi) to convert more than 80% of hydrocarbon to oxides of carbon at the outlet of the reformer. The typical composition (% by volume) of a synthesis gas at 100 psig leaving a steam-methane is as follows [2].

Component	Vol. %
H_2	74
CO	18
CO_2	6
CH_4	2
	100

After the reformer, the process gas mixture of CO and H_2 passes through a heat recovery step and is fed into a water-gas shift reactor to produce additional H_2. Normally, two stages of conversion (high temperature and low temperature) are used to reduce the CO content to a level of approximately 0.2 - 0.4 vol.%. This exothermic reaction occurs at temperatures ranging from 200° to 400°C (392 to 752°F).

The cold raw gases next pass through gas purification units to remove CO_2, the remaining CO, and other impurities to deliver the desired purified H_2 product to the battery limits of the plant. Several commercial processes for removing CO_2 exist. One of these systems uses a wet scrubbing process, such as monoethanolamine (MEA), hot potassium carbonate, and sulfinol. The residual CO and CO_2 remaining in the H_2 stream after CO_2 removal are further converted to CH_4 by a methanation reaction. The product H_2 leaves with a purity of approximately 97% to 98%.

An alternate technology to wet scrubbing is the use of the pressure swing adsorption (PSA). As can be seen in Figure 1b, this process reduces the number of unit processes and complexity of the operation by replacing the low temperature shift, CO_2 removal and methanation with a PSA process unit. In this process, the raw gas is passed through a series of beds of molecular sieves or activated carbon, where all components except H_2 are preferentially absorbed. The beds are regenerated by adiabatic depressurization at ambient temperature. The purge gas, which contains water vapor, CO_2, CO and CH_4, is then fed to the furnace for supplying heat to the reformer. The purity of H_2 from the PSA system can be 99% or higher, and can thus be used for a wide variety of chemical and petrochemical processes to considerable advantage.

Figure 1

Simplified Flow Diagram of Hydrogen Production from Methane Steam Refining

(a) Conventional Process, (b) with PSA Modification

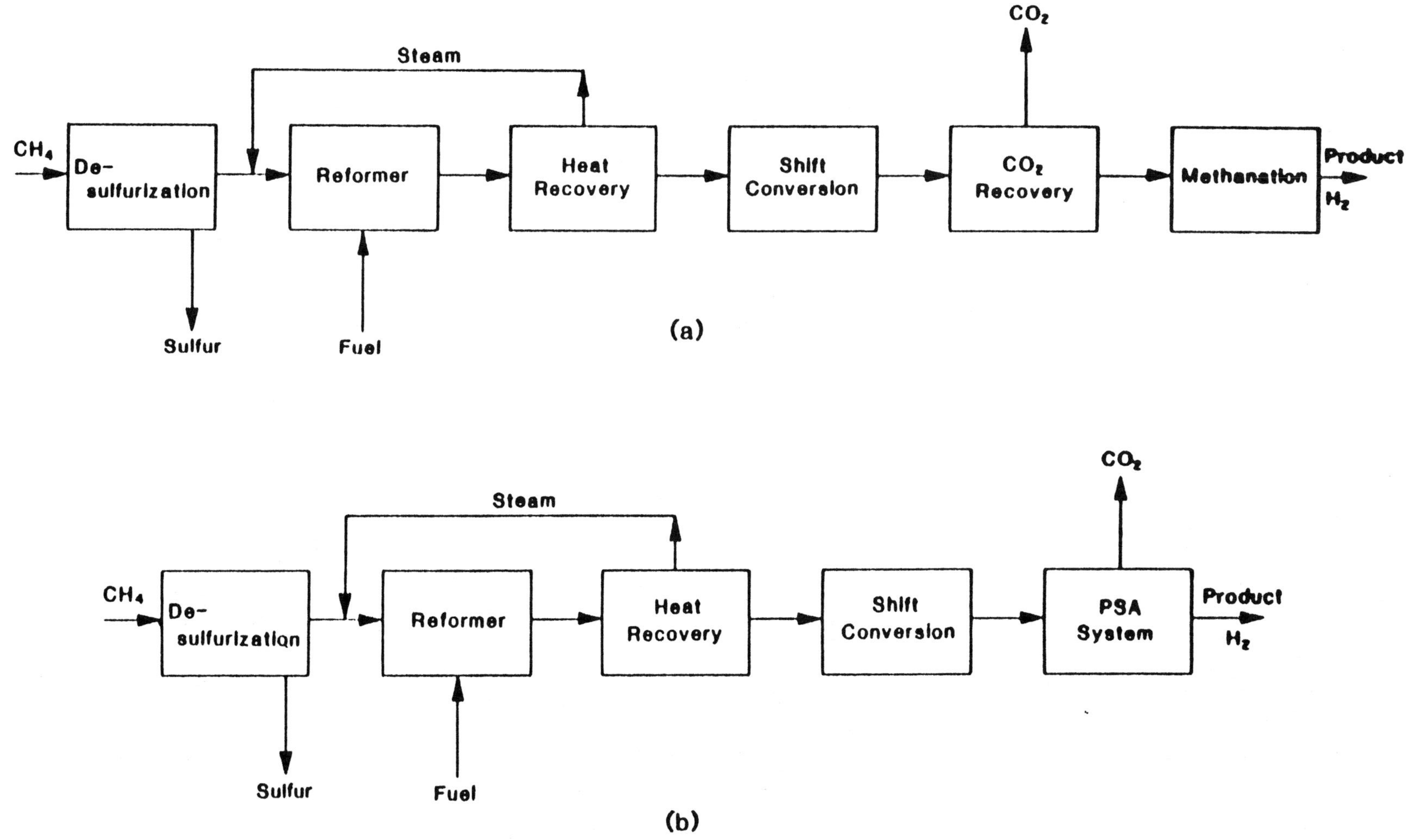

CH_4
De-sulfurization
Sulfur
Steam
Reformer
Fuel
Heat Recovery
Shift Conversion
CO_2 Recovery
CO_2
Methanation
Product H_2

(a)

CH_4
De-sulfurization
Sulfur
Steam
Reformer
Fuel
Heat Recovery
Shift Conversion
PSA System
CO_2
Product H_2

(b)

The capital investment and operation cost information for hydrogen produced from steam reforming of light hydrocarbons was obtained from References 1, 3, 4 and 5, adjusted to mid-1987 dollars. Based on the adopted economic assumptions given in Table 1, the capital investment of a 100 million SCF/D plant with PSA gas purification processes is $83.2 million as shown in Table 2. With natural gas as the feedstock at a base cost of $\$2.91/10^6$ Btu, the hydrogen production cost is $\$2.06/10^3$ SCF ($\$6.40/10^6$ Btu). By allowing credit for by-product steam at $\$5.00/10^3$ lb, the H_2 production cost is reduced to $\$1.90/10^3$ SCF ($\$5.90/10^6$ Btu).

The component costs as a percentages of the overall hydrogen production cost for steam reforming of natural gas are as follows:

	% of Production Cost
Feedstock (includes fuels)	60.7
Capital Cost	29.1
O&M Cost	10.2
	100.0

The cost of feedstock, which makes up 60% of the total production cost, has a significant effect on the hydrogen production cost. Future hydrogen prices are therefore heavily dependent on the trend in future feedstock prices.

The overall performance comparison of the PSA and conventional steam reforming of natural gas is shown in Table 3 for a 100×10^6 SCF/D hydrogen plant wtih a 90% on-line stream factor. Following the economic evaluation assumptions (Table 1), the extra capital cost of the PSA system is offset by the lower utilities and maintenance costs. Thus, the hydrogen production costs calculated for these two processes, without allowing for by-product (steam) credit, are very close (Table 3). However, by taking by-product credit, the cost of hydrogen produced from PSA is 10% lower than that of the conventional process.

The capacity of most steam reforming plants is in the range of 20-60 x 10^6 SCF/D. Single-train units are limited to 60 x 10^6 SCF/D by the size restriction of shop-fabricated construction of the reaction vessels. Economy of scale is essentially exhausted for a plant capacity of 100 x 10^6 SCF/D - the size on which we base the present cost estimates. For capacities over 60 x 10^6SCF/D, the PSA process generally requires parallel trains even using the advanced design of polybeds available from Union Carbide Corp.

Hydrogen From Partial Oxidation (POX) of Heavy Oil

The process basically involves the conversion of steam, oxygen and hydrocarbons to hydrogen and carbon oxides. The process proceeds at moderately high pressures with or without catalyst depending on the feedstock and process selected. The catalytic POX, which occurs at about 590°C (1100°F), will work with feedstocks ranging from methane to naphtha. The noncatalytic POX which occurs at 1150°-1315°C (2100-2400°F), can operate with hydrocarbons ranging

TABLE 1: BASIS FOR ECONOMIC EVALUATION

Time Frame	Mid-1987
Base Plant Capacity	100×10^6 SCF/D = 33×10^9 SCF/yr
Project Life	20 years
Operating Factor	328 D/Y (90% stream factor)
Capital Investment	
Facilities Investment (FI)	as obtained from literature and adjusted to mid-1987 basis
Interest During Construction	0.10 FI
Startup Expense	0.02 FI
Working Capital	60 days feedstock, utilities and cash supply
Total Capital Investment (TCI)	Sum of the above
Annual Operating Cost	
Feedstock and Fuel Cost	
Coal	\$40/ton (\$1.81/10^6Btu)
Natural gas	\$2.91/10^6Btu
Residual Fuel Oil	\$2.74/10^6Btu
Utilities	
Electricity	\$0.049/kWh
Process Water	\$0.5/1000 gal
Cooling Water	\$0.5/1000 gal
Cash Supply	
Operation	
Labor	\$21,750/man-year (\$11.00/man-hr)
Supervision	15% of operating labor
Maintenance	
Labor	2% of FI
Supervision	15% of maintenance labor
Material	2% of FI
Administrative & Support	
Labor	20% of all oper. & maint. labor
Payroll Extras - Fringe Benefits, etc.	20% of all labor
Insurance	2% of FI
General Administrative Expenses	2% of FI
Taxes (Fed, State, Local)	50% of net profit
Depreciation	Straight line, 5%/yr
Financing	100% equity (land cost not included)
Return on Investment (Cost of Capital)	20% of TCI before taxes
Gross Revenue Required	Sum of the above
By-Product Credits	
Sulfur	\$100/ton
Steam	\$5.00/10^3/lb
Power	\$0.03/kWh(e)
Oxygen	\$40/ton
Net Revenue Required	Gross Revenue Required - By-product Credits

Hydrogen Production Cost = Net Annual Revenue Required/Annual H_2 Production

from methane, heavy oil and coal. It is often referred to as gasification when coal is the feedstock. Currently, POX is mainly used to produce hydrogen from heavy hydrocarbons, which cannot readily be reacted over a catalyst. Due to the lower hydrogen-to-carbon ratios of the heavy feedstocks from methane, a larger portion of the hydrogen produced from POX comes from steam and thus more carbon dioxide is generated per unit of hydrogen produced. Water supplies 69% of the H_2 produced when heavy oil is used, while 83% when coal is used. Although the chemistry of both processes is similar, the engineering operations in the coal gasification is more complex than that in heavy fuel oil, since solids must be handled. Compared to steam reforming, POX requires additional facilities such as air separation plant to provide oxygen, as well as a larger shift and separation train.

A simplified basic flow diagram for hydrogen production by partial oxidation is shown in Figure 2. There are three main steps: (1) synthesis gas generation, (2) water-gas shift reaction, and (3) gas purification. The shift reaction and gas purification (either by the conventional wet scrubbing or by the PSA process), following the POX unit are similar to those of steam reforming.

In the synthesis gas generation step, the hydrocarbon feedstock is partially oxidized with oxygen and the carbon monoxide is shifted with steam to produce CO_2 and H_2. Because of the difficulty of separating nitrogen from hydrogen to produce a pure product, pure oxygen is used in the partial oxidation process. The partial oxidation reactions are typically as follows:

$$C_n H_m + n/2\ O_2 \longrightarrow nCO + m/2\ H_2 + \text{heat}$$

$$C_nH_m + nH_2O + \text{heat} \longrightarrow nCO + (n + m/2)H_2$$

$$CO + H_2O \rightleftharpoons CO_2 + H_2 + \text{heat}$$

where n = 1 and m = 1.3 for residual fuel oils and n = 1 and m = 0.8 for coal. While the hydrocarbon-oxygen reaction is exothermic, the additional energy required for the endothermic hydrocarbon-steam reaction, which is similar to steam reforming, is usually supplied by burning additional hydrocarbon feedstock. Because of high operating temperatures, 1150° to 1315°C (2100 - 2400°F), POX can reach favorable equilibrium conditions at much higher pressures than with a reformer. Typical composition (% vol.) of synthesis gas at 880 psig leaving a POX reactor with heavy oil as feedstock are as follows:

Component	Vol. %
H_2	46
CO	46
CO_2	6
CH_4	1
N_2 & Ar	1
	100

The sulphur contained in the feedstock is converted mainly into H_2S with a small portion into COS. After sulphur removal, the gas is treated in the same way as the product gas of the steam reforming process.

Figure 2

Simplified Flow Diagram Of Hydrogen Production From Residual Fuel Oil Partial Oxidation (POX)

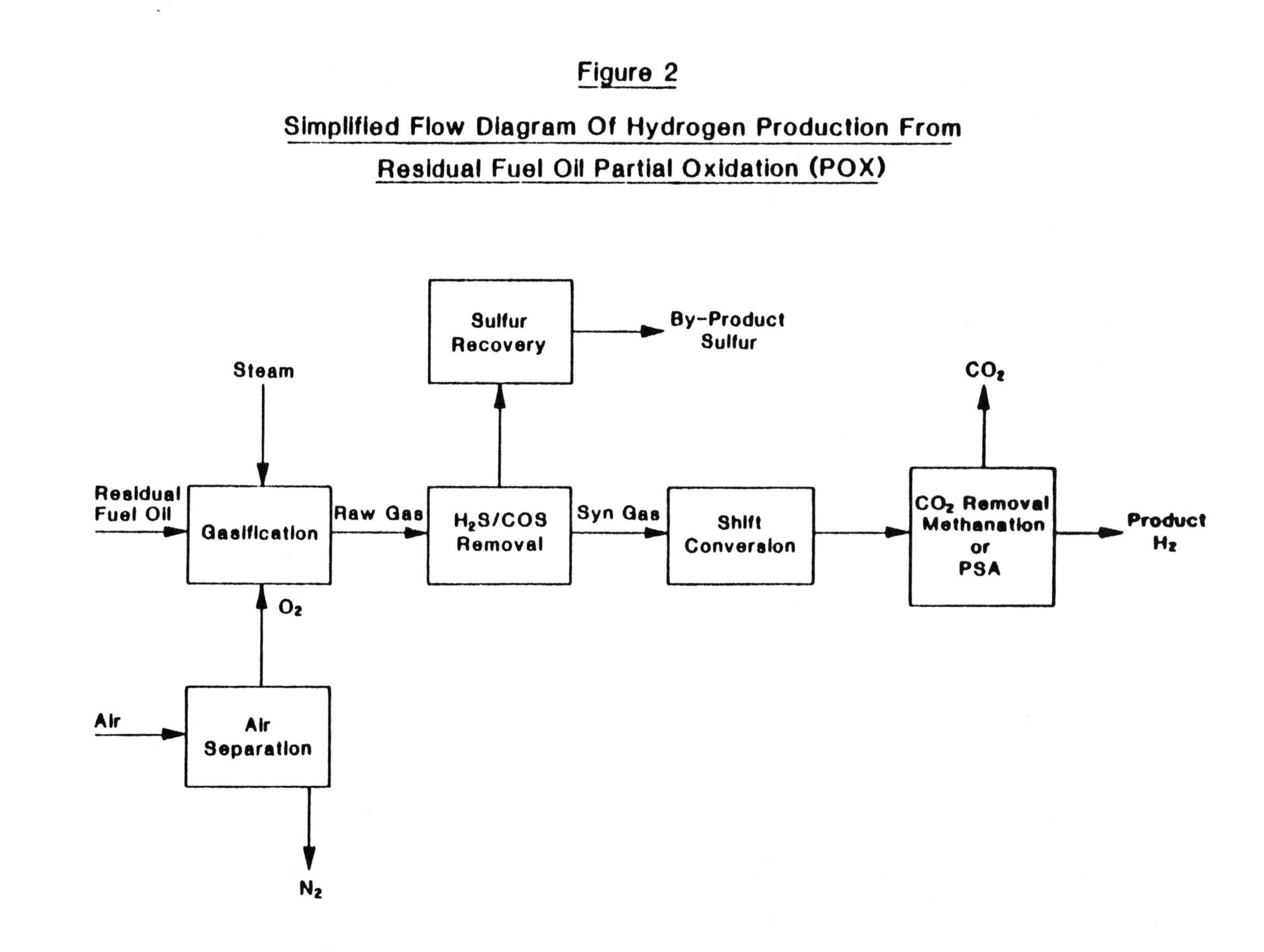

The capital investment and operating cost information for hydrogen produced from the partial oxidation of residual fuel oil was obtained from References 3, 5 and 6, adjusted to mid-1987 dollars. Following the economic evaluation basic assumption (Table 1), the capital investment of a 100 million SCF/D hydrogen plant is about $205 million as shown in Table 2. With a cost of residual oil at $2.74/10^6$ Btu, the hydrogen production cost is $3.12/10^3$ SCF ($9.70/10^6$ Btu), which is higher than for steam reforming of methane. Credit for the sulfur by-product from POX is too small to influence the overall hydrogen economics as given in Table 4.

The component costs as a percentage of the overall hydrogen production cost for the partial oxidation of residual fuel oil are as follows:

	% of Production Cost
Residual Fuel Oil Feedstock (includes fuels)	34.8
Capital Cost	47.9
O&M Cost	17.3
	100.0

As can be seen, under present economic conditions, the economic aspects of hydrogen derived from the partial oxidation of residual oil depends very much on the capital outlay involved as well as the price of residual oil. The hydrogen production cost is very sensitive to the capital cost, which accounts for 48% of the overall production cost. The detailed capital investment for various plant sections of the POX process is shown in Table 4 [6]. The POX reactor is less expensive than the steam reformer. However, the cost of the oxygen plant and the additional costs of the desulfurization steps make such a plant, capital intensive. The feedstock cost accounts for 35% of H_2 production cost in the partial oxidation of residual fuel oil, compared to 61% in steam reforming of methane (see Table 7).

For single train POX processes, the maximum plant capacity, limited by economic shop-fabrication construction is in the range of 60-75 MSCF/D. Larger plants are be built in multiple trains. For capacities over 100 MSCF/D - which is our base-line capacity, the capital costs are nearly proportional to capacity (varying with an exponential capacity factor of 0.8 - 0.9).

Hydrogen From Coal Gasification

The Koppers-Totzek and Texaco gasification processes, which use commercial or near commercial technologies, were chosen for hydrogen production in this study. The simplified block flow diagram for the two processes are shown in Figure 3.

In the Koppers-Totzek process, the pulverized coal is rapidly partially oxidized with oxygen and steam at essentially atmospheric pressure under slagging conditions. The typical composition (% vol.) of synthesis gas leaving the gasifier is as follows:

TABLE 2: COSTS FOR HYDROGEN PRODUCTION BY CONVENTIONAL PROCESSES
PRODUCTION CAPACITY - 100 X 10^6 SCF/D = 33 X 10^9 SCF/YR

Process	Steam Reforming	Partial Oxidation	Texaco Gasification
Feedstock	Natural Gas	Residual Oil	Bit. Coal
Capital Investment, $\$10^6$			
Facilities Investment (FI)	67.0	175.0	274.0
Interest During Construction	6.7	17.5	27.4
Startup Costs	1.3	3.5	5.4
Working Capital	8.2	8.5	9.6
Total Capital Required, $\$10^6$	83.2	204.5	316.4
Annual Operating Costs, $\$10^3$			
Feedstock	39,290	34,780	30,590
Utilities			5,800
Power	1,588	539	
Process Water	156	482	
Cooling Water	221	-	
Flue Gas Cleanup & Ash Disposal		525	970
Operation			
Labor	370	565	648
Supervision	55	85	97
Maintenance			
Labor	1,340	3,500	5,480
Supervision	201	525	822
Materials	1,340	3,500	5,480
Administrative & Support Labor	393	935	1,410
Payroll Extras	472	1,122	1,690
O&M for Combined Cycle By-Product Power Section	-	-	-
Insurance	1,340	3,500	5,480
General Administrative Expenses	1,340	3,500	5,480
Depreciation	3,350	8,750	13,700
Return on Investment	16,640	40,900	63,280
Gross Revenue Req'd x $\$10^3$	68,096	103,208	140,928
H_2 Prod. Cost, $\$/10^3$SCF	2.06	3.13	4.27
H_2 Prod. Cost, $\$/10^6$Btu	6.40	9.70	13.50
By-Product Credits, x $\$10^3$			
Sulfur	-	(1,000)	(2,570)
Steam	(5,500)	-	-
Power	-	-	-
Oxygen	-	-	-
Net Revenue Required, x $\$10^3$	62,596	102,208	138,358
H_2 Product Cost with By-Product Credits			
$\$/10^3$SCF	1.90	3.10	4.19
$\$/10^6$Btu	5.90	9.60	13.00

Table 2 (Cont.)

Process	Koppers-Totzek Gasification	Steam-Iron	Water Electrolysis
Feedstock	Coal	Coal	Elec. Power
Capital Investment, 10^6			
Facilities Investment (FI)	353.0	278.0	88.0
Interest During Construction	35.3	27.8	8.8
Startup Costs	7.1	5.6	1.8
Working Capital	10.2	12.2	33.2
Total Capital Required, x $\$10^6$	405.6	323.6	131.8
Annual Operating Costs, in $\$10^3$			
Feedstock	33,530	45,306	176,760
Utilities			
Power	-	-	-
Process Water	-	100	630
Cooling Water	2,765	-	-
Flue Gas Cleanup & Ash Disposal	-	-	-
Operation			
Labor	652	2,284	217
Supervision	98	342	33
Maintenance			
Labor	7,060	5,560	1,760
Supervision	1,059	834	264
Materials	7,060	5,560	1,760
Administrative & Support Labor	1,774	1,804	455
Payroll Extras	2,128	2,165	546
O&M for Combined Cycle By-Product Power Section	-	3,070	-
Insurance	7,060	5,560	1,760
General Administrative Expenses	7,060	5,560	1,760
Depreciation	17,650	13,900	4,400
Return on Investment	81,120	64,720	26,360
Gross Revenue Req'd x $\$10^3$	169,016	156,765	216,705
H_2 Prod. Cost, $\$/10^3$SCF	5.12	4.75	6.57
H_2 Prod. Cost, $\$/10^6$Btu	15.88	14.73	20.36
By-Product Credits, x $\$10^3$			
Sulfur	(640)	-	-
Steam	-	-	-
Power	-	(37,440)	-
Oxygen	-	-	(27,800)
Net Revenue Required, x $\$10^3$	168,376	119,325	188,905
H_2 Production Cost with By-Product Credits			
$\$/10^3$SCF	5.10	3.61	5.74
$\$/10^6$Btu	15.82	11.21	17.79

TABLE 3: COSTS AND PERFORMANCE DATA FOR THE CONVENTIONAL AND PSA STEAM REFORMING OF NATURAL GAS FOR 100 X 10^6SCF/D HYDROGEN PLANT*

	PSA	CONVENTIONAL
Capital Investment, $\$10^6$		
Facilities Investment (FI)	67.0	64.0
Interest During Construction	6.7	6.4
Startup Costs	1.3	1.3
Working Capital	8.2	8.3
Total Capital Required, $\$10^6$	83.2	80.0
Feed Plus Fuel, 10^6 Btu/hr	1,700	1,680
Electric Power, Kw(e)	4,000	4,330
Cooling Water, GPM	5,000	9,000
By-Product Steam, 10^3 μs/hr	140	23
H_2 Purity	99.9%	97%
Production Cost, $\$10^3$		
Feedstock	39,290	38,830
Utilities	1,980	2,210
Operation & Maintenance	6,850	7,080
Capital Cost	19,990	19,200
Gross Revenue Required	68,100	67,320
H_2 Prod. Cost, $\$/10^3$SCF	2.06	2.04
H_2 Prod. Cost, $\$/10^6$Btu	6.40	6.32
By-Product Credits, $\$10^3$		
Steam	(5,500)	(900)
Net Revenue Required	62,600	66,420
H_2 Production Cost with By-Product Credits		
$\$/10^3$SCF	1.90	2.01
$\$/10^6$Btu	5.90	6.33

*90% Stream Factor

TABLE 4: ESTIMATED CAPITAL INVESTMENT FOR VARIOUS PLANT SECTIONS OF A 100 X 10^6SCF/D HYDROGEN PLANT BY PARTIAL OXIDATION OF RESIDUAL OIL

Plant Section	Capital Cost, $\$10^6$
Gasification	38.2
Sulfinol	7.6
CO-Shift	19.8
Auxiliary Oil-Fired Boiler	10.2
CO_2-Removal & Methanation	13.3
Oxygen Plant	62.4
Total Inside Battery Limits	151.5
Total Outside Battery Limits[1]	53.0
Total Facilities Investment	204.5

[1]Estimated at 35% of Inside Battery Limits

Figure 3

Simplified Flow Diágram of Hydrogen Production from Coal Gasification
(a) Koppers-Totzek, (b) Texaco

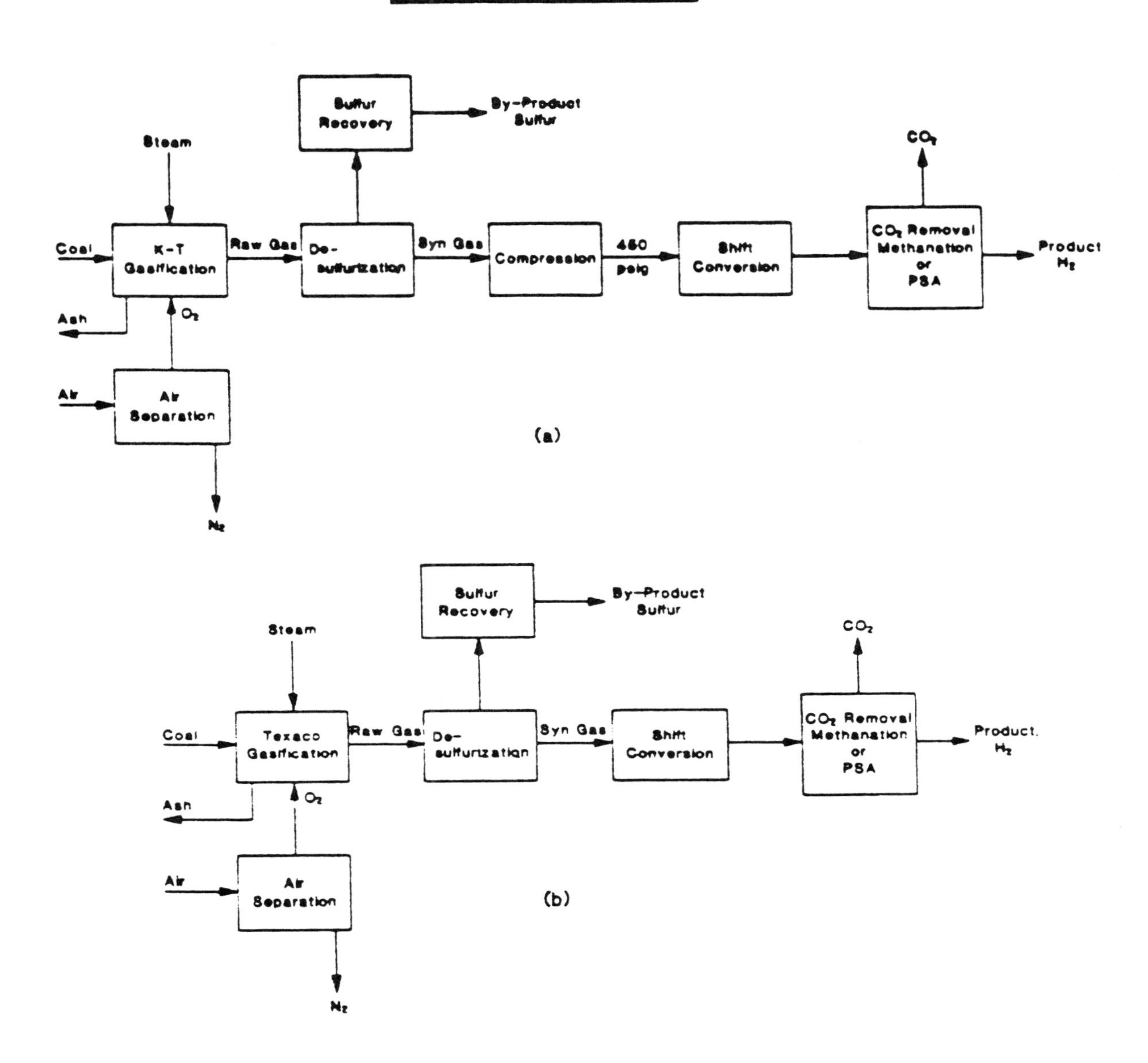

Component	Vol. %
H_2	29
CO	60
CO_2	10
N_2 & Ar	1
	100

The raw gas at 0 psig is then cooled to recover waste heat, followed by quenching with water to remove entrained ash particles before going through the steps of compression, shift and gas purification (by pressure swing adsorption or conventional wet scrubbing processes). The product hydrogen is at about 2.8 Mpa (400 psig) with a purity higher than 97.5%.

Since hydrogen compression is highly energy consuming, as is synthesis gas compression, and most of the hydrogen users today and in the future need hydrogen at elevated pressures, it is important to gasify coal at elevated pressures. By doing so, compression of the synthesis gas to the necessary pressure for future reactions is reduced or completely avoided. The Texaco gasifier is a higher pressure gasifier operating at around 5.5 Mpa (800 psi). By operating in the direct quench mode at this pressure, a high steam content in the synthesis gas is desirable for use in the shift reaction for hydrogen production. The typical composition (% vol.) of synthesis gas leaving the Texaco gasifier is as follows:

Component	Vol. %
H_2	34
CO	48
CO_2	17
N_2 & Ar	1
	100

The raw gas is then desulfurized, shifted, and purified either in a pressure swing adsorption system or a CO_2 removal system followed by methanation. The product hydrogen is at about 4.0 Mpa (600 psig) with a purity higher than 97%.

As previously mentioned, the reaction mechanisms of coal gasification resemble very much those of the partial oxidation of heavy fuel oils. However, the coal gasification processes are complicated by the necessity to handle solid fuels and to remove large amounts of ash. The solids-handling problem has a significant impact on costs and prevents much of the technology and equipment developed for petroleum from being used in the conversion of coal. The capital investment and operation cost information for hydrogen produced from coal gasification was obtained from References 3, 4, 5 and 7. Following the economic evaluation basis (Table 1), the capital investment of a 100 million SCF/D plant is $316 million for the Texaco gasification process and $405 million for the Koppers-Totzek process. With coal at a base cost of \$40/ton (\$1.81/10^6 Btu), the hydrogen production cost is \$4.27/10^3 SCF (\$13.27/10^6 Btu) for the Texaco gasification process and \$5.12/10^3 SCF (\$15.88/10^6 Btu) for the Koppers-Totzek process. Credit for the sulfur by-product from coal gasification is too small to have a significant impact on overall hydrogen economics.

The component costs as a percentage of the overall hydrogen production cost for coal gasification are as follows:

	% of Production Cost	
	Texaco	Koppers-Totzek
Feedstock (includes fuels)	25.8	21.5
Capital Cost	54.6	58.4
O&M Cost	19.6	20.1
	100.0	100.0

It clearly shows that the hydrogen production costs for coal gasification depend very much on the capital investment involved followed by the price of coal. Generally, the capital for the coal gasification section is the major portion of the capital cost of the coal gasification plant.

As can be seen in summary Table 7, (page 34), the production costs of hydrogen from coal are much higher than those for natural gas reforming based on the present cost conditions for coal and natural gas in the U.S. Even at zero cost of coal, the hydrogen production from coal would still be barely competitive with natural gas reforming. The reason for the higher costs of hydrogen from coal is the high investment needed for the gasification plant, and not the lower energy efficiency. The capital cost of producing hydrogen generally is highest for the coal gasification, next highest for partial oxidation of residual oils and lowest for natural gas reforming. It is only when we assume no capital cost at all, that hydrogen produced from coal would be competitive with hydrogen from natural gas reforming. In other words, in order to produce hydrogen from coal economically, we need coal gasification processes that allow simpler plants requiring less capital investment, rather than more sophisticated plants which generally require higher capital investment.

Compared with the optimized plant capacity of 60 x 10^6 SCF/D for steam-reforming facilities, the economical coal conversion units can be as large as 100 x 10^6 SCF/D. For capacity over 100 x 10^6 SCF/D, the capital costs are nearly proportional to capacity (i.e., the exponential capacity factor is 0.8 - 0.9).

Hydrogen From the Steam-Iron Process

The steam-iron process involves basically reacting steam with hot iron to produce a H_2-rich gas and an iron oxide. Although it is a coal-based process, hydrogen is derived from the decomposition of steam by reacting with iron oxide, rather than synthesis gas generated from coal. The synthesis gas is mainly used for the reduction of iron oxide to iron. Since hydrogen is not derived from the synthesis gas, air can be used in the gasifier. In addition, the iron oxide "barrier" prevents nitrogen from significantly contaminating the hydrogen.

A basic block flow diagram for the process is presented in Figure 4. It consists of four main steps: (1) coal gasification, (2) iron regeneration, (3) hydrogen generation, and (4) purification. Coal is gasified with steam and air to produce a synthesis gas. In the iron regenerator, the synthesis gas is mixed with iron oxide and the following reduction reactions occur:

$$Fe_3O_4 + H_2 \longrightarrow 3\ FeO + H_2O$$

$$Fe_3O_4 + CO \longrightarrow 3\ FeO + CO_2$$

$$FeO + H_2 \longrightarrow Fe + H_2O$$

$$FeO + CO \longrightarrow Fe + CO_2$$

The synthesis gas, CO and H_2, is not completely converted in reducing iron oxide. Heating value plus sensible heat at 825°C (1520°F) in the gas exiting from the iron regenerator represents about 54% of the input coal heating value. Excluding about 15% of the coal heating value in this stream used in the plant to compress air and to generate steam, most of the remaining heating value is used to generate power with a gas-steam combined cycle turbine generator. With a plant capacity of 100 x 10^6 SCF/D of hydrogen, the plant capacity of by-product electric power amounts to 158 MW(e). In a sense, this is a cogeneration plant, e.g., hydrogen and electrical power.

The regnerated iron then enters the steam-iron reactor where it is oxidized by steam to produce Fe_3O_4 and a hydrogen-rich gas. At temperatures of about 815°-870°C (1500-1600°F), the following reaction occurs in

$$3FeO + H_2O \longrightarrow Fe_3O_4 + H_2$$

The effluent from the steam-iron reactor contains 37% hydrogen and 61% steam plus small amounts of nitrogen and carbon oxides. By condensing out the steam, the hydrogen concentration increases to about 96% with 1.6% carbon oxides and 2.5% nitrogen. A methanation reaction, which follows, reduces the carbon oxides to 0.2%. There is no need for CO shift or acid gas scrubbing.

The capital investment and operating cost information for hydrogen produced by the steam-iron process was obtained from Reference 3, adjusted to mid-1987 dollars. Following the basic economic assumptions of Table 1, the capital investment for a 100 million SCF/D hydrogen plant is $324 million (Table 2). With coal at a base cost of $1.81/10^6 Btu, the hydrogen production cost is $4.75/10^3 SCF ($14.73/10^6 Btu). Credit for by-product electric power at a price of $0.03/kWh is very substantial in that the hydrogen production cost is reduced to $3.61/10^3 SCF ($11.21/10^6 Btu) (Table 2). It should be noted that this process relies heavily on by-product credit for steam or power which amounts to 24% of the gross production cost.

The component costs as a percentage of the overall hydrogen production cost for the steam-iron process are as follows:

Figure 4

Simplified Flow Diagram of Hydrogen Production from Steam-Iron Process.

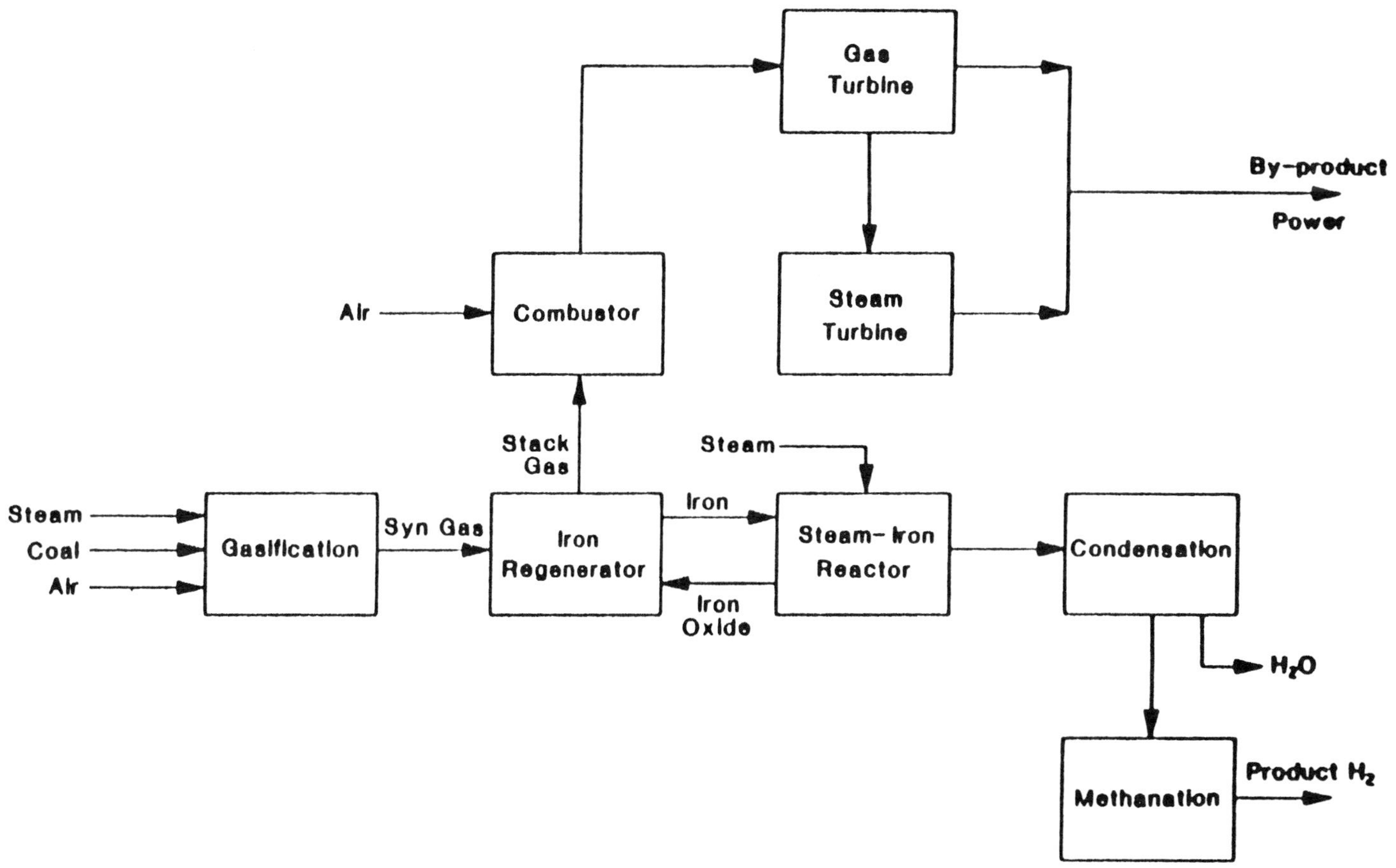

Gas Turbine
By-product Power
Air
Combustor
Steam Turbine
Stack Gas
Steam
Steam
Coal
Air
Gasification
Syn Gas
Iron Regenerator
Iron
Iron Oxide
Steam-Iron Reactor
Condensation
H_2O
Methanation
Product H_2

	% of Production Cost
Feedstock	29.1
Capital Cost	50.1
O&M Cost	20.8
	100.0

Similar to the coal gasification and to partial oxidation processes, the economic aspects of hydrogen generated from steam-iron process depends strongly on the capital investment involved followed by the price of coal.

Hydrogen From Electrolysis of Water

Water electrolysis is today the only industrial hydrogen production process which does not necessarily rely on fossil energy. The required electrical energy can derive either from fossil energy or nuclear and solar energy based. Although presently not economically competitive with hydrogen produced from hydrocarbons, electrolytic hydrogen has several advantages such as high product purity and flexibility of operation. An electrolysis plant can inherently operate over a wide range of capacity factors which is attractive for electricity load levelling either at the generation site or anywhere on the grid. In addition, the technology is convenient for small operating capacities.

Water electrolysis involves basically splitting water into hydrogen and oxygen by passing a direct electrical current through water that has been made electrically conducting by addition of hydrogen or hydroxyl ions (e.g., alkaline potassium hydroxide). The simplified block flow diagram for water electrolysis is shown in Figure 5. The process consists of a water purification plant, an electrolyzer section, gas separation, power transformer, regulator and rectifiers as well as compressor systems. The current state-of-the-art offers very reliable electrolyzers which have an energy efficiency (based on the H_2 enthalpy) of up to 80% with current densities of about 2000 A/M^2 [8]. Water electrolysis is a modular operation technology. The rate of hydrogen generation per unit electrode area is related to the current density. In general, the higher the current density, the higher is the cell voltage. Thus, the power cost per unit of hydrogen produced increases with current density, while the size of the electrolysis plant, and hence the capital investment, decreases. The designer of a new electrolysis plant must make a trade-off between capital cost and operating cost to minimize the total.

The capital investment and operating cost information for hydrogen produced by water electrolysis was obtained from References 3 and 4, adjusted to mid-1987 dollars. Based on the economics evaluation basis (Table 1), the capital investment of a 100 million SCF/D hydrogen plant is $132 million (Table 2). With electricity at a base cost of $0.049/kWh, the hydrogen production cost is $\$6.57/10^3$ SCF ($\$20.36/10^6$ Btu). Credit for by-product oxygen is substantial, amounting to about $\$0.83/10^3$ SCF hydrogen, but it is still less than the price difference between electrolytic and fossil-based hydrogen.

The component costs as a percentage of the overall hydrogen production cost for water electrolysis are as follows:

Figure 5

Simplified Flow diagram of Hydrogen Production from Water Electrolysis.

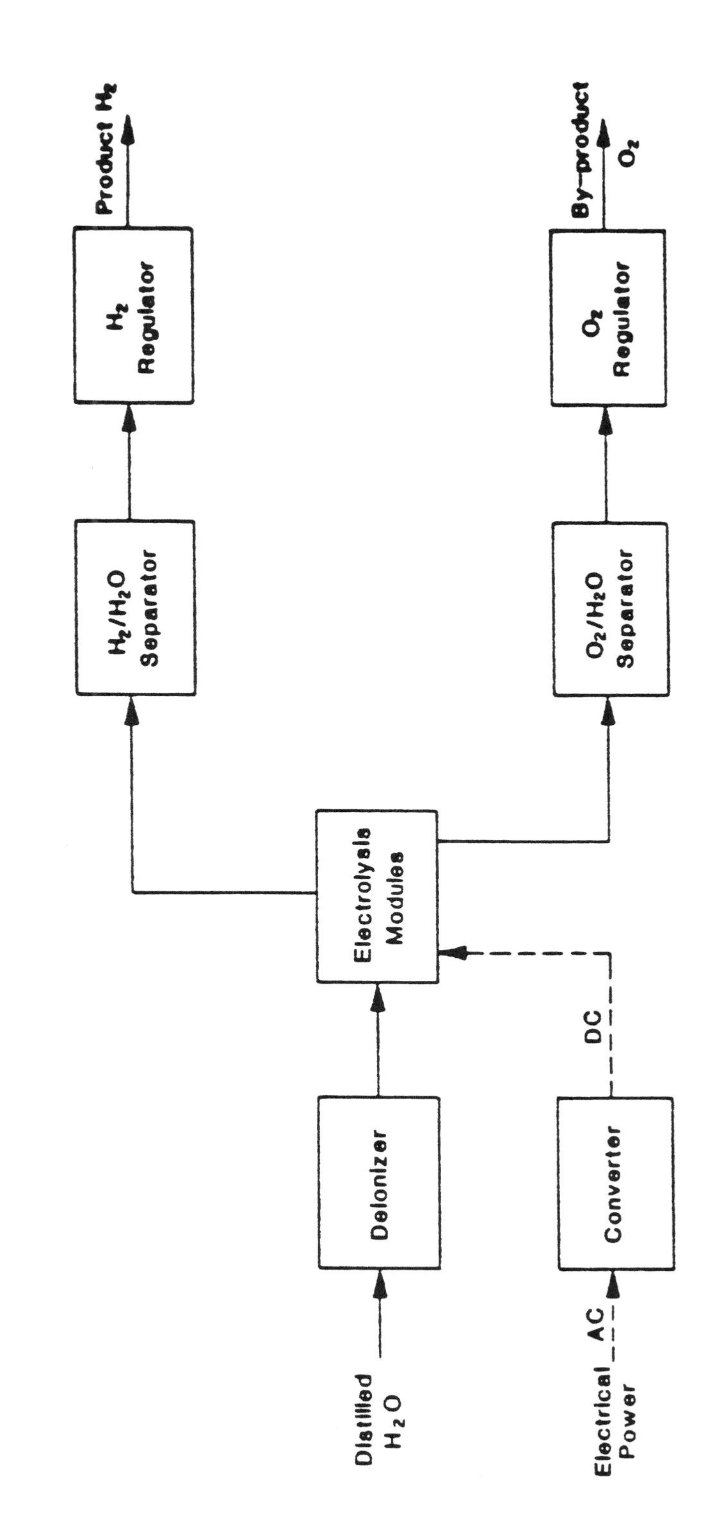

	% of Production Cost
Feedstock	81.9
Capital Cost	14.1
O&M Cost	4.0
	100.0

The cost of electricity, which amounts to 82% of the total production cost, dominates the price of electrolytic hydrogen in the U.S. Unless low-cost electricity is available, for example from hydropower or in the future possibly generated by the high-temperature nuclear reactors, energy costs as a percentage of total production cost will remain high. This puts electrolysis in a poor competitive position relative to fossil-based technologies unless or until fossil fuel costs increase significantly.

The solid polymer electrolytic (SPE) processes developed by General Electric, and high-temperature steam electrolysis (HTE) are still in the research stage and are discussed in the following sections of this report.

3. HYDROGEN AS A BY-PRODUCT FROM OTHER CONVENTIONAL PROCESSES

There are a number of processes which produce hydrogen in varying capacities as a by-product [9] some of these are mentioned as follows:

1. Hydrogen is a by-product of the production of counter-chlorine from the electrolytic chlor-alkali process.

$$2NaCl + 2H_2O = 2NaOH + Cl_2 + H_2$$

2. Hydrogen is produced as light gases from crude oil refinery processes.

3. Hydrogen is produced in the production of coke from coal in coke oven gases.

$$\underset{\text{(coal)}}{CH_{0.8}\,O_{0.2}} = 0.8C + 0.2CO + 0.4H_2$$

4. Hydrogen is emitted in chemical dehydrogenation processes (e.g. ethylene plant off-gas; ammonia dissociation, hydrodealkylation).

$$C_2H_6 = C_2H_4 + H_2$$

Even for the large scale chlor-alkali process, generally these by-product sources have not been of sufficient capacity and are too costly to warrant serious consideration as a large-scale supply of hydrogen. For example, a 1000 T/D caustic soda plant would produce only 9.5 x 10^6 SCF/D of H_2 which is more than 10 times less than the capacity of plants evaluated in this study.

Hydrogen can also be produced by the following two methods.

1. Decomposing ammonia to nitrogen and hydrogen and the hydrogen recovered

$$2NH_3 = N_2 + 3H_2$$

2. Decomposing methanol to carbon monoxide and hydrogen followed by a CO shift reaction and the total hydrogen recovered.

$$CH_3OH = CO + 2H_2$$
$$CO + H_2O = CO_2 + H_2$$

However, the above two processes are not prime sources of hydrogen. They are generally regarded as means for storing and transporting hydrogen conveniently in the form of liquid ammonia and methanol.

4. HYDROGEN PRODUCTION BY ADVANCED PROCESSES

High Temperature Electrolysis of Steam

A significant improvement in the electrolytic production of hydrogen can be achieved by the high temperature electrolysis (HTE) of steam. The low temperature water electrolysis process for hydrogen production as described previously under conventional processes depends on the electrolytic conduction of an aqueous media such as alkaline or acid solutions. Most of the energy of splitting water comes from the electrical energy fed to the cell. However, by raising the temperature of the cell an increasing amount of the energy for decomposing water can be thermal energy according to the second law of thermodynamics:

$$\Delta G = \Delta H - T\,\Delta S$$

and

$$\Delta G = nFe$$

where

ΔH is the enthalpy of water decomposition
ΔG is the free energy of water decomposition
ΔS is the entropy of water decomposition
T is the absolute temperature
e is the voltage or energy of the cell
n is the number of electron equivalents in producing a molecule of H_2.

The electronic equations for HTE at each electrode are as follows:

Cathodic Reaction $2H_2O + 4\,e^- = 2H_{2(g)} + 2O_2^=$

Anodic Reaction $2O_2^= = O_{2(g)} + 4e^-$

At temperatures where water exists as steam the conduction must occur by ionic means rather than by electrolytic as in aqueous condensed systems. Ionic conductors have been developed by a number of research groups including

Doenitz et al. [10] and Westinghouse Electric Corporation [11]. They consist of a porous metal oxide ceramic, usually zirconium oxide (ZrO_2) stabilized with yttrium oxide (Y_2O_3). Electrodes are placed against a thin slab of gas-tight ZrO_2-(Y_2O_3). Steam is supplied on one side of the ionic conductor. On applying a direct current electrical field, hydrogen forms at the cathode while oxygen ions migrate through the ionic conductor and form oxygen molecules at the anode. The ionic conductors can be arrayed in various geometrical configurations (such as in tubular form) to produce a single cell which can be placed either in parallel or series circuitry to form a battery. These in turn can be hooked up in a bank of batteries to form a large scale reactor.

Several HTE process designs have been developed based on coal as the primary energy source [12]. One lower cost system shown in Figure 6 feeds electricity and steam from a coal-fired steam power station and coal-fired steam heater. Saturated steam from a power plant is preheated to 1100°C (2012°F) in ceramic conductor tubes by a coal-fired heater before being fed to the HTE plant. The steam is electrolyzed with a temperature drop of 200°C (360°F). The process can be designed to convert 50% to 90% of the steam to hydrogen and oxygen with an overall thermal efficiency of 35 to 38% based on 35% electrical power generation from coal. For increased hydrogen productivity of the cell, current densities can vary between 3000 to 5000 amps/M^2 with cell potentials varying between 1 and 1.6 volts.

A distribution of the hydrogen production cost is as follows:

	% of Production Cost
Feedstock (includes fuels)	66.2
Capital Cost	17.6
O&M Cost	16.2
	100.0

It is noted that the feedstock cost which is primarily electricity strongly dominates the cost of H_2 production and is therefore very sensitive to the cost of electricity. Furthermore, the by-product oxygen can reduce the production cost by 16.6% (Table 5).

Electroconductive Membrane Process for the Production of Hydrogen from Gasified Coal

In the realm of advanced electrochemical techniques for hydrogen production there exists a combination process wherein gasified coal is electrochemically treated to produce a highly concentrated hydrogen gas and by-product steam [13].

The gasification of coal is generally represented by the following reaction

$$C + H_2O = CO + H_2$$

Figure 6

High Temperature Electrolysis (HTE) of Steam.

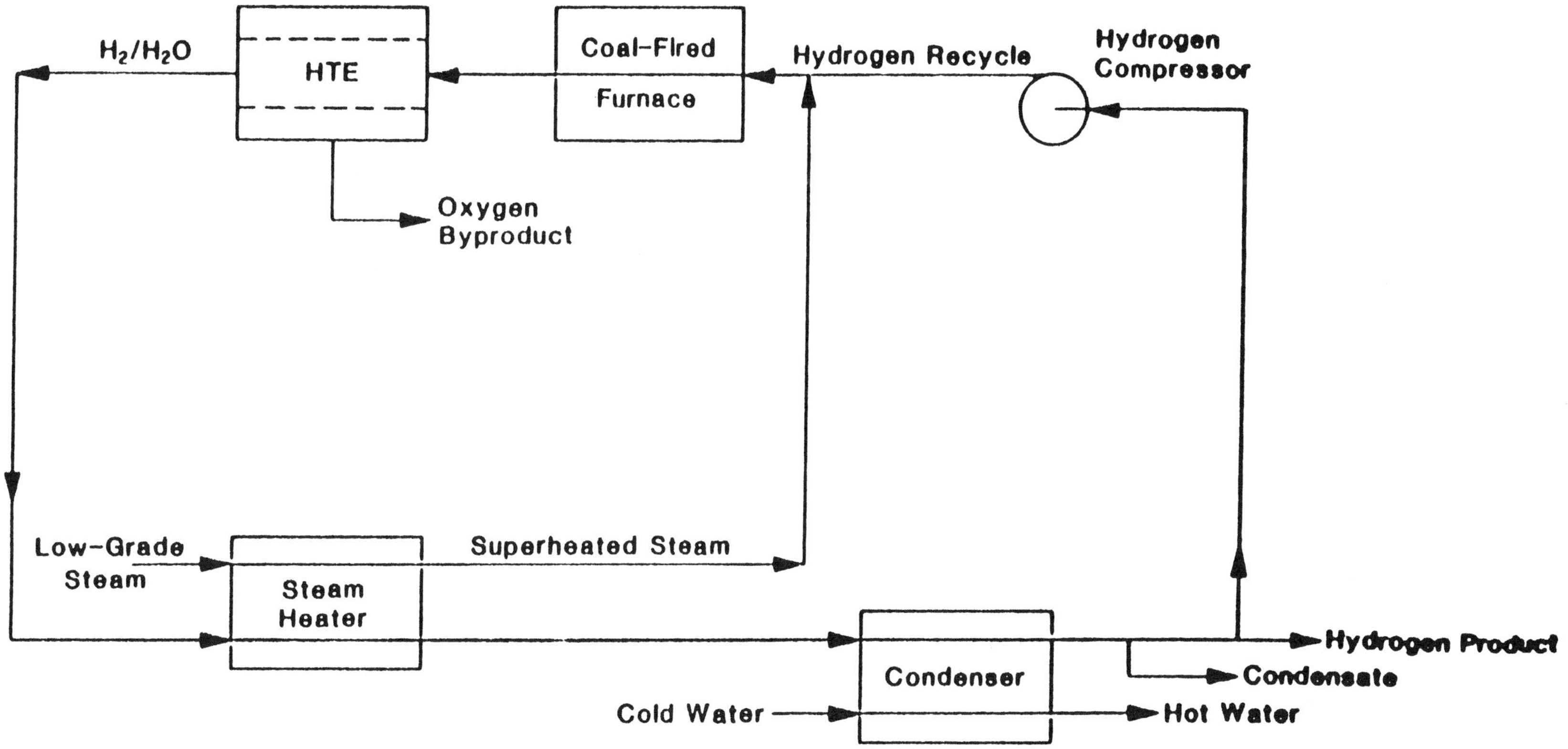

H_2/H_2O

HTE

Coal-Fired

Furnace

Hydrogen Recycle

Hydrogen

Compressor

Oxygen

Byproduct

Low-Grade

Steam

Steam

Heater

Superheated Steam

Condenser

Cold Water

Hydrogen Product

Condensate

Hot Water

TABLE 5: COST FOR PRODUCING HYDROGEN BY ADVANCED ELECTROCHEMICAL PROCESSES
PRODUCTION CAPACITY = 100 X 10^6 SCF/D = 33 X 10^9 SCF/YR

Process	High Temperature Electrolysis of Steam (HTE)	Combined Coal Gasification Electrochemical Shift (Westinghouse)	Combined Coal Gasification High Temperature Electrolysis (GC-HTE)
Feedstock	Electric Power	Coal	Bit. Coal & Electical Power
Rate	285 MW(E)	3800 T/D	95 MW(e) and 550 T/D Coal
Capital Rate Investment, $\$10^6$			
Facilities Investment	118	232	208
Engineering, Startup, Contingencies, Etc.	23	30	37
Total Capital Required	141	262	245
Annual Operating Cost, $\$10^3$			
Feedstock	110,800	50,200	57,300
Power, Water, Chemical Disposal	15,600	8,100	6,400
Operating Labor	220	3,500	2,230
Maintenance, Overhead & Insurance (10% F.I.)	11,080	23,200	20,800
Capital Charges			
Depreciation (5% on F.I.)	5,900	11,600	10,400
Return on Investment (20% on T.C.I.)	23,600	52,400	49,000
Total H_2 Production Cost	167,200	149,000	146,130
$\$/10^3$ SCF	5.07	4.52	4.43
$\$/10^6$ BTU	15.70	14.03	13.76
By-Product Credit, $\$10^3$	-27,800 (O_2)	-59,400 (steam)	-----
Net Production Cost	139,400	89,600	146,430
$\$/10^3$ SCF	4.22	2.72	4.43
$\$/10^6$ BTU	13.12	8.40	13.76

The second step in the process consists of cleaning and shifting the CO to form hydrogen and CO_2 by reacting with steam. This step is conducted in a high temperature electrolysis cell across a gas impervious mixed ionic-conducting ceramic membrane. The overall reaction for converting the CO is the chemical shift reaction,

$$CO + H_2O = CO_2 + H_2$$

However, unlike the conventional water gas shift reaction which is carried out thermocatalytically, it is unnecessary to remove sulfuric impurities to a high degree prior to shifting and to remove CO_2 after the shift reaction. This results in a savings in process cost compared to the conventional process.

The core technology for this advanced shifting process is the ability of certain ceramic materials, e.g., doped stabilized zirconia, to conduct oxygen ions at high temperature while simultaneously being electronic conductors. On one side of the membrane the following reaction occurs (cathodic)

$$H_2O + 2e^- \longrightarrow H_2 + O^+$$

and on the other side (anodic)

$$CO + O^= \longrightarrow CO_2 + 2e^-$$

The overall reaction is the shift reaction given above. The oxide membrane thickness is about 20 microns. When steam passes on one side of the membrane it will be reduced at reaction sites along the membrane by the addition of electrons from the membrane resulting in the production of hydrogen molecules and the release of oxygen ions to the membrane. On the opposite side of the membrane, at reaction sites, carbon monoxide will be oxidized by reaction with the oxygen ions, which have been transported across the membrane to form carbon dioxide and release electrons to the membrane. The oxygen ions move across the membrane due to the large difference in oxygen activity on each side of the membrane. This process will continue as long as the flow of the separate reactants continues maintaining the oxygen activity gradient across the membrane.

Typically this process would be carried out at approximately 1000°C (1832°F) and the partial pressure of oxygen in steam would be about 10^{-8} atm. Compared to the corresponding partial pressure of oxygen in carbon monoxide at this temperature, which is about 10^{-18} atm, this ten orders of magnitude gradient would result in a theoretical voltage across the membrane of about 630 mV. However, this voltage would be reduced by the resistance of the material to ionic and electronic direct current flow, as dissipated heat (IR losses). Further reduction in the voltage across the film would result from so-called irreversible polarization losses due to oxygen adsorption and desorption reactions on the surfaces. Also in an actual reactor, where the thin membrane would be supported on a porous substrate, there would be diffusional losses through the substrate, associated with getting reactants and products to and away from electrolyte surface reaction sites.

The technology for the high temperature electrolytic membrane reactor is based on the development work for the high temperature electrolysis technology. Since the free energy for the above shift reaction is exothermic, an impressed electrical potential is not required to drive the reaction. Thus no electrical power is needed for this process.

A preliminary process design and economic estimate has been attempted as follows.

A simple flow diagram of the coal-to-hydrogen plant is shown in Figure 7, consisting of three major plant sections: The Fuel Gas Generation Section; the Hydrogen Production Section; and the Steam Generation Section.

The Fuel Gas Generation Section consists of the following major components and subsystem for a 100 x 10^6 SCF/D hydrogen plant,

Coal receiving, storage, preparation, handling and feeding
Dolomite receiving, storage, preparation, handling and feeding
Coal gasifier modules (6) with in-bed desulfurization and two stages of recycle cyclones
Air blowers
Hot-gas particulate removal using a high-temperature ceramic bag house
Ash transport, processing (calcium sulfide oxidation) and disposal
Hot-gas piping

The Hydrogen Production Section consists of:

Hydrogen production reactor
Hydrogen dryer system

The Steam Generation Section consists of:

Fuel gas combustion system
Heat recovery steam generation system

The selected process design operates at near atmospheric pressure using air-blown fluidized bed gasifier technology and incorporates desulfurization with limestone or dolomite. The gasifier operates in the range of 870° to 1040°C (1600 to 1900°F). The process performance is sensitive to the selected gasifier pressure and temperature, the gasifier type and the oxidant type (air or oxygen).

The process considered here is designed so that the steam generated from the drying of the hydrogen stream and from the utilization of the lean fuel gas exhausted from the hydrogen production reactor is sufficient to provide all of the plant steam requirements and air compression requirements while minimizing steam generated for export which is substantial. This selection is based on the commodity value of hydrogen (about $8.00/MSCF) being higher than the equivalent commodity value of industrial steam or electricity (at about $5/1000 Lb or $0.039 per kWh). The plant coal feed rate is about 3800 tons per day.

Except for the high temperature electrolytic membrane, most of the plant components are commercial or near commercial items. The electrolytic membrane reactor is conceived to be of a tube-in-shell design and is based largely on the extensive development and commercialization work performed by Westinghouse on solid electrolyte fuel cells. The tubes are taken as 1.3 cm diameter by 100 cm long, with a 0.1 cm wall thickness based on presently established extrusion and coating technology. The assumed hydrogen production rate per

Figure 7

Simplified Flowsheet for Coal to Hydrogen by Gasification and Electrochemical Shift.

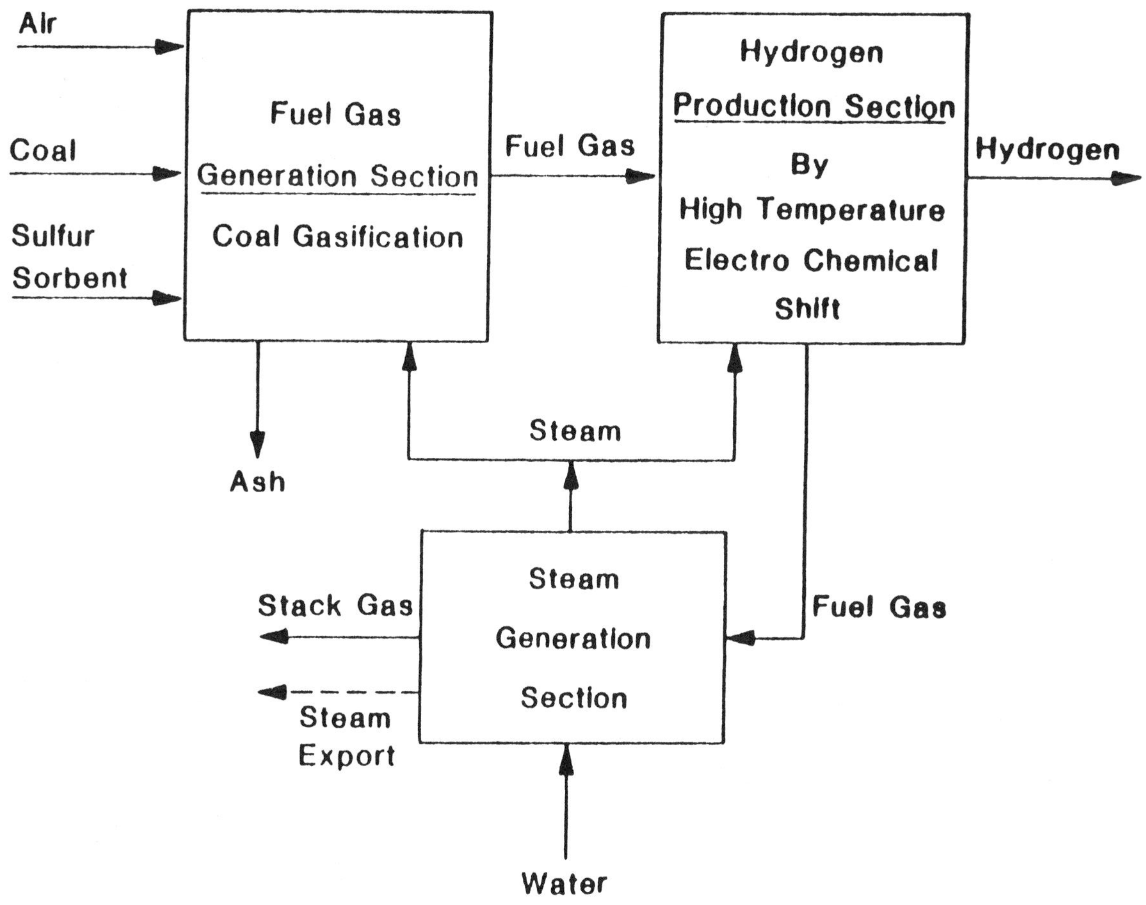

unit of tube area is 33 $ml/min-cm^2$, and the evaluation depends on how successful the technology development will be. The reactor costing is based on about 150,000 tubes, each weighing about 390 gms. With a stabilized zirconia price of $15/Kg, the resulting tube material cost is about $880,000. Based on solid electrolyte fuel cell experience, the manufactured tube cost of will be about $2,640,000. The total reactor cost for the installation is estimated to be less than $20,000,000, which is a relatively small portion of the total plant cost.

The plant consumes the following raw materials and power:

coal: 3800 tons/day
dolomite: 1050 tons/day
water: 4200 tons/day
auxiliary power: 5,000 kW

The plant produces for export about 1,500,000 lb/hr steam, with the pressure and temperature depending on the end use.

Costs are reported for a site located in western Pennsylvania, with an expected accuracy of 40%. A contingency of 15% is included on the process investment. Interest during construction is not included.

The distribution of production costs is shown as follows.

	% of Production Cost
Feedstock (includes fuels)	33.7
Capital Cost	43.0
O&M Cost	23.3
	100.0

The production cost is very sensitive to the gasifier type. By allowing credit for by-product steam at $\$5.00/10^3$ lb, the hydrogen product cost can be reduced by as much as 40% (Table 5).

Combined Coal Gasification with High Temperature Electrolysis (CG-HTE)

An advanced hybrid coal gasification - HTE process is useful because the two process systems complement each other. The coal gasification system produces synthesis gas (CO and H_2), and steam at high temperature which can be used in the HTE. The HTE produces H_2 and O_2 from the steam and the O_2 produced in the HTE, can in turn be used in the gasification of coal. A simplified flow sheet of the process is shown in Figure 8. An analysis of the system based on the contribution to hydrogen production from each section of the process can be obtained as follows.

In the gasification unit the reaction stoichiometry and energetics are:

$C + H_2O = CO + H_2$ Endothermic, $\Delta H = +42$ Kcal/mol
$0.5\ C + 0.5\ O_2 = 0.5\ CO_2$ Exothermic, $\Delta H = -47$ Kcal/mol

And for the shift reaction:

$CO + H_2O = CO_2 + H_2$ Neutral, $\Delta H = 0$ Kcal/mol

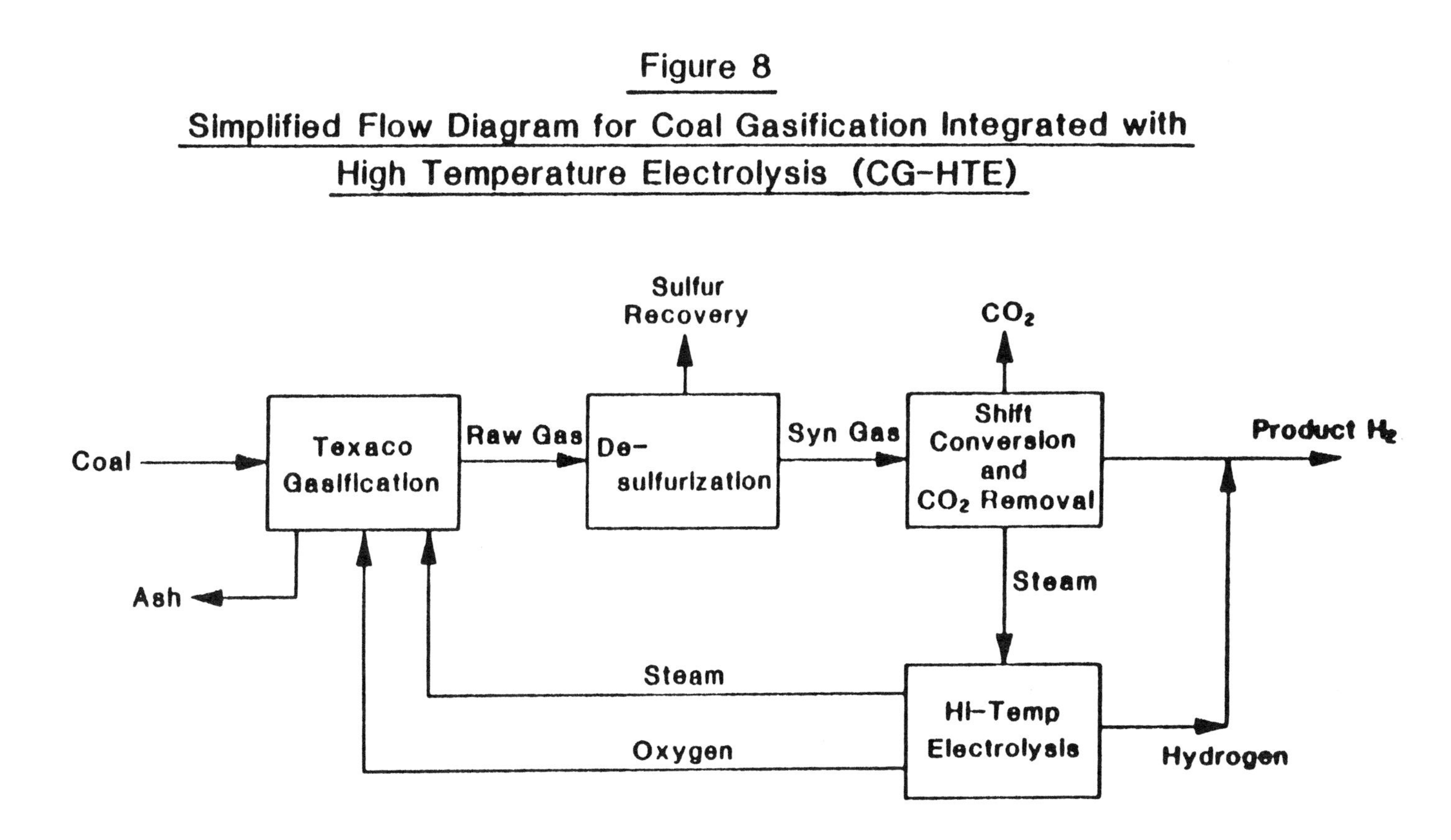

Figure 8

Simplified Flow Diagram for Coal Gasification Integrated with High Temperature Electrolysis (CG-HTE)

Overall reaction: $1.5\ C + 2H_2O = 1.5\ CO_2 + 2\ H_2$

The oxidation of 0.5 mole of carbon (coal) produces the necessary energy to sustain the steam gasification of carbon.

In the HTE unit the reaction is

$$H_2O = H_2 + 1/2\ O_2 \qquad \text{Endothermic } \Delta H = +68 \text{ Kcal/mol}$$

Two moles of H_2 are produced in the gasifier for every mole of H_2 produced in the HTE. Therefore, for a combined CG-HTE process two thirds of the hydrogen is produced in the gasifier and one third of the hydrogen in the steam electrolyzer. This production allocation is utilized in making the cost estimates shown in Table 5. Data from the earlier estimates in the conventional Texaco gasification process and the advanced HTE process were scaled to make the estimates.

The distribution of production cost is as follows:

	% of Production Cost
Coal Feedstock (includes fuels)	39.2
Capital Cost	40.7
O&M Cost	20.1
	100.0

It is noted that the feedstock cost has been significantly reduced due to the fact that a large fraction of hydrogen is produced from the coal gasification and thus a lower fraction of electrical energy requirements for HTE. On the other hand, capital investment has gone up. The total production cost, $\$4.43/10^3$ SCF for the combined GC-HTE plant is lower than for the HTE plant alone, $\$5.07/10^3$ SCF before by-product credit is taken. However, by taking by-product oxygen credit for the HTE plant, the HTE plant alone becomes more economical than the combined plant which has no by-product credit.

The Thermal Cracking of Natural Gas

The thermal decomposition of natural gas ($CH_4 = C + 2\ H_2$) has been practiced for many years for the production of carbon black for rubber tire vulcanization, for pigment and for the printing industry [14]. In the industrial process, the hydrogen produced is used to provide part of the thermal energy in the process. The process is practiced batchwise with the use of tandem furnaces at near atmospheric pressure. A methane-air flame is used to heat up firebrick to temperatures in the order of 1400°C (2550°F). The air is then turned off and the methane alone decomposes on the hot firebrick until the temperature drops to about 800°C (1472°F). The micron sized carbon particulates are collected in the effluent gas stream in bag filters. The methane-hydrogen effluent gas is then used to heat up a second furnace while the methane decomposition is continuing in the first furnace. Then the flow of gas is reversed so that while the first furnace is producing carbon black, the second furnace is being heated up. Attempts have also been made to thermally crack natural gas in a continuous fixed bed reactor [15]. It is noted that catalysts were used to increase the rate of decomposition and even carbon itself might act as a catalyst. It appears that an efficient continuous process should be possible of development for production of hydrogen with the carbon as a by-product.

The decomposition reaction is endothermic by 18 Kcal/mol so that a minimum of 9% of the heat of combustion of methane (212 Kcal/mol) is needed to drive the process. A preliminary process design and economics estimate of a continuous process has been made [16,17]. Figure 9 shows a flow sheet of the continuous process. A capital and production cost estimate of the process is given in Table 6. About 60% of the total Btu output of the plant can be recovered as H_2 and about 40% as carbon. Thus the carbon can be taken as a significant by-product to reduce the cost of the hydrogen. Carbon black is a clean ash-less, sulfur-less premium fuel which can command a higher value, however, credit has been taken conservatively at the same cost as the raw feedstock coal. Even though this is a new continuous process that has not been practiced on a commercial scale, the same economic rules as the conventional system have been applied here and are presented in Table 6. These estimates are probably not better than about ±40%.

The distribution of the production cost before credit for the by-product clean carbon fuel by-product is as follows:

	% of Production Cost
Feedstock (includes fuels)	76.2
Capital Cost	12.8
O&M Cost	11.0
	100.0

Even with the uncertainties, the net cost of hydrogen for the thermal decomposition of hydrogen comes out to be the lowest in this study at $\$1.64/10^3$ SCF or $\$5.10/10^6$ Btu after by-product credit is taken. The by-product credit in the form of carbon as a fuel amounts to only 28% of the production cost and is not heavily dependent. However, as indicated in the above table, the cost is highly dependent an the natural gas feedstock cost. At present, the price of natural gas is low because of the oil and gas glut. As the supply becomes more restricted, the cost of hydrogen by thermal decomposition of natural gas will increase relatively more rapidly than by other processes, but by allowing by-product credit, it should always be competitive with other natural-gas-based processes such as steam-gas reforming. This analysis indicates that if carbon-black can be fully used as a fuel as claimed above, an advanced continuous thermal process should be explored for the decomposition of natural gas to produce hydrogen for near-term and future use.

The HYDROCARB Thermal Conversion of Coal

An advanced process (called HYDROCARB) has been conceived for the production of carbon black from carbonaceous raw materials, the most abundant of which is coal [18,19]. A co-product of this process is a hydrogen-rich or a methane-rich gas. The process consists of the deep hydrogenation of the carbon containing raw material (coal) to produce mainly gaseous methane together with smaller amounts of CO, CO_2, and water. After removing water, the methane-rich process stream, is sent to a thermal cracker in which the methane is decomposed to carbon black particulate and hydrogen. The hydrogen-rich gas is recycled back to the coal hydropyrolyzer. Depending on the market, the excess hydrogen-rich gas or methane-rich gas is recovered as a co-product and

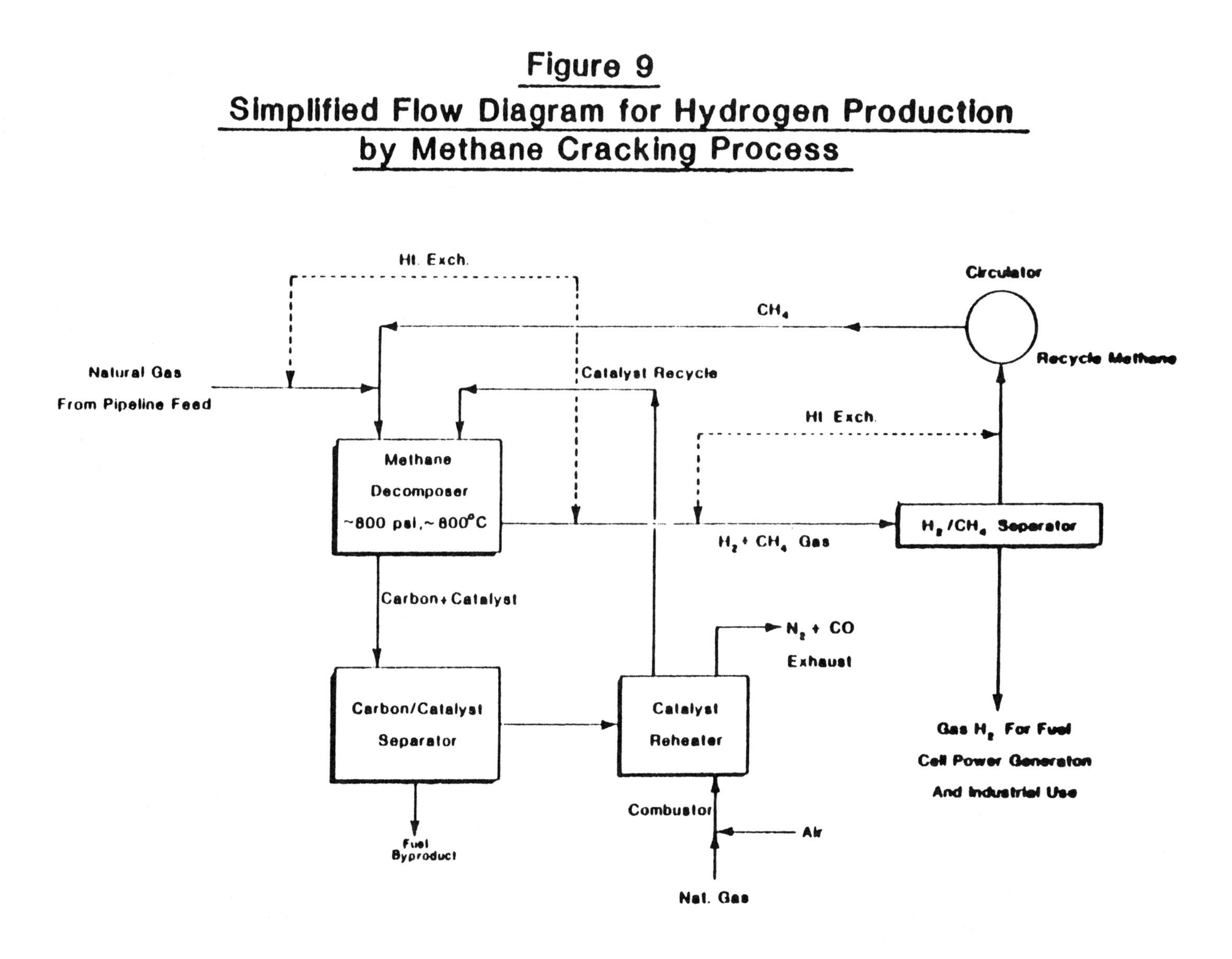

Figure 9
Simplified Flow Diagram for Hydrogen Production by Methane Cracking Process

TABLE 6: COST FOR PRODUCING HYDROGEN BY ADVANCED THERMAL PROCESSES
PRODUCTION CAPACITY = 100 X 10^6 SCF/D = 33 X 10^9 SCF/YR

Process	Methane Decomposition	Hydrocarb Process
Feedstock	Natural Gas	Bit. Coal
Rate	60 x 10^6 SCF/D	7460 T/D
Capital Investment - $\$10^6$		
Facilities Invest.	27.3	172
Interest, Startup, Working Capital	14.0	61
Total Capital Required	41.3	233
Annual Operating Cost - $\$10^3$		
Feedstock	57,600 ($\$2.91/10^3$SCF)	98,500 ($40/ton)
Power, Water, Chemicals, Disposal	5,200	20,100
Operating Labor	400	1,100
Maint., Overhead & Insurance (10% F.I.)	2,700	17,200
Capital Charges		
Depreciation (5% F.I.)	1,400	8,600
Ret. on Investment (20% T.C.I.)	8,300	46,600
Total H_2 Production Cost	75,600	192,100
$\$/10^3$ SCF	2.29	5.82
$\$/10^6$ BTU	7.11	18.09
By-Product Credit		
Carbon at $2.91/MMBTU	-21,400 (carbon)	-133,200 (carbon)
Net Production Cost	54,200	58,900
$\$/10^3$ SCF	1.64	1.78
$\$/10^6$ BTU	5.10	5.52

is therefore a source of hydrogen. A flowsheet of the conceptual process is shown in Figure 10. The reaction taking place in the hydropyrolyzer operating on a bituminous coal can be expressed as follows:

$$CH_{0.8}O_{0.08}S_{0.016}N_{0.015} + 2\ H_2 \longrightarrow CH_4 + 0.08\ H_2O + 0.32\ H_2 + 0.008\ N_2$$

The sulfur is removed by addition of limestone to form CaS which is oxidized to $CaSO_4$ for disposal. The ash remains behind. The methane, together with the equilibrium amounts of CO, CO_2, and H_2O, is sent to a regenerative condenser to remove the oxygen impurities as water and then fed to a methane pyrolysis reactor at elevated temperatures. The methane is decomposed to carbon black and hydrogen according to the following reaction which has been discussed in the previous section.

$$CH_4 \longrightarrow C + 2\ H_2$$

The hydrogen-rich gas is then sent back to the hydropyrolysis reactor to begin the cycle once again. In effect, the process cracks coal into its elements, C, H_2, and oxygen as H_2O, with the assistance of excess hydrogen as a transfer agent according to the following overall reaction.

$$CH_{0.8}O_{0.08}S_{0.016}N_{0.015} + \text{ash} + CaCO_3 \longrightarrow C + 0.32\ H_2 + 0.08\ H_2O + 0.008\ N_2 + \text{ash} + 0.016\ CaS$$

The hydropyrolysis of coal is an exothermic reaction while the methane decomposition reaction is endothermic. Process design and mass and energy balance estimates indicate that the process can be made 90% mass and energy efficient. Moving beds of granular Al_2O_3 are used in the two reactors to transfer the heat of reaction to the process gas. In one design 79.8% of the energy comes out of the process as carbon black energy and 20.2% as hydrogen energy. Thus the product from this plant is carbon black as a clean fuel for the energy market and hydrogen as a significant co-product clarified. The carbon can be burned as fuel or can be mixed with water, oil, or kerosene to produce liquid fuel mixes.

Cost estimates for the HYDROCARB process have been made [19]. These have been interpolated in terms of producing hydrogen at a production rate of 100×10^6 cu ft/day and taking the carbon as a by-product fuel with cost factors used in the conventional process mentioned earlier.

The distribution of the production cost before credit for the by-product clean carbon fuel is as follows:

	% of Production Cost
Coal Feedstock (includes fuels)	50.5
Capital Cost	30.4
O&M Cost	19.1
	100.0

It is noted that with by-product credit, the production cost of hydrogen amounts to $\$1.78/10^3$ or $\$5.52/10^6$ Btu which is second to the lowest cost process, e.g., methane decomposition. It is, however, heavily dependent on by-product carbon which accounts for 69% of the production cost of hydrogen. Development of this process must then depend on opening up a market for a clean carbon fuel. The significant advantage of the HYDROCARB process is that it makes use of one of the world's largest energy resources and that is coal. Coal in the near- and far-term future will always be available at a reasonable cost and thus the clean carbon and hydrogen co-products will be highly competitive with all other coal-based processes for hydrogen production such as coal gasification processes which are 1.74 to 2.40 times higher than the HYDROCARB process.

5. OTHER ADVANCED PROCESSES FOR HYDROGEN PRODUCTION

Thermochemical Water-Splitting Processes

A number of chemical redox systems have been investigated for the cyclical thermochemical splitting of water for hydrogen and oxygen production. Thermal sources, both fossil and non-fossil energy sources, including solar and nuclear, have been proposed [20]. The major systems devised are identified as follows:

a) The halides (bromine, chlorine, and iodine) of the metals Ca, Sr, Mn, and Fe acting as oxidation/reduction agents to split water

b) The two-component sulfur-iodine redox systems

c) Steam-iron system. This redox system has been described and discussed under the conventional processes above. It is not, however, a completely closed cyclical process since the carbon in coal is used as an energy recovery agent and is emitted as CO_2. This system is probably even more efficient than a closed cycle.

The thermochemical cycles are Carnot cycle limited which means that higher temperatures improve efficiencies of conversion. Higher temperatures cause severe materials of construction damage especially when corrosive chemical reagents are usually involved. Any loss of these costly reagents becomes a severe economic burden on the process. Hydrogen is a very low cost commodity compared to the chemicals used in thermochemical water splitting. For example, hydrogen at costs of \$2 to $\$6/10^3$ SCF H_2, amounts to \$0.38 to \$1.14/lb H_2. If iodine is used on a mol/mol basis it takes 127 lbs of I_2 to produce 1 lb H_2. At \$10/lb I_2, this means that loss of only 0.04 to 0.12 lbs of hydrogen production or a loss of one part in 1000 to 3000 parts of iodine would double the cost of hydrogen production. This is a very severe process condition.

Iron oxidation-reduction, (steam-iron) one of the lowest cost redox systems mentioned above, showed a higher cost water splitting process than the other conventional processes (Table 2). On this basis, reagents which are higher in cost would be less economical than the steam-iron process and therefore can be discounted as a nearer term economical process. The energy for reduction of the iron oxide comes directly from coal carbon instead of indirectly in a heat exchanger and thus in principle this process should be more economical than the other redox water splitting processes.

Plasma, Solar and Radiation Processes

Several high temperature and high energy processes have been suggested in the past and are listed as follows [21]:

a) Plasma-arc process
b) Photolytic laser process
c) High energy radiation process.

a) The plasma-arc essentially heats water to high temperature by means of an electric field to an extreme temperature reading up to 5000°C (9000°F). The process is equilibrium limited in that the water is cracked to seven radical and molecular components H, H_2, O, O_2, OH, HO_2 and H_2O. Up to about 50% concentration by volume of H and H_2 components is possible. In order to stabilize the hydrogen components, so that no recombination with the oxygen constitutents takes place a very rapid quench of the plasma gases must take place. This can be accomplished with direct water injection or even with a cryogenic liquid. Because of this highly irreversible quench process the process, is very wasteful of energy. The requirement of electrical energy for the arc taken together with a wasteful energy system makes the process extremely expensive.

b) The photolytic process which uses light absorption in a mercury catalyzed water vapor system at a wavelength of 3060Å is a low efficiency system in that energy must be converted from thermal to electrical to photolytic radiant energy which is then transmitted to mercury which then transmits it to water. The overall efficiency is usually less than 10%. Lasers in recent years which can emit more intense photon radiation in specific wavelengths, improves efficiencies somewhat but even these devices cannot indicate an overall efficiency greater than 10%.

c) With the advent of nuclear energy there was much effort over several decades in developing high energy radiation processes for the synthesis and production of chemicals. One system investigated was the radiation decomposition of waste. Gamma, beta and fission fragment radiation were used. The highest yield obtained was with fission fragment radiation directly in a nuclear reactor however, the energy efficiency obtained for the process, never exceeded about 18%. Furthermore the separation of highly radioactive particulates and gases as well as the separation of hydrogen from oxygen are severe process problems.

Solar Photovoltaic Water Electrolysis Process

There appears to be a growing interest in directly generating electrical energy using solar energy by means of photovoltaic (PV) devices. New solar cell materials such as amorphous silicon have reached solar to electrical energy conversion efficiencies of 12% for large area laboratory modules. Manufacturing plants are presently expected to produce cells for a cost of $1/peak watt. By directing this PV cell output to water electrolysis at 84% efficiency, it is projected that in the time frame of the next 15 or 20 years, it will be possible to produce PV cells for as low as $0.20 to $0.40/peak watt and to generate hydrogen in solar intensive areas such as in the Southwestern part of the U.S. for $3 to $5/10^3$ ft^3 at capacities of 0.5 million cubic feet/day [22].

6. COMPARATIVE ASSESSMENT DISCUSSION

The process and economic data for the six commercially available conventional processes and the five advanced unconventional processes for production of hydrogen are listed in Tables 7 and 8 respectively. Since the same basic economic factors (Table 1) were used in obtaining the cost data, all processes can be compared directly with one another. In addition, thermal efficiency of each of the processes have been calculated based on the higher heating value of hydrogen and of the by-product and expressed as a percent of the highest heating value of the feedstock including the fuel value of energy consumed in the process. A graphical presentation of the cost data for the eleven processes are shown in Figure 11.

In comparing the value of the processes one must consider the interdependence of at least three major factors: (1) the efficiency, (2) the capital investment and (3) the value of the by-product. For example, the steam reforming of natural gas is the most economical of the conventional processes because it has the highest thermal efficiency and lowest capital investment and also has a modest steam by-product credit. This is followed by the POX residual oil process with a 50% increase in net production cost due principally to a large increase in capital investment. The next highest net production cost process is the steam-iron process. The large by-product power credit is responsible for reducing the total cost and placing it third in line in this comparison. The Texaco coal gasification process is next in line which increases the total production cost by 50% above the POX oil process. This is primarily due to a significant increase in capital investment for coal gasification. Using the Koppers-Totzek (K-T) process is somewhat more expensive than the Texaco process because of the highest capital investment of the conventional data and a lower thermal efficiency. Finally, the highest production cost is exhibited by the conventional water electrolysis process. This is due to costly electrical energy required and the low thermal efficiency. The large by-product oxygen credit was not large enough (12.6%) to offset the electrical power cost.

Comparing the ranking of the advanced processes indicates that the methane cracking (decomposition) process has the lowest gross and net H_2 production cost. In this case, although the H_2 efficiency is not very high, the capital investment is very low which compensates for the lower efficiency. The by-product credit is significant and amounts to about 28% of the total cost.

The second highest ranking non-conventional process is the Hydrocarb Coal Cracking Process. In this case although the capital cost is seven times the methane cracking process and the H_2 efficiency is low (only 17.3%), the by-product carbon black credit is so high, up to 70% of the total cost, that

TABLE 7: SUMMARY OF PROCESS CHARACTERISTICS AND CAPITAL AND PRODUCTION COSTS (MID-1987 U.S. DOLLARS) COMPONENTS FOR PRODUCING HYDROGEN BY CONVENTIONAL (COMMERCIALLY AVAILABLE) PROCESSES

Basis: H_2 Production Capacity - 100 x 10^6 SCF/D Gas at 300 to 600 psig

Process	Steam Reforming (SR)	Partial Oxidation (POX)	Texaco Gasif. (TG)	K-T Gasif. (KG)	Steam-Iron	Water Electrolysis
Feedstock Requirement	Nat. Gas 41 x 10^6 SFC/D	Resid. Oil 6400 Bbl/D	Bit. Coal 2320 T/D	Bit. Coal 1550 T/D	Bit. Coal 2990 T/D	Electricity 506 MW(e)
Thermal Efficiency, %						
H_2 Product.	78.5	76.8	63.2	57.7	45.7	27.2
Byproduct	<0.1	--	--	--	5.4	—
Total	78.5	76.8	63.2	57.7	51.1	27.2
Byproduct Capacity	Steam 1.7 T/D	Sulfur 30 T/D	Sulfur 70 T/D	Sulfur 10 T/D	Power 342 MWH/D	Oxygen 695 T/D
Capital Cost, $\$10^6$	83.2	205	316	406	324	132
Prod. Costs, $\$/10^3$ SCF						
Feedstock	1.25	1.08	1.10	1.10	1.38	5.38
Capital	0.60	1.51	2.33	2.99	2.38	0.93
O&M	0.21	0.54	0.92	1.03	0.99	0.26
Total	2.06	3.12	4.35	5.12	4.75	6.57
Byproduct Credit	-0.16	-0.03	-0.08	-0.02	-1.14	-0.83
Net H_2 Prod. Cost in $\$10^3$ SCF	1.90	3.09	4.27	5.10	3.61	5.74
in $\$/10^6$ BTU	5.90	9.60	13.26	15.84	11.21	17.83
Net Production Cost Ranking	1	2	4	5	3	6

TABLE 8: SUMMARY OF PROCESS CHARACTERISTICS, ENERGY EFFICIENCIES, BYPRODUCTS, AND CAPITAL AND PRODUCTION COST COMPONENTS - (MID-1987 U.S. $) FOR PRODUCING HYDROGEN BY ADVANCED PROCESSES

Basis: H_2 Production Capacity - 100 x 10^6 SCF/D Gas at 300 to 600 psig

Process	High Temp. Electrolysis of Steam (HTE)	Coal Gasif. With Elec. Chem. Shift (Westinghouse)	Coal Gasif. With High Temp. Electrolysis (CG/HTE)	Methane Cracking	Hydrocarb Process (Coal Cracking)
Feedstock Requirement	Electricity 285 MW(e)	Bit. Coal 3800 T/D	Bit. Coal Electricity 1550 T/D, 95 MW(E)	Nat. Gas (Methane) 60 x 10^6 SCF/D	Bit. Coal 7460 T/D
Thermal Efficiency, %					
H_2 Product.	48.3	33.8	57.2	53.7	17.3
Byproduct	--	38.0	None	39.1	74.4
Total	48.3	71.8	57.2	90.8	91.7
Byproduct Capacity	Oxygen 695 T/D	Steam 18,000 T/D	None None	Carbon Black 790 T/D	Carbon Black 4920 T/D
Capital Cost, $\$10^6$	141	262	245	41	233
Prod. Costs, $\$/10^3$ SCF H_2					
Feedstock	3.35	1.52	1.74	1.75	2.98
Capital	0.89	1.94	1.80	0.25	1.17
O&M	0.82	1.05	0.89	0.29	1.67
Total	5.06	4.51	4.43	2.29	5.82
Byproduct Credit	-0.84	-1.80	None	-0.65	-4.04
Net H_2 Prod. Cost $\$/10^6$ BTU	4.22 13.12	2.71 8.40	4.43 13.76	1.64 5.10	1.78 5.52
Net Production Cost Ranking	4	3	5	1	2

the net production cost is only 8.8% higher than the methane cracking process. The Westinghouse coal gasification with electrochemical shift is the next increased production cost process exceeding Hydrocarb by another 50%, due mainly to increased capital, and O&M cost. The by-product steam credit is very significant amounting to 40% of the production cost. The next ranking process is the high temperature electrolysis followed by the hybrid coal gasification and/HTR process, the latter giving the highest net product cost which is due primarily to the fact that there is no by product credit. It should also be noted that the Hydrocarb process yields the highest thermal efficiency of 91.7%. The thermal cracking processes are much lower in cost than the advanced high temperature electrolysis processes.

Comparing conventional (available) with unconventional (developing) processes, the methane and coal cracking processes are potentially lower in net production cost than the natural gas steam-reforming process. High temperature electrolysis unconventional processes are potentially competitive with the partial oxidation and coal gasification processes. The Westinghouse HTR shift process is potentially better than the partial oxidation of resids based on net production cost. The advanced processes depend heavily on by-product credit from either steam oxygen, or carbon black.

It should be noted that the above comparisons and conclusions hold for feedstock costs as of mid-1987. Generally it is projected that oil and gas will be available at the present low cost of about $3/MM BTU in the U.S. for the next 10 to 20 years after which these fossil fuels will increase in cost again. However, coal costs, mainly because of the large U.S. indigenous natural resource, should remain reasonably stable. This means that natural gas and oil based processes can be very sensitive to feedstock cost increases and that the total and net cost can catch up or even exceed the coal cost. The electrolytic process cost should remain fairly stable, increasing slowly because electrical power cost based on coal and nuclear will tend to maintain level electrical power costs. The conclusion can be reached that for the future, effort should be directed toward supporting R&D for the processes based on coal feedstock with emphasis on the thermal processes because they show the lowest hydrogen production cost after by-product credit. Thus the carbon black processes should strongly be considered for support.

Another point which should be stressed is that hydrogen production is usually integrated with production of another end-product. For example, for synthetic fuel production from coal, synthesis gas which contains hydrogen is used to produce substitute natural gas (SNG), methanol and substitute gasolines. In ammonia production the hydrogen production process is integrated with the ammonia plant. Thus the integration of hydrogen with end-use production processes, yield some economy of an integral operation. The stand alone process system is of value where hydrogen alone is used as a fuel for transportation, automotive, aircraft and rocket engines. The purpose of using hydrogen as fuel in a so-called "hydrogen economy" is its clean burning qualities, largely eliminating acid rain, particulates emission, and CO_2 emissions which lead to the greenhouse effect. The hydrogen economy is usually related to the use of a non-fossil solar and nuclear energy source, eliminating the emissions of CO_2. However, there is a possibility of basing a pure hydrogen economy on fossil fuel when applying the processes of methane cracking and HYDROCARB coal cracking. It is only in these cases that carbon can be extracted and stored, while using only the hydrogen product as a fuel [23]. The gross cost for the HYDROCARB Process, although high, is nevertheless not that much more than conventional coal gasification processes and less than the conventional water electrolysis process.

As noted in each of the sections above, hydrogen by the exotic process of water splitting, lasers and plasma are in principal much higher in cost than the conventional and the advanced processes.

Hydrogen from by-product chemical and electrochemical processes are not economic because of the limited available production capacities.

Much government and industrial effort has already been expended on high temperature electrochemical and water splitting thermal processes. However, no significant reductions in hydrogen costs have resulted compared to conventional processes. On the other hand very little effort has been put into thermal cracking processes. It has been shown in this study that the thermal cracking processes have greater potential in reducing hydrogen costs especially with a market for the by-product carbon. It is recommended that development effort be expended to explore and advance the concept of thermally cracking fossil fuels for hydrogen production.

7. CONCLUSIONS

1. Based on present and near term costs of the conventional processes for the production of hydrogen, steam reforming of methane is the most economical followed by partial oxidation of residual oil.

2. Coal gasification, steam-iron and water electrolysis process indicate a 2 to 3-fold higher net hydrogen production cost than steam reforming of neutrals. Advanced gas separation technologies, such as pressure swing absorption has a small economic advantage over wet absorption methods in conventional gasification processes for hydrogen production.

3. Of the advanced processes, the thermal cracking of methane and coal systems are worthwhile pursuing because of the large economic potential compared to the advanced electrochemical and conventional systems.

4. In the long term, processes based on the use of coal as a feedstock for hydrogen production should not rise in cost as rapidly as with oil and gas and this leads to the conclusion that hydrogen based on advanced coal technology is well worth supporting for long-term development and commercialization.

5. The advanced thermal process for hydrogen production from coal should be developed as an integrated system with a market for by-product carbon and the use for hydrogen as a fuel.

6. The high temperature electrochemical processes suffer from high energy and high capital costs.

7. Thermochemical and high energy water splitting systems are inherently less efficient and more costly than the conventional and advanced electrochemical and thermal cracking processes. The solar photovoltaic water electrolysis process has the potential of becoming a viable hydrogen producer in solar intensive areas of the world, sometime beyond the turn of the century.

8. It is recommended that research and development effort be expended in the development of the concept of thermal cracking of fossil fuels for clean carbon and hydrogen fuel production.

8. REFERENCES

1. Minet, R. G. and Desai, K. Cost-effective methods for hydrogen production, Int. J. Hydrogen Energy, 8, 285, 1983.

2. Moore, R. B. Economic feasibility of advanced technology for hydrogen production from fossil fuels, Int. J. Hydrogen Energy, 8, 905, 1983.

3. Gregory, D. P. et al. The economics of hydrogen production, in Hydrogen: Production and Marketing, ACS Symposium Series 116, p. 3, 1980.

4. Ekman, K. R. Cost of hydrogen production from fossil and nuclear fuels, 5030-461, Jet Propulsion Laboratory, CIT, 1980.

5. Gaines, L. K. and Wolsky, A.m. Economics of hydrogen production: the next twenty-five years, in Hydrogen Energy Progress V (N. T. Veziroglu and J. B. Taylor, Ed.) Vol. 1, 259, 1984.

6. Reed, C. L. and Kuhre, C. J. Hydrogen production from partial oxidation of residual fuel oil, in Hydrogen: Production and Marketing, ACS Symposium Series 116, 95, 1980.

7. Ahn, Y. K., and Fischer, W.H. Production of hydrogen from coal and petroleum coke: technical and economic perspectives, Int. J. Hydrogen Energy 11, 783, 1986.

8. Hammerli, M. When will electrolytic hydrogen become competitive? Int. J. Hydrogen Energy, 9, 25, 1984.

9. "Faith, Keys, Clark's. Industrial Chemicals", F. A. Lowenstein, and M. K. Moran, p. 77, 4th Ed., John Wiley and Sons, Inc., NY, 1975.

10. Doenitz, W. R., Schniedberger, R., Steinheid, E. and Strucker, R. Hydrogen production by high temperature electrolysis of water vapor, Int. J. Hydrogen Energy 5, 55, 1980.

11. Westinghouse Electric Corporation. Technical Progress Report No. 1: High temperature water vapor electrolysis using solid electrolysis cells, BNL Contract No. 585847-S, Pittsburgh, PA (May 6, 1983).

12. Lipa, M.A. and Borhan, A. High Temperature steam electrolysis: technical and economic evaluation of alternative process designs, BNL 51798, Brookhaven National Laboratory, Upton, NY 11973, (September 1983).

13. Brown, S.T., Newby, R.A. and Keirns, D. Letter communication, Westinghouse Electric Corp. (December 9, 1987).

14. Donnet, J. B. Carbon Black, pp 16-18, Marcel-Dekker, NY, 1976.

15. Pohleny, J. B., and Scott, N.H. Method for hydrogen production by catalytic decompostion of a gaseous hydrocarbon stream. U.S. Patent 3,284,161 assigned to Universal Oil Products Co. (November 8, 1966).

16. Grohse, E. W. and Steinberg, M. A process to convert remote gas to a fungible fuel with capability for EOR. BNL 40424, Brookhaven National Laboratory, Upton, NY, (October 1987).

17. Steinberg, M. A proposal to develop a process for converting natural gas (methane) to premium liquid and gaseous fuel. BNL Proposal, Brookhaven National Laboratory, Upton, NY, (September 1985).

18. Grohse, E. W. and Steinberg, M. Economical clean carbon and gaseous fuels from coal and other carbonaceous materials. BNL Report 40485, Brookhaven National Laboratory, Upton, NY, (November 1987).

19. Steinberg, M. The HYDROCARB Process - Conversion of carbonaceous materials to clean carbon and gaseous. BNL Report 40731, Brookhaven National Laboratory, Upton, NY, (December 1987).

20. Dang V. D. and Steinberg, M. Hydrogen production using fusion energy and thermochemical cycles. Int. J. Hydrogen Energy, 5, 119-129 (1980).

21. Steinberg, M. Beller, M. and Powell, S.R. A survey of applications of fusion power technology to the chemical and material processing industry. BNL Report 18866, Brookhaven National Laboratory, Upton, NY, (May 1974).

22. Ogden, J. M. and Williams, R. H. Hydrogen and the revolution in amorphous silicon solar cell technology (in press), Princeton Univeristy, Princeton, NJ (April 1988).

23. Steinberg, M. and Cheng, H. C. Advanced Technologies for Reduced CO_2 Emissions. BNL Report 40730, Brookhaven National Laboratory, Upton, NY, (December 1987).

Technology Assessment of Advanced Electrolytic Hydrogen Production

S. Dutta
Florida Solar Energy Center

INTRODUCTION

Hydrogen is a transportable, storable and nonpolluting fuel that can be produced from renewable resources. It also has the highest gravimetric energy density of all known fuels. Bringing this ideal fuel into the energy network is one issue. Mass-scale production of it from renewable resources with electricity, which is itself a desirable and expensive form of energy, is another. Electrolytic hydrogen production calls for improvements in individual electrolyser cell performance and cost, and engineering analysis that will lead to the optimum plant design, operation and scale-up.

Excellent reviews are available on electrolyser technology in present practice as well as of advanced concepts [1–7]. The purpose of this paper is to present a comprehensive and critical assessment of the pros and cons of the most promising advanced technologies, and pass recommendations on the areas in need of major developments, particularly from the point of view of large-scale commercialization. A more detailed treatment can be found in Block *et al.* [8]. Only a technical assessment of advanced electrolysers is presented in this paper. The economic assessment is presented in a separate publication by Dutta *et al.* [9].

One of the most promising applications of the electrolyser technology is to harness solar energy to produce hydrogen through the integration with the photovoltaic technology. Assessment of various aspects of this important integration is available in the literature, for example by Bilgen and Bilgen [10], Hammache and Bilgen [11] and Hancock [12], and will not be discussed in this paper.

ADVANCED TECHNOLOGIES

Typical design and operating conditions of present (conventional) and advanced electrolysers under development are summarized in the Appendix. This table provides a glimpse of the wide variety of materials and operating conditions that are being used in various efforts for technology advancements.

Alkaline water electrolysis is the technology in present practice for large-scale electrolytic hydrogen production. Low efficiency, low current density and a lack of proper scale-up practice are the primary drawbacks of the present technology. At the present average grid electricity price, electrolyser electricity costs would constitute more than 75% of the cost of liquid hydrogen produced through present technologies. All advanced technologies are geared to reach improved cell efficiencies and high current densities in order to save electricity and capital costs. Significant improvements have been made possible in this regard and further improvements are anticipated. However, these improvements require more expensive materials and novel fabrication techniques, which are expected to add to the capital costs, creating new obstacles to lower hydrogen production costs.

Many advanced concepts relating to various aspects of electrolytic hydrogen production are reported in the literature. Embodiment of many of these concepts can be found in the four prominent advanced technologies, namely, advanced alkaline water electrolyser (AWE), inorganic membrane alkaline electrolyser (IME), solid polymer electrolyser (SPE) and high temperature electrolyser (HTE). Typical design and operating conditions and performance of these advanced electrolysers are compared in Table 1, keeping the conventional alkaline water electrolyser as the reference.

Many other novel concepts have been proposed and/or tested for advancement of water electrolysis. Two of these concepts have been developed in prototype and small-scale commercial units. In Life Systems Inc.'s technology [6], an ingenious static feedwater concept is used, where water to be electrolysed is fed as vapor, instead of as a liquid, to the cell electrolyte, which is in the liquid phase. Furthermore, the cell design is such that the product gases do not need to be separated from the electrolyte. The capability of using semi-pure water and

Table 1. Present and advanced electrolyser technologies at a glance

Technology	Alkaline electrolysers				
	Conventional alkaline electrolyser	Advanced alkaline electrolyser	Inorganic membrane alkaline electrolyser	Solid polymer electrolyser	High temperature (water vapor) electrolyser
Development stage	Commercial large-scale units	Laboratory-scale and prototype units	Laboratory-scale and prototype units	Laboratory-scale and prototype units	Very small laboratory-scale units
Cell voltage, V	1.84–2.25	1.5–3.0	1.6–1.9	1.4–2.0	0.95–1.3
Current density, A cm^{-2}	0.13–0.25	0.20–2.0	0.2–1.0	0.25–2.0	0.3–1.0
Temperature, °C	70–90	90–145	90–120	80–150	923–1009
Pressure	Ambient, except Lurgi—30 atm	Up to 40 atm	Up to 40 atm	Up to 40 atm	Up to 30 atm
Cathode	Steel, stainless steel or nickel	Catalytically or non-catalytically activated Ni	NiS	Porous carbon fiber paper with a layer of Pt catalyst	Ni
Anode	Nickel	Catalytically or non-catalytically activated Ni	Spinel oxide based on Co	Porous titanium with a layer of a proprietary catalyst	Ni–NiO or Perovskite ($LaNiO_3$, $LaMnO_3$, etc.)
Gas separator	Asbestos 1.2–1.7 Ω cm^{-2}	Asbestos-based below 100°C; Teflon-bonded K-titanate and polybenzimidazole (PBI) above 100°C 0.5–0.7 Ω cm^{-2}	Patented polyantimonic acid (PAM) membrane 0.2–0.3 Ω cm^{-2}	Multilayer expanded metal screens	None
Electrolyte	25–35% KOH	25–40% KOH	14–15% KOH	Nafion (Perfluorosulfonic acid) membrane 10–12 mils thick 0.46 Ω cm^{-2}	Solid Y_2O_3-stabilized ZrO_{380}
Cell efficiency	77–80%	80–90%	82–91%	85–90%	90–100%
Power consumption kWh $(Nm^3 H_2)^{-1}$	4.3–4.9	3.8–4.3	3.6–4.0		3.5

Table 2. Major advantages and disadvantages of alternative technoloigies

	Alkaline electrolyser				
Technology	Conventional alkaline electrolyser	Advanced alkaline electrolyser	Inorganic membrane alkaline electrolyser	Solid polymer electrolyser	High temperature (water vapor) electrolyser
Major advantages	*Proven technology *Simple *No specialty material required	*Advancements already demonstrated in small scales *Minimum development effort required	*Non-noble metal catalyst used *Low alkali concentration required *Lower capital cost compared to SPE *Apparently lower overall cost compared to SPE	*No corrosive electrolyte *Compact design *Most appropriate for modular design	*Has the potential to reach close to or more than 100% efficiency *No liquid and flow distribution problems *No corrosive electrolyte
Major disadvantages	*Low efficiency *Low current density *High production cost *Corrosive electrolyte	*Somewhat lower efficiency compared to other advanced technologies *Alkaline corrosion problems become severe at more desirable higher temperature operations		*Expensive membrane and noble metal catalysts add significantly to capital cost	*Severe materials and fabrication problems due to very high temperature operation *A very high temperature heat source required
Remarks		*Has immediate commercialization potential	*Technology has been demonstrated and recently commercialized	*Inexpensive membrane and catalyst developments are the key requirements of this innovative technology to bring it to widespread commercialization	*An innovative concept, but at a very early stage of development *Significant development effort required to find inexpensive materials and fabrication techniques

the elimination of the gas-electrolyte separator are the novel features of this technology. In Teledyne's 'single irriguous' electrolysis cell [6], only the anode is flooded with the electrolyte, and hydrogen is produced in the liquid-free cathode compartment. The process is thus simplified due to the need for only one gas-electrolyte separator and associated circulation equipment.

COMPARISON OF ALTERNATIVE TECHNOLOGIES AND THE KEY R & D AREAS

Major advantages and disadvantages of alternative technologies are highlighted in Table 2 and discussed in the following text.

AWE Technology has the most immediate widescale commercialization potential, since no radical development effort is required over the existing (conventional) alkaline electrolyser technology. Development of an alternative separator suitable for high-temperature operation in place of the asbestos-based material used in conventional technology is the most pressing issue for a breakthrough in this technology. Other major issues are:

- Development of inexpensive, more active and stable electrocatalysts;
- Cell design for minimum ohmic losses;
- Development of inexpensive construction and sealant materials resistant to alkaline corrosion, particularly at higher temperatures;
- Engineering analysis, process optimization and scale-up.

Replacement of conventional asbestos-based separator materials with a thin low-resistance (ohmic) PAM (polyantimonic acid membrane) membrane in *IME Technology* is a significant advancement over conventional and AWE technologies. Use of non-noble metal catalysts also makes this technology attractive when compared with SPE, where noble metal catalysts and an expensive electrolyte, the Nafion membrane, are used. The major development issues for IME are the same as those for AWE. The IME technology has been demonstrated and recently commercialized.

Compactness, simplicity of design, operation and maintenance, and the absence of corrosive liquid electrolyte have made *SPE Technology* a strong alternative for large-scale commercial electrolysis. However, requirement of a high loading of platinum catalyst, novel fabrication techniques for current collectors and electrode-membrane assemblies, and the high cost of Nafion membrane electrolyte lead to high capital costs. Development of inexpensive catalyst systems, an inexpensive alternative for Nafion and simpler fabrication techniques are the pressing issues of SPE technology. Other major issues are the following:

- Reduction of parasitic losses of current and diffusional losses of hydrogen through the membrane at higher temperatures and pressures;
- Reduction of SPE electrode junction losses by developing a better catalyst bonding technique;
- Testing of the endurance of Nafion or alternative membranes at higher temperatures and toward corrosion by water impurities.

HTE Technology is built upon a sound fundamental concept of very high cell efficiency achievable in high temperature vapor-phase electrolysis. However, availability of an inexpensive high-temperature (~1000°C) heat source is generally implied in the commercialization of this technology. Development of inexpensive ceramic electrolyte and catalyst systems, and fabrication techniques are the most pressing issues of this technology. The technology is at a very early stage of development. Many severe problems common to all high-temperature designs and operations are inherent to this technology. Some of the other major issues are the following:

- Reduction of operating voltage at economic current densities (so far, operation has not been found practical at current densities of 0.4–0.5 A cm^{-2} and below 1.3 V, which is the thermoneutral voltage of water vapor at 1000°C);
- Design optimization and scale-up.

MAJOR TECHNOLOGICAL ISSUES

Issues specific to each technology have been discussed earlier. However, there are some major issues that are of common interest to all these technologies as well as to other advanced concepts. These issues are discussed in the following subsections.

High-pressure electrolysis

All large-scale commercial electrolysers except that by Lurgi operate at close to atmospheric pressure. The Lurgi unit is designed to operate up to 30 atm pressure. The principal interest in high-pressure operation is to save the cost of compressing hydrogen after it is produced. Reduced equipment size and lower cell voltage (due to lower gas holdup or smaller gas bubbles) are claimed to be the added benefits of high-pressure operation. For liquid water electrolysis, high-pressure operation is a requirement for operating temperatures close to or above 100°C for minimizing the evaporation of the electrolyte.

However, the primary issues of concern in high-pressure operation are higher capital costs, the requirement of a more delicate pressure balance across the membrane/separator and safety. It is argued that the additional cost of a high-pressure system may outweigh the expected cost saving in compression. Whether to produce hydrogen at atmospheric pressure and compress it or to produce hydrogen at a high pressure is still debated.

Containment of the entire electrolysis module inside a single large cylindrical pressurized chamber has been an attractive approach to alleviate many problems or concerns of individual cell pressurization methods used in earlier designs.

Proper engineering analysis, process optimization, and scale-up procedure along with a proper accounting of the market demand with respect to hydrogen pressure will determine the optimum operating pressure of a large-scale commercial electrolyser.

High-temperature electrolysis

High-temperature electrolysis is favored from both thermodynamic and kinetic standpoints. Lower cell voltage and higher current density are simultaneously achievable in high-temperature operation. All advanced liquid water electrolysis techologies are geared to push electrolyser operating temperature toward as high a level as possible, preferably, in the range of 150 to 200°C. Conventional electrolysers operate at temperatures of 70–90°C. High-temperature operation, however, suffers from the following limitations:

- The requirement of higher pressure in liquid phase electrolysis for minimizing evaporation loss;
- Simultaneous increase in temperature and pressure which may add significantly to the capital cost and must be carefully weighed against the gains achievable due to higher cell efficiency and current density. This may be a major issue in the engineering design, optimization and scale-up to a very large commercial unit;
- Increased corrosion of electrodes, membrane, separator, gaskets and the metallic container and connectors;
- A significant increase in temperature calls for a radical departure from conventional technology, engineering and operating practice, and needs a substantial investment in development efforts. The search for novel materials and fabrication techniques becomes of prime importance in this situation.

Advanced separator

Development of an inexpensive gas separator that has the combined characteristics of low gas permeability (particularly at high pressure), low ohmic resistance, high mechanical strength and long life in a corrosive environment is one of the key challenges in advanced electrolysis. A significant reduction of the ohmic resistance is possible by reducing the interelectrode gap and the gas bubble fraction in the electrolyte. Most advanced technologies, therefore, use a *zero-gap* sandwich-type configuration, in which *porous electrodes* are pressed on the opposite faces of the separator/membrane, and the gases evolve predominantly on the backsides of the electrodes. Such a configuration keeps the gap between the electrodes or between the separator/membrane and electrodes virtually free from gas bubbles. A more compact cell module is also a result of this cell configuration. Table 3, adapted from Wendt and Imarisio [7], provides the status of the development of five promising separator/membrane types.

Electrocatalysts

Cathodic and anodic overvoltages constitute a substantial fraction of total cell voltage, particularly at low temperature operations. Both higher temperature and suitable catalysts favor electrode reaction kinetics and help in lowering overvoltages. Reduction of overvoltages by catalysis without a substantial increase in temperature is an ideal approach in electrolyser improvement.

Corrosive alkaline electrolyte has restricted the choice to essentially Ni-based catalysts in alkaline electrolysers. Catalysts based on noble metals such as platinum and ruthenium are, on the other hand, found to be the most active catalysts for non-alkaline electrolyte systems. A search for low-cost substitutes for noble metal catalysts or reduction of noble metal catalyst loading and improvement of the activity of nickel catalysts are the primary development goals in this field.

High surface area deposits of nickel are generally the basis for both cathodic electrocatalysis (to catalyse hydrogen evolution) and anodic electrocatalysis (to catalyse oxygen evolution), to which transition metals are added, using specific techniques. As a high surface area base, either sinter nickel or a Raney pre-alloy (Ni–Al or Ni–Zn) is used. These materials are applied to the base electrode surface and leached with potassium hydroxide to form, *in situ*, the high surface area nickel. Various methods of improvement and stabilization of the catalytic surfaces, and other methods of catalyst preparation, have been summarized by Wendt and Imarisio [7]. While a substantial reduction in the cathodic overvoltage (from 350 to 150–200 mV) has been possible by using the improved techniques, no such improvement has yet been possible on the anode side of the electrolyser.

Table 3. Characteristics of advanced separators/membranes*

Specifications	Thickness (mm)	Surface-specific resistance ($\Omega\,cm^{-2}$)	Maximum working temperature (°C)	Remarks
1. Polymer reinforced asbestos	0.2–0.3	0.15–0.20	90	Semi-commercial
2. Specific polymers for woven diaphragms	—	—	140	Laboratory
3. Polysulphone impregnated by Sb_2O_{383} polyoxides	0.2–0.3	0.16–0.20	120	Commercial (used in IME technology)
4. Teflon-bonded zirconia	0.2–0.5	0.25	160	Laboratory
5. Nickel-net backed $BaTiO_3$ diaphragm	0.3	0.15–0.20	160	Semi-commercial

* Adapted from Wendt and Imarisio [7].

Engineering analysis, process optimization and scale-up

When compared with efforts spent in various other areas of water-electrolysis technology, only minor attention seems to have been paid to engineering analysis, process optimization and scaleup. Existing or proposed designs of large-scale commercial electrolysers have been basically simple multiples of bench-scale or prototype units. Many advanced concepts and procedures of scale-up commonly used in large-scale chemical process plants have not been adequately utilized in the advancement of electrolyser technologies. Following are some of the major engineering issues to be addressed:

- Elucidation of electrochemical reaction mechanisms and kinetics on and around electrode surfaces;
- Establishing the rate-determining step/steps among various adsorption, desorption or reaction steps;
- Effects of heat and mass transfer rates around electrodes and in the bulk of the electrolyte on overpotentials and ohmic losses;
- Electrolyte flow and cell hydrodynamics;
- Shape and size of electrodes; scale-up criteria for large-scale units;
- Design of cell modules—optimum cell stacking, electrical connections and flow distribution of the electrolyte;
- Gas separator and collector design.

Process optimization would call for an analysis of an integrated system consisting of the electrolyser and all the ancillary equipment such as gas-electrolyte separators, heat exchangers and gas purifers. The optimum cell design and operating conditions will be those corresponding to the minimum overall cost of hydrogen production. Other factors that will influence the overall cost and hence the optimum design and operating conditions are the following:

- The market demand; for example, whether for gaseous or liquid hydrogen;
- The proximity of consumption sites
- Integration with an existing energy system or energy network.

A comprehensive system model that can account for the effects of all major design and operating variables needs to be developed for successful scale-up to a large-scale commercial electrolyser. The system model should address the major engineering issues stated earlier and identify the critical scale-up criteria. Application of this model, when combined with simultaneous cost analysis of alternative conceptual designs, will lead to the successful design of a commercial unit.

CONCLUSION AND RECOMMENDATIONS

Review of present and advanced water electrolysis technologies for hydrogen production leads to the following conclusions:

- Significant improvements in alkaline electrolysis have resulted from development of effective electrocatalysts, superior gas separators, improved cell design, corrosion-resistant cell components and more appropriate operating conditions. A modern large-scale commercial alkaline electrolyser, with proper incorporation of these developments, should produce hydrogen at a cost substantially lower than that possible with existing commercial units.
- Appropriate engineering analysis, process optimization, and design and scale-up practice should lead to further improvement of electrolytic hydrogen production technology and accelerated commercialization.
- Advanced alkaline electrolyser (AWE) has the most immediate wide-scale commercial potential of all the technologies currently under development.
- Inorganic membrane electrolyser (IME) and solid polyer electrolyser (SPE) are the two strongest contenders having near-term potentials for wide-scale commercialization.
- Technology improvements have promised a significant reduction of the capital cost of a modern electrolyser. However, this, in turn, has further inflated the contribution of electricity costs to the total cost of hydrogen production. Electricity cost, therefore, still remains the singular key factor in dictating electrolytic hydrogen price.

Identification of the key research issues for advancements in electrolytic hydrogen production has resulted in the following recommendations:

- Instead of predominantly fundamental research directed toward improvement of single-cell performance, a more balanced effort should be undertaken to address three parallel issues:
  —Improvement of single-cell performance with minimum additional expense
  —Engineering analysis, process optimization, and development of design and scale-up criteria for large-scale commercial plants
  —Strategy development for optimum integration with existing energy systems or energy networks.
- Development of inexpensive but improved alternatives for gas separators, electrocatalysts, cell design and cell materials should continue for improvement of single-cell performance.
- High-temperature (1000°C) water vapor electrolysis (HTE) should be advanced through development of novel materials and fabrication techniques, and substantial engineering analysis and scale-up efforts.
- While present technologies are basically concentrated around 100 and 1000°C operating temperatures, possible advantages of operating at intermediate temperatures and heat sources for such operations should be explored.
- Novel design and operation of cells in an endothermic mode should be studied for further savings in electricity costs.

Acknowledgements—The work was supported by Solar Energy Research Institute, Golden, Colorado, through US DOE subcontract No. XK707158.

REFERENCES

1. J. Bockris, B. E. Conway, E. Yeager and R. E. White, eds, *Comprehensive Treatise of Electrochemistry*, Vol. 2, Plenum Press, New York (1981).
2. R. LeRoy and A. F. Hufnagl, Progress in industrial demonstration of advanced unipolar electrolysis. *Int. J. Hydrogen Energy* **8**, 581 (1983).
3. R. L. LeRoy, Industrial water electrolysis: present and future. *Int. J. Hydrogen Energy* **8** (6), 401 (1983).
4. R. L. LeRoy, Hydrogen production by the electrolysis of water. The kinetic and thermodynamic framework. *J. Electrochem. Soc.* **130**, 2158 (1983).
5. H. Vandenborre, R. Leysen, H. Nackaerts, D. Van der Eecken, S. W. Asbroeck and J. Piepers, Advanced alkaline water electrolysis using inorganic-membrane-electrolyte (IME) technology. *Proc. 5th World Hydrogen Energy Conf.* **2**, 703 (1984).
6. B. V. Tilak, P. W. T. Lu, T. E. Colman and S. Srinivasan, In J. Bockris, B. E. Conway, E. Yeager and R. E. White, eds, *Electrolytic Production of Hydrogen, Comprehensive Treatise of Electrochemistry*, Vol. 2: Electrochemical Processing, Chap. 1, Plenum Press, New York (1981).
7. H. Wendt and G. Imarisio, Nine years of research and development on advanced water electrolysis. A review of the research programme of the Commission of the European Communities. *J. Appl. Electrochemistry* **18**, 1 (1988).
8. D. L. Block, S. Dutta, R. L. Port and A. TRaissi, Hydrogen for power applications: task 1. Electrolytic hydrogen production technology, Report # FSEC CR 18887, Florida Solar Energy Center (December 1987).
9. S. Dutta, D. L. Block and R. L. Port, Economic assessment of advanced electrolytic hydrogen production. *Int. J. Hydrogen Energy*, in press.
10. C. Bilgen and E. Bilgen, An assessment on hydrogen production using central receiver solar systems. *Int. J. Hydrogen Energy* **9**, 197 (1984).
11. A. Hammache and E. Bilgen, Performance of large scale photovoltaic electrolyzer systems. *Adv. Hydrogen Energy* **5**, 287 (1986).
12. O. G. Hancock, A photovoltaic-powered electrolyzer: its performance and economics. *Int. J. Hydrogen Energy* **11**, 153 (1986).
13. J. Fischer, H. Hoffmann, G. Luft and H. Wendt, Water electrolysis at high pressure and medium temperature, hydrogen as an energy vector. *Proc. Brussels Int. Seminar*, p. 270, Kluwer Academic, The Netherlands (February 1980).
14. H. Wendt, Electrolytic production of hydrogen, In *Hydrogen: Energy Vector for the Future*, p. 55, Graham & Trotman, London, U.K. (1981).
15. L. Helmet, H. Mezgolits, J. Prasser, A. Schall and W. Stockmans, Comparing study about the research project concerning the improvement of the electrolytic hydrogen production, Commission of the European Communities, Contract # EHB38029D(B), Final Report (1984).
16. J. Divisek, P. Malinowski, J. Mergel and H. Schmitz, Advanced techniques for alkaline water electrolysis. *Proc. 6th World Hydrogen Energy Conf.* **1**, 263 (1986).
17. J. N. Murray, Alkaline solution electrolysis advancements, *Proc. 5th World Hydrogen Energy Conf.* **2**, 583 (1984).
18. I. Abe, I. Fujimaki and M. Matsubara, Hydrogen production by high-temperature, high-pressure water electrolysis; results of test plant operation. *Proc. 4th World Hydrogen Energy Conf.* **1**, 167 (1982).
19. H. Wendt and H. Hofmann, Cermet diaphragms and integrated electrode-diaphragm units for advanced alkaline water electrolysis. *Int. J. Hydrogen Energy* **10**, 375 (1985).

APPENDIX

Typical design features of conventional and advanced electrolyser cells

Ref.	Electrolyser type	Voltage (V)/ Current Density (A cm^{-2}) [temp. (°C)/press. (atm)]	Electrode (cathode [C], anode [A]), catalyst [K] and separator [S] materials/characteristics
[6]	Conventional alkaline	1.75–2.04/0.13–0.65 [70–90/1–30]	C: steel, stainless steel or Ni A: Ni or steel activated by sulfide process or by sponge Ni-coating K: none S: asbestos (~4 mm, 1.2–1.7 Ω cm^2)
[13]	AWE	2.0–3.15/1.0–3.0 [145/30]	C: steel-blasted perforated Ni (0.6 mm holes, 45% opening) A: steel-blasted perforated Ni (1.2 mm holes) K: anode supplied with spinel or perovskite S: skeleton-free ceramic diaphragm (0.45–0.5 mm, 0.073–0.18 Ω cm^2) and metal-gauze supported ceramic diaphragm (0.3 mm, 0.027–0.075 Ω cm^2)
[14]	AWE	1.55–2.0/0.2–1.5 [90–170/30–60]	C: finely divided Ni, Co or Fe A: Ni K: perovskite spinels etc. on anode S: PTFE*bound K-titanate/metal supported porous ceramics/polymer reinforced thin asbestos
[15]	AWE	1.53–2.1/0.13–2.0 [90–150/1–40]	C and A: sintered Ni K: mixed oxides of Ni and Mo on both cathode and anode S: PTFE + Ktitanate (0.45–0.5 mm)
[16]	AWE	1.44–1.6/0.2–1.0 [80–120]	C and A: Ni K: Raney Ni on cathode; Raney Ni, spinel or perovskite on anode S: a fine-pored ceramic diaphragm on NiO basis supported on Ni-gauze
[17]	AWE	1.51–2.33/0.1–0.8 [40–100/—]	C and A: Ni screens (16–60 mesh) K: a Ni–Mo alloy on cathode, various catalysts on anode S: PBI†–K-titanate (0.038 mm)
[19]	AWE	—/1.0 [100–150/—]	C and A: porous electrode made of Ni powder K: none S: Ni-net supported $BaTiO_3$ (0.4 mm, 0.2 Ω cm^2)
[18]	AWE	1.65–1.86/0.2–0.5 [120/5–19]	C and A: Raney Ni/sintered alloy Ni/perforated Ni plate K: none S: porous PTFE impregnated with K-titanate
[15]	IME	1.8/0.4–1.0 [60–120/5–30]	C and A: perforated Ni K: Ni_3S_2 on cathode, spinel on anode S: 80% PAM‡ + 20% PSF§ (0.35–0.40 mm, 0.25–0.3 Ω cm^2)
[15]	HTE	0.95–1.33/0.1–0.7 [900–1009/—]	C: Ni cermet (40 vol% ceramic, Y_2O_3stabilized ZrO_2, +60 vol% Ni) A: Pt or Ca doped lanthanum manganate ($LaMnO_3$) K: Pt or Ca Electrolyte: solid electrolyte of ZrO_2 stabilized with 10 vol% Y_2O_3
[6]	SPE	1.4–2.0/0.25–1.1 [82–149/6.8–40)	C: porous carbon A: porous Ti K: Pt for cathode, a proprietary catalyst for anode S: multilayer expanded metal screens or the Nafion electrolyte membrane itself

* Polytetrafluoroethylene (Teflon).
† Polybenzimidazole.
‡ Polyantimonic acid membrane.
§ Polysulfone.

Economical and Technical Evaluation of UT-3 Thermochemical Hydrogen Production Process for an Industrial Scale Plant

A. Aochi and T. Tadokoro

Japan Atomic Energy Research Institute

K. Yoshida

Department of Chemical Engineering, University of Tokyo

H. Kameyama

Department of Chemical Engineering, Tokyo University of Agriculture and Technology

M. Nobue and T. Yamaguchi

Process Engineering Division, Tokyo Engineering Corporation

Abstract—The thermochemical water-decomposition cycle UT-3 has been demonstrated with the successful operation of the bench-scale model plant. A commercial size UT-3 hydrogen generating plant was conceptually designed to assess the final hydrogen production cost. The encouraging results showed that the UT-3 cycle has the potential to become a competitive process for producing hydrogen.

NOMENCLATURE

H_{H_2}	Energy content of the hydrogen produced in the plant, at HHV (high heating value) (MW)
H_{He}	Input heat to chemical process from HTGR plant by helium gas (MW)
H_R	Heat recovered (MW)
P_c	Power consumed within the plant (MW)
η_{sp}	Power generation efficiency, which means the efficiency of the conversion of recovered heat into electricity or for driving steam turbine

INTRODUCTION

It is a well known fact that hydrogen is an attractive fuel for the future, because it is renewable as an energy resource, and is clean in its energy use. The Br–Ca–Fe cycle named UT-3, which the authors have previously proposed [1, 2], was experimentally verified and the investigation was further continued to solve various problems encountered during its development and scale-up for commercialization [3]. Figure 1 shows the schematic diagram of the UT-3 cycle. This cycle consists of hydrolysis and bromination of Ca and Fe compounds, which are heterogenous endothermic or exothermic reactions proceeding in gas–solid state. This process, therefore, requires the special design and plant operation. This paper describes the results of the preliminary conceptual design and the economics of hydrogen production using the UT-3 process.

FEATURE OF UT-3 CYCLE

All the reactions of the UT-3 cycle are performed continuously in reactors by only transferring the gas-reactant from one reactor to another while keeping solid reactants in the same reactor. Thus, the solid reactants are exposed alternately to two reactions as shown in Fig. 1.

CONCEPTUAL DESIGN OF UT-3 PROCESS

The commercial size UT-3 hydrogen plant producing 20,000 $Nm^3\ h^{-1}$ was designed at a conceptual level. Figure 2 shows the bird's eye view of UT-3 Thermochemical Hydrogen Plant made based on the results of the preliminary conceptual design.

Design criteria

The design criteria for conceptual design are as follows:

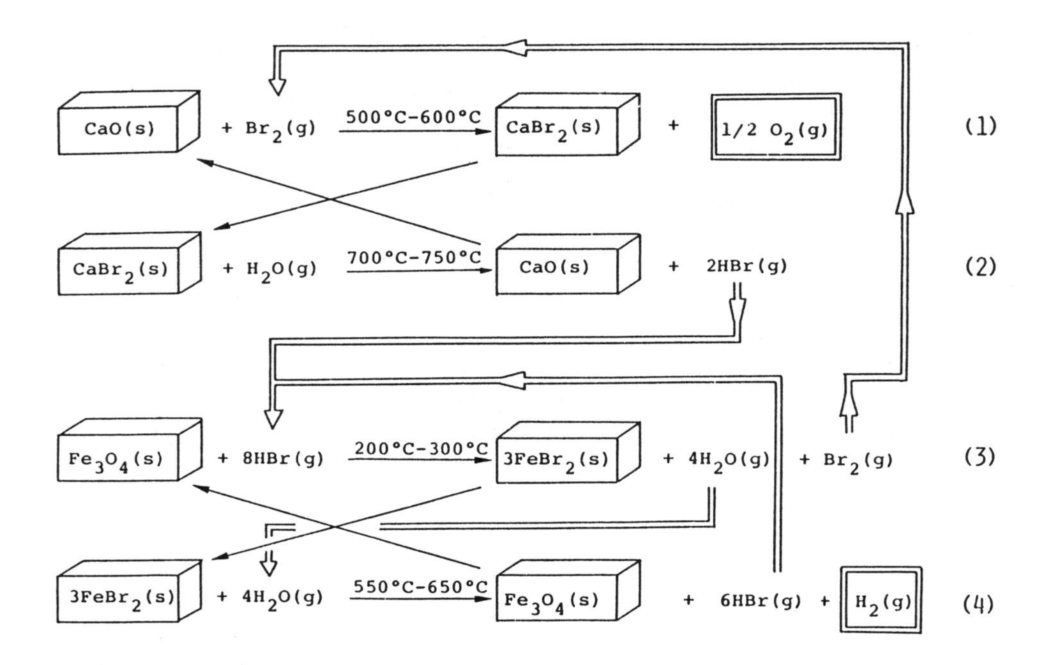

Fig. 1. Schematic diagram of UT-3 cycle.

Fig. 2. Bird's eye view of UT-3 thermochemical hydrogen plant. Assumed hydrogen production capacity: 20,000 $Nm^3\ h^{-1}\ unit^{-1}$, total 80,000 $Nm^3\ h^{-1}$.

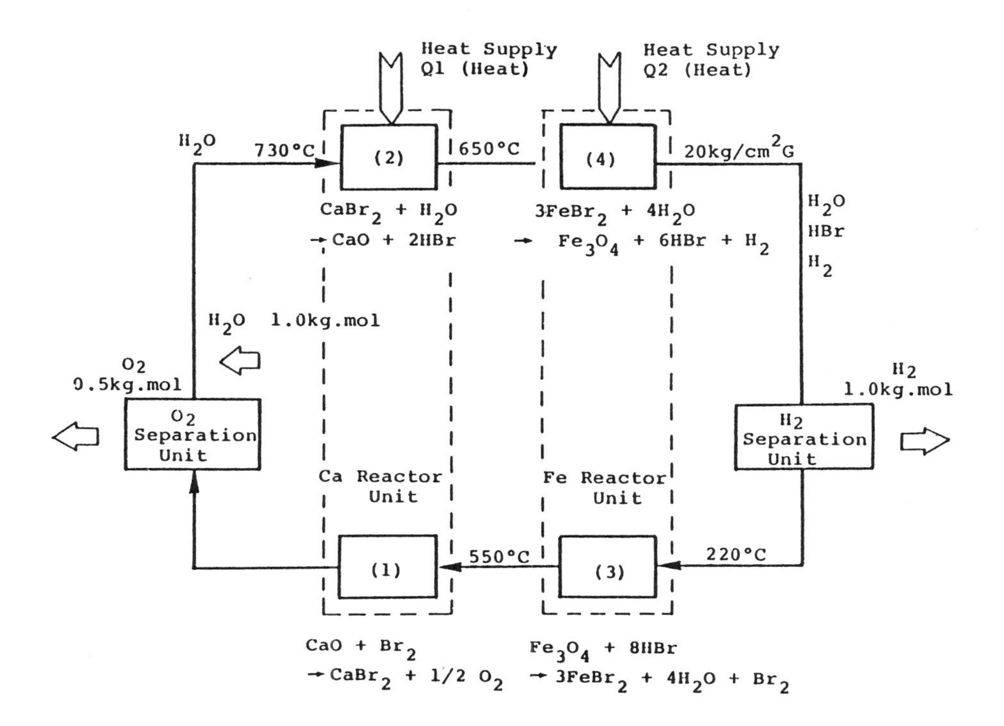

Fig. 3. Schematic system arrangement of UT-3 process (Model 1).

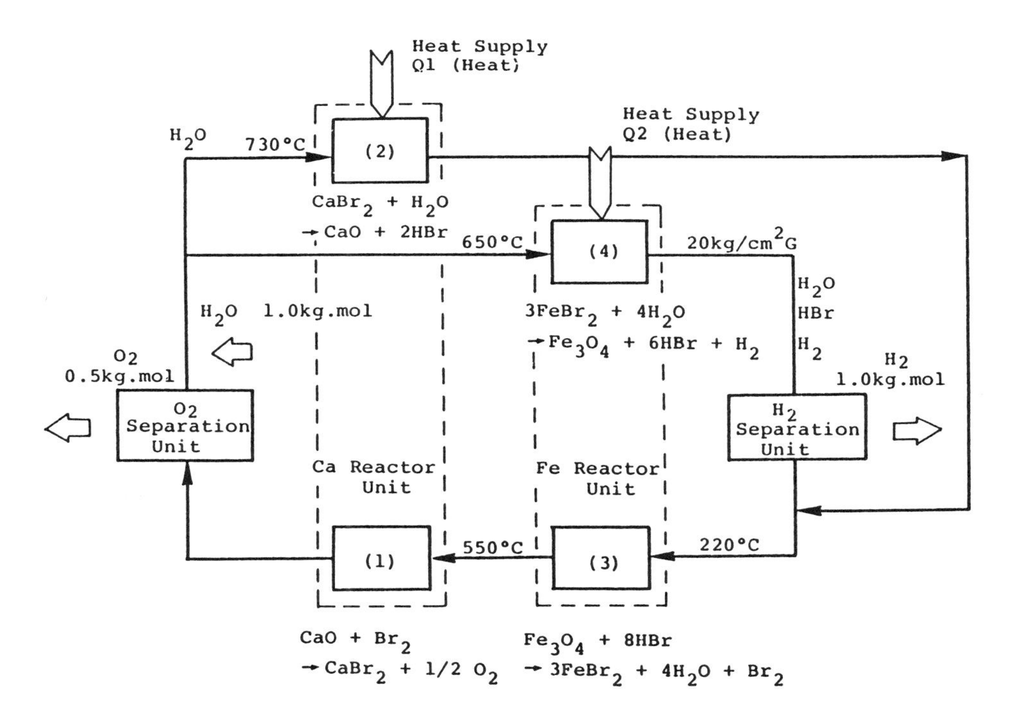

Fig. 4. Schematic system arrangement of UT-3 process (Model 2).

(1) The high temperature gas reactor (HTGR) plant supplies the process heat required for the endothermic reactions in the UT-3 cycle. The above process heat is introduced to the hydrogen plant by helium gas.

(2) Helium gas is introduced to the hydrogen plant at 850°C, 40 kg $cm^{-2}G^{-1}$ and returned at 700°C, 40 kg $cm^{-2}G^{-1}$.

(3) Hydrogen plant is located close to the HTGR plant.

(4) Hydrogen and oxygen are separated by condensation from the respective gas mixtures. This method of separation is very conservative and hence incorporating more efficient separation process will further increase the process efficiency and result in improved economics of the UT-3 process.

(5) Utilities required for the hydrogen plant are assumed to be imported from the outside.

(6) The materials for equipment are selected based on the experimental corrosion data from literature, and which are commonly used in the industry.

(7) The material balance is assumed based on the reaction data obtained in experiments conducted under atmospheric pressure. Operating pressure of each reactor is assumed to be around 20 kg $cm^{-2}G^{-1}$ as shown in Fig. 3.

(8) Pressures of product hydrogen and by-product oxygen are as follows;

Pressure of hydrogen at B.L.: 25 kg $cm^{-2}G^{-1}$
Pressure of oxygen at B.L.: 18 kg $cm^{-2}G^{-1}$

System arrangement

The UT-3 process consists of following main process units;

— Ca Reactor Unit
— Fe Reactor Unit
— H_2 Separation Unit
— O_2 Separation Unit

Model 1 system arrangement is more advantageous from energy consumption and investment cost point of view because the circulated gas flow in Model 1 is lower than Model 2. Hence, Model 1 system arrangement was applied for the conceptual design of the UT-3 process.

As reactions (2) and (4) are endothermic, heat must be supplied from an external heat source which is assumed to be from the high temperature gas reactor (HTGR). Figures 3 and 4 show the two system arrangements (Model 1 and Model 2) which were considered.

Description of the UT-3 process

Figure 5 shows the flowsheet of the UT-3 hydrogen plant. The UT-3 process mainly consists of the reactor units, separation units and gas circulation units. The HTGR is considered as the heat source. Helium gas is assumed to be introduced to the UT-3 hydrogen plant at 850°C and supplies the necessary heat for the reaction.

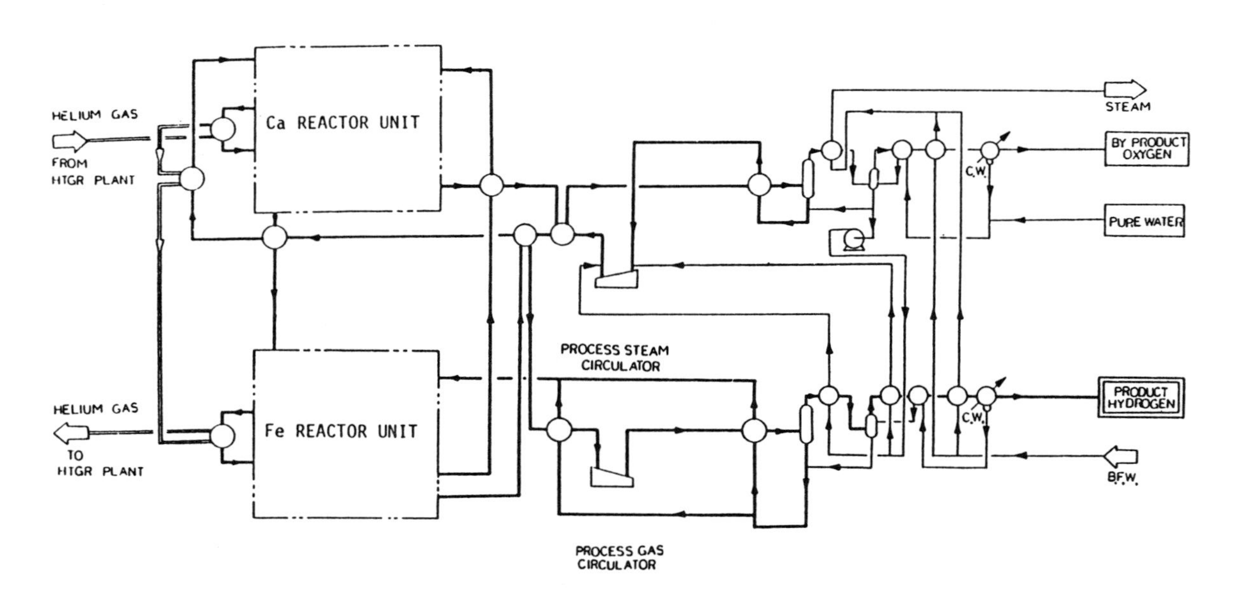

Fig. 5. Flowsheet to UT-3 hydrogen plant.

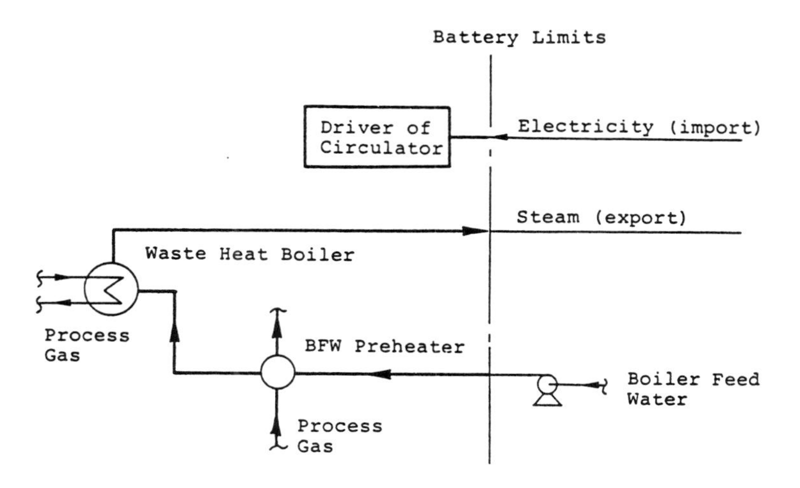

Fig. 6. Schematic flow diagram of steam and electricity for estimation of thermal efficiency.

The reactant gases must be always heated up to a certain temperature level required for each reaction. It is necessary to preheat the reactant gas up to the highest reaction temperature for each reaction, because they cannot reach that temperature by heat exchange with product gases alone. The heating as well as cooling of solid reactants is carried out by use of the sensible heat of reactant gas mixtures containing high steam content. The gaseous products from the reactions (1) and (4) must be separated. In the reaction (1), oxygen can be easily recovered from the mixture of steam and oxygen by condensation. Hydrogen produced from reaction (4) is also separated by condensation from the other two constituents.

A condensed mixture of hydrogen bromide and steam can be utilized directly as gas reactant for reaction (3). Process waste heat generated in hydrogen and oxygen separation unit is recovered as by-product steam.

Thermal efficiency of the UT-3 process

Figure 6 shows the schematic utility flow diagram of steam and electricity for estimation of thermal efficiency. In the commercialized industrial plant, recovered steam may be used directly as a power source for driving the steam turbines of circulators instead of importing the electricity from outside.

Thermal efficiency (η_T) is the ratio between H_{H_2} and H_{He} minus net recovered heat defined by:

$$\left(H_R - \frac{P_c \times 860.6}{\eta_{sp}}\right).$$

Applying the general definition of thermal efficiency and considering the efficiency of the power generation in case of utilizing the recovered heat we have,

$$\eta_T = \frac{H_{H_2}}{H_{He} - \left(H_R - \dfrac{P_c \times 860.6}{\eta_{sp}}\right)} \qquad (1)$$

Figure 7 shows the thermal efficiency (η_T) which was estimated by using the above equation (1), and also shows the possibility that more than 40% thermal efficiency is obtained if the power generation efficiency

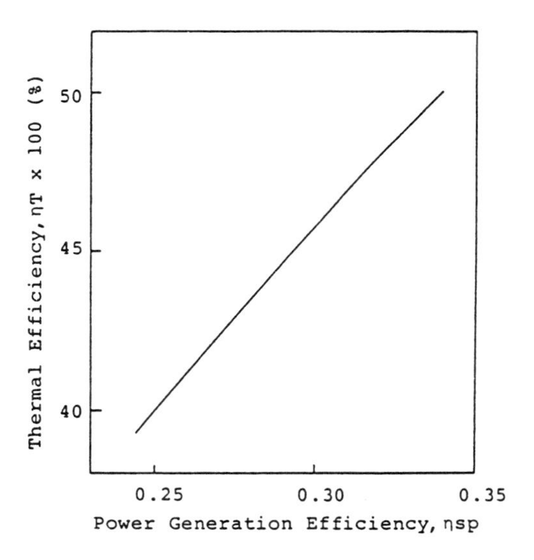

Fig. 7. Thermal efficiency of the UT-3 hydrogen plant (HHV Basis).

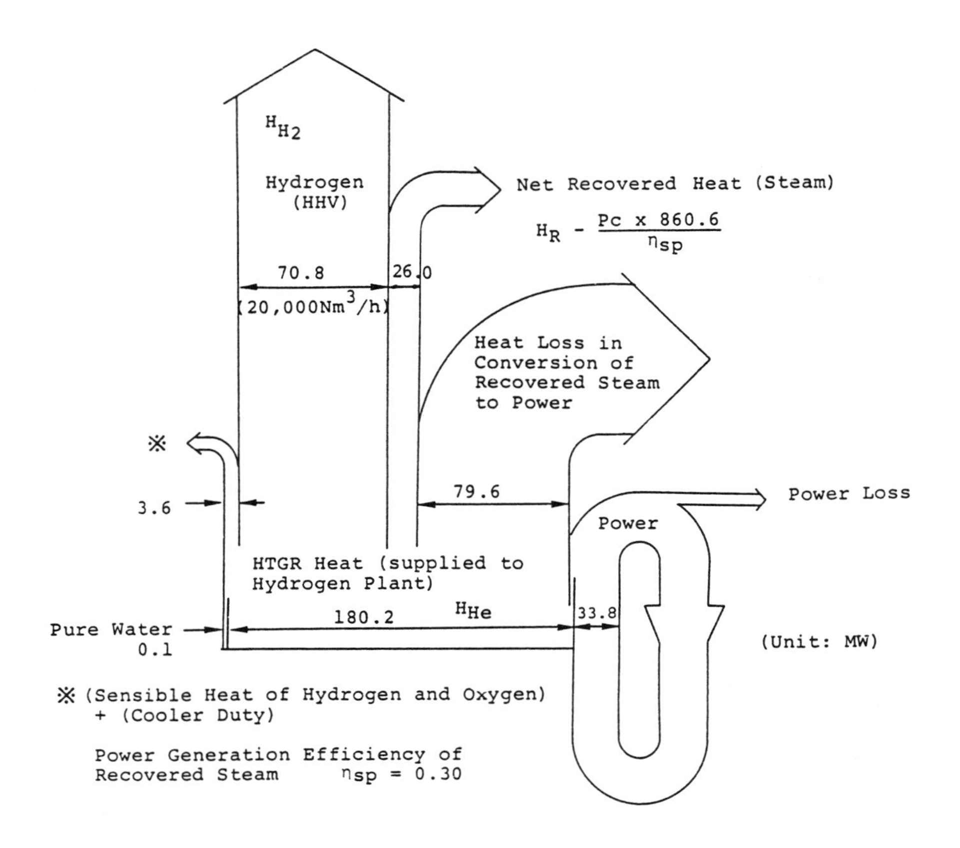

Fig. 8. Energy balance of the UT-3 hydrogen plant.

(η_{sp}) is more than 0.25. Figure 8 shows the example of the energy balance of the UT-3 thermochemical hydrogen plant. An energy balance is made using a power generation efficiency (η_{sp}) of 0.30 in equation (1). It is assumed that most of the heat recovered is utilized for generating the power for driving the motor or steam turbine.

ECONOMICS OF HYDROGEN PRODUCTION USING THE UT-3 PROCESS

Investment costs for the UT-3 hydrogen plant were calculated by application of standard chemical engineering cost estimation techniques. These techniques are based on actual industrial experience for various chemical plants. Hydrogen production cost is estimated based on the results of the preliminary conceptual design described in the previous section. The product costs are calculated using simplified discounted cash flow method with some assumptions. Financial parameters and economic assumptions are summarized in Table 1. Table 2 shows the consumption figures and unit costs of the raw material and utilities.

Table 3 shows the selling price of the hydrogen produced via UT-3 thermochemical process.

Hydrogen production cost via the water electrolysis process is estimated based on the same economic evaluation previses for the purpose of comparing the economics of the UT-3 process with a presumed competitive process in the future.

Specific figures for water electrolysis are as follows:

Plant investment	10.9×10^9 yen
Raw material and utilities consumption	
Electric power (AC)	4,870 kWh $(1{,}000\ \text{Nm}^3)^{-1}$
Pure water	0.84 ton $(1{,}000\ \text{Nm}^3)^{-1}$
Others (Chemicals, nitrogen etc.)	120 yen $(1{,}000\ \text{Nm}^3)^{-1}$

The cost in 1985 yen for hydrogen produced by water electrolysis process is estimated to be 77.2 yen $(\text{Nm}^3)^{-1}$.

To investigate the effects of the economic factor on

Table 1. Financial parameters and economic assumptions

Item	Value
Plant start-up	2010
Project life	30 yr
Plant construction period	3 yr
Plant capacity (hydrogen product)	20,000 $Nm^3\ h^{-1}$
Plant availability	7,200 h yr^{-1}
Capital cost	
Plant investment	12 × 10^9 yen
Start-up cost	10% of plant investment
Interest during construction	10% yr^{-1}
Depreciation Plant	15 yr
Start-up cost	30 yr
Tax rate	50%
Rate of return on investment	0% discounted cash flow (DCF)
Net escalation rate of products and utilities price	
Hydrogen selling price	3.9% yr^{-1}
HTGR heat cost	1.5% yr^{-1}
Electricity, steam, etc.	3.9% yr^{-1}

Table 2. Raw material and utility cost

Item	Consumption*	Unit Cost†
1. HTGR heat	7,745.8 × 10^3 kcal	3.8 yen 1,000 $kcal^{-1}$
2. Chemical (solid reactants)	1.84 l	1,800 yen
3. Utilities		
Pure water	0.804 ton	250 yen ton^{-1}
Cooling water	13.2 ton	8 yen ton^{-1}
Electricity	1,705 kWh h^{-1}	11.5 yen $kWh^{-1}\ H^{-1}$
Others‡		2% of production cost
4. By-products		
Steam	−9.3613 ton	2,600 yen ton^{-1}

* Consumption figures are shown per 1,000 Nm^3 hydrogen product.

† Figures show the projected first year (2007) price in 1985 yen.

‡ Instrument air, emergency power, etc.

Table 3. Revenue and expenditure of UT-3 thermochemical hydrogen production (Unit: 10^6 yen)

	First Year (2010)	Period Average
Sales	5,271	9,914
(Selling price, yen Nm^{-3})	(40.2)	(75.7)
Production cost		
HTGR heat cost	4,437	5,597
Solid reactants	477	477
Utilities		
Pure water, cooling water	44	44
Electricity	3,167	5,957
Others	105	105
Credit of by-product		
Steam	Δ 3,931	Δ 7,394
Variable cost (total)	4,299	4,786
Tax and insurance	218	218
Maintenance	413	413
Labour	76	76
Depreciation	959	483
Fixed cost (total)	1,666	1,190
Production cost (total)	5,963	5,976
Tax	0	1,969
Profit after tax	692	1,969

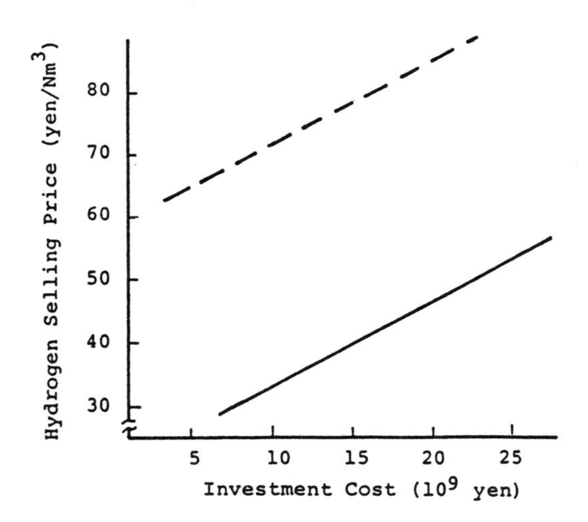

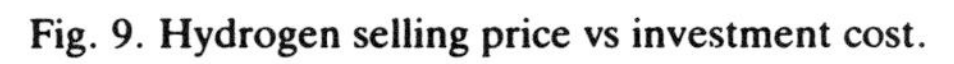

Fig. 9. Hydrogen selling price vs investment cost.

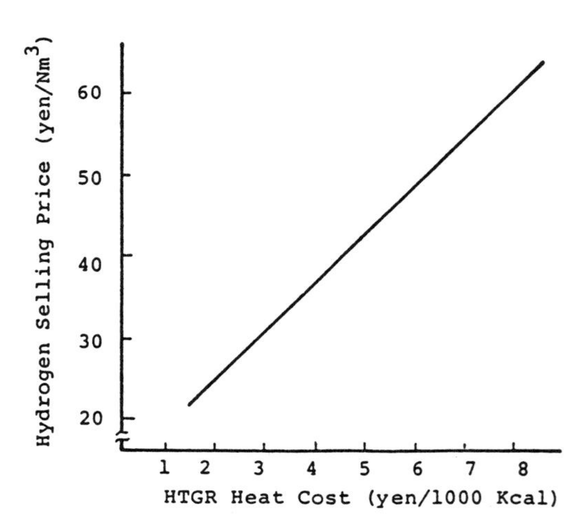

Fig. 10. Hydrogen selling price vs HTGR heat cost.

the UT-3 hydrogen production cost, a sensitivity analysis was carried out to assess the impact of variation in investment cost and cost of HTGR unclear heat. The results of the analysis are shown in Figs 9 and 10.

CONCLUSION

Preliminary conceptual design was carried out with various assumptions to evaluate the economics of an industrial scale UT-3 thermochemical hydrogen production process. The results of the preliminary conceptual design showed the possibility that the UT-3 process thermal efficiency increased to more than 40%. At this efficiency the UT-3 process economics including the hydrogen cost of production are very favourable and the UT-3 process has the great potential to become a competitive process, in comparison with the water electrolysis process. The conceptual design considered the conservative condensation method for gas separation and hence the process economics can be greatly improved if more efficient separation processes are adopted.

Acknowledgements—The authors are grateful to Mr I. Yoshida for carrying out the economic analysis.

REFERENCES

1. K. Yoshida and H. Kameyama, MASCOT—a bench-scale plant for producing hydrogen by the UT-3 thermochemical decomposition cycle. Report of Special Project Research under Grant in Aid of Scientific Research of the Ministry of Education science and Culture, Japan *SPEY* **13**, 29–34 (1985).
2. H. Kameyama and K. Yoshida, Br–Ca–Fe water-decomposition cycles for hydrogen production. *Proc. 2nd World Hydrogen Energy Conference*, Zurich, Switzerland, pp. 829–850 (1978).
3. T. Nakayama, H. Yoshida, H. Furutani, H. Kameyama and K. Yoshida, MASCOT—a bench-scale plant for producing hydrogen by the UT-3 thermochemical decomposition cycle. *Int. J. Hydrogen Energy* **9**, 187–190 (1984).

Photoelectrochemical and Photocatalytic Methods of Hydrogen Production: A Short Review

Nikola Getoff

Institute for Theoretical Chemistry and Radiation Chemistry, The University, and Ludwig Boltzmann Institute for Radiation Chemistry

Abstract—A critical overview on photoelectrochemical and photocatalytic methods for hydrogen production by water splitting is presented. First, the solar spectral irradiance and energy distribution as well as the direct water photolysis are briefly mentioned. This is followed by the liquid-junction transducer comprising semiconductor photoelectrodes, photogalvanic cells, homogeneous photoredox systems and hybrid water splitting methods utilizing solar energy. Presently the most promising technique seems to be the photoelectrochemical and the hybrid devices.

1. INTRODUCTION

The fossil fuels (coal, oil, natural gas) are presently the main energy sources of the industrial world. In some countries nuclear energy is also playing an important role in this respect. On the other hand, oil and natural gas also represent the main basic materials of the chemical and other industries. Since the resources of fossil fuels are limited (estimated world reserves of oil and natural gas are 50–60 years and for coal 450–500 years) the problem of alternative energy resources is very urgent. Hence, an important and humane obligation of our generation is to replace the present energy sources with alternative ones. Besides the hydroelectrical power, the opening up of new energy resources, such as solar, wind, ocean tides and sea streams energies is required, in addition to the thermal energy winning from hot water springs and volcanos.

It should be also mentioned that the content of CO_2, produced by combustion of the fossil fuels, increases continuously. As a sequence of this fact the earth's temperature rises (the greenhouse effect) and the climate is altering. Hence, an enlargement of the desert surfaces and melting of the polar ice caps is expected. The latter will cause a rising of the sea level [1].

Since the energy shock in 1973 a number of laboratories started an intensive research work with the aim to develop methods for hydrogen production by solar energy utilization. Because of its physical and chemical properties, hydrogen is accepted as a nearly ideal energy carrier for the future. It can be produced from water by various techniques and has advantages with respect to electricity. It can be stored and on combustion reforms water (closed cycle) with low pollution (traces of nitrous oxides only).

The photoelectrochemical and photocatalytic methods for hydrogen production are of special benefit, because solar energy is free and practically inexhaustible. The progress in this field is represented in several books [2–10], monographs [e.g. 11–13], review articles [e.g. 14–26] and a relatively great number of publications.

2. SOLAR RADIATION

The sun is located at an average distance of 1.5×10^8 km from the earth and has a surface temperature of about 6000°C. The spectral sunlight distribution is shown in Fig. 1 together with a table summarizing some spectral data. It is evident that due to scattering and absorption processes in the atmosphere the primary sun spectrum (air-mass ratio, $a/m = 0$) is modified. The vacuum-uv- ($\lambda < 200$ nm) and the far-uv-light fraction ($\lambda < 315$ nm) are cut off due to absorption by O, N, O_2, O_3, while the absorption by water vapor and CO_2 are determined by the long wavelength limit.

3. DIRECT WATER PHOTOLYSIS

Thermodynamically the decomposition of liquid water, H_2O (l), at 25°C, to gaseous hydrogen, H_2 (g), and oxygen, O_2 (g), requires a Gibbs free energy change, $G = 2.46$ eV:

$$H_2O\ (l) \rightarrow H_2(g) + 0.5\ O_2\ (g) \quad (1)$$

$$\Delta G = \Delta H - T\Delta S = 2.46\ \text{eV} \quad (2)$$

where $\Delta H = 2.96$ eV (enthalpy change), $\Delta ST = 0.50$ eV (entropy change multiplied by absolute temperature). The endothermic reaction for hydrogen production from water requires an energy of $\Delta H = 2.96$ eV (1 eV = 23 kcal mol^{-1}; 1 cal = 4.186 J or W-s), which corresponds to light of $\lambda \simeq 420$ nm. Water is, however, transparent for this light (see Fig. 2, absorption spectrum of water). Hence, appropriate systems should be

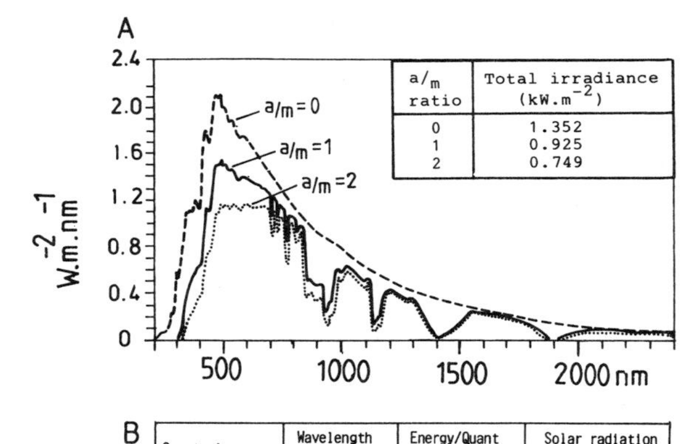

B

Spectral range	Wavelength (λ in nm)	Energy/Quant (eV/hν)	Solar radiation (%)	Watt/m²
Near u.v.	315 - 400	3.95 - 3.1	2.7	20
Blue	400 - 510	3.1 - 2.44	14.6	109
Green-yellow	510 - 610	2.44 - 2.04	16.0	120
Red	610 - 700	2.04 - 1.78	13.8	104
Near IR	700 - 920	1.78 - 1.35	23.5	176
IR-region	920 - 1400	1.35 - 0.89	21.6	162
	>1400	> 0.89	7.8	58

Fig. 1. (A) Solar spectral irradiance ($W\ m^{-2}\ nm^{-1}$) at air-mass ratio (a/m) of 0, 1 and 2 [27] ($a/m = 0$ is outside the atmosphere, $a/m = 1$ and 2 corresponds to 90° and 30° latitude, respectively). (B) Energy distribution of various ranges in the solar spectrum at $a/m = 2$.

developed, which are absorbing solar energy from 315 nm (3.96 eV $h\nu^{-1}$) upwards in order to promote water splitting.

For the sake of completeness the quantum yield (Q) of the primary products of water photolysis (H, OH, e^-_{aq}) for three wavelengths are also given in Fig. 2.

4. PHOTOELECTROCHEMICAL METHODS FOR HYDROGEN PRODUCTION

This kind of device can be divided into two groups: liquid-junction transducers (cells with at least one semiconductor–photoelectrode) and photogalvanic cells having a redox-couple as the sensitizer. The systems of the first type are very promising with respect to technical application for hydrogen production, whereas the second group is still under investigation in order to overcome instability problems. Both types of cells will be discussed separately.

4.1. *Liquid-junction transducers*

This kind of photoelectrochemical system is based on the application of semiconductors as photoelectrodes which are separated by a membrane in a cell in the presence of an aqueous electrolyte. The *n-type* semiconductors (e.g. TiO_2, $SrTiO_3$, GaAs etc.) have an excess of electrons (e^-) and hence are used as the *photoanode*. The *p-type* semiconductors (CdTe, GaP etc.) possess positive holes (p^+) excess (yielding e^- to the semiconductor–electrolyte interface) and serve as the *photocatode*. The electrodes are placed in different chambers separated by an appropriate semipermeable membrane or frit (Fig. 3A) and by immersing them into an aqueous electrolyte band-bending takes place. Hence, an interface is formed causing a charge separation similar to the solid-state Schottky-junction of a solar cell (Fig. 3B). The cell will operate even with one photoelectrode, e.g. n-TiO_2 as photoanode and metal cathode as first described by Fujishima and Honda [33]. Using a single crystal of rutile (n-TiO_2) as photoanode and a Pt-cathode these authors succeeded with a photoinduced electrolysis of water to H_2 and O_2 (Fig. 3A). The energy-level diagram of the cell is given in Fig. 3B.

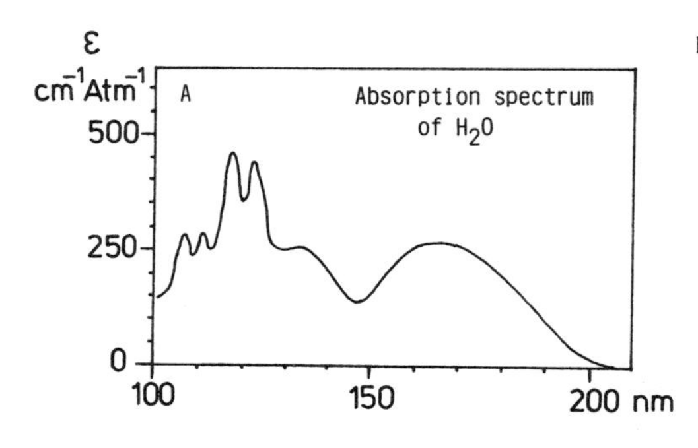

B

Q-values of primary products				
λ nm	eV/hν	Q(H,OH)	Q(e^-_{aq})	Ref.
185	6.7	0.33	< 0.04	/29,30/
147	8.4	0.72	< 0.07	/31,32/
123.6	10.0	1.0	≤ 0.10	/32/

C PHOTOLYSIS OF LIQUID WATER AND SOME PRIMARY REACTIONS

$H_2O \rightsquigarrow H_2O^*$

$H_2O^* \longrightarrow H + OH$

$2H_2O^* \longrightarrow H_2O^{**} + H_2O$

$H_2O^{**} \longrightarrow e^-_{aq} + H_2O^+_{aq}$

$H_2O^+_{aq} + H_2O \longrightarrow OH + H^+_{aq}$

$2H \longrightarrow H_2\ (k = 1x10^{10}\ M^{-1}s^{-1})$

$2e^-_{aq} \longrightarrow H_2 + 2OH^-_{aq}\ (k = 5x10^9\ M^{-1}s^{-1})$

$2OH \longrightarrow H_2O_2\ (k = 6x10^9\ M^{-1}s^{-1})$

$H^+_{aq} + e^-_{aq} \longrightarrow H\ (k = 2.3x10^{10}\ M^{-1}s^{-1})$

$H^+_{aq} + OH^-_{aq} \longrightarrow H_2O\ (k = 1.4x10^{11}\ M^{-1}s^{-1})$

Fig. 2. (A) Absorption spectrum of water (after Watanabe and Zelikoff [28]). (B) Quantum yields (Q) of H, OH and e^-_{aq} at 185, 147 and 123.6 nm. (C) Primary reaction of water photolysis by vuv-light.

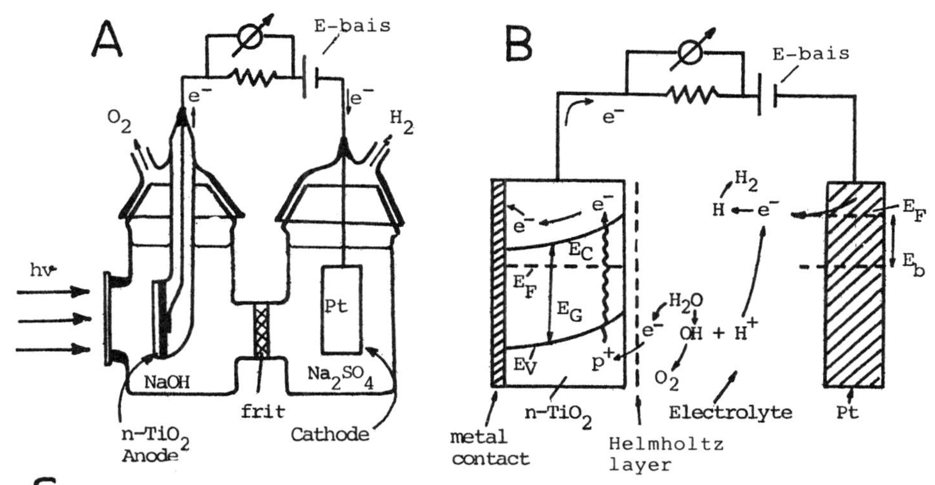

C REACTIONS ON THE ELECTRODES ON:

TiO_2 in neutral media:

$$TiO_2 \rightsquigarrow (TiO_2^+.e^-) \rightarrow TiO_2^+ + e^-$$
$$TiO_2^+ + H_2O \rightarrow TiO_2 + OH + H^+_{aq}$$
$$OH + OH \rightarrow H_2O_2 \quad (k=6x10^9 M^{-1} s^{-1})$$
$$OH + H_2O_2 \rightarrow H\dot{O}_2 + H_2O \quad (k=1.3x10^{10} M^{-1} s^{-1})$$
$$H\dot{O}_2 \rightarrow \dot{O}^-_{2aq} + H^+_{aq} \quad (pK = 4.8)$$
$$TiO_2^+ + \dot{O}^-_{2aq} \rightarrow TiO_2 + O_2$$
$$\dot{O}_2^- + \dot{O}_2^- \rightarrow O_2^{2-} + O_2 \quad (at\ pH \leq 6)$$
$$H\dot{O}_2 + \dot{O}_2^- \rightarrow HO_2^- + O_2$$
$$H\dot{O}_2 + H\dot{O}_2 \rightarrow H_2O_2 + O_2 \quad (pH \leq 3)$$

TiO_2 in basic media:

$$TiO_2^+ + OH^- \rightarrow TiO_2 + OH$$
$$OH \rightleftharpoons \dot{O}^-_{aq} + H^+_{aq} \quad (pK = 11.9)$$
$$TiO_2^+ + \dot{O}^-_{aq} \rightarrow TiO_2 + O^{\bullet}_{ads}$$
$$2O^{\bullet}_{ads} \rightarrow O_2$$

On the Pt-cathode:

$$e^- + H^+_{aq} \rightarrow H_{ads}$$
$$2H_{ads} \rightarrow H_2$$
$$e^- \rightarrow e^-_{aq} \quad (2 - 3\ ps)$$
$$e^-_{aq} + H^+_{aq} \rightarrow H \quad (k=2.3x10^{10} M^{-1} s^{-1})$$
$$2H \rightarrow H_2 \quad (2k=1.1x10^{10} M^{-1} s^{-1})$$

Fig. 3. (A) Fujishima–Honda cell with n-TiO_2 photoanode and Pt-cathode. (B) Schematic energy level diagram of the cell. E_V-valence band, E_C-conduction band, E_F-Fermi level, E_G-energy gap (for $n = TiO_2$; $E_G = 3.0$ eV). (C) Possible reactions on TiO_2 photoanode in neutral and basic media as well as on the Pt-cathode, where "solvated electrons" (e^-_{aq}) can be formed in the Helmholtz-layer.

When n-TiO_2-photoanode is illuminated (light with $\lambda = 315$–450 nm) electrons rise from the valence band (E_V) to the conductivity band (E_C) and from there to the Pt-cathode. As a result of this process a photocurrent is initiated leading to water photoelectrolysis (production of H_2 and O_2). The main reactions on the surface of n-TiO_2 in neutral and basic media, as well as on the Pt-cathode, respectively are presented in Fig. 3C.

The energy difference between both levels, $E_C - E_V = E_G$ (band-gap; eV) is a characteristic value for each type of semiconductor. In Fig. 4 the relative energy levels of various semiconductors are illustrated [34, 35]. For estimation of the optical light-wavelengths (λ in nm) for a semiconductor with a given E_G (eV) the following formula can be used:

$$\lambda(\text{nm}) = 1240/E_G. \quad (3)$$

Due to saturation with O_2 of the anode-volume, the H_2-formation on the cathode can be inhibited. This effect is usually prevented by applying a supporting voltage up to 0.7 V.

Since n-TiO_2 as single rutile crystal is rather expensive much effort has been put forward in order to replace it by a cheaper, photo-sensitive, polycrystalline n-TiO_2 layers. At the beginning of this kind of investigation a photoinduced corrosion of a n-TiO_2-film produced by anodic oxidation was observed [36, 37]. This problem was overcome by production of polycrystalline n-TiO_2-layers under controlled thermal oxidation of Ti-metal foils at 600°C for about 30 min [38–40]. The best

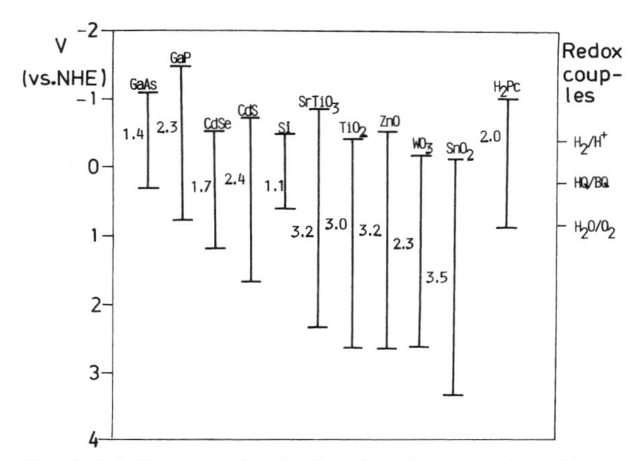

Fig. 4. Relative energy levels of semiconductors and of phthalocyanine (H_2Pc) and redox couples at pH 7, vs NHE [34, 35].

n-TiO_2-films have a thickness of 0.1–2 μm and consist of rutile. Similar results were obtained by other authors as well [41, 42]. It is of interest to mention that a strong relationship between photocurrent (mA cm^{-2}) and thickness of the n-TiO_2 photoanode exists [38–40]. The photosensitivity of these kind of layers is essentially influenced by the surface as illustrated in Fig. 5 [39, 43–46]. Obviously the best n-TiO_2-films are produced by controlled thermal oxidation of Ti-metal foils (micrograph 4 and curve 4 in Fig. 5A and B respectively).

It might be mentioned that in the last decade a relatively great number of papers have been published concerning the production and testing of various n- and p-types photoelectrodes. Special attention is given to photoelectrochemical cells comprising, e.g. n-TiO_2/p-CdTe, n-TiO_2/p-GaP, n-$SrTiO_3$/p-CdTe and n-$SrTiO_3$/p-GaP [47], as well as other systems [48–50].

The photoelectrodes possess generally only a rather limited range of light absorption. In order to enlarge their absorption band (minimizing the band-gap) a good deal of work has been performed by attachment of a suitable dye-stuff on the semiconductor surface. Such kind of sensitization of n-TiO_2, n-$SrTiO_3$ and n-SuO_2 using xanthenthiazine-, three-phenylmethane-, merocyan-dyes [e.g. 25, 51–54], as well as by use of CdSe, CdS, GaAs, GaP, ZuO phthalocyanine [55, 56] and ruthenium-based dyes etc. [57] were reported. The sensitizing mechanisms are explained by energy transfer from the excited dye to the surface of the semiconductor or/and by electron transfer from the dye-stuff into the conductivity band of the photoelectrode. It might be mentioned that excited thionine, methylene blue [58], flavines [59], proflavines and acridine orange [60] etc. in aqueous solutions are leading to the formation of solvated electrons. It is important to stress that all sensitizers from organic compounds are not stable during operation of the hydrogen production cell [57].

In order to avoid corrosion of p- or n-type silicon in an aqueous electrolyte a monomolecular layer of various metals (e.g. Pt, Co, Ni, Au, Pb, Cd, W, Mo) was used to cover the photoelectrode [61, 62]. The H_2-formation in this case is based on the charge transfer through the metal–solution layer. Very recently it was shown that by evaporation of Ni on n-Si semiconductor followed by thermal annealing (350–650°C for 30–60 min) a multilayer n-type silicon electrode is formed, namely: $NiO(OH)Ni_2Si/n^+/n$-Si, under operation in 1 M KOH electrolyte [63]. This kind of photoanode proved to be

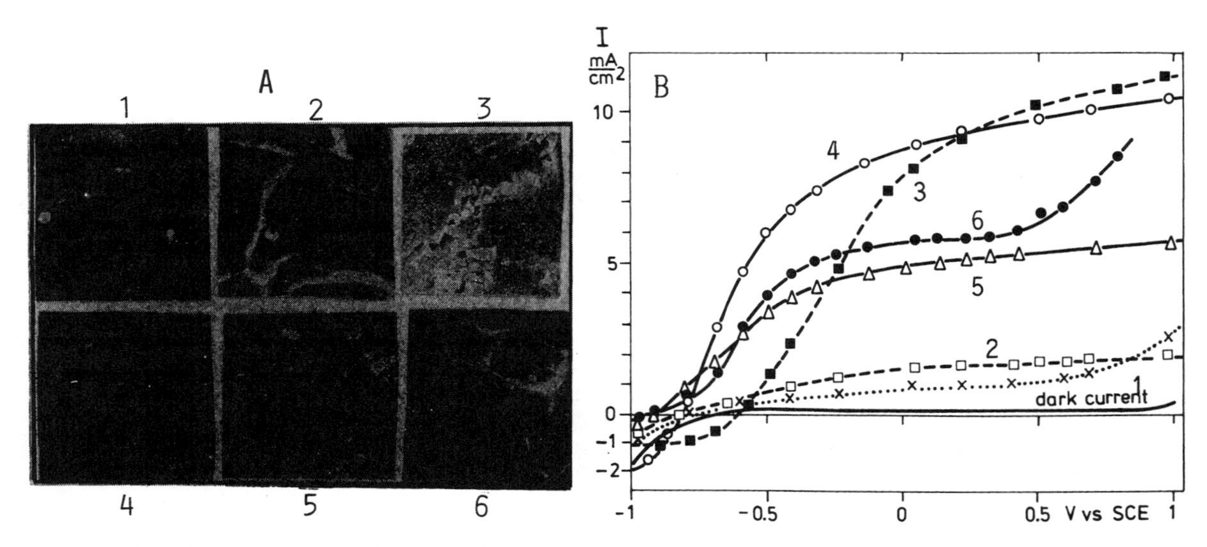

Fig. 5. (A) Scanning electron micrographs (f ~ 5000) of different n-TiO_2 layers produced by: (1) vapour deposition at 2000°C 5×10^{-3} mbar ($d = 1.32\ \mu$m), (2) anodic oxidation (60 mA, 30 V), (3) anodic oxidation followed by reduction in H_2 at 600°C, (4) thermal oxidation at 600°C, 30 min in O_2, (5) sample 4 reduced in H_2, (6) thermal oxidation at 700°C, 30 min in air. (B) Current–voltage curves for the above n-TiO_2 layers.

rather photosensitive and corrosion resistant for long operation times.

Studies in this respect have been also performed by using p-type indium phosphine electrodes covered with platinum (p-InP(Pt)), in combination with an n-type gallium arsenide (n-GaAs) photoanode in a self-driven photoelectrochemical cell for water electrolysis [64]. The photoanode possessed a thin film (~200 Å) of Mn-oxide for protection against corrosion. Using aqueous 6 M KOH as electrolyte in this PCE cell a conversion of light to H_2 with efficiency, $\eta = 8.2\%$ was achieved. An interesting PCE cell making use of bipolar semiconductor photoelectrode arrays (Cd/Se/polysulfide/COS) was investigated under various conditions (effect of light flux, redox couple concentrations and number of panels) [65 and Refs therein]. For the optimum achievement of water splitting ($\eta = 2.8\%$) there were applied 8–9 panels in the array based on n-CdSe, 0.8–1 M polysulfide concentration and a light flux of about 70 mW cm^{-2}.

An important factor in the development of semiconductor photoelectrodes is the knowledge of photovoltage and photocurrent kinetics. Such experiments were performed by using chopped light (0.1–1 s) [20, 40, 66, 67], common flash- (μs-range) and LASER flash-techniques (μs to ps range) [68–71]. At least two photovoltage and photocurrent transients were observed. Their formation is independent of excitation intensity and external resistance, but pH-dependent. An attempt has also been made recently to study the photoreactions at the semiconductor-electrolyte interface under diminishing field conditions using 10 ns laser-flash (335 nm) for excitation on n-TiO_2 photoelectrode [72]. Only at higher potentials and at relatively low light-doses was possible a transient determination by the RC-constant of the circuit. Further investigations in this respect are needed for elucidation of the reaction processes on the electrode surface.

Summarizing the results in this field it can be noticed that the liquid-junction transducers are very promising devices for hydrogen production from water utilizing solar energy. The main objectives presently are to enlarge the light-absorption range and to increase their photoefficiency and corrosion stability for long-term operation.

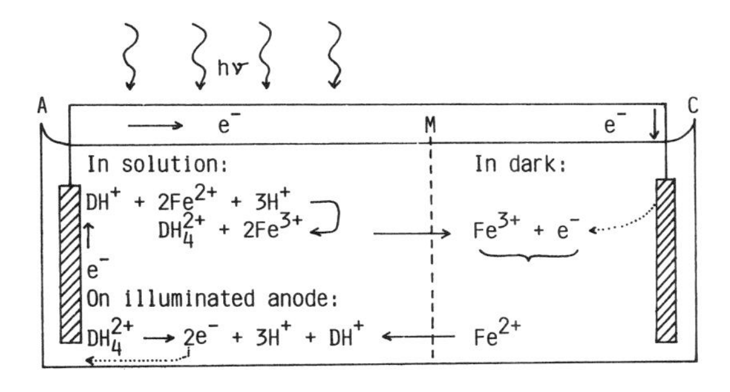

$MB^+ \rightsquigarrow {}^1MB^+$

${}^1MB + Fe^{2+} \longrightarrow$ fluorescence quenching

${}^1MB \xrightarrow{ISC} {}^3MB^+$ (triplet formation)

${}^3MB^+ + H^+ \underset{}{\overset{pK_T}{\rightleftharpoons}} {}^3MBH^{2+}$

$pK_T(TH^+) = 6.3$ and $pK_T(MB^+) = 7.5$

${}^3MB^{2+} + Fe^{2+} \longrightarrow Fe^{3+} + M\dot{B}H^+$ (semiquinone)

$2M\dot{B}H^+ \rightleftharpoons MB^+ + MBH_2^+$ (leuco-dye form)

$M\dot{B}H + H^+ + Fe^{2+} \longrightarrow MBH_2^+ + Fe^{3+}$

$M\dot{B}H^+ + Fe^{3+} \longrightarrow MB^+ + H^+ + Fe^{2+}$

$MBH_2^+ + Fe^{3+} \longrightarrow M\dot{B}H^+ + H^+ + Fe^{2+}$

Brutto reaction:

$MB^+ + 2H^+ + Fe^{2+} \underset{dark}{\overset{light}{\rightleftharpoons}} MBH_2^+ + Fe^{3+}$

Fig. 6. Schematic presentation of the Rabinowitch-cell and light induced reactions in the electrolyte consisting of methylene blue and Fe^{2+}/Fe^{3+}-ions in aqueous solution.

4.2. *Photogalvanic cells*

This kind of device for solar energy transformation into electricity are still under investigation since the first cell was proposed in 1940 by Rabinowitch [73] and in the meantime a number of improvements were performed on it [e.g. 3, 5, 6, 74–78]. The cell comprises two electrodes separated by a membrane and an electrolyte of acid aqueous solution containing Fe^{2+}/Fe^{3+} couple and a dye as sensitizer (thionine, methylene blue, acridine orange etc., Fig. 6). By illumination of the solution in the anodic compartment, several photoredox reactions, resulting in various intermediates, are taking place: formation of singlet and triplet states, electron transfer processes causing the formation of dye semiquinone (e.g. MBH^+), leuco-dye (MBH_2^+) and Fe^{3+} ions. In the dark the process is reversible gaining a potential of about 0.3 to 0.5 V. A simplified reaction mechanism is presented in Fig. 6, where methylene blue (MB^+) is acting as sensitizer.

Theoretically no replenishment of cell-components is expected. However, this is not the case. The instability of the cell is certainly based on several processes among which the energy transfer from single excited dye, ${}^1MB^+$, to Fe^{2+} (reaction 4) in addition to the formation of Fe^{3+} ions (reaction 5) [79, 80] are the most important:

$${}^1MB^+ + Fe^{2+} \longrightarrow MB^+ + {}^*Fe^{2+} \text{ (energy transfer)} \quad (4)$$

$$\xrightarrow{+H^+} MBH^+ + Fe^{3+} \text{ (charge transfer)} \quad (5)$$

$${}^*Fe^{2+}_{aq} \rightarrow Fe^{3+}_{aq} + e^-_{aq} \quad (6)$$

$$e^-_{aq} + H^+_{aq} \rightarrow H \; (k = 2.3 \times 10^{10}\,M^{-1}s^{-1}). \quad (7)$$

Reaction (4) occurs at $\lambda \leqslant 440$ nm and initiates the formation of "solvated electrons" (e^-_{aq}, reaction 6) [79, 81, 82]), which are converted into H-atoms (reaction 7). It was further found [58] that e^-_{aq} are also formed by illumination of thiazine dyes (thionine and methylene blue) in aqueous solutions. The same observation was made for a number of flavines [83]. Hence, it can be concluded that the instability of the photogalvanic cell is effected in addition to other processes [76, 77] also by the H-atoms, formed by side reactions [23].

Reaction (5) takes place at $\lambda \geqslant 400$ nm and is analogous to triplet dye reduction by Fe^{2+} ions.

In order to elucidate the role of H-atoms in the photogalvanic cell, specially designed pulse radiolysis investigations on thiazine dyes were performed [84, 85] in combination with a semi-linear computation method [86]. It was found that 86% of H atoms are leading to the formation of desired dye semiquinone (MBH^+), while the rest of the H-atoms decompose the sensitizer. Similar observations were also made for acridine orange [87] and proflavin [88]. The reaction mechanism and the resolved kinetics of the H-attack on methylene blue is presented as an example:

(MB^+) + H $\xrightarrow{k_8}$ (semiquinone) (8)

$\xrightarrow{k_9}$ (H-adduct) (9)

$\xrightarrow{k_{10}}$ (scission of the chromophore group) (10)

$(MBH_2^+$, leucobase)

$k_8 = 9.5 \times 10^9\ dm^3 mol^{-1} s^{-1}$
$k_9 = 0.6 \times 10^9\ dm^3 mol^{-1} s^{-1}$
$k_{10} = 0.95 \times 10^9\ dm^3 mol^{-1} s^{-1}$

It might be mentioned that a photogalvanic cell using thionine (Th^+) as a photosensitizer EDTA (instead of Fe^{2+} ions) as a reductant is recently reported [90]. The effect of various experimental parameters (pH, $[Th^+]$, [EDTA] etc.) have been studied in respect to the electrical output of the cell. Using 8×10^{-4} M EDTA, 1.47×10^{-4} M Th^+ at pH 11.3 a photocurrent of 80 μA and a photopotential of 900 mV were observed.

5. PHOTOCATALYTIC METHODS

A relatively great number of publications are concerned with this kind of system, which are presented in a broad variety of components [e.g. 3, 5, 6, 12–14].

A system having a redox potential $E^0 \geqslant +0.82$ V is thermodynamically capable to oxidize water (O_2-production) and contrary, with $E^0 \leqslant -0.41$ V, to reduce it (H_2 formation). This fact is based on the following processes (pH 7):

$$2H^+ + 2e^- \rightarrow H_2\ (E^0 = -0.41\ V) \quad (11)$$

$$2H_2O \rightarrow O_2 + 4H^+ + 4e^-\ (E^0 = +0.82\ V) \quad (12)$$

$$H_2O \rightarrow H_2 + 0.5\ O_2\ (E^0 = 1.23\ V). \quad (13)$$

Hence, systems with appropriate E^0 value can mediate the formation of H_2 or O_2 from water by utilization of solar energy. The photocatalytic methods are founded on these basic redox-processes.

The proposed devices for water splitting can be divided into *homogeneous* and *heterogeneous* systems. A relatively large number of model systems belonging to the first group have been proposed [e.g. 12, 13, 19, 22, 78 and ref. therein]. A more advanced coupled photocatalytic system of this kind is presented schematically in Fig. 7(A). It consists of: a *photosensitizer*, e.g. tris-(2,2′-bipyridine)-ruthenium-(II) dication, $Ru(bipy)_3^{2+}$, a *H_2-producing* catalyst (acting as an electron pool; e.g. colloidal Pt, platinized TiO_2 powder etc.), an *electron relay* (e.g. methyl viologen, MV^{2+}) and an *O_2-producing catalyst*, e.g. RuO_2. On illumination with visible light (400–600 nm) the sensitizer becomes electronically excited (life time, $\tau = 0.6\ \mu s$) and reduces MV^{2+} to the radical cation, $M\dot{V}^+$. The products being $Ru(bipy)_3^{3+}$ and $M\dot{V}^+$ are tending to react back ($k = 2.4 \times 10^9\ M^{-1} s^{-1}$), but this is suppressed by both catalysts. Thereby the redox-potentials of the couples are: E^0 $(MV^{2+}/M\dot{V}^+) = -0.45$ V and E^0 $(Ru(bipy)_3^{3+}/Ru(bipy)_3^{2+}) = 1.26$ V [6, 12, 13, 90, 91]. In the sequel $M\dot{V}^+$ transfers one electron on the surface of the Pt-catalyst and oxidizes to MV^{2+}, where H_2 is produced; $Ru(bipy)_3^{3+}$ is reduced on RuO_2-surface and O_2 is formed. At prolonged operation of the system a decomposition of $Ru(bipy)_3^{2+}$ [92] and of MV^{2+} [93] was observed, which leads to an instability of the system. The probable reactions taking place on the surface of both catalysts are given in Fig. 7(B) and (C). It is obvious that simultaneously with the production of H_2 and O_2 the sensitizer and the electron relay are also decomposed.

By application of the pulse radiolysis technique in combination with a semilinear computation procedure [86] it was possible to elucidate the reaction mechanisms of the MV^2 decomposition by attack of H-atoms [94, 95]:

$$MV^{2+} + H \xrightarrow{k_{14}} H_3C\text{-}\overset{+}{N}(C_5H_4)\text{-}(\dot{C}_5H_4)N\langle^{H}_{CH_3}\ (M\dot{V}^{+}H^{+}) \xrightarrow{k_{14'}} H_3C\text{-}\overset{+}{N}(C_5H_4)\text{-}(\dot{C}_5H_4)N\text{-}CH_3 + H^{+}\ (M\dot{V}^{+}) \quad (14)$$

$$\xrightarrow{k_{15}} \text{e.g. } H_3C\text{-}\overset{+}{N}(C_5H_4)\text{-}(\dot{C}_5H_4(H)(H))\overset{+}{N}\text{-}CH_3\ (MV^{2+}H) \xrightarrow{k_{15a}} \text{Products} \quad (15)$$

$k_{14} = 3.1 \times 10^8\, dm^3 mol^{-1} s^{-1}$

$k_{14} = 2.0 \times 10^4\, s^{-1}$

$k_{15} = 2.9 \times 10^8\, dm^3\, mol^{-1}\, s^{-1}$

$k_{15a} = 6.0 \times 10^8\, dm^3\, mol^{-1}\, s^{-1}$

$$MV^{2+} + e^-_{aq} \rightarrow M\dot{V}^+ \ (k_{16} = 7.5 \times 10^{10}\, M^{-1}\, s^{-1}\ [94]). \quad (16)$$

The OH-radicals occurring on RuO_2 surface and on its vicinity (see Fig. 7C) can also attack MV^{2+} leading to various products [96]:

$$MV^{2+} + OH \rightarrow \text{transients} \quad (17)$$
$$(k = 1.3 \times 10^8\, M^{-1}\, s^{-1}\ [96]).$$

The sensitizer, $Ru(bipy)_3^{2+}$, can principally be involved in similar reactions. As a model substrate 2,2′-bipy-H^+ was used for pulse radiolysis experiments. The H-attack on its results in the formation of two transients, which lead to degradation of the compound [97]:

$$(2,2'\text{-bipy-}H^+) + H \rightarrow (2,2'\text{-bipy-}H_2^{+})\ (R_1)\ (k_{18} = 1.4 \times 10^8\, M^{-1}s^{-1}) \quad (18)$$

$$(2,2'\text{-bipy-}H^+) + H \rightarrow \text{(adduct on ring carbon)}\ (k_{19} = 0.6 \times 10^8\, M^{-1}s^{-1}) \quad (19)$$

As a conclusion it can be stated, that homogeneous photo-redox systems of this kind are unstable in operation and hence they are presently of academic interest only. Worth mentioning is the fact that both, H_2 and O_2, are formed simultaneously (detonating gas), therefore special techniques for their separation are needed.

A catalytic photodissociation of water can be also achieved only in the presence of semiconductor suspension or colloids. It has been also demonstrated that by illumination ($\lambda > 300$ nm) of a fine TiO_2-suspension (grain size $\leqslant 0.06$ mm, 3–15 mg ml^{-1}) in diluted sulfuric acid, H_2 and O_2 were produced in a ratio of about 2:1 [98]. In the presence of 10^{-3} M Ce^{4+}-ions an enhanced evolution of O_2 was observed. In both systems n-TiO_2 is acting as an efficient electron donor.

By irradiation of aqueous CdS–dispersion in the presence of sulfide ions or H_2S with visible light a water cleavage is observed [99]. When RuO_2 is deposited on the CdS–particles an increased H_2–formation occurs reaching a quantum yield, $Q(H_2) = 0.35$. Further, suspensions of CdS and mixtures of CdS with TiO_2, Al_2O_3, RuO_2, Pt or CdS + TiO_2 + RuO_2 etc. when illuminated in alkaline aqueous solutions containing SH^- ions show an increased H_2-formation in the given succession and H_2S is decomposing [100]. Thereby an electron transfer from the conduction band of CdS to that of TiO_2 etc. is taking place. It has been also shown that a mixture of catalysts, e.g. Pt(RuO_2)/TiO_2 used as an aqueous suspension can mediate water splitting [101]. The best combination of physical characteristics was achieved with high surface area and small size of agglomerates containing Pt on TiO_2. Such kind of simple systems are opening a new branch in field of solar energy utilization. They are relatively stable against photocorrosion and hence show promising aspects for the future.

Relatively less work is performed for developing of heterogeneous photoredox systems. It might be mentioned the photocatalytic production of H_2 and O_2 from water in separate reaction vessels utilizing Eu^{2+} as photosensitizer and a bifunctional catalyst (Pt and manganese-oxichloride) [102]. A part of the gained energy is used for pumping the solution from the one into the other vessel. More research work is recommended in this field.

6. HYBRID WATER SPLITTING SYSTEMS

Various multi-stage hybrid systems (Yokohama cycles) for water decomposition by solar energy utilization have been developed by Ohta *et al.* [5, 103–107]. They are based on an appropriate combination of photochemical and electrochemical reactions (Yokohama Mark 5; Fig. 8) or on photochemical, electrochemical and thermochemical processes (Yokohama Mark 6). Both systems are making use of iodine as sensitizer in acid aqueous solution (presented as I_2 and I_3^-). A diagram of the Yokohama Mark 5 system is pictured in Fig. 8 together with the photoinduced brutto-reactions.

According to the scheme the aqueous solution containing iodine and ferrous sulfate ($FeSO_4$) is passed as thin layer in front of the Fresnel lense. The formed HI and Fe^{3+} ions are subsequently reduced electrochemically in two separated cells under formation of H_2 and O_2 and rebuilding of I_3^- and Fe^{2+} ions. The unabsorbed light fraction is collected by a Fresnel lense and the obtained heat is partly converted into electricity by a thermoelectric device (p- and n-type semiconductors, Fig. 8). Additional electric power is, however, needed for covering the energy demand in Mark 5 system. According to the authors [107] the overall efficiency of the absorbed light is about 20%. The energy necessary in the improved Mark 6 system is shared by thermochemical and electrochemical processes (Fig. 8). The efficiency of the photochemical conversion process depends on the light intensity, concentration and kind of the reactance and amounts to about 15–20%.

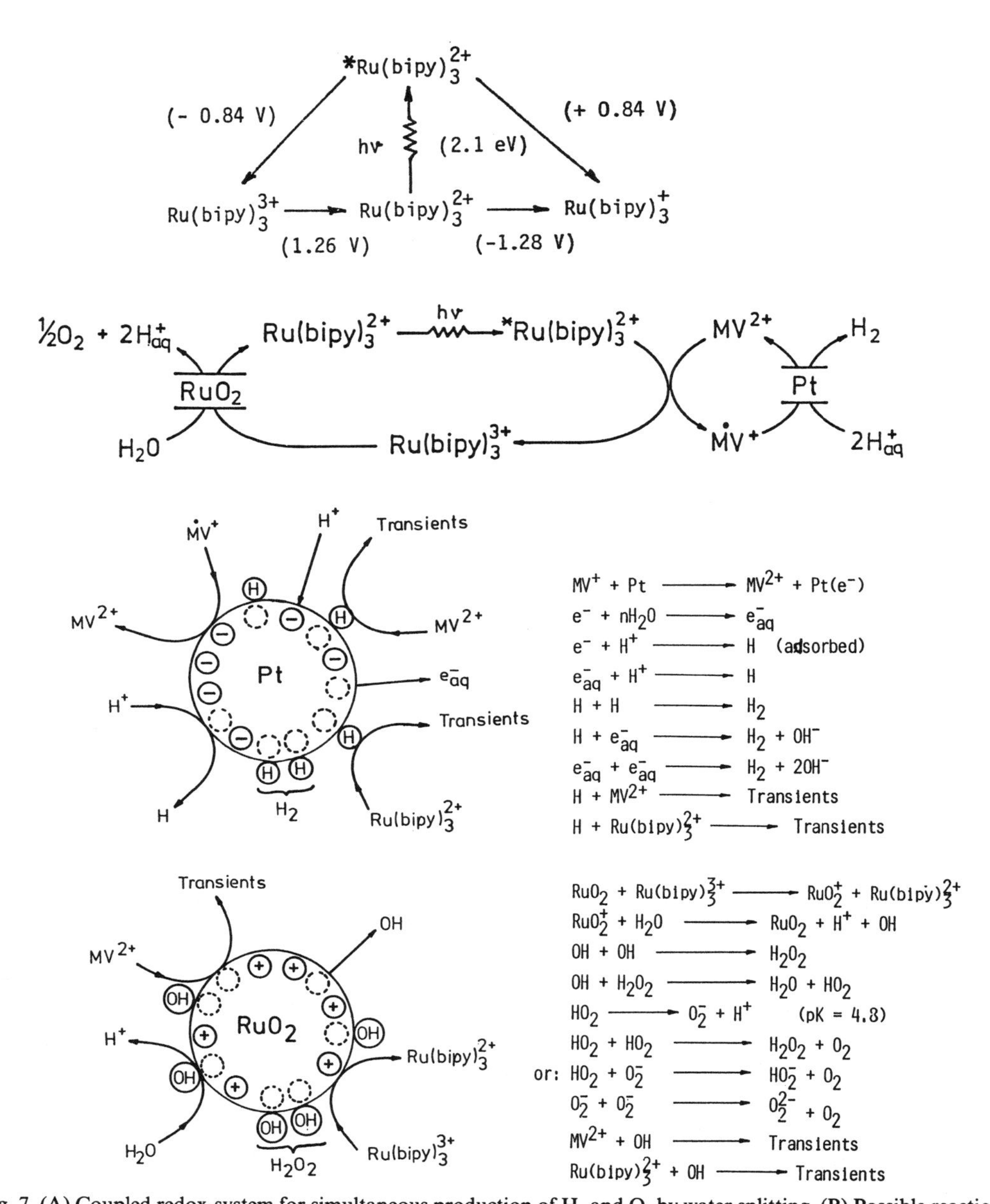

Fig. 7. (A) Coupled redox-system for simultaneous production of H_2 and O_2 by water splitting. (B) Possible reactions on the Pt-surface and its vicinity (reduction processes). (C) Probable reactions on the RuO_2-surface and its vicinity (oxidation processes).

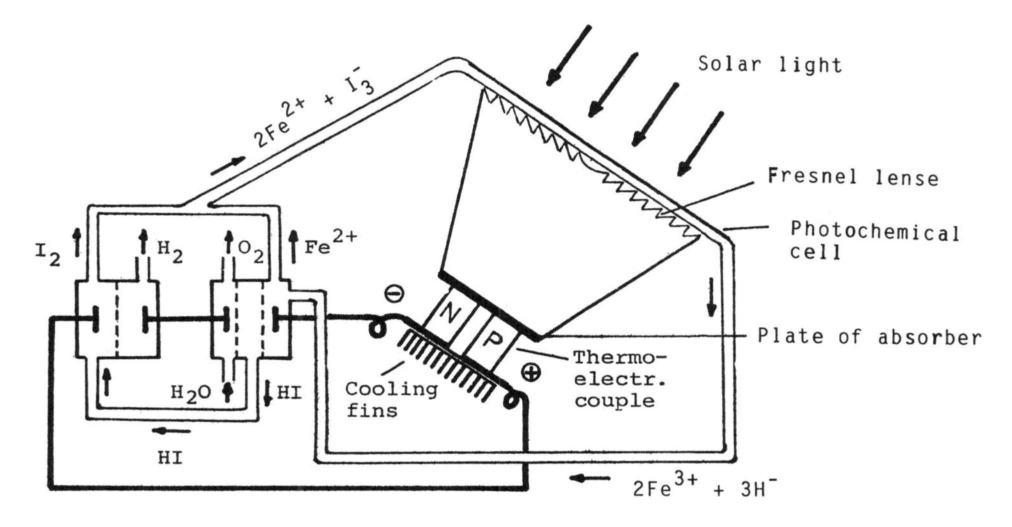

Overall-reactions for production of H_2 and O_2 by Yokohama Mark 5 and Mark 6 systems:

$$I_2 + 2FeSO_4 + H_2SO_4 \xrightarrow[(\lambda < 600\ nm)]{h\nu} 2HI + Fe_2(SO_4)_3$$

$$2HI \xrightarrow[(Mark\ 6\ :\ thermochem.)]{(Mark\ 5\ :\ electrochem.)} H_2 + I_2$$

$$Fe_2(SO_4)_3 + H_2O \xrightarrow{(electrochem.)} 2FeSO_4 + H_2SO_4 + \frac{1}{2}O_2$$

Fig. 8. Diagram of hybrid water splitting system (Yokohama Mark 5) and Brutto-reactions for H_2 and O_2-production.

The hybrid systems are rather promising devices for a technical solar energy utilization, although nothing is known about eventual corrosion problems, stability etc. under prolonged operations.

7. CONCLUSION

In the last years the research activity in the field of photoelectrochemical and photocatalytic methods of hydrogen production including the hybrid systems has made good progress. The most promising devices for conversion and storage of solar energy are the liquid-junction transducers, having semiconductor photo-electrodes and the hybrid systems. The photocatalytic methods are not stable because several undesired side reactions are leading to transients (e^-_{aq}, H, OH), which result in a decomposition of the system. Hence, these systems are presently of academic interest only.

Acknowledgements—The author thanks the Austrian Federal Ministry for Science and Research, as well as the Jubilee Fund of the Austrian National Bank and the Fund for Furtherance of Scientific Research in Austria for financial support, which made possible various studies cited in this review.

REFERENCES

1. T. N. Veziroğlu, *Int. J. Hydrogen Energy* **11,** 1 (1986).
2. J. O'M. Bockris, *Energy—The Solar Hydrogen Alternative.* The Architectural Press, London (1976).
3. N. Getoff, K. J. Hartig, G. Kittel, G. A. Peschek and S. Solar, *Hydrogen as Energy Carrier: Production, Storage and Transportation* (German). Springer, Wien–New York (1977).
4. J. R. Bolton (ed.), *Solar Power and Fuels.* Acad. Press, New York (1977).
5. T. Ohta (ed.), *Solar–Hydrogen Energy Systems.* Pergamon Press, Oxford (1979).
6. R. R. Hantala, R. B. King and C. Kutal (eds), *Solar Energy—Chemical Conversion and Storage.* Humana Press Clifton, New Jersey (1979).
7. Gerischer and J. J. Katz (eds) *Light-Induced Separation in Biology and Chemistry.* Verlag Chemie, Weinheim and New York (1979).

8. Yu. Ya. Gurevich, Vu. V. Preskov and Z. A. Rothenberg, *Photoelectrochemistry*. Consultants Bureau New York and London (1980).
9. N. N. Semenov (ed.) *Solar Energy Conversion* (Russian). Z. H. F., Academy of Sciences USSR, Chernogolovka (1981).
10. F. Cardon, W. P. Gomes and W. Dekeyser (eds) *Photovoltaic and Photoelectrochemical Solar Energy Conversion*. Plenum Press, New York and London (1981).
11. S. Cleason and Holmström (eds) *Solar Energy, Photochemical Processes Available for Energy Conversion*. National Swedish Board for Energy Source Development, Stockholm (1982).
12. Grätzel M. (ed.) *Energy Resources Through Photochemistry and Catalysis*. Acad. Press, New York (1983).
13. K. I. Zamaraev and R. F. Khairutdinov, *Soviet. Sci. Rev. Section B*, **2,** 357 (1980).
14. G. Porter and M. D. Arthur, *Interdiscip. Sci. Rev.* **1,** 119 (1976).
15. T. Ohta and T. N. Veziroğlu, *Int. J. Hydrogen Energy*, **1,** 255 (1976).
16. N. Getoff, *Österr. Chem. Zeitschr.* **79,** 243 (1978).
17. M. Calvin, *Photochem. Photobiol.* **37,** 349 (1983).
18. A. Harriman, *Platinum Metal Rev.* **27,** 109 (1983).
19. M. Grätzel, in J. O'M. Bockris and B. E. Convay (eds) *Modern Aspects of Electrochemistry*, Vol. **15,** pp. 81. Plenum Press, New York (1983).
20. K. J. Hartig and N. Getoff, *Proc. 5th WHEC, Toronto*, Vol. **3,** p. 1085 (1984).
21. N. Getoff, *Int. J. Hydrogen Energy* **9,** 997 (1984).
22. K. Kalyanasundaram and M. Grätzel, *Photochem. Photobiol.* **40,** 807 (1984).
23. N. Getoff, in *Proceed. Summer School at Igls, Austria* p. 105, ESA SP-240 (1985).
24. J. L. Abrahams, L. G. Casagrande, N. D. Rosenblum, M. L. Rosenbluth, P. G. Santangelo, B. J. Tufts and N. S. Lewis, *Nouveau J. Chim.* **11,** 157 (1987).
25. A. Heller, *Nouveau J. Chim.* **11,** 187 (1987).
26. N. Getoff, *Proc. Symp. Hydrogen-Energy Carrier of the Future*. Vulkan, Essen, F.R.G.
27. M. P. Thekaekara, NASA-Goddard Space Flight Centre, U.S.A. (1971).
28. K. Watanabe and M. Zelikoff, *J. Opt. Soc. Am.* **43,** 753 (1953).
29. F. S. Dainton and P. Fowles, *Proc. R. Chem. Soc.* **A287,** 295 (1965).
30. N. Getoff, *Monatsh. Chem.* **99,** 136 (1968).
31. U. Sokolov and G. Stein, *J. Phys. Chem.* **44,** 3329 (1966).
32. N. Getoff and G. O. Schenck, *Photochem. Photobiol.* **8,** 167 (1968).
33. A. Fujishima and K. Honda, *Nature*, **238,** 37 (1982).
34. A. J. Nozik, *Annu. Rev. Phys. Chem. Phil. Trans. R. Soc., London* **A295,** 453 (1980).
35. C. D. Jaeger, F. Fu-Ren Fan and A. J. Bard, *J. Am. Chem. Soc.* **102,** 2592 (1980).
36. F. DiQuatro, K. Doblhofer and H. Gerischer, *Electrochim. Acta* **23,** 51 (1981).
37. N. Getoff, S. Solar and M. Gohn, *Naturwiss.* **67,** 7 (1980).
38. K. J. Hartig, J. Lichtscheidl and N. Getoff, *Z. Naturforsch.* **35a,** 51 (1981).
39. J. Lichtscheidl, K. J. Hartig, N. Getoff, Ch. Tauschitz and G. Nauer, *Z. Naturforsch.* **36a,** 727 (1981).
40. K. J. Hartig and N. Getoff, *Int. J. Hydrogen Energy*, **11,** 773 (1986).
41. V. Antonucci, N. Giordano and J. C. J. Barth, *Int. J. Hydrogen Energy* **7,** 769 (1982).
42. V. Antonucci, N. Giordano and J. C. J. Barth, *Int. J. Hydrogen Energy* **7,** 867 (1982).
43. K. J. Hartig and N. Getoff, *Proc. 5th WHEC*, Vol. **3,** 1085 (1984).
44. K. J. Hartig and N. Getoff, *Monatsh. Chem.* **116,** 453 (1985).
45. K. J. Hartig and N. Getoff, *Hydrogen Systems*, in T. N. Veziroğlu, Zhn Yajie and Ban Deyon (eds), Vol. **1,** p. 81. Pergamon Press–Academic, Beijing, China (1985).
46. K. J. Hartig, N. Getoff, Rumpelmayer, G. Popkirov, K. Kotchev and St. Kanev, in T. N. Veziroğlu, N. Getoff and P. Weinzierl (eds), *Hydrogen Energy Progress VI*, Vol. **2,** 546. Pergamon Press, New York (1986).
47. K. Ohashi, J. McCann and M. O. M. Bockris, *Energy Res.* **1,** 259 (1977).
48. K. D. Legg, A. B. Ellis, J. M. Boltz and M. S. Wrighton, *Proc. Nat. Acad. Sci.* **74,** 4416 (1977).
49. T. Inone, T. Watanabe, A. Fujishima and K. Honda, *Bull. Chem. Soc. Japan* **52,** 1243 (1979).
50. A. J. Nozig, *Phil. Trans. R. Soc.* (London) **A295,** 453 (1980).
51. H. Tributsch and H. Gerischer, *Z. Naturforsch.* **23B,** 736 (1969).
52. H. Gerischer, *Ber. Bunsenges. Phys. Chem.* **77,** 771 (1973).
53. H. Gerischer, *Faraday Disc.* **58,** 219 (1974).
54. T. Osa and M. Fujihira, *Nature* **264,** 349 (1976).
55. C. D. Jaeger, F. R. F. Fan and A. J. Bard, *J. Am. Chem. Soc.* **102,** 2592 (1980).
56. M. P. Dare-Edwards, J. B. Doodenough, A. Hamnett, K. K. Seddon and R. D. Wight, *Faraday Disc.* **70,** 285 (1980).
57. W. Schwarz and N. Getoff, unpublished results.
58. W. Vonach and N. Getoff, *J. Photochem.* **23,** 233 (1983).
59. N. Getoff, S. Solar and D. B. McCormick, *Science* **201,** 616 (1978).
60. W. Vonach and N. Getoff, to be published.
61. J. O'M Bockris and R. C. Kainthla, in T. N. Veziroğlu, N. Getoff and P. Weinzierl (eds), *Hydrogen Energy Progress VI*, p. 449. Pergamon Press, New York (1986).
62. H. Tsubomura and Y. Nakato, *Nouveau J. Chim.* **11,** 167 (1987).
63. G. Li, Sh. Wang and N. Getoff, *Z. Naturforsch.* **43a,** 248 (1988).
64. R. C. Kainthla, B. Zelenay and J. O'. M. Bockris, *J. Electrochem.* **134,** 841 (1987).
65. S. Cervera-March, E. S. Smotkin, A. J. Bard, M. A. Fox, T. Mallouk, S. E. Webber and J. M. White, *J. Electrochem.* **135,** 567 (1988).
66. K. J. Hartig and N. Getoff, in T. N. Veziroğlu, Z. Yajie and B. Deyon (eds), *Hydrogen Systems*, p. 81, China Acad. Publ., Beijing and Pergamon Press, Oxford (1986).
67. A. Schwarz, K. J. Hartig and N. Getoff, *Int. J. Hydrogen Energy* **13,** 81 (1988).
68. F. Willing, K. Bitterling, K.-P. Charle and F. Decker, *Ber. Bunsenges. Phys. Chem.* **88,** 374 (1984).
69. D. Bahnemann, A. Henglein, J. Lilie and L. Spanhel, *J. Phys. Chem.* **88,** 709 (1984).
70. M. Evenor, S. Gottesfeld, Z. Harzion, D. Huppert and S. W. Feldberg, *J. Phys. Chem.* **88,** 6213 (1984).
71. K. J. Hartig, G. Grabner, N. Getoff, G. Popkirov and St. Kanev, *Ber. Bunsenges. Phys. Chem.* **89,** 835 (1985).
72. M. Neumann-Spallart, A. Schwarz and G. Grabner, *J. Phys. Chem.* **93,** 1984 (1989).
73. E. Rabinowitch, *J. Chem. Phys.* **8,** 551 (1940).
74. A. E. Porter and L. H. Thalter, *Solar Energy* **3,** 1 (1957).

75. K. G. Mathai and E. Rabinowitch, *J. Phys. Chem.* **66,** 663 (1962).
76. N. N. Lichtin, in J. R. Bolton (ed.), *Solar Power and Fuels*, p. 19. Academic Press Inc. New York (1977).
77. W. J. Albery, P. N. Bartlett, J. P. Davis, A. W. Foulds, A. R. Hillman and F. S. Bachiller, *Faraday Disc.* **76,** 341 (1980).
78. J. Rabani (ed.) *Photochemical Conversion and Storage of Solar Energy*. The Weizmann Sci. Press, Israel (1982).
79. S. Solar and N. Getoff, *Int. J. Hydrogen Energy* **4,** 403 (1979).
80. G. Köhler, S. Solar and N. Getoff, *Z. Naturforsch.* **35a,** 1201 (1980).
81. N. Getoff, *Z. Naturforsch.* **17b,** 87 (1962).
82. P. L. Airey and F. S. Dainton, *Proc. R. Soc.* **A291,** 340 (1966).
83. N. Getoff, S. Solar and D. B. McCormick, *Science* **201,** 616 (1978).
84. S. Solar, W. Solar and N. Getoff, *Radiat. Phys. Chem.* **20,** 165 (1982).
85. S. Solar, W. Solar and N. Getoff, in J. Dobo, P. Hedvig and R. Schiller (eds), *Proc. 5th Tihany Symp. Rad. Chem.*, Vol. 3, p. 279. Akad. Kiado, Budapest (1983).
86. S. Solar, W. Solar and N. Getoff, *J. C. S. Faraday Trans. II*, **79,** 123 (1983).
87. S. Solar, W. Solar and N. Getoff, *Z. Naturforsch.* **37a,** 78 (1982).
88. S. Solar, W. Solar and N. Getoff, *Z. Naturforsch.* **37a,** 1077 (1982).
89. S. C. Ameta, S. Khamesra, S. Lodha and R. Ameta, *J. Photochem. Photobiol. A*, **48,** 81 (1989).
90. J. Kiwi and M. Grätzel, *Angew. Chem. Int. Ed. Engl.* **17,** 860 (1978).
91. M. Kirch, J. M. Lehn and J. P. Sauvage, *Helv. Chim. Acta* **62,** 2745 (1979).
92. M. Gohn and N. Getoff, *Z. Naturforsch.* **34a,** 1135 (1979).
93. O. Johansen, A. Launikonis, J. M. Loder, A. W.-H. Man, W. H. F. Sasse, J. D. Swift and D. Wells, *Austr. J. Chem.* **34,** 981 (1981).
94. S. Solar, W. Solar, N. Getoff, J. Holcman and K. Sehested, *J.C.S. Faraday Trans. I.* **78,** 2476 (1982).
95. S. Solar, W. Solar, N. Getoff, J. Holcman and K. Sehested, *J.C.S. Faraday Trans. I*, **80,** 2929 (1984).
96. S. Solar, W. Solar, N. Getoff, J. Holcman and K. Sehested, *J.C.S. Faraday Trans. I*, **81,** 1101 (1985).
97. S. Solar, *Radiat. Phys. Chem.* **26,** 109 (1985).
98. W. Vonach and N. Getoff, *Z. Naturforsch.* **36a,** 876 (1981).
99. E. Borgarello, K. Kalayanasundaram and M. Grätzel, *Helv. Chim. Acta* **65,** 243 (1982).
100. N. Serpone, E. Borgarello and M. Grätzel, *J.C.S. Chem. Comm.* **342,** (1984).
101. J. C. Escudero, J. Gimenez, R. Simarro and S. Cervera-March, *Solar Energy Mater.* **17,** 151 (1988).
102. P. R. Rayson, USA-Patent, NASA-Case-NPO-13657-1 from 13 Feb. 1976.
103. T. Ohta and N. Kamiya, *Proc. 9th IECEC*, p. 317 San Francisco (1974).
104. T. Ohta, S. Asakura, M. Yamaguchi, M. Kamiya and T. Otagawa, *Int. J. Hydrogen Energy* **1,** 113 (1976).
105. T. Ohta, N. Kamiya, T. Otagawa, M. Suzuki, S. Kurita and A. Suzuki, *Proc. 2nd WHEC*, Zürich (1978).
106. T. Ohta, N. Kamiya, M. Yamaguchi, N. Gotoh, T. Otagawa and S. Asakura, *Int. J. Hydrogen Energy* **3,** 203 (1978).
107. T. Ohta, N. Kamiya and T. Otagawa, *Int. J. Hydrogen Energy* **4,** 55 (1979).

Hydrogen Production by Cyanobacteria

G. D. Smith, G. D. Ewart and W. Tucker
Department of Biochemistry, Faculty of Science
The Australian National University

Abstract

Cyanobacteria are capable of biophotolysis, the light-driven splitting of water into hydrogen and oxygen, in reactions which involve the enzymes nitrogenase and hydrogenase. *Anabaena cylindrica* possesses an integral membrane hydrogenase and a soluble hydrogenase, both of which are nickel-dependent. The release of nitrogenase - catalyzed hydrogen can be controlled chemically and also by manipulating nickel levels in the growth medium, suggesting that selection of uptake hydrogenase-deficient mutants is an important objective for maximizing biophotolysis. We have cloned and sequenced the hydrogenase genes from *A.cylindrica* with a view to genetically manipulating the hydrogen metabolism of this and other cyanobacteria.

1. INTRODUCTION

Cyanobacteria in general possess three hydrogen-metabolizing enzymes: nitrogenase, membrane-bound uptake hydrogenase and soluble hydrogenase [1,2]. The former two enzymes interact in the hydrogen metabolism in the manner indicated in Fig.1, as is the case with many nitrogen-fixing bacteria. Attempts to maximize the amount of hydrogen produced by cyanobacteria have usually involved manipulation of the metabolic scheme depicted in Fig.1, namely to maximize the hydrogen *produced* by nitrogenase and minimize that *consumed* by the so-called uptake hydrogenase. Until recently little has been known of the role of the soluble hydrogenase of cyanobacteria although there is now evidence that it plays a role in their dark fermentative metabolism, with hydrogen being produced as an electron sink [3,5]. As such it offers little as a practical source of hydrogen gas. Two possible approaches to using soluble hydrogenase-mediated hydrogen formation are based on the observations that photobleached *Anabaena cylindrica* appears to link its photosystems to the hydrogenase [6,7] in a light-dependent hydrogen evolution, and that anoxygenic formation of hydrogen from sulfide occurs in certain organisms [8,9]. Attempts have also been made to link cyanobacterial photosystems artificially to external hydrogenases by chemical means [10].

At this stage, however, the most promising approach to using cyanobacteria has involved manipulating the metabolic system shown in Fig.1. This paper describes results from our laboratory concerning three types of biochemical manipulation. The first involved the use of enzyme inhibitors (carbon monoxide and acetylene), the second involved variation of the nickel concentration in the cellular growth medium and the third involves genetic manipulation of the hydrogenase enzymes.

2. MANIPULATION OF HYDROGEN FORMATION IN CYANOBACTERIA

As indicated in Fig.1 the light-dependent formation of hydrogen by cyanobacteria represents a balance between that produced by the nitrogenase and that consumed by hydrogenase. The results of three approaches to optimizing this production by manipulation of the activities of the respective enzymes (chemical, nickel ion concentration and genetic) are described below.

Chemical

The nitrogenase enzyme has, during normal catalysis, two activities which function together, these being nitrogen fixation and proton reduction to form hydrogen. If cultures of this organism are illuminated in a gaseous environment free of dinitrogen the nitrogenase simply catalyzes proton reduction and releases hydrogen [11]. If oxygen is present, however, the hydrogen is consumed, despite an absence of dinitrogen fixation [12]. It has been shown that carbon monoxide, an inhibitor of N_2 reduction by nitrogenase but not of proton reduction, present together with acetylene, an inhibitor of hydrogenase, can be used to optimize the hydrogen formation even in air [13,14]. This approach has been tested with a number of marine cyanobacteria [15] and has been optimized for long term hydrogen production by *Anabaena cylindrica* [16]. This approach was combined with the use of glass-immobilized organisms [17] to develop an outdoor biophotolytic system [18].

Nickel ion concentration

The demonstration that the hydrogenase activities of *A.cylindrica* required nickel ions as an essential cofactor [19,20] suggested that hydrogen formation could be controlled by controlling the nickel ion concentration of the medium used to grow cells. As shown in Fig.2 the amount of hydrogen released by *A. cylindrica* is strongly influenced by the nickel ion concentration in which the cells were grown. The cells represented in Fig.2 were incubated in argon but it was also shown that, in the absence of nickel, hydrogen was released even in air [19]. Interestingly, as shown in Fig.3 the logarithmic growth rate of cells of *A.cylindrica* was unaffected by whether nickel was present or absent, suggesting that an absence of uptake hydrogenase does not impair in any way growth of the organisms under the growth conditions which were used in this work.

Genetic

We have been studying the hydrogenase enzymes of *A.cylindrica* with a view to understanding their properties and functions and also to genetically manipulating them to optimize cellular hydrogen formation.

The soluble hydrogenase has been purified to homogenity and shown to consist of two subunits, one catalyzing tritium-exchange activity and the other conferring reductive hydrogenase activity [21]. The N-terminal amino acid sequences of both subunits have been determined and used to synthesize oligonucleotide probes which have in turn been used to clone the respective genes. The complete nucleotide sequence of the putative gene encoding the tritium-exchange subunit has been determined [22] and that of the second subunit partially so. Sequence analysis of the gene encoding the tritium exchange subunit showed no homology with any of the hydrogenase genes already known in other bacteria whose sequences have been determined [22].

As yet the membrane hydrogenase has not been purified and so we have attempted to clone its gene using heterologous hybridization. Although others have attempted this approach without success we employed a modified Southern blotting procedure reported to have improved sensitivity and resolution [23]. With this method, hydridization was detected on Southern blots between *A.cylindrica* DNA, digested with various restriction enzymes, and three plasmids containing cloned hydrogenase genes from *Bradyrhizobium japonicum* [24], *Desulfovibrio vulgaris* [25] and *Alcaligenes eutrophus* [26]. The recombinant clones from these sources were pHU1, pGH1 and pGM7.3, respectively. The *B.japonicum* gene has now been sequenced and shows that the enzyme has characteristics of hydrophobicity which are presumed to contribute to its location in the bacterial membrane [27]. It has a high degree of homology to the hydrogenases of other bacteria, including two species of Desulfovibrio [27]

The hybridization between each of these probes and *A.cylindrica* DNA was weak and only detectable with the alkaline Southern transfer procedure. In the absence of strong hybridization with any single probe, it was considered that the DNA fragments detected by more than one of the probes were most likely to represent hydrogenase genes. It was observed that hybridization between the three probes and restriction endonuclease*Bgl*II - digested *A.cylindrica* DNA was predominantly to fragments larger than 5 kb.

A partial library of *A. cylindrica* DNA, produced from 5-12 kb *Bgl*II restriction fragments cloned into pGEM-1, was screened with the pGH1 and pHU1 probes. One clone hybridizing to both probes contains a single *Bgl*II fragment of 5.4 kb which is presumed to include the relevant *A. cylindrica* hydrogenase gene(s). Sequencing and characterization is in progress.

A major aim of such genetic work is to produce hydrogenase-deficient (Hup^-) strains of cyanobacteria, which would be expected, on the basis of the work described above (see Fig.1), to produce hydrogen spontaneously.

Acknowledgement

This work was supported by the Australian Research Council.

REFERENCES

1. Lambert, G.R. and Smith, G.D.(1981) Biological Reviews *56*, 589-660.

2. Houchins, J.P. (1984) Biochim. Biophys. Acta *768*, 227-255.

3. van der Oost, J., Kanneworff, W.A., Krab, K. and Kraayenhof, R. (1987) FEMS Microbiol. Lett. *48*, 41-45.

4. Asada, Y. and Kawamura, S. (1985) Report of the Fermentation Research Institute No.63, 39-54. Yatabe-machi Higashi, Ibaraki, Japan 305.

5. Howarth, D.C. and Codd, G.A. (1985) J.Gen. Microbiol. *131*, 1561-1569.

6. Laczko, I. (1985) Physiol. Plant. *63*, 221-224.

7. Laczko, I. and Barabas, K. (1981) Planta *153*, 312-316.

8. Belkin, S. and Padan, E. (1978) FEBS Lett. *94*, 291-294.

9. Fry, I, Robinson, E., Spath, S. and Packer, L. (1984). Biochem. Biophys. Res. Comm. *123*, 1138-1143.

10. Smith, G.D., Muallem, A. and Hall, D.O. (1982) Photobiochem. Photobiophys. *4*, 307-319.

11. Benemann, J.R. and Weare, N.M.(1974) Science *184*, 174-175.

12. Lambert, G.R. and Smith, G.D. (1981) Arch. Biochem. Biophys. *211*, 360-367

13. Bothe, H., Tennigkeit, J., Eisbrenner, G. and Yates, M.G.(1977) Planta *133*, 237-242.

14. Daday, A., Platz, R.A. and Smith, G.D. (1977) Appl. Environ. Microbiol. *34*, 478-483.

15. Lambert, G.R. and Smith, G.D. (1977) FEBS Lett. *83*, 159-162.

16. Lambert, G.R., Daday, A. and Smith, G.D. (1979) Appl. Environ. Microbiol. *38*, 530-536.

17. Lambert, G.R., Daday, A. and Smith, G.D. (1979) FEBS Lett. *101*, 125-128.

18. Smith, G.D. and Lambert, G.R. (1981) Biotechnol. Bioeng. *23*, 213-220.

19. Daday, A. and Smith, G.D. (1983) FEMS Microbiol. Lett. *20*, 327-330.

20. Daday, A., Mackerras, A.H. and Smith, G.D. (1985) J.Gen.Microbiol. *131*, 231-238.

21. Ewart, G.D. and Smith, G.D. (1989) Arch. Biochem. Biophys. *268*, 327-337.

22. Ewart. G.D., Reed, K.C. and Smith, G.D. (1990) Eur. J. Biochem. *187*, 215-223.

23. Reed, K.C. and Mann, D.A. (1985) Nucleic Acids Res. *13*, 7207-7221.

24. Cantrell, M.A., Haugland, R.A. and Evans, H.J. (1983) Proc. Nat. Acad. Sci. U.S.A. *80*, 131-135.

25. Voordouw, G. and Brenner, S. (1985) Eur. J. Biochem. *148,* 515-520.

26. Chow, W.Y. and Atherly, R.G. (1984) Abstr. Annu. Meeting. Am.Soc. Microbiol. *84*, H31.

27. Sayavedra-Soto, L.A., Powell, G.K., Evans, H.J. and Morris, R.O. (1988) Proc. Nat. Acad. Sci. U.S.A. *85*, 8395-8399.

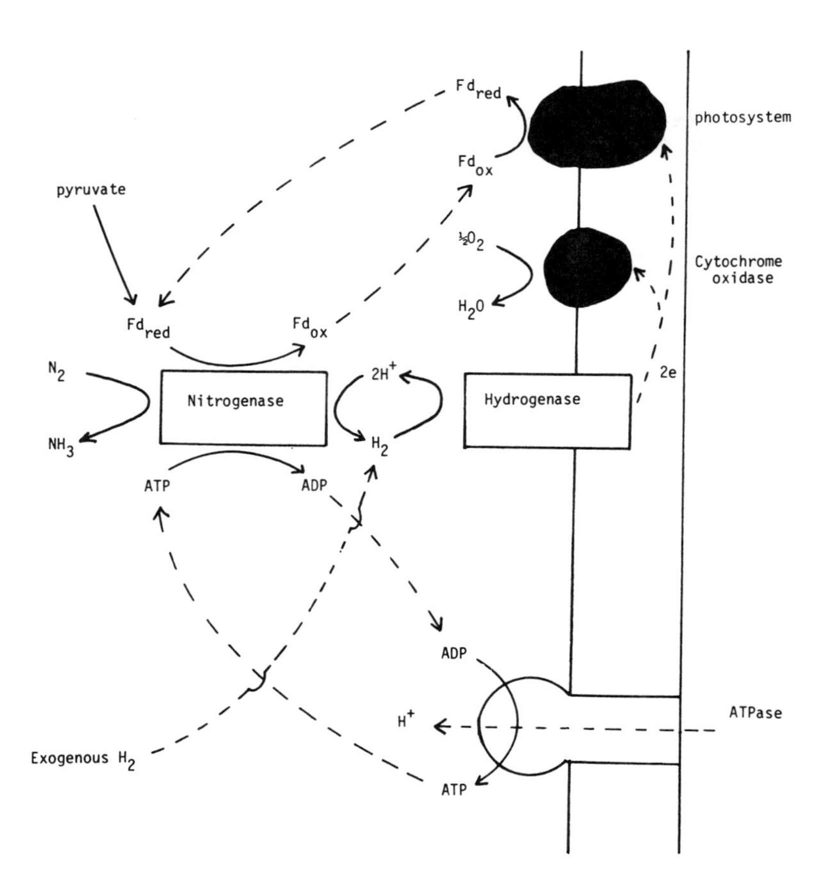

Fig.1 -The Relationship Between Nitrogenase-catalyzed Hydrogen-Formation and Hydrogenase-calalyzed Hydrogen Uptake in Cyanobacteria

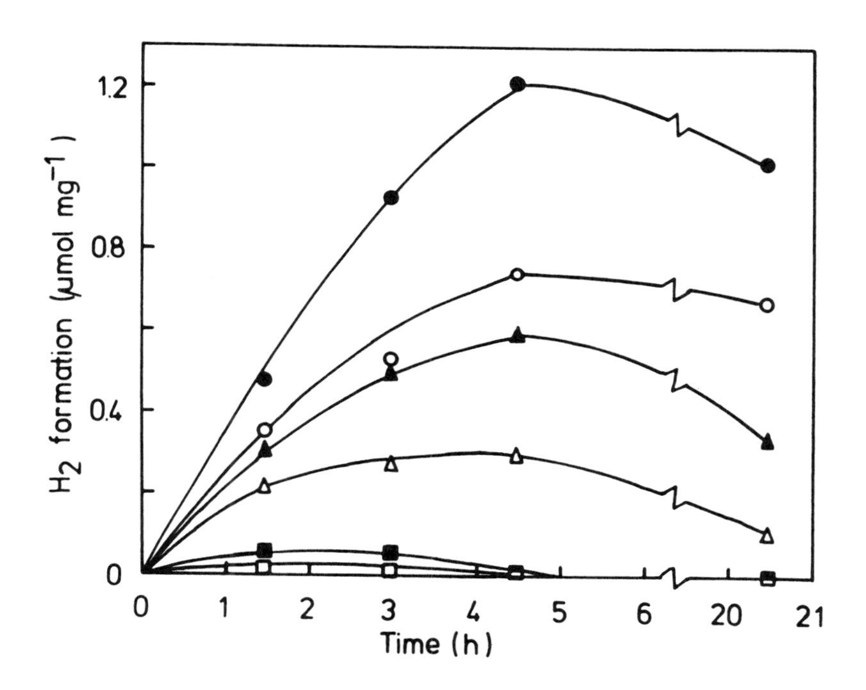

Fig.2 -The Effect of Nickel on Hydrogen Formation by *A.cylindrica*

Cells were incubated under argon after growth in the presence of 0μM (O), 0.021 μM (O), 0.043μM (▲), 0.17μM (Δ), 0.34 μM (■) or 0.68 μM (□) $NiSO_4$.

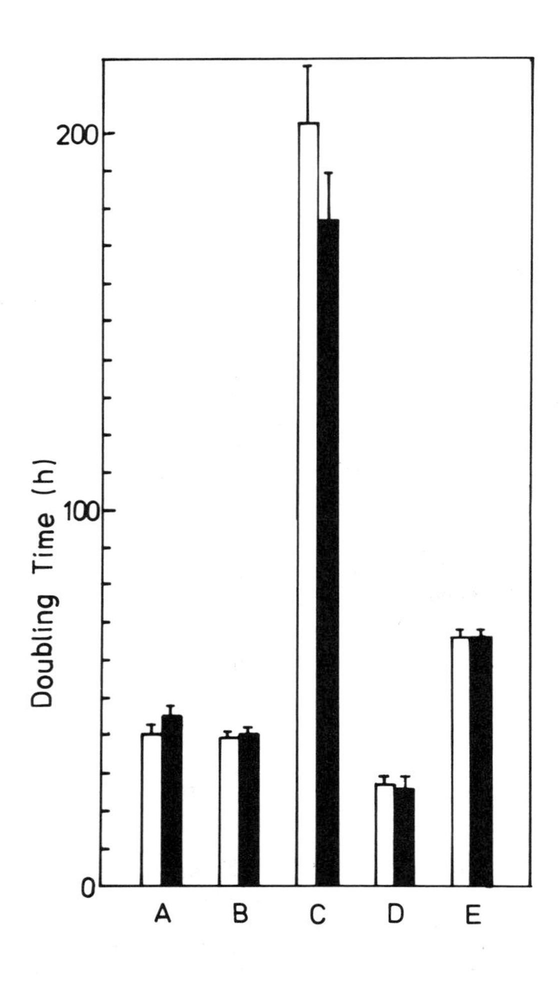

Fig. 3 -The Effect of Nickel on the Exponential Growth Rate of *A.cylindrica* grown under Different Conditions.

The white bars represent cells grown without nickel and the black bars those grown in 0.68 µM $NiSO_4$. The growth conditions were: A, 12/12 dark/light cycle and gassed with Air/4% H_2/0.3% CO_2; B, 12/12 dark/light cycle and gassed with 0.3% CO_2 in air; C, as with B but with a light intensity reduced to 15 $\mu Es^{-1}m^{-2}$ from the usual 120 $\mu Es^{-1}m^{-2}$; D, continuous light and gassed with N_2/4% H_2/0.3% CO_2; E, 12/12 dark/light cycle and gassed with air.

Pine and Willow as Carbon Sources in the Reaction Between Carbon and Steam to Produce Hydrogen Gas

R. C. Timpe, R. E. Sears, and T. J. Malterer

University of North Dakota Energy and Environmental Research Center

ABSTRACT

The depletion of the fossil fuels supply in an energy-dependent society may eventually spell disaster if supplemental fuel forms are not developed. Even with present oil conservation practices and the availability of coal and peat, tar sands, oil shale and solar and hydro energy, satisfactory means are needed for producing and efficiently converting renewable resources (biomass) with a net energy yield. The substrate selected for such energy-producing processes should be readily replaced in nature, easily harvested, require minimal tending, be as rich as possible in reactive atomic specie, and provide minimal harmful disturbance to the natural environment. At the University of North Dakota Energy and Environmental Research Center (UNDEERC), a series of bench-scale experiments were carried out using western U.S. Ponderosa pine (Pinus ponderosa) and northern Minnesota hybrid willow (Salix viminalis) as carbon sources for a reaction with steam to produce hydrogen. The kinetics of the reaction between the carbon and steam in the production of hydrogen gas at ambient pressure was investigated at temperatures of 700°, 750°, and 800°C. This paper discusses the results of the experiments in terms of reaction kinetics, product gas, and potential for scale-up to pilot scale testing.

BACKGROUND AND INTRODUCTION

The energy needs of the world's population are currently being met by the fossil fuel industry. In 1974 Cheney (2) pointed out that the world's consumption of energy was doubling every decade and, without effective energy conservation of non-renewable fuels, the lifetime of known reserves was predicted to be relatively short. The supplies of oil, gas, and tar were being depleted at a rate which would make them a rare commodity, even within the lifetime of a large percentage of the current users. The petroleum reserves were projected to be substantially consumed by 2025, while the readily obtainable coal reserves were predicted to support energy needs for three or more centuries.(2) Since the early 1970's consumption rates have declined somewhat, but additional energy consuming industries and goods are making their appearance, causing additional demand for fuels. Conservation and new reserve discoveries will extend the projected lifetime of total reserves but even with the most optimistic of conservation plans, the fact remains that a non-renewable resource will eventually be exhausted.

Extensive efforts have been made to convert energy-rich materials such as coal and peat to products whose properties conform to those of petroleum-based fuel supplies. In addition to consuming these nonrenewable resources, processing coal and peat is itself an energy consuming operation, thus putting additional drain on the limited fuel supply. Another resource, nuclear energy, is used to supplement fossil fuels in electrical generation and marine propulsion but has its inherent waste disposal and control problems.

Renewable resources have not been ignored and are slowly becoming an increasingly larger part of the energy picture. The waterways, tides, sun, plants and plant products, wood and wood waste, groundwater supplies including geothermal sources, and garbage have all been investigated with varying degrees of success as to their possible contributions as supplements to the fossil fuel supplies (10). It becomes imperative that, if we are to maintain living conditions that draw on large energy supplies, as fossil fuel supplies decrease, there must be an increase in the proportion of energy provided by renewable fuels such as these. In addition, in developing and using these fuels there exists the caveat which requires that there be minimal deleterious impact on the environment and that the renewable fuel have widespread availability.

With the exception of tides, which are localized on the margins of continents, the other renewable resources mentioned above meet the availability requirement. One of the most likely candidate fuels is the biomass resource which has as one of its most attractive features a very low sulfur content, and thus reduced potential as a pollutant. Biomass production can provide food, building materials, and wildlife habitat and ground cover for preventing erosion. Carbon dioxide emissions from combustion processes are utilized by the growing plants thus lessening potential for the "greenhouse effect" and closing the cycle of energy production and utilization. This resource can be generated as a primary energy source, a secondary source in the form of horticultural residues, or a tertiary source as waste materials. (6) In poor, underdeveloped nations as well as in countries that have a more industrial base, cultivation of fast-growing plants of high carbon content could be used not only as fuel for combustion but also as sources of distillable liquid products and reactants for production of alternate fuels. One such alternate, and the subject of this paper, is hydrogen. Besides being a clean-burning attractive fuel providing 343 Btu/sfc (5), hydrogen has additional uses in the chemical, aerospace, and smelting industries. Bench-scale studies of hydrogen production using coal as feedstock for the char-steam reaction have been successful (4). The production of hydrogen from the reaction of biomass char and steam should also be successful.

In order for woody plants to be a feasible energy feedstock, the plant must be fast-growing in the soil type of the prospective geographic region and must be capable of growing in dense stands with the available water and nutrient supply. Manpower requirements would be similar to surface coal mining and involves ground preparation, seeding, and tending the growing crop (irrigating and thinning if necessary), this being comparable to handling overburden and returning the topsoil to its original condition. Harvesting and transportation can be roughly compared to the agricultural handling of corn ensilage. Tillman (9) addresses the question of the efficiency of such a process by predicting a trajectory efficiency, that is, the net energy yield of 20.3% for growth, extraction, conversion, transport, and combustion. As in utilization of low-rank coal where the most economical method is in situating the coal-fired or coal-conversion plant as near the mine as possible ("mine-mouth" operation), the crop of biomass should be grown as near as possible to the point of utilization and in large enough quantity to support the plant.

Recent research sponsored by the U.S. Department of Energy's Short Rotation Woody Crops Program suggests that there are many locally adapted tree species that make good candidates for renewable fuels. A hybrid willow species (Salix viminalis), grown in Minnesota, shows promise as a renewable fuel that would be grown exclusively for fuel use. It is well adapted to northern climates and it produces large amounts of good quality biomass yearly. Research efforts on this species are being conducted by the University of Minnesota, under the auspices of the USDOE (1).

Besides trees being grown exclusively for fuel purposes, wood wastes and wood products are good candidates for renewable fuels. Ponderosa pine (Pinus ponderosa) is a popular commercial tree used in the building industry. Substantial amounts of wood waste result during cutting and milling of the logs. Fuel use of these wood wastes has considerable potential for use as a feedstock for an hydrogen-production process.

PURPOSE OF THE STUDY

Hydrogen is an important fuel that can be produced in commercial quantities from reacting the carbonaceous material with steam. The reaction can be carried out on the char remaining after removal of valuable distillates and pyrolysates from the feedstock. The purpose of this study was to investigate the bench-scale kinetics of the hydrogen production reaction involving chars from Minnesota hybrid willow and waste wood from Ponderosa pine. This involves reaction of wood with steam over a temperature range of 700°-800°C at atmospheric pressure. Catalysis of the reaction was also investigated with each wood.

THEORY

In hydrogen production studies of this general type, the hydrogen is produced in an unpressurized reaction chamber according to two elementary reactions. The first reaction, termed the gasification reaction, occurs between steam and the charred remains of carbonaceous substrate as shown in reaction (i).

$$C + H_2O + 32.23 \text{ kcal/mole} \text{ -----> } CO + H_2 \quad \text{(i)}$$

Char is produced by heating a carbonaceous substrate under argon at a selected temperature in order to dry the material and remove volatile matter. Preheated steam is reacted with the char to produce one mole carbon monoxide and one mole H/mole C (synthesis gas). The gasification reaction is endothermic, as shown in (i), and must occur in the heated zone of the reactor.

The second hydrogen-producing reaction occurs as the gaseous product containing the CO and steam begins to cool. This reversible "water-gas shift" reaction ideally results in the production of a second mole of H_2 and a mole of CO_2 as the synthesis gas and steam react (ii).

$$CO + H_2O \text{ -----> } CO_2 + H_2 + 9.19 \text{ kcal/mole} \quad \text{(ii)}$$

These two reactions yield an equilibrium dry H_2 concentration from the wood char-steam reaction predicted by an engineering computer model to be > 60 mole % at 750°C.

The gasification reaction follows pseudo first order kinetics when excess steam is used and the rate equation used to calculate the reactivity which is expressed in terms of the reactivity parameter, k, given by the equation: (4)

$$r = \frac{d[C]}{dt} = k[1-C]^n \quad \text{where} \quad r = \text{rate} = \frac{d[C]}{dt}$$

n = reaction order = 1

k = reactivity parameter

C = conversion

This bench-scale experimental approach enabled the investigators to study the kinetics of the reaction by measuring the change in weight of the char with time. The bench-scale work is the precursor to further work with a 20 lb/hr fluid-bed gasifier at the University of North Dakota Energy and Environmental Research Center (UNDEERC).

EXPERIMENTAL

Hybrid willow stems were harvested from experimental plots near Virginia, Minnesota. The University of Minnesota Natural Resources Research Institute maintains the plots as part of a seven year research project that deals with evaluation of the USDOE Rotation Woody Crops Program. The hybrid willow is one of the more promising woody fuels. Samples were supplied by researchers from the University of Minnesota. Samples of Ponderosa pine were collected as waste material of irregular size particles from the sawdust pile of a local lumber company. Both willow and pine were ground to -60 mesh ($< 250\ \mu m$) for use in the experiments. The proximate and ultimate analyses of the wood samples are shown in Table 1.

Alkali metals in the form of salts have been shown to be excellent catalysts for coal char-steam reactions. Potassium carbonate (K_2CO_3) was one of the best (11) and therefore was chosen for use to catalyze the wood char-steam reaction. The catalyzed reaction involving the K_2CO_3 (56.6 wt % K) was carried out as above using an admix of 90 wt% wood and 10 wt% K_2CO_3. Ten weight percent of dry catalyst was added to the ground wood and was dispersed as uniformly as possible by thorough mixing.

The reactivity studies of uncatalyzed and K_2CO_3-catalyzed pine and hybrid willow were carried out at ambient pressure on a DuPont 951 thermogravimetric analyzer

TABLE 1.

Proximate/Ultimate Analysis of Pine and Hybrid Willow

	As-Run wt %	
	Pine	Willow
Proximate		
Moisture	6.81	6.38
Volatile Matter	78.15	75.16
Fixed Carbon	13.83	14.52
Ash	1.21	3.91
Ultimate		
Carbon	48.06	43.49
Hydrogen	6.71	6.26
Nitrogen	0.02	1.11
Sulfur	0.00	0.00
Oxygen (diff)	44.00	45.17
Oxygen, mf	37.95	39.12

TABLE 2

Raw and Normalized Elemental Composition of Hybrid Willow and Pine Ash in Mole Element/Unit Weight Wood.

	WILLOW		PINE	
ELEMENT	(MOL/G)*E4	NORM	MOL/G)*E4	NORM
Aluminum	0.117	0.022	0.075	0.112
Silicon	0.189	0.036	0.180	0.266
Sodium	0.000	0.000	0.000	0.000
Potassium	2.008	0.380	0.061	0.090
Calcium	2.272	0.430	0.187	0.278
Magnesium	0.288	0.054	0.058	0.086
Iron	0.061	0.012	0.043	0.064
Titanium	0.009	0.002	0.003	0.005
Phosphoru	0.249	0.047	0.030	0.045
Sulfur	0.087	0.016	0.036	0.053

(TGA) module coupled to a DuPont 1090 Thermal Analyzer data station. The reaction chamber consisted of a quartz tube with a side-arm for inlet of steam from a steam generator (See Figure 1). Twenty-five to forty mg of -60 mesh wood was loaded onto the balance pan and a flow of 160 cc/min of argon was introduced into the chamber. The chamber was inserted into the furnace and heated at 100°C/min to a predetermined target temperature, i.e., 700°, 750°, or 800°C, and held at that temperature until devolatilization was nearly complete as indicated by small weight change with time. With the temperature held constant, the argon flow was reduced to approximately 60 cc/min, and the steam flow into the chamber was started at 2-5 mg/min. Weight and temperature were recorded every two seconds over the duration of the experiment. Weight percent carbon remaining versus time for each experiment, as shown in Figure 2, was plotted and a table of time, temperature, weight %, and first derivative of weight % with time was printed. The total product gas samples were collected in 5-liter gas bags and analyzed by gas chromatography. All experiments were carried out at a minimum of two times each.

Several of the experiments that were performed on the small TGA were repeated on a larger capacity unit that was built at the UNDEERC. Sample size using this instrument was limited by the large bulk of the low-density ground wood but typically ranged up to 1.5 grams. During the experiments steam flow was set at ~300 mg/min and argon carrier at ~200 mL/min. This enabled the collection of larger gas samples for analysis, thus reducing the uncertainty in the gas chromatographic analytical method.

Ash was prepared from the hybrid willow by heating it to 500°C under argon and then removing remaining carbon by combustion at that temperature. Analysis of the ash was by X-Ray Fluorescence Analysis (XRFA). The results are shown in Table 2.

Fifty MHz solid ^{13}C CP/MAS Nuclear Magnetic Resonance (NMR) spectra of the ground wood and char were obtained using a Varian XL200 NMR spectrometer with a superconducting magnet fitted with a Doty broadband probe. The spectra of the pine and hybrid willow are shown in Figure 2.

RESULTS AND DISCUSSION

As is shown in reaction (i), the gasification of carbonaceous material is an endothermic reaction requiring a compromise between a reaction temperature high enough to

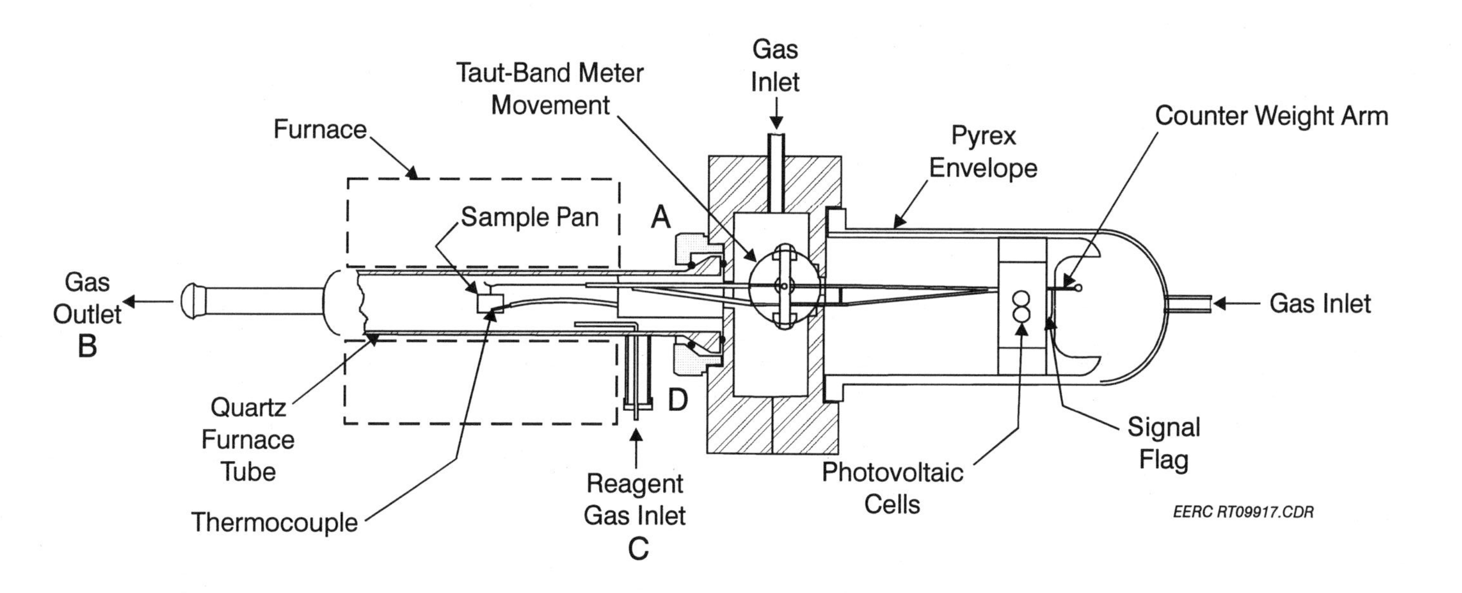

Figure 1. Char-steam reaction chamber for DuPont 951 Thermogravimetric Analyzer.

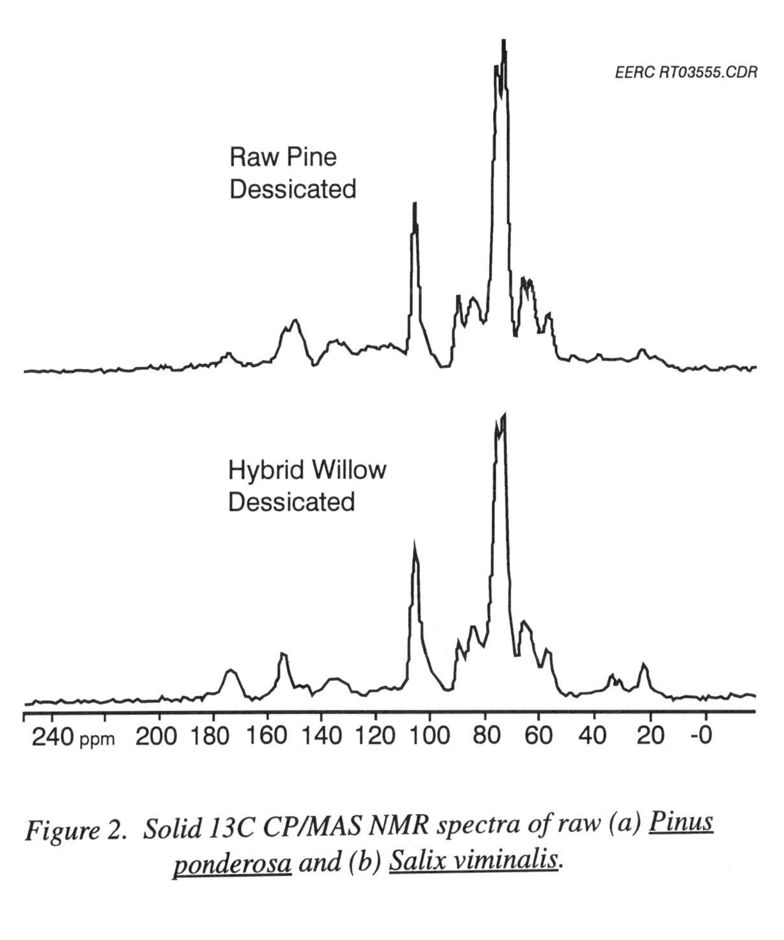

Figure 2. Solid 13C CP/MAS NMR spectra of raw (a) Pinus ponderosa and (b) Salix viminalis.

provide carbon conversion rapidly yet not so high as to decrease the water-gas shift reaction and to reflect lower costs for capital equipment, preheated steam and carrier gas, and product handling. Reaction (ii) is an exothermic equilibrium reaction requiring reduced temperature for maximizing hydrogen production from this step. Theoretically, optimization of reaction conditions would result in >60 mole % hydrogen.

Recently there has been interest in both of the aforementioned reactions, with emphasis on the low-cost production of hydrogen. All ranks of coal with and without reaction catalysts have been investigated as carbon sources and shown to be reasonable substrates for the process (8). However, they are a non-renewable resource and need to be conserved. Wood and wood products, with more rapid growth and increased production could become a substitute for coal in some applications. Research on raising of hybrid poplar and willow that have the necessary growth characteristics for rapid, abundant production has been successful in Minnesota (1).

The results of proximate and ultimate analyses of the wood samples are shown in Table 1. As can be seen by the proximate analysis, up to 88 wt % of the wood samples were determined to be moisture and volatiles. Of the remaining 12%-15%, 70-75 wt % was considered to be gasifiable according to the above reaction (i). The solid ^{13}C NMR spectra of the two woods are shown in Figure 2. The spectra are similar in that both contain the expected large cellulose carbon peaks at 64, 89, 72-74, and 105 ppm. The differences occur primarily between the larger regions of the more stable aromatics and phenolics in the pine spectrum and the larger regions of the more reactive carbonyl and anomeric (C1) aliphatic carbons of the willow spectrum. A comparison of Figure 2 with spectra of the respective wood chars (not shown) indicates a loss of most of the cellulose and hemicellulose from both woods as a result of the char production. No effort was made to analyze the volatiles produced during the charring process. It was assumed the volatiles could be condensed as a liquid side-stream and be used as a chemical feedstock or fed back into the reactor as either additional hydrogen-producing substrate or gasifier fuel.

To demonstrate that hydrogen was the principal gaseous product of the reaction, after the char had been prepared and steam was introduced into the reaction chamber, the total off-gas was collected and analyzed. As a result of the small size of the samples reacted in the small TGA, the product gas was greatly diluted with carrier and the uncertainty in the analysis was correspondingly large. However, reactions carried out in the large TGA allowed for reduction of the dilution effect by decreasing carrier gas flow while increasing the amount of H_2, CO_2, and CO produced. Typical results of analysis of gases produced on both the large and small TGA systems are shown in Table 3. A computer model developed at UNDEERC to predict the equilibrium gas concentrations in mole % of coal char-steam hydrogen production reactions indicated 63-65% H_2, 20-23% CO_2, and 12-15% CO yields on a dry basis could be the expected in the temperature range of this study. Results close to these values in a reasonable char residence time would be considered successful. The gas analyses of the product from the experiments carried out on the large TGA showed values near those predicted by the computer model for H_2. In addition, the production of methane during the process was kept low due to low severity operating conditions.

The kinetics of the gasification reaction were determined from weight change as a function of time. Plots of wt % versus time in Figures 3-6 illustrate the data obtained from the TGA and used to calculate k. For the pine and K_2CO_3-catalyzed pine and willow samples the reactivity was calculated for 50 wt % conversion. The reactivities for the willow were obtained for 25 wt % conversion. The reactivity parameter, k, was calculated assuming first-order kinetics (7) with respect to carbon conversion according to the equation mentioned above:

$$\frac{dC}{dt} = k(1-C)^n$$

where C is the carbon conversion and is defined as the following:

$$C = 1 - \frac{M - Mf}{Mi - Mf}$$

Mi = initial mass
M = mass at time t
Mf = final mass

As seen in Figures 7 and 8 the reactivity parameter varies with the temperature. This allows the calculation of the apparent energy of activation (E_a) from Arrhenius theory according to the equation:

$$k = A \exp(-E_a/RT)$$

A = preexponential factor
E_a = apparent energy of activation
R = ideal gas constant
T = absolute temperature
k = reactivity parameter

TABLE 3.

Product Gas from Hybrid Willow-Steam Reaction-Small TGA

SAMPLE	TEMP	MOLE % H_2	CO_2	CO	CH_4	C_2H_6
Hyb.Willow	800°C	58.0	32.1	9.9	--	--
Hyb.Willow	750°C	50.0*	50.0	ND	--	--
Hyb.Willow	700°C	80*	20	ND	--	--

Product Gas from Hybrid Willow-Steam Reaction-Large TGA

SAMPLE	TEMP	MOLE % H_2	CO_2	CO	CH_4	C_2H_6
Hyb.Willow	800°C	61.5	32.2	5.5	0.8	0.02
Hyb.Willow	750°C	61.5	32.9	4.4	1.3	--
Hyb.Willow	700°C	67.5	27.8	3.1	1.6	0.03

*Gas samples were diluted with excessive amounts of carrier.

TABLE 4.

Apparent Arrhenius Energy of Activation (E_a) and Correlation Coefficient (r^2) for 700-800°C Pine and Willow Char-Steam Reaction.

SAMPLE	E_a (KCAL/MOLE)	r^2
Pine	23.02	1.00
Pine/K2CO3	13.50	0.87
Willow	50.03	1.00
Willow/K2CO3	19.61	0.98

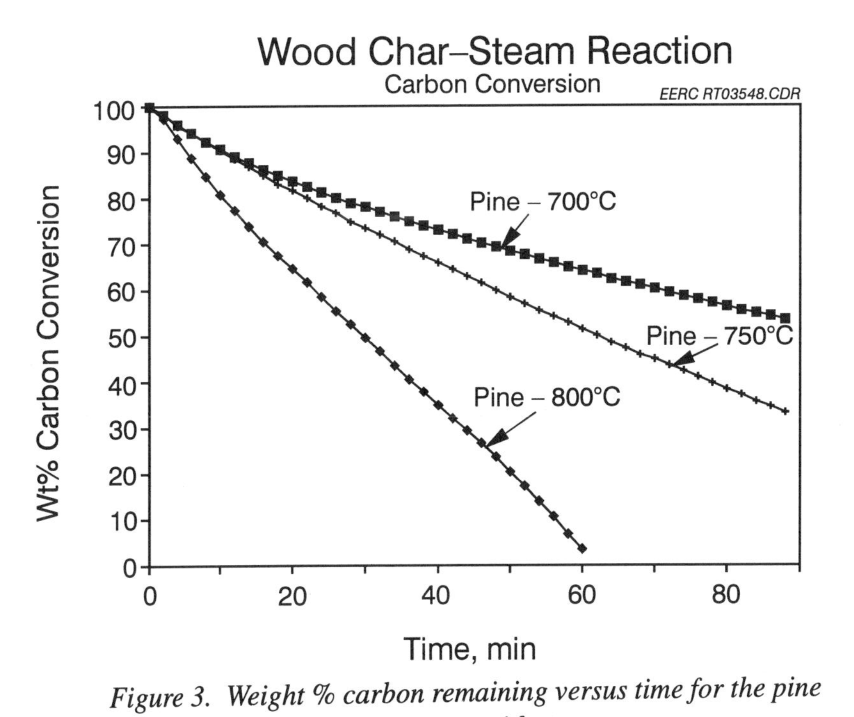

Figure 3. Weight % carbon remaining versus time for the pine char reaction with steam.

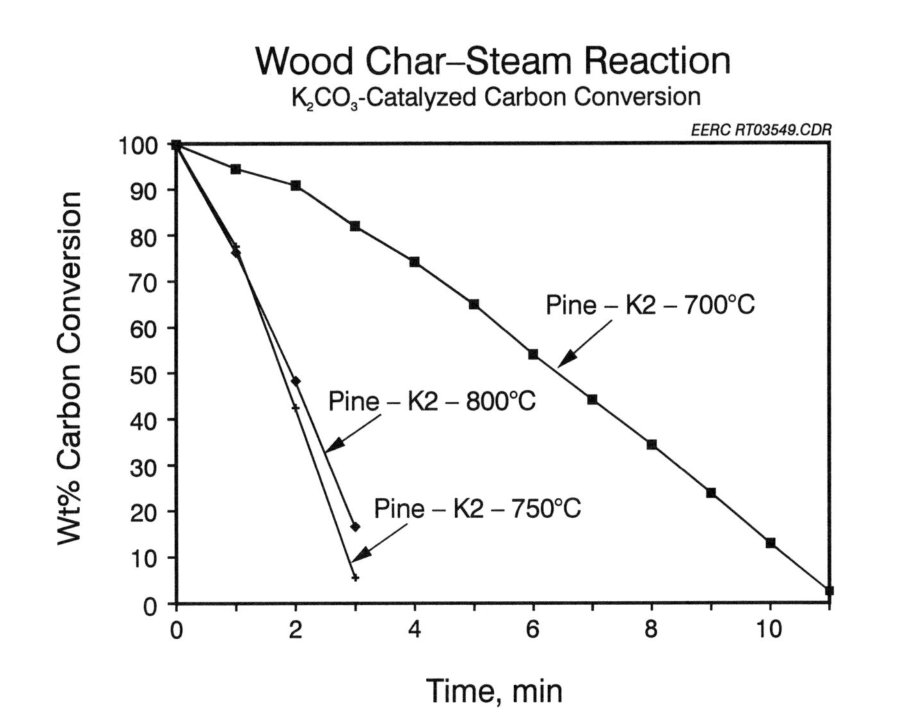

Figure 4. Weight % carbon remaining versus tie for the K_2CO_3-catalyzed pine char-steam reaction.

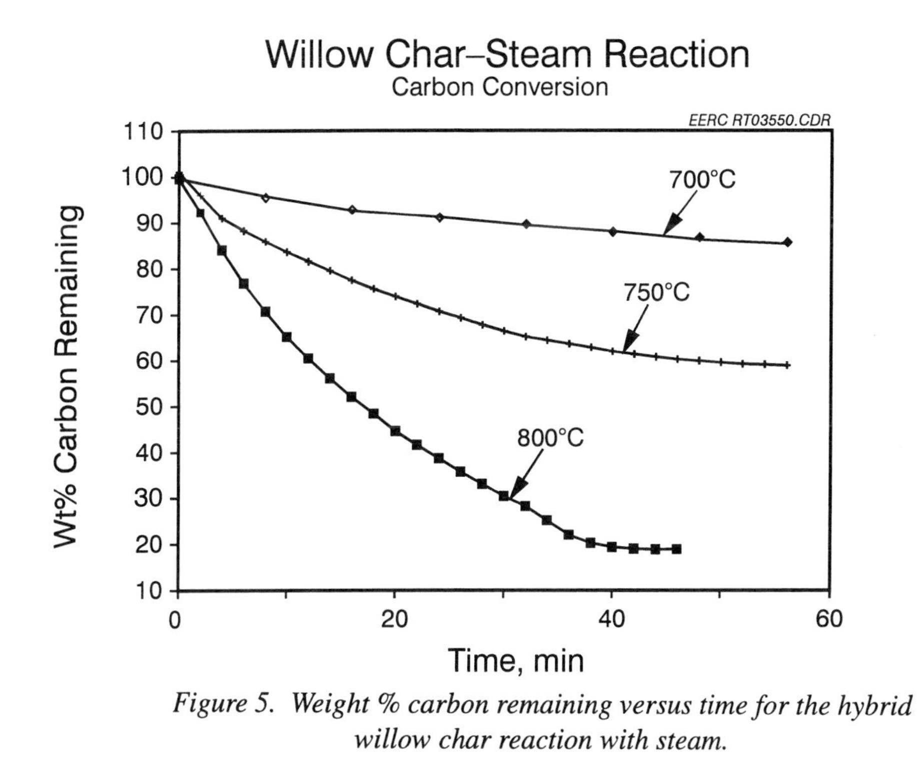

Figure 5. Weight % carbon remaining versus time for the hybrid willow char reaction with steam.

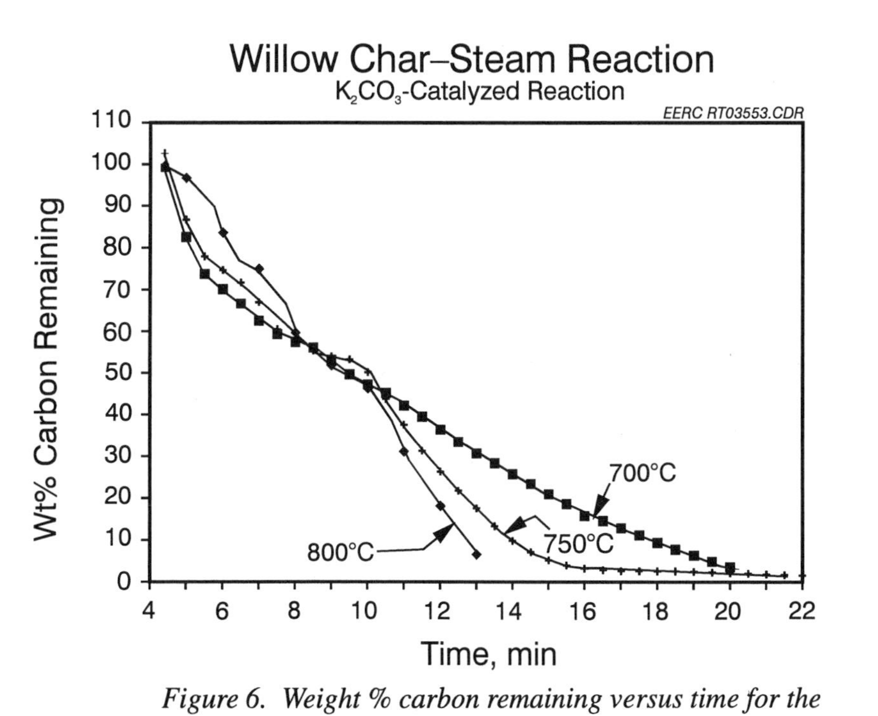

Figure 6. Weight % carbon remaining versus time for the K_2CO_3-catalyzed hybrid willow char-steam reaction.

E_a is calculated from the slope of the Arrhenius plot of ln k versus 1/T and ln A is the intercept. The E_a values shown in Table 4, although not absolute, are useful in determining the affect on the reaction of changes in temperature in terms of such occurrences as chemical versus diffusional control and the affect of added catalysts. The A values are most useful in determining a better estimate of the order of the reaction according to the method of Duvvuri et al when n = 1 is obviously not an acceptable choice for order of the reaction (3). However, for this study the choice of n = 1 was considered an acceptable value and a better estimate was not required. The energies of activation and the correlation coefficients (r^2) for the pine char-steam and K_2CO_3 catalyzed pine char-steam reactions are shown in Table 4. They indicate the effect of catalyst in lowering the energy of activation thus accounting for the increased rate of reaction, and a possible change in mechanism is implied by significant change in r^2. The addition of catalyst to the pine char-steam reaction at 800°C increased the reactivity by a factor of 4 as shown in Figure 7. A more significant finding, however, was that the reactivity at 750°C was as great as that at 800°C, thus indicating that the reaction could be run at milder conditions without sacrificing production. Similarly, catalysis of the willow char-steam reaction with K2CO3 resulted in similar behavior with a substantial lowering of apparent energy of activation as shown in Table 3 and increased reactivity at lower temperatures. This translates to operational savings in this type of hydrogen production process.

Table 2 shows the elemental analysis of ash prepared from the willow and pine used in this study. The major components of the willow ash are calcium (Ca) and potassium (K), of which carbonates and hydroxides of the latter are known from extensive studies of coal char-steam reactions to be excellent catalysts, and similar compounds of Ca are also known to exhibit some catalytic effects (12). The major constituents of the pine ash were silicon (Si) and calcium (Table 2). Although both wood samples had satisfactory reactivity for a hydrogen production process, the pine char, with its lower content of known catalyts, exhibited slightly lower reactivity than the willow. The reactivities, in addition to the absence of such catalyst-killers and environmental hazards as chlorine and sulfur, would appear from a chemical standpoint to make wood chars excellent candidates for hydrogen production reactions.

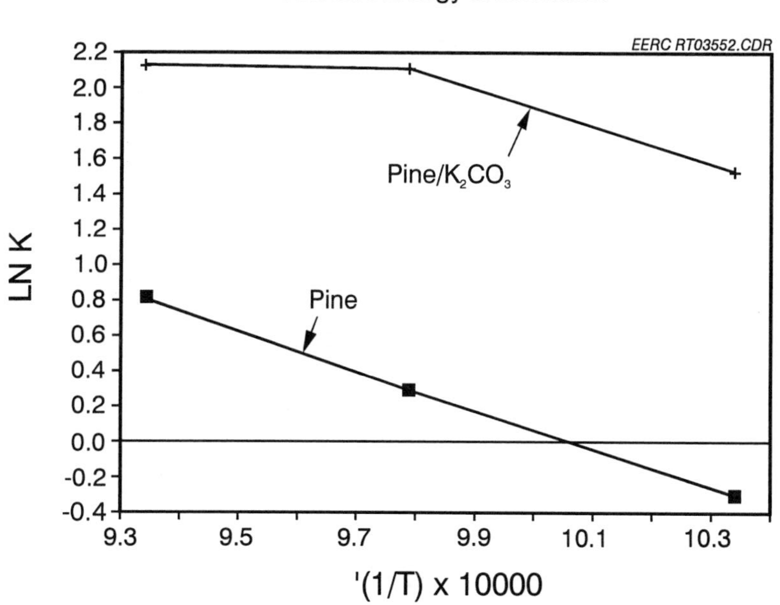

Figure 7. Arrhenius plot of the natural log of the reactivity parameter (ln k) of raw pine and pine/K_2CO_3 versus the reciprocal of the absolute temperature [(1/T) × 10000].

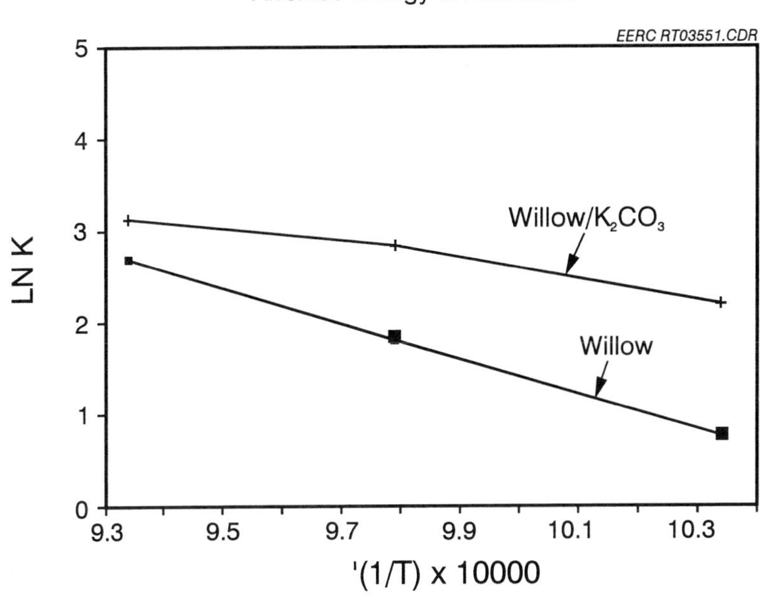

Figure 8. Arrhenius plot of the natural log of the reactivity parameter (ln k) of raw hybrid willow and hybrid willow/K_2CO_3 versus the reciprocal of the absolute temperature [(1/T) × 10000].

CONCLUSION

The reactivity of the hybrid willow char was comparable to that of the pine char tested. Catalysis with K_2CO_3 increased the reactivity of both chars by a factor of ~4 over those of the uncatalyzed woods, respectively. The catalyzed willow reaction had reactivities twice as great as those of the catalyzed pine.

A high hydrogen-content product gas can be produced via the gasification of wood char. Product gas compositions of >60 mol % H_2 were achieved for the willow char samples in the temperature range of 700°-800°C at atmospheric pressure.

ACKNOWLEDGEMENTS

The authors wish acknowledge Art Ruud for NMR spectra and Jan Lucht and Ron Kulas for assistance in the TGA experiments and gas analyses. Thanks are also due to W.A. Berguson of the Natural Resources Research Institute at the University of Minnesota-Duluth for providing hybrid willow samples.

DISCLAIMER

Mention of specific brand names or models of equipment is for information only and does not imply endorsement of any particular brand.

REFERENCES CITED

1. Berguson, W.A., Connolly, B.J., Farnham, R.F., Garton, S., Gugol, D.F., Larson, W.E., LeVar, T.E., Lewis, K.A., Read, P.E., Schmitt, M.D.C., and Sherf, D.B., Agroforestry Research on Minnesota Peatlands: Status Report 1983, USDOE, University of Minnesota, 1983.

2. Cheney, E.S., "U.S. Energy Resources: Limits and Future Outlook", American Scientist, 62, 14, 1974, Jan-Feb.

3. Duvvuri, M.S., Muhlenkamp, S.P., Iqbal, K.Z., Welker, J.R., "The Pyrolysis of Natural Fuels", J. Fire Flammability, 6, 468, 1975, October.

4. Galegher, S.J., Timpe, R.C., Willson, W.W., Farnum, S.A., Kinetics of Catalyzed Steam Gasification of Low-Rank Coals to Produce Hydrogen, Final Report, Office of Scientific and Technical Information, United States Department of Energy, 1986.

5. Lange, N.A., Handbook of Chemistry, 820-821, Ohio: Handbook Publishers, Inc., 1956.

6. Narayanaswami, V., Krishnamoorthy, S., Sekaran, P.M.C., Kalyanaraman, L., "Biomethanation of Leucaena leucocephala: A Potential Biomass Substrate", Fuel, 65, 1129, 1986, August.

7. Serageldin, M.A., and Pan, W.P., "Coal: Kinetic Analysis of Thermogravimetric Data", Thermochimica Acta, 71, 1, 1983.

8. Takarada, T., Tamai, Y., Tomita, A., "Reactivities of 34 Coals Under Steam Gasification", Fuel, 64, 1438-1442, 1985, October.

9. Tillman, D.A., Wood as an Energy Source, 207-214, New York: Academic Press, Inc., 1978.

10. Tillman, D.A., Wood as an Energy Source, 233-236, New York: Academic Press, Inc, 1978.

11. Veraa, M.J., and Bell A.T., "Effect of Alkali Metal Catalysts on Gasification of Coal Char", Fuel, 57, 194-200, 1978, April.

12. Wood, B.J. and Sancier, K.M., "The Mechanism of the Catalytic Gasification of Coal Char: A Critical Review", Cat. Rev.-Sci. Eng., 26(2), 233-279, 1984.

Chapter 5

Hydrogen Safety

5.1 Introduction

The word hydrogen generally carries negative connotations among the general public. Hydrogen fuel is often associated with either the Hindenburg or Challenger disasters, or even the hydrogen bomb. For the most part, however, hydrogen's bad reputation is unjustified. Hydrogen is a potentially dangerous and highly explosive fuel, but other fuels such as gasoline and natural gas pose similar dangers. While some of the accidents involving hydrogen have been dramatic, some have also probably been overly sensationalized [1]. As for hydrogen bombs, there is simply no relationship between the H-bomb and hydrogen as a fuel for surface transportation.

Hydrogen actually has a good overall safety record due to strict adherence to regulations and procedures, and good training of the personnel who handle hydrogen. Most people are unaware that hydrogen has been used safely as a fuel for decades. Today, it is used worldwide in space programs and a number of industrial applications [2]. Hydrogen has been shipped through extensive pipeline networks in Germany since the 1940s [3]. Mercedes Benz and BMW have both operated demonstration fleets in Germany for a number of years without a mishap. Other small scale or single hydrogen vehicle demonstrations have also proved to be quite safe. NASA also has considerable experience with handling hydrogen as part of their space program.

The fact that hydrogen is a potentially dangerous fuel must not be discounted, however, in the development of a hydrogen vehicle and distribution system. It cannot simply be assumed that the experience of trained professionals and strict adherence to regulations will extrapolate to more widespread public use. The safety of hydrogen vehicles and storage and distribution systems must be well documented before they are widely introduced. Precautions must be taken to ensure safe day-to-day operation and minimize the dangers of severe accidents. The hasty introduction of hydrogen into a marketplace without adequate demonstration of safe operations could have negative consequences if the public's initial image of hydrogen vehicles is tarnished. In this chapter, procedures and techniques for safely handling hydrogen will be reviewed along with the results of safety tests.

5.2 Accidents and Safety Studies

Accidents

Although the use of hydrogen has resulted in some accidents, hydrogen's unique characteristics have often prevented accidents from being fatal. The National Aeronautics and Space Administration (NASA) is one of the largest users of liquid hydrogen (LH_2) in the U.S., and they frequently transport LH_2 using highway tankers which carry between 11,000 and 60,500 ℓ of LH_2. NASA has experienced a number of accidents involving hydrogen systems throughout its history. These accidents have occurred both during off-loading procedures at the test site and on the highway [4].

In one of these highway accidents, the driver was traveling too fast around a curve and rolled over an embankment. Although the entire truck rolled about 12 m downhill, the tank was still in good condition and functioned normally. In another accident, the tank unhitched from the cab and slid about 24 m down a ditch. In this case, the safety discs ruptured, as they were designed to do, and the hydrogen merely evaporated. In a third accident, a delivery truck traveling about 80 km/h (50 mi/h) was hit by another truck. The tank was not damaged, however, and the vacuum remained intact [4].

Probably the most famous of all accidents involving hydrogen is the Hindenburg incident, which received widespread media coverage. Although the cause of the accident is not officially known, some suspect that it was started when hydrogen being vented was ignited by electrostatic charges present in the atmosphere as the result of a thunderstorm. Some have also pointed to the possibility of sabotage. It should be noted, however, that two-thirds of the Hindenburg passengers survived that accident. Most of the 35 who died perished because they jumped from the vessel at too great a height. Although some of the passengers also died on-board from diesel burns, no one died from hydrogen burns [5]. It is also little known that the Hindenburg made ten successful round-trip transatlantic flights before its infamous end. Also noteworthy is the fact that the Hindenburg incident was the result of a fire rather than a full-scale explosion [4].

Safety Evaluations

A series of safety experiments comparing hydrogen and jet fuel were conducted by the U.S. Air Force at Wright-Patterson Air Base. For these tests, several Styrofoam-lined, aluminum containers were set up, some of which contained LH_2, while the rest contained jet fuel. In one experiment, incendiary (explosive) and fragment simulator (non-explosive) bullets were fired at the containers. When struck by fragment simulator bullets the hydrogen simply spilled out through the bullet holes and evaporated. The fragmentation bullet shot into the jet fuel container, on the other hand, caused a shock wave which ruptured the container. These differences can be attributed to the greater fluid density of the jet fuel, as a traveling bullet will build up more pressure in a denser fluid. The incendiary bullets ignited both fuels, but neither container exploded and it was the jet fuel fire that persisted longer. In another experiment, a six million volt generator was used to simulate lightning strikes on the containers. Although fires resulted in both fuel containers, no explosions occurred [4].

In a study done at the University of Miami, the relative dangers of hydrogen, methane, and propane gas leaks were investigated. Ten leaks of different geometries were tested using kitchen models with gas stoves. Four different kitchen models were used, each with a simple vent over the stove but no forced convection (i.e., no fans operating). Computer models were generated to show cloud formation in the kitchen four and twelve minutes after the leak began. In each case, propane was found to present the greatest danger, while hydrogen was found to present the least danger. Hydrogen simply diffused straight up through the vent each time without forming a combustible cloud. Methane formed a small combustible cloud, but the bulk of it diffused through the vent as well. Propane, on the other hand, formed a large combustible cloud throughout the kitchen during each test [6].

BMW is currently involved in a four-year research program (1992-1995) to examine in detail the safety of hydrogen systems. Working with BMW on this program are tank manufacturers (Messer Griesheim GmbH and Linde AG) licensing authorities, universities, and the German army. This program is funded by the Commission of European Communities as part of the Euro-Quebec Hydro-Hydrogen Pilot Project (EQHHPP). The program focus is on the safety aspects of hydrogen storage in LH_2 tanks. Tank bursting process itself and the potential mechanisms for causing the tank to burst will be examined during the program. The first stage of the program is focusing on situations that can occur as a result of a sudden and complete failure of the LH_2 tank [7]. More details on this project can be found in the reprint by Pehr at the end of this chapter.

5.3 Hazards

Some of the important characteristics of hydrogen as compared to those of methane, propane, and gasoline are listed in Table 5.1.

Physical Properties

Similar to methane, hydrogen is invisible and odorless and thus can not be easily detected. This problem can be alleviated by adding small amounts of a colorant or odorant to the gas as is done for natural gas.

Fire and Explosion

Probably the greatest hazards in dealing with hydrogen are fire and detonation. For any fire to occur, the proper mix of hydrogen and oxygen must be present and the minimum ignition energy must be supplied. Of the fuels compared here, hydrogen represents the greatest fire and detonation hazard based on its wide flammability range and low ignition energy. Thus, safety guidelines should be based on the assumption that a hydrogen fire will occur if it can.

In some practical situations, however, other fuels can present a comparable fire threat. Where leaks are concerned, for example, it is the lower limits of flammability and detonability that are generally the distinguishing factors in ascertaining whether a fire will occur. For the fuels presented here the differences between the lower flamma-

Table 5.1 Properties of H_2, CH_4, C_3H_8, and Gasoline

Property	H_2	CH_4	C_3H_8	Gasoline
Molecular Wt. (g)	2.016+	16.04+	44.094+	107.0†
Density of NTP gas (g/cm^3)	0.0838+	0.6512+	1.87+	4.4*
Limits of Flammability in air, vol. %	4 - 75+	5.3-15.0+	2.1- 9.5+	1.0 - 7.6*
Limits of detonability in air, vol. %	18.3 - 59+	6.3 - 14+	—	1.1 - 3.3*
Minimum ignition energy in air (mJ)	0.02*	0.3*	0.26*	0.24*
Autoignition Temperature (K)	858*	813*	760*	501- 744*
Adiabatic Flame Temperature (K)	2318+	2148+	2385+	2470†
Burning Velocity in air (cm/s)	265- 325†	37 - 45†	46 - 47.2*	37 - 43†
Detonation velocity in air (km/s)	2.0*	1.8*	1.85*	1.4 - 1.7*
Buoyant Velocity in air (m/s)	1.2 - 9†	0.8 - 6†	—	non buoyant
Diffusion Velocity in air (cm/s)	≤2.00†	≤0.51†	—	≤0.17†
Detonation Induction Distance (L/D)	100†	—	—	—
Explosion Energy (g TNT/g fuel)	24+	11+	10+	10†
Explosion Energy (kg TNT/m^3 NTP gas)	2.02+	7.03+	20.5+	44.22†

* Data taken from reference [8].
\+ Data taken from reference [9].
† Data taken from reference [2].

bility limits are relatively minor. With regard to the minimum ignition energy, this number is most important in the context of weak ignition sources, such as sparks from electrical equipment. Studies have shown that a weak electrostatic discharge from the human body can release about 10 mJ, which is more energy than is needed to ignite a combustible mixture of any of the fuels discussed here [2].

One consequence of hydrogen's high burning velocity is that hydrogen has a higher probability of undergoing a deflagration-to-detonation transition (DDt), i.e., a transition from simple combustion to an explosion. Although the detonation velocities for all four of the fuels listed in Table 5.1 (gasoline, propane, methane, and hydrogen) are comparable, the normal burning velocity for hydrogen is much higher compared to that of the other three fuels. Thus, hydrogen can accelerate to its detonation velocity in a much shorter distance than the other fuels.

Some of the properties of hydrogen can be beneficial in terms of fire prevention. Hydrogen has the highest diffusion velocity and the highest buoyant velocity. Thus, in unconfined areas, hydrogen leaks will rise and diffuse rapidly to non-combustible proportions. A gasoline spill will remain a fire hazard much longer than an LH_2 spill, which will rapidly rise and disperse in the air to concentrations below the lower flammability limit. In enclosed areas, however, hydrogen can quickly spread to all areas of the room if leaks develop, forming combustible mixtures. In order to minimize this danger, hydrogen should be handled outdoors whenever possible, and in confined areas only when adequate ventilation is possible.

Hydrogen fires have several distinguishing features. They are typically more difficult to locate, extinguish, or avoid than other fires since they are invisible (i.e., nearly impossible to detect during the day with the naked eye) and radiate little heat to their surroundings. The flames are visible at night, however, and ultraviolet and infrared technology has been developed to detect hydrogen fires in daylight [2]. Hydrogen fires do burn very rapidly, and thus are relatively short-lived. Hydrogen fires and methane fires also do not generate toxic smoke like gasoline fires.

Overpressure

A detonation hazard consists of several key components: fire, a pressure wave resulting from a dramatic increase in pressure, and shrapnel. Of these components, the pressure wave causes the greatest structural damage. While unconfined overpressures are usually less than 7 kPa, confined overpressures can exceed 800 kPa and have devastating effects. In general, the wider the flammability limits, the greater the chances of a detonation that can cause extensive structural damage and injury. Because hydrogen has the widest combustion and detonation range, it has the greatest danger of detonating [2].

Asphyxiation and Frostbite

Hydrogen itself is non-toxic, and it emits non-toxic combustion products when it is burned with oxygen. As with any gas other than oxygen, however, it can act as a simple asphyxiant. Asphyxiation should not be taken lightly, as it can cause dizziness, vomiting, unconsciousness, and even death. In enclosed areas this can be a problem if large leaks develop in a hydrogen containment vessel. The problem is considerably less for small leaks, as the danger from a small leak is usually eliminated by a standard ventilation system [10]. Since H_2 has high diffusion and buoyant velocities (i.e., it diffuses and rises rapidly in air), this problem is also almost negligible in unconfined areas. If an asphyxiation danger does exist, self-contained breathing apparatus should be kept on hand.

When dealing with LH_2, frostbite is another hazard of concern. Human skin can easily be frozen or torn if it comes into contact with cryogenic surfaces which maintain hydrogen at its liquefaction temperature of 20 K. All pipes containing cryogenic hydrogen must be insulated and operators must insulate body parts which may come into contact with these cold surfaces. Insulation of pipes is also necessary to prevent the surrounding air from condensing on the pipes [11], as an explosion hazard can develop if air condenses on pipes and then drips onto nearby combustibles. Asphalt pavement, for example, is composed of combustible petroleum products which can ignite in the presence of liquid air. Thus, areas where LH_2 is used should be paved with concrete instead of asphalt [10].

5.4 Preventive and Safety Measures

Leaks and Spills

A study done by Zalosh and Short reported that the highest percentage of hydrogen accidents are caused by undetected leaks [12]. Hydrogen can diffuse readily through joints or connections, even those that are considered 'air-tight,' due to its small size. Thus, it must be assumed that leaks can occur in any design. Leaks can be detected by running inert gases, such as helium, through the system. Soapy water should be used to test any spot where a leak is suspected. In order to reduce the potential of forming leaks, the number of joints and connections in a system should be reduced and welded joints should be used wherever possible.

Some plants use flare stacks to vent leaks to the atmosphere (see Fig. 5.1), or to relieve boil-off pressure from LH_2 dewars (see Fig 5.2). Flare stacks allow excess hydrogen to be burned without damaging the system. Valves within the stack prevent air from flowing back into the system, thus preventing detonable mixtures from forming within the system. Flare stacks, however, are generally only used when a large volume of hydrogen must be quickly ventilated. Most plants simply vent the boil-off to the atmosphere.

Since leaks will occur, it is necessary to eliminate ignition sources wherever possible. All electrical equipment must be securely grounded to eliminate electrostatic sparks. Equipment should also be kept at safe distances from

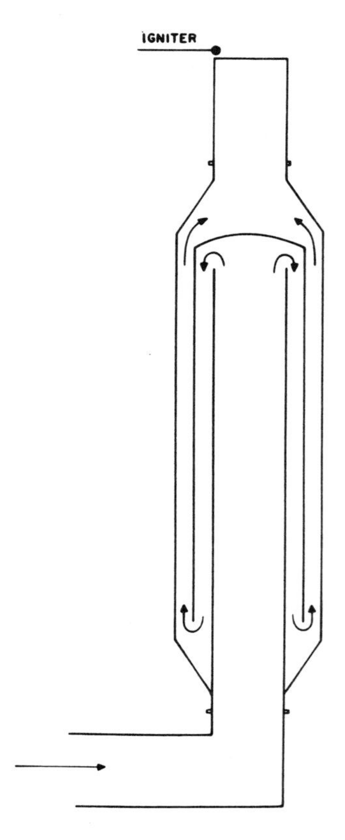

Figure 5.1. Hydrogen flare stack. (Source: ref. [10])

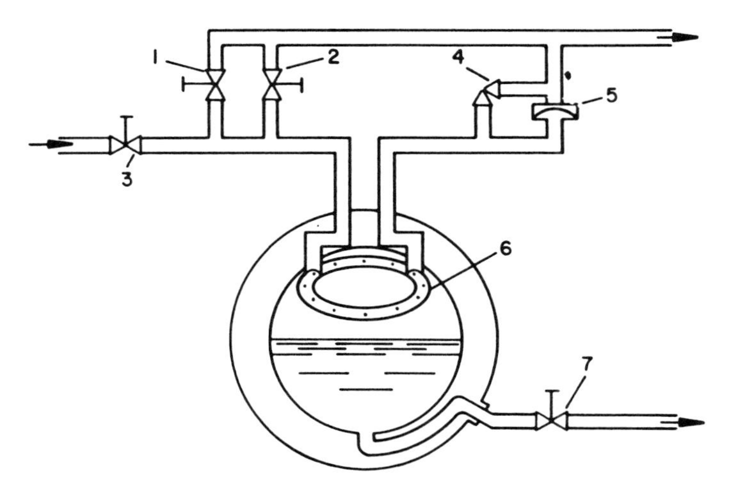

Figure 5.2. LH_2 storage dewar vent system. (1) Normal boil-off vent valve. (2) High rate vent valve. (3) Valve to admit dewar pressurization gas. (4) Safety pressure relief valve. (5) Rupture disc. (6) Dewar pressurization and vent ring. (7) Liquid hydrogen discharge valve. (Source: ref. [10])

hydrogen sources, as friction or mechanical sparks can arise during normal operation of equipment. Of course, smoking or match striking should never be permitted in an area where hydrogen is present. Even precautions such as lightning rods, to protect against lightning strikes, are used [13]. As an additional safety measure, all other combustibles should be removed from the site, as situations can occur as a result of a sudden and complete failure of the LH_2 tank [7]. More details on this project can be found in the reprint by Pehr at the end of this chapter.

Purging

To remove air from the system so that combustible mixtures cannot develop within a hydrogen containment vessel, the system must be purged with an inert gas, such as nitrogen. The purging procedure is important and must be done properly. In fact, it was found that approximately 8% of hydrogen accidents are a result of inadequate purging [12]. After a purge, the system must be tested for residual oxygen before hydrogen is allowed to enter the system. Gas chromatography is an accurate method of testing for trace amounts of oxygen [10].

An accident once occurred at the Los Alamos National Laboratory that demonstrated the consequences of not following correct purging procedures. As a result of procedure violations, a tube containing hydrogen and a tube containing oxygen were inadvertently connected to the same manifold. An inadequate purging procedure allowed oxygen to enter a cylinder containing hydrogen, creating a near stoichiometric mixture. The mixture was accidentally ignited and the flame front sped back to the hydrogen tube where a detonation occurred, rupturing the tube. Shrapnel from the overpressure was thrown more than 400 m. While the tube was designed to withstand a pressure increase of about 20 times that of the normal system pressure, the pressure that resulted from this detonation may have been 40 times greater than the ambient system pressure [1].

Pressure Relief Devices

Pressure relief devices are designed to vent excess gas when the pressure in a vessel exceeds a certain design value. Two of the most common pressure relief devices are spring-loaded safety valves and rupture discs. A spring-loaded safety valve is designed to remain closed until the pressure in the vessel exceeds the nominal pressure (P) by a certain amount (∅P). If the pressure in the tank exceeds P + ∅P, the force on the valve then exceeds the spring force, and the spring will extend, bleeding off excess hydrogen until the pressure is reduced to below P. When the force on the valve is less than the spring force, the spring will contract again, discontinuing the venting.

Rupture discs consist of a ring covered with a membrane of predetermined thickness. The thickness is designed to withstand a maximum pressure, similar to the spring in the safety valve. When the design pressure is exceeded, the membrane bursts, releasing the gas. Once this membrane breaks, however, the venting cannot be stopped until all of the gas has been released. At this point, a new rupture disc must be installed on the system before it is operational again. Rupture disks are generally not used alone, but rather as a back-up device in case the spring loaded safety valve fails.

Fire Fighting Techniques

Several techniques can be used to prevent hydrogen fires from evolving into detonations. Before anything is done, however, all personnel should be evacuated from the area. If it is safe to do so, the fuel flow should also be disconnected. Other hydrogen containers should be removed from the area, because if these containers become too hot, i.e., above 52° C (125° F), pressure in the system can build to a point where the gas will be vented by the relief system [14].

It is sometimes advisable to allow spilled fuel to burn off, once the source is disconnected, as an explosive hazard can develop when trying to extinguish a hydrogen fire [2]. In such cases, water can be used to cool nearby flammable objects. This will reduce the possibility of explosive reignition, which can occur after the hydrogen flame is extinguished if nearby surfaces were not sufficiently cooled with water [15]. It should be noted that only trained fire fighting professionals should attempt to extinguish larger hydrogen fires. Methods used by fire fighters to extinguish hydrogen flames include using water (about 80% effective in reducing structural damage), dry chemical extinguishers (about 86% effective), production shutdown (about 81% effective), and explosion venting (about 91% effective).

Preventing Explosions

The geometry of the containment vessel can affect the probability for a transition from deflagration to detonation. One way to minimize the probability of this transition is to decrease the length of the pipe and increase the diameter. The benefits of such a geometry are related to a property known as the minimum induction distance, which is the distance necessary for the flame to reach the detonation velocity. The induction distance is defined as the distance from the point of ignition to the point where the flame front reaches detonation velocity. Pipes with a high length to diameter ratio (L/D) are more likely to propagate a DDt than enclosures with a low L/D ratio [2]. Table 5.2 lists the corresponding pipe lengths necessary for a DDt to occur in pipes of different diameter.

Another way of inhibiting explosions is by using a "lean" mixture of hydrogen and air, or a mixture toward the lower end of the flammability range. This lowers the burning velocity substantially, bringing the flame into laminar flow regions. Under these conditions, the flame cannot reach detonation velocity unless a strong mechanism, such as a mixing fan, is available to induce turbulence. A lean mixture of hydrogen is actually safer than lean mixtures of the other fuels, as lean hydrogen mixtures, which can have as little as 4% hydrogen by volume, are well outside the detonable range (the lower limit of detonability for hydrogen is 18.3% hydrogen by volume) [8]. For other fuels, however, their lower limits of flammability and detonability are similar. These limits, respectively, are 5.3% and 6.3% for methane, and 1.0% and 1.1% for gasoline, thus even lean mixtures of these fuels are still close to the detonable range.

Shrapnel Protection

Detonations can send shrapnel fragments flying for several hundred meters, thus minimum safety distances are necessary to protect personnel from shrapnel (See Fig. 5.3). In addition, shrapnel protection shields should be used whenever possible. Shrapnel protection shields are difficult to design because it is hard to predict the volume, mass, or velocity of the fragments, as these quantities vary with the strength of the detonation. It has, however, been shown experimentally that materials with a high Young's Modulus of Elasticity, such as structural steel, high-strength/low alloy steel, and stainless steel, are effective as shrapnel shields [2].

Embrittlement

Due to hydrogen's unique molecular properties, it is capable of diffusing into materials which are impermeable to most fluids or gases. Diatomic H_2 is the most stable form of hydrogen. But when provided with enough energy, diatomic hydrogen can dissociate into hydrogen atoms, which are small enough to permeate a container surface (see Fig. 5.4). This permeation can cause structural damage to the container, especially when using metal containers. This phenomenon is known as hydrogen embrittlement. The actual degradation of the metal can occur via several processes, which differ primarily in how the hydrogen interacts with the metal lattice and the degree of degradation they induce. The net effect of any

Table 5.2. Required Pipe Length (m) for a Deflagration to Detonation Transition

Gas	Pipe Diameter (mm)		
	100	200	400
Methane	12.5	18.5	30.0
Propane	12.5	17.5	22.5
Hydrogen	7.5	12.5	12.5

(Source: ref. [13])

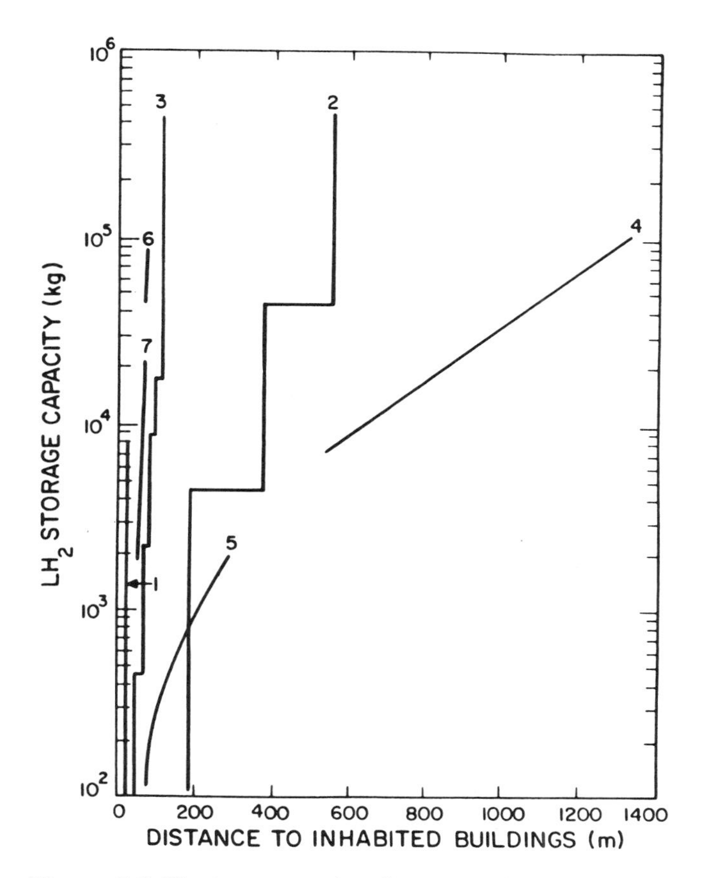

Figure 5.3. Various quantity-distance relationships suggested for inhabited buildings to LH_2 storage. (Sources: ref. [10] and references therein)

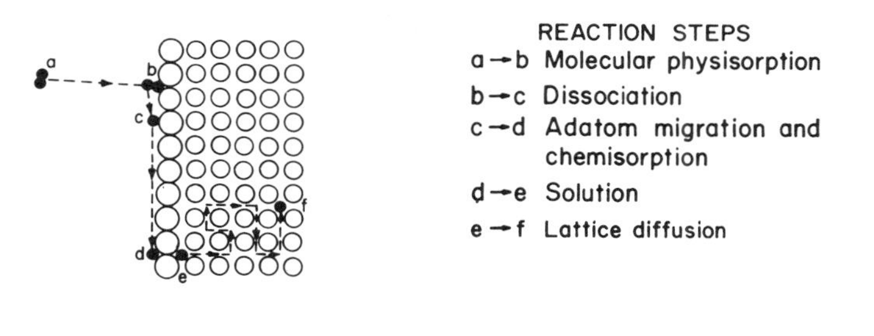

Figure 5.4. Possible reaction steps involved in the transport of hydrogen from the molecular hydrogen environment. (Source: ref. [16])

of these mechanisms, however, is to weaken the metal lattice, causing it to lose ductility and eventually fracture [16]. Cryogenic embrittlement can also occur when LH_2 is used. This happens when the cold temperatures of liquid hydrogen cause the metal to lose ductility under stress and eventually crack [11].

Certain materials are more susceptible to embrittlement than others, and some have proven quite resistant to embrittlement. Martensitic (body centered tetrahedral structure) and ferritic (body centered cubic) steels are more susceptible to embrittlement than highly alloyed austenitic (face centered cubic [fcc]) stainless steels because hydrogen can be more readily transported in martensitic and ferritic materials. The fcc structure of austenitic steels, on the other hand, inhibits hydrogen diffusion. As far as specific compounds are concerned, nickel alloys exhibit all of the degradation mechanisms and aluminum alloys exhibit significant surface cracking. Copper alloys, however, show almost no degradation except at high temperatures [16]. Carbonaceous steel cannot be used as the hydrogen can react with carbon in the steel to form CH_4 bubbles in the metal, though this usually only occurs at high temperatures [13].

Hydrogen diffusion into the lattice is more likely to occur in the gas phase than in the liquid phase, as hydrogen transport is much slower at cryogenic temperatures than at ambient temperatures. Thus, materials that exhibit high degrees of degradation at ambient temperatures can sometimes be used to store liquid hydrogen. Gas-phase diffusion is still possible in cryogenic systems, however, as any cryogenic system will lose some liquid hydrogen as boil-off [16]. LH_2 is safely stored in double-walled dewars made of an inner, stainless steel core and a carbon steel vacuum jacket.

5.5 Conclusion

Safety considerations are of the utmost importance in large-scale handling of hydrogen. In view of hydrogen's excellent safety record, however, hydrogen need not be feared. Hydrogen has been safely produced and used around the world for decades. While hydrogen poses different hazards than other fuels, they are not necessarily any greater. Before hydrogen will be widely accepted by the public as a viable transportation fuel, the potential dangers of using hydrogen in transportation applications must be fully understood. Thus, studies in the area of safety should continue, in coordination with the development of safety standards and regulations.

References

1. Edeskuty, F.J., Haugh, J.J., and Thompson, R.T., "Safety Aspects of Large-Scale Combustion of Hydrogen," in *Hydrogen Energy Progress VI*, Pergamon Press, Elmsford, NY, p. 147, 1986.

2. Hord, J., "Is Hydrogen a Safe Fuel?" *Int. J. Hydrogen Energy*, 3, p. 157, 1978.

3. Harkin, T., *Proposal for a Sustainable Energy Future Based on Renewable Hydrogen*, 1993.

4. Hoffmann, P., *The Forever Fuel: The Story of Hydrogen*, Westview Press, Boulder, Colorado, 1981.

5. Zweig, R.M., Video.

6. Swain, M.R. and Swain, M.N., "A Comparison of H_2, CH_4 and C_3H_8 Fuel Leakage in Residential Settings," in *Hydrogen Energy Progress IX*, International Association for Hydrogen Energy, Coral Gables, FL, p. 1121, 1992.

7. Pehr, K., "Aspects of Safety and Acceptance of LH_2 Tank Systems in Passenger Cars," in *Hydrogen Energy Progress X*, International Association for Hydrogen Energy, Coral Gables, FL, p. 1399, 1994.

8. Eichert, H. and Fischer, M., "Hydrogen Safety in Energy Application Compared with Natural Gas," in *Hydrogen Energy Progress V*, Pergamon Press, Elmsford, NY, p. 1869, 1984.

9. Fischer, M., "Safety Aspects of Hydrogen Combustion in Energy Systems," *Int. J. Hydrogen Energy*, 11, p. 593, 1986.

10. Edeskuty, F.J., "Safety," in *Hydrogen: Its Technology and Implications*, Edited by K.E. Cox and K.D.J. Williamson, CRC Press, Inc., Boca Raton, FL p. 208, 1979.

11. Union Carbide—LINDE, *Constant Vigilance: The Safe Handling of Hydrogen Gas.*

12. Zalosh, R.G. and T.P. Short, *Comparative Analysis of Hydrogen Fire and Explosive Incidents*, Factory Mutual Research Corp., March 1978.

13. Balthasar, W. and Schödel, J.P., *Hydrogen Safety Manual*, Commission of the European Communities, Directorate General for Science, Research and Development, Luxembourg, 1983.

14. Union Carbide—LINDE, *Material Safety Data Sheet*, Union Carbide-LINDE Division, 1986.

15. National Fire Protection Association, *Gaseous Hydrogen Systems*, 1986.

16. Nelson, H.G. *A Review—Materials and Safety Problems Associated with Hydrogen Containment in Hydrogen Energy Progress V*, Pergamon Press, Elmsford, NY, p. 1841, 1984.

Aspects of Safety and Acceptance of LH_2 Tank Systems in Passenger Cars

K. Pehr
BMW AG, Germany

ABSTRACT

So far safety concepts for hydrogen-drive cars have been derived mainly from existing rules and regulations for pressure vessels and gas plants. Whether these rules are sufficient for the strict requirements of hydrogen supply systems in passenger cars, however, is still unclear and therefore requires verification.

Unlike industrial plants, it would not be appropriate in the case of passenger cars to assume that only trained personnel will be involved with the new energy carrier, as vehicles with hydrogen drive will be used for all kinds of purposes - and also abused in some cases, with various damage scenarios resulting in the process.

In the light of this situation, it is essential to start out by systematically determining and assessing possible hazards, effective improvements of the automobiles involved and their surrounding aspects only being possible after such a fact-finding procedure. This in turn calls for thorough cooperation between the automobile industry, on the one hand, and suppliers, energy utilities, the gas industry, and licensing authorities, on the other.

Only a clear and open discussion of the approaches taken and results achieved in safety tests can pave the way for general acceptance by the public at large of hydrogen as a possible energy carrier.

Conducting research on the safety of liquefied hydrogen (LH_2), BMW is currently cooperating with Messer Griesheim GmbH and Linde AG in a series of research operations and tests serving first to determine the hazard potential of LH_2 tanks used in road vehicles, taking into consideration in this process the various parameters influencing the effects of this hazard, and second to determine further requirements to be made of such vehicle tanks in their basic concept. This experimental research is funded by the Commission of the European Communities (CEC) as part of the EQHHPP (Euro-Québec Hydro-Hydrogen Pilot Project).

The following report presents the initial results of these tests.

1. INTRODUCTION

Whenever human beings assess risks and hazards, their perception is based not only on objective circumstances weighted against one another, but also on personal feelings or experiences. The subjective assessment of a given hazard often differs significantly from the actual risk determined by way of statistical material or precise, objective analysis [1,2,3], and defined as the product of the actual scope of damage and the probability of damage occurring. It is important to note in this context that even risk analyses may involve certain subjective features, albeit based on good engineering judgment.

Factors leading to subjective assessment of a risk are for example the degree to which a specific hazard is already known, the extent to which a risk is assumed voluntarily, the possibility of influencing a specific risk or danger, the possible consequences of damage actually occurring, and the time at which damage occurs. These factors are assessed individually by the parties concerned as a function of their specific features and character [4].

In paving the way for a hydrogen infrastructure in the future, it would not be admissible, given the advanced state of the art, to base the safety technologies applied on experience gained with accidents or damage in the past, as was done above all in the early days of technology. Instead, any lack of historical experience must be compensated for by systematic, analytical forecasts revealing the hazards to be expected.

2. STORAGE OF LIQUEFIED HYDROGEN IN THE PASSENGER CAR

Benefitting from a high level of energy density, the LH_2 tank - as opposed to other hydrogen storage systems - is particularly suitable for the passenger car [5].

In this case hydrogen is stored in cylindrical, double-wall, vacuum super-insulated tanks at a temperature of about 20 K and a maximum overpressure of 5 bars [6,7]. Weighing 60 kg or 132 lb, the prototype tanks currently fitted in test vehicles have an inner volume of approximately 120 litres (26.4 Imp gals) providing a cruising range of more than 300 km. Special vehicles designed from the start for liquid hydrogen might well be able to achieve a range of more than 500 km [8].

A hydrogen vehicle may well be parked for several days without any loss of energy. Then, in the event of a longer standstill period, one must expect daily evaporation losses amounting to 2 per cent of the tank volume at most.

Still made elaborately by hand today, hydrogen tanks have a significant potential for cost reduction when manufactured by suitable production methods.

3. HIGHLIGHTS OF BMW'S HYDROGEN SAFETY RESEARCH

The high standard of safety and reliability required these days of vehicles with conventional drive, also applies to hydrogen drive.

Results achieved by BMW in working on liquefied hydrogen safety concepts have for many years been applied step-by-step in our test vehicles.

With hydrogen - like natural gas and liquid petroleum gas - being a gaseous fuel under normal conditions, the fuel storage and supply system must meet particularly stringent quality requirements. As with stationary tanks and facilities [9], it is very important to avoid any leakage of gas. The various components and the overall system must therefore remain absolutely tight and function properly throughout the entire service life of the vehicle, withstanding extreme environmental conditions as well as vibrations, sudden impacts, and accidents.

The basic rule in designing the hydrogen storage and supply system is to keep the number of potential leakage points - pipe connections, seals on components, evacuation valves, and pressure relief vent ports - as small as possible. Any gas discharged from the system must be able to escape from the vehicle through defined channels and as safely as possible, that is without entering the passenger compartment or other sealed-off areas. A methodical failure analysis [10] will systematically reveal the design modifications and improvements required in each case.

With vehicles being subject to the risk of collisions on the road, all components storing or conveying hydrogen must be located at points protected from damage to the greatest possible extent. Components inevitably installed in more critical areas - such as the tank filler coupling and pipe - must be connected to the fuel system as such by flexible units in order to minimise the risk of leakage caused by pipes breaking loose or becoming damaged.

With hydrogen being non-toxic and odourless, pressure surveillance of all relevant components and gas monitors inside the vehicle are required to detect leakage as early as possible, thus providing ample time for automatically initiated counter-measures. As an example, forced air extraction from the interior of the car upon detection of even very small, quite uncritical leakage ensures risk-free disposal of hydrogen whenever necessary.

Transmission of data exceeding a specific threshold from the system monitors through modern information channels to the headquarters of a closely-knit, mobile service network would allow rapid and uncomplicated help by trained specialists in the event of a deficiency. BMW`s mobile service already has nearly 300 service vehicles worldwide and is being constantly improved [11].

Apart from precautions ensuring that the vehicle is absolutely safe in operation, allowance must also be made for the risk of very severe accidents destroying the very structure of the car and enabling hydrogen to escape. Assessment of the physical and chemical properties of hydrogen already discussed in detail in the literature would not provide any meaningful new information on the risks involved under such circumstances, since the behaviour of liquid hydrogen tank systems in situations of this kind is still unknown.

Studies must set out by examining the hydrogen tank in which a major share of the energy in the vehicle is stored and which is crucial in its underlying concept to the overall design package of the car. One must find out how much damage the liquid hydrogen tank may potentially cause, how it will respond in very severe accidents, and how its concept may be improved in the interest of maximum safety.

These are the tasks and questions currently being considered in a four-year research programme (1992 - 1995) carried out by BMW in cooperation with the tank manufacturers, Messer Griesheim GmbH and Linde AG, and including licensing authorities (Technical Inspection Authority in Bavaria/Saxony), universities and scientific institutes (Federal Institute for Materials Research and Testing in Berlin, Fraunhofer Institute of Chemical Technology in Berghausen, Stuttgart Materials Testing Institute, Munich Technical University), as well as the German army (Defence Technology Division 52). In addition, the programme is funded by the Commission of European Communities as part of the EQHHPP (Euro-Quebec Hydro-Hydrogen Pilot Project) [12].

An experimental examination of the overall vehicle system as such would only appear meaningful when the liquefied hydrogen storage and supply concept for a passenger car has achieved a defined, very high safety standard (all the more because the vehicle as a whole will still be subject to change up to that point in time).

4. RESEARCH PROGRAMME ON THE SAFETY OF LH_2 TANKS SUITABLE FOR USE IN VEHICLES

The first consideration is that customers and users of LH_2-drive automobiles cannot be expected to meet any greater demands and requirements than "regular" motorists today. Accordingly, one must proceed from the same standard of behaviour on the road as in present-day traffic, which means that the hydrogen-drive car may be used under all kinds of conditions, abused in various ways, and subjected to defects and damage scenarios of all types.

Taking a methodical approach - for example a fault tree analysis - we are able to identify the main loss events affecting LH_2 tanks in automobiles, their causes, and the way such damage is brought about. Analyses to this effect are being carried out at the moment, the current research programme being based for the time being on the most significant types of damage and incidents generally valid quite irrespective of the design and structure of the LH_2 tank. Any modification of the tank concept resulting from the research programme and subsequent work would only affect the likelihood of such principal types of damage occurring and their effects in practice.

The worst case scenario proceeds from sudden failure of the tank releasing all the mechanical, thermal and chemical energy stored in an instantaneous shock. In an accident such sudden failure may result directly from mechanical damage to the outer and inner tank, excessive pressure build-up in the inner tank, or the vehicle catching fire. Further causes of such damage may result from the development, production, or handling of the tank - but since this might lead via the same loss mechanisms to the same event at the top of the fault tree, these circumstances need not be considered in detail in the context of this research programme. Proceeding from the top event on the fault tree (Fig 1), BMW has developed a two-stage research programme providing information on the degree of damage (but not the risk) possibly emanating from LH_2 vehicle tanks and allowing the engineer to enhance such tanks to the highest conceivable level of safety by optimising their technical parameters and the design of the tanks as such.

The first step is to examine the actual process of a tank bursting. One must consider what consequences such an event may have, what parameters influencing the process of bursting have to be taken into account, and how their effects must be judged. Apart from the quantity of energy stored, one of the most important parameters directly affecting such incidents is tank pressure not only crucial to the design of an LH_2 tank, but also determining the condition of hydrogen in the tank and characterising to a large extent both the actual occurrence of damage and the subsequent effects.

Stage 2 of the research programme serves to examine the mechanisms potentially making the tank burst. The objective is to understand the process of damage in a double-wall vacuum super-insulated LH_2 tank, to determine weak points and make suitable improvements.

In the event of mechanical damage inflicted from outside, we may regard a tank as "benign" in its behaviour if leakage can be avoided by the stable design of the tank and by using ductile materials or if at least cracks allowing leakage do not grow larger and make the tank burst, spontaneously releasing all the energy stored.

This principle of "leak before rupture" also applies to excessive pressure building up inside the inner tank. With damage spreading from inside to outside, the outer tank may in this case fulfil a specific safety function by damping the effects of the inner tank failure. Such an excessive, inadmissible pressure may build up within the inner tank if the effects of an accident, for example, destroy the insulating vacuum, the safety valves are blocked at the same time, and pressure relief is rendered impossible by the control unit either being inaccessible or inoperative.

In the event of the vehicle catching fire the safety valves must be sufficiently large in order to avoid pressure building up within the tank. If a fire continues over a lengthy period a suitable pressure relief strategy must release pressure from the inner tank before it can rupture.

In the following the report will present and discuss the results obtained in the first stage of examinations.

5. EXPERIMENTAL EXAMINATION OF THE BURSTING TANK SCENARIO

Regardless of the likelihood of a tank bursting, which may by reduced almost at random by suitable technical modifications and refinement, it is essential to carefully examine the effects of such a worst case scenario. Assessing the effect of specific parameters and their influence, we can determine a suitable approach for improving the actual standard of safety.

The most significant parameters influencing the extent of damage are assumed to be the amount and condition of hydrogen stored, its flow behaviour, the time of ignition, ignition energy, and surrounding atmospheric conditions.

Since the pressure range required for running the vehicle depends on the specific type of energy converter and LH_2 tank system used, particular consideration was given in the tests to the respective condition of the hydrogen as a parameter influencing the extent of damage.

Choosing a suitable tank failure mechanism, we have been able to keep the actual tank rupture process and the conditions under which the escaping hydrogen ignited as reproducible as possible: All tanks were cut open around the entire circumference of the cylinder within a time interval of approximately 0.2 ms by means of a cutting charge as shown in Fig 2. The cutting jet generated by the exploding charge and facing in a specific, predetermined direction spontaneously ignited the hydrogen/air mixture forming in the process. The amount of explosives used was however so small that its effect was not significant compared to the bursting process as such.

Weather conditions remained relatively stable during the tests and the test location protected from wind helped to ensure reproducible test conditions.

To limit the amount of technical equipment required in the tests, the bursting process was studied on single-wall tanks equivalent to the inner tank in an LH_2-driven vehicle and insulated with a layer of foam. To determine the current state of the hydrogen, the test tanks were equipped with a level sensor, pressure sensor, and temperature sensor. In all, 10 tanks were ruptured with their hydrogen contents in various conditions.

The tank bursting process was recorded visually, effects such as flames, thermal radiation and pressure waves being measured by suitable equipment.

While the results obtained from the tests are certainly very helpful, they have not been verified statistically and cannot be simply transferred to other damage scenarios.

6. DEVELOPMENT OF THE FIREBALL

Within fractions of a second after ignition of the cutting charge the hydrogen escaping from the tank mixed with the ambient air, the energy released by the cutting charge making this mixture ignite. Indeed, even when a tank bursts without the additional effect of explosives it is quite likely that the gas/air mixture will ignite immediately, for example on account of the energy released by the material rupturing.

With the burst process continuing, the next phenomenon was the development of a fireball emitting not only ultraviolet radiation in the 250-340 nm wavelength range (OH^- radicals) characteristic of the hydrogen flame, but also infrared and visible radiation due to particles flying through the air and atomised in the process.

Fig 3.2 shows a characteristic example of a fireball 250 ms after ignition of the cutting charge. Just one second later, the fireball had already expanded to almost its full size and started to rise up (Fig 3.3). On ground level (between 0 and 3 metres elevation) the maximum diameter of the fireball measured horizontally ranged from 6 to 15 metres, depending on tank pressure and the amount of hydrogen stored in the tank. At the latest 1.8 seconds after ignition the lower end of the fireball rising up into the air had reached a height of three metres and was continuing to move up (Fig 3.4). The maximum diameter reached by the fireball in this process was 20 metres. The maximum height achieved by the fireball prior to extinction after four seconds was between 16 and 20 metres above ground.

The fireball itself was an inhomogenous mixture of radiating particles and gas molecules expanding three-dimensionally. The radiation emitted by the solid particles involved in this process can be described by and large as black-body radiation observable as a continuous infrared spectrum with a characteristic wavelength and temperature dependent intensity distribution expressed by Planck's law of radiation. The radiation bands characteristic of H_2O molecules (a product generated in the combustion of hydrogen) are also to be found in the infrared range (1300-1500 nm, 1700-2100 nm, 2400-3100 nm, and 6000-8000 nm). The emission bands between

4200 and 4500 nm, in turn, involve CO_2 molecules resulting inter alia from the combustion of small amounts of insulating material.

Radiation spectra were analysed by two filter wheel spectrometers developed by the Fraunhofer Institute of Chemical Technology (FhG-ICT), working in the range from 1.2 - 14 microns and applying a time resolution factor of 100 spectra per second. A filter wheel spectrometer is able to select the spectrum of infrared radiation by means of a wheel broken down into segments made of different filters rotating in front of the detector. We were not able to use other radiation meters either because their temporal resolution is inadequate or because they are not in a position to determine the radiation intensity of an emitter with an unknown spectrum.

Applying Planck`s law of radiation, we are able to determine from the spectra measured (Fig 4 shows an extract of one spectrum) that the maximum temperature reached by black-radiation particles ranges from 1500 to 1800 K. It is important to note in this context that the level of continuous radiation at 2200 nm is slightly increased by low intensity branches of a water band system.

These data were confirmed by additional measurements using a two-colour pyrometer determining the intensities emitted in the spectra from 900 to 1100 nm and 1000 to 1260 nm. Again applying Planck`s law of radiation, the temperature may be determined from the ratio between the various radiation intensities.

Integration of the data as a function of wavelength then enables us to calculate the specific radiation capacity, the estimated share of the continuum ranging from 40-65 per cent, depending on the test. Considering the particle temperature measured between 1500 K and 1800 K, the relatively high integrated radiation capacity, and the ratio between the intensities of different molecular band systems, one may reasonably assume that at some points gas temperature briefly came close to the stochiometric combustion temperature of hydrogen in air (2318 K).

Since the distribution of particles, the concentration of gas and temperatures in the fireball expanding very quickly is unknown, it is impossible to make any exact statements on the optical depth of the combustion geometry. This is also why it was not possible to develop an experimentally verified calculational model for the fireball allowing a determination of the absolute thermal radiation load in the vicinity of the fireball generated. Proceeding from the results obtained, further research in the area of radiation measurement would be required in order to make any statements of this kind.

7. PRESSURE WAVES AND TANK FRAGMENTS

Pressure sensors were arranged perpendicular to the propagation path of the pressure wave in order to determine the waves generated by the bursting LH_2 tank. This is the measuring procedure generally applied for assessing pressure waves (side-on overpressure), the pressure measured allowing to determine the load conditions actually generated depending, inter alia, on the angle of contact, the mass and area of the object exposed, and its distance from the blast [13].

The pressure phenomena measured were attributable to the following causes:

- The explosion of the cutting charge generated a brief, sudden surge of pressure initiating the entire bursting procedure as such.

- As soon as the tank was opened a certain amount of the liquid hydrogen stored therein was able to spontaneously evaporate due to the sudden drop in pressure. This created a pressure wave, as caused by the sudden expansion of gaseous hydrogen from the tank.

- Additional pressure events followed from the acceleration of flames and the expansion of exhaust gas behind the progressing flame front of the ignited hydrogen/air mixture.

These processes cause a different pattern of pressure waves varying from one test to another. Fig 5 presents three examples, showing how the energy converted into pressure waves is concentrated in one, two or three pressure peaks. The significant peaks are followed by a phase of partial vacuum. Considering the different, non-controllable temporal course of the individual processes, it is obvious that the tests cannot be reproduced in full.

Expanding radially, all the pressure waves measured in the range between three and 36 metres from the centrepoint of the tank decreased in intensity to about 1/10th of their respective intensity level at three metres.

The pressure peaks measured three metres away from the middle of the test tank are shown in Fig 6, thus presenting the pressure curves obtained in all the meaningful tests evaluated. These values range over a wide area, the lower limit of which increases with increasing tank pressure. This, in turn, is attributable to the greater gas-stored compression with increasing tank pressure, and possibly also to the larger contribution of hydrogen evaporating spontaneously. Another factor is that escape flow velocity and turbulences are greater in the first few fractions of a second after ignition of the cutting charge, which improves the mixture of air and gas. Finally, faster combustion of the hydrogen/air mixture is able to generate a greater pressure load.

The results of tests 3 and 7 provide an over-proportional increase in the maximum overpressure measured with increasing tank pressure, as shown in Fig 6. The high pressure load measured in test 7 may be attributed to the rapid succession and, as a result, the aggregate effect of the cutting charge explosion, spontaneous evaporation, expansion of gas, and combustion of hydrogen. A further factor in test 3 was that due to defective insulation and the low temperature of the wall there was probably an enrichment of oxygen on the tank wall at the time of ignition contributing to the violent reaction.

It was not possible to establish any further substantial effect of factors such as the amount of hydrogen, which varied from 1.8 to 5.4 kg, or the traction of the gaseous phase which could not be measured accurately because the capacity-related level sensor worked imprecisely at high pressures.

The ignition process, one of the most important factors influencing the effects of bursting LH_2 tanks, was kept virtually unchanged by using the same cutting charge in all tests. Explosives were placed at the same spots and the mixture was ignited in all cases within the first two milliseconds after ignition of the cutting charge. Examinations [13] with LH_2/LO_2 mixtures show that ignition delayed between 200 and 800 milliseconds after the media have been able to escape has the greatest explosive effect.

The effect of pressure waves on the human body is described in the literature [14,15]. Apart from the maximum pressure occurring in an explosion, one must also consider the duration of exposure, another factor being specific frequencies of pressure curves causing damage to human organs in special cases. This latter point may nevertheless be neglected in the present context, since there were no characteristic frequencies in the tank burst tests conducted.

The maximum admissible strain tolerated by the human lungs and hearing system serves as the main factor for judging pressure waves. The maximum admissible limit for the human eardrum exposed for a period between 3 and 400 ms is stated in literature to be 345 mbar [14]. The maximum strain tolerated by the lungs, in turn, increases very significantly with a decreasing period of exposure. Fig 7 shows the threshold curves for lung damage and survival with a 70 kg man hit by a pressure wave at right angles to the longitudinal axis of his body either in an open space or - worse - directly next to a reflecting wall [15]. Substantial pressure peaks measured during the tank burst tests at a distance of three metres from the middle of the tank are shown in Fig 7.

Given the scenario applied here, any direct risk to a 70 kg man attributable to pressure effects at a distance of three metres may be ruled out almost completely. It is however somewhat problematic to directly convey the results obtained to realistic accident scenarios actually involving vehicles on the road. In such cases the effect of pressure waves might increase several times due to the blocked position of the tank in the vehicle. In addition fragments flying at high speed through the air might cause severe injuries.

All of the test tanks ruptured with a burst pressure beneath the critical pressure level of H_2 (12.9 bars) remained intact in terms of the actual structure of the two half-shells forming the tank. A tank ruptured at 14.8 bars, on the other hand, broke up into several fragments some of which were subsequently found at a distance of more than 15 metres from the original location of the tank.

The formation of fragments is attributable to the crack propagation speed increasing as a function of tank pressure.

8. CONCLUSION

The first stage of the research programme currently being conducted on the safety of LH_2 tanks for road vehicles examined the sudden, complete failure of an LH_2 tank. Considering the great stability of hydrogen tanks and the safety concept applied, such sudden and complete failure would appear to be unlikely, but should not be excluded altogether. In the immediate vicinity of such an extreme accident, destruction might be quite severe and human beings in the area could well suffer fatal injury.

The effect of damage - and in particular the effect of pressure waves - may be reduced by minimising the pressure level within the tank. Other parameters such as ignition circumstances cannot be influenced in the event of an accident or, respectively, are of lesser significance.

Involving suitable experiments, the second stage of the research programme will examine the mechanisms (internal build-up of pressure, fire, damage from outside) potentially making a tank burst, the objective being to determine the safety of modern LH_2 tanks in realistic accident scenarios and to capitalise on the concept and design-related potentials for further improvement.

REFERENCES

1. Slovic, P, Fischhoff, B, Lichtenstein, S
   Regulation of Risk: A psychological perspective, decision research report 82-11, April 1982

2. Winterfeldt, D V, John, R S, Borcherding, K:
   Cognitive components of risk ratings, risk analysis, 1, 1981, 277-287

3. Prescott-Clarke, P: Public attitudes towards industrial, work-related and other risks, social and community planning research,
   Survey Research Centre Publication, London, p 595, March 1982

4. Fritzsche, A F
   Wie sicher leben wir? (How safely do we live?)
   Rhineland Technical Inspection Authority Publishing Centre, 1986

5. Regar, K-N, Fickel, H-C, Pehr, K:
Der neue BMW 735i mit Wasserstoffantrieb
(The new hydrogen-drive BMW 735i)
VDI Reports No 725, pp 187/98, VDI Verlag, Düsseldorf 1989

6. Rüdiger, H:
Design characteristics and performance of a liquid hydrogen tank system for motor cars
Cryogenics 1992, Vol 32, No 3, pp 327-329

7. Ewald, R:
Liquid hydrogen fuelled automobiles: on-board and stationary cryogenic installations
Cryogenics 1992, Vol 30, September Supplement

8. Reister, D, Strobl, W:
Current development and outlook for the hydrogen-fuelled car
Proceedings of the 9th World Hydrogen Energy Conference
Paris, France, 22-25 June, 1992

9. Pehr, K:
Safety concept of an engine test rig with liquid hydrogen supply
Int Journal of Hydrogen Energy, Vol 18, No 9, pp 773-781, 1993

10. Vu-Han, V, Appel,H:
Zuverlässigkeit eines Airbag-Systems
(The reliability of airbag systems)
ATZ, 87, 1985

11. Anwenderbericht über den Bereitschaftsdienst der BMW AG
(User report on BMW AG`s mobile service)
Business Computer, December 1993

12. Gretz, J, Drolet, B, Kluyskens, D, Sandmann, F, Ullmann, O:
Phase II and III, 0 of the 100 MW Euro-Quebec Hydro-Hydrogen Pilot Project EQHHPP
Proceedings of the 9th World Hydrogen Energy Conference
Paris, France, 22-25 June, 1992

13. Baker, W E, Kulesz, J J, Ricker, R E, Bessey, R L, Westine, P S, Parr, V B, Oldham, G A:
Workbook for predicting pressure wave and fragment effects of exploding propellant tanks and gas storage vessels
NASA Contractor Report 134 906, 1977

14. White, C S:
The scope of blast and shock biology and problem areas in relating physical and biological parameters
Annals of the New York Academy of Sciences 152, pp 89-102, 1968

15. Bowen, I G, Fletcher, E R, Richmond, D:

Estimate of man`s tolerance to the direct effects of air blast

DASA Report 2113, 1968

FIGURES

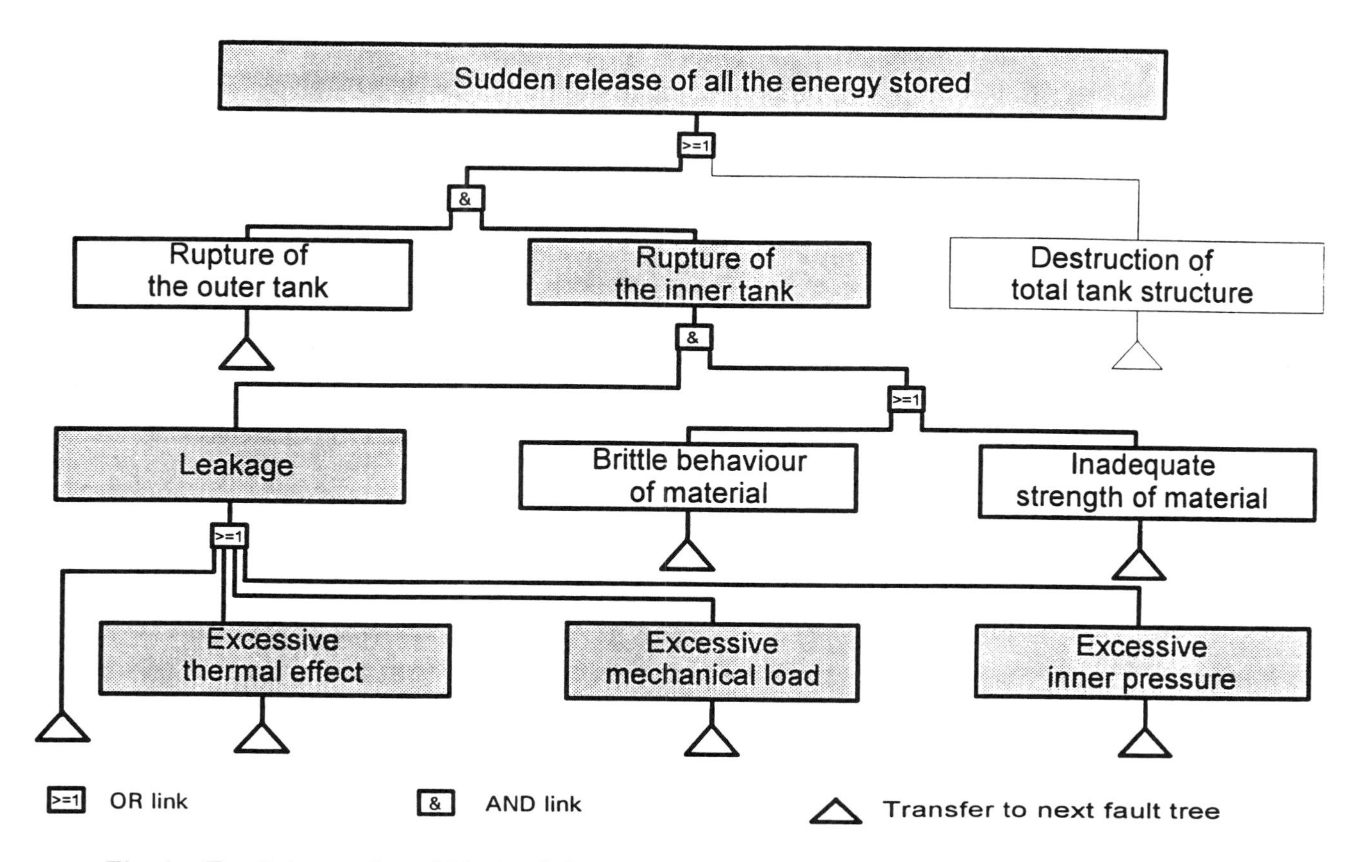

Fig.1. Fault tree of an LH_2 tank in a passenger car with "tank rupture" as the main event

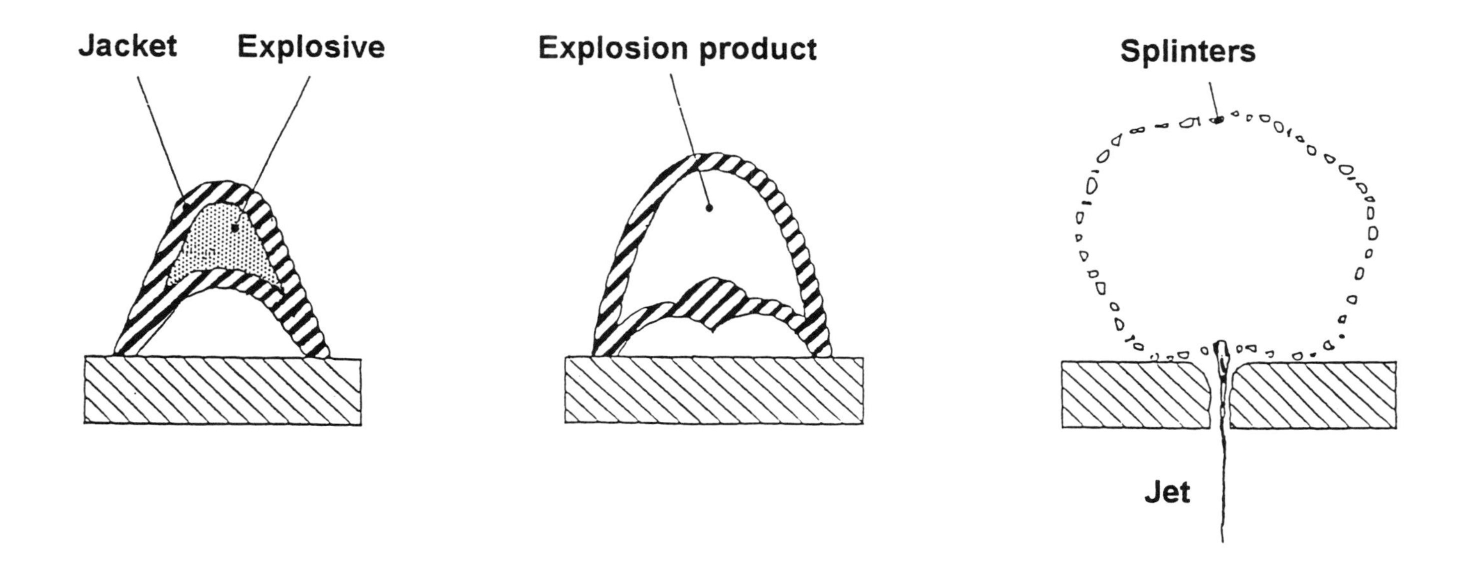

Fig.2. Jet development of a cutting charge

1. Ignition

2. 250 ms after ignition

3. 1250 ms after ignition

4. 1800 ms after ignition

Fig.3. Development of a fireball

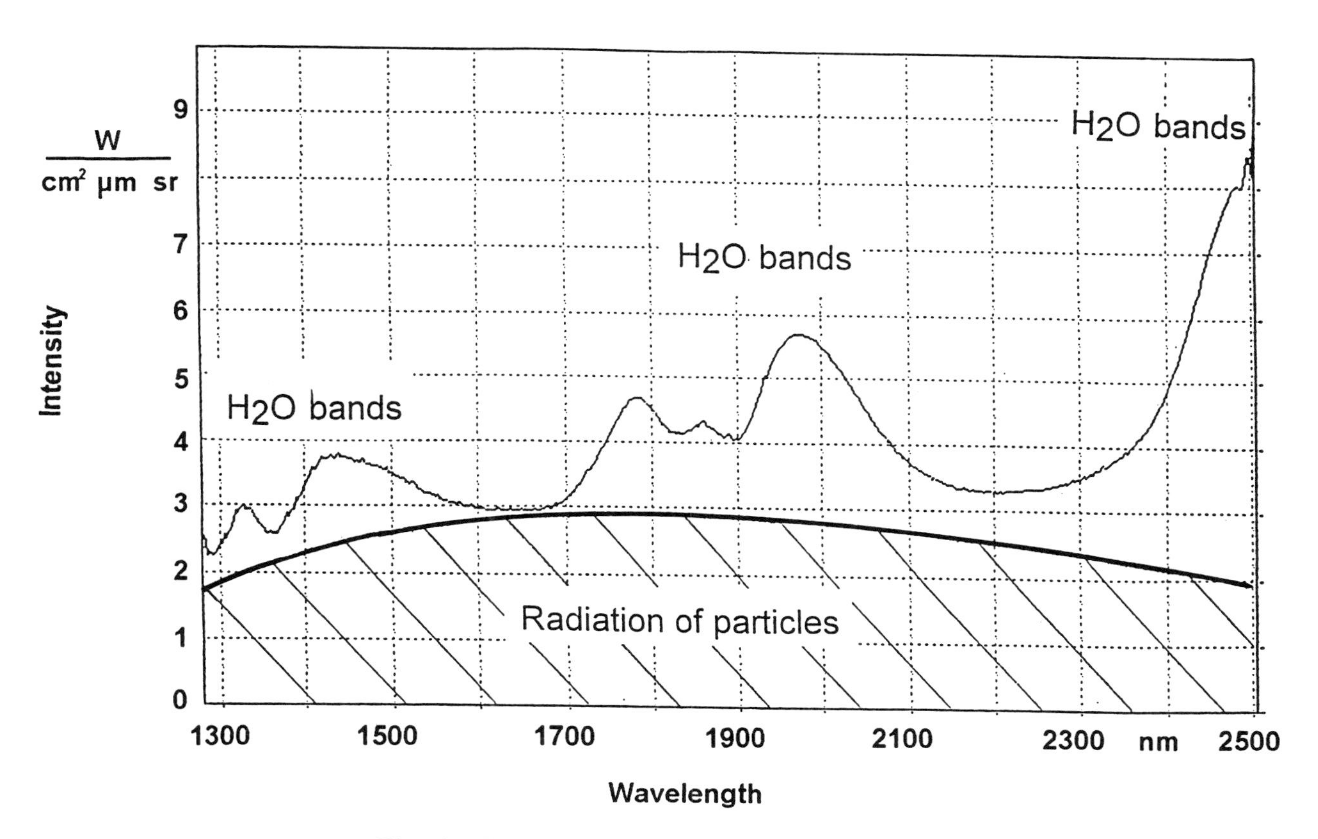

Fig.4. Extract of IR spectrum measured

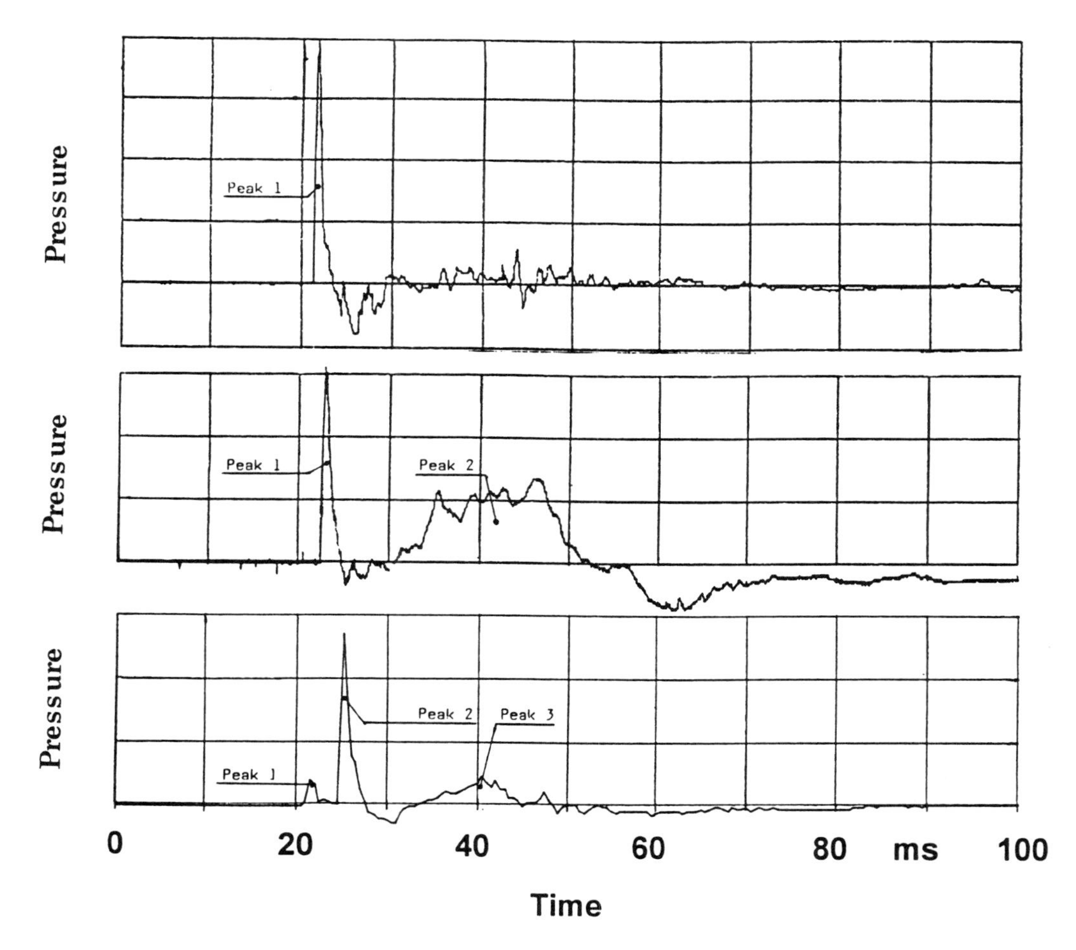

Fig.5. Energy conversion concentrated in one, two or three pressure peaks

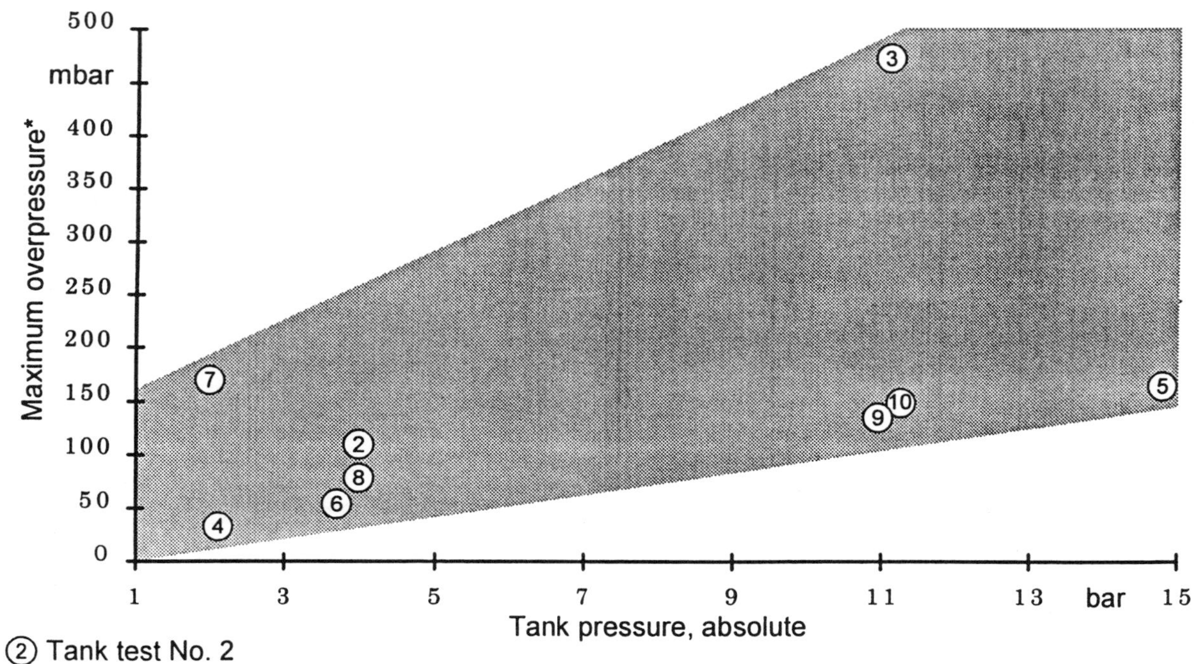

② Tank test No. 2
* Mean value of three measuring points 3 metres from the middle of the test tank

Fig.6. Amplitudes of pressure waves of tank tests No. 2 to 10

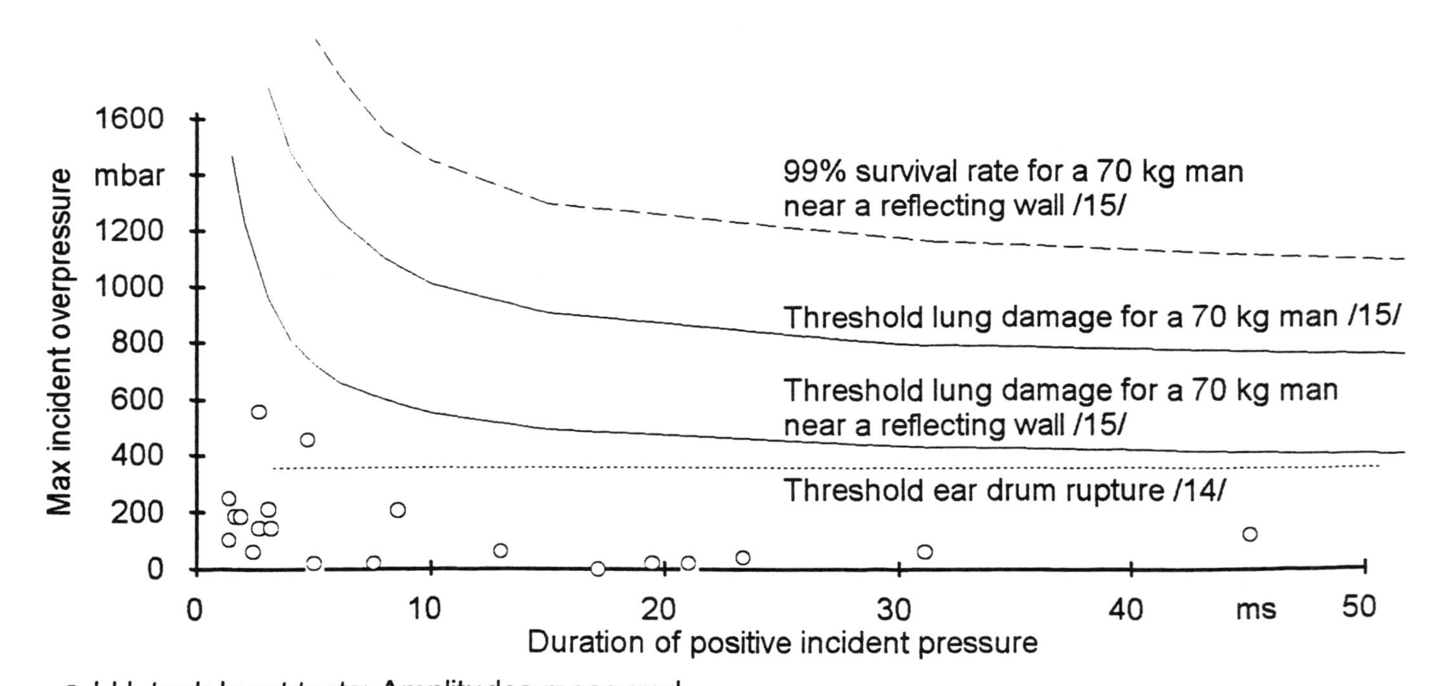

○ LH_2 tank burst tests: Amplitudes measured at a distance of 3 metres

Fig.7. Tolerance of human beings to direct effects of air blast

Is Hydrogen a Safe Fuel?

J. Hord

Cryogenics Division, Institute for Basic Standards, National Bureau of Standards

Abstract—The safety aspects of hydrogen are systematically examined and compared with those of methane and gasoline. Physical and chemical property data for all three fuels are compiled and used to provide a basis for comparing their various safety features. Each fuel is examined to evaluate its fire hazard, fire damage, explosive hazard and explosive damage characteristics. The fire characteristics of hydrogen, methane and gasoline, while different, do not largely favor the preferred use of any one of the three fuels; however, the threat of fuel-air explosions in confined spaces is greatest for hydrogen. Safety criteria for the storage of liquid hydrogen, liquefied natural gas (LNG) and gasoline are compiled and presented. Gasoline is believed to be the easiest and perhaps the safest fuel to store because of its lower volatility and narrower flammable and detonable limits. It is concluded that all three fuels can be safely stored and used; however, the comparative safety and level of risk for each fuel will vary from one application to another. Generalized safety comparisons are made herein but detailed safety analyses will be required to establish the relative safety of different fuels for each specific fuel application and stipulated accident. The technical data supplied in this paper will provide much of the framework for such analyses.

INTRODUCTION

For many years hydrogen has been considered a suitable, if not ideal, synthetic fuel for future generations. Its clean-burning, rapid-recycling characteristics are lauded by hydrogen advocates and its explosive characteristics are emphasized by hydrogen opponents. The safety aspects of hydrogen are many times uppermost in the mind of the man-on-the-street when he is first introduced to the hydrogen fuel concept. This spontaneous fear reaction is probably attributable to the automatic association of hydrogen fuel with the "hydrogen-bomb" and the Hindenburg fire. Subsequent to the Hindenburg incident the U.S.A. has experienced more tragic and devastating fires involving natural gas, yet natural gas is a fuel that is commonly accepted by the general public. Also, there is no connection whatsoever between the chemical explosive potential of hydrogen fuel and the thermonuclear explosive potential of the hydrogen isotopes as they relate to "hydrogen-bombs". Thus, even the novice will recognize that the wide-eyed fear of hydrogen is unjustified. Simultaneously we must realize that hydrogen is one of the most flammable and explosive fuels available to us and it must be handled with appropriate respect and safeguards.

Evaluation of the safety hazards of a particular fuel is a highly complex task requiring interpretation of specific technical data and intercomparisons with other fuels. Fire and explosion hazards must be carefully assessed to determine the relative safety of a fuel in each potential application. Therefore, hydrogen can be safer than conventional fuels in some applications and more hazardous in other applications. Because of the complexity and depth of this topic, it is treated rather superficially in page-limited technical articles dealing with hydrogen-energy concepts. In this paper we present the condensed results of a comprehensive document† prepared to systematically examine the safety aspects of hydrogen and to determine if hydrogen is sufficiently safe for use as a future fuel. It will be demonstrated herein that the answer is overwhelmingly, YES, although its use may be restricted in some future applications.

GENERAL PHYSICAL AND CHEMICAL PROPERTIES

To permit insight and provide perspective for the relative safety of hydrogen, comparative data are given in Table 1 for methane and gasoline. Methane is the major constituent of most compressed natural gases and of liquefied natural gas (LNG); therefore, methane and LNG are used interchangeably in this document. The significance of the technical data, listed in Table 1, is discussed in considerable detail throughout the remainder of this paper. These data were obtained from numerous

† See Reference [65].

TABLE 1. Properties of hydrogen*, methane and gasoline**

Property	Hydrogen	Methane	Gasoline
Molecular weight	2.016 [1]	16.043 [5]	~107.0 [7]
Triple point pressure, atm	0.0695 [1]	0.1159 [5]	---
Triple point temperature, K	13.803 [1]	90.680 [5]	180 to 220[1] [7]
Normal boiling point (NBP) temperature, K	20.268 [1]	111.632 [1]	310 to 478 [7,8]
Critical pressure, atm	12.759 [1]	45.387 [5]	24.5 to 27 [7]
Critical temperature, K	32.976 [1]	190.56 [5]	540 to 569 [7]
Density at critical point, g/cm^3	0.0314 [1]	0.1604 [5]	0.23 [9]
Density of liquid at triple point, g/cm^3	0.0770 [1]	0.4516 [5]	---
Density of solid at triple point, g/cm^3	0.0865 [1]	0.4872 [1]	---
Density of vapor at triple point, g/m^3	125.597 [1]	251.53 [5]	---
Density of liquid at NBP, g/cm^3	0.0708 [1]	0.4226 [1]	~.70[2] [7]
Density of vapor at NBP, g/cm^3	0.00134 [1]	0.00182 [1]	~0.0045[2] [9]
Density of gas at NTP, g/m^3	83.764 [1]	651.19 [1]	~4400 [9]
Density ratio: NBP liquid-to-NTP gas	845 [1]	649 [1]	156[3] [7,9]
Heat of fusion, J/g	58.23 [1]	58.47 [1]	161 [12]
Heat of vaporization, J/g	445.59 [1]	509.88 [1]	309 [7,9]
Heat of sublimation, J/g	507.39 [1]	602.44 [1]	---
Heat of combustion (low), kJ/g	119.93 [2]	50.02 [2]	44.5 [2,4,9]
Heat of combustion (high), kJ/g	141.86 [2]	55.53 [2]	48 [2,4,9]
Specific heat (C_p) of NTP gas, J/g-K	14.89 [1]	2.22 [1]	1.62[2] [9]
Specific heat (C_p) of NBP liquid, J/g-K	9.69 [1]	3.50 [1]	2.20[2] [9]
Specific heat ratio (C_p/C_v) of NTP gas	1.383 [1]	1.308 [1]	1.05[2] [7]
Specific heat ratio (C_p/C_v) of NBP liquid	1.688 [1]	1.676 [1]	---
Viscosity of NTP gas, g/cm-s	0.0000875 [1]	0.000110 [1]	0.000052 [9]
Viscosity of NBP liquid, g/cm-s	0.000133 [1]	0.001130 [1]	0.002 [9]
Thermal conductivity of NTP gas, mW/cm-K	1.897 [1]	0.330 [1]	0.112 [9]
Thermal conductivity of NBP liquid, mW/cm-K	1.00 [1]	1.86 [1]	1.31 [9]
Surface tension of NBP liquid, N/m	0.00193 [1]	0.01294 [1]	0.0122 [9]
Dielectric constant of NTP gas	1.00026 [3]	1.00079 [1]	1.0035[5] [10]
Dielectric constant of NBP liquid	1.233 [3]	1.6227 [1]	1.93[4] [9]
Index of refraction of NTP gas	1.00012 [1]	1.0004 [1]	1.0017[5] [10]
Index of refraction of NBP liquid	1.110	1.2739 [1]	1.39[4] [9]
Adiabatic sound velocity in NTP gas, m/s	1294 [1]	448 [1]	154
Adiabatic sound velocity in NBP liquid, m/s	1093 [1]	1331 [1]	1155[6] [12]
Compressibility factor (Z) in NTP gas	1.0006 [1]	1.0243 [1]	1.0069
Compressibility factor (Z) in NBP liquid	0.01712 [1]	0.004145 [1]	0.00643[2]
Gas constant (R), cm^3-atm/g-K	40.7037 [1]	5.11477 [1]	0.77
Isothermal bulk modulus (α) of NBP liquid, MN/m^2	50.13 [3]	456.16 [6]	763[6] [11]
Volume expansivity (β) of NBP liquid, K^{-1}	0.01658 [3]	0.00346 [6]	0.0012[2] [9]

TABLE 1. Properties of hydrogen, methane and gasoline (*continued*)

Property	Hydrogen	Methane	Gasoline
Limits of flammability in air, vol. %	4.0 to 75.0 [13,14]	5.3 to 15.0 [13,14]	1.0 to 7.6 [8,18]
Limits of detonability in air, vol. %	18.3 to 59.0 [15]	6.3 to 13.5 [19]	1.1 to 3.3 [20]
Stoichiometric composition in air, vol. %	29.53	9.48	1.76
Minimum energy for ignition in air, mJ	0.02 [16]	0.29 [16]	0.24 [16,21]
Autoignition temperature, K	858 [17]	813 [8]	501 to 744 [8,18]
Hot air-jet ignition temperature, K	943 [22]	1493 [22]	1313[7] [22]
Flame temperature in air, K	2318 [23]	2148 [23]	2470 [24]
Percentage of thermal energy radiated from flame to surroundings, %	17 to 25 [25,26]	23 to 33 [25,26]	30 to 42 [25,26]
Burning velocity in NTP air, cm/s	265 to 325 [27,28]	37 to 45 [18,19]	37 to 43 [18,24]
Detonation velocity in NTP air, km/s	1.48 to 2.15 [29,35]	1.39 to 1.64 [19]	1.4 to 1.7[8] [30]
Diffusion coefficient in NTP air, cm^2/s	0.61	0.16	0.05
Diffusion velocity in NTP air, cm/s	$\lesssim$2.00	$\lesssim$0.51	$\lesssim$0.17
Buoyant velocity in NTP air, m/s	1.2 to 9	0.8 to 6	nonbuoyant
Maximum experimental safe gap in NTP air, cm	0.008 [31]	0.12 [32]	0.07 [31]
Quenching gap in NTP air, cm	0.064 [16,33]	0.203 [16,33]	0.2 [33]
Detonation induction distance in NTP air	$L/D \approx 100$ [35,36]	---	---
Limiting oxygen index, vol. %	5.0 [34]	12.1 [34]	11.6[9] [34]
Vaporization rates (steady state) of liquid pools without burning, cm/min	2.5 to 5.0 [26]	0.05 to 0.5 [26]	0.005 to 0.02
Burning rates of spilled liquid pools, cm/min	3.0 to 6.6 [25,26]	0.3 to 1.2 [25,26]	0.2 to 0.9 [25,26]
Flash point, K	gaseous	gaseous	~230 [8]
Toxicity	nontoxic [37] (asphyxiant)	nontoxic [37] (asphyxiant)	slight [37] (asphyxiant)
Energy[10] of explosion, (g TNT)/(g fuel)	~24	~11	~10
Energy[10] of explosion, (g TNT)/(cm^3 NBP liquid fuel)	1.71	4.56	7.04
Energy[10] of explosion, (kg TNT)/(m^3 NTP gaseous fuel)	2.02	7.03	44.22

NBP = Normal boiling point.

NTP = 1 atm and 20 C (293.15 K).

* Thermophysical properties listed are those of parahydrogen.

** Property values are the arithmetic average of normal heptane and octane in those cases where "gasoline" values could not be found (unless otherwise noted).

[1] Freezing temperatures for gasoline @ 1 atm.

[2] @ 1 atm and 15.5 C.

[3] Density ratio @ 1 atm and 15.5 C.

[4] @ 1 atm and 20 C.

[5] @ 1 atm and 100 C.

[6] @ 1 atm and 25 C.

[7] Based on the properties of butane.

[8] Based on the properties of n-pentane and benzene.

[9] Average value for a mixture of methane, ethane, propane, benzene, butane and higher hydrocarbons.

[10] Theoretical explosive yields.

sources and are believed to be the best available. Some of the data given in Table 1 were obtained by giving weighted consideration to several sources of data and by performing appropriate computations.

Most of the properties listed in Table 1 will be familiar to the average reader; however, some of the combustion properties will be briefly described in the following paragraphs because of their importance to this safety analysis or because they are not commonly used properties. The thermophysical properties conform with the conventionally accepted definitions; however, explanatory notes are provided at the bottom of Table 1 to explain the bases for some of the "gasoline" properties. The heats of fusion and sublimation are taken at the triple point (at the freezing point for gasoline) and the heats of vaporization are taken at the normal boiling point.

Hot air-jet ignition temperature

The temperature of a jet of hot air as it enters pure fuel vapors or a combustible fuel–air mixture at N.T.P. and causes ignition to occur. The data given in Table 1 represent the jet temperature of hot air as it enters pure fuel vapors at N.T.P. The jet diameter for these data is 0.4 cm. This ignition temperature decreases with increasing jet diameter [22, 38, 39] and for a given jet diameter the hot gas jet ignition temperature increases if hot jets of nitrogen gas (rather than air) are squirted into combustible fuel–air mixtures [38]. The hot gas jet ignition temperature is dependent upon the composition of the combustible mixture and the velocity of the jet of hot gas.

Percentage of thermal energy radiated from flame to surroundings

The percentage of the heat of combustion (high) that is radiated from the combustion zone to its surroundings. The higher heating value of every flame is eventually dissipated by radiative processes. The data given in Table 1 are for flames fueled by vaporization of pools of liquid fuels in an air environment. These data are similar to those obtained in laboratory experiments with stationary gaseous diffusion flames [25, 26]. Atmospheric moisture absorbs thermal energy radiated from a fire and can reduce the values recorded in Table 1. Hydrogen fires benefit most from this absorption effect [25], e.g. it is estimated that 45% of radiant hydrogen flame energy is absorbed within a distance of 8 m in 25°C air containing water vapor at 15 mmHg partial pressure.

Diffusion velocity in N.T.P. air

The velocity at which a gaseous fuel diffuses through air. For a specified fuel concentration gradient the diffusion velocity is proportional to the diffusion coefficient and can be estimated from Stefan's equation [40]. Diffusion velocity varies with temperature according to $T^{3/2}$ and consequently low temperature gases produced by cryogenic liquid fuel spills will diffuse more slowly than N.T.P. fuel gases. The values recorded in Table 1 are based on N.T.P. fuel gas and N.T.P. air densities and fuel concentrations that vary from 99.99% to 0.0% over path lengths of 3 cm to 30 m.

Buoyant velocity in N.T.P. air

The velocity at which gaseous fuels rise in air under the influence of buoyant forces. This velocity cannot be determined in a direct manner as it is dependent upon drag and friction forces that oppose buoyant forces acting on the rising volume of gaseous fuel. Atmospheric turbulence as well as shape and size of the rising volume of gas can affect the terminal velocity of the buoyant gas. Buoyant forces are related to the difference in air and fuel densities; therefore, cold, dense fuel gases produced by cryogenic fuel spills will rise more slowly than NTP fuel gases. The buoyant velocities recorded in Table 1 were estimated from fundamental principles of dynamics and with the aid of empirical data [41, 42]; it was assumed that the radii of the buoyant masses of N.T.P. fuel gas varied from 3 cm to 1.5 m.

Maximum experimental safe gap (MESG) in N.T.P. air

The maximum permissible clearance, between flat parallel steel surfaces, that prevents the propagation of dangerous flames or sparks through the gap. The MESG is measured by igniting a combustible fuel air mixture inside of a test enclosure and observing a similar combustible mixture surrounding the enclosure to detect its ignition. The MESG is the largest gap size that does not permit ignition outside of the test enclosure and is of vital importance to the design and manufacture of explosion-proof equipment.

Quenching gap in N.T.P. air

The spark gap between two flat parallel plate electrodes at which ignition of combustible fuel–air mixtures is suppressed, i.e. smaller gaps have the effect of totally suppressing spark ignition and flame propagation.

Detonation induction distance in N.T.P. air

The distance required for a deflagration to transit to a detonation in a detonable fuel–air mixture. This distance is usually experimentally determined in a long cylindrical tube with a spark or hot wire ignitor on one end of the tube. The tube is instrumented along its length to sense the velocity of the flame front as it propagates through the detonable mixture of gases. The distance from the ignitor to the axial position in the tube where the flame front first attains the detonation velocity is reported as the induction distance. This distance is dependent upon the combustible mixture constituents, the pressure, temperature and concentration of the gaseous mixture, the enclosure geometry [43] and strength of the ignition source [18]. A deflagration is a low order explosion resulting from subsonic flame speed, relative to the unburned gas. It is conventionally defined as a propagating reaction in which the energy transfer from the reaction zone to the unreacted zone is achieved through ordinary rate-limiting transport processes such as heat and mass transfer. A detonation is a high order explosion resulting from supersonic flame speed, relative to the unburned gas. It may be defined as a propagating reaction in which energy is transferred from the reaction zone to the unreacted zone on a reactive shock wave.

Limiting oxygen index

The minimum concentration of oxygen that will support flame propagation in an unknown mixture of fuel, air and nitrogen, e.g. no mixture of hydrogen, air and nitrogen at N.T.P. conditions will propagate flame if the mixture contains less than 5.0 vol. % oxygen [14]. Use of diluents other than nitrogen results in different values for the limiting oxygen index of each fuel [14, 34].

Vaporization rates of liquid pools without burning

The rate at which the liquid level decreases after a pool of liquid fuel has been formed by spilling fuel onto a warm surface such as sand or soil. These evaporation rates are measured after subsidence of the violent boiling that accompanies the initial liquid spill. Vaporization rates of the cryogenic fuels can be expected to vary widely with the conductivity and heat capacity of the soil or other material confining the spilled liquid fuel. In the case of gasoline, vaporization rates will vary with the volatility of constituents (blend), age, fuel temperature, ground surface texture and temperature, etc. Wind velocity influences the vaporization rate of all fuels considered herein.

Burning rates of spilled liquid pools

The rate at which the liquid level decreases after a pool of liquid fuel has been formed by spilling fuel onto a warm surface and the resultant vapor–air mixture has been ignited. Again, these burning rates are measured after the initial-spill violent boiling has subsided and the vapor is mixing and burning in air above the pool of spilled fuel. These burning rates may also be obtained by adding the vaporization rate (without burning) and the liquid level regression rate attributable to the burning of vapors in the open air over liquid fuels that are contained in open-mouthed insulated vessels. Steady state burning rates increase with liquid pool diameter while vaporization rates continuously decrease with time, irrespective of pool size. Burning rates can be expected to vary with pool diameter and wind velocity [25, 26, 44].

Energy of explosion

The theoretical maximum energy available from a chemical explosion. This maximum energy release is determined by computing the isothermal decrease in the Helmholtz free-energy function. Explosive energies listed in Table 1 are expressed in terms of equivalent quantities of TNT (symmetrical trinitrotoluene) and may be converted directly to energy units by multiplying by 4602 J/(g TNT).

FIRE HAZARDS

By considering fire hazards, fire damage, explosive hazards and explosive damage attributable to each fuel we can expose the relative safety merits of each fuel for potential fuel applications. By definition, a hazard identifies a pending risk or peril. Of course, the existence of a hazard does not assure the occurrence of a fire or explosion or that damage will be sustained if a fire or explosion does occur. Thus, the safety hazards and damage potential of candidate fuels must be compared to provide the requisite safety criteria for fuel selection. The degree or extent of hazard and damage potential are frequently difficult to express in indisputable scientific terms. Consequently, scientific data are usually tempered with experienced judgment to formulate safety evaluations for different fuels. Hence, the perpetual controversy over the relative safety of fuels.

By referring to the combustion properties listed in Table 1 we can systematically step through the fire hazards of hydrogen, methane and gasoline. First we will consider the accidental or inadvertent means of obtaining flammable mixtures of fuel in air. Usually, such mixtures are the result of fuel leakage or spillage which may be attributable to mechanical failure of equipment, material failure, erosion, physical abuse, improper maintenance, collision, etc.

The rate at which the fuel vapors mix with air is indicated by their diffusion velocities and their buoyant velocities. The buoyant effect is dominant for hydrogen and methane and from the data listed in Table 1 it is apparent that hydrogen can be expected to mix with air more rapidly than methane or gasoline—the latter is obviously the slowest mixing fuel of the three fuels considered. In the event of a fuel spill, one could expect hydrogen to form combustible mixtures more rapidly than methane because hydrogen has a higher buoyant velocity and a slightly lower flammable limit. Again, gasoline would be orders of magnitude slower than hydrogen or methane in forming combustible mixtures in air. In some fuel applications these relative mixing times may be important while in others they are meaningless, e.g. an instantaneous fire hazard exists for the impact rupture of an auto fuel tank irrespective of the type of fuel carried.

Because of their higher buoyant velocities, hydrogen and methane can also be expected to disperse more rapidly than gasoline and thus shorten the duration of the flammable hazard. Even though the upper flammable limit (UFL) of hydrogen is much higher than that of methane the higher buoyant velocity of hydrogen permits it to disperse to concentrations below the lower flammable limit (LFL) more rapidly [45] than methane. Thus, one could expect a fire hazard to exist most readily with hydrogen, methane and gasoline, respectively, and to persist in the inverse order.

We must exercise some caution in analyzing fuel mixing and dispersion rates by comparing relative buoyant and diffusion velocities of the N.T.P. fuel gases. In large cryogenic liquid fuel spills, the vaporization of liquid and warming of the vapor can cool large masses of air. In addition, NBP hydrogen vapor density approaches that of N.T.P. air while NBP methane vapor density is greater than the density of N.T.P. air. Consequently, for some finite period of time these cold vapor–air mixtures are nonbuoyant and may extend to appreciable distances from the spill. Therefore, both the range and duration of the fire hazard may be extended somewhat when cryogenic liquid fuels are spilled. More definitive experimental data are needed in this area to supplement existing knowledge [25, 26, 46]. Some experimental data and analyses for LNG spills are available [44, 46]; these data treat the dispersion and drift characteristics of vapor clouds that form over LNG spills.

The low value of limiting oxygen index reflects the high value of the UFL for hydrogen–air mixtures. The wide flammability limits of hydrogen are of practical significance only when fuel leakage into enclosed spaces is a major concern. In this case the flammable limits of hydrogen are sufficiently wide to enhance the probability of combustion from a random ignition source. This flammability characteristic should not preclude the use of hydrogen because the LFL is the vital one in most applications. The LFL is important because ignition sources are nearly always present when a leaking fuel first reaches combustible proportions in air.

The rate of vapor generation and burning over spilled liquid pools is of interest for the various fuels. As indicated in Table 1 the volumetric vaporization rates and burning rates are highest for hydrogen, methane and gasoline, respectively. Consequently, for a given liquid spillage volume, gasoline fires will last the longest and hydrogen fires the shortest while all of the fuels burn at about the same flame temperature. The thermal energy radiated from these pool-fed fires may be computed by multiplying the appropriate (burning rate) × (NBP liquid density) × (high heat of combustion) × (percentage of thermal energy radiated from the flame to its surroundings) using the data given in Table 1. These computations indicate that the radiated thermal energy should not exceed 276 W/cm^2 of pool liquid–vapor surface area for hydrogen, 155 W/cm^2 for methane or 212 W/cm^2 for gasoline. Then, the scene of a hydrogen fire may be hotter than that of a hydrocarbon fire, but the hydrocarbon fires will last five to ten times longer than hydrogen fires (for equivalent fuel spillage volumes). Additional data for LNG vaporization and burning rates and for radiant heat flux emitted from fires over LNG pools are presented in [44].

Another matter of concern in evaluating fire hazards is the rate of leakage flow of liquids or gases through leakage paths of varying geometries, e.g. leakage through cracked welds, improperly mated flanges, threaded fittings, damaged seals, etc. Previous work [47] has shown that volumetric leakage rates are either inversely proportional to the square root of the density, or inversely proportional to the absolute viscosity of the leaking fluid. Since care is usually taken to minimize fuel leaks the viscous leakage flows are considered to have the greatest practical significance. Using the data given in Table 1 we can estimate the relative volumetric leakage rates of the NBP liquid fuels and of the gases at N.T.P. We find that liquid hydrogen is much more difficult to contain than liquid methane or liquid gasoline and that N.T.P. gasoline vapors are more difficult to contain than either N.T.P. hydrogen or N.T.P. methane. Industry has proven that all of these fuels can be safely and easily contained.

Techniques for detection of hydrogen leakage are effectively summarized by two documents [48, 49] and general procedures for leak testing are detailed in the handbook by Marr [48].

The minimum spark energy required for ignition of hydrogen in air is about an order of magnitude less than that for methane or gasoline; however, the ignition energy for all three fuels is sufficiently low that ignition is relatively assured in the presence of thermal (weak) ignition sources, e.g. sparks, matches, hot surfaces or open flames. Even a weak spark due to the discharge of static electricity from a human body may be sufficient to ignite any of these fuels in air—10 mJ sparks may be produced in such electrostatic discharges [50].

Although hydrogen has a higher auto-ignition temperature than methane or gasoline, its low ignition energy characteristic makes it more readily ignitable than either of the hydrocarbon fuels. The hot air-jet ignition temperature is highest for methane and lowest for hydrogen; therefore, hydrogen is easiest to ignite by jets of hot combustion products emitted from an adjacent enclosure. The flash point is meaningless for the cryogenic fuels, hydrogen and methane, within the temperature range of interest because these fuels will flash at all temperatures above their normal boiling points. The boiling points of the cryogenic liquid fuels are so low that these fuels are considered to behave like gases. The flash point of gasoline is also well below room temperature; therefore, all three fuels must be considered volatile and will generate sufficient vapor to create a fire hazard at earth surface temperatures. Then, all three fuels are relatively easy to ignite. Hydrogen is most susceptible while methane and gasoline appear to be equally susceptible to ignition.

The burning velocity is a fundamental property of a combustible gas mixture and should not be confused with the flame speed [51]. The burning velocity influences the severity of the explosion and along with quenching gap is important in the design of flame arresters [51]. Higher burning velocities indicate a greater tendency for the combustible gas mixture to support the transition from deflagration to detonation in long tunnels or pipes. In general, faster-burning gases have smaller quenching gaps and flame arresters for faster-burning gases must have smaller apertures [51]. The quenching gap is the passage gap dimension required to prevent propagation of an *open flame* through a flammable fuel–air mixture that fills the passage and it is clearly distinguishable from the MESG. The latter is the maximum permissible clearance between flanges to assure that an *explosion* does not propagate from within an enclosure to a flammable mixture surrounding the enclosure. Because of the high explosion pressures the MESG is always smaller than the quenching distance.

Available data [52] indicate that the pressure rise ratios for adiabatic combustion of stoichiometric mixtures of hydrogen–air and methane–air in closed vessels are nearly identical. Similar data [31] produced pressure rise ratios for hydrogen–air that were 20 to 40% higher than those for gasoline–air. In long tubes or tunnels, hydrogen–air mixtures will transit to detonation more rapidly than methane–air or gasoline–air mixtures; therefore, overpressure hazards in confined spaces are enhanced in hydrogen systems. Thus, the high burning velocity of hydrogen is an indication of its high explosive potential and of the difficulty of confining or arresting hydrogen flames and explosions. Industrial equipment is currently available to safely confine hydrogen explosions but hydrogen pipeline flame arresters are not yet considered reliable [53].

FIRE DAMAGE

Some of the combustion characteristics discussed in the previous section are also applicable to comparisons of fire damage by the different fuels. Similarly, many of these characteristics are applicable to explosive hazard and explosive damage criteria, as will be revealed in subsequent sections of this paper. In the previous section we discussed the risk or likelihood of having a fire. In this section we examine the damage potential of a fire.

The main fire damage parameters are thermal radiation, flame engulfment (fireball), smoke inhalation, fire detection and extinguishment. Explosion overpressure, impulse and shrapnel damage are related parameters that are reserved for detailed discussion under the section entitled explosive damage. Thermal radiation characteristics of the three fuels under consideration have already been discussed so we turn our attention to fireballs.

Fireball damage is the direct result of combustion of materials initiated through contact or engulfment by flames that are consuming fuels. A fireball may result from the ignition of fuel–air mixtures or from the explosion of solid or chemical explosives. The explosion fireball is short-lived but flames will persist until all of the fuel is consumed in fuel–air fires. Ignition of fuel–air mixtures above pools of spilled liquid fuels produces flames with dimensions that vary with the volume of spilled liquid, rate of spillage, nature of spillage containment surface, wind velocity, location of ignition source and time delay before ignition. A simple mathematical expression [54] seems to adequately predict the maximum equivalent spherical radii of fireballs for a wide variety of explosives including hydrogen–air and rocket bipropellants. The diameter of the fireball is given by

$$D \approx 7.93\ W_f^{1/3},$$

where D = diameter in meters and W_f = weight of fuel in kilograms. The fireball duration may be estimated from

$$t \approx 0.47\ W_f^{1/3},$$

where t is in seconds and W_f is the weight of fuel in kilograms. See [54] for limitations of these approximations. Various fireball and fire radiation criteria are summarized along with overpressure criteria on Figs 1 and 2. The development and discussion of these figures has been treated elsewhere [54]

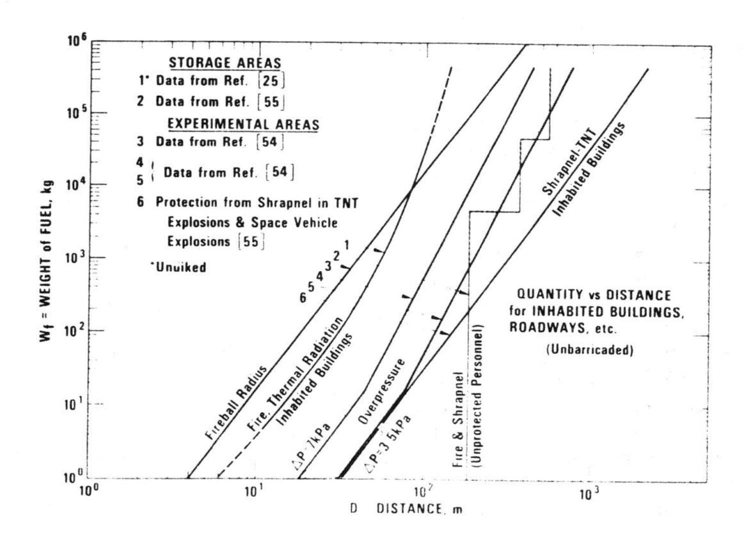

FIG. 1. Quantity distance relationships for the protection of personnel near liquid hydrogen storage and experimental areas (unbarricaded).

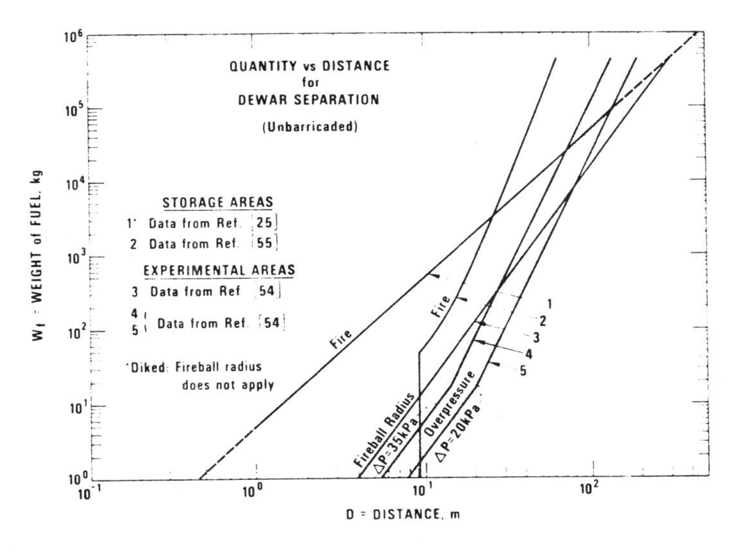

Fig. 2. Quantity-distance relationships for liquid hydrogen dewar separation in storage and experimental areas (unbarricaded).

and will not be repeated here. Although these figures were developed specifically for hydrogen they are believed to provide conservative safety criteria for methane and gasoline. It must be emphasized that these figures are based upon potential fire and explosion hazards in unbarricaded storage and experimental areas and are not indicative of industrial storage standards for fuels. Industrial hydrogen storage standards are much less stringent—see Fig. 3.

Smoke inhalation is one of the major causes of injury and death in any fire. When fuel–air fires cause buildings and other combustible materials to burn, smoke inhalation is of concern for hydrogen, methane and gasoline. When fuels burn in the open air only gasoline can cause severe smoke inhala-

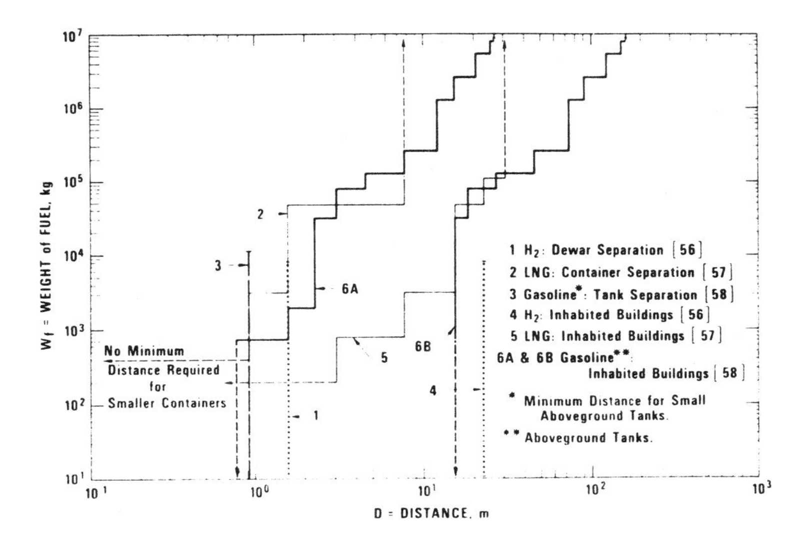

Fig. 3. Industrial storage standards (quantity–distance) for hydrogen, liquefied natural gas (LNG) and gasoline.

tion damage as both hydrogen and methane are clean burning fuels. Inhalation of the combustion products from hydrogen–air or methane–air fires is considered less serious because both of these fuels are buoyant and require a large influx of fresh air to sustain the fire. Also, the combustion products of hydrogen–air fires (mainly nitrogen and water vapor) and methane–air fires (mainly nitrogen, carbon dioxide and water vapor) are not foreign to the lungs as are the sooty combustion products of gasoline–air fires. Of course, the lungs can be seared by breathing hot combustion gases produced by burning any of these fuels. Breathing any of these fuels or their combustion products in sufficiently rich concentrations can also cause asphyxiation. The physiological effects of breathing aerosols, toxic gases (such as CO and CO_2), hot gases and oxygen-deficient gases are reviewed by Custer and Bright [59].

Current fire detection technology is summarized in a recent document [59] and hydrogen fire detection is reviewed in an older publication [49]. Hydrogen flames are nearly invisible in daylight but their visibility is improved by the presence of moisture and/or impurities in the air. Hydrogen fires are readily visible in the dark or in subdued light and large hydrogen fires are quite detectable in daylight because convective "heat ripples" are visible in the air at near-range distances and thermal radiation heats the skin. Small hydrogen fires are more difficult to detect and require that certain precautions be taken to avoid personnel and equipment damage. Methane flames, though clean-burning, are yellowish in color and quite visible in daylight. Gasoline flames are similar to those of methane but are mixed with large volumes of soot and smoke so that fire detection is obvious.

Two types of sensors, thermal and optical, are used to detect hydrogen fires. Thermal sensors are the conventional type and are fully discussed by Custer and Bright [59]. These conventional sensors (including smoke and ionization detectors), coupled with flame visibility, are adequate for detection of methane or gasoline fires and not quite adequate for detection of all hydrogen fires. The aerospace industry has advanced the use of optical sensors for detecting hydrogen fires in bright-field environments. The most common optical sensors detect ultraviolet or infrared radiation and several detection schemes exist [49, 59]. Closed-circuit infrared and ultraviolet television sets, equipped with appropriate filters, have been successfully used to detect hydrogen fires on rocket engine test stands [49]. Intumescent paints have also been used to detect hydrogen fires. These paints char and swell at low temperature ($\sim 200\,°C$) and emit pungent gases. Hydrogen fires are obviously more difficult to detect than methane or gasoline fires; however, the availability of modern detection equipment makes it possible to quickly and reliably detect the flames of all three fuels.

A brief review of fire extinguishment methods and recommended fire extinguishing procedures for various combustible materials has been published by the NFPA [8]. Class B extinguishing agents [8] are generally suited for gasoline fires and water deluge, or water spray, are usually useful in fighting gasoline, methane or hydrogen fires. The water is used to cool and protect adjacent exposed combustibles and may not extinguish the fire unless it is used in a prescribed manner by skilled personnel. Water may be particularly ineffective in extinguishing liquid gasoline fires as the gasoline is less dense than water and will float on top of the water and continue to burn.

It is sometimes advisable to permit hydrogen and methane fires to burn until gas-flow is stopped or liquid-spills are consumed because of the potential explosive hazard created by extinguishing such flames. If the fuel source is neither depleted nor shut off, an explosive fuel–air mixture may be formed with far greater damage potential (if ignited) than the original fire. In those instances where extinguishment is judged imprudent, a water deluge may be used to cool surrounding combustibles and control fire damage.

Recent experiments [44] have evaluated the effectiveness of commercially available dry chemical agents and high-expansion foams in controlling and extinguishing LNG pool fires. These experimental results show that dry chemicals can be used to extinguish LNG fires and foams applied to the pool surface reduce the radiant heat flux to surroundings while reducing vapor evolution. Thus, fire control and fire extinguishment methods and equipment are commercially available to combat LNG fires.

It is anticipated that these same fire-fighting procedures would be effective in controlling liquid hydrogen fires but no such data exist. Inert gas-flooding and CO_2 extinguishers have been successfully used to extinguish gaseous hydrogen fires.

Thus, we may conclude that: (1) water deluge or water spray is useful in fighting hydrogen, methane or gasoline fires, (2) gasoline or methane (LNG) fires may be controlled or extinguished using commercial dry chemical or high-expansion foam agents and (3) the effectiveness of dry chemicals or foams in controlling or extinguishing liquid hydrogen fires has not been evaluated.

EXPLOSIVE HAZARD

The ignition of a combustible mixture of fuel–air can result in a fire or an explosion. An explosion is always accompanied by a fireball and a pressure wave (overpressure). The fireball may ignite surrounding combustibles or fuel released by the explosion so that a fire may follow an explosion. If the fuel–air mixture is partially or totally confined the explosion may propel fragments of the enclosure material at high velocities over great distances. By its nature an explosion hazard constitutes fire, overpressure, impulse and shrapnel hazards. The extent of overpressure, impulse (overpressure-force multiplied by the time interval of explosive overpressure) and shrapnel hazards is dependent upon the severity of the explosion. Detonations cause more damage than deflagrations.

The same combustion parameters that influence fire hazards also influence the explosive hazards associated with each fuel; therefore the discussion on fire hazards is equally applicable to this section on explosive hazards. Two of the combustion properties previously reviewed, MESG and hot air-jet ignition temperature, are vitally important to the containment of an explosion and to prevent propagation of the explosion to explosive mixtures of fuel–air surrounding the enclosure. The containment vessel must be sufficiently strong to withstand the explosion pressure without emitting jets of combustion products that are larger or hotter than those specified by the "hot air-jet ignition temperature" and simultaneously the MESG must not be exceeded.

The limits of detonability are important in an evaluation of the explosive hazards of fuel–air mixtures. The wider these limits, the greater the probability of a high-order explosion with attendant high overpressures and severe shrapnel hazards—detonation. The flammable limits define the fuel–air concentrations that will burn and low-order explosions may occur within these limits. Such explosions are called deflagrations and they result in lower overpressures and less shrapnel hazard than those associated with detonations. In order to have a fire or an explosion there must exist in combination an oxidant, a fuel and an ignition source. The fuel and oxidizer are supplied and mixed by the release of fuel into the air. Hydrogen has by far the widest limits of detonability of the three fuels considered herein; therefore, it presents the greatest hazard to explosion damage. The explosive potential of all three fuels is discussed in the next section.

The ignition source may be a mechanical or electrostatic spark, flame, impact, heat by kinetic effects, friction, chemical reaction, etc. The strength of the ignition source influences whether a detonable mixture deflagrates or detonates. Weak (thermal) ignition sources initiate deflagrations in open and closed systems; however, a deflagration may develop into a detonation in a closed system due to the influence of the confining walls. Strong (shock-wave) ignition sources tend to initiate detonations in open or closed fuel–air systems. Matches, sparks, hot surfaces and open flames are considered to be thermal (weak) ignition sources while shock-wave (strong) ignition sources are blasting caps, bursting vessels, TNT, high voltage-capacity shorts (exploding wires), lightning and other explosive charges.

The geometry of an enclosure has a strong effect on the transition from deflagration to detonation. Experimental data indicate that a U-shaped enclosure plus the ground comprise sufficient confinement to support "strong" explosions in detonable hydrogen–air mixtures that are ignited by thermal ignition sources. Geometrical changes in the confining walls that induce turbulence also enhance transition to detonation. The distance in a tunnel, pipeline, heating duct, or hallway that it takes for a reaction to progress from a deflagration to a detonation is related to the detonation induction distance. Hydrogen is a rapid burning fuel and the flame front has a tendency to accelerate in long enclosures. Consequently, detonation induction distances have been experimentally observed using hydrogen–air mixtures but no such data exist for the slower burning gasoline–air or methane–air mixtures. Transition to detonation occurs because compression of the unburned fuel–air mixture by deflagration increases the mixture temperature and pressure, both of which increase the burning velocity of the mixture. Recent experiments [53] indicate that it is difficult to design flashback arrestors that successfully disrupt deflagrative or detonative combustion in hydrogen-rich mixtures of hydrogen, methane and air that are contained within or flowing in cylindrical pipelines. Burgess [20], Zabetakis [60] and Carhart [30] agree that methane–air and gasoline–air mixtures will transit from deflagration to detonation if the pipe is long enough and its diameter is large enough. Experimental apparatus used to determine detonation induction distance is usually small and consequently these data are yet to be determined for methane and gasoline. The largest detonation experiments conducted to date were performed by Kogarko [19] and Burgess *et al.* [61]. Kogarko used a 30.5 cm inside-diameter tube and Burgess *et al.* used a 61 cm diameter pipe and earthen tunnels with cross-sectional areas ranging from 0.025 to 1.39 m^2. We have no assurance that gasoline–air or methane–air mixtures will not detonate from spark ignition in long tunnels or corridors; however, we can be quite sure that hydrogen–air mixtures will detonate under the proper circumstances ($L/D \gtrsim 100$).

We are belaboring the fuel detonation characteristics as they are believed to be of vital importance in future fuel applications. The tendency of hydrogen to detonate from spark ignition is perhaps the most significant deterrent to its widespread use. The pressure rise ratio of a detonation may easily be an order of magnitude higher than that of a deflagration—see detailed discussion in reference [54]. Overpressures due to deflagrations in open air are usually considered negligible; however, open-air deflagrations can cause structural damage if they are close to the structure and are of sufficiently large volume.

Shrapnel hazards relate directly to explosion overpressures so that all of the foregoing arguments concerning overpressure apply to the evaluation of shrapnel hazards for the different fuels. Thus, we see that overpressure shrapnel hazards associated with ordinary enclosures ($L/D \lesssim 30$) are about the same for hydrogen–air and methane–air and somewhat less stringent for gasoline–air. In long tunnels, etc., hydrogen is a greater explosion threat than either of the other two fuels because it has a greater tendency to transit to detonation. The wider flammable limits and detonable limits of hydrogen also tend to make hydrogen a greater explosive threat than methane or gasoline.

A number of preventive measures can be enacted to minimize the explosive threat of all three fuels and are particularly helpful for the rapid-dispersion fuels, hydrogen and methane. Roof vents and forced ventilation, where practical, are accepted methods of minimizing accumulations of gaseous fuel within enclosures. In some applications the quantity of fuel permitted within an enclosure can be restricted. Ignition sources can be minimized but seldom are they eliminated. Frangible (weak) walls can be used to relieve deflagration overpressures within enclosures—rupture discs can be used to provide the same protection for pressure vessels. Frangible walls and discs are of little value in relieving detonation overpressures, although in some instances they may prevent or lessen the effects of detonation [54]. It appears that weak but pressure-wave reflecting walls will support transition from deflagration to detonation—the use of elastic membranes (plastic curtains) may inhibit or prevent transition to detonation. Fuel storage tanks can be buried or storage areas can be diked (for fuel containment), barricaded and confining structures minimized. Major spillage can be avoided by using storage vessels constructed of ductile materials and by adherence to established safety procedures.

EXPLOSIVE DAMAGE

The elements of explosive damage, fireball, ensuing fires, overpressure and shrapnel, have already been discussed and only overpressure and shrapnel warrant additional attention.

Explosions that create overpressures and shrapnel may be rated in terms of the amount of energy that is released. This energy release may be evaluated directly in energy units such as kJ although it is commonly expressed as an equivalent quantity of TNT. The explosive strength of TNT is well-known and reproducible and it is a good standard for rating the explosive potential of various substances. Expressing explosive potential in terms of an equivalent mass of TNT is a good technique for evaluating damage potential at distances well-removed from the explosion; however, at distances inside or near the reaction zone, this procedure is less accurate because of the differences in shape and peak magnitude of the impulse diagrams for TNT and fuel–air mixtures. A fuel–air explosion may deliver a considerably lower overpressure, relative to TNT, over a longer time interval and thus have less crushing effect on some structures, but a greater overturning moment. Although there is general dissatisfaction [62] with the TNT concept, it will continue to be used until non-ideal explosions can be characterized more definitively. To provide conservative results the TNT equivalent concept can be used [54] to evaluate impulse and overpressure effects at distances far from the explosion source, and to evaluate impulse effects in the near-combustion zone.

The theoretical TNT equivalent, of various fuels, can be determined by using the decrease in Helmholtz free energy to compute the maximum energy available for explosive yield. Following this procedure we obtain the theoretical limiting values of explosive potential for hydrogen, methane and gasoline as recorded in Table 1. Note that hydrogen is the most potent on a mass basis and the least potent on a volumetric basis. The explosive potential per kJ of stored heating value (based on the high heat of combustion) is 0.17 (g TNT)/kJ for hydrogen, 0.19 (g TNT)/kJ for methane and 0.21 (g TNT)/kJ for gasoline. Thus, for equivalent energy storage, hydrogen has the least theoretical explosive potential of the three fuels.

It must be emphasized that only a fraction of this theoretical explosive yield can be realized in an actual open-air mishap because it is virtually impossible to spill or release a large quantity of fuel and have all of it mixed in proper proportions with air prior to ignition. Experimental data and computations indicate that the fraction [46, 54, 62] of fuel within the combustible range at any time following a massive or continuous fuel spillage will be less than 10% of the quantity spilled. Such explosive yield data are meager for all three fuels; however, the vapor or gas phase mixing limitations are equally applicable to all fuels. Hydrogen disperses much more rapidly than methane or gasoline, but it also has much wider flammable and detonable limits, etc. Thus, in the absence of more definitive experimental data it is impossible to accurately assess the probable explosive yield attributable to accidental release and ignition of hydrogen, methane or gasoline in air. The "energy of explosion" values listed in Table 1 should be considered theoretical maxima and yield factors of 10% are considered reasonable for fuel–air explosions.

Overpressure damage is highly dependent upon the nature of the explosion. A confined and unvented deflagration [52] of hydrogen–air or methane–air will produce a static pressure rise ratio of less than 8:1. Explosion pressures [31] for confined deflagrations of gasoline–air are about 70–80% of those for hydrogen–air. Unconfined deflagration overpressures are usually less than 7 kPa; however, 3–4 kPa is sufficient side-on pressure to cause structural damage [63] to buildings and unconfined large volume gas-phase explosions can be destructive. Ordinary glass window panes fracture under pressures of 3–7 kPa, non-reinforced masonry walls fail at pressures below 55 kPa, and human eardrums rupture at pressures of approximately 35 kPa. Thus, it is apparent that confined deflagrations (even if relieved) can be very devastating—up to 8 atm (811 kPa) of explosion pressure—and unconfined deflagrations can cause slight to moderate structural damage and injure people via fire, window-glass shrapnel, etc.

Detonations, whether confined or unconfined, can be expected to severely damage or totally destroy ordinary buildings in the near vicinity of the explosion. TNT pressure–distance data [55] can be used [55, 61] to estimate overpressures resulting from fuel–air detonations. The applicability of TNT equivalence to vapor or gas-cloud explosions is fully reviewed in [55, 62]. The pressure accompanying detonation of any fuel is approximately double that obtained by adiabatic combustion of a stoichiometric mixture of the fuel in air at constant volume. Consequently, we could expect static pressure rise ratios of ~15:1 for hydrogen–air or methane–air detonations and a ratio of ~12:1 for a gasoline–air detonation. Much higher reflected pressure rises can be attained if the explosion transits from deflagration to detonation because the deflagration compresses the unburned fuel–air mixture prior to transition to detonation, e.g. a reflected pressure rise ratio of 120:1 (8:1 × 15:1) could be achieved where a hydrogen or methane deflagration transits to detonation. Such transitions are more easily accomplished with hydrogen than with methane or gasoline.

The impulse created by explosion overpressures is of concern in evaluating explosion damage and in the design of barricades or structures to withstand explosions. Although the overpressure created by a gas-phase explosion is of lower peak magnitude and longer duration [64] than the overpressure due to an equivalent quantity of TNT, the extensive TNT data may be used [55, 61, 64] for design purposes. The fundamentals of dynamic blast loads and structural response to shock waves is treated elsewhere [28] as is the applicability [55, 61, 62, 64] of TNT explosive data to the design of structures to withstand gas-phase explosions.

Unbarricaded distances required for the protection of personnel in inhabited buildings that are exposed to shrapnel from TNT explosions are indicated by curve 6 of Fig. 1. These data also predict the maximum observed fragment distances for space vehicle explosions [55] and are more restrictive than the unbarricaded distances required for shrapnel protection of personnel on roadways (see curve 2 of Fig. 1. Fletcher [64] has suggested that TNT shrapnel hazard data may be used to estimate propellant-explosion shrapnel hazards if the appropriate TNT equivalent is used. There is evidence [64] that large low-velocity fragments emitted from such explosions may exceed the TNT shrapnel scatter limits. This situation results because propellant explosions endure longer and can impart more impulse to the projectile; however, the range of high population-density projectile scatter is normally greatest [64] for an equivalent quantity of TNT. As an interim measure the Department of Defense and NASA have adopted [55] the TNT shrapnel hazard data for propellant explosives at range launch pads and rocket engine test stands. These TNT data adequately predict [55] maximum fragment distances for known incidents involving propellant explosions.

Design for shrapnel protection is difficult because it is necessary to estimate the size, mass, and velocity of fragments emitted from explosions of varied type, strength, and location. In brief, shrapnel shields made of materials with large modulus of elasticity (Young's) are the most effective: for example, steel is more shrapnel resistant than copper and copper is more resistant than aluminum, etc. The opposite is true for shrapnel projectiles themselves. A projectile having a given size and momentum will penetrate deeper if it has a lower density—aluminum will penetrate deeper than copper and copper deeper than steel, etc. Recall that momentum is mass × velocity and a constant momentum and size require a higher velocity for the lower density projectiles. Selected references on this subject are given in the parent-document [65] of this paper.

Figures 1 and 2 summarize much of the fire and explosive hazard data presented herein. These figures illustrate the variation in conservatism of various authorities that generate safety criteria. Obviously, when in doubt, the more conservative criteria should be used. The overpressure band on Fig. 1 corresponds to breakage of ordinary window glass (3.5–7 kPa) and the overpressure band on Fig. 2 relates to the estimated external pressure capability of liquid hydrogen storage dewars. The derivation and use of data shown on these figures (safe unbarricaded distances) are fully discussed in a summary document [54] treating the explosive hazards of hydrogen. Safe barricaded distances for TNT and fuel–air explosions may also be estimated from data made available in references [54, 55]. A single series of documents [55] offer comprehensive treatment of overpressure, impulse, fireballs, shrapnel, barricades, structural response and physiological effects, as they relate to propellant explosions. The author feels that these documents are applicable to fuel–air explosions where the TNT equivalent is properly estimated. Care should be exercised when attempting to assess the damage potential of large-volume gas-phase explosions because line-of-sight from such explosions to vulnerable targets may pass over or around barricades that were erected to provide protection from concentrated explosives.

Figure 3 provides a ready comparison of industrially accepted fuel storage standards for hydrogen, LNG and gasoline. By comparing curves 1 and 4 of Fig. 3 with the data given on Figs 1 and 2, we find that the industrial quantity–distance standards (Fig. 3) for hydrogen are less demanding than those suggested for experimental areas (Figs 1 and 2). Also, by comparing curves 1, 2, 3 and curves 4, 5, 6 of Fig. 3 we observe that industrial storage standards are more restrictive for hydrogen, methane and gasoline, respectively.

Curve 3 of Fig. 3 represents the minimal distance for separation of two adjacent above-ground gasoline storage tanks. The distance [58] between such tanks shall not be less than 3 ft and not less than one-sixth the sum of the diameters of two adjacent tanks. When the diameter of one tank is less than one-half the diameter of the adjacent tank, the distance [58] between the two tanks shall not be less than one-half the diameter of the smaller tank. Curves 6A and 6B of Fig. 3 represent the variation in quantity–distance standards for the protection of personnel in buildings adjacent to gasoline storage tanks. These curves bound standards [58] that vary with type of tank construction, fire control measures and protection for exposures, tank operating pressure and emergency venting equipment. The distance [58] from any part of an underground tank (storing gasoline) to the nearest wall of any basement or pit shall be not less than 1 ft, and not less than 3 ft from inhabited buildings.

It is apparent that industrial storage standards are least restrictive for gasoline; however, the industrial storage standards for LNG and hydrogen fuels are not prohibitive and should not limit their use.

COMPARISON OF FUEL STORAGE METHODS

Hydrogen may be stored in the compressed gas, liquid or hydride forms. The relative costs [66] of storage in these various forms are dependent upon the quantity of hydrogen stored and upon desired storage pressure and storage duration.

Currently, the most promising alloys for metallic hydride storage contain titanium and magnesium. Storage data [67] for magnesium–nickel (Mg–Ni) and iron–titanium (Fe–Ti) hydrides are available and hydride storage system evaluations are in progress [68]. Safety standards for the production, processing, handling and storage of titanium and magnesium are well documented [69]; however, the safety hazards are much less certain for the candidate Fe–Ti and Mg–Ni ores in combination with stored hydrogen. A recent experimental evaluation [70] indicates that Fe–Ti hydride can be considered a safe method of storing hydrogen and hydride storage appears attractive [68] for certain short-term storage applications.

Hydrogen is routinely stored as a compressed gas in industry and this practice must be considered safe. It is usually stored in metal containers at pressures ranging from near-ambient to more than 20 MPa. It may also be possible to store compressed hydrogen in abandoned natural gas fields, caverns, aquifers, etc. Metal storage containers are normally constructed of a hydrogen-compatible ductile steel and are not generally susceptible to catastrophic failures. Such containers are normally equipped with pressure and thermally-induced pressure relief devices. Vessel fracture would most likely be accompanied by autoignition of the released hydrogen and air mixture with an ensuing conflagration lasting until the contents of the ruptured vessel are consumed. Considerable experimental data substantiate this statement—see [54, 71]. Adjacent metal storage vessels are relatively shrapnel and fire resistant and may be water-cooled or buried in sand for additional fire and blast resistance.

Hydrogen is also routinely stored as a liquid in industry and in university and government laboratories supporting the U.S. space exploration program. Again, catastrophic failures are not technically plausible as dewars are constructed of ductile metals and are diked to confine the liquid contents of the dewar in the event of spillage. The purpose of the dike is to reduce the liquid evaporation rate and to confine the potential conflagration to the vicinity of the defective dewar. The double wall construction of liquid storage dewars provides good protection against fire and shrapnel and additional fire resistance for adjacent storage dewars can be provided with a water deluge system. Liquid hydrogen storage dewars are usually built with carbon steel vacuum jackets and aluminum or stainless steel inner vessels. Liquid hydrogen has been safely stored in large metal dewars for nearly 20 years.

Hydrogen continues to be stored and used in both the compressed gas and liquid forms in an industrially safe manner. Hydride storage should prove to be equally safe.

Gasoline is normally stored in simple steel containers at near-ambient pressures both above and below ground. The principles of diking and water-cooling to protect adjacent buildings or storage vessels are equally applicable to above-ground gasoline, liquefied natural gas (LNG) and liquid hydrogen storage. The use of ductile steel tanks virtually precludes catastrophic failures and diking confines potential conflagrations. Gasoline has been safely stored in large quantities for over half a century.

Methane is commonly and safely stored in large quantities as a constituent of compressed gas in natural gas fields, caverns, abandoned coal mines, etc. The safety record of the natural gas industry is exceptionally good.

Liquefied natural gas, whose primary constituent is methane, has been stored in large quantities in the U.S. since 1941. The early tanks were of double-walled steel construction (3.5 % nickel steel inner shell and carbon steel outer shell) and a disastrous fire accompanied the failure of one of these tanks in 1944. Changing the inner shell material from 3.5 % nickel steel to 9 % nickel steel or aluminum has cured the early storage problems associated with LNG and it has been safely stored in large quantities since the mid-sixties. Storage vessels with concrete inner shells and carbon steel outer shells have also been placed in service within the last several years. Liquefied natural gas is stored in above-ground and below-ground containers and the space between the double walls is usually filled with a foam, powder, or fibrous insulation material and gas-filled with nitrogen or natural gas. All of the foregoing arguments concerning the storage of gasoline and liquid hydrogen apply to the storage of LNG. In fact, a comprehensive fire and explosion hazard study [26] by the Bureau of Mines concluded that LNG could be safely stored in much the same manner as gasoline.

Of the three fuels examined, gasoline is certainly the easiest and perhaps the safest fuel to store because of its higher boiling point, lower volatility and narrower flammable and detonable limits. All of the previous discussion in this paper concerning fire and explosion hazards support this generalized conclusion; however, we must recognize that hydrogen and methane (or LNG) can also be safely stored. The degree of risk associated with the storage of each fuel cannot be specified at this time but industrial experience indicates that all three fuels can be safely stored using current technology.

FUTURE HYDROGEN APPLICATIONS

Hydrogen is being considered as a replacement fuel [72] in all of the major fuel markets and will continue to penetrate these markets as time passes. It is anticipated that all of the major fuel markets (industrial, commercial, transportation and residential) will ultimately rely on electricity or hydrogen. A detailed safety analysis covering the use of hydrogen fuel in each of these market areas is needed but well beyond the scope of this paper. Credible accidents must be postulated and detailed fault trees must be developed to provide meaningful hazards data. A resourceful analyst is required because some of the experimental data required for thorough hazards evaluation are nonexistent, e.g. the effects of partial confinement on explosion overpressure and transition to detonation, the effects of elastic weak-walls on deflagration overpressures and the effects of atmospheric dispersion rates on vapor or gas-cloud accumulations—see [46, 54, 62]. A current review article [62] summarizes the state-of-the-art concerning characterization and evaluation of accidental explosions and is a useful guide to current research efforts in this area. The interested reader or analyst may find additional information and assistance with hydrogen safety problems by referring to the recent compilations by Ludtke [73] and to the safety manual [74] prepared by the NASA.

The data and discussions presented in this paper are intended to aid the hazards analyst in preparing safety analysis for each specific application of hydrogen fuel as the applications arise. Without resorting to specific applications and specific accident criteria we can categorically analyze some of the safety hazards associated with the various fuel markets.

Hydrogen is a feedstock in the process industries [72] and has been sucessfully handled for decades. Thus, the use of hydrogen in industry is essentially routine and no major safety problems are anticipated with its increased use in industrial processes. New processes and new uses of hydrogen may pose new safety hazards that must be dealt with as they arise.

Hydrogen has not been used extensively in commercial applications although it has been successfully piped cross-country as a compressed gas in a few locales. A demonstration project [68] is currently in progress to evaluate the feasibility of using hydrogen in electrical utility load-leveling operations. The electrical utility industry has also successfully used hydrogen gas to cool the rotor and stator coils in large turbine-generators for more than 40 years. No significant hydrogen safety problems are foreseen in the commercial sector.

Hydrogen is a potential replacement fuel in the voracious transportation market. Conceptually, and technically, hydrogen can be used to fuel aircraft, ships, trains, trucks, buses and automobiles; however, the economics, logistics and safety of supplying and distributing hydrogen to fuel these vehicles are as yet undetermined. Excluding a few demonstration projects with automobiles, trucks, buses and airplanes, we find that hydrogen is a relatively untried fuel in the transportation market. Hydrogen was tried briefly as an auto fuel in the mid-thirties and as an inflatant for balloons and dirigibles prior to the ill-fated Hindenburg fire of 1937. Considerable safety analysis is in order prior to the widespread use of hydrogen fuel in transportation—particularly in highway vehicles where personnel exposure is maximum and fuel handling procedures are most difficult or impossible to enforce. Hydrogen is already the accepted fuel of the aerospace industry and is safely handled in large quantities. Bowen [75] and Lippert [76] recently completed reviews of some of the hazards associated with the use of hydrogen as a military fuel.

Hydrogen can also be used to supply residential fuel needs because appliances, furnaces, etc. can be made to operate on hydrogen gas. From a technical viewpoint this application can be readily satisfied, but from a safety viewpoint there are a number of significant concerns. The major concerns relate to the problems of gas leakage, detection and the potential severity of explosions of hydrogen–air mixtures in confined or partially-confined spaces. More experimental data and safety analyses are needed to fully resolve these questions and to determine the comparative risks of hydrogen and natural gas as residential fuels; however, hydrogen-enriched gases (coal gas, town gas, producer gas, etc.) have been successfully used in European countries to satisfy residential fuel needs during the last century. Thus, there also appears to be ample precedent for acceptably safe use of hydrogen in the residential sector.

SUMMARY

The safety aspects of any fuel are intimately related to the fuel application and to the postulated accident criteria. Thus, specific conclusions await hard comparisons of competing fuels in applications where credible accidents can be specified; however, generalized conclusions and judgments may be drawn from the comparative technical data and discussions presented herein.

A comprehensive list of thermophysical and combustion properties of hydrogen, methane and gasoline was compiled and presented in this paper. These data provide the bases for future safety analysis and for direct safety comparisons of hydrogen, methane and gasoline.

Liquid hydrogen is more difficult to contain than either liquid methane or liquid gasoline and N.T.P. gasoline vapors are more difficult to contain than either N.T.P. hydrogen or N.T.P. methane. Industry has proven that all three fuels can be safely and easily contained in both gaseous and liquid phases.

In the event of a fuel spill, we can expect a fire hazard to develop most rapidly with hydrogen, methane and gasoline, respectively, and the fire hazard should persist in the inverse order. For a specified liquid spillage volume and ensuing fire we can expect gasoline fires to last the longest and hydrogen fires to be the shortest lived, while all three fuels burn at nearly the same flame temperature. The scene of a hydrogen fire may be hotter (1.3 × to 1.8 ×) than that of a hydrocarbon fire, but the hydrocarbon fires will endure five to ten times longer than hydrogen fires (for spillage of identical liquid fuel volumes).

All three fuels are easily ignited by weak ignition sources such as matches. Even a weak spark generated by the discharge of static electricity from a human body may be sufficient to ignite any of these fuels in air. Hydrogen is more readily ignitable than either of the hydrocarbon fuels which appear to be equally susceptible to ignition.

Hydrogen fires are more difficult to detect than methane or gasoline fires but modern detection equipment makes it possible to quickly and reliably detect the flames of all three fuels. In some applications hydrogen and methane fires should be allowed to burn until gas flow is stopped or until liquid spills are consumed because of the potential explosive hazard created by extinguishing such flames: however, the fire should be controlled in all situations and in many cases it is advisable to extinguish the fire. Water may be used to fight fires of all three fuels and commercial dry chemicals and high-expansion foams can be used to extinguish LNG and gasoline fires.

The potential for smoke inhalation damage is judged to be most severe in gasoline, methane and hydrogen fires, respectively.

The wider flammable limits and detonable limits of hydrogen coupled with its rapid burning velocity tend to make hydrogen a greater explosive threat than methane or gasoline. Unconfined fuel–air explosions are not normally very destructive; however, confined fuel–air explosions can be devastating and hydrogen presents the greatest confined-explosion threat of the three fuels considered.

For equivalent energy storage or for equivalent volume storage, hydrogen has the least theoretical explosive potential of the three fuels considered—even though it has the highest heat of combustion (and explosive potential) on a mass basis.

Hydrogen is currently being safely stored and used in industry in both the compressed gas and liquid forms and it is anticipated that metal hydride storage will be equally safe. Of the three fuels examined, gasoline is the easiest and perhaps the safest fuel to store because of its lower volatility and narrower flammable and detonable limits.

Personnel and equipment safety criteria (fuel quantity–distance) are concisely charted on figures presented herein. It is believed that these figures provide safe exposure distances for all three fuels.

Consideration of future hydrogen applications reveals no safety problems in the industrial and commercial markets. Hydrogen safety problems may exist in the transportation and residential fuel markets and additional safety analyses are needed in these areas. Lower risk (or lower cost) fuels will most likely be used to satisfy many of these markets over the next few decades; however, hydrogen cannot currently be considered unsafe and cannot be excluded from consideration in any of these applications on the grounds of safety. It is the author's belief that fuel availability and cost will outweigh fuel safety in the selection of fuels in the future and hydrogen must be considered a contender in the chemical fuel market.

REFERENCES

1. A. F. Schmidt, Recommended materials and practices for use with cryogenic propellants, Aerospace Information Report, AIR 839B, Soc. of Auto. Engrs, 68 pp. (April 1969).
2. *Mechanical Engineers' Handbook*, 5th edn, p. 342, Edited by L. S. Marks, McGraw-Hill, New York (1951).
3. R. D. McCarty and L. A. Weber, Thermophysical properties of parahydrogen from the freezing liquid line to 5000 R for pressures to 10,000 psia, Nat. Bur. Stand. (U.S.) Tech. Note TN-617 (April 1972).
4. B. Lewis and G. von Elbe, *Combustion, Flames and Explosions of Gases*, 2nd edn, p. 685, Academic Press, New York (1961).

5. R. D. Goodwin, The thermophysical properties of methane, from 90 to 500 K at pressures to 700 bar, Nat. Bur. Stand. (U.S.) Tech. Note TN-653 (April 1974).
6. R. D. McCarty, Cryogenics Division, Nat. Bur. Stand., Boulder, Colorado, private communication (1975).
7. *Standard Table of Physical Constants of Paraffin Hydrocarbons and Other Components of Natural Gas*, GPA Publication 2145–75 (1975); (available from Gas Processors Assoc., 1812 First Place, Tulsa, OK 74103).
8. *Fire Hazard Properties of Flammable Liquids, Gases, Volatile Solids* 1969, NFPA Pamphlet No. 325M (1969); (available from National Fire Protection Assoc., 470 Atlantic Ave., Boston, MA 02210).
9. *Engineering Data Book*, 9th edn, Section 16, *Physical Properties*, Compiled and Edited by Gas Processors Assoc. (1972); (available from Gas Processors Suppliers Assoc., 1812 First Place, Tulsa, OK 74103).
10. *Handbook of Chemistry and Physics*, 39th edn, p. 2343, Edited by C. D. Hodgman, R. C. Weast and S. M. Selby, The Chemical Rubber Co., Cleveland, Ohio (1957–58).
11. *Ibid.* p. 2011.
12. *Handbook of Tables for Applied Engineering Science*, p. 68, Edited by R. E. Bolz and G. L. Tuve, The Chemical Rubber Co., Cleveland, Ohio, (1970).
13. Lewis and von Elbe, *op. cit.* pp. 692–693.
14. H. F. Coward and G. W. Jones, Limits of flammability of gases and vapors, Bureau of Mines Bulletin 503 (1952); (available from Superintendent of Documents, U.S. Gov't. Printing Office, Washington, D.C. 20402).
15. Lewis and von Elbe, *op. cit.* p. 535.
16. Lewis and von Elbe, *op. cit.* pp. 329–335.
17. *Handbook of Chemistry and Physics, op. cit.* p. 1790.
18. M. G. Zabetakis, Flammability characteristics of combustible gases and vapors, Bureau of Mines Bulletin 627 (1965); (available from Superintendent of Documents, U.S. Gov't. Printing Office, Washington, D.C. 20402).
19. S. M. Kogarko, Detonation of methane–air mixtures and the detonation limits of hydrocarbon–air mixtures in a large-diameter pipe, *Sov. Phys. tech. Phys.* **28**, (3) (1959).
20. D. S. Burgess, Bureau of Mines, Pittsburgh, Pennsylvania, private communication (1975).
21. E. C. Magison, *Electrical Instruments in Hazardous Locations*, 2nd edn, p. 74, Instrument Society of America, Pittsburgh, Pennsylvania, (1972).
22. *Ibid.* p. 304.
23. Lewis and von Elbe, *op. cit.* p. 706.
24. *Handbook of Tables for Applied Engineering Science, op. cit.* p. 303.
25. M. G. Zabetakis and D. S. Burgess, Research on the hazards associated with the production and handling of liquid hydrogen, Bureau of Mines Report of Investigations 5707 (1961).
26. D. S. Burgess and M. G. Zabetakis, Fire and explosion hazards associated with liquefied natural gas, Bureau of Mines Report of Investigations 6099 (1962); (available from NASA Scientific and Technical Information Facility, N63-18682).
27. Lewis and von Elbe, *op. cit.* pp. 381–382.
28. M. G. Zabetakis, *Safety with Cryogenic Fluids*, p. 57, Plenum Press, New York (1967).
29. Lewis and von Elbe, *op. cit.* p. 530.
30. H. Carhart, Naval Research Laboratory, Washington, D.C., private communication (1975).
31. An Investigation of Fifteen Flammable Gases or Vapors with Respect to Explosion-proof Electrical Equipment, Underwriters' Laboratories Bulletin of Research No. 58 (April 1970); (available from Underwriters' Laboratories, Publication Stock, 333 Pfingsten Road, Northbrook, IL 60062).
32. Magison, *op. cit.* p. 116.
33. *Ibid.* pp. 26–27.
34. Lewis and von Elbe, *op. cit.* p. 698.
35. L. E. Bollinger, Experimental detonation velocities and induction distances in hydrogen–air mixtures, *AIAA J.* **2** (1) 131–133 (Jan. 1964).
36. E. W. Cousins and P. E. Cotton, Design closed vessels to withstand internal explosions, *Chem. Engr, N.Y.* 133–137 (August 1951).
37. N. I. Sax, *Dangerous Properties of Industrial Materials*, 4th edn, pp. 785, 817, 903, Van Nostrand Reinhold, New York (1975).
38. M. Vanpee and H. G. Wolfhard, Ignition by hot gases, Bureau of Mines Report of Investigations 5627 (1960).
39. J. M. Kuchta, R. J. Cato, G. H. Martindill and W. H. Gilbert, Ignition characteristics of fuels and lubricants. Bureau of Mines Technical Report AFAPL-TR-66-21 (March 1966); (available from Clearinghouse for Federal Scientific and Technical Information, AD632730).
40. W. M. Rohsenow and H. Choi, *Heat, Mass and Momentum Transfer*, pp. 397–401, Prentice-Hall, Englewood Cliffs, New Jersey (1961).
41. R. B. Bird, W. E. Stewart and E. N. Lightfoot, *Transport Phenomena*, pp. 60, 182, Wiley, New York (1960).
42. *Mechanical Engineers' Handbook*, 5th edn, *op. cit.* p. 1483.
43. L. E. Bollinger, M. C. Fong, J. A. Laughrey and R. Edse, Experimental and theoretical studies on the formation of detonation waves in variable geometry tubes, NASA Technical Note D-1983 (June 1963).

44. LNG Safety Program, Phase II, Consequences of LNG Spills on Land, American Gas Association Project IS-3-1 (Nov. 1973); see also, H. R. WESSON, L. E. BROWN and J. R. WELKER, Vapor dispersion, fire control, and fire extinguishment of high evaporation rate LNG spills, paper 74-D-36 in 1974 Operating Section Proceedings of the American Gas Association, 1515 Wilson Boulevard, Arlington, VA 22209 (1974).
45. J. M. ARVIDSON, J. HORD and D. B. MANN, Efflux of gaseous hydrogen or methane fuels from the interior of an automobile, Nat. Bur. Stand. (U.S.) Tech.Note TN-666 (March 1975).
46. D. S. BURGESS, J. N. MURPHY, M. G. ZABETAKIS and H. E. PERLEE, Volume of flammable mixtures resulting from the atmospheric dispersion of a leak or spill, *15th International Symposium on Combustion*, pp. 289–303, The Combustion Institute, Union Trust Building, Pittsburgh, PA (1974).
47. J. HORD, Correlations for predicting leakage through closed valves, Nat. Bur. Stand. (U.S.) Tech. Note TN-355 (August 1967).
48. J. W. MARR, Leakage Testing Handbook, prepared for Jet Propulsion Laboratory, NASA, Pasadena, Califonia, under contract NAS 7-396, S-67-1014 (June 1967); (also available as NASA CR-952, April 1968).
49. B. ROSEN, V. H. DAYAN and R. L. PROFFIT, Hydrogen leak and fire detection, a survey, NASA SP-5092 (1970).
50. Static Electricity 1972, NFPA Pamphlet No. 77 (1972); (available from National Fire Protection Assoc., 470 Atlantic Ave., Boston, MA 02210).
51. Guide to the Use of Flame Arresters and Explosion Reliefs, New Series, No. 34, Ministry of Labour, H.M.S.O., England.
52. ZABETAKIS, *Safety with Cryogenic Fluids, op cit.* p. 49.
53. W. B. HOWARD, C. W. RODEHORST and G. E. SMALL, Flame arresters for high-hydrogen fuel–air mixtures, *Loss Prevention* 9 (American Institute of Chemical Engineers, 345 East 47th Street, New York, NY 10017 (1975)).
54. J. HORD, Explosion criteria for liquid hydrogen test facilities, Nat. Bur. Stand. (U.S.) NBS Report (Feb. 1972).
55. *General Safety Engineering Design Criteria*. Vol. 1, CPIA publication 194, (Oct. 1971); also, *Liquid Propellant Handling, Storage and Transportation*, Vol. 3, CPIA publication 194 (May 1972)—documents prepared by the JANNAF propulsion committee of the JANNAF Hazards Working Group and are available from the Chemical Propulsion Information Agency (CPIA) of the Johns Hopkins University Applied Physics Laboratory, 8621 Georgia Ave., Silver Spring, MD 20910; (also available to the public through National Technical Information Service, Springfield, VA 22151: Vol. 1 Accession No. AD 889 763, Vol. 3 Accession No. AD 870 259).
56. Standard for Liquefied Hydrogen Systems at Consumer Sites, NFPA Pamphlet No. 50B (ANSI Z292.3).
57. Standard for the Production, Storage and Handling of Liquefied Natural Gas (LNG), NFPA Pamphlet No. 59A (ANSI Z225.1).
58. Flammable and Combustible Liquids Code, NFPA Pamphlet No. 30 (ANSI Z288.1).
59. R. L. P. CUSTER and R. G. BRIGHT, Fire detection: the state-of-the-art, NASA CR-134642, NBS TN-839 (June 1974); (available from National Technical Information Service, Springfield, VA 22151).
60. M. G. ZABETAKIS, U.S. Bureau of Mines, private communication (1975).
61. D. S. BURGESS, J. N. MURPHY, N. E. HANNA and R. W. VAN DOLAH, Large-scale studies of gas detonations, Bureau of Mines Report of Investigations 7196 (1968).
62. R. A. STREHLOW and W. E. BAKER, The characterization and evaluation of accidental explosions, NASA CR-134779 (June 1975); (available from National Technical Information Service, Springfield, VA 22151).
63. R. REIDER, H. J. OTWAY and H. T. KNIGHT, An unconfined large-volume hydrogen/air explosion, *Pyrodynamics* 2, 249–261 (1965).
64. R. F. FLETCHER, Characteristics of liquid propellant explosions, *Ann. N.Y. Acad. Sci.* **152**, 432–440 (Oct. 1968); see also, R. F. FLETCHER, Liquid-propellant explosions, *J. Spacecraft Rockets* **5**, (10) 1227–1229 (Oct. 1968).
65. J. HORD, Is hydrogen safe? Nat. Bur. Stand. (U.S.) Tech. Note TN-690 (Oct. 1976).
66. J. HORD and W. R. PARRISH, Economics of hydrogen (to be published as Chapter 1 of Volume 5, *Implications of Hydrogen*, in the book: *Hydrogen, Its Technology and Implications*, CRC Press).
67. R. H. WISWALL, JR and J. J. REILLY, Metal hydrides for energy storage, Proc. 7th Intersociety Energy Conversion Engr. Conf., ACS, Washington, D.C., pp. 1342–1348 (1972).
68. F. J. SALZANO, C. BRAUN, A. BEAUFRERE, S. SRINIVASAN, G. STRICKLAND and J. J. REILLY, Hydrogen for energy storage: a progress report of technical developments and possible applications, presented at the Energy Storage Conf., Asilomar Conf. Grounds, Pacific Grove, California, Feb. 8–13, 1976 (Jan. 1976).
69. *NFPA National Fire Codes*, Volume 3 (1975); (available from National Fire Protection Assoc., 470 Atlantic Ave., Boston, MA 02210).
70. C. E. LUNDIN and F. E. LYNCH, The safety characteristics of FeTi hydride, Proc. 10th Intersociety Energy Conversion Engr. Conf., pp. 1386–1390 (August 1975).
71. P. M. ORDIN, Review of hydrogen accidents and incidents in NASA operations, Proc. 9th Intersociety Energy Conversion Engr. Conf., San Francisco, California, pp. 442–453 (August 1974).
72. W. R. PARRISH, R. O. VOTH, J. G. HUST, T. M. FLYNN, C. F. SINDT and N. A. OLIEN, Selected topics on hydrogen fuel, Edited by J. Hord, Nat. Bur. Stand. (U.S.) SP 419 (May 1975).
73. P. R. LUDTKE, Register of specialized sources for information on selected fuels and oxidizers, NASA CR-

134807 (March 1975); see also, P. R. LUDTKE, Register of hydrogen technology experts, NASA CR-2624 (Oct. 1975).
74. *Hydrogen Safety Manual*, NASA Lewis Research Center, NASA Tech. Memo TM-X-52454 (1968).
75. T. L. BOWEN, Investigation of hazards associated with using hydrogen as a military fuel, Report 4541, Naval Ship Research and Development Center, Bethesda, MD 20084 (August 1975).
76. J. R. LIPPERT, Vulnerability of advanced aircraft fuel to ballistic and simulated lightning threats, *Int. J. Hydrogen Energy* **1**, 321–330 (1976).

Safety Aspects of a Hydrogen-Fuelled Engine System Development

L. M. Das
Indian Institute of Technology, New Delhi

Abstract—In view of the performance and emission characteristics, hydrogen can be considered an excellent fuel for internal combustion engines. However, its temperamental combustion behaviour often raises problems of safety. This paper describes the development of an overall effective gaseous hydrogen fuel supply system for an engine test cell. Various safety measures adopted to combat the symptoms of undesirable combustion phenomena have been discussed.

INTRODUCTION

Hydrogen is probably a unique, versatile fuel which possesses the potential of providing an ultimate solution to the twin problems of the energy crisis and environmental pollution. Unfortunately this fuel cannot be routinely handled exactly in the same manner as the conventional petroleum-based fuel, primarily because of the wide difference in combustion characteristics of both fuels.

Hydrogen is a renewable, recyclable fuel which can be generated from an infinite source potential using practically any non-fossil energy. Upon combustion, it produces almost no harmful pollutants. Tables 1 and 2 list some important properties of hydrogen which must be carefully reviewed in planning out the development of a safe hydrogen-fuelled engine system. Initially, at present a hydrogen engine has to be basically a converted system, from one which has been originally designed to operate on petroleum fuels. Any "converted system", in general is bound to exhibit some reliability and safety problems. Evaluation of the safety hazards of hydrogen fuel operating in a petroleum-fuel based engine system is obviously a highly complex task. Perhaps the magnitude of these safety related problems could be drastically reduced when used in a hydrogen-specific engine system. However, in the present context, a hydrogen-operated engine system should be designed for a high degree of inherent safety features. The main safety consideration, as clearly demonstrated in Table 1 should be based on adequate ventilation, prevention of possible leakage and elimination of undesirable ignition sources. In most parts of the system and the surroundings, additional safety devices must be installed to protect the equipment and the personnel. A detailed discussion on these aspects is described below.

There exists an inherent difference between the use of hydrogen and hydrocarbon fuel in engines irrespective of in whatever form the former is used, i.e. whether gas (medium pressure or high pressure), cryogenic liquid or metallic hydrides. Therefore, care must be exercised throughout the entire sequence of creating a convenient situation for adopting a facility so that this new fuel exhibits a standard of safety currently accepted for existing conventional fuels which have achieved such standards after long years of widespread use.

Hydrogen as a fuel and as a chemical, has been used for more than two centuries. This shows that a vast experience exists in handling hydrogen. Today, production, storage, transport and utilization of hydrogen are essentially routine procedures in chemical industries. Generally new safety problems are not expected for the increase in hydrogen utilization in these industrial sectors. However, a large scale introduction of hydrogen into any daily energy supply system raises questions of safety. In the early part of eighteenth century, Lemory found that a mixture of hydrogen and air explodes in the presence of a flame and later in 1781, Cavendish discovered that hydrogen burns in air and ultimately produces water. Thus the historical experience of operating hydrogen stresses that it must be handled carefully.

DESIGN CONSIDERATIONS

As emphasized earlier, hydrogen has a few properties which set it apart from conventional fuels. The properties which require special treatment to lessen the chances of risk are broad combustible range, low minimum ignition energy and rapid rate of diffusion. It will not be out of context to emphasize at this point that hydrogen has been classified by the NFPA (National Fire Protection Association) in its most hazardous group of flammable liquids, gases and volatile solids [1]. It is thus essential that specific care be taken in selecting the material as well as location for hydrogen fuel utilization.

Table 1. Thermodynamic properties of hydrogen, methane, and gasoline (generally accepted values from the literature)

Property	Hydrogen	Methane	Gasoline
Molecular weight	2.016	16.043	107.0
Density of gas at NTP (g m^{-3})	83.764	651.19	4400
Heat of combustion (low), kJ g^{-1}	119.93	50.02	44.5
Heat of combustion (High), kJ g^{-1}	141.86	55.53	48
Specific heat (C_P) of NTP gas, J g^{-1} K^{-1}	14.89	2.22	1.62
Viscosity of NTP gas, g cm^{-1} S^{-1}	0.0000875	0.000110	0.000052
Specific heat ratio (v) of NTP gas	1.383	1.308	1.05
Gas constant (R) cm^3 atm g^{-1} K^{-1}	40.7030	5.11477	0.77
Diffusion coefficient in NTP air cm^2 S^{-1}	0.61	0.16	0.005

Hydrogen has a specific gravity of 0.0695 at 20°C and 101.3 kPa. It is the smallest and the lightest element in nature. Thus, it can diffuse rapidly through air. Hence, if hydrogen is used in a confined area, there is a possibility of diffused hydrogen being amassed at the top. In an engine system, cast parts should be avoided as far as possible. Cast parts, being porous, could help diffuse hydrogen under certain circumstances and create unanticipated problems.

Hydrogen operations should preferably be conducted out-of-doors where, in the event of any possible leakage hydrogen diffuses to the surroundings and is diluted to non-combustible proportions. It should be borne in mind that it is very important to detect the hydrogen once it is released from the system. Gas detectors only effectively function indoors because of hydrogen's high buoyancy. They only function outdoors with difficulty. It has been demonstrated that an excessive hydrogen leakage into open atmosphere has not resulted in any explosion, but has dispersed without causing any harm. Hence, it is often suggested that open-air hydrogen-installations must be built on the basis of a successful preventive measure.

The low molecular weight of hydrogen results in a high diffusivity such that hydrogen diffuses 3–8 times faster than air. The properties such as high diffusivity and buoyancy are highly advantageous in the event of a spill. It has been estimated that a spill of about 2000 l of liquid hydrogen dissipates to an extremely low concentration (below the explosive limit) after only 1 min. Mechanical components such as flanges, connections or joints in transfer lines must be properly designed as they are quite vulnerable to leakage. Even though hydrogen is basically non-toxic, there is a definite disadvantage in hydrogen operation being carried out in a confined space. If a sufficient amount of hydrogen is released in a confined area, it can alter the air composition by displacing oxygen and thus introducing an asphyxiation risk. When oxygen levels in the air fall below 17%, thinking and coordination becomes difficult and below 6% breathing becomes practically impossible. However, since hydrogen is physiologically inert, fairly large quantities can be tolerated up to a point where oxygen gets displaced to as low as 13% at 0.1 MPa.

Hydrogen is colourless, odourless and tasteless. Most of the gaseous fuels, in general, contain odorants. So, odorized warning agent will work well as an additive to gaseous hydrogen. At stoichiometric composition, the minimum spark energy of hydrogen is an order of magnitude lower than that of hydrocarbons. It is because of this property, hydrogen–air mixture can be ignited by a number of ignition sources some of which may be relatively weak also. Sometimes, the discharge or static electricity is found to be unidentified initiator of hydrogen fires. In fact this is suspected to be at the root of Hindenberg disaster [2].

In addition to mixture composition, the flammability limits of any fuel also depend on the energy of ignition. Hydrogen fuel does possess a broad range of flammability which, in principle, can be highly disadvantageous

Table 2. Combustion of properties of hydrogen, methane, gasoline (generally accepted values from the literature)

Property	Hydrogen	Methane	Gasoline
Limits of flammability in air, vol%	4.0–75.0	5.3–15.0	1.0–7.6
Stoichiometric composition in air, vol%	29.53	9.48	1.76
Minimum energy for ignition in air, MJ	0.02	0.29	0.24
Autoignition temperature, K	858	813	501–744
Flame temperature in air K	2318	2148	2470
Burning velocity in NTP air, cm s^{-1}	265–325	37–45	37–43
Quenching gap in NTP air, cm	0.064	0.203	0.2
Percentage of thermal energy radiated from flame to surrounding, %	17–25	23–32	30–42
Diffusivity in air, cm^2 s^{-1}	0.63	0.2	0.08
Normalized flame emissivity 2000°K, 1 atm	1.00	1.7	1.7
Limits of flammability (equivalence ratio)	0.1–7.1	0.53–1.7	0.7–3.8

with respect to potential risks. In most of the practical accidents; it is the "lower flammability limits" that do play a very important role.

Hydrogen fuel possesses a high normal burning velocity which is a very important safety-related property of a combustible fuel. This, however, is different from the flame speed, which is the sum of burning velocity and displacement velocity of the unburnt gas mixtures. The stoichiometric value of hydrogen burning velocity is almost five times that of C_3H_8.

The minimum ignition temperature of hydrogen–air mixture is reported to be about 50°C at atmospheric pressure, whereas it is about 432°C for propane, 440°C for gasoline and 482–632°C for natural gas depending upon its composition. Thus hydrogen does possess a comparable minimum ignition temperature with natural gas and in fact, this value is actually higher than many hydrocarbon fuels. However, the low minimum ignition energy of hydrogen makes it more readily ignitable than these fuels. For hydrogen, ignition of explosive mixtures occurs with very low energy input: it is almost 1/10th that of gasoline–air mixture. Hence, an invisible spark can also cause explosion. The radiation flux of hydrogen flame decreases rapidly with distance. This may be beneficial to nearby combustible gases, but this almost invisible flame makes it easier for a person to contact the flame inadvertently thus increasing the possibility of burn-danger. Such a situation is quite likely to occur in practice since the heat radiating from the flame envelopes does not provide sufficient warning. Sometimes it so happens that even if the location of the fire is known, the flame envelope is not easily seen.

CURATIVE AND PREVENTIVE MAINTENANCE

It is advisable to use special, effective hydrogen-specific sensors to monitor this combustible gas in a hydrogen environment. In the event of an accidental leakage the combustible gas monitoring system should automatically come into operation thereby shutting off the fuel supply. The location of the detecting device is a critical consideration. Sometimes the response and sensitivity of the detector may be influenced by the presence of moisture or by a mixture of gases such as nitrogen, carbon dioxide and helium.

The need for reliable ventilation of the hydrogen-system-surrounding has already been adequately emphasized. In some operating conditions, it is not possible to permit sufficient ventilation to some test chambers. In such cases, the potential hazards inside the chambers can be rendered non-hazardous by building an atmosphere of inert gases. Nitrogen, carbon dioxide or helium can be used for this purpose.

Besides using nitrogen, carbon dioxide, etc., extinguishing powders such as ammonium phosphate or potassium chloride can also bring about inertness. These have been found to be quite successful as a means for temporary inertization of hydrogen–air mixtures, thereby preventing explosions from progressing through the systems. These are, therefore functionally effective explosion suppressants.

It has been reported that a liquid hydrogen splash does not damage the tissue immediately. This is so because the blood supply to the tissue acts as a heat source which causes the formation of a protective gas film between the cryogen of the warm tissue [3]. The extent of damage seems to be larger in the case of touching surfaces which have been cooled by liquid hydrogen or cold hydrogen vapours. The period of exposure, in all such cases essentially determines the magnitude of loss caused due to frostbite.

Flame arrestors are a set of systems for suppressing explosions inside a hydrogen-containing system. However, these arrestors are difficult to design reliably mainly because of the fact that hydrogen has a very small limiting diameter through which explosions can progress. These flame arrestors operate on the basic principle that the flame gets quenched if sufficient heat can be removed from the gas by the arrestors. In this respect, the quenching distance does play a decisive role. The quenching distance, by definition, is the spacing between parallel walls which just permits a flame to pass. The quenching distance for hydrogen is of the order of 1/4 that of other fuels. It depends on the pressure. As the pressure increases, the quenching distance also increases. It will be important to note in this context that a flame arrestor designed for a hydrocarbon flame will not be able to stop hydrogen flame. As far as metallurgical production design considerations are concerned, sintered bronze flames are effective in stopping a hydrogen flame.

Besides flame arrestors and passivity mechanisms, there are certain other protective measures, which are quite effective in preventing occurrence of high magnitude explosions in hydrogen-containing systems. These steps do mainly comprise of explosion relief systems. In a hydrogen engine system development, flame traps have been observed to work extremely satisfactorily in overcoming the undesirable combustion phenomena, especially backfire.

Installation of a non-return valve in the fuel line prevents a reverse flow to the system. Such a possibility of reverse flow can occur sometimes in a hydrogen-injected engine, particularly in the latter part of injection duration.

In view of the small size of hydrogen atoms and the molecules they can be readily dissolved into and diffuse through many materials, thus resulting in considerable degradation of the mechanical properties of most metals. Such a phenomenon is referred to as embrittlement. Hydrogen embrittlement often causes premature structure failure by brittle fracture [4]. The failures of aerojet pressure vessels [5] and Los Alamos pipeline [6] have already shown that the extent of damage due to embrittlement could be quite severe at welds. Thus safety aspects in material selection form an important factor in hydrogen service. All structures essentially are susceptible to varying extents of hydrogen embrittlement. It has been found out that ferric steels and nickel-based superalloys are highly susceptible, whereas austenitic stainless steels are only moderately affected and low aluminium alloys and copper alloys are relatively unaffected.

The process of welding used, the stresses (both residual and applied) and the microstructure are some of the important parameters which determine the susceptibility of certain welds to hydrogen utilization. Therefore, selection of the welding process and the adopted procedures of testing must be clearly defined in order to achieve a higher level of weld integrity. It has been observed in this context that the joints be also scrupulously designed. It has been generally observed that leaks from gas joints and valve packing account for about 65% of the fires.

Hydrogen is soluble in many materials including low-alloy steels and stainless steels. The solubility of hydrogen increases with temperature. However, even if the gas is at room temperature, hydrogen embrittlement can still occur since it depends on pressure.

The selection of the chemical for hydrogen use is a very important safety consideration because hydrogen reacts with a number of chemicals. For example, it explodes with chlorine in light.

ENGINE SPECIFIC SYSTEM

Certain inherent reliability problems of the system such as leakage, abrasion, wear, corrosion and material incompatibility must be thoroughly investigated and mechanical properties such as strength, hardness and machinability must be studied before adopting any material for hydrogen use. The choice of proper material is important. Permeability of steel to hydrogen is low at room temperature whereas the diffusion coefficient of hydrogen in steels is relatively high.

While conducting experiments with a hydrogen operated engine inside a laboratory, it is suggested that the hydrogen supply line be preferably routed through the roof thereby minimizing the dangerous effects of possible unintentional leakages. Additionally seals are to be provided at all vulnerable parts to ensure that no leakage of hydrogen occurs. All pipings and equipments should be internally cleaned before installation.

Engine operation should preferably be carried out with well dispersed water sprays in the exhaust. This helps suppress the detonation pressure and thus reduces the number and temperature of ignition sources in the exhaust system. Sometimes CO_2 may be used along with the water sprays to reduce the effects of potential hazards.

It is advisable to purge the fuel supply system with nitrogen; because filling out and opening the system are dangerous periods when there exists a bigger possibility of hydrogen and air coming in contact in the pipeline. Besides, an emergency fuel shut-off system should be incorporated at the point just previous to the point of fuel induction.

The fuel supply system should preferably work with precision and no improper adjustment or alignment problems should be permitted between different moving and connecting parts. In a technical report, Zalosh and Short [7] have brought forth some generic causes of hydrogen accidents as given in Table 3.

Table 3.

Category	% of Incidents
Undetected leaks	22
Hydrogen–oxygen off-gas explosion	17
Piping and pressure vessel rupture	14
Inadequate inert gas purging	8
Vent and exhaust system incidents	7
Hydrogen chlorine incidents	7
Other incidents	25

The incidents due to undetected leaks demonstrate hydrogen's propensity to leak past normally air-tight seals. Depending upon the pressure difference and the leak area, the volumetric leak rate is governed either by gas density or viscosity. A comparative evaluation between air and hydrogen indicate that hydrogen has larger viscous and inviscid leaks than air. So it will be erroneous to assume that an air-tight equipment is also "gas-tight".

It is better that a written procedure for system operation along with the relevant drawings be kept near the equipment to provide constant guidelines and precautionary measures to the operating staff. Personnel involved in the production, transportation and storage of hydrogen would require some training in handling.

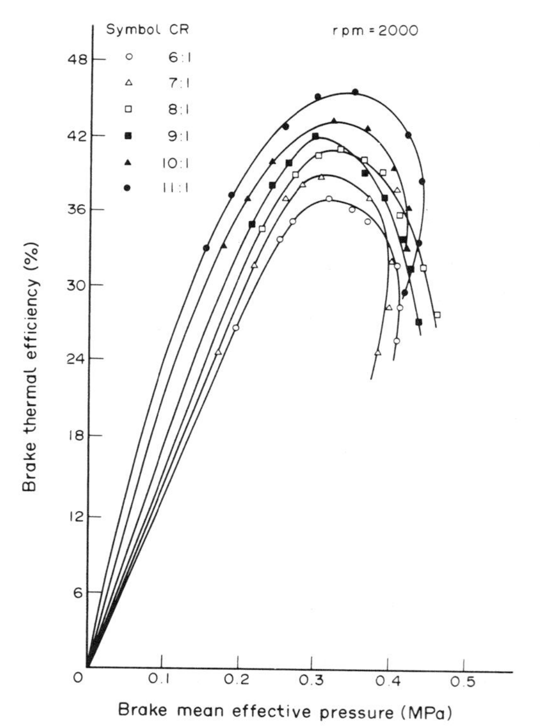

Fig. 1. Brake thermal efficiency as a function of brake mean effective pressure (BMEP) at constant speed for various compression ratios.

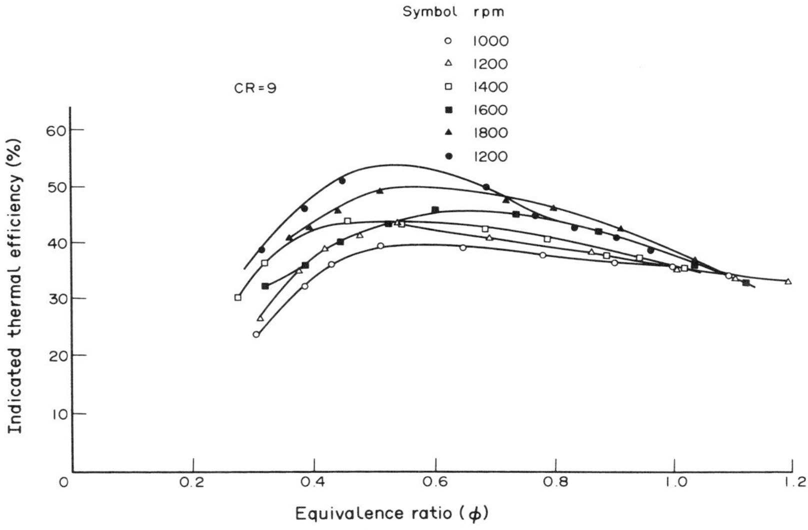

Fig. 2. Indicated thermal efficiency as a function ratio at constant compression ratio for various speeds.

As has been discussed earlier, the conversion of an existing engine system to hydrogen operation is not radically a new concept. The efforts carried out over the years in this direction range from a simple laboratory curiosity to a strong desire of meeting the twin challenges of the fuel crisis and environmental degradation. A detailed account of such efforts has been given elsewhere [8, 9].

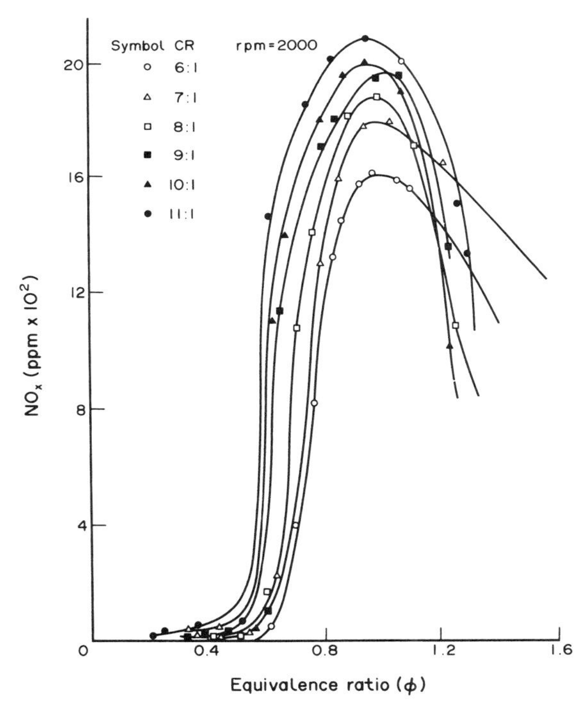

Fig. 3. Oxides of nitrogen (NO_x) emissions as a function of equivalence ratio at constant speeds for various compression ratios.

Extensive research work embracing all aspects of these studies is being pursued in IIT Delhi for the past several years. Both carburetion [10] and timed manifold injection [8] have been widely investigated as fuelling modes of an engine. The merits of timed manifold injection system have been brought out in terms of the engine's performance [11, 12], exhaust emission [13] and smoother combustion characteristics [14]. Results of these studies are demonstrated in Figs 1, 2 and 3. A technologically superior approach that might evolve in future may be the design of an engine specifically tailored to the needs of hydrogen fuel.

Delayed port admission, reported by Watson and Coworkers [15] has been, like TMI, a very safe method of hydrogen fuel induction. It overcomes the problem of residual gas ignition and ensures a backfire free operation. Other prominent causes of backfiring due to hot deposit ignition and hot surface ignition have also been omitted from this system by several other measures such as stringent oil control measures, selection of a synthetic lubricating oil and water injection.

Fuel induction techniques do play a sensitive role as far as safety aspects are concerned. The timed manifold injection system [8] has been found to be superior to carburetion in that in the former the fuel injection can be delayed somewhat after the intake of air has begun with a view to facilitate reduction of temperature levels of the potential hot spots responsible for backfire and other undesirable combustion phenomena. However, development of a suitable injector demands certain rigorous conditions. In view of the wide flammability range of hydrogen and its invisible odourless flame, the injector, under no circumstances should leak hydrogen either into the engine cylinder or in the intake manifold at any unscheduled point in the cycle. In this context, apart from precision surface finish, choice of proper sealing material is extremely important.

Some abnormal conditions could occur during the engine operation. Leak detection and necessary remedial action form a very important design aspect of a safe hydrogen-engine-system development. A combustible gas detector should be mounted at the highest level which should actuate an alarm indicating that a significant quantity of hydrogen has leaked off thereby posing problems of safety. However, in a well-ventilated test cell; if the quantity of hydrogen leakage is very small, normally no safety problem arises.

There occurs sometimes the possibility of engine misfire. Under such circumstances, any unburned hydrogen should be conducted to the outside through a closed enhaust system. Similarly, in the event of the engine stopping abruptly due to some reasons hydrogen supply system should be cut-off immediately. These design precautions will ensure the possibility of no hydrogen being amassed in the event of a sudden accidental failure of engine operation.

CONCLUSION

There is no room for doubt that a well-educated person should be able to use hydrogen fuel safely. However, the hydrogen automobile need not be hastily introduced in a large scale without adequately demonstrating its safe operation. An initial bad image from premature introduction in public, can ultimately cause serious damage. Hydrogen-specific vehicles should be as routinely accepted as gasoline-powered vehicles probably after an initial run-in period in which demonstration automobiles would exhibit ample reliability and safety.

REFERENCES

1. National Fire Protection Association (NFFPA), Standard 325M, Fire hazard properties of flammable liquids, gases and volatile solids.
2. D. P. Gregory, *Scientific American*, p. 381 (1973).
3. T. L. Bowen, Hazards associated with hydrogen fuel, *Proc. 11th Intersoc. Energy Conversion Engng Conf.*, p. 997 (1976).
4. J. G. Hust, Survey for materials for hydrogen service. NBS Special Publication No. 419, Chap. 4 (1975).
5. J. S. Laws, V. Frick and J. McConnel, Hydrogen gas pressure vessel problems in the M-1 facilities. NASA CR-1305, Washington D.C. (1969).
6. D. A. Mathis, *Hydrogen Technology for Energy*. Noyes Data Corporation, Park Ridge, NJ (1977).
7. R. G. Zalosh and T. P. Short, Comparative analysis of hydrogen fire and explosion incidents. Factory Mutual Research Corporation (1978).
8. L. M. Das, Studies on timed manifold injection in hydrogen operated spark ignition engine: Performance, combustion and exhaust emission characteristics. Ph.D. Thesis, I.I.T. Delhi, India (1987).
9. L. M. Das, Hydrogen engines: a view of the past and a look into the future. *Int. J. Hydrogen Energy* **15**, 425–443 (1990).
10. P. R. Khajuria, Hydrogen fueled spark ignition engine—its performance and exhaust emission characteristics. Ph.D. Thesis, IIT Delhi, India (1981).
11. L. M. Das, Fuel induction techniques for a hydrogen operated engine. *Int. J. Hydrogen Energy* **15**, 833–845 (1990).
12. H. B. Mathur and L. M. Das, Performance characteristics of hydrogen fuelled S.I. engine using timed manifold injection. *Int. J. Hydrogen Energy* **16**, 115–127 (1991).
13. H. B. Mathur and L. M. Das, Automobile exhaust pollution control through hydrogen fuel substitution. *Proc. 8th World Clean Air Congress*, Vol. 4, pp. 429–439, Netherlands (1989).
14. H. B. Mathur and L. M. Das, Combustion related studies on a timed manifold injection hydrogen engine. *9th Miami Int. Cong. Energy and Environment* (1989).
15. H. C. Watson, E. E. Milkins, W. R. B. Martin and J. Edsell, An Australian hydrogen car. *Proc. 5th World Hydrogen Energy Conf.*, pp. 1549–1562, Canada (1984).

Safety Aspects of Hydrogen Combustion in Hydrogen Energy Systems

M. Fischer

D.F.V.L.R., Institute for Technical Thermodynamics, F.R.G.

Abstract—Safety aspects will become essential for the introduction and acceptance of gaseous and liquid hydrogen as an energy carrier and fuel in energy supply systems. Prevention and control of accidental formation and ignition of large volumes of fuel–air mixtures are of primary importance when safety aspects of released gaseous hydrogen are discussed. Detailed knowledge of the overpressure in an accidental situation is essential for the protection of the public as well as for the corresponding plants and safety installations. Considerable progress has been made in the last few years concerning the understanding of the complex phenomena involved in combustion processes of gaseous mixtures. This holds in particular for flame acceleration and maximum turbulent flame speeds in unconfined and confined geometries. Fast turbulent deflagrations often transit spontaneously to detonations if flame speeds are high enough, depending on the combustible and boundary conditions. This paper discusses the potential hazards of hydrogen in the energy market as compared with other and already familiar energy carriers like natural gas and propane.

INTRODUCTION

Hydrogen is a feedstock in the chemical industry and has been handled safely for decades in large-scale chemical process engineering. Production, storage, transport and utilization of hydrogen are essentially routine in chemical industry. In the Federal Republic of Germany the world's largest industrial pipeline network for gaseous hydrogen supply has operated since 1940 with a total length of 220 km and 18 industrial connections [1]. With this network very high operational reliability and safety have been demonstrated. Replacement of parts in the hydrogen pipeline network have shown that, in contrast to gas pipelines, only negligible deposition or corrosion damage exists [2]. However, this hydrogen supply pipeline is at relatively low pressure (11–25 bar) and of small diameter (150–300 mm); the resulting low hoop stress in the walls of the network makes any material degradation due to hydrogen embrittlement highly unlikely. Such degradation, however, could be very prominent in future large-scale hydrogen transmission lines which may be operated at pressures of 100 bar or greater and thus experience high stresses.

Liquid hydrogen is being used, and also safely handled and controlled, in large quantities as a fuel for spacecraft propulsion. In power plants, hydrogen is applied in cooling big electrical generators. No new safety problems are expected in industrial applications of hydrogen, even assuming a strong increase of hydrogen utilization in industrial processes. However, the envisaged large-scale introduction of hydrogen as an energy carrier into the energy supply system raises questions concerning specific risks and necessary safety measures of a much greater significance than the solely industrial use of hydrogen.

SAFETY GOALS—GENERAL CONSIDERATIONS

The use of hydrogen as an energy carrier and as a fuel inevitably results, as with the use of any other combustible, in risks for the public. When used in public energy supply systems, hydrogen must be handled safely not only by hydrogen experts, but in addition and even primarily by lay persons. Untrained people must be able to use hydrogen as a fuel with the same degree of safety and with no more involved risks than with conventional gaseous and liquid fuels today. The term risk is defined as the product of the entrance probability of an incident or accident and the possible extent of damage as a result of an occurred incident or accident. In this perspective, risk evaluations are used by safety experts to assess a well-balanced and rational safety technology. Two aspects are, therefore, of overriding importance with respect to safety design in the field of energy technology. First, prevention of accidental situations (i.e. reduction of entrance probabilities of incidents and accidents where this is useful and meaningful) and second, mitigation of postulated accidents (i.e. reduction of damage or consequence of assumed accidents).

Safety aspects and considerations will be essential for the acceptance and introduction of a new energy carrier like hydrogen. First of all, hydrogen must be compared with conventional gaseous and liquid fuels with respect to the basic safety characteristics. Of fundamental importance for all safety-related considerations concerning production, storage, transport and utilization of large quantities of combustibles and fuels, is the investigation and assessment of the ignition, combustion and eventually detonation behavior including the resulting pressure development during accidents.

Triggering incidents for most accidental situations with flammable combustible mixtures are fuel leakages caused by material defects, material embrittlement, corrosion, mechanical overload of construction materials, design weaknesses, collision (in the case of vehicles) or insufficient maintenance of equipment.

In the course of an uncontrolled coincidence of (a) released fuel with (b) air or an oxidant and (c) ignition source (the classical fire triangle) one must thoroughly distinguish between combustion processes in confined or even partially confined areas and combustion processes in the open atmosphere.

In the case of an ignition, the possibility of a flame front acceleration to fast deflagration and the transition to detonation are decisive for the damage potential of uncontrolled released combustibles. Experiments with released gaseous and liquid hydrogen in unconfined and completely or partially confined areas show that in unconfined situations, neglecting some very special initial and boundary conditions, no deflagration-to-detonation transition (DDT) processes have been observed [3–14]. To trigger a detonation in a hydrogen–air mixture in the open atmosphere, an initiation charge, e.g. Tetryl, of critical energy content is required [5].

However, in confined or partially confined compartments there are realistic possibilities that deflagrations in hydrogen–air mixtures, with corresponding composition, may lead to detonation due to turbulence-promoting structures and interaction of shock waves with the flame front. Such processes result in high pressure peaks.

Among the combustibles considered in this paper (H_2, CH_4, C_3H_8), hydrogen is most sensitive to DDT. Therefore, the prevention of DDT in containments, compartments, pressure vessels, pipeline networks and storage systems is one of the most important specific safety goals in hydrogen energy technology. Inherently working mechanisms resulting in a reduction of the entrance probability of DDT processes may be of outstanding importance in hydrogen energy safety technology.

BASIC SAFETY CHARACTERISTICS

To assess potential risks and related safety problems of the energy carrier hydrogen, a comparison of its basic safety characteristics with other combustibles used on a large scale is illuminating. In Table 1 the most important data are summarized. A comprehensive set of data is given in [15].

Limits of flammability and ignition mechanism

Besides the mixture composition, the limits of flammability are in principle also dependent on energy of ignition, pre-pressure and temperature and moisture content in the unburnt mixture [16, 17]. Concerning realistic accidental situations, the influence of the ignition energy on flammability limits is very small, in particular with respect to a self-sustaining flame propagation. The range of concentration for flammability increases as the pressure in the unburnt mixture is increased above normal pressure (101.3 kPa). At low pressures (below normal pressure) the range of flamm-

Table 1. Basic safety relevant characteristics of hydrogen, methane (natural gas) and propane [15].

Characteristics	Hydrogen H_2	Methane (nat. gas) CH_4	Propane C_3H_8
Density gas at NTP ($kg\ m^{-3}$)	0.0838	0.6512	1.8700
Autoignition temperature (K)	858	813	760
Minimum energy for ignition in air (mJ)	0.02	0.29	0.26
Limits of flammability in air (vol. %)	4–75	5.3–15.0	2.1–9.5
Flame temperature in air (K)	2318	2148	2385
Limits of detonability in air (vol. %)	18.3–59	6.3–14	
Detonation velocity in air ($km\ s^{-1}$)*	2.0	1.8	1.85
Detonation overpressure (kPa)*	1470	1680	1825
Low heating value ($kJ\ g^{-1}$)	119.93	50.02	46.35
High heating value ($kJ\ g^{-1}$)	141.86	55.53	50.41
Specific heat c_p NTP gas ($J\ gK^{-1}$)	14.89	2.22	1.67
Sound velocity NTP gas ($m\ s^{-1}$)	1294	448	260
Stoichiometric mixture in air (vol. %)	29.53	9.48	4.03
Diffusion coefficient in NTP air ($cm^2\ s^{-1}$)	0.61	0.16	0.12
Explosion energy ($kg\ TNT\ m^{-3}$ NTP gas)†	2.02	7.03	20.5
Explosion energy ($g\ TNT\ g^{-1}$ Fuel)†	24	11	10
Explosion energy ($g\ TNT\ kJ^{-1}$)†	0.17	0.19	0.20

* Stoichiometric mixture

† Theoretical explosion yields (realistic value about 10% of theoretical maximum)

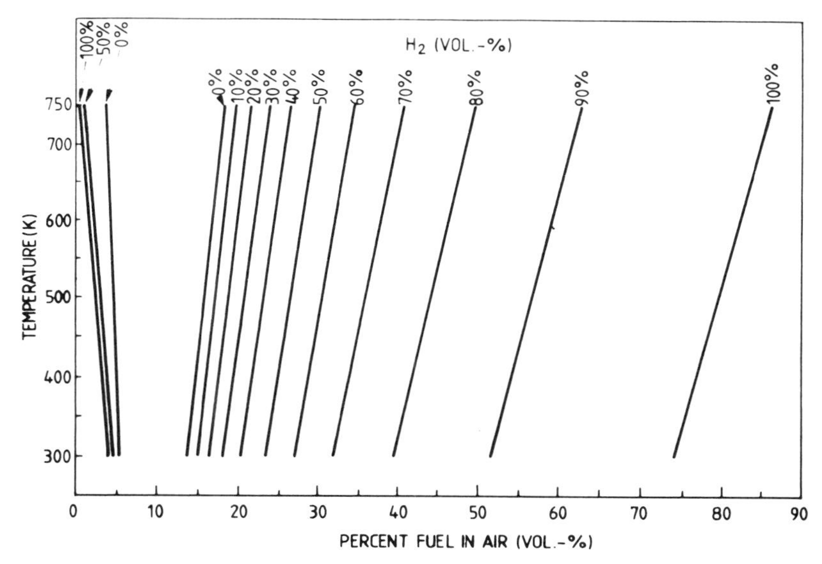

Fig. 1. Limits of flammability of H_2–air, CH_4–air and H_2–CH_4–air mixtures vs initial temperature of the mixture (upward flame propagation) [18].

ability narrows down until a self-sustaining flame propagation becomes possible.

Figure 1 shows the influence of the initial temperature of the mixture on the limits of flammability of H_2–air mixtures, CH_4–air mixtures and H_2CH_4–air mixtures for upward flame propagation [18].

The wide range of flammability of hydrogen–air mixtures as compared to other combustibles is in principle a disadvantage with respect to potential risks. On the other hand, there are only minor differences between hydrogen and methane concerning the lower limit. The lower limit of propane is even lower. However, in most accidental situations, the lower flammability limit is of particular importance. This is due to the fact that in realistic accident sequences, ignition sources with sufficient energy are nearly always present, once leaking fuels and combustibles have reached flammability concentrations in air.

For hydrogen, the minimum energy for ignition in air is about an order of magnitude less than for other combustibles [19]. However, the decisive aspect is the fact that so-called weak ignition sources, like sparks of electrical equipment (e.g. switch gears, relays, motors) or even electrostatic sparks and sparks from striking objects, already involve more energy than necessary to ignite even methane (natural gas), propane and other fuels [20]. A weak electrostatic spark discharged from a human body releases about 10 mJ [21], sufficient to ignite methane–air mixtures (minimum ignition energy 0.29 mJ). Surfaces with catalytic effects should also be considered in evaluating ignition hazards because they will reduce ignition temperature considerably.

Figure 2 shows the minimum ignition energy for H_2–air and CH_4–air mixtures. On both sides of the stoichiometric composition the minimum energies for ignition increase steeply for both, H_2– and CH_4–air mixtures, and are about equal at the lower flammability limits [22], but still far below the energy of weak ignition sources (about 10 mJ).

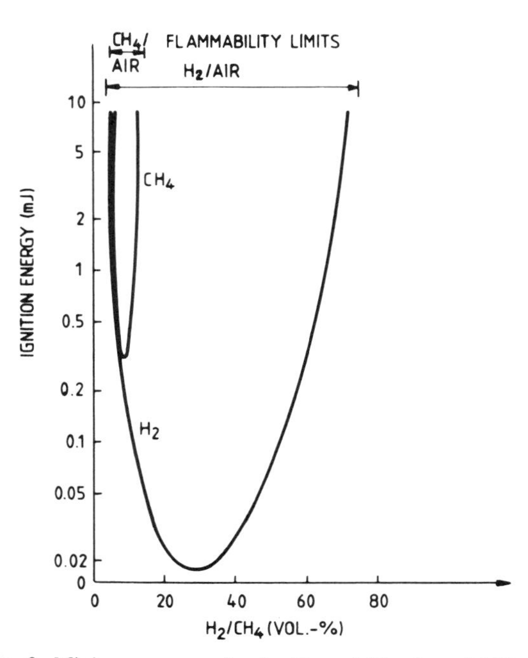

Fig. 2. Minimum energy for ignition of H_2–air and CH_4–air mixtures [22].

Table 2. Comparison of normal burning velocities.

Fuel	V_{max} (cm s^{-1})	c_{max} (vol. %)	V_{sto} (cm s^{-1})	c_{sto} (vol. %)
H_2	346	42.5	237.0	29.58
CH_4	43.0	10.17	42.0	9.50
C_3H_8	47.2	4.27	46.0	4.07

Burning velocity

The burning velocity is a fundamental safety-related property of combustibles and must not be confused with the flame speed. The flame speed is the sum of the burning velocity and the displacement velocity of the unburnt gas mixture. The higher the normal burning velocity of a combustible at NTP*, the stronger the tendency towards a DDT process. The normal burning velocity is the movement of a laminar flame front relative to the unburnt gas mixture and can be calculated theoretically as a function of the mixture composition. One of the specific characteristics of hydrogen is its high normal burning velocity. Table 2 shows a comparison of stoichiometric and maximum values of the normal burning velocities at NTP.

*NTP: Normal temperature and pressure is 293.14 K and 101.3 kPa.

Figure 3 shows the influence of the mixture composition on the normal (laminar) burning velocity at NTP [23]. In Fig. 4 the dependence of the normal (laminar) burning velocity on the initial temperature of stoichiometric H_2–air and H_2–O_2 mixtures is shown. The dependence of the burning velocity on pressure and temperature of the unburnt gas mixture, $V = V(T, p)$, is the flame propagation law. For not too high pressures and temperatures the flame propagation law may be empirically approximated by $V = V_o(p/p_o)^{0.2}(T/T_o)^2$. The very high normal (laminar) burning velocity V_o of hydrogen represents, in combination with the strong temperature dependence of V, one of the major reasons for the acceleration from laminar to turbulent hydrogen flame fronts and for the possibility of DDT. Other major influences on DDT processes are turbulence-promoting structures and interactions of the flame front with pressure and shock waves.

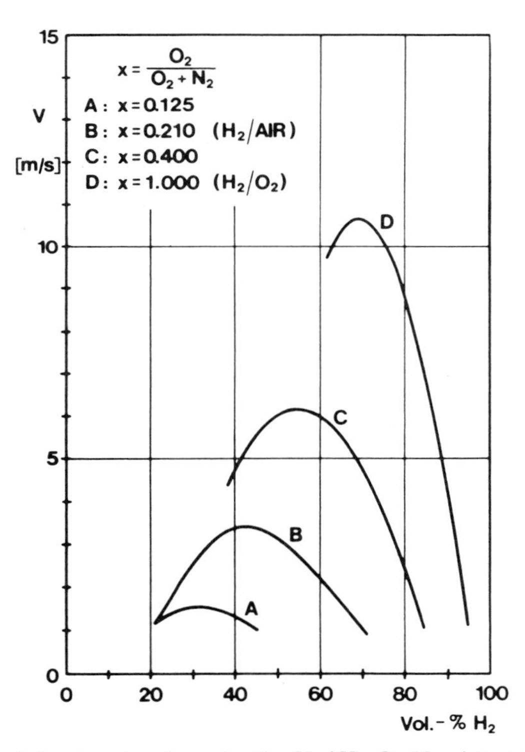

Fig. 3. Laminar burning velocities V of H_2–O_2–N_2 mixtures vs mixture composition [23].

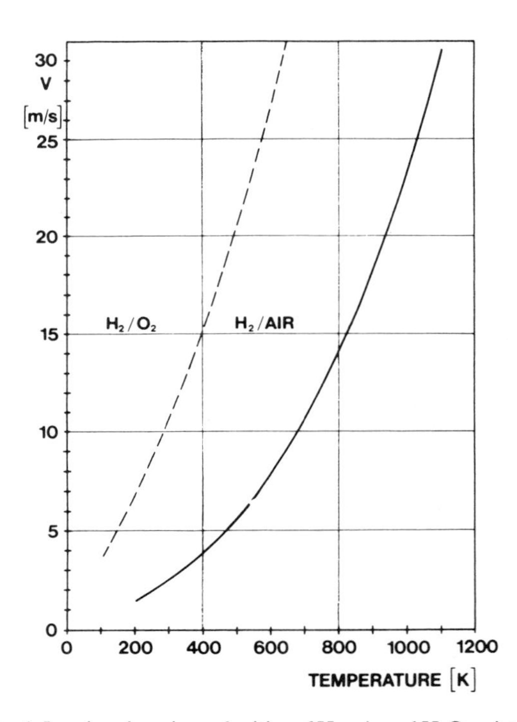

Fig. 4. Laminar burning velocities of H_2–air and H_2O_2 mixtures at stoichiometric composition vs initial temperature.

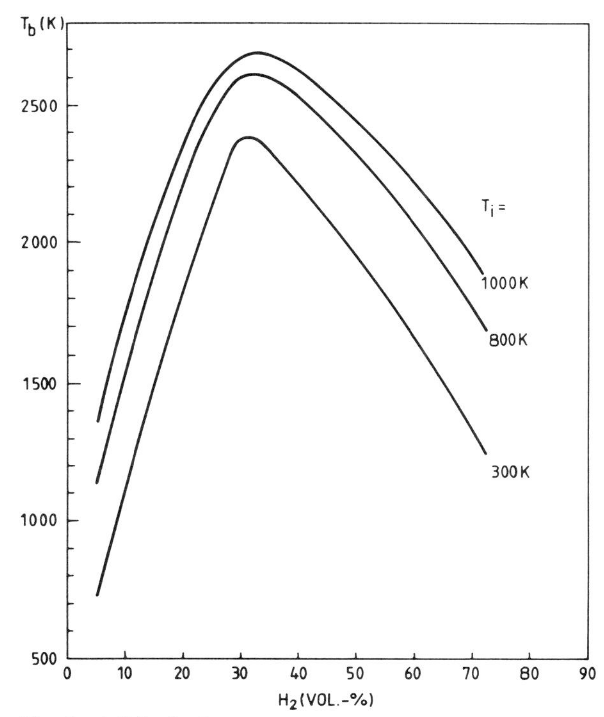

Fig. 5. Adiabatic flame temperature of H_2–air mixtures vs mixture composition and initial temperature.

Flame temperature

The adiabatic flame temperatures of H_2, CH_4, C_3H_8 and gasoline in stoichiometric mixtures with air are not very different. The strong dependence of the adiabatic H_2–air flame temperature on the mixture composition is shown in Fig. 5. Burning hydrogen exhibits very low emissivity, only a fraction of that of burning natural gas.

Detonation velocity and overpressure

Figure 6 shows the detonation velocity of H_2–air mixtures as a function of the mixture composition [19]. The corresponding detonation overpressure may be calculated approximately from $p_2 = (\rho_1 D^2)/1 + \gamma_2)$, where D is the detonation velocity, ρ_1 the density of the unburnt gas mixture and γ_2 the adiabatic exponent of the burnt gases.

Overpressures caused by deflagration and in particular by detonation are the main hazards of accidentally ignited combustibles and fuels. The existing boundary conditions determine the amount of overpressure. In confined and unvented cubic or spherical volumes, deflagrations of stoichiometric mixtures of H_2, CH_4, C_3H_8 and most fuel vapors in air result in a maximum pressure rise which is around 8:1. The rate of pressure rise due to deflagration, however, may be very different for different combustibles and fuels subject to the same initial and boundary conditions. In pipelines, the maximum pressure rise during deflagration may be considerably higher than 8:1, depending on the geometrical boundary conditions. Therefore, the geometry of the containment must be taken into account. Detonations of stoichiometric combustibles in closed volumes and containments lead to a maximum pressure rise of about 20:1, i.e. to more than twice the value of an adiabatic deflagration [24]. However, considerably higher pressure rises will result if a fast deflagration transits to detonation due to the pressure build-up in the unburnt combustible in front of the deflagration. As already mentioned, hydrogen is the most sensitive to DDT processes among the combustibles H_2, CH_4 and C_3H_8 considered in this paper.

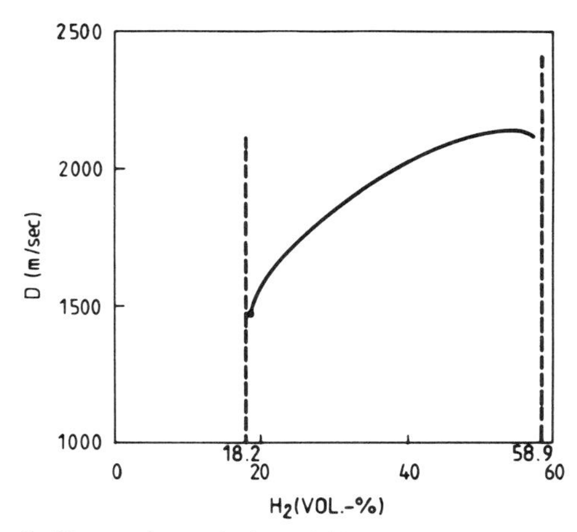

Fig. 6. Detonation velocity of H_2–air mixtures vs mixture composition [19].

Explosion energy

Despite basic shortcomings [25], the TNT†-equivalent concept will be further used until nonideal explosions, e.g. those with combustibles, can be characterized more definitely. The theoretical TNT-equivalent of combustibles and fuels can be calculated [26]. In Table 1, these theoretical explosion yields are quoted for H_2, CH_4, and C_3H_8. Hydrogen has the lowest theoretical explosion potential on a volumetric basis and the highest on a mass basis. For practical safety purposes, the volume-based explosion energy is relevant. For equivalent energy storage, related to the upper heating value, hydrogen has about the same theoretical explosion potential as methane and propane. However, experimental and computation investigations show that only a fraction of about 10% of the theoretical explosion energy is released in fuel–air explosions, because in the course of an accident only small fractions of released fuels are in the flammability range at any time [25, 27].

†Symmetrical Trinitrotoluene.

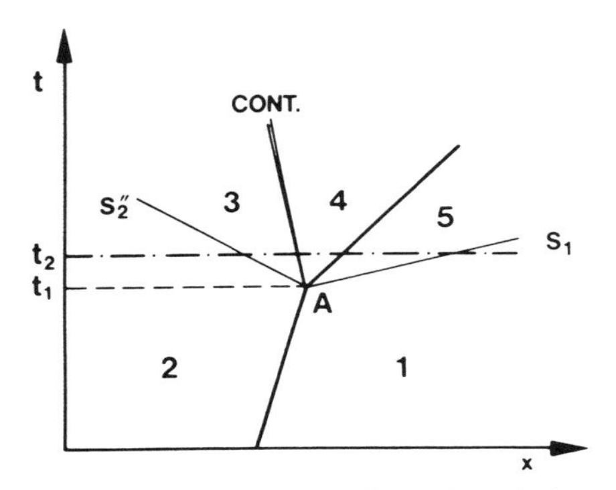

Fig. 7. Sudden acceleration of a flame front (schematic).

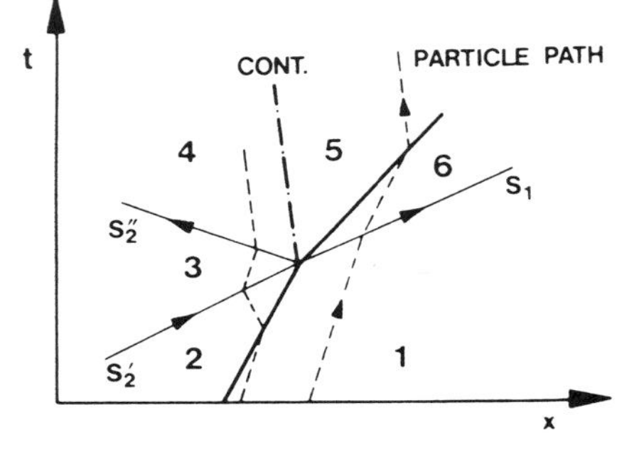

Fig. 8. Shock wave flame front interaction (schematic).

DEFLAGRATION AND DETONATION BEHAVIOR

Precise knowledge of acceleration processes of laminar and turbulent flame fronts during deflagration of accidentally released combustibles and fuels is of central significance for the control of such accident sequences. Dependent on initial and boundary conditions, the same combustible–air mixture can adopt burning velocities which may differ by orders of magnitude.

Laminar burning velocities of stoichiometric mixtures of conventional combustibles, e.g. CH_4 and C_3H_8, with air are about 0.5 m s^{-1} at NTP; the corresponding H_2–air mixture has a laminar burning velocity of about 2.4 m s^{-1} (Table 2). The related laminar flame velocities (burning velocity plus displacement velocity of unburnt gas) amount to a few m s^{-1} for CH_4 and C_3H_8, and to more than 10 m s^{-1} for H_2–air mixtures. Due to a change of the flame front character from laminar to turbulent turbulence-promoting structures, interactions of reflected pressure and shock waves with the flame front and prepressurization, as well as preheating of the unburnt gas mixture by precursory pressure or shock waves, flame speeds of combustible–air mixtures may be accelerated to several 100 m s^{-1}, and under certain conditions up to critical turbulent flame speeds which may transit to detonation processes. Such acceleration processes are shown schematically in Figs. 7–9. According to Fig. 7, the combustion process starts with ignition on the left hand side at the closed end of a combustion tube. At point A, the flame front accelerates due to a change from laminar to turbulent status. Shock waves moving upstream and downstream (S_1, S_2'') and a contact discontinuity (cont) are produced. The unburnt gas mixture in front of the flame is now preheated by the shock wave S_1 which results in further acceleration of the flame front according to Fig. 4.

An example for acceleration processes due to interaction of reflected pressure and shock waves with the flame front is shown in Fig. 8. A fast moving shock wave S_2' enters the flame front from behind resulting in the transmitted shock S_1 and the reflected shock S''. The acceleration of the flame front is partly due to the impact of the shock wave and partly to preheating of the unburnt gas mixture.

Figure 9 shows the case of an interacting shock wave of critical strength entering the flame front from behind, resulting in DDT. Such highly nonsteady combustion processes were modeled and compared with experimental results [28].

In Fig. 10 a DDT process in a tube closed at the ignition end is shown. In this case, the calculated flame acceleration process leads to DDT at a certain detonation induction distance and a corresponding induction time which are in good agreement with experimental data [28].

Figure 11 shows an example of the pressure as a function of time and tube distance for a DDT process. The figure shows (1) deflagration, (2) transition to detonation, (3) and (4) transition process to Chapman–Jouquet detonation.

A detonation shock wave propagates with supersonic speed relative to the unburnt gas mixture ahead of the wave. Typical detonation velocities are of the order of a few km s^{-1}. Detonation velocities of H_2–air mixtures are shown in Fig. 6 as a function of mixture composition. The temperature rise associated with the compression in a detonation shock wave triggers autoignition of the

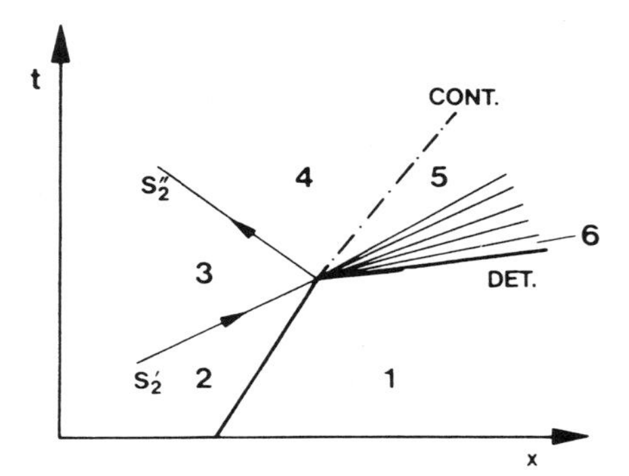

Fig. 9. Shock wave flame front interaction with DDT (schematic).

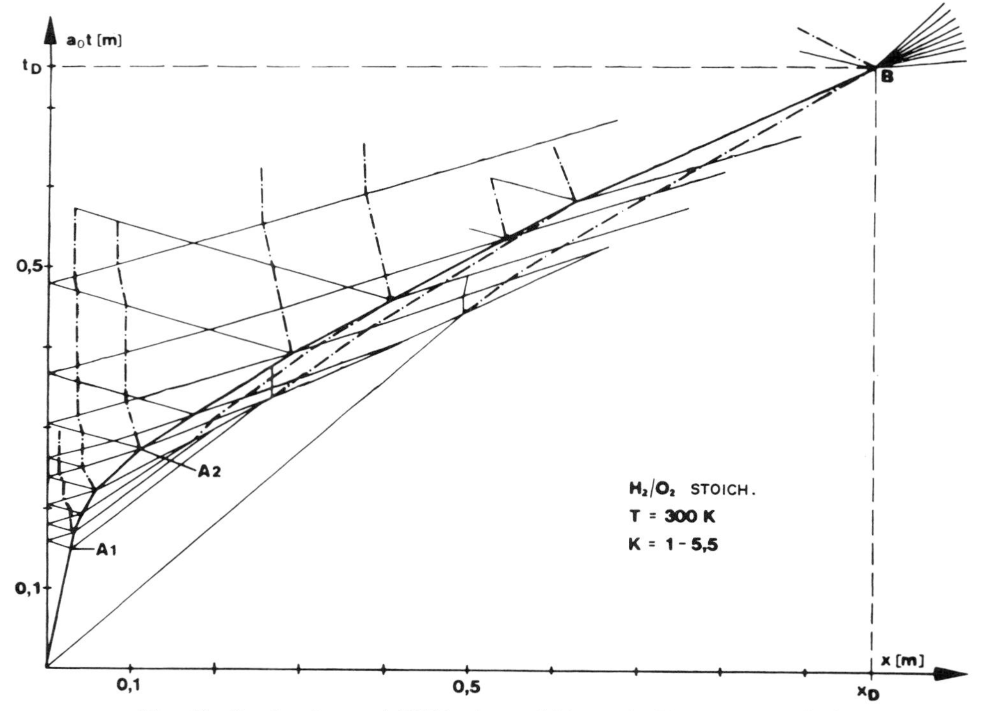

Fig. 10. Combustion and DDT of a stoichiometric H_2–O_2 mixture [28].

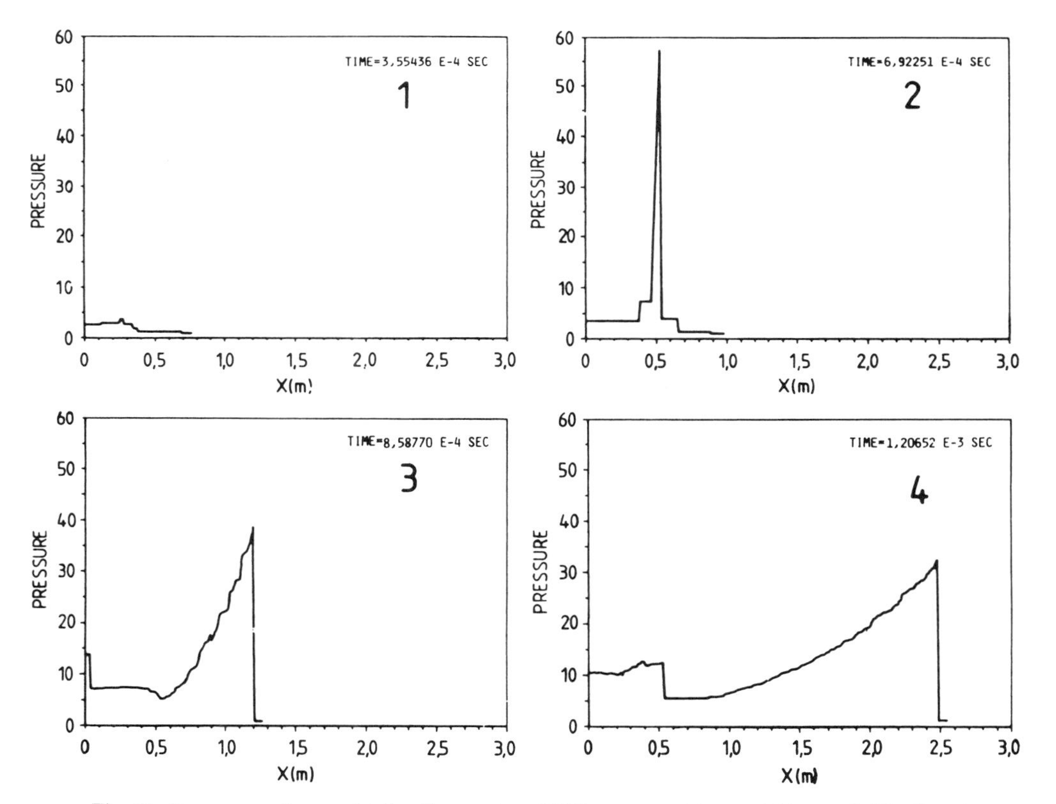

Fig. 11. Pressure vs time and tube distance in a DDT process with a stoichiometric H_2–O_2 mixture.

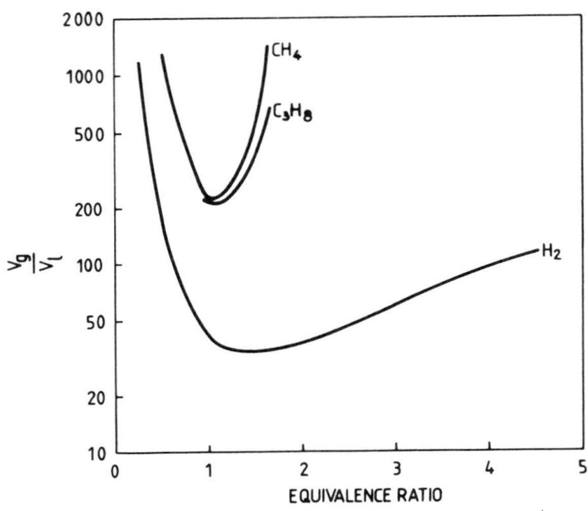

Fig. 12. Ratio of critical to laminar burning velocities v_g/v_l for CH_4–, C_3H_8–, H_2–air mixtures vs equivalence ratio.

combustible–air mixture in the shock front. Therefore, the reaction front and the detonation shock wave are strongly coupled and propagate as a unit with detonation velocity.

Considerable progress has been made in recent years concerning the understanding of the complex phenomena involved in turbulent deflagration and detonation processes [28–40]. Provided that corresponding initial and boundary conditions exist, the discussed acceleration processes of flame fronts have the potential to lead to DDT in all combustibles. This transition phenomenon is unlikely to occur if deflagrations take place in the open atmosphere. However, in confined geometries, e.g. pipelines, containments, compartments, and vessels, critical turbulent burning velocities can be predicted at which the DDT takes place without any interaction with reflected shock waves and gas dynamic discontinuities. Figure 12 shows critical burning velocities of methane–, propane– and hydrogen–air mixtures, related to the normal (laminar) burning velocity, as a function of the equivalence ratio.

For stoichiometric and overstoichiometric fuel–air mixtures this ratio is much lower for hydrogen compared to the other combustibles. This is due to the already discussed high normal (laminar) burning velocity of hydrogen (Table 2). The lower this ratio, the easier it is for DDT to take place. However, if understoichiometric fuel–air mixtures are considered, i.e. lean mixtures, the differences between the combustibles diminish. Similar conclusions may be drawn concerning the aforementioned critical shock strength. A shock wave having critical strength, e.g. reflected from confinement walls, leads directly to DDT in the interaction process with the flame front (Fig. 9). For each ratio of the actual turbulent to the normal (laminar) burning velocity, v_t/v_l, a critical shock strength may be predicted, dependent on the fuel–air composition.

In Fig. 13 for methane–, propane– and H_2–air mixtures the critical shock strength $\pi_k = p_2 p_1 - 1$ is shown as a function of the equivalence ratio and for $v_t/v_l = 15$. In this example, the interacting critical shock enters the flame front from behind. Hydrogen within the limits of detonability is most sensitive to DDT, compared to the other combustibles. However, from a safety point of

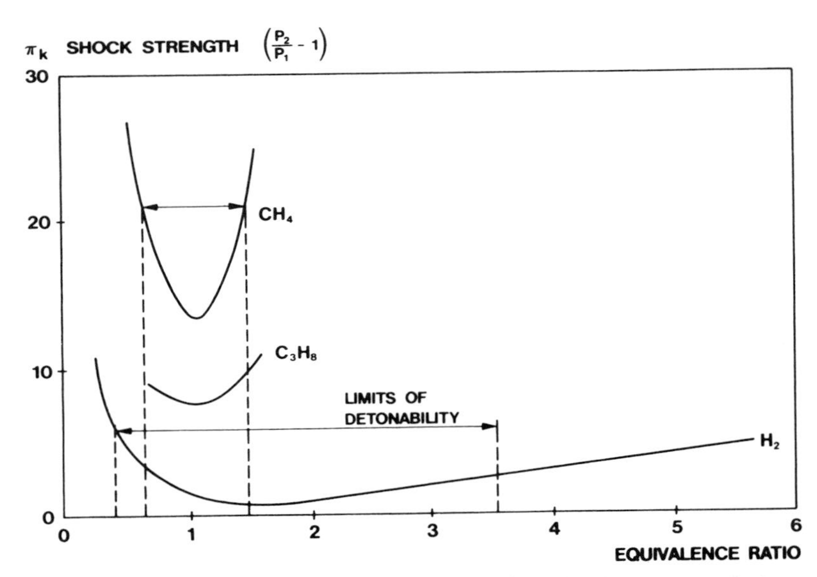

Fig. 13. Critical shock strength π_k for H_2–, C_3H_8– and CH_3–air mixtures vs equivalence ratio; $v_t/v_l = 15$.

view, the lower limit of detonability has to be considered again: for hydrogen in air this is 18.3 vol%, which is much higher than for methane in air (6.3 vol.%). Nevertheless, hydrogen safety research must still find ways to prevent or suppress DDT in the whole range of hydrogen–air mixtures. Possible DDT processes in confined areas and the associated pressure loadings represent one of the most important safety problems of all combustibles, including the use of hydrogen as an energy carrier and fuel.

Acknowledgement—The author expresses his sincere thanks to H. Eichert for his assistance and valuable contributions and discussions.

REFERENCES

1. C. Isting, Efrahrungen mit einem Pipeline-Verbundnetz für Wasserstoff. *3 R International* **6** (1974).
2. G. Haske, Petrochemischer Wasserstoff–Herstellung, Transport und Verwendung. *Chemie-Ingr-Tech.* **48** (1976).
3. D. S. Liu, R. MacFarlane and L. J. Clegg, Some results of WNRE experiments on hydrogen combustion, *Proc. Workshop Impact Hydrogen Water Reactor Safety*, Albuquerque, U.S.A., Vol. III 72–91 (1981).
4. W. Bartknecht, *Explosionen*. Springer, Berlin (1978).
5. J. H. Lee, Overview of gas explosions and recent results in the study of turbulent deflagrations and detonations. Int. Symp. Control Prevent. Gas Explos., London (1983).
6. W. Bartknecht, Ablauf von Staub- und Gasexplosionen und deren Bekämpfung. Jahrg Inst. Chem. Treib- und Explosivstoffe, Pfinzthal-Berghausen (1973).
7. W. E. Baker, Blast waves emitted from gas explosions. Int. Symp. Control Prevent. Gas Explos., London (1983).
8. L. B. Thompson *et al.*, Hydrogen combustion and control in nuclear reactor containment buildings. *Nucl. Safety* **25**, 3 (1984).
9. M. Berman and J. C. Cummings, Hydrogen behaviour in light water reactors. *Nucl. Safety* **25**, 1 (1984).
10. G. D. Brewer, G. Wittlin, E. F. Versaw, R. Parmley, R. Cima and E. G. Walther, Assessment of crash fire hazards of LH_2-fueled aircraft, NASA-Report CR-165525 (1981).
11. L. H. Cassut, F. E. Maddocks and W. A. Sawyer, A study of the hazards in the storage and handling of LH_2. *Adv. Cryogen. Engng* **5**, 55–61 (1959).
12. M. G. Zabetakis, A. L. Furno and G. H. Martindill, Explosion hazards of LH_2. *Adv. Cryogen. Engng* **6**, 185–194 (1960).
13. D. L. Ward, D. G. Pearce, D. J. Merret, Liquid hydrogen explosions in closed vessels. *Adv. Cryogen. Engng* **9**, 390–400 (1963).
14. R. D. Witkofski and J. E. Chirivella, Experimental and analytical analysis of the mechanisms governing dispersion of flammable clouds formed by liquid hydrogen spills. *Proc. 4th WHEC*, **4**, 1659–1674 (1982).
15. J. Hord, Is hydrogen safe? Inst. Basic Standards, National Bureau of Standards, Boulder, Colorado, NBS Tech. Note 690 (1976).
16. H. Christner, Experimentelle Bestimmungen der Zündgrenzen von Mehrstoffgemischen in Abhängigkeit von Anfangs- Temperatur-, Druck und Zündenergie. Dechema-Ausschuß Gas- und Flammenreaktionen (1974).
17. H. F. Coward and G. W. Jones, Limits of flammability of gases and vapors. National Bureau of Standards No. 503 (1952).
18. N. R. Baker, W. D. Van Vorst, Mixture properties for hydrogen supplementation of natural gas. In *Hydrogen Energy Systems*, Pt III, Zürich. Pergamon Press, Oxford (1978).
19. B. Lewis and G. von Elbe, *Combustion Flames and Explosions of Gases* (2nd edn). Academic Press, New York (1961).
20. C. McKinley, Safe handling of hydrogen. *Am. Chem. Soc., Symp. Series* **116** (1980).
21. National Fire Protection Association, Boston. Static electricity NFPA No. 77 (1972).
22. Sicherheitsaspekte einer künftigen europäischen Wasserstofftechnologie. Dornier Systemtechnik (1979).
23. J. Warnatz, Concentration, pressure and temperature dependence of the flame velocity in hydrogen–oxygen–nitrogen mixtures. *Combust. Sci. Tech.* **26** 203–213 (1981).
24. Y. Zeldovich and A. S. Kompaneets, *Theory of Detonations*. Academic Press, New York (1960).
25. R. A. Strehlow and W. E. Baker, The characterization and evaluation of accidental explosions, NASA CR-134799 (1975).
26. G. F. Kinney, *Explosive Shocks in Air*. Macmillan, New York (1962).
27. D. S. Burgess, J. N. Murphy, M. G. Zabetakis and H. E. Perlee, Volume of flammable mixtures resulting from atmospheric dispersion of a leak or spill. 15th Int. Symp. Combust., Combust. Inst., Pittsburgh (1974).
28. H. Eichert and M. Fischer, Hydrogen safety in energy applications compared with natural gas. *Proc. 5th WHEC* **4**, Toronto, 1869–1880 (1984).
29. J. H. Lee and I. O. Moen, The mechanism of transition from deflagration to detonation in vapor cloud explosions. *Progr. Energy Combust. Sci.* **6** (1980).
30. F. Bartlmä, *Gasdynamik der Verbrennung*. Springer, Berlin (1975).
31. F. Bartlmä, The transition from slow burning to detonation. *Acta Astronautica* **6**, 435–447 (1979).
32. C. K. Westbrook, Hydrogen oxidation kinetics in gaseous detonations. *Combust. Sci. Tech.* **29**, 67–81 (1982).
33. T. V. Bazhenova, Y. U. Lobastov, J. Brossard, T. Bonne, B. Brion and N. Charpentier, Influence of nature of confinement of gaseous detonation. 7th ICOGER, Göttingen (1979).
34. G. K. Adams and D. C. Pack, Some observations on the problem of transition between deflagration and detonation. Proc. 7th Symp. Combust. (1959).
35. R. Edse and J. R. Lawrence, Detonation induction phenomena and flame propagation rates in low temperature hydrogen–oxygen mixtures. *Combust. Flame* **13** (1969).
36. F. E. Belles, Detonability and chemical kinetics: prediction of limits of detonability of hydrogen. Proc. 7th Int. Symp. Combust. (1959).
37. J. H. S. Lee, Initiation of gaseous detonation. *A. Rev. Phys. Chem.* **28**, 75–104 (1977).
38. D. Pawel, P. J. van Tiggelen, H. Vasatko and H. Gg. Wagner, Initiation of detonation in various gas mixtures. *Combust. Flame* **15**, 173–177 (1970).
39. A. A. Boni, M. Chapman, J. L. Cook and G. P. Scheyer, Transition of detonation in an unconfined turbulent medium. AIAA 15th Aerospace Sci. Meet., Los Angeles (1977).
40. D. S. Burgess, J. N. Murphy, N. E. Hanna and R. W. van Dolah, Large scale studies of gas detonations, Bureau of Mines. Investigations Report 7196 (1968).

Chapter 6

The Hydrogen Economy

6.1 Introduction

In light of the problems associated with excessive fossil fuel use, many people have advocated the use of hydrogen on a global scale as a universal fuel. A major barrier to this goal is the establishment of the necessary infrastructure—an expensive proposition that could take several decades to complete. Given these obstacles, hydrogen vehicles will probably have to be initially introduced on a smaller scale, into niche markets such as fleet operations with central refueling sites. A flow chart showing some of the infrastructure requirements for a major "transition to hydrogen" is shown in Fig. 6.1. This chapter examines different aspects of the hydrogen infrastructure—including production, storage, transportation, and distribution—from a feasibility and cost-effectiveness perspective. The technical aspects of hydrogen production are discussed in Chapter 4.

6.2 Production

Producing hydrogen at rates and costs that will be comparable to those of other transportation fuels will be a major challenge. Hydrogen currently costs on the order of 3 to 15 times more to produce than natural gas and 1.5 to 9 times more to produce than gasoline, depending on the method of hydrogen production used [1]. Today, the principal, and least expensive, method of producing hydrogen is from fossil fuels such as natural gas, oil, or coal. Unfortunately, fossil fuels are nonrenewable and emit pollutants when burned. Considering the environmental consequences of using fossil fuels, it is unlikely that they will be used to meet the long range needs of hydrogen production if these needs are significantly expanded. Although fossil fuel production of hydrogen could play a more critical role in the short-term introduction of hydrogen vehicles, the focus of this section will be on the economic prospects of hydrogen production from renewable sources, in particular from water electrolysis and biomass.

Water Electrolysis

From an environmental perspective, water electrolysis is one of the best methods of producing hydrogen, provided the electricity used in the process is generated from clean, renewable sources. For electrolysis, the generation of electricity is the primary process expense (as opposed to capital costs), and hence the primary consideration for pricing scenarios [3]. To date, the high costs of renewable energy sources have precluded their more widespread use, and will likely continue to remain an obstacle in the near term. Currently, there are a number of technologically feasible methods available to generate electricity without burning fossil fuels, including solar photovoltaics, solar thermal electricity, wind power, or off-peak hydropower.

Solar Photovoltaics

Although solar photovoltaics (PV) are a well developed and proven technology, they are currently only cost-effective, on a lifecycle basis, for remote sites located where

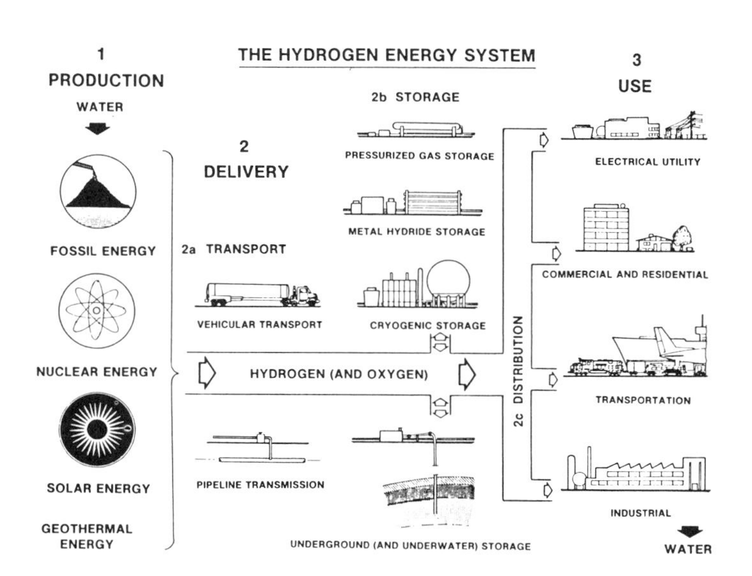

Figure 6.1. The hydrogen energy system. (Source: ref. [2])

grid power is unavailable. The first generation of PV cells was based on thick film, single crystal silicon cells with high conversion efficiencies. Thin film amorphous silicon cells, which have lower efficiencies but are less costly, now occupy about one-third of the market share. Efficiencies of solar cells currently range from 3-17 percent for field applications, and from 6-34 percent for laboratory applications [4].

Unit prices for solar cells have declined by nearly two orders of magnitude in the last 20 years. While solar cells cost over $300,000/kWp in the early 1970s [4], large (> 5 MW) PV systems today cost about $4000-9000/kW installed [5]. Today's large PV systems can provide electricity at a rate of about 14 to 35 cents/kWh [5]. Presently, over 40 major manufacturers are investing in the technology, and several are gearing up for a second stage of larger-scale production, in anticipation of larger markets and cost reductions [4]. It has been estimated that the cost of PV systems could drop to $1500-3500/kW during the 1990s, with DC electricity costs of 6-14 cents/kWh. By the early part of the next century, thin-film solar-cell or concentrator systems could potentially bring costs down to $500-1100/kW, with DC electricity costs of 2.2-4.4 cents/kWhDC [5].

Solar Thermal Electric

Another method of capturing the energy of sunlight is to use solar thermal techniques. In solar thermal systems, solar radiation is collected over a large area "collector" and focused onto a smaller area receiver. The heat collected at the receiver is then used to run an electric generator. There are three main types of solar thermal collectors: parabolic trough collectors, central-receiver collectors, and parabolic dishes.

<u>Parabolic Troughs</u>

Parabolic troughs are the most developed solar thermal technology. This type of collector focuses sunlight onto a central pipe which contains oil. The hot oil in the pipe, at 300-400° C, is then passed through a heat exchanger which produces steam that is used to run a steam turbine generator. The overall efficiency of converting sunlight to electricity for these systems is about 13-17% [5]. A majority of the solar thermal-electric parabolic trough systems are located in California, which has a capacity of 354 MW [4]. Parabolic trough systems are now commercially available at $2800-3500/kW, and can produce electricity at 12-26 cents/kWh. With improvements in capital costs, prices could drop to $2000-2400/kW resulting in electricity costs of about 8-12 cents/kW [5].

<u>Parabolic Dish</u>

Parabolic dish systems consist of an array of parabolic dishes, each of which tracks the sun in two axes and focuses light onto a receiver at the focal point of the dish. The thermal energy is converted to electricity either by a small engine-generator located at each dish, or by having the receiver heat a working fluid which is then piped to a central location to produce steam and electricity. Stirling engines have received the most attention for use with parabolic dishes. These engines utilize external heat to drive a working gas, typically hydrogen or helium at high pressures, to generate energy. The expansion and contraction of the working gas moves a piston which is mechanically linked to a generator. The parabolic dishes themselves are highly efficient since they have high concentration ratios and always point toward the sun. Efficiencies for these systems are currently about 16-24% and could reach 20-28% [5]. One of the major obstacles for this technology is the cost of developing an adequate Stirling engine [6]. Systems with circulating fluids have also encountered difficulties in the heat transfer process. In the near term, parabolic dish systems are projected to cost $3000-5000/kW, with electricity costs of 17-38 cents/kWh. In the longer term, costs of $1250-2000/kW and 6-12 cents/kWh are projected [5].

<u>Central-Receiver Systems</u>

In central-receiver systems, an array of moveable flat plate heliostats focus sunlight onto a central-receiver tower, heating a working fluid, which in turn drives a steam turbine to generate electricity. Central-receiver systems can operate at temperatures from 500-1500° C and at efficiencies of about 8-15%. In the future, efficiencies are expected to reach levels of 10-16% [5]. This technology has been demonstrated in pilot projects ranging in size from <1 MW to 10 MW [7]. The largest of these demonstrations was the 10 MW Solar One test site located in Barstow, CA, which was operated successfully for more than six years. This facility is currently being redesigned under the Solar Two project to utilize molten salt rather than water as the operating fluid. The retrofit is expected to be completed in July of 1995 and the Solar Two plant will operate from January 1996 through 1998 [8].

With the current technology, central-receiving systems are expected to cost between $3000-4000/kW, yielding electricity costs of about 10-20 cents/kWh. Capital costs could drop to $2000-3000/kW in the near term, resulting in electricity costs of 7-12 cents/kWh. Over the longer term this number could drop to between 5.4-7.6 cents/kWh [5]. With research and development, the Department of Energy hopes to eventually achieve costs of 5 cents/kWh [7].

Hydroelectric Power

Hydropower is a mature, commercial electricity generation technology that is used around the world. Hydropower currently supplies 6.7% of the world's energy [9], and can be an inexpensive source of electricity at sites where excess off-peak power is available. At such sites, electricity for hydrogen production can be made available for 2-4 cents/kWhAC for eight hours [5]. Many of the best sites for hydroelectric power facilities in the United States and Western Europe are already exploited [10], however, and the development of further sites could be difficult due to environmental and land issues. A consortium of Canadian and European interests (the Euro Quebec Hydro Hydrogen Pilot Project) is currently investigating the potential of using hydrogen produced from off-peak Canadian hydropower in Europe as a transportation fuel [11]. Some of the demonstration vehicles that have been developed under this project are discussed in Chapter 7.

Wind Power

Wind power technology has improved dramatically over the past ten years and today nearly 1500 MW of wind power are installed in the U.S. Today's state-of-the-art technologies, which are capable of operating at variable rotor speeds, are expected to be able to produce electricity at a cost of about 5 cents/kWh, based on an average annual wind speed of 16 miles per hour [12]. Over the longer term, prices are expected to reach between 2.6 to 4 cents/kWh [5]. While much of the United States' wind power resources are currently in California, the prime locations for wind farms are largely in the central, Great Plains region. Under "moderate" land-use constraints, i.e., excluding 100% of environmentally sensitive and urban areas plus 50% of forest areas, 30% of agricultural areas, and 10% of range areas, 0.6% of the U.S. land area can be used for wind power applications with today's technologies. This number is expected to increase to 13% as the technology advances [13]. It is estimated that approximately 2% of the total U.S. land area would be needed to supply enough energy to power all the cars and light-duty trucks in the U.S. With such large areas involved, land-use issues and the effects of wind power on the local environment including visual impact, noise, and electromagnetic interference are concerns [14]. Dual land use for grazing or agriculture is possible, however. The potential for birds to fly into the rotor blades and be killed has also raised some attention [15].

Biomass

Another potential source of renewable hydrogen production is biomass (see Section 4.6). Biomass is currently the most widely used form of renewable energy, supplying about 14% of the world's energy, and about 35% of the energy needs in developing countries [16]. Biomass has the potential of producing enough energy to displace a major fraction of future oil use. The U.S. currently produces about 3.6 Quads of energy annually from biomass, or about 4.2% of the U.S. supply [17], and it is estimated that up to 14.6 Quads/yr of energy could be derived from biomass. From a cost perspective, biomass could also be the most competitive source of renewable hydrogen (see below). The primary problem with biomass hydrogen production is that it is land and water intensive. To produce enough hydrogen to fuel all the cars and light trucks in the U.S., 3% of the U.S. land area would have to be devoted to biomass production [5]. On the other hand, these land requirements could be met by using approximately 70% of idled cropland.

Comparison with Gasoline

As technological advances continue, the costs of renewable energy systems are expected to go down, especially when they are built on larger scales. Some projections of the present and future costs of hydrogen production from different sources are given in Table 6.1. For comparison, the wholesale price of gasoline at 63 cents per gallon is equivalent to $4.85/GJ. The projections in Table 6.1 show that while the costs of renewable electricity sources are currently greater than those of gasoline, they are projected to drop considerably over the next 10-20 years.

When comparing the price per gallon of gasoline to that of hydrogen, there are several factors that should be considered. The first is that hydrogen engines and, to a greater extent, fuel cells have the potential of achieving higher fuel efficiencies than gasoline engines. Thus, if a fuel cell vehicle is twice as efficient as a gasoline vehicle, the price of gasoline on a per mile basis would have to be doubled to provide an appropriate comparison to the fuel cell vehi-

Table 6.1 Current and Projected Production Costs of Hydrogen ($/GJ)

	1991	Near Term	Post 2000
Renewable Resources			
Electrolytic Hydrogen from:			
Solar PV	54-121	29-57	12-19
Wind (630 W/m^2)			13
(500 W/m^2)	37	25	17
(350 W/m^2)	53	38	21
Solar Thermal	45-60	37-63	22-30
Off Peak Hydroelectricity	12-19	12-19	12-19
Hydrogen from Bio Gasification			
Large plant (50 million scf/day)			6.2-8.8
Fossil Sources			
Hydrogen from Steam Reforming of Natural Gas			
Large Plant (100 million scf/day)	6.1-8.1	6.1-8.1	8.1-10.1
Small Plant (0.5 million scf/day)	11-14	11-14	14-17
Hydrogen from Coal Gasification			
Large Plant (100 million scf/day)	8	8	8
Medium Plant (25 million scf/day)	13	13	13

(Source: ref. [5])

cle, i.e., hydrogen produced at $9.70/GJ would give the same cost per mile as gasoline at $4.85/GJ. Fuel cell vehicles also have the potential for longer lifetimes and lower maintenance costs than gasoline vehicles (see Chapter 3), and hence some researchers have projected that, on a total life cycle cost basis, hydrogen fuel cell vehicles could be economically competitive with gasoline powered vehicles [18]. The prospects of renewable hydrogen production methods are discussed further in the reprint by Odgen and DeLuchi at the end of this chapter and in their more detailed report in ref. [5].

External Costs

Some researchers have pointed out that there are external costs associated with using fossil fuels that should be considered when comparing the economics of fuels. Among these costs for fossil fuel use are the costs of damage to the environment, and expenditures necessary to protect the supply of fossil fuels (e.g., military expenditures) [19].

6.3 Storage

Regardless of whether hydrogen is produced from fossil fuels or renewable sources, an effective method of storage will be needed to develop an adequate hydrogen infrastructure. For renewable energy sources in particular, an energy storage system will be necessary to keep up with oscillating energy demands, and variable (sometimes seasonal, as in the case of the sun) energy production rates. Hydrogen can be stored in three primary ways: as a gas, liquid, or as a solid combined chemically with a metal. This section will focus on large-scale storage of hydrogen at the production site, in each of the three forms mentioned above. On-board vehicular storage of hydrogen is discussed in Section 2.4.

Gaseous Storage

Storage in Tanks or Pipelines

Tank storage of hydrogen as a compressed gas is an old and well-established technology. Gaseous hydrogen is conventionally stored in banks of "K" bottles or "tube" tanks at pressures as high as 41 MPa (6000 psig). For special applications, tube tanks have been fabricated from a low carbon steel, and utilized at pressures as high as 70 MPa (10,000 psig) [20]. Larger quantities of gaseous hydrogen are stored in welded, laminated-wall vessels at pressure levels as high as 103 MPa (15,000 psig). These vessels are double-walled pressure containers with an inner layer that is compatible with hydrogen at the resultant stress levels, and an outer wall, made of high-strength welded steel, that is designed to contain the pressure. When a large volume of hydrogen is needed that can be discharged at a desired rate, tanks or vessels can be manifolded together [21]. In times of excess hydrogen production, hydrogen can also be stored in the pipeline system at elevated pressures.

Underground Storage

Gases can also be stored in underground storage facilities. Natural gas, in particular, has been stored underground since 1916 [22]. Facilities for underground storage of gases fall into two general categories: (1.) porous media storage (such as depleted oil and gas fields or aquifers) where the gas occupies the naturally occurring pore space between mineral grains, or crystals in the sandstones, and (2.) cavern storage, in which the gas is contained in excavated or solution-mined cavities in dense rock. At present, the first category accounts for the majority of all underground storage of natural gas [21]. The areas in the United States that are suitable for underground storage of gas in porous rock are shown in Fig. 6.2.

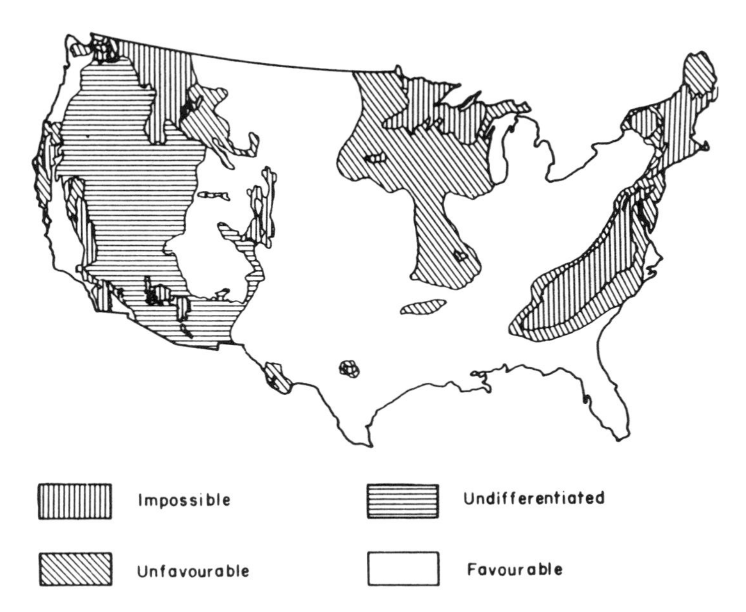

Figure 6.2. Areas in the United States suitable for underground storage of gas in porous rock. (Source: ref. [23])

Underground storage of gaseous hydrogen in natural formations has so far been limited. Hydrogen has been successfully stored in solution-mined salt caverns in England by Imperial Chemical Industries (ICI) at Teeside. The Teeside storage facility utilizes three brine-compensated caverns to store hydrogen at 750 psi at a depth of 1200 ft. For a time, Gaz de France was injecting hydrogen-rich gas into a storage aquifer near Beyenes, France. In 1973, the field was successfully converted to natural gas storage [20, 21].

Underground storage of hydrogen can be a relatively low cost option. Depleted oil and gas fields can be utilized for underground gas storage with a minimal capital investment [24]. Although capital expenditures can be considerable when a salt cavern must be created and compressors installed, this method can also be cost-effective overall.

The reprint by Taylor, et al explores the potential of underground storage of hydrogen in more detail.

Microsphere

Another technique for hydrogen storage that is still being developed is storing hydrogen in small glass microspheres. This technology is based on the ability of hydrogen to diffuse through small glass beads at elevated temperatures. This technique is still in the experimental stages, but if developed to the point of commercialization, microspheres could provide hydrogen storage equivalent to that of hydrides but at a lower overall weight [21].

Liquid Storage

Cryogenic Liquid

Storage of hydrogen as a cryogenic liquid is presently the only large-scale method being employed by industry. On a weight basis, liquid hydrogen (LH_2) has the highest energy density available in a chemical fuel. This is why LH_2 is the principal fuel for all of the national space programs. Spherical, vacuum-jacketed containers of up to 850,000 gallons (3.2 million ℓ) capacity have been built and are operated by NASA to ensure a continuous supply of liquid hydrogen fuel for the space programs [21, 23].

One of the problems with using liquid hydrogen is that the liquefaction process is energy intensive and involves several steps, including: (1.) the compression of gaseous hydrogen with reciprocating compressors; (2.) the pre-cooling of the compressed gas to liquid nitrogen temperature (78K); followed by (3.) expansion through turbines. An additional process is also needed to convert H_2 from ortho- to para-hydrogen spin isomers. This extra process is necessary because normal hydrogen at room temperature consists of a mixture of 75% ortho-hydrogen (nuclear spins parallel [s = 1]) and 25% para-hydrogen (nuclear spins anti-parallel [s = 0]), whereas the equilibrium mixture at 20K is almost entirely para-hydrogen. Thus, if hydrogen were cooled in its normal state, large boil-off losses would result since the ortho-hydrogen would slowly, and exothermically, transform to para form during storage to reach its equilibrium state [25, 26]. In total, the energy required for the liquefaction process is the equivalent of 40% of the combustion energy of the hydrogen itself [27].

A certain percentage of liquid hydrogen is also lost as it boils off. The larger the size of the dewar, the lower the percentage boil-off, as shown in Fig. 6.3 [26].

Intermediate Liquids (Open-loop Cycle)

A number of different intermediate liquids can be used to store hydrogen. Some of the carriers are once-through or open-loop carriers, which generate an end-product that is not recyclable. Ammonia (NH_3) and methanol (CH_3OH) are examples of open-loop carriers that can be stored as liquids and used for hydrogen production as needed. Ammonia has long been used as an in-plant source of hydrogen. To produce hydrogen from ammonia, NH_3 vapor is exposed to an iron oxide type catalyst at temperatures in excess of 700° C yielding hydrogen via the following reaction [26]:

$$2NH_3 \Leftrightarrow N_2 + 3H_2$$

Similarly, steam reformation techniques, used to produce hydrogen in large-scale plants, can be used to release hydrogen that is stored in methanol or other hydrocarbons. For methanol, the reaction proceeds as follows:

$$CH_3OH + H_2 \Leftrightarrow CO_2 + 3H_2$$

Unfortunately, this storage method requires energy intensive processing. Additionally, the gases used to produce ammonia and methanol, namely nitrogen and carbon dioxide, must also be stored on the production site. All things considered, it is also probably more effective to use methanol and ammonia in other applications [28]. Methanol, for example, can be used directly as a fuel for the IC engine. For these reasons, intermediate (open-loop) hydrogen carriers would probably not play a significant role in a large-scale hydrogen economy.

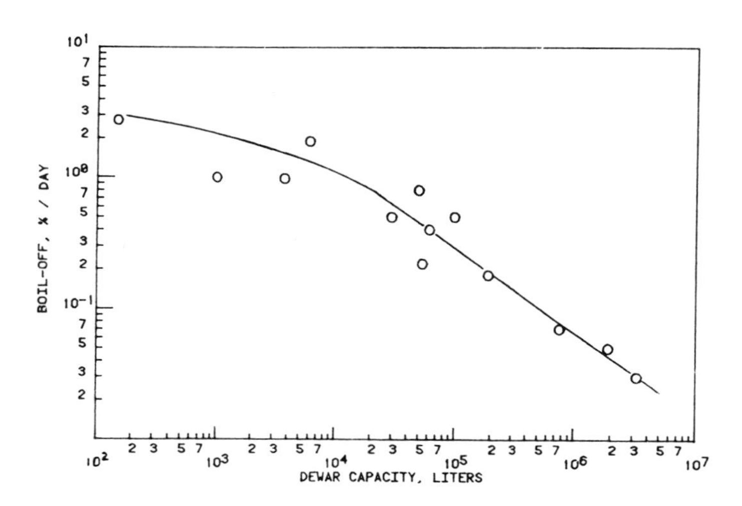

Figure 6.3. LH_2 Dewar boil-off losses. (Source: ref. [26])

Reversible Intermediates (Closed-loop Cycle)

Unlike once-through intermediates or open-loop cycles, reversible systems consist of hydrogen-containing liquids that can be stripped of some (or all) of their hydrogen and the product(s) saved for subsequent recharge with hydrogen (hydrogenation) at the production site. This liquid closed-loop cycle storage system is compatible with an energy system in which hydrogen is produced from renewable sources (e.g., water electrolysis using PVA electricity). After production, hydrogen could be stored in hydrogen-containing liquids (hydrogenation step), and then transported to central stations where the intermediate storage liquids are dehydrogenated (stripped of the hydrogen).

The CNR Institute in Messina, Italy has studied the following potential (de)hydrogenation reaction [28]:

$$3H_2 + C_6H_6 \text{ (Benzene)} \Leftrightarrow C_6H_{12} \text{ (Cyclohexane)}$$

In other work by scientists in Switzerland, the use of a Methylcyclohexane, Toluene, Hydrogen system (MTH-system) has been investigated. This (de)hydrogenation proceeds as follows [29-31]:

$$\text{Toluene} + \text{Hydrogen} \Leftrightarrow \text{Methylcyclohexane}$$
$$C_7H_8 + 3H_2 \Leftrightarrow C_7H_{14}$$

The general scheme of the MTH storage and transportation system is shown in Fig. 6.4.

For both of the above reactions, the hydrogenation reaction proceeds from left to right while the dehydration reac-

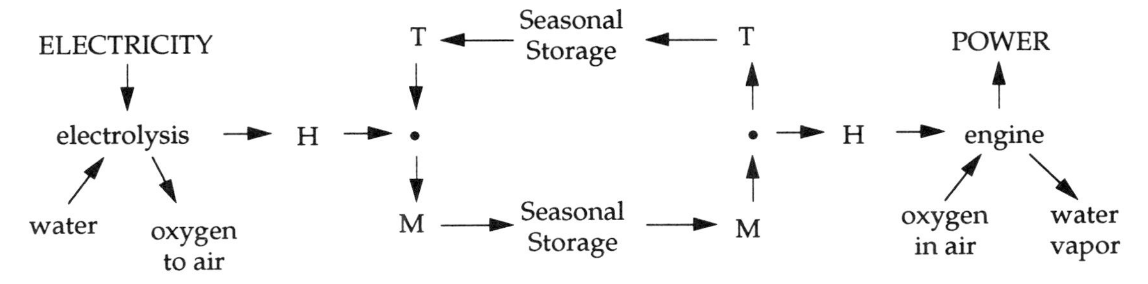

Figure 6.4. General scheme of the MTH-system. (Source: ref. [29])

tion proceeds from right to left. Details on how a complete (de)hydrogenation system would operate are provided in Section 6.3.

Solid Storage

Metal-Hydride

Hydrogen can also be stored chemically by combining it with metals and metal alloys to form hydrides. Hydride storage offers several advantages. Hydrogen can be stored with a relatively high volumetric storage density in a metal hydride, while eliminating the need to use high-pressure gas, thus increasing system safety.

Although there are some elemental metals (Mg, Ti, V, Pd) that may be used for hydride hydrogen storage, almost all practical compositions of rechargeable hydrides are intermetallic compounds [26]. These intermetallic compounds consist of combinations of A elements (the hydride formers) and B elements (the additions which alter thermodynamics and enhance kinetics). Compound types and typical compositions are as follows [26, 32]:

$$AB = TiFe,\ TiFe_{0.9}Mn_{0.1},\ TiFe_{0.8}Ni_{0.2}$$
$$AB_2 = ZrFe_2,\ TiCr_2,\ TiV_{0.6}Fe_{0.15}Mn_{1.25}$$
$$AB_5 = LaNi_5,\ MmNi_5^*,\ MmNi_{4.15}Fe_{0.85}$$
$$A_2B = Mg_2Ni,\ Mg_2Cu$$

* Mm = Mischmetal, an alloy of rare earth elements typically consisting of 48-50% Ce, 32-34% La, 13-14% Nd, 4-5% Pr, and about 1.5% other rare earth elements.

AB, AB_2, and AB_5 compounds can be used in hydrides with operating temperatures up to about 120° C, while A_2B are used for higher temperature hydrides operating at around 300° C. The hydride formers are titanium for the AB and AB_2 compounds, the rare-earth metals for the AB_5 compounds, and Mg for the higher temperature A_2B materials. The main alloying elements are Fe, Ni, and Mn [32].

Materials costs, storage capacity, and operating parameters are key factors in the overall hydride storage system cost. When hydrides are used at temperatures around 300° C, A_2B compounds are the most cost-effective, while for ambient temperature operations AB (FeTi-type) compounds are the lowest cost alloys.

6.4 Transportation and Distribution of Hydrogen

Pipeline Distribution

Hydrogen can be distributed as either a gas or a liquid through a pipeline. Liquid hydrogen has been transported for short distances through well-insulated, vacuum-jacketed pipelines in liquid hydrogen production plants and for the space program. Insulating longer pipelines for liquid hydrogen would not be practical, however. Ideally, the present natural gas infrastructure, including a 1.3 million mile pipeline system, could be used for the initial stages of the transition to hydrogen, although it probably would not be adequate for a full-scale hydrogen economy [33]. Although hydrogen has unique physical and chemical characteristics, the general distribution techniques needed for hydrogen are similar to those used for natural gas. A schematic drawing of components of a representative natural gas distribution system that could potentially be converted to hydrogen use is given in Fig. 6.5 [21].

One of the most important issues with using hydrogen in natural gas pipelines is the problem of embrittlement, or the tendency of hydrogen to react with metals and promote cracking [34, 35]. If this problem is not resolved, a new dedicated pipeline system will have to be installed—this could be a major obstacle to the implementation of a large-scale hydrogen infrastructure. It is possible that the development of a dedicated hydrogen pipeline system could prove more cost-effective in the long run, however [36].

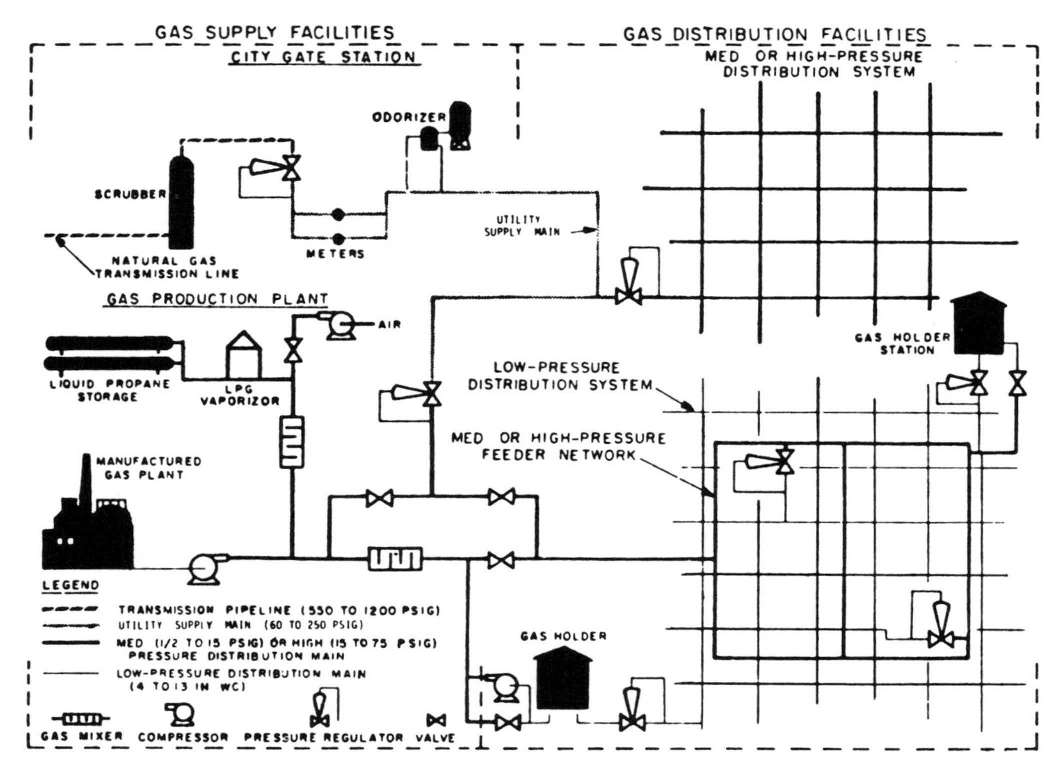

Figure 6.5. Various natural gas distribution system layouts with representative supply. (Source: ref. [21])

The low volumetric energy density of hydrogen presents another problem of compatibility with the existing natural gas system. On a volumetric basis, the energy density of hydrogen has only one-third the energy content of natural gas, hence three times the amount of hydrogen gas must be pumped through a pipeline to transmit an equivalent amount of energy. Thus, a conversion to hydrogen could require the installation of pumps and compressors which operate at higher pressure capacities than those presently used. Additionally, all the equipment would have to be modified so that it is compatible with hydrogen's unique characteristics. These changes could require a major investment.

The utilization of hydrogen or mixtures of hydrogen and other gases as a substitute for natural gas in residences and industries has been the subject of several preliminary investigations [20]. The Public Service Electric and Gas Company of Newark, New Jersey, has determined that present-day appliance burners are satisfactory for an 80/20 mixture of natural gas and hydrogen. For higher percentages of hydrogen, the present burners would have to be replaced with ones that would function properly with the anticipated hydrogen level [20.]

The transmission of hydrogen via pipelines has been successfully demonstrated in several countries. The oldest hydrogen pipeline is a 210 km (130 mile) system which has been operated in the Ruhr Valley of Germany by Hüls Company since 1938 [20]. The layout of the Ruhr Valley network is shown in Fig. 6.6. L'Air Liquide operates a 400 km (250 mile) hydrogen pipeline between Belgium and points in Northern France, the longest in the world, along with other shorter pipelines in Europe. There are also several hydrogen pipelines in Canada. In the United States, there are 725 km (450 miles) of hydrogen pipelines. Praxair operates three hydrogen pipelines in Texas, Whiting, Indiana, and Carney's Point, New Jersey with a combined length of 258 km (160 miles). Air Products and Chemicals, Inc. operates two hydrogen pipelines in La Porte, Texas and Plaquemine, Louisiana, which combine for 160 km (100 miles) [37].

Closed-Loop Cycle

Another possibility is to transmit hydrogen through a pipeline using reversible intermediate hydrogen carriers such as benzene-cyclohexane or toluene-methylcyclohexane. For such a transmission system, a closed-loop cycle would be used, where a compound is hydrogenated at the

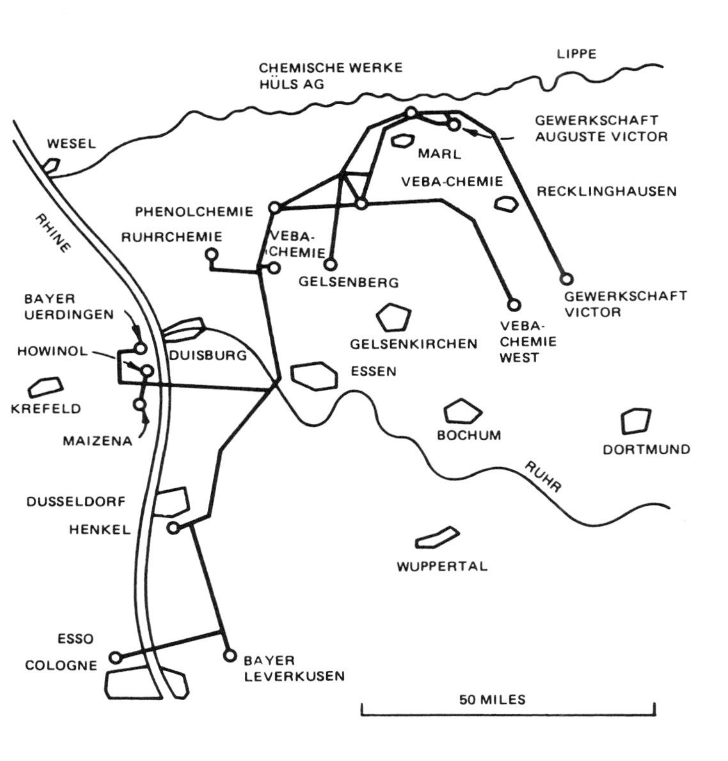

Figure 6.6. Hydrogen pipeline grid of Chemische Werke Hüls AG, Germany, 1970. (Source: ref. [20])

hydrogen production site, then transferred by means of pipelines to points of utilization, and finally dehydrogenated for use. Such a system requires two pipelines, one for the hydrogenated compound and one for the dehydrogenated compound [28].

Although this system is more complex than some of the others discussed, this does not necessarily mean there will be a proportional increase in cost. The closed-loop cycle pipelines can have smaller diameters than those required for gaseous hydrogen transport. Also, the energy needed to pump a liquid is less than that required by compressors of gaseous hydrogen. In the long run, these advantages could prevail over the increased complexity of the system to lower the final costs [28].

As discussed previously, Cacciola and Giordano have investigated (de)hydrogenation reactions using benzene and cyclohexane for possible use in long-range transport [28]. This system offers several advantages for large-scale, long-distance transport of hydrogen. The hydrogen density per volume of cyclohexane is 56 g H_2/ℓ as compared to 18 g H_2/ℓ in the case of gaseous hydrogen at 20 MPa. Moreover, the energy requirement of this reaction is compatible with an overall system where hydrogen is produced by electrolysis and used in a fuel cell. In this case, the excess heat produced by the fuel cell system could be used to increase the efficiency of the electrolysis [28]. A schematic of the whole system is shown in Fig. 6.7.

As shown in Fig. 6.7, cyclohexane, the product of the hydrogenation reaction, is pumped through pipelines at ambient temperature to the utilization points where it is dehydrogenated. For the dehydrogenation reaction, cyclohexane is transformed by an endothermic catalytic reactor into hydrogen, which is sent to a fuel cell to generate useful power (electricity), and benzene. Benzene is then pumped through another set of pipelines back to the pro-

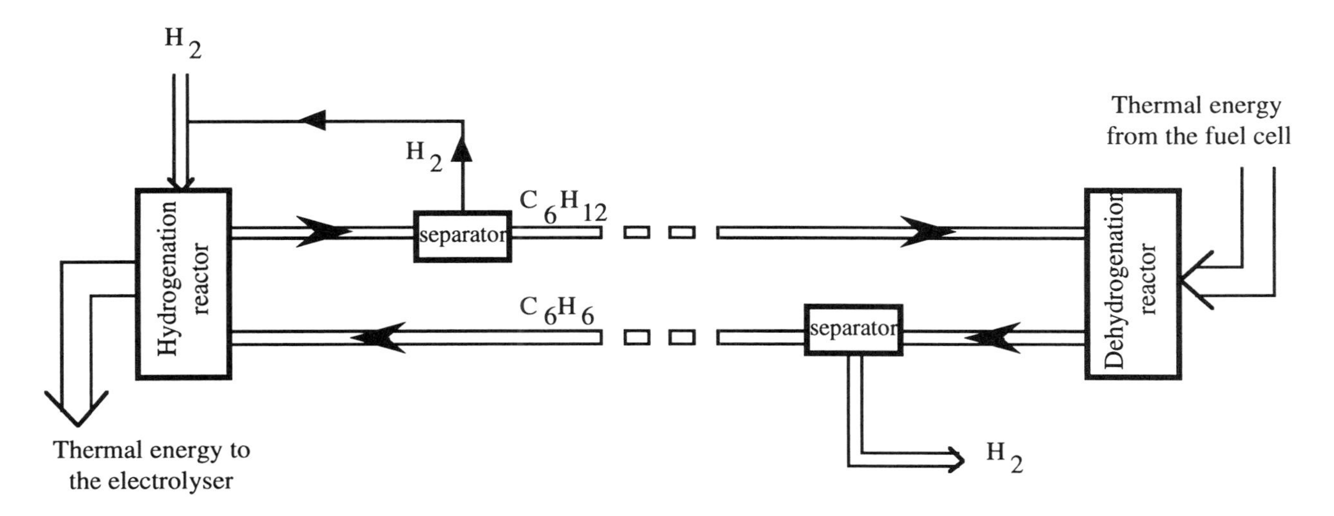

Figure 6.7. Closed-loop cycle. (Source: ref. [28])

duction site. One disadvantage of this system is that both cyclohexane and benzene have high freezing points (279.5 and 278.5 K, respectively) and could be difficult to transport in medium-cold winters. Cacciola and Giordano believe that this problem can be overcome by either adding antifreeze agents, or simply using another (de)hydrogenation reaction [28].

The methylcyclohexane-toluene-hydrogen system, discussed in the previous section, is a possible alternative to the cyclohexane-benzene reaction [29-31]. The freezing points of methylcyclohexane and toluene are both low (146.4 and 178 K, respectively) and the transmission of toluene is safer than that of the highly toxic benzene. Methylcyclohexane has a slightly lower hydrogen density (47.4 g H_2/l) than that of cyclohexane, but the density is still greater than the 18 g H_2/ℓ for the transport of gaseous hydrogen at 20 MPa (200 bar).

Gas Transportation by Truck

Hydrogen can be transported by truck as a high-pressure gas in cylinders at pressures ranging from 15 to 40 MPa [38]. The tube trailer consists of a number of large, high-strength and high-pressure steel tubes which have been manifolded together and mounted on a truck trailer. While this technique is well developed, it is very inefficient and can only be used to haul small volumes of hydrogen gas. A steel cylinder weighing 20 to 30 kg will hold only 1 kg of hydrogen at 15.2 MPa and 2.5 kg of hydrogen at 40.5 MPa. As a result of the large container weights involved, the hydrogen typically constitutes only two to four percent of the cargo weight [38].

Liquid Transportation by Truck

Presently, semi-trailers transport liquid hydrogen in batch quantities over public roadways from the producer to the user directly. A liquid hydrogen tube trailer is shown in Fig. 6.8. Typically, the cryogenic liquid hydrogen tank used by semi-trailers consists of concentric inner and outer tanks separated by an evacuated space filled with perlite. Liquid hydrogen is carried in the inner tank and effectively insulated from the environment by the annular, evacuated space. Maintenance costs are much higher for liquid as opposed to gaseous transportation [21].

Liquid Transportation by Rail

Liquid hydrogen is transported by rail in tank cars of 9500 to 28,000 gal. (36,000 to 106,000 ℓ) capacity. These tank cars are equipped with insulation layers that consist of vacuum jackets and multi-layer radiation shielding to keep evaporation losses to below 0.5% per day [21].

Figure 6.8. Praxair, Inc. liquid hydrogen truck. (Source: Praxair, Inc.)

Marine Transportation

At present, dedicated liquid hydrogen ships do not exist. Three liquid hydrogen barges are used as part of the space program, each with a capacity of 250,000 gallons (~950,000 ℓ). Liquid hydrogen tankers are also being designed under the Euro Quebec Hydro-Hydrogen Pilot Project. The liquid hydrogen tanks for these ships are being designed so that fuel boil-off will not occur for a period of at least 50 days. It should be noted that there are more than 100 liquid natural gas tanker ships, which provide for all commercial natural gas transport between Europe, North and South America, and the Pacific Rim countries [37].

6.5 Conclusion

While hydrogen could be an important fuel source in the future, there are still a number of obstacles that must be overcome before a major hydrogen infrastructure can be developed. Hydrogen production (from sources other than fossil fuels), storage, transportation, and distribution are still in the early stages of development, and the monetary costs involved in using hydrogen to fuel our economy currently exceed the costs of using fossil fuels. While much development work is still needed to establish a large-scale hydrogen economy, there is some potential for introducing hydrogen at a smaller scale in niche markets, such as fleet operations with centralized refueling facilities. Increased efforts to reduce air pollution, however, could promote greater use of alternative fuels and stimulate developments which could lead to a larger penetration of hydrogen-fueled vehicles into the transportation marketplace.

References

1. Kukkonen, C.A. and Shelef, M., "Hydrogen as an Alternative Automobile Fuel: 1993 Update," SAE Technical Paper No. 940766, Society of Automotive Engineers, Warrendale, PA, 1994.

2. Veziroglu, T.N., "Hydrogen Technology for Energy Needs of Human Settlements," *Int. J. Hydrogen Energy*, Vol. 12, p. 99, 1987.

3. Dutta, S., Block, D.L., and Port, R.L., "Economic Assessment of Advanced Electrolytic Hydrogen Production," *Int. J. Hydrogen Energy*, Vol. 15, p. 387, 1990.

4. Anderson, D. and Ahmed, K., "Where We Stand with Renewable Energy," *Finance & Development*, p. 40, June 1993.

5. DeLuchi, M.A. and Ogden, J.M., "Solar hydrogen Transportation Fuels," PU/CEES Report, Center for Energy and Environmental Studies, Princeton University, Princeton, NJ, February 1992.

6. De Laquil III, P., Kearney, D., Geyer, M., and Diver, R., "Solar-Thermal Electric Technology," in *Renewable Energy: Sources for Fuels and Electricity*, Edited by T. B. Johansson, H. Kelly, A. K. N. Reddy, R. H. Williams, and L. Burnham, Island Press, Washington, D.C., p. 213, 1993.

7. "Central Receiver Technology: Status and Assessment," Department of Energy, Report No. SERI/SP-220-3314, 1989.

8. Department of Energy, "The Solar Two Challenge," Information Pamphlet.

9. Jackson, T., "Renewable Energy: Great Hope or False Promise," *Energy Policy*, Vol. 19, p. 2, 1991.

10. Sims, G.P., "Hydroelectric Energy," *Energy Policy*, Vol. 19, p. 776, 1991.

11. Drolet, B., Gretz, J., Kluyskens, D., Sandmann, F., and Wurster, R., "The Euro-Quebec Hydro-Hydrogen Pilot Project (EQHHPP)," in *Hydrogen Energy Progress X*, Edited by D. L. Block and T. N. Veziroglu, International Association for Hydrogen Energy, Coral Gables, FL, p. 23, 1994.

12. "A Growth Market in Wind Power," *EPRI Journal*, p. 4, December 1992.

13. Grubb, M.J. and Meyer, N.I., "Wind Energy: Resources, systems and regional strategies," in *Renewable Energy: Sources for Fuels and Electricity*, Edited by T. B. Johansson, H. Kelly, A. K. N. Reddy, R. H. Williams, and L. Burnham, Island Press, Washington, D.C., p. 157, 1993.

14. Clarke, A., "Wind Energy: Progress and Potential," *Energy Policy*, Vol. 19, p. 742, 1991.

15. Tamkins, T., "Tilting at Wind Power," *Audubon*, p. 24, Sept.-Oct. 1993.

16. Hall, D.O., "Biomass Energy," *Energy Policy*, Vol. 19, p. 711, 1991.

17. Pimentel, D., Rodrigues, G., Wang, T., Abrams, R., Goldberg, K., Staecker, H., Ma, E., Brueckner, L., Trovato, L., Chow, C., Govindarajulu, U., and Boerke, S., "Renewable Energy: Economic and Environmental Issues," *BioScience*, Vol. 44, p. 536, 1994.

18. DeLuchi, M.A. and Ogden, J.M., "Solar-Hydrogen Fuel-Cell Vehicles," *Transportation Research, Part A, Policy and Practice*, Vol. 27A, p. 255, 1993.

19. Barbir, F. and Veziroglu, T.N., "Effective Costs of The Future Energy Systems," *Int. J. Hydrogen Energy*, Vol. 17, p. 299, 1992.

20. Kelley, J.H. and Hagler, R., "Storage, Transmission and Distribution of Hydrogen," *Int. J. Hydrogen Energy*, Vol. 5, p. 35, 1980.

21. Blazek, C.F., Daniels, E.J., Donakowski, T.D., and Novil, M., "Economics of Hydrogen in the '80s and Beyond," in *Recent Developments in Hydrogen Technology*, Edited by K. D. J. Williamson, Jr. and F. J. Edeskuty, CRC Press, Inc., Boca Raton, FL, p. 1, 1986.

22. Taylor, J.B., Alderson, J.E.A., Kalyanam, K.M., Lyle, A.B., and Phillips, L.A., "Technical and Economic Assessment of Methods for the Storage of Large Quantities of Hydrogen," *Int. J. Hydrogen Energy*, Vol. 11, p. 5, 1986.

23. McAuliffe, C.A., *Hydrogen and Energy*, Gulf Book Division, Houston, TX, 1980.

24. Styrikovich, M.A. and Malyshenko, S.P., "Bulk Storage and Transmission of Hydrogen," in *Hydrogen Energy Progress VI*, Edited by T. N. Veziroglu, N. Getoff, and P. Weinzierl, Pergamon Press, Elmsford, NY, p. 765, 1986.

25. Sarangi, S., "Cryogenic Storage of Hydrogen, D. Reidel Publishing Company," presented at the Progress in Hydrogen Energy Conference, New Delhi, India, 1985.

26. Huston, E.L., "Liquid and Solid Storage of Hydrogen," in *Hydrogen Energy Progress V*, Edited by T. N. Veziroglu and J. B. Taylor, Pergamon Press, Elmsford, NY, p. 1171, 1984.

27. Yamane, K., Hiruma, M., Watanabe, T., Kondo, T., Hikino, K., Hashimoto, T., and Furuhama, S., "Some performance of engine and cooling system on LH_2 refrigerator van Musashi-9," in *Hydrogen Energy Progress X*, Edited by D. L. Block and T. N. Veziroglu, International Association for Hydrogen Energy, Coral Gables, FL, p. 1825, 1994.

28. Cacciola, G. and Giordano, N., "Hydrogen Transportation by a Chemical Closed-Loop Cycle, an Alternative Option for Long Distance Transport of Energy," in *Hydrogen Energy Progress V*, Edited by T. N. Veziroglu and J. B. Taylor, Pergamon Press, Elmsford, NY, p. 1237, 1984.

29. Grünenfelder, N.F. and Schucan, T.H., "Seasonal Storage of Hydrogen in Liquid Organic Hydrides: Description of the Second Prototype Vehicle," *Int. J. Hydrogen Energy*, Vol. 14, p. 579, 1989.

30. Taube, M., Rippin, W.T., Cresswell, D.L., and Knecht, W., "A System of Hydrogen-Powered Vehicles with Liquid Organic Hydrides," *Int. J. Hydrogen Energy*, Vol. 8, p. 213, 1983.

31. Taube, M., Rippin, D., Knecht, W., Hakimifard, D., Milisavlejevic, B., and Grunenfelder, N., "A Prototype Truck Powered by Hydrogen from Organic Liquid Hydrides," *Int. J. Hydrogen Energy*, Vol. 10, p. 595, 1985.

32. Angus, H.C., "Storage, Distribution and Compression of Hydrogen," *Chemistry and Industry (London)*, N. 2, p. 68, January 16, 1984.

33. Bockris, J.O'M., *Energy Option: Real Economics and the Solar-Hydrogen System*, Australia & New Zealand Book Company, Sydney, 1980.

34. Lewis, F.A., "Storage and Distribution of Hydrogen: Materials and Reliability," in *Hydrogen Energy Progress VI*, Edited by T. N. Veziroglu, N. Getoff, and P. Weinzierl, Pergamon Press, Elmsford, NY, p. 979, 1986.

35. Cialone, H.J., Scott, P.M., Holbrook, J.H. et al, "Hydrogen Effects on Conventional Pipeline Steels," in *Hydrogen Energy Progress V*, Edited by T. N. Veziroglu and J. B. Taylor, Pergamon Press, Elmsford, NY, p. 1855, 1984.

36. DeLuchi, M.A., "Hydrogen Vehicles: An Evaluation of Fuel Storage, Performance, Safety, Environmental Impacts, and Cost," *Int. J. Hydrogen Energy*, Vol. 14, p. 81, 1989.

37. Cannon, J.S., *Harnessing Hydrogen: The Key to Sustainable Transportation*, INFORM, Inc., New York, NY, 1995.

38. Williams, L.O., *Hydrogen Power: An Introduction to Hydrogen Energy and its Applications*, Pergamon Press, Oxford, U.K., 1980.

Economics of Hydrogen as a Fuel for Surface Transportation

H. J. Plass, Jr., F. Barbir, H. P. Miller and T. N. Veziroglu
University of Miami

Abstract—This paper presents an analysis of the systems which produce hydrogen from renewable energy sources (solar and wind), and which use hydrogen as a fuel for surface transportation. A steady state model has been developed which analyses process performances, the costs of each of the stages in hydrogen production, storage and distribution, and estimates material and energy requirements for building and operating the Solar Hydrogen Energy System. Analysis of the solar hydrogen production, storage and distribution has shown that the price of delivered hydrogen depends mostly on the cost of available solar technologies (photovoltaic cells and wind turbines) and their efficiencies. The real cost of transportation has been calculated for both gasoline, as the existing fuel for surface transportation, and hydrogen as an alternative in the future. The real cost includes environmental and societal damage caused either by burning the fuel or by processing it. The unit costs and efficiencies of photovoltaics and wind turbines, which enable electrolytic hydrogen production at competitive prices, have been calculated.

1. INTRODUCTION

The transportation sector uses almost one-third of the total energy consumed in the U.S.A. (22.3 EJ out of 80 EJ in 1987) and it is completely based on fossil fuels—petroleum products. Motor gasoline accounts for 63% of transportation fuels and 18% of the total energy consumption in the U.S.A. [1]. Furthermore, the prices of transportation fuels, because they are highly refined, are the highest prices paid for fluid fuels.

Hydrogen, because of its physical and chemical properties, can be easily used as a fuel for transportation, particularly for air transportation (subsonic and supersonic) and for surface transportation, i.e. in internal combustion engines and probably in fuel cells. However, hydrogen is very expensive to produce, particularly from renewable energy sources. Technologies for conversion and utilization of renewable energy sources (solar, wind) are not developed enough and it is expected that further development will significantly improve the efficiencies and reduce the costs.

On the other hand, the prices of fossil fuels, particularly petroleum products, will rise due to depletion and due to detrimental environmental effects, which more and more seem to become the matter of public and legislative concern.

Taking these two trends into account it is possible to determine when and under which conditions hydrogen can become a competitive fuel for surface transportation.

2. HYDROGEN SYSTEM ANALYSIS

Hydrogen can be produced directly or indirectly from renewable energy sources, using different methods, such as direct thermal decomposition—thermolysis, thermochemical cycles, electrolysis (electricity from photovoltaic cells or from wind turbines), and photolysis which includes photo-electrolysis, catalytic photolysis and bio-photolysis. Most of these methods are still available only in laboratory scale, and therefore few data are available on their technical feasibility and economics, particularly for large scale applications. Nevertheless, wind electricity is already feasible at some locations [2]. Photovoltaics technology is rapidly developing and seems to be very promising, even in the near future [3–5]. Expected development could enable significant price reduction of produced electricity and consequently the price of hydrogen. Electrolysis is a developed and mature technology, although not for massive production of hydrogen.

Because of these reasons the analysis of hydrogen systems was limited only to photovoltaics and wind electrolytic hydrogen production. In addition, large scale underground hydrogen storage and regional and local transport and distribution of hydrogen have been included in this analysis.

Process analysis

The analysed solar hydrogen energy system may be described by a sequence of process steps: electricity production (PV or wind), electrolysis, hydrogen storage including compression, hydrogen transportation (regional and local), and finally, hydrogen delivery to the end use. As a fuel for surface transportation hydrogen can be used as a gas or as a liquid; therefore high pressure compression or liquefaction must be considered as the last step. Each of these steps involves energy losses and has associated with it an efficiency. Knowing or assuming the efficiency of each step it is possible to calculate the energy value of produced and delivered hydrogen as follows:

For the PV system:

$$\text{HD (GJ yr}^{-1}\text{ kW}_p^{-1}) = 0.0036\,(I_a/I_p)\eta_{BOS}\eta_T\eta_C\eta_{EL}\eta_{ST} \quad (1)$$

where:

I_a = average annual solar availability (2400 kWh m^{-2} for fixed flat plate [3])
I_p = peak insolation (1.1 kW m^{-2}, [3])
η_{BOS} = balance-of-system efficiency (0.85, [3])
η_T = temperature correction (0.9, [3])
η_C = PV-electrolysis coupling efficiency (0.93, [6])
η_{EL} = electrolysis efficiency (0.8–0.9, [4])
η_{ST} = gaseous hydrogen storage efficiency (0.97, [7, 8]).

For the wind system:

$$\text{HD (GJ yr}^{-1}\text{ kW}_p^{-1}) = 0.0036\,(W_a/W_p)\eta_{WT}\eta_{BOS}\eta_{EL}\eta_{ST} \quad (2)$$

where:

W_a = average annual wind energy potential available (3000 kWh m^{-2})
W_p = peak specific power—depends on design wind speed (kW m^{-2})
η_{WT} = average annual wind turbine efficiency (0.3)
η_{BOS} = balance-of-system efficiency (0.9).

With these assumptions it has been calculated that with an advanced photovoltaic–electrolysis plant, hydrogen can be delivered at the rate of 4.6–6.4 GJ yr^{-1} per peak kW of installed photovoltaic panels. A wind–electrolysis plant can produce and deliver hydrogen at the rate of 5.0–6.0 GJ yr^{-1} per kW of installed wind turbines.

However, to produce and to transport hydrogen, a certain amount of energy is required to run pumps, compressors and other auxiliary equipment. Net produced energy is the difference between delivered hydrogen and used energy:

$$E_{NET} = \text{HD} - [E_{AUX} + E_{COM} + E_{HCOM}] \quad (3)$$

or

$$E_{NET} = \text{HD} - [E_{AUX} + E_{COM} + E_{LIQ}] \quad (4)$$

where:

E_{AUX} = energy for auxiliary equipment (E_{AUX} = 0.01 HD [9, 10])
E_{COM} = energy for compressors for storage and transportation (E_{COM} = 0.095 HD [10])
E_{HCOM} = energy for ultra-high compression up to 200 or 400 bar (E_{HCOM} = 0.035 HD for 200 bar and E_{HCOM} = 0.050 HD for 400 bar [10])
E_{LIQ} = energy for liquefaction (E_{LIQ} = 0.30 HD [11]).

The amount of energy which must be invested accounts for 14–16% of delivered hydrogen energy if hydrogen is delivered as a high pressurized gaseous fuel and up to 40% for liquid hydrogen.

Materials and energy requirement

The solar hydrogen energy system should not be analysed in isolation, without interactions with its surroundings, i.e. the main economy. The main economy uses the product of the energy system—fuel for transportation, as the driving force which contributes to the growth of the main economy. On the other hand, an energy system, such as the solar hydrogen energy system, requires some goods and services from the main economy, namely equipment and labor. In the case of an entirely new energy system, such as the solar hydrogen energy system, most of the equipment is not available on the market. It has to be designed and manufactured. Materials required for manufacturing the equipment must be processed first. To build the solar hydrogen energy system, energy is required, which can be obtained from the main economy. Of course, labor, services and information (knowledge) are also very important in the manufacturing process. These services can be represented by money flow, because money paid for goods and services always goes to the people for services provided [12].

Some of the analyses performed previously [9, 13] have shown that the materials requirement, although substantial, should not be a barrier for the realization of a solar hydrogen energy system. A 1 GW solar hydrogen power plant, including PV modules, module supports, electrolysis, auxiliary equipment, storage and transportation system, requires about 1.1 million tons of different construction materials, namely iron, nonferrous metals (copper, aluminum, nickel, zinc), concrete, glass, plastics, and of course silicon. The materials requirement for a PV plant strongly depends on the area required, which means that the module efficiency is very important in such considerations.

A similar analysis performed for the wind power systems has shown similar material requirements (more iron and less concrete, glass and plastics, and no silicon).

Energy required for the construction includes the energy required to process the raw materials, the energy needed for equipment manufacture, final processing, transport to the site and the energy needed at the site for construction. In this analysis, the data from Voigt [13] and Spreng [14] were used. The energy requirement for a solar hydrogen plant has been found to be 19 GJ kW^{-1}, and for a wind power plant 10 GJ kW^{-1}. The difference is mainly due to high energy intensity of silicon in PV modules.

The energy payback time, defined as the time in which a power plant produces the amount of energy which was necessary to construct it, is a measure of its usefulness. Shorter energy payback time means that the total net energy delivered to society will be greater.

A solar hydrogen plant, with fixed plate PV modules, can produce and deliver to the consumers 4.6 GJ yr^{-1} of compressed hydrogen per kW of installed power [10]. However, the amount of energy required to operate the system (water treatment, cooling, auxiliary power, and compressors) is estimated to be approximately 14% of the hydrogen energy delivered. Therefore, the net energy is 4 GJ yr^{-1} kWe^{-1}. The energy payback time is 19/4 = 4.75 years. For liquid hydrogen, the energy payback is 6.9 years.

A wind power hydrogen plant can produce and deliver to the consumers 5.4 GJ yr^{-1} of hydrogen per kW of installed power [10], with a net energy of 4.64 GJ yr^{-1} kWe^{-1}. The energy payback time is 10/4.64 = 2.1 years.

Cost analysis

For the cost analysis, the following equations were derived [10]:

- The cost of PV electricity:

$$C_E(\$\,kWh^{-1}) = [0.053(C_{PV} + C_{BOS}) + 0.6]/(1000\,\eta_{PV}). \quad (5)$$

- The cost of wind electricity:

$$C_E\,(\$\,kWh^{-1}) = 0.06(C_{WT} + C_{BOS})/1000. \quad (6)$$

- The cost of electrolytic produced hydrogen:

$$C_P\,(\$\,GJ^{-1}) = 0.0045\,C_{EL}/CF + 351\,C_E. \quad (7)$$

- The cost of hydrogen storage:

$$C_{ST}\,(\$\,GJ^{-1}) = 0.0025\,C_{STV} + 0.0015\,C_{STP} + 20.2\,C_E + 0.03\,C_P. \quad (8)$$

- Total cost of gaseous hydrogen as a fuel:

$$C_{HD}\,(\$\,GJ^{-1}) = C_P + C_{ST} + C_{TR} + C_{COM}. \quad (9)$$

- Total cost of liquid hydrogen as a fuel:

$$C_{HD}\,(\$\,GJ^{-1}) = C_P + C_{ST} + C_{TR} + C_{LIQ}. \quad (10)$$

In these equations the following assumptions have been included:

—annual discount rate: 0.061 [3, 4]
—lifetime: 20 years for wind turbines, compressors, pumps, electrolysis equipment and 30 years for PV modules, balance-of-system, pipes, etc. [4]
—indirect cost factor: 0.25 [3, 4]
—operation and maintenance costs: 1.1 \$ m^{-2} for PV plant, 2% of capital cost for wind systems and electrolysis plant, and 5% of capital cost for storage and distribution of hydrogen [3, 4].

The parameters which have been chosen as variables are unit costs and efficiencies:

- The unit PV module cost (C_{PV}), today is approximately \$300 m^{-2}, but long term technical goals aim to achieve \$45–80 m^{-2} for flat plate systems and \$60–100 m^{-2} for concentrator systems [3], which corresponds to \$0.3–0.5 Wp^{-1}.
- The balance-of-system (C_{BOS}) cost includes all subsystems and components, exclusive of photovoltaic cells and modules, needed for a fully functional power system. Area related BOS costs are those for the central station, the module supporting structure, land, site preparation, module installation, roads, fences, and other civil engineering, instrumentation, grounding, wiring, and surge protection. In addition, for tracking systems and concentrators, BOS also includes tracking costs. Minimum achievable area related BOS costs for fixed flat-plate systems are expected to be about \$50 m^{-2} [3], although there are some estimates that these costs can be as low as \$30 m^{-2} [4]. For 1-axis and 2-axis tracking systems, and for concentrator systems, BOS costs are expected to be \$75 m^{-2}, \$100 m^{-2} and \$125 m^{-2} respectively [3].
- The module efficiency η_{PV} is very important for the costs because it determines the area required (the higher efficiency the less area required). Module efficiency at 25°C for flat-plate systems is assumed to be 0.1 to 0.25, and for concentrator systems 0.2 to 0.35 [3].
- The unit cost of wind turbines (C_{WT}) is presently about \$1000–1200 kW^{-1} [2]. Projected costs may be assumed to be \$500–750 kW^{-1}, and C_{BOS} may add another 30%, or $(C_{WT} + C_{BOS})$ = \$650–1000 kW^{-1}.
- The unit capital cost of the electrolyser (C_{EL}) depends on the type and size of the plant. If a modular type of electrolyser is going to be developed for large scale production, there will be no significant scale economies for hydrogen production capacity above 5–10 MW [4]. In that case unit costs may be assumed to be \$170–250 kW^{-1} of installed or rated power.
- The capacity factor (CF) depends on the source of electricity, and in the case of solar or wind produced electricity it depends on the availability of the source. For the analysed photovoltaic systems the capacity factor is in the range from 0.20 for fixed flat plate systems up to 0.27 for concentrator systems, and for the wind electricity it is about 0.23 [10], and it may vary at different locations and for different types of the wind turbines.
- Production costs of hydrogen (C_P) have been calculated to be \$11.93–32.21 GJ^{-1} for photovoltaic systems and \$17.17–26.17 GJ^{-1} for wind systems.
- Volume related unit cost of hydrogen storage (C_{STV}) includes costs for site preparation, cavern adaptation, piping, control, etc. These costs are estimated to be about \$7 kg^{-1} of H_2 stored, or \$50 GJ^{-1} [9, 15]. Power related costs (C_{STP}) primarily include the compressor capital costs, which are about \$600 kW^{-1} for large compressors (> 3000 kW) [4, 15]. The cost of large scale underground hydrogen storage is calculated to be \$1.84–3.51 GJ^{-1}.
- Regional and local transport and distribution cost (C_{TR}) have been estimated to be \$0.54 GJ^{-1} [4, 10]. Since hydrogen must be compressed or liquefied to be suitable as a fuel for surface transportation, the cost of compression (C_{COM} = \$0.50 GJ^{-1} for 200 bar [10]) or the cost of liquefaction (C_{LIQ} = \$4.50 GJ^{-1} [11] must be added to the price of delivered hydrogen.
- The costs of delivered hydrogen (C_{HD}) have been calculated to be \$14.81–36.76 for gaseous hydrogen and \$18.81–40.76 for liquid hydrogen.

The costs mainly depend on the unit cost and efficiencies of photovoltaic and wind power systems, and such a wide range is due to uncertainties of projected unit costs and efficiencies.

3. HYDROGEN AS A FUEL FOR SURFACE TRANSPORTATION

Hydrogen has good properties as a fuel for I.C. engines in automobiles. Some of the characteristic properties of hydrogen/air mixture which can definitely influence the engine design and performance are: low ignition energy, low density, wide range of ignition limits, high diffusion speed, and high flame speed. Because of these properties hydrogen can be considered more thermally efficient than gasoline, primarily because it burns better in excess air, and permits the use of a higher compression ratio [16]. Data from engine and automibile tests [17, 18] indicate that hydrogen's thermal efficiency is 15–50% more than gasoline. The average value of 22% better overall efficiency has been suggested [16, 19]. Therefore:

$$e = \frac{\text{efficiency of hydrogen engine}}{\text{efficiency of gasoline engine}} = \frac{\eta_{H_2}}{\eta_G} = 1.22. \quad (11)$$

A generally accepted measure of car economy is mpg, i.e. how many miles a car can run with one gallon of gasoline. In energy units it can be expressed as:

$$F_E = \frac{82{,}000}{\text{mpg}} \quad (12)$$

where F_E is fuel economy in kJ km^{-1}.

Today, an average car has a fuel economy of approximately 20 mpg (4100 kJ km^{-1}), but new designs have much better fuel economy, up to 98 mpg, such as a prototype Toyota AXV [4].

The main disadvantages of using hydrogen as a fuel for automobiles are huge on-board storage tanks which are required because of hydrogen's extremely low density. Hydrogen may be stored on board a vehicle as compressed gas in ultra-high-pressure vessels, as a liquid in cryogenic containers, or as a gas bound with certain metals in metal hydrides.

Another obstacle in large scale hydrogen applications is its price. Hydrogen produced from renewable sources is rather expensive fuel when compared with today's fuels. However, the market price of fossil fuels does not include the costs of environmental damage. The use of fossil fuels have harmful impacts on the environment. During the extraction, transportation, refinement and storage of petroleum and petroleum products, spills and leakages occur, which causes water and air pollution. Most of the fossil fuel environmental impact occurs during the end use—combustion when tremendous amounts of various gases (CO_2, CO, SO_x, NO_x, CH), soot and ash, are produced and released into the atmosphere. In the atmosphere they cause air pollution, especially in the interaction with water vapor, sunlight, and other atmospheric processes. Air pollution causes damage to human health, animals, crops and reduces visibility, etc. Most of the air pollutants fall down as acid deposition, causing water pollution, soil and lakes acidification, affecting humans, animals, vegetation and man-made structures. The remaining products of combustion in the atmosphere, mainly carbon dioxide, together with other so-called greenhouse gases result in thermal changes by absorbing the infrared energy the Earth radiates back into the atmosphere, causing global temperature increase. The effects of the temperature increase are melting of the ice caps, sea level rise and climatic changes, which include heat waves, droughts, floods, stronger storms, more wildfires, etc. The cost of these effects have been estimated to be $8.10 per GJ of fossil fuel used in the U.S., and particularly for petroleum based fuels $8.47 per GJ [20].

On the other hand, hydrogen can be considered as a clean fuel, because the main combustion product is water. Hydrogen vehicles would not produce significant amounts of CO, HCs, SO_x, particulates, sulfur-acid deposition, ozone and other oxidants, benzene and other aromatic hydrocarbons, aldehydes, lead and other toxic metals, smoke, and particularly CO_2 and other greenhouse gases. In properly designed and maintained hydrogen engines, only a small fraction of the lubricants is burned producing trace amounts of HCs and CO. The only pollutant of concern would be NO_x [16]. However, the results of several hydrogen engine and vehicle emissions tests [17, 18], have shown that optimized hydrogen vehicles can emit much less NO_x than do comparable, optimized gasoline vehicles, under the same test conditions, with the same pollution control equipment, primarily because they can be run leaner and cooler.

If all the advantages of hydrogen (better efficiency and no pollution) are taken into account it should be possible to calculate the real transportation costs for hydrogen and gasoline vehicles, to compare them and to determine under which conditions hydrogen could become competitive.

A transportation cost (in $ km^{-1}) should include capital cost (investment in an automobile, including insurance, registration, taxes etc.), operation and maintenance costs, and fuel costs.

DeLuchi [16] has estimated that a hydrogen engine would be $50–200 more expensive than a comparable gasoline engine. However there is a cost credit for removed pollution control, which amounts to $350–450. The equipment for hydrogen storage, either as a liquid or as a gas, is considerably more expensive than a gasoline tank; it can add up to $700–1700.

Fuel cost should include not only the cost of fuel production and distribution, but also the cost of environmental and societal damage caused either directly (by burning the fuel) or indirectly (by processing it):

$$C_F = F_E[C_G + C_{ENV} + aC'_{ENV}] \quad (13)$$

where:

C_F = fuel cost ($ km^{-1})
F_E = fuel economy (GJ of gasoline km^{-1})
C_G = market fuel cost ($ GJ^{-1})

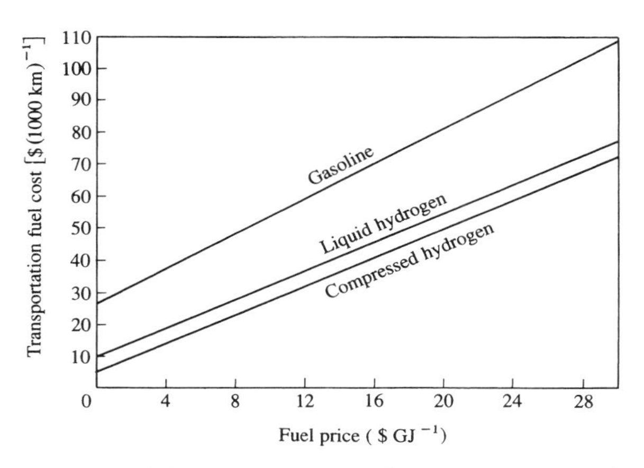

Fig. 1. Real fuel cost as a part of transportation cost for different fuels (fuel economy 30 mpg of gasoline).

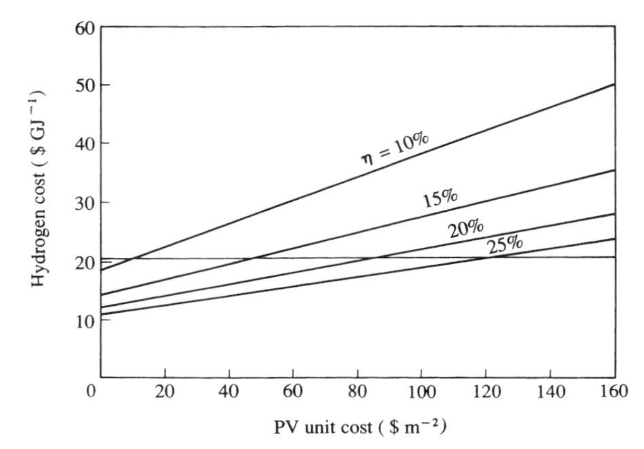

Fig. 2. Price of delivered compressed hydrogen as a function of PV unit price and efficiency.

C_{ENV} = cost of the direct environmental damage (for gasoline it is 8.47 \$ GJ^{-1} [20])

a = fraction of auxiliary energy required (for gasoline it is 0.15)*

C'_{ENV} = cost of the indirect environmental damage (for fossil fuels it is 8.1 \$ GJ^{-1} [20]).*

For hydrogen, this equation should be slightly different, taking into consideration that hydrogen does not cause any direct damage and that hydrogen engines have better efficiencies and better fuel economy:

$$C_F = (F_E/e)\,[C_{HD} + aC'_{ENV}] \tag{14}$$

where:

e = ratio of efficiencies of hydrogen to gasoline engines = 1.22 [16, 19]

C_{HD} = cost of delivered hydrogen (\$ GJ^{-1})

a = fraction of auxiliary energy required ($a = 0.28$ for compressed hydrogen and $a = 0.54$ for liquid hydrogen).*

Figure 1 shows fuel costs, as a part of the real transportation costs, for different fuels (gasoline, liquid hydrogen and compressed hydrogen).

It is possible to calculate at which price hydrogen would become competitive with gasoline, i.e. when it would have the same transportation cost (\$ km^{-1}) as gasoline. Since a hydrogen vehicle would be \$300–1500 more expensive than a gasoline vehicle (because of the storage tanks), one must take this difference into account when estimating transportation cost. For the projected price of gasoline in the year 2000 [21], which is in the range \$10.5–12.5 GJ^{-1}, the competitive price of compressed hydrogen would be \$17.2–23.8 GJ^{-1}, or on average \$20.5 GJ^{-1}, and competitive price of liquid hydrogen would be \$15.0–21.5 GJ^{-1}, or on average \$18.2 GJ^{-1}.

The price of hydrogen depends mostly on the unit costs of the PV modules and wind turbines and their efficiencies. Figure 2 shows that if projected costs of the PV modules, \$45–80 m^{-2}, and projected efficiency, 15–20%, are achieved, compressed hydrogen could become competitive with gasoline. About the same are the results for the wind power (Fig. 3): compressed hydrogen could become competitive only for the lower limit of projected unit costs, which means that only the most advanced wind turbines located at the most windy areas (with the highest capacity factor) would be applicable for the large scale hydrogen production. Liquid hydrogen would still be more expensive in both cases; therefore

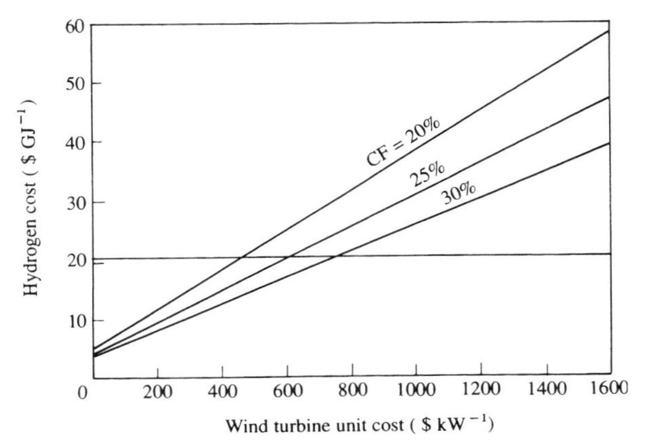

Fig. 3. Price of delivered compressed hydrogen as a function of wind turbine unit price and capacity factor.

*As shown before, there is a significant amount of energy required for hydrogen storage, transportation and distribution which accounts for approximately 14% of delivered hydrogen energy for compressed hydrogen and up to 40% for liquid hydrogen. In addition, material and energy requirement analysis has shown that another 14% is required to build up the plants for hydrogen production. Since only hydrogen as a fuel for surface transportation has been considered in this study, it is assumed that this "energy subsidy" would be supplied by fossil fuels, at least during the transition period until the establishment of the overall hydrogen energy system. Energy required for gasoline production, transportation, storage and distribution has been estimated to be 15%.

research efforts are required to improve the liquefaction processes and to reduce the costs.

4. CONCLUSION

The results presented in this study indicate that hydrogen produced from renewable energy sources (solar and wind) can become cost competitive with gasoline by the year 2000, as a fuel for surface transportation, provided that the cost of hydrogen (compressed gas) can be reduced to approximately $\$21\ GJ^{-1}$. This is possible for solar photovoltaic hydrogen production if the PV module cost is low enough. For example, a 15% efficient PV module must cost less than $\$47\ m^{-2}$. This cost is thought to be possible by the year 2000, based on progress presently being made on PV silicon technology. About the same costs could be achieved by wind energy conversion progress. Competitiveness in this study has been based on the real costs which include all the internal (capital cost for the automobile, fuel cost) and external costs related to transportation. Internal costs include the purchase of the car, maintenance and fuel. External costs include environmental damage due to fossil fuel use. Automotive air pollution in the urban environment is perhaps the most important factor motivating a shift to renewable hydrogen. However, for this shift, a further development and improvement of all components of the hydrogen system is required to improve efficiency and to reduce costs.

REFERENCES

1. *Annual Energy Review* 1988, Energy Information Administration, Washington, D.C. (1989).
2. J. M. Ohi and L. R. Brown, Wind-powered electricity for California, *Solar Today* **3,** 16–18 (1989).
3. National Photovoltaics Plan, Five Year Research Plan 1987–1991. Photovoltaics: USA's Energy Opportunity, U.S. Department of Energy, (DOE/CH10093-7) (1987).
4. J. M. Ogden and R. H. Williams, Hydrogen and the Revolution in Amorphous Silicon Solar Cell Technology. Center for Energy and Environmental Studies, Princeton University, Report No. 231, Princeton, NJ (1989).
5. K. Zweibel and H. S. Ullal, Thin Film Photovoltaics, SERI/TP-211-3501, Prepared under Task No. PV940301, Golden, CO (1989).
6. C. Carpetis, An assessment of electrolytic hydrogen production by means of photovoltaic energy conversion, *Int. J. Hydrogen Energy* **9,** 969–991 (1984).
7. T. N. Veziroğlu, Hydrogen Energy Technology for Developing Countries. Report prepared for UNIDO, Clean Energy Research Institute, University of Miami, Coral Gables, FL (1989).
8. E. W. Justi, *A Solar–Hydrogen Energy System.* Plenum Press, New York (1987).
9. C.-J. Winter and J. Nitsch, *Hydrogen as an Energy Carrier.* Springer, Berlin (1988).
10. H. J. Plass, Jr, F. Barbir and H. Miller, Hydrogen Systems Application Analysis. Report to DOE/SERI #XL-9-18168-1, Clean Energy Research Institute, University of Miami, Coral Gables, FL (1989).
11. D. L. Block, S. Dutta and A. T-Raissi, Hydrogen for Power Applications, Task 2: Storage of Hydrogen in Solid, Liquid and Gaseous Forms. Report No. FSEC-CR-204-88, Florida Solar Energy Center, Cape Canaveral, FL (1988).
12. H. T. Odum *et al.*, Environmental Systems and Public Policy. Center for Wetlands, University of Florida, Gainesville, FL (1988).
13. C. Voigt, Material and energy requirements of solar hydrogen plants, *Int. J. Hydrogen Energy* **9,** 491–500 (1984).
14. D. T. Spreng, *Net-Energy Analysis and the Energy Requirements of Energy Systems.* Praeger Publishers, New York (1988).
15. J. B. Taylor *et al.*, Technical and economic assessment of methods for the storage of large quantities of hydrogen. *Int. J. Hydrogen Energy* **11,** 5–22 (1986).
16. M. A. DeLuchi, Hydrogen vehicles: an evaluation of fuel storage, performance, safety, environmental impacts and cost. *Int. J. Hydrogen Energy* **14,** 81–130 (1989).
17. K. Feucht *et al.*, Hydrogen drive for road vehicles—results from the test run in Berlin. *Int. J. Hydrogen Energy* **13,** 243–250 (1988).
18. G. Withalm and W. Gelse, in T. N. Veziroğlu, N. Getoff and P. Weinzierl (eds), The Mercedes Benz hydrogen engine for application in a fleet vehicle, *Hydrogen Energy Progress VI, Proc. 6th WHEC,* Vol. 3, pp. 1161–1173. Pergamon Press, New York (1986).
19. A. H. Awad and T. N. Veziroğlu, Hydrogen versus synthetic fossil fuels. *Int. J. Hydrogen Energy* **9,** 355–366 (1984).
20. F. Barbir, T. N. Veziroğlu and H. J. Plass, Jr, Environmental damage due to fossil fuels use. Submitted to *Int. J. of Hydrogen Energy* (1990).
21. *Annual Energy Outlook, With Projections to* 2000, Energy Information Administration, Washington, D.C. (1987).

Renewable Hydrogen Transportation Fuels

Joan M. Ogden

Center for Energy and Environmental Studies, Princeton University

Mark A. DeLuchi

Institute of Transportation Studies, University of California

Abstract

In this paper we assess the prospects for producing hydrogen from renewable resources for use as a low polluting transportation fuel. First, we review current and projected costs and potential resources for renewable hydrogen production from solar, wind, hydro electricity and biomass. Next, cost and performance estimates are presented for a hydrogen fuel cell automobile based on a proton exchange membrane (PEM) fuel cell. Using post 2000 projections for fuel cell, advanced battery and renewable hydrogen production technologies, we find that the lifecycle cost of owning and operating a fuel cell vehicle fueled with renewable hydrogen would be comparable to that of a gasoline vehicle, and less than that of an electric battery powered vehicle. This suggests that hydrogen fuel cell vehicles could play an important role in emerging "zero emissions vehicle" markets, starting around the year 2000.

1. INTRODUCTION

Concerns about global warming, urban air quality, acid deposition, and energy supply security are motivating increased interest in low polluting alternative transportation fuels. Hydrogen is a high quality, exceptionally clean fuel, which can be produced from a variety of non-petroleum feedstocks, and has been demonstrated in prototype cars, buses, trucks and airplanes. With today's internal combustion engine hydrogen vehicles the only significant pollutants are nitrogen oxides (NOx). With hydrogen fuel cell vehicles which could be developed over the next several years, even NOx emissions would be eliminated. If hydrogen is produced from renewable resources, via electrolysis powered by solar photovoltaic, solar thermal electric, wind or hydro electricity or via gasification of renewably grown biomass, it would be possible in principle to produce and use transport fuel on a large scale with greatly reduced greenhouse gas emissions and very little local pollution. Here we assess the technical and economic prospects for renewable hydrogen as a transportation fuel for automobiles [1].

---

[a] This paper was adapted from Ref. [1].Please see [1] for details of the calculations and extensive references for Tables A-H.

2. PRODUCTION OF HYDROGEN FROM RENEWABLE RESOURCES

Renewable Hydrogen Production Methods

In solar electrolytic hydrogen systems, a source of renewable electricity [hydro, wind, solar thermal electric or solar photovoltaic (PV)] is connected to an electrolyzer, splitting water into hydrogen and oxygen [2,3]. The hydrogen can be used onsite or compressed for storage or pipeline transmission to offsite users. Electrolysis is a commercially available technology, which is now being optimized for use with intermittent power sources such as solar and wind [4]. Rapid progress is being made in wind, solar thermal electric and solar photovoltaic technologies [5], and low intermittent electricity costs are projected for wind and PV by the early part of the next century (Table B). The cost and performance of wind and PV electrolysis systems are shown in Table A, based on post 2000 projections [5].

Hydrogen can also be produced by gasifying at high temperatures biomass feedstocks such as wood chips and forest and agricultural residues [6,7]. The gasifier output, mainly hydrogen, carbon monoxide and methane, can then be reformed and shifted to produce a mixture of hydrogen and carbon dioxide. The carbon dioxide is removed leaving hydrogen. Biomass gasifiers have been demonstrated at the laboratory scale, and several types of biomass gasifiers now under development for methanol production could probably be used for hydrogen production as well. The rest of the equipment needed for converting the gasifier output to hydrogen (methane reformers, shift reactors, CO_2 removal, and pressure swing adsorption for hydrogen purification) is commercially available and widely used in chemical process industries. Cost and performance estimates are given in Table A for a biomass hydrogen plant[7].

The cost of producing hydrogen from renewable resources is estimated in Table C, for present (1991) technology, for the "near term" (1990s) and for the "long term" (post 2000, with mature technologies widely employed, so that economies of scale are fully exploited). Levelized costs (in constant 1989 US dollars) are calculated using a common set of economic assumptions (6.1% discount rate, annual insurance rate is 0.5% of the installed capital cost and taxes are 1.5% of the installed capital cost). At large scale (17,000 GJ/day or 50 million scf/day), biomass hydrogen would cost about $6.2-8.8/GJ to produce, making it the least expensive method of renewable hydrogen production (all hydrogen costs and efficiencies are given in terms of the higher heating value of hydrogen). Electrolytic hydrogen from wind or solar PV would cost about twice as much, $12-20/GJ. However, electrolytic hydrogen systems could be employed at much smaller scale (200 GJ/day), and would probably be less expensive than biomass hydrogen plants up to capacities of several 1000 GJ/day. Where it is available hydropower could also be used to produce hydrogen at costs of $10-20/GJ. For comparison, the cost of producing hydrogen from fossil feedstocks is also given. At large scale, hydrogen from steam reforming of natural gas would cost $6.1-10.1/GJ (assuming natural gas prices of $2-6/GJ), comparable to hydrogen from biomass. At smaller scale (200 GJ/day), steam reforming would be more expensive, approximately competitive with solar or wind electrolysis. Coal gasification plants would also exhibit strong scale economies. For large plant sizes, hydrogen from coal would cost about $8-13/GJ. At a given plant size, hydrogen from biomass gasification would probably be less expensive than hydrogen from coal gasification, because the plant would be less complex.

Potential Resources For Renewable Hydrogen Production

As shown in Table D, ample resources for renewable hydrogen production are available in most areas of the world. Table E illustrates the specific land and water use for various methods of hydrogen production and estimates the land needed to produce an amount of energy equivalent to: US oil use in 1988; world fossil fuel use; and projected non-electric fuel use in 2025 and 2050 [1,5]. From Tables D and E, it is apparent that PV hydrogen would require by far the least land and water use, and would be the least constrained geographically. Using only about 0.5% of the global land area (or 2% of the world's desert area), it would be possible to produce PV hydrogen equivalent in energy to current fossil fuel use (300 EJ/year). Wind resources are also large (an amount of energy equivalent to current fossil fuel use could be produced using about 1/4 of the potentially available wind resource). Although wind hydrogen would require about 10 times the land area of PV, the land could perhaps be used simultaneously for other purposes. Biomass hydrogen equivalent in energy to current oil use (115 EJ/yr) could be produced on 10% of forest, woodland and cropland area (4% of global land area). Biomass would require about 20-30 times the land needed for PV, and several thousand times as much water, so that resource constraints might ultimately limit biomass hydrogen development. Hydropower, although much smaller than solar, wind or biomass resources could make local contributions as well.

3. HYDROGEN AS A TRANSPORT FUEL

Many prototype hydrogen internal combustion engine vehicles (ICEVs) have been demonstrated [8]. Recently, several companies including Ballard Power Systems (Canada), Elenco (Netherlands), Energy Partners (USA), Roger Billings (USA), H-Power (USA), Siemens (Germany), and Daimler Benz (Germany) have begun developing hydrogen fuel cell vehicles [9,10]. The potential benefits of fuel cell vehicles are: elimination of all harmful tailpipe emissions; and higher efficiency, which makes onboard storage requirements more tractable and requires less supply capacity (this is particularly important for renewable energy supplies, which tend to be capital and/or land intensive).

In a fuel cell electric vehicle (FCEV), a hydrogen-air fuel cell provides electricity to an electric drive train similar to those used in battery powered electric vehicles. Hydrogen fuel is stored directly (as a compressed gas or hydride) or in the form of methanol, which is reformed onboard to produce hydrogen. In some designs peak power demands are met by a small supplemental battery or perhaps an ultra-capacitor. In our study, we have chosen a proton exchange membrane (PEM) fuel cell, which offers high power density, quick start-up time, modest operating temperature (100°C) and the potential to reach low costs in mass production. PEM fuel cells are now being developed and should be available within a few years. The supplemental peak power battery is assumed to be a bipolar lithium iron disulfide battery, a promising technology now under development, which has the high power density needed for peak power. For hydrogen storage have chosen high pressure compressed hydrogen gas cylinders because of they are simple, commercially available, safe and can be refilled in only 3 minutes.

To assess the prospects for hydrogen fuel cell vehicles, we compare them to gasoline ICEVs, methanol fuel cell vehicles and battery powered electric vehicles. The characteristics of the vehicles used in our comparison are given in Table F. To facilitate comparison, the weight, range and performance of the vehicles has been chosen to be comparable. The fuel efficiency of the fuel cell vehicles is about 2 to 3 times that of the gasoline vehicle. Moreover, the lifetime is assumed to be about 50% longer than a gasoline vehicle and the maintenance cost is less, assumptions which are based on experience with electric battery vehicles. The initial cost of the fuel cell vehicles is about $8000-9000 higher than that of the gasoline vehicle. The lifecycle cost of transportation is then computed assuming that hydrogen is produced from solar or wind (at a delivered cost of $23.4/GJ without taxes) or from biomass (at a delivered cost of $13.2/GJ). Methanol from biomass is estimated to cost $13.0/GJ delivered, electricity 7 cents/kWh, and gasoline $1.21/gallon (a price projected for the year 2000). Table G shows the lifecycle cost, and also the breakeven gasoline price, which would make the total lifecycle cost of the gasoline vehicle equal to that of the alternative vehicle.

Rather surprisingly, we find that fuel cell vehicles fueled with hydrogen from solar or wind would have a lifecycle cost comparable to that of a gasoline vehicle. With biomass hydrogen or methanol the lifecycle cost would be even lower. Even though the initial vehicle cost and fuel costs are higher for renewable hydrogen than those for gasoline, the lifecycle cost is about the same (or slightly lower for biomass hydrogen) because: 1) hydrogen can be used 2 to 3 times as efficiently as gasoline, so that the fuel cost per km is less; 2) the lifetime of the fuel cell vehicle is 50% longer so that the contribution of the vehicle cost to the lifecycle cost is only slightly higher than for gasoline; 3) maintenance costs are lower for FCEVs than for ICEVs.

4. A STRATEGY FOR DEVELOPING RENEWABLE HYDROGEN AS A TRANSPORT FUEL

Large and rapidly expanding markets for zero emissions vehicles (ZEVs) are expected to open over the next ten to twenty years. The California Air Resources Board has mandated that by 2003 10% of the cars sold in California must be ZEVs,a market of several hundred thousand vehicles per year. Because of their much faster refueling time and potential to reach lower cost, hydrogen FCEVs could eventually serve a much larger fraction of the passenger car market than battery powered electric vehicles. To compete economically with BPEVs, however, hydrogen FCEVs would have to be developed, tested and commercialized on a large enough scale to significantly reduce fuel cell costs.

The first prototype hydrogen FCEVs could be developed within the next few years. Testing in small experimental fleets is the first step toward gaining experience with vehicle and refueling technology and evaluating consumer acceptance. Beyond this, the next step might be the introduction of modest sized fleets of several hundred to a few thousand vehicles, which would be centrally refueled. With a committment from industry, it is possible that these fleets could be ready around the year 2000 in response to the California market. If hydrogen FCEVs proved successful in fleet service, the hydrogen distribution network might be expanded to general consumers. During the early decades of the next century, FCEVs might come to capture a large share of the

ZEV market.

How would hydrogen be produced to satisfy these potential markets? In the 1990s, experimental fleets would probably use truck delivered industrial hydrogen derived from natural gas. After the year 2000, larger government or utility owned fleets might be introduced. A 1000 car fleet would require about 75,000 GJ of hydrogen per year or 0.5 million scf/day. At this scale, it would be less costly to produce hydrogen at the filling station, either from small scale steam reforming of natural gas or via electrolysis. [By the early part of the next century, wind and PV hydrogen would become roughly cost competitive with small scale steam reforming of natural gas (assuming natural gas costs $4-6/GJ in this time frame). At this small scale biomass hydrogen would probably be too expensive.] In the longer term, large numbers of FCEVs might be introduced as general purpose passenger cars. Hydrogen fuel would be produced in a centralized large plant and made available to urban consumers at local filling stations. To supply a city with hundreds of thousands of FCEVs, biomass hydrogen would be the least expensive renewable source. Even though many regions of the world are close to good solar, wind or biomass resources, it might be neccessary in areas such as Northern Europe to transmit hydrogen long distances. For a large (0.5 EJ/yr) 1600 km pipeline, the delivered cost of PV hydrogen would be about the same as for a local city supply, because increased transmission and distribution costs would be offset by savings in large scale storage and compression costs [1,5].

5. CONCLUSIONS

With projected advances in the technology of fuel cells, solar-electric power generation and biomass gasification, hydrogen from renewable resources could become attractive as a low polluting transport fuel in the early part of the next century. The first renewable hydrogen systems might be introduced around the year 2000, with small PV or wind electrolysis systems supplying hydrogen for small (1000 car) fleets of fuel cell vehicles. The use of hydrogen might then be expanded to the general public. At this scale, biomasss hydrogen would become economically attractive, although the amount of biomass hydrogen produced might be limited by environmental constraints and the availability of land. By contrast, wind and especially solar resources are huge and could meet all potential demand for transport fuel. In the long term, hydrogen would be produced from the best available resource, balancing environmental with traditional economic criteria. Hydrogen fuel cell vehicles could be an important part of a strategy for reducing greenhouse gas emissions and improving urban air quality.

Acknowledgements

The authors would like to thank Eric Larson and Robert Williams for useful discussions during this work. This research was supported by the USDOE National Renewable Energy Laboratory, and the US Environmental Protection Agency.

REFERENCES

1. M.A.DeLuchi and J.M. Ogden, "Solar Hydrogen Transportation Fuels," Princeton University Center for Energy and Enviromental Studies Report, forthcoming 1992.

2. J.M.Ogden and R.H. Williams, Solar Hydrogen: Moving Beyond Fossil Fuels, Washington DC:World Resources Institute. 123 pp, 1989.

3. C.-J. Winter and J. Nitsch, (eds.) Hydrogen as an Energy Carrier. Berlin, New York:Springer-Verlag. 377 pp, 1989.

4. S.N.Pirani and A.T.B. Stuart, "Testing and Evaluation of Advanced Water Electrolysis Equipment and Components," Proceedings of the 2nd International Energy Agency Hydrogen Production Workshop, Julich, Germany, September 4-6, 1991. Stucki, S. Operation of Membrel Electrolyzers Under Varying Load, Proceedings of the 2nd International Energy Agency Hydrogen Production Workshop, Julich, Germany, September 4-6, 1991

5. J.M. Ogden and J.Nitsch, "Solar Hydrogen," in T. Johannson, H. Kelly, A.K.N. Reddy, R.H. Williams, eds., Renewable Energy: Fuels and Electricity from Renewable Resources, Island Press, Washington DC, forthcoming 1992.

6. M.A.DeLuchi, E.D.Larson, and R.H. Williams, "Biomass Methanol and Hydrogen for Transportation," Princeton University Center for Energy and Environmental Studies Report, August 1991.

7. E.D. Larson and R. Katofsky, Biomass methanol and hydrogen production technology, Princeton University Center for Energy and Environmental Studies Report, forthcoming 1992.

8. M.A.DeLuchi, "Hydrogen vehicles: An evaluation of fuel storage, performance, safety, environmental impacts and cost," International Journal of Hydrogen Energy, 14:81-130, 1989.

9. Society for Automotive Engineers, Proceedings, Topical Technical Conference on Fuel cells for Transportation, Arlington, Virginia, November 4-5, 1991.

10. R.A. Lemmons, "Fuel Cells for Transportation," Journal of Power Sources, v.29, pp. 251-264, 1990.

TABLE A: POST 2000 PROJECTIONS FOR RENEWABLE HYDROGEN PRODUCTION SYSTEMS

Electrolytic Hydrogen Systems [1]

Atmospheric pressure unipolar electrolyzer (>10 MW)

Rated voltage	1.74 Volts
Rated current density	250 mA/cm^2
Max. operating current density	333 mA/cm^2
Electrolyzer annual O&M cost	2% of capital cost
Electrolyzer lifetime	20 years
Efficiency at max. op. voltage	81% (AC Plant), 85% (DC Plant)
Installed plant capital cost @ max.op. current density	$371/kWACin , $231/kWDCin

Wind Hydrogen System		Solar Photovoltaic Hydrogen System	
Horizontal axis wind turbine		Thin film Solar PV Modules	
Turbine capacity	1000 kWAC	Tilted, fixed flat plate array	
Turbine diameter	52 m	PV Module efficiency	12-18%
Hub height	50 m	PV system efficiency	10.7-16.0%
System availability	95%	PV system capacity	> 10 MWDC
Hectares/MWe	16	Hectares/MWe	1.25-1.87
Total installed system cost	$750/kWpeak	PV System Cost	$522-1077/kWp
Annual O&M cost	$0.008/kWhAC	O&M Cost	$0.5/m^2/yr
System lifetime	30 years	System Lifetime	30 years
Electrolyzer coupling efficiency	94%	Coupling Efficiency	93%

Cost of wind hydrogen			Cost of PV hydrogen	
Annual ave. wind power density W/m^2 (power per unit of rotor area)	630	350	Ann.ave.insolation	271 W/m^2
Annual average capacity factor	0.49	0.27		0.27
Levelized electricity cost	$0.026	0.040/kWhAC		$0.022-0.044/kWhDC
Levelized hydrogen cost ($/GJ)	13.0	20.6		11.6-19.1

Biomass Gasifier Hydrogen Plant [7]

Battelle Columbus Laboratory Gasifier	
Biomass input capacity	1650 dry tonne biomass/day (1382 GJ/h)
Plant lifetime (years)	25
Total investment cost (10^6 $)	138
Working capital (10^6 $)	10.2
Land (10^6 $)	2.07
Variable operating costs excl. biomass (10^6 $/year)	9.90
Biomass costs (10^6 $/year)	21.8-43.7 ($2-4/GJ)
Fixed operating cost (10^6 $/year)	7.75

Levelized hydrogen costs ($/GJ)	
Capital	1.71
Labor, maintenance, chemicals	1.96
Biomass	2.57-5.15
Total	6.24-8.82

TABLE B: LEVELIZED PRODUCTION COST OF SOLAR ELECTRICITY (CENTS/KWH) [1]

Technology	1991	Near term	Post 2000
Wind (630 W/m^2)	-	4.5	2.6
(500 W/m^2)	8.3	5.2	3.3
(350 W/m^2)	14.3	7.0	4.0
Solar thermal electric (SW US)	11-16	11-16	5.5-7.8
Solar PV (SW US)	14-35	7-16	2.2-4.4 (DC) 3.2-5.4 (AC)
Hydropower (Off-peak)	1-4	1-4	1-4

Levelized costs are given for intermittent electricity at the plant site without storage. For wind power the annual average wind power density at hub height is shown in parentheses.

TABLE C: CURRENT AND PROJECTED COSTS OF HYDROGEN PRODUCTION ($/GJ) [1]

	1991	Near Term	Post 2000
Renewable sources			
Electrolytic hydrogen (for plants producing 14,160 Nm^3 (180 GJ) H2/day) from:			
Solar PV (SW US)	54-121	29-57	12-19
Wind (630 W/m^2)			13
(500 W/m^2)	37	25	17
(350 W/m^2)	53	38	21
Solar thermal (SW US)	45-60	37-63	22-30
Off peak hydroelectricity	10-20	10-20	10-20
Hydrogen from biomass gasification			
Large plant (17360 GJ H2/day)			6.2-8.8
Fossil sources			
Hydrogen from steam reforming of natural gas			
Large plant (36170 GJ H2/day)	6.1-10.1	6.1-10.1	6.1-10.1
Small plant (180 GJ H2/day)	11-17	11-17	11-17
Hydrogen from coal gasification			
Large plant (36170 GJ H2/day)	8	8	8
Medium plant (9041 GJ H2/day)	13	13	13

The levelized hydrogen production cost is shown in constant 1989 US$. Compression and storage (which would be required for wind or solar electrolysis) would add about $2-3/GJ, local distribution about $0.5/GJ and filling station costs about $5.2/GJ, so that the delivered cost of hydrogen for transportation would be $8-9/GJ higher than the production cost. For biomass, coal or natural gas hydrogen, compression and storage would not be needed, so that the delivered cost would be about $6/GJ higher. Electrolytic hydrogen costs were calculated assuming the electricity production costs in Table B.

TABLE D: POTENTIAL RESOURCES FOR RENEWABLE HYDROGEN PRODUCTION [1,5]

Region	Land area (10^6 km^2)	Electrolytic Hydrogen From: Technically Useable Hydro (EJ H_2/yr)	Total Wind Potential EJ H_2/yr	10^6 km^2	PV on 1% Land Area EJ H_2/yr	10^6 km^2	Biomass H2 Produced on 10% of Forest Woodland, crop EJ H_2/yr	10^6 km^2
Africa	29.7	9.1	257	7.1	128	0.30	19	0.89
Asia	24.4	15.5	68	2.2	103	0.24	22	1.01
Australia	10.6	1.1	75	1.8	47	0.11	5	0.21
N America	21.9	9.1	308	7.7	94	0.22	18	0.83
S/C America	17.8	11.0	122	3.2	77	0.18	25	1.17
Europ+USSR	30.3	10.6	366	9.4	130	0.30	25	1.45
World	137.7	56.3	1196	31.7	579	1.38	118	5.56

TABLE E: LAND AND WATER REQUIREMENTS FOR RENEWABLE HYDROGEN PRODUCTION [1,5]

	Land requirements hectares/MWe,peak	Land requirements MJ/yr/m^2	Water requirements liters/GJ (HHV)
Electrolytic hydrogen from:			
PV	1.3	530	63
Solar Thermal Electric	4.0	175	63
Wind	4.7-16	30-160	63
Hydroelectric	16-900	2-90	>>63
Biomass Hydrogen	-	21	35,000-81,000

Land requirements (10^6 km^2) to produce hydrogen equivalent in energy to:

from:	Present US Oil (34 EJ)	Present World Oil (115 EJ)	World Fossil Fuel (300 EJ)	Projected World Non-Electric Fuel Demand (IPCC) 2025 (286 EJ)	2050 (289 EJ)
PV	0.079	0.268	0.700	0.667	0.674
Wind	0.87	2.9	7.7	7.3	7.4
Biomass	2.0	6.8	17.8	17.0	17.1

TABLE F: COST AND PERFORMANCE OF ALTERNATIVE FUELED VEHICLES [1]

	Vehicle type				
	Battery EV	H2 FCEV	H2 FCEV	MeOH FCV	Gasoline ICEV
Fuel Storage System	-	Comp.gas @ 55 MPa	Adv. System	Metal tank	Metal tank
Battery	----Bipolar lithium iron disulfide----				
Driving range (km)	400	400	400	480	640
Power to wheels (kW)	81	72	71	74	98.4
Delivered fuel price (excl. taxes)		--PV/wind	Biomass--		
($/gallon gasoline equiv)	2.54	3.06	1.73	1.69	1.21
($/GJ)	19.4	23.4	13.2	13.0	9.3
Refueling time (min)	60-360	2-3	2-5	2-3	2-3
Gasoline-equivalent fuel economy (l/100 km)	2.18	3.08	3.11	3.63	9.08
(mpg)	108	76.5	75.6	64.8	25.9
Curb weight (1000 kg)	1.44	1.24	1.26	1.27	1.37
Initial price (1000 $)	29.6	30.4	28.3	29.3	17.3
Vehicle life (1000 km)	289	289	289	289	193
Annual maintenance cost($)	358	401	401	416	516

TABLE G: LIFECYCLE COST OF TRANSPORTATION FOR ALTERNATIVE VEHICLES(CENTS/KM) [1]

Cost Component	Battery EV	Solar/wind H2 FCV	Biomass H2 FCV	Biomass MeOH FCV	Gasoline ICEV
Purchased electricity	1.47				
Vehicle (excl.fuel cell, battery, storage)	7.09	6.72	6.72	6.73	11.17
Battery	6.71	2.67	2.67	2.52	
Fuel storage system		0.83	0.83	0.02	
		(comp.H2 gas @55 MPa)			
Fuel cell system		2.25	2.25	2.65	
Fuel for vehicle (excl. taxes)		2.45	1.42	1.73	2.89
Maintenance	1.70	1.90	1.90	1.97	2.89
Misc. other costs	5.00	4.72	4.72	4.66	4.56
Total cost (cents/km)	21.99	21.55	20.54	20.93	21.51
Breakeven gasoline price incl. tax ($/gal)	1.71	1.52	1.09	1.26	-
w/ Advanced H2 Storage System		1.26	0.82	-	-

Technical and Economic Assessment of Methods for the Storage of Large Quantities of Hydrogen

J. B. Taylor,* J. E. A. Alderson,* K. M. Kalyanam,† A. B. Lyle,‡ and L. A. Phillips†

**Division of Chemistry, National Research Council, †Canatom Inc. Canada and ‡Monenco Engineers & Constructors Inc., Canada*

Abstract—The storage of large quantities of hydrogen is analysed on the basis of five conceptual scenarios. These scenarios are representative of possible practical situations. They involve storage above ground as a cryogenic liquid as well as high-pressure storage both above and below ground. The latter includes natural and specially created cavities. No comparison is made between these various methods since in practice one would rarely be free to make a choice among various alternatives. Each scenario is defined and then costed in detail. The incremental cost of storage is calculated and the sensitivity determined for variations of throughput, capital cost and electricity cost. Also included in the data are the costs of electrolytic hydrogen and of hydrogen liquefaction.

The cost of storage can add between 30 and 300% to the cost of hydrogen so that large offsetting benefits have to exist in order to give economic viability. There is a considerable interplay between the magnitude of the capital charges and the system throughput or utilization. In the case of the electrolyser alone, non-continuous operation results in a considerable increase in the unit hydrogen costs. For storage systems where the unit capital charges are high then the operating costs are very sensitive to changes in both throughput and capital costs. A hidden capital charge occurs in some cases of underground storage where a large inventory of cushion gas is inaccessible to the user, but is necessary to define the minimum size of the reservoir.

The results clearly show the need to consider many factors in order to optimize a storage system. In practice, this has to be done for each specific situation and general consideration can be very misleading.

INTRODUCTION

Electrical energy is a secondary resource. In its industrial applications, it is normally generated at the resource site and is either distributed to customers or is used on site. In the latter case, typified by the aluminum industry, raw materials are moved to the site and the product is shipped to markets. Numerous suggestions have been made which involve the interconversion of electricity and hydrogen. The incentives for such action can include the requirements for hydrogen as a chemical commodity or as a special fuel such as in space application or the creation of a commodity which can be used to store energy. Storage is an essential element of almost all ideas on electric to hydrogen conversion. The storability of hydrogen may be the motivation in the first place (i.e. use of intermittent energy sources such as tidal, wind, etc.) or it may be an imposed requirement derived from the end-use (e.g. liquid hydrogen for aircraft).

The insertion of a storage buffer in an electrical energy supply sequence creates an interesting new set of conditions which may be advantageous in a practical sense. However, the use of storage will be dictated by economic factors. The costs for bulk hydrogen storage are poorly documented and there has been very limited experience with the storage of hydrogen on such a scale. The work reported here was part of a larger study which besides the costing aspects also addressed such topics as site specific requirements, geology, safety and regulatory considerations.

AVAILABLE TECHNOLOGIES

For the purposes of this study, it was decided to confine the analysis to storage technologies which involved no conversion of the hydrogen by chemical combination (this excluded reversible organic hydrogenation/dehydrogenation, hydrides and ammonia). All the technologies considered are currently available and there is no technical obstacle to their use. Engineering and cost data can be generated but site and applications specificity dominate the final considerations. One would rarely be in a position to select from a menu of technologies for a specific application and some overriding considerations would usually force the choice. A recent very detailed analytical treatment [1] takes a rather different view of the problem and defines a procedure for selection of the storage method for economic optimization of the whole system.

Methods exist to store hydrogen in large quantities as a gas or as a liquid. The choice between the two forms is largely dictated by the end-use. Since there are few end-uses for liquid hydrogen, it has rarely been used as a method for bulk storage. It would of course be the only feasible storage for off-shore shipment comparable to the present LNG (liquid natural gas) activity. However, the time frame for such market development was viewed to be outside the interests of the present analysis.

A major study of underground storage of gaseous hydrogen was conducted by the Institute of Gas Technology in the U.S.A. [2]. The report of this study provided much background material for the present analysis. It also confirmed the economic and technical feasibility of such large-scale hydrogen storage. Significant differences exist between this work and the IGT study. In particular, this work uses different site specificity, different cycle rates, above-ground storage options, and the application is entirely to the use of electrolytic hydrogen.

STATE-OF-THE-ART

Pressure vessels

Suppliers of these tube-type vessels include the U.S. Steel Corporation and Mannesmann Pipe (F.R.G.). The steel is typically seamless tempered material (C ~ 0.5%; Mn ~ 1.4–1.8%) with a tensile strength of 120,000 psi (586,000 kg m^{-2}). Specifications generally meet ASTM A372-Type IV or Type V. For certification, the vessels are hydraulically tested to 1.5 times their rating which is normally 2500 psi (12,200 kg m^{-2}). Manifolding is done with brass and copper with a safety relief set at about 10% less than the allowed pressure set by the vessel code. In the U.S. this is the ASME Boiler and Pressure Vessel Code for integrally forged vessels (SA372 section VII, Division I). The vessels can be certified for hauling on road trailers. Since the system is modular, the capital costs are insensitive to scale. However, the use of high-pressure tube storage is confined to quite small gas volumes (500,000 scf or 14,000 Nm^3) for reasons that will become apparent from the results of this study.

Gas storage underground

There are four underground formations in which gas can be stored under pressure:

- Depleted oil or gas field
- Aquifers
- Excavated rock caverns
- Solution mined salt caverns.

Natural gas has been stored underground since 1916 and much of the experience is directly applicable to hydrogen. Originally underground storage was confined to depleted oil and gas fields. They are porous rock structures not unlike those associated with aquifers. Such storage facilities tend to be extremely large; volumes of gas stored exceed 10^9 Nm^3. Pressures can be up to 40 atm. Any underground storage in porous media requires the following:

- A stratum of water-bearing porous rock, usually sand or sandstone, at a depth of 500–3000 ft (150–900 m) below the surface.
- An impervious caprock of adequate thickness.
- A suitable geological structure such as an anticline, which usually forms a dome-shaped geologic structure.

The porosity of the sandstone or sand must be sufficiently high to provide a reasonable void space in an aggregate sense to yield an economically acceptable storage volume. The permeability must be high enough to provide an adequate rate of inlet flow (injection) and outlet transmission injection (withdrawal). On the other hand, the caprock structure must be reasonably impermeable if it is to contain the gas.

An important point to note regarding the performance of the caprock structure is the mechanism involved in 'sealing' the top of the underground reservoir, either a depleted field or an aquifer. This sealing occurs because of water capillary action, in which water fills all the voids of the caprock structure and must be expelled by sufficiently high pressure to overcome the capillary resistances (the threshold pressure of the caprock). Below this incremental pressure, the caprock will act as an effective barrier to the passage of any gas. This incremental or threshold pressure and the effectiveness of the sealing action is independent of the nature of the gas because it is a water–rock capillary effect. This is an important observation in that it indicates that hydrogen in underground storage will behave much like natural gas insofar as integrity against leakage is concerned.

The gas stored in a field is divided into active working gas and cushion gas. The latter is inactive base gas not recoverable at acceptable withdrawal rates, but necessary to define the storage volume reservoir pressure conditions to place a field into storage service. The ratio of working to cushion gas varies widely with a ratio greater than 2:1 generally being preferred. The cost of the cushion gas is a very significant part of the capital charges for such large storage reservoirs.

Where previously mined cavities are available which are, or can be made, gas tight, an opportunity to use these as artificial underground structures for storage is presented. A special case is the use of salt caverns, particularly in salt domes [3, 4]. Unlike depleted field and aquifer storage systems, which store gas in porous rock formation, cavern storage involves large open, void spaces to be filled with gas. A more complex structural analysis is therefore required to establish feasibility. For example, if the pressure in the cavity is allowed to drop significantly below ambient pressure, a collapsing stress situation is created which might result in loss of structural integrity of the storage volume. One approach considered is to replace the gas drawn off with water so that the cavity pressure is maintained.

There is of course no cushion gas requirement in such a hydraulically compensated scheme and the delivery pressure is constant—equivalent to the hydraulic head. The disadvantage for storage in salt caverns is that the working fluid must be saturated brine. Surface storage of this is complicated by many environmental constraints. Single salt caverns can be 50×10^6 m^3 whereas hard rock caverns would be one tenth of this size. However, modern mining machinery and techniques make the costs of mined caverns comparable to other underground options.

There are two pertinent experiences with under-

Table 1. Cryogenic properties of gases

	Boiling point (K)	Melting point (K)	Critical temperature (K)	Heat of vaporization at NBP, ($J\ m^{-3}$)	Relative loss index*
Hydrogen	20.3	13.9	33.1	3.15×10^7	1.0
Methane	111.5	90.5	190.5	2.16×10^8	0.1
Oxygen	90.0	54.6	154.2	2.44×10^8	0.1
Nitrogen	77.2	63.2	125.9	1.62×10^8	0.15

* Loss index is the difference between room temperature and normal boiling point (NBP) divided by volumetric heat of vaporization.

ground storage of hydrogen. At Beynes in France, from 1957 to 1974, Gaz de France stored town gas containing 50% hydrogen. An aquifier of capacity 330×10^6 Nm^3 was used with no special problems at all. Current experience is at Imperial Chemical Industries on Teeside in England. Here, there are three brine compensated salt caverns at 1200 ft (366 m) storing hydrogen at 50 atm (5×10^6 Pa) pressure for an industrial chemical complex.

Cryogenic liquid storage

Table 1 shows some important parameters for a number of gases which are often stored as cryogenic liquids.

Liquid hydrogen presents numerous engineering challenges. Its low heat of vaporization results in a very high loss index. Even the exothermic conversion of *ortho* to *para* hydrogen which occurs at low temperature is sufficient to vaporize almost all the hydrogen if it is left in the normal manufactured state of 75% *ortho*.

Liquid hydrogen cannot be stored in cylindrical tanks of the type used for LNG because the boil-off would be too high. Spherical tanks are always used for large-scale applications and the technology has largely been driven by the needs of space programs. The U.S. programs of the National Aeronautics and Space Administration (NASA) uses tanks up to 3.8×10^3 m^3 (10^6 U.S. gallons) which are about 22 m in diameter. The practical volume limit would be about five times this current usage and economies would result because of the approx. 2/3 power relationship shown in Fig. 1. The tanks costed in this figure are double walled, vacuum insulated with a cavity of about 1 m.

In costing a practical system the boil-off and transfer losses must be considered. For the largest tanks, the boil-off will be about 0.06% per day. At about 10^3 m^3

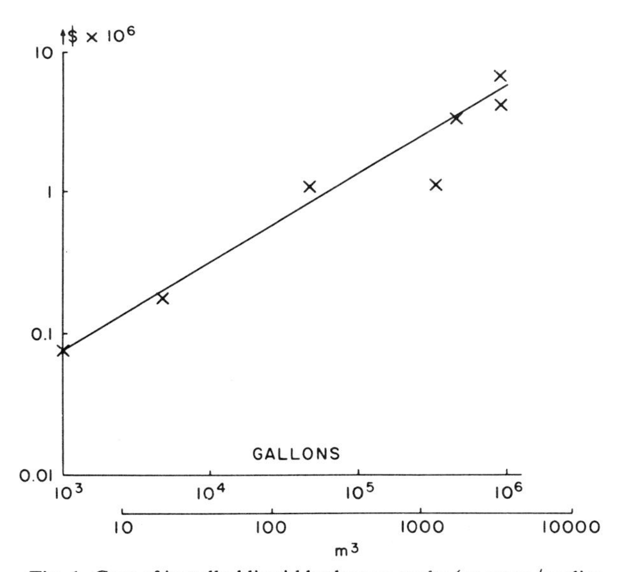

Fig. 1. Cost of installed liquid hydrogen tanks (vacuum/perlite insulated spheres—data from various published sources).

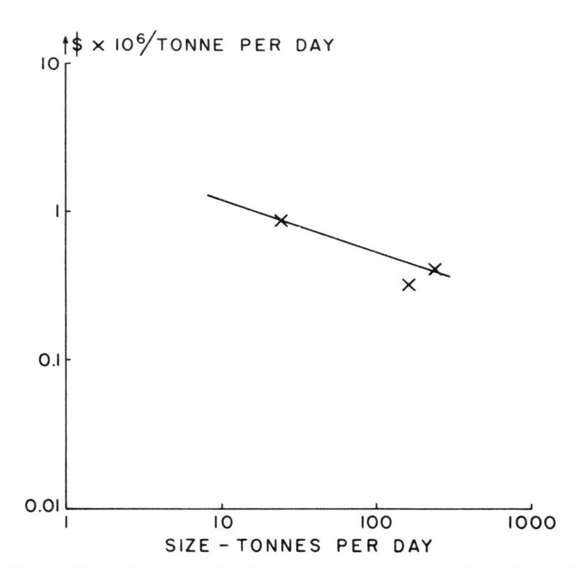

Fig. 2. Cost of installed hydrogen liquefaction facilities (excluding storage tanks—data points from various published sources).

sizes this figure will be 0.2% per day. Transfer losses may be extremely high (50%) when small amounts are involved and when a cold gas recovery loop cannot be integrated with the liquefaction facility.

It is debatable whether the cost of liquefaction is to be considered part of a liquid hydrogen storage facility since it is presumed that the end use demands liquid and therefore should bear the costs of obtaining that form. However, we have chosen to include the liquefaction costs in our analysis.

Figure 2 plots the cost of liquefaction plants based on the Claude Cycle including the plant to make liquid nitrogen for pre-cooling but excluding liquid hydrogen tankage. Capital costs are extremely high and are at least three times greater those of high pressure tube costs for a comparable daily throughput.

ENERGY RECOVERY

Energy can potentially be recovered from the cold when a cryogenic liquid is vaporized. Similarly, energy can be potentially recovered from expansion when gas is withdrawn from high pressure storage. Both concepts have been exploited to some extent.

The enthalpy change of hydrogen in going from a liquid to gas at room temperature is about 3700 J/g (1600 BTU/lb). This is 10% of the practical energy required to liquefy and convert to the *para* form. Thus a theoretical maximum of 10% of the liquefaction energy can be recovered. In addition to this limitation, the demand must coincide with the availability. In the case of LNG, cold recovery is achieved in France and Japan where the evaporation occurs at large terminals. Liquid air production is carried out at both sites. Also low-grade cooling is used for food warehousing in Tokyo. Both countries integrate LNG evaporators with power station condenser cooling water to increase power production efficiencies. All these things can be accomplished with liquid hydrogen in place of LNG. Because of the lower temperature, a simple system of direct liquefaction of air is possible. The real problem is to integrate the facilities and here the anticipated smaller scale may have a negative influence. If the liquid hydrogen is subsequently used to produce hydrogen to power an engine, there are many opportunities for energy recovery. These could include blade cooling in turbines, charge cooling and charge compression (high pressure hydrogen from evaporation of liquid) for internal combustion engines. The incorporation of these features are a downstream engineering consideration separate from the storage system. They do constitute a form of energy recovery, but it cannot be quantified in the present analysis.

Compression energy recovery is possible when high-pressure gas is withdrawn from storage. However, recovery systems would only be practicable on very large systems. The necessary turboexpanders are commercially available to generate electricity [5].

For the case of natural gas the production of useful work requires the addition of thermal energy to the gas either before or after expansion. There is no expansion scheme which will recover all the energy spent in compression. The resulting expansion work comes from the heat added to the system and the efficiency of conversion is about double that normally experienced in a thermal power plant (4×10^6–5.8×10^6 J/kWh as compared to 11×10^6 J/kWh). A pay-back time of 1.5–3 y would be normal.

Because of a negative Joule–Thompson coefficient for hydrogen at ambient temperatures there is not the requirement of thermal input that was indicated above for the case of natural gas. However, the low molecular weight causes problems in obtaining power from expansion of high-pressure hydrogen. Detailed calculations were not made for energy recovery on high-pressure hydrogen storage. Based on the difference in the compression requirements we estimate that the maximum recoverable energy from hydrogen is only 25% of that available from an equivalent volume of natural gas. Pay-back would not be expected in under 6 y.

DEFINITION OF PROCEDURE AND SCENARIOS

Economic analyses were performed on five site-specific examples or scenarios. Each example is developed in sufficient detail to establish the quantities of hydrogen to be stored and the injection/withdrawal cycle. In each case, it is assumed that the hydrogen is generated by electrolysis. Capital and operating costs are obtained for the storage facilities and based on these data, unit costs of storing hydrogen are calculated. The sensitivities of these costs to variations in throughput, capital costs and electric power costs are also analysed.

The five conceptual scenarios are:

(i) Pressure vessel storage for the edible oil and fat industry.
(ii) Liquid hydrogen for aircraft.
(iii) Salt cavern storage for retiming tidal energy.
(iv) Mined cavern for utility peakshaving.
(v) Underground porous media storage for heavy oil upgrading plant.

A capsule definition of each scenario follows but readers should note that a very complete analysis and definition formed the basis of this study. Space does not permit inclusion of such data here.

(i) *Pressure vessel storage for the edible oil and fat industry*

- Plant output 10^5 kg/day hydrogenated product.
- Hydrogen required—4550 m^3/day at 17 atm pressure.
- Electrolytic plant power requirement—815 kW.
- Compression power—40 kW.
- It is assumed that requirement is met by part of output of 80 MW electrolysis plant located 80 km away.
- Hydrogen storage to accommodate seasonal variations in demand and provide security of supply.

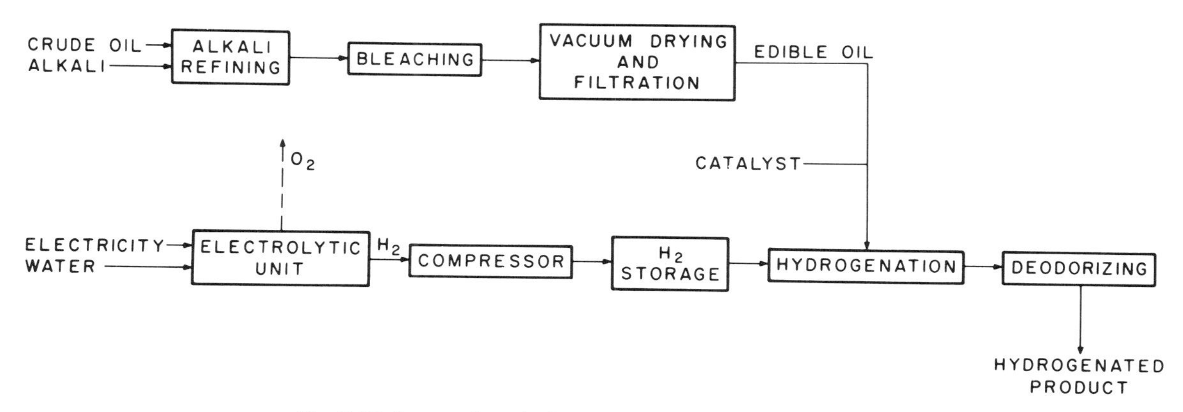

Fig. 3. Hydrogenation of edible oil using electrolytic hydrogen.

- Three 'JUMBO' tube trailers—5200 m^3 capacity. One delivery per day.
- Two trailers always at user site.

The conceptual plant is shown in Fig. 3.

(ii) *Liquid hydrogen for aircraft*

- Flight schedule—one 4800 km flight every day.
- Fuel load—25,000 kg liquid hydrogen.
- Supply required with allowance for boil-off and transfer losses—30,000 kg/day.
- Liquefaction plant—36 × 10^3 kg/day using integrated production, liquefaction and storage facility.
- Input electrical power—40 MW.
- Hydrogen storage buffer—14,000 m^3.
- Three storage spheres for hydrogen—10^6 kg total capacity.
- Minimum of 10 days supply will be in storage at all times.
- Pipeline transfer system with cold gas recovery.
- Anticipated fuelling boil-off rate is 1000 kg h^{-1} ≡ liquefaction rate.
- Fuelling time 38 min.
- The conceptual installation is shown in Fig. 4.

(iii) *Salt cavern storage for retiming tidal energy*

- Power input to electrolytic plant—1000 MW.
- Availability—4.5 h on each tidal cycle (once per 12.5 h).
- Maximum hydrogen production rate 0.22 × 10^6 m^3/h of available energy.
- Maximum production per tidal period—10^6 Nm^3.
- Storage capacity—2 × 10^6 Nm^3.

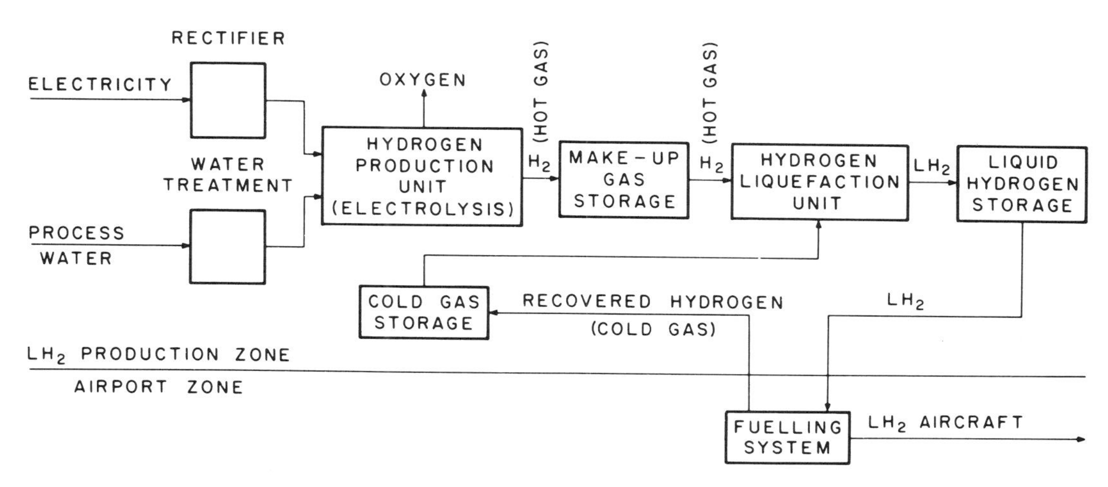

Fig. 4. Liquid hydrogen production and fuelling for an airport.

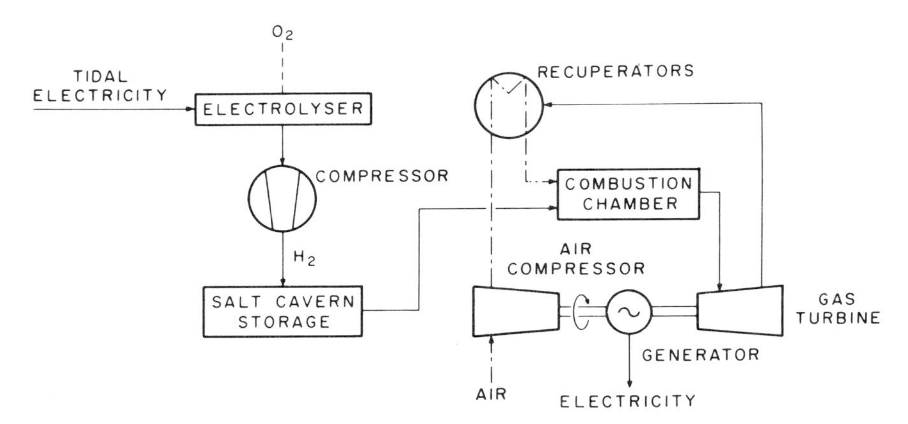

Fig. 5. Salt cavern storage of hydrogen for re-timing energy from tides.

- Two storage scenarios are considered:

	Dry	Brine compensated
Maximum pressure (atm)	25	57
Minimum pressure (atm)	17	0
Cavern spherical diam. (m)	79	40
Base gas volume (m^3)	4.5×10^6	0
Compressors	7 × 6000 HP	3 × 3500 HP
	Depth ~ 500 m	

- Energy reconversion not costed. Sizing of storage is for peak power demand of 500 MW for 3.6 h this is obtained from eight GE 7821 gas turbines (60 MW each) with expansion turbines for energy recovery.

The conceptual installation is shown in Fig. 5.

(iv) *Mined cavern for utility peakshaving*

To assist comparisons, there are many similarities between this scenario and the previous one.

- Power input from surplus capacity—1000 MW.
- Availability—9 h each day.
- Hydrogen production—2×10^6 Nm^3/day.
- Maximum pressure—25 atm.
- Minimum pressure—17 atm.
- Base gas volume—4.5×10^6 m^3.
- Compressors—7 × 6000 HP.

The conceptual installation does not differ significantly from that of Fig. 5.

(v) *Underground porous media storage for heavy oil upgrading plant*

- Plant output 148,500 bbl/day of synthetic crude.
- Hydrogen requirement for hydrocracker—4×10^6 Nm^3 per day.
- Hydrogen requirement for hydrotreater—3–9 × 10^6 Nm^3 per day.
- Half of hydrogen supplied from partial oxidation of residuum using oxygen from electrolytic plant.

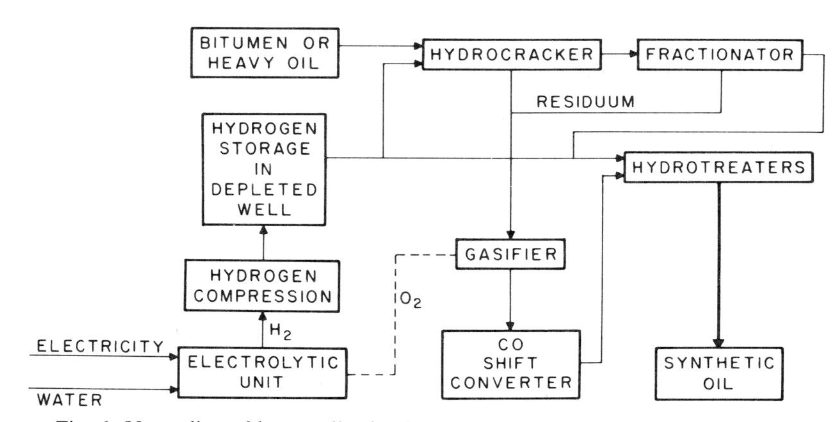

Fig. 6. Upgrading of heavy oil using hydrogen and oxygen from electrolysis.

- Electrolytic hydrogen requirement—4×10^6 Nm^3 per day.
- Electricity available year round—473 MW with additional 473 MW available for six months.
- Storage in depleted gas well operated on six month cycle.
- Pressure range—1.9–7.5 atm.
- Working gas—0.24×10^9 Nm^3.
- The particular formation studied was the Whiteside/Victoria gas pool in Saskatchewan Canada. It has three wells at present for natural gas withdrawal and is 87% depleted. To satisfy the hydrogen delivery requirement, a further 27 wells would be needed. The costs for these wells are included in the calculations.

The conceptual plant is shown in Fig. 6.

ECONOMIC ASSESSMENTS

Standard procedures have been followed in each case to determine the total erected capital costs and the operating expenses. From this, the cost of gas delivered is calculated on a unit basis and the incremental cost of

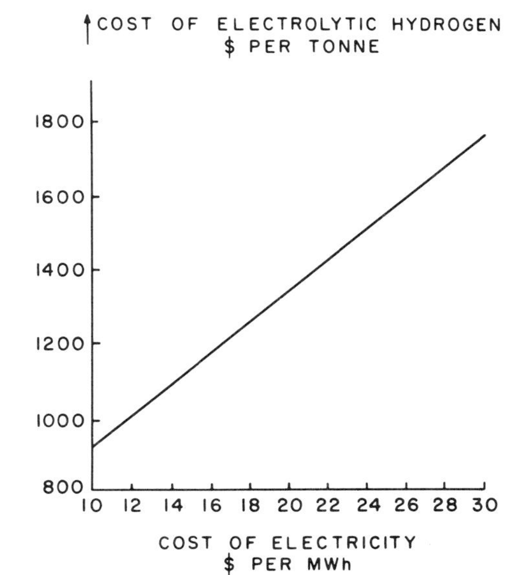

Fig. 8. Hydrogen cost as a function of electricity cost.

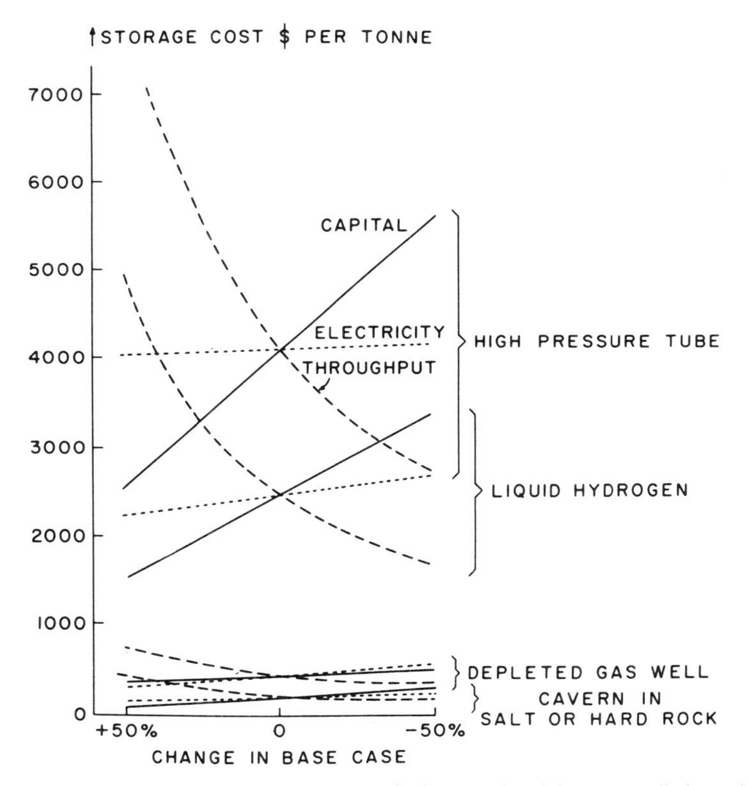

Fig. 7. Sensitivity of hydrogen storage costs to capital cost, electricity cost and throughput for five systems.

storage is then obtained. The base case data are given in Appendices I–V. Appendix VI contains the data for electrolytic hydrogen production. Finally the sensitivity is presented graphically in Fig. 7 which represents the variation of incremental storage costs with throughput, capital cost and electricity cost as a percentage of the base case situation. Figure 8 shows the effect of electricity costs on the cost of electrolytic hydrogen.

All accounting is in $U.S. (1982).

OBSERVATIONS

There is a wide variation in cost between the various modes chosen for storing hydrogen. Before drawing any quick conclusions, or making any comparisons, it is essential to bear in mind that each storage mode has been tailored to suit a specific scenario, and hence it is not unreasonable to expect storage costs which are widely different. In fact, any direct comparison between modes is of little significance. Contributing to this wide variation is the fact that each scenario has its own cycle of operation and annual throughput. Both of these factors not only affect the costs of the storage system but also influence the cost of electrolytic hydrogen supplied to the scenario. This can be observed by reference to the tables in Appendix I, and is further discussed here.

Cost of electrolytic hydrogen

The sensitivity of the cost of electrolytic hydrogen to variations in the cost of electricity is illustrated in Fig. 8. Typically, the cost of power comprises between 45 and 70% of the total cost; however, another not-too-obvious influence is the effect of the service factor imposed by the consuming scenario. This effect can be seen by comparing the cost of the electrolytic hydrogen to be supplied to the scenarios utilizing a salt cavern and a depleted gas pool, see Appendices III and V, respectively. Both scenarios employ the same size of electrolysis plant and the same power cost, yet the cost of product hydrogen is significantly different between the two. The cyclic nature of tidal power supply imposes a service factor of about 36% on the one electrolysis plant, thereby restricting its annual production. The other plant can, in effect, run continuously for nine months of the year and can produce more than twice as much hydrogen per year. This increase in production more than compensates for the increased annual operating costs.

High-pressure tube storage

The most significant item comprising the cost of storage is the capital charge which, for the base case, represents almost 61% of the total. This is due to the high value of investment per unit of product for this mode, and explains the high sensitivity of storage cost to variations in both capital cost and throughput.

It should be noted that this scenario is based on a demand which can be satisfied with the delivery each day of a single full trailer load of hydrogen. It is envisaged that the delivery and return of the empty trailer will be made during the normal working day and that the refilling operation will be performed during the remainder of the 24-h day by one of the two compressors. By operating both compressors together, an additional half delivery can be fitted into each 24-h period at a minimal incremental cost to cover only the additional operating expenses. Acquisition of a fourth tube trailer would permit up to three deliveries per day, i.e. three times the base case throughput, without requiring any additional compressor capacity. It is estimated that delivery of 14,000 Nm^3/day of hydrogen would result in a base case delivered price of $3600/tonne and a storage cost of $2250/tonne. This would produce a significant lowering of the storage cost and result in the maximum utilization of the capital involved.

There is considerable confusion in the literature about the cost of electrolytic hydrogen produced in small quantities on site in electrolysers of about 1 MW size. Thus, the Electric Power Research Institute in the U.S. concluded [6] that bottled gas was cheaper. Our own figures indicate that break-even would occur in the application considered if on-site electrolysers were installed and supplied with electricity at 35 mils/kWh. Energy at this price is readily available in many parts of Canada and elsewhere.

Liquid hydrogen storage

The most significant item in the storage cost is that due to capital charges, which represent 62.6% of the total. This incidentally is the same ratio as for the high-pressure tube storage, i.e. both scenarios 1 and 2 have the same ratio of capital investment to annual throughput. The fact that this scenario has a lower storage cost is due to its lower unit charges for labour and maintenance resulting from its 70 times greater annual throughput. However, it should be noted that storage as a liquid in this scenario is dictated by the end use requirements and not by the cost of storage.

The high value of capital investment per unit of product per annum also contributes to the high sensitivity of storage cost to variations in both capital cost and throughput as shown in Fig. 8.

Salt cavern storage

Even though this mode of storage involves a considerable amount of capital expenditure in the creation of a cavern and the installation of compressors, this scenario results in the lowest cost for storing hydrogen. Examination of the unit cost figures in Appendix III and comparison with the same data for the depleted gas well scenario, in Appendix V, shows that the low cost for this scenario is attributable to its high annual throughput. Given the same throughput as the depleted gas well scenario, the storage cost for this salt cavern scenario would be almost three times that shown.

Rock cavern storage

The increased cost of excavating a cavern in hard rock as opposed to solution mining in a salt deposit is reflected in an increased investment of $7.39 million and an increased storage cost of hydrogen. Although the range of cost estimates for hard rock mining has been found to be quite wide ($28–$70/m depending upon depth and volume), the cost of solution mining at $19.50/m favours this latter method of cavern creation. A suitable credit for the excavated rock could however, alter the balance in favour of the excavated cavern.

Depleted gas field storage

The most significant feature of this scenario is that almost 80% of the total investment is represented by stored hydrogen, either as cushion gas or the initial six months supply of working gas. The annual capital charge on this hydrogen inventory comprises 62% of the storage costs.

Another effect is to make this scenario exhibit a higher degree of sensitivity to the cost of power than the other scenarios. For this scenario only, the cost of power shows a greater influence than does the capital cost.

The cost of the gas ex-storage from this scenario is the lowest of all the scenarios studied. The reason can be traced to the lower cost of the feed hydrogen which more than compensates for the increased storage charges incurred by the base gas and inventory.

REFERENCES

1. C. Carpetis, A system consideration of alternative hydrogen storage facilities for estimation of storage costs, *Int. J. Hydrogen Energy* **5**, 423–437 (1980).
2. S. Foh, M. Novil, E. Rockar and P. Randolph, Underground hydrogen storage, Final report, Institute of Gas Technology, Chicago, Issued as BNL 51275 (December 1979).
3. L. T. Pasiechnyk, Solution-mined salt cavern storage of fluid hydrocarbons, *Can. Inst. Min. Bull.* **71**, 91–97 (1978).
4. M. H. Allan, Natural gas storage in caverns in Saskatchewan, in *Symposium on Salt*, Vol. 2, pp. 412–421. Northern Ohio Geological Society Cleveland, Ohio (1966).
5. J. M. Coliss, D. E. Jones, Battelle Columbus Lab, issued by Gas Research Institute, Chicago as GRI-79/0107 (April 1981).
6. F. Kalhammer, R & D status report—energy management and utilization division, *EPRI J.* September, 50 (1984).

APPENDIX I

Capital cost of pressure vessel storage

	Description	Cost ($000)
(3)	'JUMBO' tube trailers, each with 16 tubes manifolded and installed @ $280,000	840
(1)	Truck-cab for the trailer	90
(2)	Four-stage horizontal reciprocating compressors c/w 100 HP explosion-proof motor @ $150,000	300
(1)	Regenerative desiccant dryer and prefilter package	30
	Foundations, installation	120
	Interconnecting piping, valves, electrical systems, controls	200
(1)	Gas holder, 2800 Nm^3 capacity	150
	Total installed cost	1730

Unit cost of pressure vessel storage

Basic parameters:	
Delivery rate	: 4550 m^3/day 0.39 tonnes/day
Power $/MWh	: 20.00
Service factor	: 0.9863 (360 days/y)
Date	: 1983
Annual capital charge	: 15% (incl. depreciation, interest and profit)

Operating expenses	Unit cost ($)	Annual cost ($000)
Electrolytic hydrogen, 0.39 tonnes/day	1350.29	188.87
Power, 90.00 kWh/h	20.00/MWh	15.55
Cooling water, 40 tonnes/day	0.06	0.86
Tube trailers and cab (fuel, taxes, insurance, etc.)		30.00
Operating personnel, two @	26,000 $/y	52.00
Supervision and fringe benefits @ 55% of labour		28.60
Administration @ 20% of labour and supervision		16.12
Maintenance materials and labour @ 2% of capital cost		34.60
Other @ 2.5% of capital cost		43.25
Total operating expenses		409.85

Investment	($000)
Total erected capital cost	1730.00
Land cost, 0.2 ha	5.00
Engineering costs, 10% of erected cost	173.00
Total plant cost	1908.00
Cushion gas, 0.165 tonnes @ 1350.29 $/tonne	0.22
Working capital, 30 days of operating expenses	33.68
Start-up cost, 2% of total plant cost	38.16
Interest before start-up @ 12% per y	
Year 1 40% of total plant cost, 194.16	
Year 2 60% of total plant cost, 137.38	331.53
Total investment	2311.42

Cost of annual gas delivered

	Annual charge ($)	Cost/tonne ($)
Feed gas	188.87	1350.26
Power, utilities and water	46.42	331.84
Labour, supervision, adminstration	96.72	691.47
Maintenance and miscellaneous	77.85	556.57
Capital charges @ 15% ROI	346.71	2478.74
Total	756.57	5408.88
Incremental cost of storage	567.70	4058.62

APPENDIX II

Capital cost of liquid hydrogen storage

	Description	Cost ($ × 10^6)
(1)	40 t/day (36 × 10^3 kg/day) hydrogen liquefaction unit	32.3
(3)	Double-walled, vacuum-perlite insulated spherical tanks, each capable of storing 300,000 kg LH_2	13.5
(1)	Cold hydrogen gas storage tank—1250 Nm^3	1.6
(1)	Hydrogen gas buffer tank—14,000 Nm^3	0.7
	LH_2 Fuelling system, including pumps, piping, valves, etc	23.0
	Interconnecting piping, valves, etc., for the storage tanks	2.0
	Foundations and installation for the tanks	2.0
	Sub-total	75.1
	Contingency allowance (10%)	7.5
	Total installed cost	82.6

Unit cost of liquid hydrogen storage

Basic parameters:

Delivery rate	: 29.48 tonnes/day power $/MWh
Power $/MWh	: 20.00
Service factor	: 0.9726 (355 days/y)
Date	: 1983
Annual capital charge	: 15% (incl. depreciation, interest and profit)

Operating expenses	Unit cost ($)	Annual cost ($ × 10^6)
Electrolytic hydrogen, 30.31 tonnes/day	1350.29	14.53
Power, 25.00 MWh/h	20.00/MWh	4.26
Utilities (chemicals, nitrogen, water, etc)		1.47
Operating personnel, nine @ 26000 $/y		0.23
Supervision and fringe benefits @ 55% of labour		0.13
Administration @ 20% of labour and supervision		0.07
Maintenance materials and labour @ 2% of capital cost		1.65
Other @ 2.5% of capital cost		2.07
Total operating expenses		24.41

Investment	($ × 10^6)
Total erected capital cost	82.60
Land cost, 4 ha	0.10
Engineering costs, 6% of erected cost	4.96
Total plant cost	87.66
Non-recoverable liquid, 100 tonnes @ 1350.29 $/tonne	0.14
Working capital, 30 days of operating expenses	2.01
Start-up cost, 3% of total plant cost	2.63
Interest before start-up @ 12% per y	
Year 1 60% of total plant cost, 13.38	
Year 2 40% of total plant cost, 4.21	17.59
Total investment	110.03

Cost of annual gas delivered

	Annual charge ($S × 10^6)	Cost/tonne ($)
Feed gas	14.53	1349.99
Power, utilities and water	5.73	532.45
Labour, supervision, administration	0.44	40.44
Maintenance and miscellaneous	3.72	345.40
Capital charges @ 15% ROI	16.51	1534.44
Total	40.93	3802.72
Incremental cost of storage	26.40	2452.73

APPENDIX III

Capital cost of salt caverns ('dry mode') storage

Description	Cost ($000)
Cost of cavern development and well-head equipment	5500
Three-stage reciprocating compressor, c/w 6000 HP electric drive	
(Total for seven operating units and one spare)	18,400
Foundations and erection	1840
20 MW transformer and primary breaker	
(Total for 2 units, installed)	2660
Building, c/w heating, ventilation, lighting and gas monitoring and alarm system	1360
Gas holder, 14,000 Nm^3 capacity	700
Interconnecting pipework	900
Total installed cost	31,360

Unit cost of salt cavern ('dry mode') storage

Basic parameters:

- Injection rate : 159.71 tonnes/day
- Power $/MWh : 20.00
- Service factor : 0.36 (two tidal periods of 4.5 h per 25 h)
- Date : 1983
- Annual ROI % : 15

Operating expenses	Unit cost ($)	Annual cost ($ × 10^6)
Electrolytic hydrogen, 159.71 tonnes/day	2053.11	119.68
Power, 28.00 MWh/h	20.00	1.74
Cooling water, 7474 tonne/day	0.06	0.06
Utilities		0.25
Operating personnel (shift), 16 @ 26000 $/y		0.42
Supervision and fringe benefits @ 55% of labour		0.23
Administration @ 20% of labour and supervision		0.13
Maintenance materials and labour @ 2% of capital cost		0.63
Other @ 2.5% of capital cost		0.78
Total operating expenses		123.91

Investment:	($ × 10^6)
Total erected capital cost	31.36
Land cost, 2 ha	0.05
Engineering costs, 10% of erected cost	3.14
Total plant cost	34.55
Cushion gas, 386.60 tonnes @ 2053.11 $/tonne	0.79
Working capital, 30 days of operating expenses	10.18
Start-up cost, 2% of total plant cost	0.69
Interest before start-up @ 12% per y	
Year 1 40% of total plant cost, 3.52 M$	
Year 2 60% of total plant cost, 2.49 M$	6.00
Total investment	52.21

Cost of annual working gas throughput

	Annual charge ($ × 10^6)	Cost/tonne ($)
Feed gas	119.68	2053.01
Power, utilities and water	2.05	35.10
Labour, supervision, administration	0.77	13.27
Maintenance and miscellaneous	1.41	24.21
Capital charges @ 15% ROI	7.83	134.21
Total	131.74	2259.90
Incremental cost of storage	12.06	206.89

APPENDIX III

Capital cost of salt cavern ('wet mode') storage

Description	Cost ($000)
Cost of cavern development and well-head equipment	1000
Cost of brine lagoons and handling equipment	1000
Four-stage reciprocating compressor, c/w 3600 HP electric drive	
(Total for 13 operating units and one spare)	25,200
Foundations and erection	2520
20 MW transformer and primary breaker,	
(Total for 2 units, installed)	2660
Building, c/w heating, ventilation, lighting and gas monitoring and alarm system	2380
Gas holder, 14,000 Nm^3 capacity	700
Interconnecting pipework	1575
Total installed cost	37,035

Unit cost of salt cavern ('wet mode') storage

Basic parameters:

Injection rate	: 159.71 tonnes/day (calendar)
Power $/MW	: 20.00
Service factor	: 0.36 (two tidal periods of 4.5 h per 25 h)
Date	: 1983
Annual ROI %	: 15

Operating expenses	Unit cost ($)	Annual cost ($ × 10^6)
Electrolytic hydrogen, 159.71 tonnes/day (calendar)	2053.11	119.68
Power, 31.00 MWh/h	20.00	1.93
Cooling water, 8186 tonnes/day (calendar)	0.06	0.06
Utilities		0.25
Operating personnel (shift), 16 @ 26000 $/y		0.42
Supervision and fringe benefits @ 55% of labour		0.23
Administration @ 20% of labour and supervision		0.13
Maintenance materials and labour @ 2% of capital cost		0.74
Other @ 2.5% of capital cost		0.93
Total operating expenses		124.36

Investment	($ × 10^6)
Total erected capital cost	37.04
Land cost, 2 ha	0.10
Engineering costs, 10% of erected cost	3.70
Total plant cost	40.84
Working capital, 30 days of operating expenses	10.22
Start-up cost, 2% of total plant cost	0.82
Interest before start-up @ 12% per y	
Year 1 40% of total plant cost, 4.16 M$	
Year 2 60% of total plant cost, 2.94 M$	7.10
Total investment	58.98

Cost of annual working gas throughput

	Annual charge ($ × 10^6)	Cost/tonne ($)
Feed gas	119.68	2053.01
Power, utilities and water	2.24	38.50
Labour, supervision, administration	0.77	13.27
Maintenance and miscellaneous	1.67	28.59
Capital charges @ 15% ROI	8.85	151.78
Total	133.21	2285.15
Incremental cost of storage	13.53	232.14

APPENDIX IV

Capital cost of mined cavern storage

Description	Cost ($000)
Cost of cavern development and well-head equipment	11,040
Three-stage reciprocating compressor, c/w 6000 HP electric drive	
(Total for seven operating units and one spare)	18,400
Foundations and erection	1840
20 MW transformer and primary breaker,	
(Total for 2 units, installed)	2660
Building, c/w heating, ventilation, lighting and gas monitoring and alarm system	1360
Gas holder, 14,000 Nm^3 capacity	700
Interconnecting pipework	900
Total installed cost	36,900

Unit cost of mined hard rock cavern storage

Basic parameters:

Injection rate	: 159.71 tonnes/day
Power $/MWh	: 20.00
Service factor	: 0.36
Date	: 1983
Annual ROI %	: 15

Operating expenses	Unit cost ($)	Annual cost ($ × 10^6)
Electrolytic hydrogen, 159.71 tonnes/day	2053.11	119.68
Power, 28.00 MWh/h	20.00	1.74
Cooling water, 7474 tonnes/day	0.06	0.06
Utilities		0.25
Operating personnel (shift), 16 @ 26000 $/y		0.42
Supervision and fringe benefits @ 55% of labour		0.23
Administration @ 20% of labour and supervision		0.13
Maintenance, materials and labour @ 2% of capital cost		0.74
Other @ 2.5% of capital cost		0.92
Total operating expenses		124.16

Investment	($ × 10^6)
Total erected capital cost	36.90
Land cost, 2 h	0.05
Engineering costs, 10% of erected cost	3.69
Total plant cost	40.64
Cushion gas, 425.71 tonnes @ 2053.11 $/tonne	0.87
Working capital, 30 days of operating expenses	10.20
Start-up cost, 2% of total plant cost	0.81
Interest before start-up @ 12% per y	
Year 1 40% of total plant cost, 4.14 M$	
Year 2 60% of total plant cost, 2.93 M$	7.06
Total investment	59.58

Cost of annual working gas throughput

	Annual charge ($ × 10^6)	Cost/tonne ($)
Feed gas	119.68	2053.01
Power, utilities and water	2.05	35.10
Labour, supervision, administration	0.77	13.27
Maintenance and miscellaneous	1.66	28.49
Capital charges @ 15% ROI	8.94	153.33
Total	133.10	2283.32
Incremental cost of storage	13.42	230.19

APPENDIX V

Capital cost of depleted gas well storage

Description	Cost ($000)
Three-stage reciprocating compressor, C/W 5000 HP electric drive (total for two operating units)	4000
Foundations and erections	400
8 MW transformer and primary breaker, installed	700
Building, c/w heating, ventilation, lighting and gas monitoring, and alarm system	400
Gas holder 14,000 Nm^3 capacity	700
Interconnecting pipework to compressors	300
New wells (27 total), complete with well-head equipment and interconnecting pipework	2000
Total installed cost	8500

Unit cost of depleted gas well storage

Basic parameters:

Injection rate	: 112.76 tonnes/day
Power $/MWh	: 20.00
Service factor	: 0.5
Date	: 1983
Annual ROI	: 15

Operating expenses	Unit cost ($)	Annual cost ($ × 10^6)
Electrolytic hydrogen, 112.76 tonnes/day	1463.98	30.13
Power, 7.00 MWh/h	20.00	0.61
Cooling water, 1900 tonnes/day	0.06	0.02
Utilities		0.25
Labour, 7 @ 26,000 $/y (Oper. & Maint.)		0.18
Supervision and fringe benefits @ 55% of labour		0.10
Administration @ 20% of labour and supervision		0.06
Maintenance materials and labour @ 2% of capital cost		0.17
Other @ 2.5% of capital cost		0.21
Total operating expenses		31.73

Investment	($ × 10^6)
Total erected capital cost	8.50
Land cost, 2 ha	0.05
Engineering costs, 10% of erected cost	0.85
Total plant cost	9.40
Cushion gas, 13217 tonnes @ 1463.98 $/tonne	19.35
Working capital, 182.5 days of operating expenses	15.87
Start-up cost, 2% of total plant cost	0.19
Interest before start-up @ 12% per y	
Year 1 60% of total plant cost, 0.96 M$	
Year 2 40% of total plant cost, 0.68 M$	1.63
Total investment	46.44

Cost of working gas inventory

	Annual charge ($ × 10^6)	Cost/tonne ($)
Feed gas	30.13	1464.03
Power, utilities and water	0.88	42.96
Labour, supervision, administration	0.34	16.45
Maintenance and miscellaneous	0.38	18.59
Capital charges @ 15% ROI	6.97	338.82
Total	38.70	1880.80
Incremental cost of storage	8.57	416.77

APPENDIX VI

Capital cost of electrolytic hydrogen (100 MW module)

Description	Cost ($\$ \times 10^6$)
For the supply of all electrolysis cells, gas headers and seal pots, mixed-bed ion exchange demineralization system including regeneration equipment and piping, closed-cycle cooling system including pump and cooling tower, electrical transformer, rectifier, switchgear, controls (including computer), initial caustic potash supply, engineering and erection.	29.0
Building to house cells, water treatment equipment and controls: 5300 m^2 floor area complete with lighting, foundation, grading, etc.	4.2
Total erected cost of 100 MW unit	33.2

Unit cost of hydrogen production by electrolysis (100 MW module)

Basic parameters:

Rated capacity MW	: 100
Hydrogen, tonnes/day	: 47.66
Power cost $/MWh	: 20.00
Service factor	: 0.9863 (360 days/y)
Date	: 1983
Annual % ROI	: 15

Operating expenses	Unit cost ($)	Annual cost ($\$ \times 10^6$)
Power, 100 MWh/h	20.00	17.28
Raw feed water, 496 tonnes/day	0.056	0.01
Cooling water, 25,174 tonnes/day	0.1433	1.30
Chemicals, etc.	2391.00	0.24
Operating personnel (shift), nine @ 26,000 $/y		0.23
Supervision and fringe benefits @ 55% of labour		0.13
Administration @ 20% of labour and supervision		0.07
Maintenance materials and labour @ 1.5% of capital cost		0.60
Other @ 0.5% of capital cost		0.20
Total operating expenses		20.06

Investment	($\$ \times 10^6$)
Capital cost installed (excl. compressors for O_2, H_2)	40.00
Land cost, 2.5 ha	0.06
Owner's engineering cost @ 1% of capital cost (other engineering costs already included)	0.40
Total plant cost	40.46
Working capital, 30 days of operating expenses	1.65
Start-up cost, 2% of total plant cost	0.81
Interest before start-up @ 12% per y	
Year 1 70% of total plant cost, 7.21 M$	
Year 2 30% of total plant cost, 1.46 M$	8.66
Total investment	51.58

(All above costs in $ Canadian. Table below in $U.S. (1982). Conversion 1$ Canadian = 0.83 $U.S.)

Products: Hydrogen 47.66 tonnes/day

	Annual charge ($\$ \times 10^6$)	Cost/tonne ($)
Hydrogen cost:		
Power, utilities and water	15.63	910.79
Labour, supervision, administration	0.37	21.06
Maintenance and miscellaneous	0.66	38.70
Return on investment (15%)	6.42	374.28
Total	23.07	1344.83
Oxygen credit: nil		

Chapter 7

Hydrogen Vehicles: Past and Present

7.1 Introduction

Efforts to convert vehicles to hydrogen operation began in the mid-1920s with the work of German engineer Rudolf Erren, who promoted hydrogen engines for a variety of applications including trucks, buses, submarines, and internal combustion engines. Kurt Weil, who was Erren's technical director in the 1930s, reported that Erren's engines could operate on either hydrogen or gasoline and his vehicles could be switched back and forth from one fuel or mixture to any other [1]. Weil states that late in 1938, when he left for the United States, there were scores of "errenized" trucks traveling between Berlin and the city of Ruhr switching on the way from one fuel to another. World War II, however, ended these large scale tests, and allied bombing of the Berlin offices of the "Deutsche Erren Studien Gesellschaft" burned the only files containing a complete description of Erren's pilot operations in Germany.

During the two decades following World War II, there was little effort in the development of hydrogen vehicles. The hydrogen movement was renewed in the 1970s, however, as oil prices surged and pollution became a more prominent issue. Pioneering programs began in the early 1970s, with many continuing to the present. Automakers, most notably Daimler-Benz and BMW in Germany, also joined the hydrogen movement. Today, there are a number of programs worldwide that are actively developing and demonstrating hydrogen vehicles. In this chapter, the worldwide development of hydrogen vehicles from the early 1970s to the present state of the art will be profiled.

7.2 Hydrogen Vehicle Projects

United States

Roger Billings and Co-workers

Roger Billings has been a pioneer of the hydrogen movement in the United States since the mid-1960s. Billings made his first vehicle conversion when he was only 16 years old with the help of his father. The resulting converted Model A truck was the first hydrogen-fueled internal combustion engine vehicle in the United States [2]. In his earlier days, Billings also experimented with using hydrogen to fuel a lawnmower. Following the General Motors-sponsored 1972 Urban Vehicle Design competition, Billings began a more concerted effort to build hydrogen vehicles by teaming up with Frank Lynch, who was a member of the student team that designed the UCLA hydrogen car (see below), to form the Billings Energy Corporation [3].

The Billings Energy Corporation converted a number of vehicles to hydrogen operation throughout the 1970s and 1980s. The corporation was also involved in developing metal hydrides for on-board hydrogen storage. Early conversions by this group include a Wankel rotary engine, a Ford Falcon, and a Chevrolet Monte Carlo [3]. The Monte Carlo was equipped with both cryogenic hydrogen and metal hydride storage. The metal hydride tank was an iron-titanium container that utilized hot water from the engine's cooling system as a source of heat for releasing hydrogen gas [2]. The Billings Corporation also converted two buses to run on hydrogen. One of these buses was delivered to Riverside, California where it was operated in regular service by the local transit authority. In another early project, Billings provided a dual-fuel converted 1977 Cadillac Seville as part of a program to demonstrate a complete hydrogen homestead [4].

Other projects by Billings and co-workers included the conversion of a Peugeot sedan [5] and a U.S. Post Office Jeep [6] to hydrogen operation. The Jeep was outfitted with a carburetor modified to accommodate hydrogen and a water injection system to suppress backfire. The hydrogen-fueled Jeep was able to obtain 21% greater fuel efficiency than a gasoline-fueled jeep when compared on an equal energy basis. The Peugeot engine was outfitted with direct cylinder injection controlled by a microprocessor.

The engine was said to have operated successfully without backfiring and attained reasonable output power at fuel consumption levels comparable to those of a gasoline engine.

More recently, Billings and co-workers have constructed a prototype hydrogen fuel cell vehicle [7]. The vehicle is a Ford Fiesta which was previously retrofitted as an electric vehicle and operated by the U.S. Postal Service. A solid polymer electrolyte fuel cell from the Asahi Chemical Company of Japan is used to power the vehicle, and hydrogen is stored on-board the vehicle in a metal hydride. The vehicle is called a LaserCell vehicle—this name is derived from the fact that a YAG laser is used in the fabrication of the anodes and cathodes in the fuel cell.

Clean Air Now

Clean Air Now (CAN) has converted three Ford Ranger trucks to hydrogen operation [8]. For this conversion, the engines were bored out and the stroke increased to yield a 2.9 ℓ displacement. A Rootes type supercharger was also added to increase the mass air flow into the engine. The engine uses Constant Volume Injection (CVI) technology to provide hydrogen to the engine. Hydrogen is stored on the vehicle in 3600 psi composite tanks kept in the truck bed. The trucks are used as part of the Xerox and the City of West Hollywood fleets. They are refueled at the Xerox Solar Hydrogen Facility in El Segundo.

Energy Partners

Energy Partners of West Palm Beach, Florida has designed and built a prototype Fuel Cell Electric Vehicle (FCEV) called "The Green Car." The vehicle's power system includes three Nafion Proton Exchange Membrane (PEM) fuel cell stacks, batteries for cold starts and extra acceleration, and a 3000 psi hydrogen storage cylinder. The total vehicle weight is about 3000 pounds. The car has a range of about 60 miles for city driving, accelerates from 0 to 30 mph in 10 seconds, and has a top speed of 60 mph [9]. This vehicle is profiled in the reprint by Nadal and Barbir at the end of this chapter.

Hydrogen Consultants, Inc.

Hydrogen Consultants, Inc. (HCI) has been modifying vehicles for hydrogen operation for well over a decade. The company was founded by long-time hydrogen advocate Frank Lynch, whose previous experience includes work at UCLA, Billings Energy Corporation, and the Desert Research Institute. In the early 1980s, this group investigated the use of hydrogen as an alternative fuel for underground mining machinery. For these tests, a Caterpillar diesel engine was converted to a hydrogen-fueled spark ignition engine. A turbocharger with aftercooling was also added to boost power. The fuel was delivered using the parallel injection technique, which is similar to inlet port injection (see Section 2.3). The engine was able to achieve low NO_x emissions, without a loss in power or backfiring problems. A similar conversion was also performed on an 18-passenger Mitsubishi Rosa bus. This bus was tested using the Federal Test Procedure (FTP) normally designed for light-duty vehicles. Emissions of hydrocarbons and carbon monoxide were minimal and those of NO_x fell below the federal standards of the time. This was despite the fact that the test mass (rolling inertia) of the Rosa bus was about four times that of a regular car [10].

HCI has also played a major role in the development of the hydrogen programs in Riverside, California. Vehicles converted for these projects include a Dodge D-50 pickup truck and, more recently, a Ford Ranger pickup truck for the University of California at Riverside. HCI has also studied the potential of a fuel mixture known as hythane that combines hydrogen with natural gas (see Section 2.3 and the reprint by Raman, et al at the end of Chapter 2).

H-Power

H-Power is the prime contractor for the Department of Energy phosphoric acid fuel cell bus project. This project calls for the development of three fuel cell buses, the first of which was unveiled in April 1994. More details on this project can be found in the reprint by Kaufman at the end of Chapter 3. H-Power is also developing another bus based on proton-exchange membrane fuel cell technology [11]. Commercial production is scheduled for 1996. This bus is designed as a retrofit for existing diesel buses. The bus will utilize a reformer to convert diesel fuel to hydrogen, and thus can be refueled using diesel fuel. The bus is being designed such that the weight of the power pack (including the fuel cell, reformer, batteries, and motor) can be reduced to 2,500 lb, about equal to the weight of conventional diesel engines and transmissions. H-Power is also the primary developer of the sponge-iron hydrogen storage technology in the United States, as discussed in Chapter 2.

Ben Jordan

Ben Jordan has played an active part in the hydrogen movement. Mr. Jordan claims to have run at least a dozen engines on some form of hydrogen since he became interested in cars in 1932. A 1924 turbocharged Model "T" Ford is among the hydrogen conversions done by Mr. Jordan. Mr. Jordan and his wife Barbara have also played an important role in organizing races for hydrogen-powered vehicles at the Bonneville Salt Flats in Utah [12].

Los Alamos National Laboratory

The Los Alamos National Laboratory (LANL) has been involved in several hydrogen-related vehicle projects and has an extensive fuel cell research program (see Chapter 3). The first vehicle modified for hydrogen operation at LANL was a 1972 Dodge 1/2 ton pickup truck with a 4-speed manual transmission [13]. The truck was modified by replacing the carburetor with an IMPCO gas mixer originally designed for use with LPG and propane. The vehicle was equipped with a 190 ℓ LH_2 dewar designed and fabricated by the Minnesota Valley Engineering Company.

Figure 7.1. View of the liquid hydrogen tank installed in the trunk of the LANL Buick. (Source: ref. [14])

Later, LANL evaluated the performance of a converted LH_2 1979 Buick Century (see Fig. 7.1) over a 2 1/2 year project period [14]. Over this period, the car was driven 3633 km (2252 miles) and operated for 138 hours. The car, which was equipped with a carburetion fuel delivery system, had less power on hydrogen than gasoline and some backfire problems, but in general operated without any major difficulties. This project was done in cooperation with the German Aerospace Research Establishment (DFVLR) in Germany, which provided a liquid hydrogen fuel tank and a liquid hydrogen refueling station. Two liquid hydrogen tanks were tested during this period. The first had a 110 ℓ (29 gal) capacity, corresponding to an operating range of 274 km (170 miles). The second, which was used toward the end of the study, had a capacity of 155 ℓ (40 gal), corresponding to an operating range of 362 km (224 miles). The vehicle was refueled at least 65 times at the semiautomatic refueling station provided by DFVLR, with a refueling time of about nine minutes.

LANL has also been involved in projects to develop FCEVs. In early tests, a golf cart was used to evaluate different fuel cell technologies, including a fuel cell with a phosphoric acid electrolyte and one with a solid polymer electrolyte [14]. Although the vehicle was initially fueled by hydrogen and oxygen from gas cylinders, this system was later replaced by a methanol reformer. The performance results for the golf cart were used to assess the possibility of using a fuel cell in a more conventional vehicle. The results of the assessment indicated that a fuel cell vehicle with greater efficiency and comparable performance to that of a gasoline vehicle was feasible.

Researchers at LANL have recently begun a new project to develop a hydrogen refueling station [15]. The proposed refueling station will be designed to accommodate liquid hydrogen but allow hydrogen to be distributed to vehicles as either a liquid, a low-pressure gas (i.e., for a metal hydride), or a medium-pressure gas. Researchers at LANL are currently procuring equipment for the project.

Middle Tennessee State University

Alternative fuel projects have been ongoing at Middle Tennessee State University (MTSU) since 1979. Prior to investigating hydrogen, Dr. Cliff Rickets, et al were successful in running engines on both ethanol and natural gas fuels and built an ethanol vehicle. The first successful attempt to run a hydrogen engine at MTSU was in 1987. Since that time, these researchers have run engines, cars, and tractors using hydrogen. In 1991, the MTSU team had the fastest timing for a hydrogen-fueled vehicle, 88 miles/hour (142 km/hour), in the Bonneville Speed Trials at the Great Salt Flats in Wendover, Utah. The team ran their hydrogen vehicle in a time-only trial since there is no official division for hydrogen-powered vehicles.

The MTSU team recently improved on their previous clocking at the 1994 World of Speed trials at the Great Salt Flats. Their new top speed of 108 miles/hour (174 km/ hour) eclipses the old mark by 20 miles/hour (32 km/

hour). The vehicle used for this trial was powered by a standard Nissan KA24E engine equipped with three valves per cylinder. Two of these valves are used to supply hydrogen and air separately to the combustion chamber so that the engine can be operated without premixing the hydrogen and oxygen. The engine is also equipped with water injection. The vehicle is pictured in Figure 7.2 [16].

Figure 7.2. Hydrogen-fueled vehicle designed and built at Middle Tennessee State University. (Source: ref. [16])

Riverside, California

The hydrogen movement has had strong support in the Riverside, California area for several decades. Dr. Robert Zweig, and the Clean Fuels Institute (CFI) that he founded, have been among the most important influences in this area. Early efforts by the CFI led to acquisition of a 19-passenger bus by the City of Riverside. The bus was built by the Billings Corporation and featured a ten-vessel FeTi hydride storage system which was heated by circulating engine coolant through a water jacket on the tanks [17]. The bus was also equipped with propane fuel and could be switched to propane operation with the flip of a switch. Several problems with the bus were identified during this test period, however, including persistent backfiring and inadequate power.

CFI next became involved in a project to convert a Dodge D-50 pickup truck to hydrogen fuel. To eliminate the problem of backfire, the truck was equipped with a parallel induction fuel delivery system, developed by Frank Lynch and co-workers at the Denver Research Institute. A turbocharger and an intercooler were also added to the vehicle to increase power output. Much of the development work and testing for this vehicle was carried out at Hydrogen Consultants, Inc. (HCI) in Colorado. The truck has been fueled by both metal hydride and high-pressure hydrogen storage.

More recently, a program to develop and evaluate hydrogen-fueled transportation vehicles has been initiated at the University of California at Riverside's College of Engineering—Center for Environmental Research and Technology (CE-CERT). This program currently includes two hydrogen vehicles. One of the vehicles is a Ford Ranger which was converted to hydrogen operation by HCI. The vehicle has a 2.3 ℓ 4-cylinder engine with a modified head and camshaft, a turbocharger, an intercooler, and a hydrogen fuel injection system. Hydrogen is stored on-board in high pressure composite gas cylinders. The hydrogen fuel injection system is a form of sequential multi-port fuel injection called Constant Volume Injection (CVI), which was developed at HCI. The hydrogen itself is produced at an on-site solar electrolysis production facility. The UC Riverside hydrogen truck is shown in Fig. 7.3 and discussed in further detail in the reprint by Heffel and Norbeck at the end of this chapter. CE-CERT is presently developing a new electronic fuel injection system to replace the current system on the truck.

Student researchers at CE-CERT have designed and built a small electric vehicle with an on-board hydrogen fuel cell and storage system as part of a hybrid vehicle development program. The hybrid electric vehicle (HEV) utilizes the electricity from a fuel cell to keep its batteries charged, enabling it to have a much greater range than a pure electric vehicle. This vehicle will serve as a test bed for researchers to experiment with various types of fuel cells and become familiar with their operation. The current power unit being utilized is a 1.5 kW phosphoric acid fuel cell. The hydrogen is stored at 4,500 psi in lightweight composite pressure vessels provided by Structural Composites Industries.

CE-CERT is presently developing a second HEV test bed utilizing a Ford Ranger chassis. This vehicle will include a lean burn hydrogen engine with an electronic fuel injectio system. An energy management system is also being developed in conjunction with this vehicle which will be used to ensure that the energy required for demanding driving conditions is always available. Operational and emissions models are also being developed to allow quantitative analysis of the system usage and the emissions benefits of HEVs under real-world driving conditions.

Figure 7.3. U.C. Riverside's Ford Ranger truck at the on-site vehicle refueling station.

Perris, California

Among the earliest advocates of hydrogen fuel in the post-war movement in the United States was a small group of residents from Perris, California. The group, composed of three engineers, Patrick Lee Underwood, Frederick F. Nardecchia, and Paul B. Dieges, and a newspaper reporter, Dwight B. Minnich, were known as the Perris Smogless Automobile Association. The group successfully operated their first test vehicle, a 1950 Studebaker, by remote control from behind a large rock, until the engine failed to restart after being overfueled with hydrogen. The vehicle was purchased for only $10 from a local junk dealer. The group next converted an antique 1930 Model A Ford pickup to cryogenic hydrogen and oxygen operation. The vehicle featured a system in which steam in the exhaust vapor could be condensed to water and excess hydrogen in the exhaust could be recirculated. The group also converted a second Ford truck to hydrogen in preparation for the 1970 Clean Air Race from Cambridge, MA to Pasadena, CA. Although this vehicle was run in some preliminary tests, the group was unable to participate in the race due to logistical and other problems [3].

University of California, Los Angeles

The University of California, Los Angeles (UCLA) was involved in a number of projects to develop and test hydrogen vehicles throughout the 1970s. Several undergraduate students, under the direction of Professor William Van Vorst, initiated this program in October of 1970 when they began converting a American Motors Gremlin to cryogenic hydrogen operation [18]. After modification, the Gremlin was easily able to meet the early emissions standards of the time. At the 1972 Urban Vehicle Design Competition the vehicle claimed the overall prize and was the runner up in the antipollution category [3].

Van Vorst et al also performed a series of tests on a liquid hydrogen-fueled U.S. Postal Service Jeep. This vehicle was configured with both port and direct injection to study the prospects of using different fuel delivery systems to control backfire problems commonly found in hydrogen engines [19, 20]. These researchers found that the direct injection system was prone to incomplete combustion due to the formation of an inhomogeneous fuel-air mixture. The performance of the port injection system, however, was more favorable in that it did not suffer from any problems related to poor fuel-air mixing. At one point, this vehicle was involved in an accident while mounted on a trailer being towed, causing it to flip over. There were no injuries and the vehicle was driven away from the scene [21].

University of Miami

The University of Miami has one of the most important and longstanding hydrogen engine development programs in the nation. Researchers at the University of Miami were among the first to investigate advanced delivery systems as a means to alleviate premature ignition in hydrogen engines. These investigations led to the development of the Hydrogen Induction Technique [22]. Later, the University of Miami in conjunction with Hawthorne Research and Testing, Inc. conducted an extensive study of different engine configurations from 1977 to 1982. The objective of this study was to provide a design database for light-duty hydrogen-fueled engines, with an emphasis on exploring as many different engine configurations as possible. In all, 19 configurations were tested including models with a normally aspirated gas-mixer, intake port injection, and early and late direct cylinder injection. Engine performance and emissions were evaluated over a maximal load range using both four- and single-cylinder operating modes. Exhaust gas recirculation and water injection were also used in certain cases where the fuel injection took place before the intake valve was closed. The results of this comprehensive study are reported in detail in ref. [23] and summarized in [24].

Since this study, the University of Miami has been involved in a number of different projects related to hydrogen engine development. These researchers have studied the performance of engines utilizing hythane, a mixture of natural gas and hydrogen [25]. They have also investigated the potential of redesigning the combustion chamber, as opposed to the fuel delivery system, to accommodate hydrogen fuel [26]. A study of the safety of using hydrogen in kitchen stoves was also performed by these researchers (see Section 5.2) [27].

Australia

Researchers at the University of Melbourne, Australia have been investigating the possibility of using hydrogen as a fuel for transportation since the 1970s [28]. The group developed a hydrogen-fueled vehicle, as part of this work [29]. The vehicle was powered by a 2-ℓ, 4-cylinder Ford engine which was equipped with a delayed port admission fuel delivery system. The test car demonstrated low emissions and a 63% improvement in fuel economy compared with gasoline operation. The acceleration of the vehicle was hampered, however, by reduced power output and increased vehicle mass. This group is still active in hydrogen research. More recent work includes computer model studies of combustion phenomena in dual-fuel hydrogen engines [30] and studies of the economics of operating diesel generator sets on waste hydrogen [31]. In other work in Australia, the Hydro-Electric Commissions of Tasmania, Australia converted a 6-cylinder, General Motors Model HQ Holdem utility vehicle to operate on hydrogen [14].

Belgium

The first of four buses to be constructed under the Euro Quebec Hydro Hydrogen Pilot Project (EQHHPP) is currently being demonstrated and tested in Belgium [32, 33]. This project is designed to show the feasibility of hydrogen transportation and promote the use of hydrogen in Europe. The 88-passenger Van Hool bus ("Greenbus") was converted to run on liquid hydrogen by VCST-Hydrogen Systems of Belgium. It is powered by a 6-cylinder M.A.N. diesel engine and is equipped with a 125 ℓ liquid hydrogen tank built and designed by Messer-Griesheim. No CO, CO_2, or HC emissions were detected in emissions tests of the vehicle, and it is estimated that NO_x emissions were reduced from 7g/kWh, for a standard advanced diesel, to 0.25g/kWh. The vehicle has a range in excess of 60 km (37.5 miles) and a top speed of 63 km (39.3 miles) per hour.

Brazil

UNICAMP

The UNICAMP laboratory in Brazil has been involved in hydrogen technology research for more than 15 years, including electrolytic generation, purification, storage, and other applications [34]. This group has produced two hydrogen vehicles, a Toyota and a Kombi-VW van. The Toyota is run on a mixture of hydrogen and diesel fuels, with approximately 20% of the energy value of the fuel mixture supplied by hydrogen. The Kombi-VW van operates on pure hydrogen, supplied by FeTi and MgNi tanks. While the Toyota showed no operational difficulties or power loss, the van experienced large power losses and backfire problems. This group is currently involved in the construction of a fuel cell hybrid electric vehicle with metal hydride storage tanks.

Canada

Ballard Power Systems Inc.

Ballard Power Systems Inc. is currently demonstrating a transit bus powered by PEM fuel cells. The 32-foot long bus uses 24 Ballard PEM 5 kW stacks arranged in three parallel strings which are in turn arranged in a series of eight to provide a gross power of 120 kW. In phase one of the project (November 1990 to March 1993), it was demonstrated that the Ballard PEM fuel cell was capable of delivering all the power necessary to run the bus. Dynamometer testing of the completed bus began in December 1992 and the bus, carrying a full load of passengers, ran on the road for the first time on January 27, 1993 [35]. The bus has been driven over 2000 km in Vancouver, Los Angeles, and Sacramento, CA. Additional information on this Ballard bus can be found in the reprint by Prater at the end of this chapter.

Ballard is converting a second transit bus to fuel cell operation. The bus will be powered by a more advanced fuel cell stack that delivers twice as much power as the earlier stack. The 250 kW fuel cell will also fit in the same amount of space allocated for the diesel engine on an equivalent diesel bus. The bus is expected to have a range of 400 km provided by hydrogen which will be stored on the roof. Ballard plans to introduce a commercial fuel cell bus which will have a range of 550 km by 1998 [36].

The EUREKA Project

The EUREKA Project is a fuel cell demonstration program which combines the efforts of four different companies in four different European countries [37]. The organizations participating include Elecno of Belgium (a fuel cell manufacturer), Airproducts of the Netherlands (a producer of hydrogen), Saft of France (a battery manufacturer), and Ansaldo Richerche of Italy (a bus manufacturer). The project was initiated in 1992 with the goal of demonstrating an alkaline hydrogen-air fuel cell system as part of a hybrid electric traction system in a city bus. An alkaline hydrogen-air fuel cell is to be the primary power source for this bus, with Ni-Cd batteries available as a secondary power source. Hydrogen is stored in liquid hydrogen tanks located on the rear end of the bus. Installation on the bus is supposedly near completion and driving tests are expected to begin shortly.

Figure 7.4. Mercedes-Benz T model with hydride storage unit. Refueling connections are shown in the lower photograph. (Source: Daimler-Benz)

Germany

Daimler-Benz

Daimler-Benz (DB) has been studying the prospects of hydrogen as a fuel for surface transportation since 1973. The early work at DB culminated with a fleet test of hydrogen vehicles in Berlin which began in 1984. Two types of vehicles were used for this test: (1.) five model 280 TE passenger cars which used mixed fuel (hydrogen/gasoline), and (2.) five model 310 delivery vans converted to pure hydrogen operation. Both types of vehicles were powered by hydrogen stored on-board as a metal hydride. Vehicles were fitted with a tachograph to monitor driving time, mileage, etc. The vehicles operated reliably and safely over the four-year test period. The primary problems encountered in these tests were mechanical, such as incorrect assembly and durability of newly developed parts. It was also noted that the tires, brakes, and springs on the car showed signs of increased wear due to the heavy weight of the hydride tank [38]. The station wagon with its metal hydride storage unit and refueling connections is shown in Fig. 7.4.

A more advanced Mercedes 230E was developed in 1989 [11]. This vehicle included a modified fuel intake system to mix jets of hydrogen and air prior to injection into the engine cylinder. Small quantities of water were also directly injected into the cylinder to help eliminate backfire. The vehicle was equipped with a metal hydride system which weighed 700 lb (317 kg) and provided for an 80 mile (129 km) range.

Daimler-Benz is currently involved in the Swiss-German cooperative "hydrogen powered applications using seasonal and weekly surplus of electricity" (or HYPASSE) program [39, 40]. DB is in the process of building a bus for the project which is expected to be ready for testing and demonstration by 1995. The bus will be fueled by pressurized hydrogen tanks which will be mounted on the roof. A 12 ℓ, 6-cylinder OM447hLA engine with a turbocharger and aftercooling is slated to be used in the bus. Some sources have hinted, however, that a new engine specifically developed for hydrogen operation will be used in the vehicle. In conjunction with this project, Swiss agencies are working on new methods of hydrogen storage, including the methylcyclohexane-toluene-hydrogen system discussed in Chapter 6. More details about the DB role in the HYPASSE program can be found in the reprint by Zieger at the end of this chapter. Daimler-Benz is also involved in an effort to develop a production model fuel cell that would be small enough to use in a standard passenger car. DB has recently unveiled a Mercedes van which is powered by a fuel cell. This van was developed jointly with Ballard Power Systems of Canada [41].

Bavarian Motor Works (BMW) and the German Aerospace Research Establishment (DFVLR, or currently DLR)

A considerable amount of work has been done in Germany through the cooperation of the DFVLR (now the DLR), BMW, and several other institutions. These efforts have been primarily directed at the development of LH_2 prototype vehicles and refueling systems [42]. One of the earliest projects was a joint effort between the DFVLR and the University of Stuttgart to convert a BMW 518 to LH_2 operation [43]. This was the first European liquid hydrogen-fueled automobile. This car was equipped with timed individual port injection, water injection, and an exhaust gas turbocharger for improved power output, and could be operated on either gasoline, hydrogen, or in a dual-fuel operation mode. This vehicle was completely taken over by the DFVLR in 1980 and has been modified several times since then [13].

BMW and the DFVLR have worked together to produce several LH_2 vehicles. One of the early conversions was a BMW 745i which was converted to operate on LH_2. It was equipped with a turbocharged 3.5 ℓ engine and was fueled using timed individual port injection [42]. The mixture control, spark advance, and other engine parameters for this vehicle were monitored by a digital electronic system. In another joint project, a 745i was configured for direct cylinder injection of hydrogen. This was the first direct cylinder injection hydrogen test vehicle in Europe [44].

Aside from its 7-series hydrogen vehicles, BMW is also conducting basic engine research using one-cylinder and modified engines both with and without supercharging. These engines range in capacity from 2.5 to 5.0 liters and cover an output range from 80 to 150 kW. BMW has also teamed up with Robert Bosch GmbH on the development of a sequential hydrogen metering system for the intake air [45]. Additionally, BMW is involved in a four-year program with Messer Griesheim GmbH, Linde AG, and others to investigate the safety aspects of LH_2 tank systems in passenger cars [46]. Additional information about these programs can be found in the reprint by Peschka and Escher at the end of Chapter 2, the reprint by Pehr at the end of Chapter 5, and the reprint by Reister and Strobl at the end of this chapter.

Italy

Another bus for the EQHHPP is currently being developed by Italy's Ansaldo Richerche, of Genoa [47]. This bus will be powered by eight 5 kW PEM fuel cells that were developed by Italy's De Nora Permelec. Backup power is provided by a battery pack. Three liquid hydrogen fuel tanks will be used to store the hydrogen. The tanks were contributed by Messer Griesheim and will be stored on the roof. The vehicle is expected to be ready for the first road tests in Brescia, northern Italy, by the end of the 1994.

Several Italian institutes are also involved in an ongoing program to investigate the feasibility of hydrogen as a fuel for transportation [48, 49]. This effort involves the cooperative effort of ENEA (The Italian Agency for New Technologies, Energy and Ambient) of Rome, the Department of Energetics of the University of Pisa, the Weber-Marelli of Bologna, and the VM Motori of Centro. This group has converted a Fiat Ducato van to dual-fuel gasoline hydrogen operation. An electronically-controlled timed port injection system was adopted to deliver hydrogen to the engine. Electronic signals from sensors are used to optimize the fuel injection parameters. The hydrogen is stored in aluminum pressure vessels. Road tests are currently being performed to study the performance of the vehicle.

Japan

Musashi Institute of Technology (MIT)

The Musashi Institute of Technology began its study of hydrogen engines in 1970. Since then, a comprehensive program of engine and hydrogen storage research has resulted in a series of hydrogen vehicles bearing the Musashi name. The first Musashi vehicle (Musashi-1) was a converted one-ton truck equipped with high-pressure hydrogen tanks. Musashi-2 was a Datsun B-210 car converted to hydrogen operation in 1975. This vehicle was equipped with a 230 ℓ LH_2 storage dewar installed in the rear of the car [50]. The vehicle was designed to run in the Student Engineered Economy Design (SEED) Rally in the United States and is shown in Figure 7.5. Other Musashi vehicles are discussed in ref. [14] and references therein.

Mushashi-8 and -9 are the latest vehicles produced by this group. Musashi-8 is a converted Nissan Fairlady sportscar which is described in greater detail in the reprint by Furuhama at the end of Chapter 2 [51]. Musashi-9 is a refrigerator van which was modified to run on LH_2. The van is designed to efficiently utilize the cold energy of LH_2 which is typically lost during operation. This van demonstrates that LH_2 can be used to cool the refrigerator unit to 0° C, in addition to fueling the engine, resulting in a vehicle which has a greater overall efficiency. Further details on this vehicle are given in the reprint Yamane, et al at the end of this chapter [52].

Figure 7.5. LH_2 tank installed in the rear of Musashi-2 (Source:Musashi Institute of Technology)

Mazda

The Mazda Corporation has recently developed several prototype vehicles based on rotary engines. This engine is, in some ways, ideal for use with hydrogen since the operational chamber moves so that the intake, compression, combustion, and exhaust strokes occur at different places [53, 54]. Backfire and preignition can, thus, be reduced since the temperature during the intake stroke is low and there is no direct contact between the air/fuel mixture and the exhaust gases. Lower combustion temperatures can also lead to a reduction in the exhaust emission levels.

Mazda introduced its first concept car, the HR-X, at the Tokyo Motor Show in 1991 [55, 56]. Special features on this vehicle, aside from the rotary engine, included a metal hydride storage tank and an Active Torque Control System (ATCS) to provide extra power. The ATCS essentially utilizes a battery-powered electric motor to add additional torque to the engine during acceleration periods. When braking, the system converts excess momentum into electricity which is stored for later use in a nickel-metal hydride battery. The modified hydrogen rotary engine is discussed in further detail in the reprint by Morimoto, et al at the end of this chapter.

Mazda has since released an updated version of the HR-X, the HR-X2, which is pictured in Fig. 7.6. [57]. The latter vehicle incorporates a larger displacement engine with a peripheral intake port for enhanced power. The engine achieves a maximum power of 130 hp (97 kW) and a maximum torque of 17 kg-m. The HR-X2 also utilizes an improved metal hydride tank composed of 18 individual

Figure 7.6. Mazda's hydrogen-fueled HR-X2. (Source: Mazda)

long rectangular cells. The temperature of the hydride can be controlled by passing heat through coolant passages in the walls between separated cells. A liquid crystal polymer fiber reinforced plastic which is recyclable was also used in the design of this vehicle.

Other vehicles released by Mazda include a hydrogen-fueled Miata and a 626 station wagon. The rotary engine used in the Miata generates 118 hp (88 kW) and 121 lb/ft of torque, about the same figures as for the production model [58]. The Miata also incorporates the cell-type metal hydride utilized in the HR-X2. The hydrogen-fueled 626 station wagon has been loaned to the Nippon Steel Corporation for extended testing [59]. Mazda has also demonstrated a fuel cell powered golf cart [36].

Romania

Researchers in Romania have investigated the advantages of dual-fuel hydrogen/gasoline operation in tests using a single-cylinder engine, a four-cylinder engine, and a DACIA 1300 car [60]. Both emissions reductions and an increase in the thermal efficiency of the engine were found with dual-fuel operation. When the vehicle was tested over the ECE City Cycle test, CO levels were found to decrease while NO_x levels remained nearly the same under dual-fuel as opposed to pure gasoline operation. Gasoline consumption for the test vehicle was also found to be reduced from 9.37 ℓ per 100 km for pure gasoline fueling to 7.75 ℓ per 100 km for dual fueling.

Former Soviet Union

Two major institutions have been the main centers of hydrogen research in the former Soviet Union: the Institute for Problems in Machinery, Ukrainian Academy of Sciences; and the Research Institute of Automobiles, Ministry of Transportation. Researchers at these institutes have experimented with dual-fuel, gasoline/hydrogen operation in passenger cars and vans, medium- and heavy-duty trucks, and truck loaders. In vehicle tests designed to simulate city traffic, it was found that gasoline consumption could be reduced by 38% and that fuel economy could be increased by 21% when vehicles were operated in a dual-fuel mode as compared to pure gasoline operation [61]. Significant reductions in the emissions of CO, HC, and NO_X were also found under dual-fuel operation. Reduced emissions and more efficient operation were also observed for a petrol/hydrogen truck studied by the Institute for Problems in Machinery [62]. The Leningrad Agricultural Institute has also reported the conversion of a tractor to dual-fuel, diesel/hydrogen operation [63].

7.3 Conclusion

The development of hydrogen vehicles over several generations has gradually produced more powerful and efficient vehicles leading towards more commercial viability. Fuel cell electric vehicles, in particular, have advanced rapidly in the last five years. Efforts in this area must continue, however, if a commercially viable hydrogen-fueled vehicle acceptable for mass marketing is to be produced. Cost, power density, and on-board storage are among the most important issues to address. Government incentives and mandates to phase in low- and zero-emission vehicles and the growing concern over pollution and energy security should continue to boost efforts to develop more marketable hydrogen vehicles.

References

1. Weil, K.H., "The hydrogen I.C. engine—its origin and future in the emerging-transportation environment system," in *7th Intersociety Energy Conversion Engineering Conference*, Edited by American Chemical Society, Washington, D.C., p. 1355, 1972.

2. Escher, W.J.D., "Hydrogen-Fueled Internal Combustion Engine, A Technical Survey of Contemporary U.S. Projects," Escher Technology Associates, Report No. TEC-75/005, 1975.

3. Hoffmann, P., *The Forever Fuel: The Story of Hydrogen*, Westview Press, Boulder, Colorado, 1981.

4. Billings, R.E., "Hydrogen Homestead," in *2nd World Hydrogen Energy Conference*, Edited by T. N. Veziroglu and W. Seifritz, Pergamon Press, Elmsford, NY, p. 1709, 1978.

5. Billings, R.E., "Advances in hydrogen engine conversion technology," *Int. J. Hydrogen Energy*, Vol. 8, p. 939, 1983.

6. Billings, R.E., Hatch, S.M., and DiVacky, R.J., "Conversion and testing of hydrogen-powered post office vehicle," *Int. J. Hydrogen Energy*, Vol. 8, p. 943, 1983.

7. Billings, R.E., Sanchez, M., Cherry, P., and Eyre, D.B., "LaserCell Prototype Vehicle," *Int. J. Hydrogen Energy*, Vol. 16, p. 829, 1991.

8. Kaiser, W., Provenzano, J., Scott, P.B., Staples, P., and Zweig, R., "Supercharged hydrogen fuel powered

trucks," presented at the 1996 World Car Conference, Riverside, CA, 1996.

9. Nadal, M. and Barbir, F., "Development of a Hydride Fuel Cell/Battery Powered Electric Vehicle," in *Hydrogen Energy Progress X*, Edited by D. L. Block and T. N. Veziroglu, International Association for Hydrogen Energy, Coral Gables, FL, p. 1427, 1994.

10. Lynch, F.E., "Parallel induction: a simple fuel control method for hydrogen engines," *Int. J. Hydrogen Energy*, Vol. 8, p. 721, 1983.

11. Cannon, J.S., *Harnessing Hydrogen: The Key to Sustainable Transportation*, INFORM, Inc. New York, NY, 1995.

12. "Official Program of the 43rd Annual Bonneville Speed Week."

13. Peschka, W., *Liquid Hydrogen-Fuel of the Future*, Springer-Verlag, Wein, New York, 1992.

14. Stewart, W.F., "Hydrogen as a Vehicular Fuel," in *Recent Developments in Hydrogen Technology*, Edited by K. D. J. Williamson, Jr. and F. J. Edeskuty, CRC Press, Inc., Boca Raton, FL, p. 69, 1986.

15. Personal Communication Edeskuty, F.J., Los Alamos National Laboratory.

16. Personal Communication Rickets, C., Middle Tennessee State University.

17. Woolley, R.L., "Design Considerations for the Riverside Hydrogen Bus," in *2nd World Hydrogen Energy Conference*, Edited by T. N. Veziroglu and W. Seifritz, Pergamon Press, Elmsford, NY, p. 1978.

18. Finegold, J.G., Lynch, F.E. and Bush, A.F., "The UCLA Hydrogen Car: Design, Construction, and Performance," SAE Technical Paper No. 730507, Society of Automotive Engineers, Warrendale, PA, 1973.

19. MacCarley, C.A., "Electronic Fuel Injection Techniques for Hydrogen Fueled I.C. Engines," M.S. Thesis in Engineering, University of California, Los Angeles, 1978.

20. MacCarley, C.A. and Van Vorst, W.D., "Electronic fuel injection techniques for hydrogen powered I.C. engines," *Int. J. Hydrogen Energy*, Vol. 5, p. 179, 1980.

21. Finegold, J.G. and Van Vorst, W.D., "Crash test of liquid hydrogen automobile," in 1st World Hydrogen Energy Conference, University of Miami, Miami Beach, FL, p. 6, 1976.

22. Swain, M.R. and Adt, Jr., R.R., "The Hydrogen-Air Fueled Automobile," in *7th Intersociety Energy Conversion Engineering Conference*, Edited by American Chemical Society, Washington, D.C., p. 1382, 1972.

23. Swain, M.R., Adt, Jr., R.R., and Pappas, J.M., Experimental Hydrogen-fueled Automotive Engine Design Database Project, U.S. Department of Energy, Washington D. C., Report No. DOE/CS/31212-1, 1983.

24. Swain, M.R., Pappas, J.M., Adt, Jr., R.R., and Escher, W.J.D., "Hydrogen engine design database summary," in *18th Intersociety Energy Conversion Engineering Conference*, American Institute of Chemical Engineers, New York, NY, p. 536, 1983.

25. Swain, M.R., Yusuf, M.J., Dulger, Z., and Swain, M.N., "The Effects of Hydrogen Addition on Natural Gas Engine Operation," SAE Technical Paper No. 932775, Society of Automotive Engineers, Warrendale, PA, 1993.

26. Swain, M.R., Swain, M.N., and Adt, Jr., R.R., "Considerations in the Design of an Expensive Hydrogen-Fueled Engine," SAE Technical Paper No. 881630, Society of Automotive Engineers, Warrendale, PA, 1988.

27. Swain, M.R. and Swain, M.N., A Comparison of H_2, CH_4 and C_3H_8 Fuel Leakage in Residential Settings, International Association for Hydrogen Energy, Coral Gables, FL, 1992.

28. Dozier, K., "Aussies Tout Hydrogen Alternative Watson: Tests Show Significant CO, CO_2 and HC Reductions," *Environment Week*, November 22, 1990.

29. Watson, H.C., Milkins, E.E., Martin, W.R.B., and Edsell, J., "An Australian Hydrogen Car," in *Hydro-*

gen Energy Progress V, Edited by T. N. Veziroglu and J. B. Taylor, Pergamon Press, Elmsford, NY, p. 1549, 1984.

30. Watson, H.C. and Lambe, S.M., "A study of ultra fast combustion in a hydrogen engine with pilot diesel fuel ignition," in *Hydrogen Energy Progress IX*, Edited by T. N. Veziroglu, C. Derive, and J. Pottier, International Association for Hydrogen Energy, Coral Gables, FL, p. 1271, 1992.

31. Lambe, S. and Watson, H., "Economics of Diesel Generator Sets Utilizing Waste Hydrogen Fuel," in *Hydrogen Energy Progress X*, Edited by D. L. Block and T. N. Veziroglu, International Association for Hydrogen Energy, Coral Gables, FL, p. 1815, 1994.

32. Vandenborre, H. and Sierens, R., "Greenbus: A Hydrogen Fuelled City Bus," in *Hydrogen Energy Progress X*, Edited by D. L. Block and T. N. Veziroglu, International Association of Hydrogen Energy, Coral Gables, FL, p. 1959, 1994.

33. Hoffmann, P., "Europe unveils proof of concept hydrogen bus," *The Hydrogen Letter*, Vol. 9, 1994.

34. Silva, E.D., Gallo, W., Szajner, J., Amaral, E.D., and Bezerra, C., "A Solar/Hydrogen/Electricity Hybrid Vehicle," in *Hydrogen Energy Progress X*, Edited by D. L. Block and T. N. Veziroglu, International Association for Hydrogen Energy, Coral Gables, FL, p. 1441, 1994.

35. Howard, P., "Ballard Fuel Cell Powered ZEV Bus," in *World Car 2001 Conference*, Edited by College of Engineering Center for Environmental Research and Technology, University of California, Riverside, Riverside, California, 1993.

36. Prater, K.B., "Polymer Electrolyte Fuel Cells: A Review of Recent Developments," *Journal of Power Sources*, Vol. 51, p. 129, 1994.

37. DeGeeter, E., Broeck, H.V.D., Bout, P., Woortmann, M., Cornu, J., Peski, V., Dufour, A., and Marcenaro, B., "Eureka Fuel Cell Bus Demonstration Project," in *Hydrogen Energy Progress X*, Edited by D. L. Block and T. N. Veziroglu, Coral Gables, FL, p. 1457, 1994.

38. Feucht, K., Hurich, W., Komoschinski, N., and Povel, R., "Hydrogen Drive for Road Vehicles—Results from the Fleet Test Run in Berlin," in *Hydrogen Energy Progress VI*, Edited by T. N. Veziroglu, N. Getoff, and P. Weinzierl, Pergamon Press, Elmsford, NY, p. 1079, 1986.

39. Hoffmann, P., "HYPASSE-Bus to operate next year, maybe with new H-engine," *The Hydrogen Letter*, 1994.

40. Zieger, J., "HYPASSE-Hydrogen powered automobiles using seasonal and weekly surplus of electricity," in *Hydrogen Energy Progress X*, Edited by D. L. Block and T. N. Veziroglu, International Association for Hydrogen Energy, Coral Gables, FL, p. 1367, 1994.

41. "Daimler Unveils Electric Vehicle Using Fuel Cells," *H_2 Digest*, September/October 1994.

42. Peschka, W., "Liquid hydrogen fueled automotive vehicles in Germany—Status and Development," *Int. J. Hydrogen Energy*, Vol. 11, p. 721, 1986.

43. Peschka, W., "Operating characteristics of a LH_2-fuelled automotive vehicle and of a semi-automatic LH_2-refuelling station," *Int. J. Hydrogen Energy*, Vol. 7, p. 661, 1982.

44. Peschka, W. and Escher, W.J.D., "Germany's Contribution to the Demonstrated Technical Feasibility of the Liquid-Hydrogen Fueled Passenger Automobile," SAE Technical Paper No. 931812, Society of Automotive Engineers, Warrendale, PA, 1993.

45. Reister, D. and Strobl, W., "Current Development and Outlook for the Hydrogen-Fuelled Car," in *Hydrogen Energy Progress IX*, Edited by T. N. Veziroglu and C. D.-J. Pottier, International Association for Hydrogen Energy, Coral Gables, FL, p. 1201, 1992.

46. Pehr, K., "Aspects of Safety and Acceptance of LH_2 Tank Systems in Passenger Cars," in *Hydrogen Energy Progress X*, Edited by D. L. Block and T. N. Veziroglu, International Association for Hydrogen Energy, Coral Gables, FL, p. 1399, 1994.

47. Marcenaro, B., "EQHHPP FC BUS: Status of the Project and the Presentation of the First Experimental Results," in *Hydrogen Energy Progress X*, Edited by D. L. Block and T. N. Veziroglu, International Association for Hydrogen Energy, Coral Gables, FL, p. 1447, 1994.

48. Dini, D., Botarelli, C., Nardi, G., Pagni, G., Sglavo, V., Sterzo, B.L., and Mazza, D., "Conversion to Hydrogen Operation of an Electronically Controlled Gasoline Supply System," in *Hydrogen Energy Progress X*, Edited by D. L. Block and T. N. Veziroglu, International Association for Hydrogen Energy, Coral Gables, FL, p. 1389, 1994.

49. Ciancia, A., Pede, G., Brighigna, M., and Perrone, V., "A Compressed Hydrogen Fuelled Vehicle at ENEA: Status and Development," in *Hydrogen Energy Progress X*, Edited by D. L. Block and T. N. Veziroglu, International Association for Hydrogen Energy, Coral Gables, FL, p. 1415, 1994.

50. Furuhama, S., Hiruma, M., and Enomoto, Y., "Development of a liquid hydrogen car," *Int. J. Hydrogen Energy*, Vol. 3, p. 61, 1978.

51. Furuhama, S., "Trend of Social Requirements and Technological Development of Hydrogen-Fueled Automobiles," *JSAE review*, Vol. 13, p. 4, 1992.

52. Yamane, K., Hiruma, M., Watanabe, T., Kondo, T., Hikino, K., Hashimoto, T., and Furuhama, S., "Some performance of engine and cooling system on LH_2 refrigerator van Musashi-9," in *Hydrogen Energy Progress X*, Edited by D. L. Block and T. N. Veziroglu, International Association for Hydrogen Energy, Coral Gables, FL, p. 1825, 1994.

53. Morimoto, K., Teramoto, T., and Takamori, Y., "Combustion Characteristics in Hydrogen Fueled Rotary Engine," SAE Technical Paper No. 920302, Society of Automotive Engineers, Warrendale, PA, 1992.

54. Teramoto, T., Takamori, Y., and Morimoto, K., "Hydrogen Fuelled Rotary Engine," SAE Technical Paper No. 925011, Society of Automotive Engineers, Warrendale, PA, 1992.

55. Various Mazda News Releases.

56. Winfield, B., "Hydrogen: It's come a long way since the Hindenburg," *Automobile Magazine*, p. 60, April 1992.

57. Mazda press release information from the 30th Tokyo Motor Show, 1993.

58. Normile, D., "Mazda's Hydrogen Miata," *Popular Science*, p. 40, October 1993.

59. Kludjian, V.Z., "Hydrogen for Vehicles—Mazda's Hydrogen Vehicle Development Program," present at the National Hydrogen Association's 5th Annual U.S. Hydrogen Meeting, Washington, D.C., 1994.

60. Sfinteanu, D. and Apostolescu, N., "Efficiency, Pollution Control, and Performances of Hydrogen-Fueled Passenger Cars," *Int. J. Hydrogen Energy*, Vol. 17, p. 539, 1992.

61. Mishchenko, A.I., Belogub, A.V., Talda, G.B., Savistsky, V.D., and Baikov, V.A., "Hydrogen as a Fuel for Road Vehicles," in *Hydrogen Energy Progress VII*, Edited by T. N. Veziroglu and A. N. Protsenko, Pergamon Press, Elmsford, NY, p. 2037, 1988.

62. Belogub, A.V. and Talda, G.B., "Petrol-Hydrogen Truck with Load-carrying Capacity 5 Tons," *Int. J. Hydrogen Energy*, Vol. 16, p. 423, 1991.

63. Kolbenev, I.L., Kolpakov, V.E., and Soldatkin, A.V., "Hydrogen Diesel Fuel Engine for Application on a Universal Tractor," in *Hydrogen Energy Progress VII*, Edited by T. N. Veziroglu and A. N. Protsenko, Pergamon Press, Elmsford, NY, p. 2095, 1988.

Development of a Hybrid Fuel Cell/Battery Powered Electric Vehicle

M. Nadal and F. Barbir
Energy Partners, Inc.

Abstract

This paper describes the design and performance of a prototype zero emission electric vehicle, powered primarily by air breathing Proton Exchange Membrane (PEM) fuel cells using gaseous hydrogen as fuel. The fuel cell system is composed of the fuel cell stacks, hydrogen tank, air compressor, solenoid valves, pressure regulators, water pump, water tank, heat exchangers, sensors, programmable controller, and voltage regulator. The battery system provides power to the vehicle during periods of peak power demand such as vehicle acceleration or traveling at a high constant speed. The batteries also provide the power to initiate fuel cell startup. The vehicle has been designed, assembled, tested, and has traveled over 100 miles solely on fuel cell power with satisfactorily performance. It has successfully proved the concept of a fuel cell powered zero emission vehicle. Further enhancements will improve the performance in terms of increased speed, acceleration, fuel efficiency, range and reliability.

1. INTRODUCTION

Automobiles are one of the major sources of air pollution in urban areas. As air pollution in heavily populated areas becomes unbearable, the search for cleaner alternatives emerges as an imperative.

The goal of this project is to design, develop and demonstrate a zero emission prototype vehicle powered by PEM fuel cells, referred to as the Green Car. PEM fuel cells are very efficient, compact and low weight, operate at almost ambient temperature, and use hydrogen - the cleanest fuel. As such the PEM fuel cells offer promise as the best future replacement for internal combustion engines in transportation applications.

Although the fuel cell was invented more than 150 years ago, it has not left the research laboratories yet, except for special applications such as the space program. Only recently, mainly because of environmental concerns, the fuel cell technology in general, and PEM fuel cells in particular, have come closer to commercialization. The technology exist, although it still needs some improvements, but it needs market and infrastructure development. The ultimate goal of this project is to bring fuel cell technology out of the laboratory into the marketplace for use in transportation as well as in distributed stationary power generation.

The project included development and manufacturing of the PEM fuel cell system and its integration into an existing lightweight vehicle, referred to as the Green Car, followed by testing of the vehicle's performance.

2. GREEN CAR DESIGN

The Green Car was designed to incorporate a lightweight body with low aerodynamic drag and small frontal area (Figure 1). It is a hybrid electric vehicle powered primarily by PEM fuel cells. Figure 2 illustrates the functional block diagram of the hybrid propulsion system showing all of the major components, which include, fuel cell stacks, fuel cell support equipment, fuel cell voltage regulator, electric drive motor, battery bank, and microprocessor-based programmable controller (MPC). The motor is a highly efficient DC brushless motor capable of providing regenerative braking which can be used to charge the auxiliary battery bank. The vehicle operates primarily on fuel cell power and draws power from the batteries only for peak power requirements, such as acceleration. The fuel cell power system works in parallel with a battery bank. The fuel cells use hydrogen as the fuel and ambient air as the oxidant. The support components of the fuel cells consist of the high pressure gaseous hydrogen storage tank, air compressor, solenoid valves, pressure regulators, sensors, water pump, water tank, heat exchangers and piping system. The voltage regulator is required to condition the voltage output from the fuel cell stacks. It provides a constant voltage output independent of the large fuel cell voltage swing. Any excess power from the fuel cell stacks can be used to charge the batteries. Two DC/DC converters provide 12 VDC to power the vehicle accessories, instrumentation, and fuel cell support equipment. A microprocessor-based programmable controller starts, monitors, and shuts down the fuel cell system. It monitors all system parameters, which include temperatures, pressures, voltages, currents, and flow rates.

2.1 Vehicle Body

The vehicle body was obtained from Consulier Automotive, West Palm Beach, Florida. The high-strength, low weight, monocoque construction is made of foam core polycarbonate advanced composite material. The material is impervious to corrosion, does not rust, and will last indefinitely. Advanced composite materials are fully recyclable and can actually be ground up and reused. The weight of the body used for the Green Car is approximately 185 lbs.

2.2 Electric Motor and Motor Controller System

The electric motor and controller that were selected for the Green Car were designed and built by Unique Mobility, Inc. The 35 hp motor is an air cooled, DC, brushless type motor and can deliver a constant torque of 650 in-lbs from 0 to 3,300 rpm. The maximum speed is 4,000 rpm, and maximum efficiency is 95%. This motor was selected because of its lightweight (approx. 40 lbs) and compactness. The motor controller weighs another 40 lbs, and is required for both commutation and control functions. The controller is capable of driving the motor both in the forward and reverse directions. Regenerative braking is also provided by the controller and the level of regenerative braking can be set by an external signal to the controller. The motor speed is controlled by the accelerator pedal which provides an analog voltage (-10 to +10 volts full reverse to full forward). Hall effect sensors in the motor are used for determining the timing information needed for motor commutation. These signals are processed by the controller and compared with the desired speed. In the event that the controller or motor overheats, the controller will automatically limit current by itself.

2.3 Transmission System

A variable speed transmission has been used in the design because of the motor's inability to operate at speeds in excess of 4,000 rpm without overheating. The transmission is a fully synchronized 5 speed manual transaxle with combined gear reduction, ratio selection and differential functions in one unit housed in a die-cast aluminum case. A reversing gear, although available, is not required since the electric drive motor is reversible. The motor is mounted directly on the transmission with a flexible coupling installed on the input shaft.

2.4 Battery System

The battery system consists of ten 12 V batteries installed to provide both start-up power to the fuel cell system and auxiliary power for acceleration. The batteries are lead acid, sealed, maintenance-free, and gelled electrolyte type. Each battery weights 66 lbs, and according to manufacturer specifications, can deliver 25 Amps for 200 minutes (1 kW-hr of energy) when totally charged.

2.5 Fuel Cell Power System

The Green Car fuel cell power system consists of the fuel cell stacks and their supporting subsystems, namely:

- Fuel storage and management subsystem,
- Oxidant management subsystem,
- Water management subsystem,
- Voltage regulation subsystem,
- Control & monitoring subsystem.

The fuel cell stacks are connected electrically in series and mechanically in parallel. Figure 2, the hybrid propulsion system functional block diagram, shows how the fuel cell power system is integrated into the electric vehicle.

2.5.1 Fuel Cell Stacks

The three proton exchange membrane (PEM) fuel cell stacks used for this project were designed and manufactured by Energy Partners. The PEM fuel cell uses a solid, fixed electrolyte and exhibits low temperature and pressure operation, rapid start-up, and rapid load response.

A fuel cell stack is composed of a series of cells separated by bi-polar current collector plates. The collector plates in the Green Car stacks have an active area of 0.84 ft^2 and are made of graphite with a polymer binder, which produces a corrosion resistant material. The collector plates contain internal manifolds and provide a proper flow field for fuel and oxidant, as well as providing pathways for water removal. Each cell consists of gas chambers and electrodes separated by Nafion™ 115, a proton conducting polymer membrane. Electrodes are made of water repellent carbon fiber paper with platinum based catalyst layer deposited on one side.

The fuel cell stack assembly hardware includes the fluid and compression endplates, which hold the stack together by sandwiching properly ordered cells and components with the use of tie-rods, nuts, and washers. The bus plates are made of copper and are designed so that wire cables can be attached easily.

Each of the stacks in the Green Car contains 60 cells for a total of 180 cells. Originally, the fuel cell system was designed to consist of the two stacks, 86 cells each. However, difficulties were encountered in building the stack with more than 60 cells, and the decision was made to build and install three 60 cell stacks. Cooling of the stack is provided by water flowing through specially designed cells strategically distributed through the active section of the stack. The stack also contains an internal humidification section which provides the proper conditions for the electrochemical reaction and prevents dehydration of the membrane. Table 1 lists the Green Car fuel cell stack specifications.

2.5.2 Fuel Storage and Management Subsystem

The fuel cells require pure hydrogen (at least 99.9% purity). The hydrogen storage system chosen for the Green Car is compressed gaseous hydrogen at a moderate storage pressure of 3,000 psig in a composite pressure vessel. This method was chosen because it is simple in design, the technology is well advanced, and when combined with a suitable pressure regulating device provides a passive, simple load following system. The compressed gas fuel storage system is designed to hold approximately 413 standard cubic feet (11,700 standard liters) of pure hydrogen. The vessel is made of an inner 6061-T6 aluminum liner with hemispherical ends, wrapped with a composite spiral wound E-glass/epoxy resin. This design results in a lightweight structurally sound cylinder. The cylinder is designed to the latest applicable DOT specifications for compressed gas fuel storage for automotive applications.

The fuel management subsystem receives hydrogen from the pressure vessel and regulates the pressure down to system pressure of 20 psig. The pre-regulated pressure is measured and supplied to the MPC to provide information on remaining fuel stores; the post regulated pressure signal is supplied to the MPC to ensure proper system pressure. Hydrogen is directly fed to the stacks in a dead-ended mode, with intermittent purging for removal of contaminants.

2.5.3 Oxidant Management Subsystem

In the Green Car air is used as the oxidant. The ambient air is drawn through a particulate filter by a compressor and raised in pressure to approximately 15 psig. A rotary vane type compressor has been selected, rated at 28 scfm and 20 psig. Power requirement at rated conditions is 3.5 kW. The air flow rate is kept constant at a level that supports the electrochemical reaction at rated power, a stoichiometric ratio of about 1.3. The stoichiometric ratio at lower power levels is therefore always higher than 1.3. The pressure and flow rate values for the air are supplied to the MPC by suitable sensors. The air is then directed through the heat exchanger and water/air separator and into the stacks. Oxygen in the air stream is used to support the fuel cell reaction while the rest of the stream carries the product water from the stack.

2.5.4 Water Management Subsystem

Water is used for cooling the fuel cell stacks and for humidification of the reactant gases within the stack. Water is drawn from a reservoir by a pump and directed through a particulate filter and into the fuel cell. The water moves through the dedicated cooling passages within the stack where it removes the waste heat. After flowing through the cooling passages, the cooling water moves through the humidification section of the stack where through hydrated membranes humidifies the reactant gas streams. The remainder of the cooling water exits the stack where its temperature and flow rate are measured. Depending on the temperature of the exiting water, a thermal control valve directs the water either through a suitable heat exchanger to release the heat or directly back to the water reservoir.

2.5.5 Fuel Cell Voltage Regulator System

The purpose of the fuel cell voltage regulator is to provide a constant voltage output to the drive train from the fuel cell stack by conditioning the large dynamic voltage range of the fuel cell. Under no-load condition, the fuel cell voltage can be as high as 180 volts, and under full load condition the voltage can drop below 120 volts. The regulator is adjustable in the range of 110 to 130 VDC and can be set at a point slightly above battery nominal voltage to maintain the batteries in a float charge condition without overcharging them. This satisfies the requirement of low power demand being supplied only by the fuel cell to the limit of the regulator output. Thus, the primary power of the vehicle is drawn from the fuel cell stack. Any excess current from the fuel cell is used to charge the battery bank. The capability of adjusting regulator output voltage is important because it limits the amount of maximum charge that the fuel cell stacks can deliver to the batteries. The efficiency of the voltage regulator is about 95%.

Since the regulator is connected between the fuel cell and battery bank, and the battery bank is directly connected to the motor drive system, the absence of fuel cell power will not interfere with the operation of the vehicle on battery power alone. Also, there is no possibility of reverse current flow from the battery bank or drive motor (in regenerative braking mode) to the fuel cell through the regulator. The regulator has internal circuitry to prevent operation at too low an input voltage and can be enabled or disabled by an external signal. In the event of a system shutdown by the controller, this last feature allows the load to be disconnected from the fuel cell.

2.5.6 Microprocessor-Based Programmable Controller

The microprocessor-based programmable controller (MPC) that is installed in the Green Car is capable of monitoring the sensor inputs for specific limit conditions (temperatures, pressures, flows and water level) and taking appropriate action in the event of a fuel cell malfunction. An additional dedicated analog multiplexer with 180 differential channels monitors the individual fuel cell voltages. The MPC also provides analog outputs for control of external devices, such as pumps and solenoids, and a display output capable of interfacing with a remote LCD display located on the vehicle dashboard. The MPC displays the fuel cell system status information on the instrument panel. In the event of a shutdown, the MPC stores the current system parameters and these can be read directly from the LCD display or downloaded to any IBM compatible computer through the RS232 communication port on the MPC.

The microprocessor is an Intel 8052H-BASIC chip operating at 12 MHz. The instruction set to implement the control logic is stored in an EPROM, where any subsequent changes to the control strategy can be made easily. A "watchdog" timer circuit is employed to prevent any uncontrolled operation in the event of a controller failure. All programming was done in BASIC and in assembly languages. However, only the fixed hardware dependent drivers were written in assembly language for optimum speed and minimum memory use.

The capability to monitor individual cell voltages is one of the most important functions of the MPC. The monitoring of total stack voltage alone does not provide sufficient information to detect a malfunction of an individual cell within the stack. A cell with a low voltage reading is an indication of a potential hazardous condition, such as a gas leak in that cell, a perforated membrane, or a reversal in cell potential. For air operation, the individual cell voltages are usually between 0.5 and 1.0 V. A shutdown is performed if any cell voltage is below 0.4 V or if the voltage difference between two adjacent cells is greater than 0.2 V. The differential voltage test is important because a defective or erratic cell can be detected before the fuel cell is subjected to a large load.

2.5.7 Instrumentation

The instrument panels on the vehicle dashboard and center console contains the digital meters displays, status and warning indicator lights, and switches for the driver to start and monitor the fuel cell. Meter displays include fuel cell amps, volts, and temperature; battery amps and volts; motor amps, temperature, and rpm; inlet air pressure; and hydrogen pressure. An LCD display with 2 rows of 20 characters located on the front dashboard provides the driver with status information on fuel cell performance. In the event of a fail-safe shutdown, the display will output the sensor reading that caused the shutdown. The center console contains status and warning lights for the stacks and fuel cell voltage regulator.

3. GREEN CAR PERFORMANCE

After all the components were installed in the Green Car, a series of tests were conducted in order to determine the performance of the Green Car. First, the performance of each component and subsystem was tested independently. After necessary modifications, driving tests were conducted. The performance of the Green Car is described below, and summarized in Table 2.

3.1 Vehicle Speed and Acceleration

Only limited driving tests have been performed due to unavailability of a suitable test track. The car has only been driven on the parking lot around the Energy Partners facilities. The Green Car has successfully completed over 100 miles solely on fuel cell power. Average driving speed was about 30 mph. Maximum speed, however, is determined by the maximum power of the main propulsion motor (35 HP), vehicle weight, drag and rolling resistance. A computer program has been developed to predict the Green Car performance based on these constraints. Figure 3 shows vehicle speed as a function of required power from the fuel cell system to the electric motor and auxiliary (hotel) loads. For 35 HP (26 kW) the maximum attainable speed is 60 mph. During the test drive it was not possible to develop maximum speed, due to limitations of the test track. According to Figure 3, the fuel cells can provide power for constant speeds up to 46 mph, which means that excess power has to be supplied by the battery bank. The battery bank starts to supply electric power as soon as the fuel cell voltage regulator output drops below battery nominal voltage (120 V).

Similarly, the acceleration performance is also determined by the main propulsion motor's capabilities. The Green Car needs approximately 10 seconds to accelerate from 0 to 30 mph. Acceleration needs more power than cruising at constant speed. Again, the fuel cell system is capable of providing up to 15 kW, and the remaining power required for acceleration has to be supplied by the battery bank.

3.2 Fuel Consumption and Range

Fuel consumption in the fuel cell stacks (in standard cubic feet per hour) is a function of current, according to the following equation, originally derived from the Faraday's Law:

$$FC = 0.0159\,(I + I_d)\,N_{cell}$$

where:

I = current (Amps),
I_d = current losses due to diffusion of hydrogen and oxygen through the membrane; $I_d = 6.5$ Amps (calculated from Faraday's Law based on experimentally determined diffusion rates).
N_{cell} = number of cells; $N_{cell} = 180$.

Approximately 10 kW are needed for constant cruising speed of 30 mph only on fuel cell power (assuming that the batteries are fully charged, i.e., power from fuel cells is used only to run the main electric motor and hotel load). The fuel cell stacks generate 10 kW at 77 Amps and 130 V. Therefore, the calculated fuel consumption is 239 scfh. Specific fuel consumption per mile traveled at constant speed of 30 mph is 8 scf/mile. In energy terms this corresponds to approximately 56 mpg for a gasoline powered vehicle. Since the amount of hydrogen stored on-board is 413 scf, the maximum vehicle range at constant 30 mph speed is 51.5 mile if the vehicle runs solely on fuel cell power. In addition, the battery bank has a practical storage capacity of 35 Ah, which is enough for at least an additional 18.5 miles, and makes the total vehicle range of 70 miles for a constant speed of 30 mph. For city driving, the predicted range obtained from the computer model is about 60 miles.

3.3 Other Performance Characteristics

Responsivity. PEM fuel cells have demonstrated very fast response to load changes, even at cold start-up. The start-up procedure takes about 15 seconds, measured from the moment the power switch is on until the fuel cell system is ready to supply power. This time is an arbitrary value programmed into the controller, that gives the controller time to check all the operating parameters. Regular shut-down takes less than one second, but the heat exchanger fans may continue to run until the cooling water temperature drops below a certain pre-set value (110 F).

Emissions. The Green Car is truly a zero emission vehicle. The exhaust is actually depleted air from the fuel cell stacks, which also contains the product water from the electrochemical reaction. Water is released both in liquid form (about 0.15 lit/mile) and as water vapor (100% saturated exhaust gas). There are no pollutants of any kind present, whatsoever.

Noise. Although an electric vehicle is supposed to be very quiet, the Green Car generates an unexpectedly loud "buzz." The moving parts in the vehicle include an electric motor with integral cooling, air compressor, compressor electric motor and cooling blower, water pump and heat exchanger fans. By far, the "noisiest" component is the rotary vane type air compressor, which generates a noise level of about 85 dB at 1 m distance.

Safety and reliability. Safety of hydrogen powered vehicles is usually a matter of concern. However, during the assembly and testing of the Green Car no accidents or any hazardous or potentially hazardous situations were experienced. Regular leak checks are required for all hydrogen piping and fittings, including the fuel cell stacks.

The fuel cell is monitored and controlled by a programmable controller (MPC), which automatically performs a shut-down whenever it detects a parameter out of a prescribed range. During the test drives, intermittent shut-offs were experienced. In case of a shut-down, the vehicle continues to operate on battery power. The three main causes for these shut-downs were overheating of the compressor electric motor, fuel cell problems indicated by low cell voltage, and false-alarms caused by inaccurate voltage readings in the MPC. These causes were constantly corrected, and the time between two shut-downs has dramatically increased.

4. FACTORS AFFECTING PERFORMANCE

Overall performance was affected either by the component suppliers not being able to meet Energy Partners' specifications and/or by the fuel cell stacks not meeting the design goals when integrated into the vehicle. The major issues concerning performance are discussed below.

4.1 Drive Motor/Controller

The limitations of the motor/controller system are the low power level of 35 hp (originally a 67 hp motor was designed), and the inability to operate at speeds in excess of 4,000 rpm without overheating the motor. Because of this limitation, a variable speed transmission has been used. However, since this motor was purchased (July, 1991), Unique Mobility and other motor manufactures have developed lightweight electric vehicle motors with power ratings over 80 hp and speeds over 7,500 rpm. The car performance, in terms of maximum speed and acceleration would dramatically change with a more powerful motor. The goal to eliminate the variable speed transmission in favor of directly matching the motor to the differential is possible with the new generation motors.

4.2 Air Compressor

The compressor was selected to deliver 28 scfm at 20 psig (which corresponds to a stoichiometric ratio of 1.3 at 20 kW design point). Power required for the compression is 3.5 kW (efficiency of the selected compressor is therefore less than 50%). However, during the tests, the air compressor motor was constantly overheating, and it was necessary to reduce power requirements below 3 kW. This was possible by lowering the air compressor motor's rpm setpoint, which in turn limited the output pressure to 12-14 psig and air flow to 24 scfm. The performance of the fuel cell stacks was affected by this change.

4.3 Fuel Cell Stacks

Manufacturing of the fuel cell stacks is still in an experimental stage, which explains the differences in performance between the three 60-cell stacks installed in the Green Car. Figure 4 shows the actual performance profiles of the three Green Car stacks under air operation obtained in the test stand. The solid line was generated from a computer model and shows the performance goal. One of the stacks was able to deliver 212 Amps at 40 Volts, but the maximum attainable current from each of the other two stacks was only 160 Amps. Since the stacks are connected in series, the stack with the lowest current rating set the system performance. Referring to Figure 4, at 160 Amps combined stack voltage is approximately 120 Volts, which implies a total power output of 19.2 kW. Although this is near the design goal of 21 kW, this performance was not achieved after installing the stacks in the Green Car, because the Green Car system lacks the capability of maintaining and controlling ideal operating conditions, such as temperature, pressure and flows, as compared with the test stand. The maximum power output of the Green Car fuel cell power system that has been achieved is approximately 15 kW (120 Amps at 125 V). The performance of the entire stack is usually held down by a few cells that experience water management problems. Since the stacks performance is monitored and controlled by an automatic programmable controller (MPC), low voltage in only one cell creates a shut-down of the entire fuel cell system, although the remaining 179 cells are capable of delivering much more power.

4.4 Microprocessor-Based Programmable Controller

The MPC performs three basic functions:

- startup of the fuel cell power system;
- monitoring of the fuel cell power system to assure that it is within a predetermined operational envelop; and
- shut-down of the fuel cell power system in case of any system's failure.

A fail-safe shut-down operation is the result of high stack temperature, high or low stack pressure, low stack voltage, low cell voltage, high differential voltage between adjacent cells, low fluid flow, or low cooling water reservoir level. Startup and shutdown include the enabling/disabling of pumps and solenoids.

Although the MPC performs satisfactorily in carrying out its control functions, it has experienced difficulties in accurately reading the individual cell voltages, resulting in occasional fuel cell fail-safe shut-downs due to false-alarms. In-house testing of the Green Car MPC system has shown that the individual cell voltage readings by the controller are not accurate since they deviate by as much as 15% from actual values, which has caused unnecessary system shut-downs. These inaccurate readings have been caused primarily by the lack of electrical isolation between the sensed analog voltage signals and the MPC, resulting in a common mode voltage problem in the multiplexer controller boards. Note that under no-load condition, cell number 180 is about 180 Volts above fuel cell ground. Thus, all 180 input voltage channels to the multiplexer's differential amplifier are scaled so that the voltage readings are within its common mode voltage limit (±15 V). Both fuel cell and controller are tied to a common ground, and there is no electrical isolation between fuel cell and controller. The readings become more inaccurate for the higher numbered cells. Energy Partners is working with the manufacturer to correct these inaccurate readings.

5. FUTURE IMPROVEMENTS

Although the Green Car has successfully proved the concept of a fuel cell powered zero emission vehicle, its performance can be further enhanced. The following is a list of areas that could be improved:

- fuel cell stacks: more efficient, more reliable, improved heat and water management, lower pressure drop, capability to operate at low air pressure.
- drive motor: more power, higher rpm, better efficiency,
- air compressor: more efficient, variable speed motor, lower noise level
- vehicle weight reduction: reduce number of stacks, reduce number of batteries, eliminate transmission requirement, optimize the supporting equipment
- hydrogen storage: increase the size and/or improve storage density (without compromising safety).
- MPC: more reliable, isolate all fuel cell analog voltage readings from controller hardware, decrease complexity by reducing the number of monitored (input) signals without sacrificing safety.

6. CONCLUSIONS

The goal of this project was to power a lightweight passenger vehicle using a PEM hydrogen/air fuel cell system. The proof-of-concept vehicle, known as the Green Car™, was the result of substantially modifying an existing composite two-seater vehicle by installing three stacks producing 15 kW of pollution-free power, along with appropriate instrumentation, programmable controller, electric drive motor, auxiliary battery bank, and support hardware. With a curb weight of 3,000 lbs, the demonstration vehicle has a top speed of 60 mph, a range of 60 miles in city driving, and can accelerate from 0 to 30 mph in 10 seconds with its 35 hp electric motor. The vehicle has been successfully driven more than 100 miles on fuel cell power during the initial test period. The concept has been proved.

It is now clear that PEM fuel cells are serious candidates for the propulsion of zero emission vehicles. Although the performance of the Green Car is not spectacular, it exceeds the internal combustion engine powered vehicles in two important characteristics, namely fuel efficiency, and emissions. The realization of the technology to build the Green Car has clearly demonstrated the feasibility of the fuel cell powered propulsion systems for transportation vehicles. The potential benefits are enormous, including reduced impacts on environmental quality, reduced dependency on petroleum imports, and the advancement of fuel cell technology into other application areas.

Phase I of the Green Car Project has been completed and Phase II is ready to begin. Its primary objective will be to improve the fuel cell stacks and all support systems. The final goal is to promote PEM fuel cell technology and accelerate its commercialization, making it accessible to all applications requiring a clean and efficient source of energy.

Acknowledgments

This project has been made possible by generosity and visionariness of Mr. John H. Perry, Jr., the Chairman of the Board of Energy Partners. A support by South Coast Air Quality Management District, Los Angeles, California, is greatly appreciated. The authors would also like to thank all the employees of Energy Partners who contributed to the success of this project.

TABLES

TABLE 1 FUEL CELL STACKS SPECIFICATIONS

Manufacturer	Energy Partners
Type	PEM
Membrane	Nafion™ 115
Collector Material	Graphite with binder
Reactants	Hydrogen/Air
Rated Power Output	21 kW
Voltage @ Rated Output	126 Volts
Current Density @ Rated Output	200 Amps/ft^2
Cell Voltage @ Rated Output	0.7 Volts
Stack Efficiency @ Rated Output	54 % (LHV)
Active Area	0.84 sq. ft.
Average Operating Temperature	125 degrees F
Reactant Pressure	20 psig
Cell active area	0.84 ft^2
Number of Cells	60 (per stack)
Number of Stacks	3
Stack Dimensions	14 x 14 x 24 inches
Stack Weight	230 lbs each (690 lbs total)

TABLE 2 GREEN CAR PERFORMANCE

PARAMETER	PERFORMANCE
Average cruising speed	30 mph
Electric power required for cruising speed	10 kW
Vehicle maximum speed	60 mph
Acceleration 0 to 30 mph	10 sec
Fuel consumption at cruising speed	8 scf/mile
Vehicle range at constant speed	70 miles
Vehicle range for city driving cycle	60 miles
Refueling time*	5-10 min
Start-up/shut-down time	15 sec / <1 sec
Emissions (gaseous)	11% O_2, 84% N_2, and 5% H_2O
Emissions (liquid)	H_2O: 0.15 lit/mile
Noise	85 dB at 1 meter
Safety	no accidents or any hazardous situations

* this is not a vehicle feature, but rather a function of available fuel pressure and refueling equipment

FIGURES

Figure 1. Energy Partners' Green Car

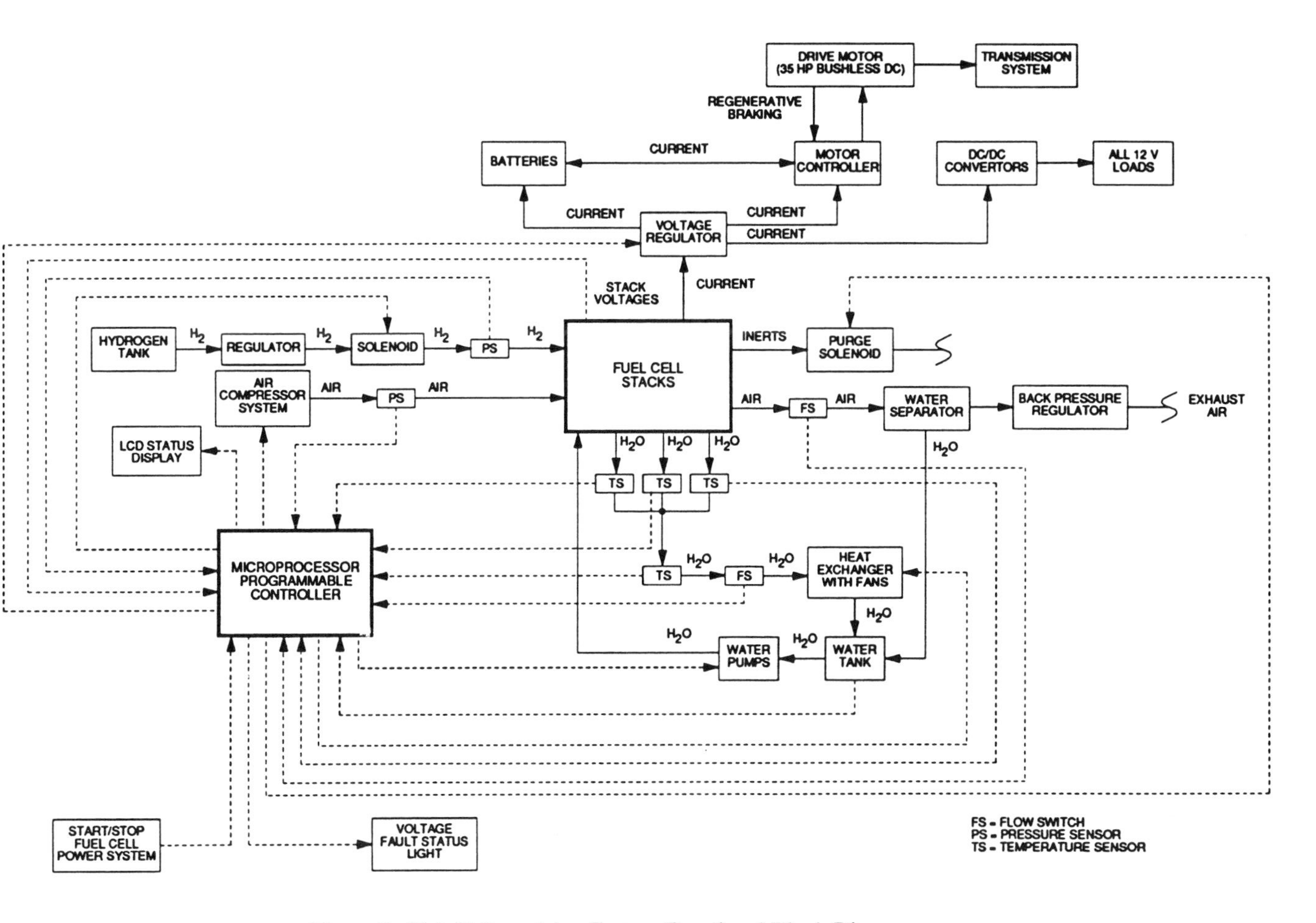

Figure 2. Hybrid Propulsion System Functional Block Diagram

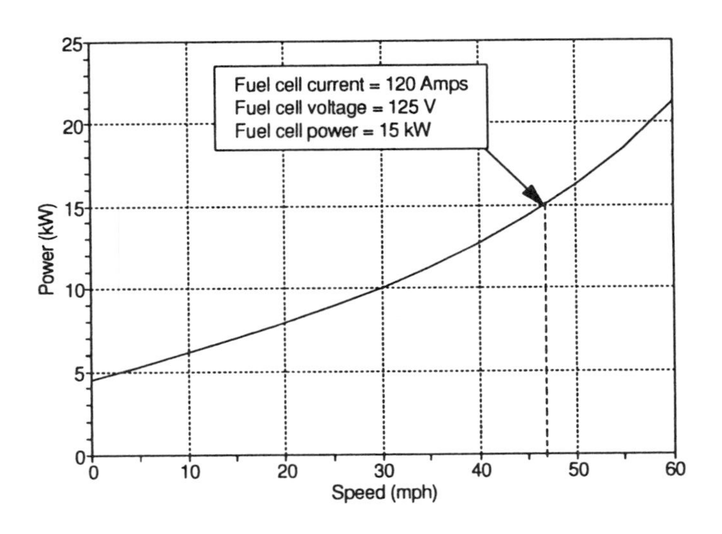

Figure 3. Fuel Cell Power Requirement for the Green Car Travelling at Constant Speed

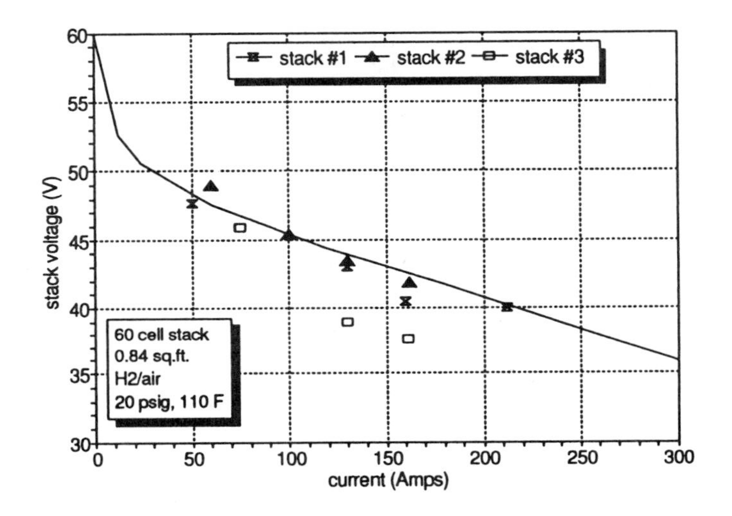

Figure 4. EP Green Car Fuel Cells Performance
(solid line = performance goal, markers = actual test data)

Preliminary Evaluation of UC Riverside's Hydrogen Powered Truck

James W. Heffel and Joseph M. Norbeck
University of California, Riverside
Bourns College of Engineering

Abstract

The South Coast Air Quality Management District (SCAQMD) sponsored CE-CERT to review and evaluate hydrogen as a fuel for transportation, develop and evaluate a state-of-the-art hydrogen fuel vehicle, and install, operate, and evaluate a solar hydrogen production and vehicle refueling station. As part of the vehicle demonstration project, a Ford Ranger truck was converted to run on hydrogen.

This project demonstrated the lean burn operation of a hydrogen fuel internal combustion engine using a Constant Volume Injection (CVI) system with closed-loop control and exhaust oxygen feedback. Preliminary tests have indicated that the vehicle is capable of meeting a 100 mile range and has operated without any pre-ignition problems during both chassis dynamometer and on-road testing. The vehicle exhibits similar performance characteristics, at speeds above 40 mph, as that of the gasoline version, but needs further improvements at the lower speeds.

1. INTRODUCTION

Over the past two decades, federal, state, and local institutions have been actively cooperating to develop the technologies to reduce air pollution. CE-CERT and the SCAQMD are currently working together to research, develop, and demonstrate the latest hydrogen vehicle and production technologies in an effort to evaluate hydrogen as a future fuel for surface transportation. As part of the implementation phase of the hydrogen program at CE-CERT, a project was conducted to review and evaluate hydrogen as a fuel for transportation; develop and evaluate a state-of-the-art hydrogen fuel vehicle; and install, operate, and evaluate a solar hydrogen production and vehicle refueling station.

For the vehicle demonstration project, a Ford Ranger truck was converted to operate on hydrogen. The three principal objectives of this project were to (1) demonstrate the lean burn operation of an internal combustion (IC) engine using hydrogen fuel (with no pre-ignition problems), (2) achieve a range of at least 100 miles, and (3) verify the vehicle exhibited similar driving characteristics as the gasoline version of the truck. This paper summarizes the preliminary results and evaluation of the hydrogen fuel truck.

2. BACKGROUND

Research on using hydrogen in an IC engine has been conducted for more than 100 years. Reverend Cecil was the first to utilize hydrogen in the 1820s[1]. Otto used a synthetic gas for fuel in the 1870s which had a hydrogen content of over 50% [2]. The idea of using hydrogen fuel was extended in the 1930s by Rudolf Erren, an engineer who worked in England and Germany. Erren reportedly converted over 1000 vehicles to run on hydrogen or hydrogen/gasoline[3]. During WWII a German scientist named Oehmichen reported efficiencies of over 50% from an engine running solely on hydrogen[4]. Although interest in hydrogen fuel waned immediately following the end of World II, some research into hydrogen vehicles continued, most notably that of R. Q. King in Canada. Then, in the 1970s there was a resurgence in research into the possibilities of hydrogen fueled transportation when programs were initiated in Japan, West Germany, and the United States. Today, research is being conducted by auto manufacturers such as Madza, BMW, Mercedes Benz, and small companies and research organizations.

The primary problem that has been encountered in developing an operational hydrogen vehicle is premature ignition. If the premature ignition occurs near the fuel intake valve and the resultant flame travels back into the induction system, a backfire condition exists and can lead to the total destruction of the fuel system. Since pre-ignition has been a major obstacle to the development of hydrogen vehicles, there has been a considerable amount of research dedicated to determining the causes of, and finding the solutions for, this problem.

Much of the progress made toward eliminating pre-ignition in hydrogen systems has been through the development of more advanced fuel delivery systems. One of the most common techniques of delivering hydrogen to the engine is inlet port injection. With inlet port injection, the fuel and air are injected into the engine from separate ports into the combustion chamber during the intake stroke and, as such, are not premixed in the intake manifold. In the 1980s, Lynch, et al. developed a technique similar to that of inlet port injection which they called parallel induction[5]. The engines used in the tests of Lynch and co-workers also incorporated turbochargers with after-cooling to increase the power output of the engine to a level more comparable to that of a standard gasoline engine. The parallel induction scheme is the precursor to the Constant Volume Injection (CVI) system that is utilized on the hydrogen truck discussed here.

3. VEHICLE AND ENGINE MODIFICATIONS

The stock Ford Ranger truck used in this project was delivered to Advanced Machine Dynamic (AMD) in Highland, California in October of 1992, where the vehicle conversion was performed. In parallel with truck modifications, the CVI was developed at Hydrogen Consultants Inc. (HCI) in Littleton, Colorado.

3.1 Vehicle Description Prior to Conversion

The vehicle used for this project was a new 1992 Ford Ranger truck. The vehicle came with a four speed automatic transmission, a 4.11:1 rear differential, a 2.3 liter four cylinder engine and power steering (no air conditioning).

The specifications of the stock engine with this vehicle are as follows:

- Naturally Aspirated
- 9:1 Compression Ratio
- Dual Spark Plug Distributorless Ignition
- Multiport Electronic Fuel Injection
- Exhaust Gas Oxygen Sensor (EGO)
- Exhaust Gas Recirculation (EGR)
- Three-way Catalyst

3.2 Engine Modifications

Before pulling the engine from the vehicle, all the gasoline related components were removed from the engine so that the new retrofit hardware (turbocharger, intercooler, exhaust and air induction system, etc.) could be sized for the engine compartment. The engine was then removed from the vehicle and the following modifications performed:

1. Irregularities (potential "hot spots") in the intake and exhaust ports and combustion chamber were removed. These modifications increased the flow capacity of the heads by 30% and also reduced the stock compression ratio of the engine from 9:1 to 8:1.

2. An Engle TCS210H camshaft with a 0.495" lift and 214° duration was installed.

3. The stock dual spark plug distributorless ignition system was replaced with an Engine Electronics Compu-fire distributorless ignition system with fixed timing.

4. Champion G-81 spark plugs (10 mm) and shroud chambers in the spark plug ports on the intake side of the cylinder head were installed (see Figure 1). The shroud chambers were used to provide assurance that the spark plug would not cause ignition due to a hot electrode.

5. The spark plugs on the exhaust side of the cylinder head were removed and replaced with "blank" plugs. The dual spark plugs (per cylinder) on the gasoline version of the engine were deemed unnecessary for hydrogen operation.

6. A special high volume crankcase ventilation system was installed to purge unburned air-fuel mixtures from the crankcase.

7. A pressure relief valve on the rocker arm cover was installed to prevent engine damage in the event of ignition of gases in the crankcase.

8. A Turbonetics Model T-2 turbocharger with 0.48 a/r ratio turbine and integral wastegate (set at 12 psi) was installed.

9. An air-to-air intercooler was installed.

10. A HCI Constant Volume Injection (CVI) system for delivering timed aliquots of hydrogen to the ports in phase with the intake stroke of the engine (i.e., sequential multiport injection) was installed[6].

11. An exhaust gas oxygen sensor was installed.

12. A new Engine Control Module (ECM) design by HCI (the original ECM remained in the vehicle, but was only used to provide control of the automatic transmission) was installed.

13. A new air induction and exhaust system was installed.

3.3 Vehicle Modifications

Modifications to the truck include the following:

1. The removal of the gasoline tank.

2. The installation of three hydrogen storage vessels in the bed of truck. Storage vessels were provided by Comdyne industries and have the following specifications:

   a. Aluminum liner with a full fiberglass wrapping.
   b. DOT approved for CNG (DOT exemption for use with hydrogen).
   c. Maximum working pressure of 3,600 psi.
   d. Each vessel has its own thermal/pressure relief valve and isolation valve.

e. Two of the three vessels have a water volume of 4.8 cubic feet for a hydrogen storage capacity of 940 scf (4.8 lbs. of hydrogen at 3,600 psi).
f. The third vessel has water volume of 1.6 cubic feet for a hydrogen storage capacity of 313 scf (1.6 lbs. of hydrogen at 3,600 psi).
g. Total hydrogen storage capacity of 2193 scf (11.2 lbs. of hydrogen at 3,600 psi). At total water volume of 11.2 cubic feet).
h. Total energy storage capacity of 616,00 Btu's (lower heating value) or a gasoline equivalent of approximately 5.1 gallons.

3. The installation of the hydrogen ancillary system.

4. The rear differential gear ratio was changed from 4.11:1 to 4.56:1.

5. The stall speed of the torque converter was increased from 600 rpm to 1400 rpm.

3.4 Additional Modifications

One of the most important goals of the program was to test and demonstrate the vehicles driveability. To this end, the vehicle was entered in the Speed Week at the Bonneville Salt Flats, Utah. Although, there is not currently a class for hydrogen fueled vehicles at Bonneville, in the past a few hydrogen vehicles have run in a "time only" category. We believe that as more hydrogen vehicles enter this event, a hydrogen fuel class will eventually be added (They have just recently added a category for electric vehicles). In order to participate at Bonneville, a number of safety modifications were incorporated into the vehicle. In addition, some minor modifications were made to improve the vehicle's aerodynamic profile. The following modifications were made in preparation for Bonneville:

1. A roll bar was installed.

2. A fire suppression system was installed in the cab and engine compartment.

3. A driveshaft hoop was installed.

4. A bucket seat with a five point safety harness was installed.

5. The vehicle was lowered three inches.

6. Air dams and a tonneau cover were installed.

Unfortunately, due to heavy rains the event had to be canceled.

4. INSTRUMENTATION

After the vehicle modifications were completed, instrumentation was installed to measure the following parameters:

1. Manifold air temperature and pressure

2. Intercooler air inlet and outlet temperature

3. Engine speed

4. Exhaust temperature

5. Throttle position

6. Exhaust Gas Oxygen Sensor (EGO) signal

7. Hydrogen storage pressure

8. Hydrogen inlet pressure

9. Air flow rate

The possibility of adding the following items are currently being investigated:

1. In-cylinder combustion pressures

2. Hydrogen flow rate

3. Hydrogen storage temperature

4. Ignition timing

5. PRELIMINARY VEHICLE EVALUATION

Following the completion of the vehicle and engine conversion in August 1994, the truck was evaluated at CE-CERT. The results of this evaluation are described below.

5.1 Baseline Emission Testing

Following the completion of the vehicle and engine modifications, and prior to any attempts to optimize engine operation, a US 75 Federal Test Procedure (FTP) emissions test was performed at the Automobile Club of Southern California's test facility in Los Angeles, California. The results of this baseline test were has follows:

CO 0.0 g/mile (meets CARB's ZEV requirements)
CO_2 3.03 g/mile (currently not regulated by CARB)
HC 0.0 g/mile (meets CARB's ZEV requirements)
NO_x 0.37 g/mile (exceeds CARB's LEV requirements)

Again, these values were for an "un-optimized" vehicle, without a three-way catalytic converter.

5.2 Fuel Economy

Fuel economy calculations for US 75 FTP are based on a carbon count of the exhaust. Because only a very small amount carbon is present in the exhaust of the vehicle (from oil seeping past the rings), and the computer program does not adjust for the fact that the vehicle was burning hydrogen instead of gasoline, the fuel economy was calculated to be 7915.74 mpg (unfortunately this is not the correct fuel economy).

Using the procedure for the EPA city/highway fuel economy test as a guideline, a procedure for evaluating the fuel economy of the hydrogen fuel vehicle was developed. This test has not been conducted as of yet, but a simple fuel economy test, based on a 67 mile round trip from CE-CERT to SCAQMD, indicates that the truck averages a little over 20 miles per equivalent gallon of gasoline. Because the on-board hydrogen storage capacity of the vehicle is equivalent to 5.1 gallons of gasoline, the vehicle range is expected to be approximately 102 miles.

5.3 Engine Starting, Idling, and Shutdown

There is a slight delay, as expected, in engine start-up as a result of a delay programmed into the Engine Control Module (ECM) to restrain firing the spark plugs until the engine has cranked over a few revolutions. This delay allows any residual gas that may have accumulated in the combustion chambers to be vented. After a few turns of the engine, the ECM commands the ignition system to fire the spark plugs and engine start-up occurs immediately thereafter. During the first 20 to 30 seconds of engine idle, the engine operates in an open-loop (high idle) configuration, thus allowing the Exhaust Gas Oxygen (EGO) sensor to warm up. After the EGO sensor warms up, the control system returns to a closed-loop configuration (normal idle). For engine shutdown, the ECM sends a signal which closes the fuel inlet solenoid valve, but continues to fire the spark plugs until the engine runs out of fuel, at which point the engine stops. This occurs with in a second of turning off the ignition. This procedure is used to ensure that no hydrogen is left in the combustion chamber. All these systems perform as designed.

5.4 Driveability

To compensate for the low volumetric energy density of hydrogen, a turbocharger was added to the engine. Unfortunately, the turbocharger provided little benefit at low engine speeds. As a result, the vehicle's driveability (acceleration) in the city suffers significantly. Highway driveability, however , is very comparable to the gasoline version of the truck since the engine can utilize the assistance of the turbo boost.

5.5 Chassis Dynamometer Testing

Laboratory dynamometer testing was conducted using a Clayton 200 horsepower water brake chassis dynamometer at CE-CERT. The resulting plot of power output versus engine speed is shown in Figure 2. Also included on this plot is the temperature of the air as it enters and exits the intercooler, the exhaust gas temperature, manifold air pressure, and the power output of a gasoline version Ford Ranger truck. From this plot it can be seen that turbocharger, under full boost, increases the air temperature by about 200°F and the intercooler effectively reduces the temperature back to ambient temperatures. The lower power output between the hydrogen engine and the gasoline engine can also be seen.

6. ANOMALIES

6.1 Turbo Boost Bypass Valve

There are two methods typically used to install a turbocharger on an engine. These two methods are called the "draw through" configuration or the "blow through" configuration. In the "draw through" configuration, the turbocharger is installed between the intake manifold and the fuel metering system. In this configuration the fuel metering system is never exposed to pressures above ambient. In the "blow through" configuration, however, the fuel metering system is installed between the turbocharger and the intake manifold, therefore subjected to the turbo boost pressure, which is above ambient. This configuration is used on the Ford Ranger truck. For this type of configuration, a device called a turbo boost bypass valve is required between the outlet of the turbocharger and the butterfly valve of the air/fuel metering system. The purpose of this device is to relieve (vent) the boost pressure in the line when the butterfly is closed. An Audi 034145710A Turbo Boost Bypass Valve was applied to the UC Riverside truck. On October 31, 1994, this device failed in the open position (typical mode of failure), causing the boost pressure to be vented prior to its entering the engine. Failure of this device did not inflict any damage on the engine, but did cause a reduction in power due to the lost in turbo boost. After repairing this bypass valve, it failed once more on November 15, 1994 and was replaced with a Turbonetics 30270. The main difference between these two bypass valves is that the first one had a plastic housing (that kept cracking) while the second utilizes an aluminum housing. So far we have not had a problem with the new device.

6.2 Noisy Regulator

The pressure regulator, located between the hydrogen storage vessels and Buzmatic regulator, is used to reduce the high pressure hydrogen from the gas storage cylinder to approximately 100 psi. This regulator is a diaphragm type and it produces an audible howling noise during medium to hard acceleration. This regulator has been flow tested in a laboratory to verify its operation. During these bench tests, no noise was noticed. Thus, the howling sound detected when it is in the vehicle is probably due to the feedback from the Buzmatic regulator which is located in close proximity to the regulator's outlet. To test this theory, the regulator will be installed further from the Buzmatic regulator and as close as possible to the storage vessels.

6.3 Water on the Spark Plugs

On two occasions the vehicle failed to start as a result of water collecting on the spark plugs. It was determined that the water accumulation was the outcome of shutting the vehicle off shortly after it was started and allowing it to sit. It is believed that because the engine did not have a chance to warm up, the combustion product (steam) rapidly condensed, forming water deposits within the combustion chamber. Although this does not occur every time, we have recently implemented a standard operating procedure to allow a cold engine to warm up (run 3 to 5 minutes) before shutting it off.

7. CONCLUSIONS

The lean burn operation of a hydrogen fuel internal combustion engine using a Constant Volume Injection (CVI) system with closed-loop control and exhaust oxygen feedback has been demonstrated. Preliminary tests indicate that the vehicle has a greater than 100 mile range between refueling stops, and that it can be operated without any pre-ignition problems. The vehicle exhibits similar performance characteristics to that of the gasoline version at speeds above 40 mph, but performance suffers at lower speeds.

The two main reasons for the poor low-end performance can be attributed to the following:

1. *The physical properties of hydrogen.* Hydrogen has a lower volumetric energy density than gasoline and because it is injected as a gas as opposed to a liquid, it displaces approximately one third of the air in the combustion chamber at a stoichiometric air/fuel ratio (gasoline only displaces one to two percent of the air). This problem was further compounded by operating the engine under lean burn conditions (excess air). As a result, an IC engine using hydrogen fuel will inherently produce 20 to 50 percent (depending on the air/fuel ratio) less power than the same size engine using gasoline.

2. *Insufficient turbo boost at low engine speeds.* To compensate for the deficiencies mentioned in (1) above, a turbocharger was used to increase the amount of air (and thus fuel) that could be inducted into the engine. The turbocharger in this case effectively increases the displacement of the engine by more than 80 percent at full boost. This modification worked quite successfully at higher engine speeds (2,000 rpm and above), but not very well at lower engine speed (from idle to 2,000 rpm). A total of five different

turbochargers were tried in an attempt to increase the turbo boost at lower engine speeds. The engine currently is using the turbocharger that gave the best low-end performance (boost at 2,000 rpm). The poor performance of the turbocharger is primarily a result of the density and temperature of the gas being pumped through the turbocharger as too low to provide for adequate operation. This is a particular problem when the engine is being run under lean conditions which leads to lower exhaust gas temperatures and densities. An additional contribution to the problem is turbo lag.

A number other things can be done to improve the turbocharging performance, such as utilizing dual turbochargers (where a small turbocharger is used for low engine speeds and a second, larger turbocharger is used for higher engine speeds) or using an electric motor to assist in boosting the turbo. Another option is to utilize a supercharger instead of a turbocharger. This approach is currently being implemented in the next generation hydrogen truck (S/N 002) being built for Xerox/Clean Air Now and will be compared with the performance of CE-CERT's hydrogen truck (S/N 001). A third option, which is currently being investigated by CE-CERT, is to not use a turbocharger or supercharger, but just increase the displacement of the engine. Preliminary analysis has indicated that to get similar performance from a hydrogen engine operating at an equivalence ratio of approximately 0.4, an engine with a displacement approximately twice that of a gasoline engine will be needed. For our truck, this would translate to a displacement of approximately 4.6 to 5 liters (a common displacement for V-8 engines).

8. RECOMMENDATIONS FOR FURTHER RESEARCH

1. Determine the vehicle's city and highway fuel economy using a modified EPA MPG test procedure.

2. Install a hydrogen flow meter. Use the output from this device, in conjunction with output of the air flow meter currently installed on the vehicle, to determine a "real time" A/F ratio.

3. Investigate methods to reduce the chances of water accumulating on the spark plugs.

4. Modify (or replace) the current ignition system so that the timing can be advanced during engine operation. Develop a timing map for the engine under various operating conditions (i.e., A/F ratios, engine speed, load, etc.)

5. Relocate the pressure reduction regulator, mentioned in Section 6.2, from the engine compartment to the bed of the truck where the hydrogen storage vessels are located.

6. Install a high pressure, electrically actuated solenoid valve at the outlet of the hydrogen storage tanks manifold. This valve will be used as a redundant shutoff valve and will be configured to close when it loses power (fail/safe position).

7. Install a thermocouple (TC) in the one of the hydrogen storage vessels. This TC will be used to measure the hydrogen temperature increase during refueling. In addition, the signal from the TC will also be used in conjunction with the output signal of the hydrogen storage pressure transducer to determine the mass of hydrogen on-board the vehicle (i.e., fuel level gauge).

8. Incrementally increase the compression ratio of the engine from 8:1 to the highest compression ratio prior to on-set of knock.

9. ACKNOWLEDGMENTS

CE-CERT would like to thank the South Coast Air Quality Management District for sponsoring this project, and its partners Hydrogen Consultants, Incorporated, and Advanced Machine Dynamic for achieving significant results. CE-CERT would also like to thank the following organizations for their help in the project: Praxair, The Electrolyser Corporation, Burke-Porter, Bourns Incorporated, Comdyne, Advantage Suspension Works, and the Automobile Club of Southern California.

10. REFERENCES

1. Cecil, W., "On the application of hydrogen gas to produce a moving power in machinery; with a description of an engine which is moved by the pressure of the atmosphere, upon a vacuum caused by explosions of hydrogen and atmospheric air", *Trans. Cambridge Philos. Soc.*, 1, p. 217, 1882.

2. Van Vorst, W.D. and Woolley, R.L. "Hydrogen-fueled surface transportation", in Hydrogen: Its Technology and Implications, Edited by K.E. Cox and D.K. Williamson, Jr., CRC Press, Boca Raton, Fla., 1979.

3. Hoffmann, P., The forever fuel: The story of hydrogen, Westview Press, Boulder, Colorado, 1981.

4. Oehmichen, M., "Wasserstoff als Motorteib-mittel", in Verein Deutsche Ingenieur, Deutsche Kraftfahrtforschung, Edited by Verlag GMBH, Berlin, 1942.

5. Lynch, F.E., "Parallel induction: A simple fuel control method for hydrogen engines", *Int. J. Hydrogen Energy*. 8, p. 721, 1983.

6. Lynch, F.E., Kaiser, W.J., and Fulton J., "UCR hydrogen fueled pickup truck with Constant Volume Injection (CVI)" Final Report., July 28, 1994.

Figure 1. Spark Plug and Shroud Chamber

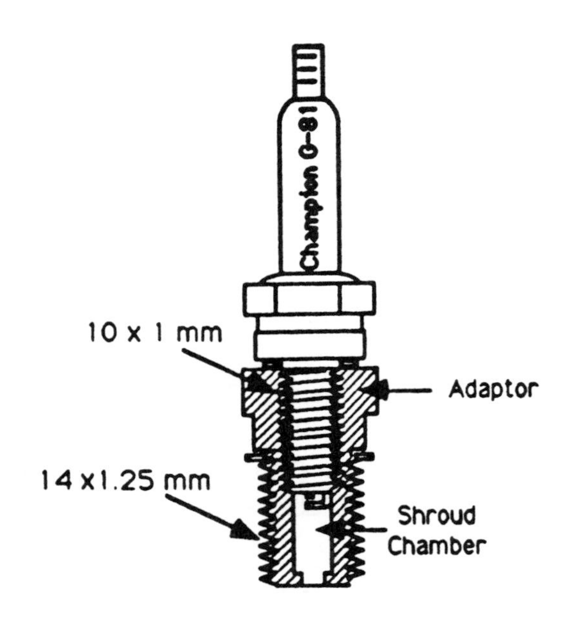

Figure 2. Graph of Chassis Dynamometer Test Results

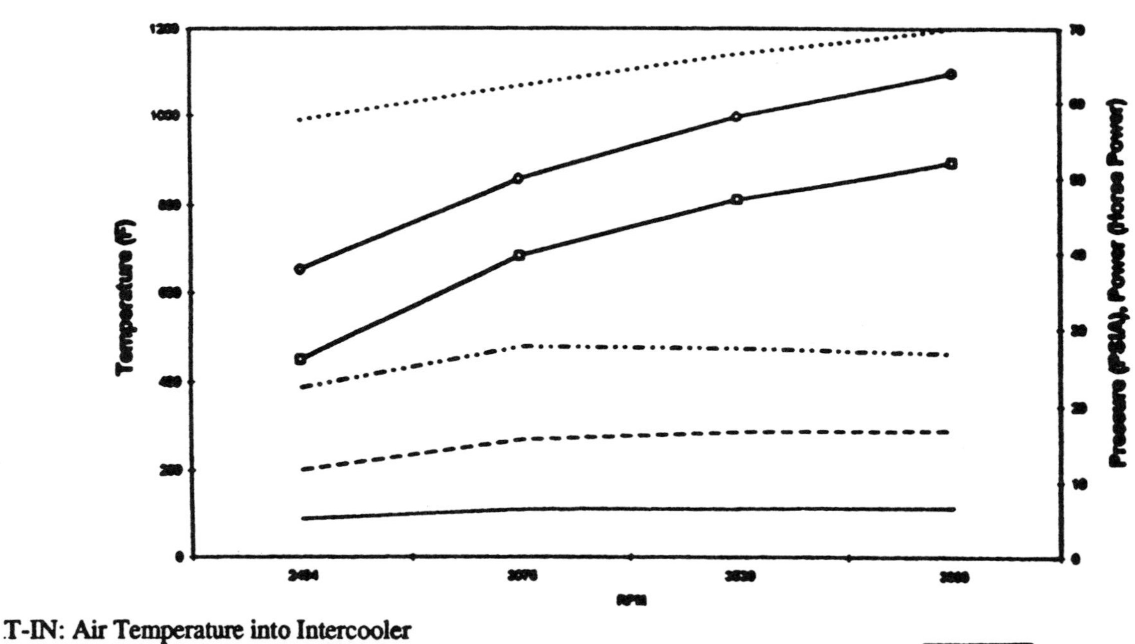

T-IN: Air Temperature into Intercooler
T-OUT: Air Temperature Out of Intercooler
T-EXH: Exhaust Temperature
MAP: Manifold Air Pressure
HP: Power Output for H2 Truck
SR-HP*: Power Output for Stock Ranger
*Data taken from a gasoline powered 1992 Ford Ranger with a 2.3L engine

Solid Polymer Fuel Cell Developments at Ballard

Keith B. Prater
Ballard Power Systems, Inc.

Abstract

Over the last several years, Ballard Power Systems has significantly advanced the state of solid polymer fuel cell (also known as PEM fuel cell) technology, having demonstrated substantial increases in power density, effective operation on air as the oxidant, and satisfactory operation on synthetic gas streams analogous to those obtained from the reforming of hydrocarbon fuels. More recently, Ballard has focussed its attention on the development of fuel cell systems and the demonstration of fuel cells in practical applications. Ballard has developed a fuel cell system which accepts pressurized hydrogen and air and produces d.c. power. The system contains the fuel cell stack, a fuel/air management sub-system, a heat management system, a water management sub-system and a control system. The unit is started by pushing a single button and delivers power within five seconds. The 3–5 kW units have been shipped to several organizations around the world for evaluation. Ballard has also begun a program to demonstrate the operation of a 30 ft transit bus powered entirely by a 105 kW fuel cell system, fueled by gaseous hydrogen. The bus is to be operational within one year. Ballard has also demonstrated the operation of a brassboard fuel cell system operating on methanol as the fuel. The system contains a reformer and an air compressor, as well as heat and water management sub-systems and an integrated control system.

Introduction

Recent advances in the performance [1] of the solid polymer fuel cell (known generically as the proton exchange membrane or PEM fuel cell) have stimulated significant interest in this technology throughout the world. Workers in the field have reported significant reductions in catalyst loading [2–4] and increased understanding of the importance of water transport in electrolyte membranes and fuel cell structures [5–8].

At Ballard Power Systems, the activities have been largely focused on translating this increased understanding of the fuel cell into useable fuel cell stacks and power systems. A further focus of these activities has been the demonstration of the potential of this technology in real applications. This paper presents an overview of recent developments in the fuel cell stack, hydrogen/air systems, and in methanol/air systems, as well as a description of a program now underway to develop and demonstrate a transit bus fully powered by solid polymer fuel cells.

Fuel cell stack developments

The performance breakthrough in this technology resulted from a combination of developments in stack hardware by Ballard [1] and the availability of a new developmental membrane from Dow Chemical [9]. The initial version of the Dow membrane was very thin and somewhat difficult to handle. Later versions of the membrane exhibit substantially improved handling properties while providing somewhat diminished performance [1].

Ballard has incorporated the latest version of the Dow membrane into a production fuel cell stack, which consists of 35 active cells and an integral gas humidification section. The dimensions of the stack are 10 in $\times$ 10 in $\times$ 18 in (25.4 cm $\times$ 25.4 cm $\times$ 45.7 cm) and the active electrode area is 36 in^2 (232 cm^2) per cell. A photograph of that stack is presented in Fig. 1. Typical performance data for the production stack operating on hydrogen and air are presented in Figs. 2 and 3. Figure 2 shows the current/voltage behaviour of the stack operating on hydrogen and air each at a pressure of 30 psig (3 atm.). At an air pressure of 50 psig (4.4 atm.) the stack will produce slightly more than 5 kW. If air is replaced by oxygen at 30 psig (3 atm.), the stack will produce over 10 kW of power. Figure 3 shows the uniformity of the performance of the 35 cells making up the stack at two different currents.

The physical properties of the Dow membrane are somewhat different from those of Nafion, the membrane electrolyte which had been the *de facto* standard for this technology. This required that the fuel cell stack be redesigned to assure proper gas sealing and to provide an operating environment conducive to long operating life time with the Dow membrane. Both of those objectives have now been achieved.

The gas permeability of the membrane/electrode assembly has been found to be a good indication of the reliability of the membrane in operation. A significant increase in the gas permeability with time suggests that the membrane is suffering some progressive damage which will ultimately result in failure. Ballard has now demonstrated stack operation in excess of 2000 h with the Dow membrane with no increase in the gas permeability over that which is characteristic of the fresh membrane. At the same

Fig. 1. A 35-cell solid polymer fuel cell stack.

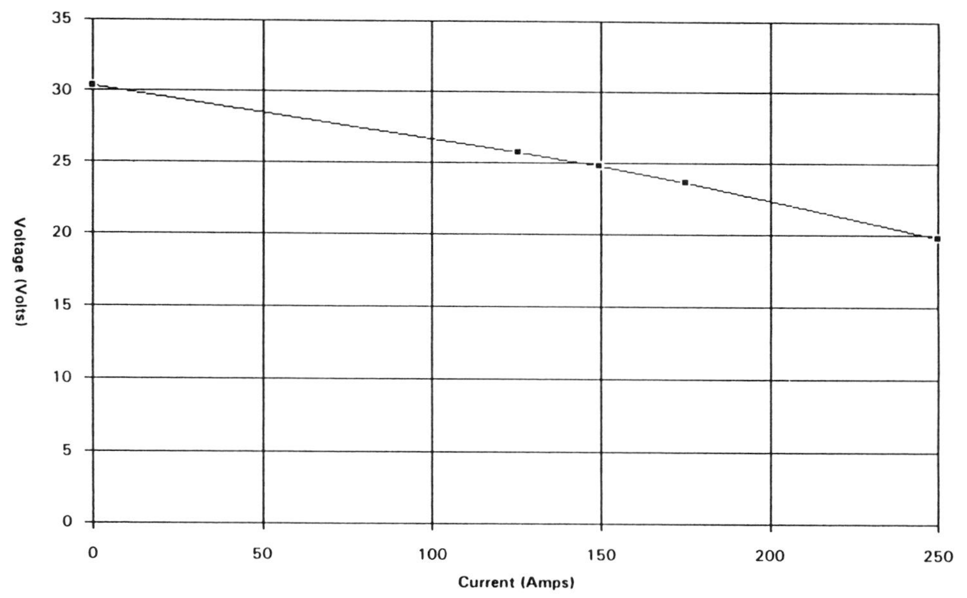

Fig. 2. Polarization curve for 35-cell stack, 232 cm^2 electrode area, H_2/air at 3 atm./3 atm. and 1.5/2.0 stoichiometry, 70 °C.

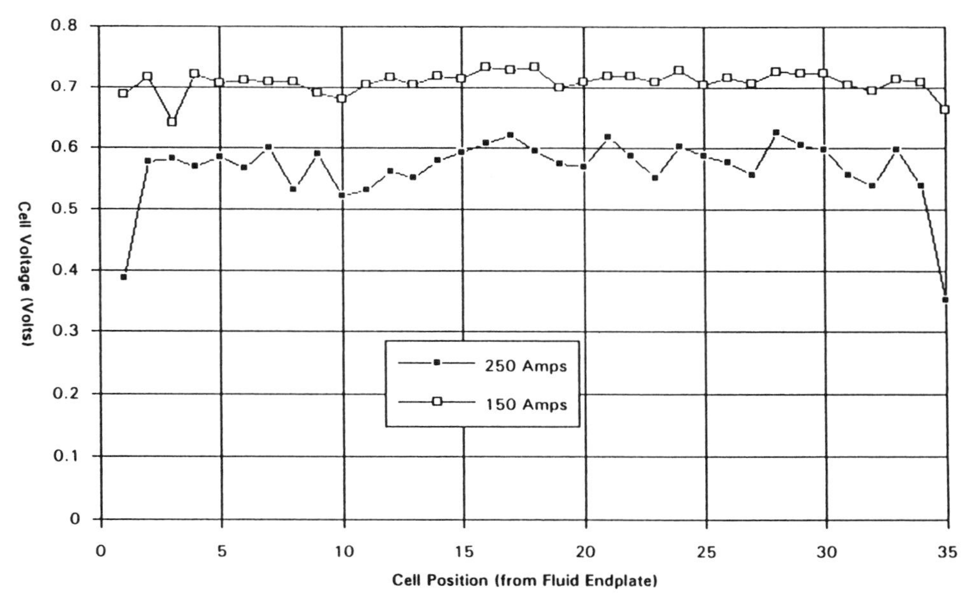

Fig. 3. Cell-to-cell consistency in 35-cell stack, 232 cm^2 electrode area, H_2/air at 3 atm./3 atm. and 1.5/2.0 stoichiometry, 70 °C.

time, Ballard has recently developed sealing technology which eliminates external gas leakage to a detection limit of 1 cc/min for the entire stack.

As part of its bus development program, which will be discussed below, Ballard has recently subjected one of its fuel cell stacks to the vibration protocol specified by MIL-STD-810E. This test simulated the vibration associated with operating the fuel cell in a city bus for 5000 miles (8000 km). The 35-cell production stack was vibrated in the longitudinal and vertical axes while the fuel cell was in operation. No change in fuel cell operation, internal gas transfer or external gas leakage was observed during these tests.

Hydrogen/air power system

In order to facilitate the testing of the fuel cell stack, Ballard has developed an integrated fuel cell system, based upon the 35-cell production stack. The system operates on pressurized hydrogen and oxidant (either air or oxygen) and includes sub-systems for gas control, product water collection, stack cooling, and system control. The parasitic electrical load due to these sub-systems is about 250 W.

A single push-button causes the control system to start the power system by opening the gas control valves, starting the coolant circulation and hydrogen recirculation pumps and monitoring the stack temperature and voltage. Within five seconds, the system connects the fuel cell power system to the external load. The control system automatically shuts the unit down when the operator touches the 'off' button or in the event of system overload or malfunction. The unit is shown in an exposed top view in Fig. 4.

While this unit was designed for room temperature operation, it was recently used to test stack operation at more extreme temperatures, as part of the bus

Fig. 4. Hydrogen/air system, exposed top view.

demonstration program. The system was cold soaked overnight at 3 °C in an environmental chamber. The system started immediately and by the time full load had been applied, 45 s after start-up, the system produced 50% of its rated power. In the 3 °C environment, it took the system about 15 min to achieve full operating temperature and power. That time could certainly have been shortened had the system been designed for low temperature operation. The system was subsequently operated in a 40 °C environment, as well.

Methanol/air fuel cell system

Ballard has now demonstrated a complete methanol/air laboratory brassboard fuel cell system. This system, which is largely assembled from components purchased from various suppliers, consists of a methanol reformer, a selective oxidizer to remove CO from the reformate gas stream, an air compressor, two fuel cell stacks operated in parallel, a voltage regulator, and control systems for the reformer and the fuel cell system.

To our knowledge, this is the first time a solid polymer fuel cell has been demonstrated to operate in conjunction with a methanol reformer and the first demonstration of a self-contained system demonstrating methanol in and regulated d.c. power out. The system was designed to produce a net 4 kW, but the purchased selective oxidizer was unable to sustain that power level. The purchased selective oxidizer was replaced with a prototype developed by Ballard which had been designed for a 1 kW throughput. With that unit in place, the system has been run for a number of three-hour tests at the 1 kW level, with no indication of performance decay.

Ballard is presently involved in the development of an improved reformer and system. The reformer will be significantly smaller than the unit purchased for the 4 kW brassboard and the selective oxidizer will be scaled to meet the needs of the system. The progress of this program will be reported elsewhere.

Fuel cell bus demonstration program

Ballard has begun a program to develop a commercial transit bus based upon solid polymer fuel cell technology. Phase 1 of this program is funded at a level of Cdn $4.84 million by the Province of British Columbia and the Government of Canada. This phase is 30 months in duration and will be completed in Mar. 1993.

The objective of Phase 1 is the demonstration of a commercial transit bus, fully powered by solid polymer fuel cells, which provides the same performance and driver acceptance as the diesel version of the same bus. The Phase 1 bus is to be based upon existing fuel cell hardware, to be fueled by hydrogen stored on-board as the compressed gas, and is to achieve a range of at least 150 km. The specific bus performance objectives are shown in Table 1.

The direction of this program is somewhat unusual in that, although it is largely funded by governments, control of the program and the allocation of funds is in the hands of a steering committee, chaired by a representative of the local transit company which will test the bus. The committee, with representatives from Canada, the US, and the UK, includes potential users, representatives of regulatory agencies, and academics. Ballard's objective in establishing such a steering committee was to create

TABLE 1

Bus performance specifications
Requirements (all with full seated passenger load)

Gradability	30 kph on 8% grade
Acceleration	1–50 kph in 20 s accelerate to and maintain 70 kph on level road
Start	start on 20% grade
Gross weight	9752 kg (21 500 lbs)

a program which would be driven by the demands of the market, rather than by technical targets.

Input from the committee has prompted three specific program choices:

(1) to select, as the test bed, a standard commercial bus, available as a diesel and licensed for transit applications;

(2) to power the bus entirely with fuel cells;

(3) to fuel the bus with hydrogen, rather than a hydrocarbon fuel such as methanol.

The commercial bus option was chosen to permit a direct performance comparison with the diesel version, to ensure that the bus could be licensed for transit applications, and to preclude any criticism that the bus chassis had been specially designed to accommodate the fuel cell system. The decision to use only fuel cells to provide motive power, rather than to use a hybrid battery–fuel cell system, was taken to eliminate any doubt that the fuel cell was, in fact, powering the bus and to explore the technical issues with such an option.

Compressed hydrogen gas had originally been selected as the fuel for Phase 1 to simplify the systems considerations and to focus the demonstration on the operation of the fuel cells rather than on a reformer. Discussions with potential users and regulators have demonstrated a clear preference for hydrogen as the on-board fuel rather than a hydrocarbon fuel such as methanol. The regulators see this choice as the only reasonable option for eliminating CO_2 emissions, whereas operators wish to avoid the capital cost, weight, and maintenance expense associated with having a reformer on-board each bus.

A sixteen-passenger, 21 500 lb (9752 kg) bus from National Coach has been selected as the test bed and is on order. Delivery of the bus is scheduled before the end of the year. The fuel cell power system will be based upon 21 of Ballard's standard nominal 5 kW stacks. These fuel cells are now being delivered to the bus program for integration into a brassboard power plant consisting of three 7-stack strings in parallel with the necessary fuel and oxidant delivery systems, cooling system, and control system. The fuel cell power system will ultimately deliver 75 kW to the wheels.

The power system brassboard is to be assembled by the end of Oct. and will be tested until the power system is installed in the bus in mid-1992. Testing of the bus is to begin in Aug. 1992 with Phase 1 being completed in Mar. 1993.

The layout of the bus is presented in Fig. 5. Note that the fuel cells are located at the rear of the bus, while the fuel tanks are located beneath the bus floor. The batteries located along one side of the bus provide starting power, as well as power for a 12 V control system, but provide no motive power. These batteries will be recharged by the fuel cell system and are completely analogous to the SLI battery in an automobile. The power budget for the bus system is presented in Table 2.

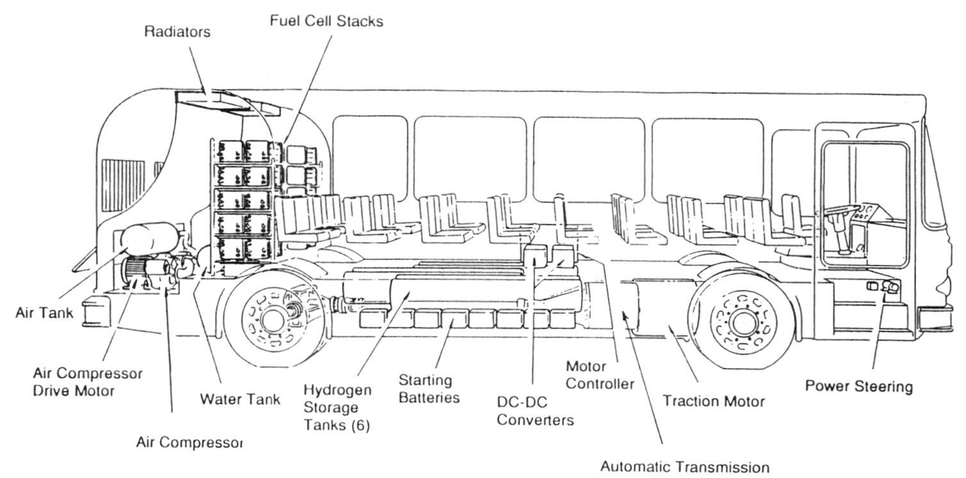

Fig. 5. Fuel cell bus system layout.

TABLE 2

Bus power budget

	Load	Duty cycle	Average	Fuel cell	12 V system
Traction	84	1	84	84	
Process air compressor	14.5	1	14.5	14.5	
Power steering and braking	1	0.5	0.5	0.5	
Lights, wipers, driver control	1.2	1	1.2		1.2
Cooling water pump	1.85	1	1.85	1.85	
Radiator fan	0.5	1	0.5		0.5
Lube oil pump	0.1	1	0.1	0.1	
Traction motor blower	0.75	1	0.75	0.75	
Control system	0.3	1	0.3		0.3
Safety system	0.02	1	0.02		0.02
Ethylene glycol pump	1	1	1	1	
			104.72	102.7	2.07

Total fuel cell power: 105.29 kW
Bus range at top speed (94 km/h) on level ground: 244 km
Bus range on simulated UMTA route: 150 km

On the assumption that the Phase 1 bus is successful, Phase 2 will involve improvements in range, based upon the incorporation of improved hydrogen storage technology and the use of more compact fuel cells and ancillaries. The size of the bus will also be increased to 40 ft.

Summary

Significant progress has been made in the development of solid polymer fuel cell stacks and systems and programs are now underway to demonstrate the potential commercial value of this technology for motive applications.

Acknowledgements

The achievements reported here are the results of the combined efforts of all of the many employees of Ballard Power Systems. The author also wishes to acknowledge the contributions of those employees of SAIC involved in the integration of the bus systems under sub-contract to Ballard.

References

1 K. B. Prater, *J. Power Sources, 29* (1990) 239–250.
2 S. Srinivasan, E. A. Ticianelli, C. R. Derouin and A. Redondo, *J. Power Sources, 22* (1988) 359–375.
3 E. A. Ticianelli, C. R. Derouin and S. Srinivasan, *J. Electroanal. Chem., 251* (1988) 275–295.
4 S. Srinivasan, D. J. Manko, H. Koch, M. A. Enayetullah and A. J. Appleby, *J. Power Sources, 29* (1990) 367–387.
5 D. M. Bernardi, *J. Electrochem. Soc., 137* (1990) 3344–3350.
6 E. A. Ticianelli, C. R. Derouin, A. Redondo and S. Srinivasan, *J. Electrochem. Soc., 135* (1988) 2209–2214.
7 P. Aldebert, F. Novel-Cattin, M. Pineri, P. Millet, C. Doumain and R. Durand, *Solid State Ionics, 35* (1989) 3–9.
8 M. W. Verbrugge and R. F. Hill, *J. Electrochem. Soc., 137* (1990) 3770–3777.
9 G. A. Eisman, *Ext. Abstr., Fuel Cell Technology and Applications Int. Seminar, The Netherlands, Oct. 1987,* p. 287ff.

HYPASSE - Hydrogen Powered Automobiles Using Seasonal and Weekly Surplus of Electricity

J. Zieger
Daimler-Benz AG, Germany

Abstract:

HYPASSE is a Swiss-German project to study and test how summertime and weekend surpluses of electricity from hydropower plants can be utilised to power busses. The surplus electricity is used for the electrolytic production of non-fossil hydrogen, which becomes a fuel for city busses. The hydrogen is stored in the bus in pressure tanks at 300 bar. These tanks are placed on the roof of the bus. The bus is powered by a hydrogen engine with internal mixture formation which is based on a standard-production diesel engine. The exhaust gas emission of that engine are about 20 % of the limits expected for the year 2000 with the same energy consumption as for a conventional engine.

1. INTRODUCTION

The automobile has been widely discussed in our industrial society. Although there are continuos improvements in fuel consumption and exhaust gas emission, the negative impacts on the environment can no longer be neglected. Above all, there is the greenhouse effect - mainly caused by the CO_2-emissions of the combustion of hydrocarbons and the immission impact (formation of smog) in urban areas - caused by the exhaust gas emissions of vehicles.

The use of hydrogen as an energy carrier would allow CO_2-free road traffic, but it goes without saying that the emissions which occur during hydrogen production must be taken into consideration. Hydrogen as a secondary energy carrier has the further advantage that well-known engine technology can be employed to a great extent. Daimler-Benz has been intensively dealing with the use of hydrogen for powering motor vehicles since 1973. The progress achieved and the experience made allowed a fleet test to be carried out by customers in Berlin between 1984 and 1988 (References 1, 2). Until today our hydrogen vehicles have covered more than 800.000 km. It was proved by all these tests that hydrogen is suitable for use as a fuel. The safety standard achieved can be compared to that of conventional vehicles.

2. HYDROGEN IN TRAFFIC

The low density of hydrogen, its low temperature of liquefaction and its broad band of ignition have an essential influence on its use in road traffic.

Currently, hydrogen can be stored on board a vehicle in a technically proven manner, either in liquid or gaseous form, or adsorped in metal hydrides. A bus for city transport has a fuel consumption of approx. 100 liters of diesel fuel for a range of 200 km. A fuel tank with a capacity of 100 l diesel fuel has a weight of about 100 kg. To reach the same range with hydrogen, storage in liquid form is most favourable (volume 560 l, weight 200 kg), but liquefaction itself takes one third of the hydrogen energy content. Gaseous storage at 300 bar (volume 2380 l, weight 1040 kg) has an advantage in weight as compared to storage in metal hydrides (volume 830 l, weight 2500 kg); however, a larger volume is required for gaseous storage.

The low ignition energy and the better ignitability in comparison with conventional hydrocarbons have a considerable influence on the combustion quality of hydrogen/air mixtures. Backfiring occurs in engine concepts in which hydrogen is injected into the intake manifold. Hot spot ignition and pre-ignition in the combustion chamber result in an irregular combustion. These problems can be solved either by adding a ballast gas, or by lean operation - both with the disadvantage of a lower power output - or by an internal mixture formation, which is technically more complex.

Based on our experience with hydrogen-powered vehicles, it becomes obvious that customers have to accept restrictions and disadvantages as compared to conventional gasoline and diesel vehicles (Reference 3). These mainly relate to range, payload, comfort, economy and refuelling times. In addition, no infrastructure for hydrogen supply is available as yet.

Therefore, it is difficult to make predictions on a future application of hydrogen in road transport. For the reasons mentioned above, hydrogen is most likely to be used in public transport (busses) and for the distribution of goods (vans), because in these vehicles the storage-related disadvantages are relatively low. The vehicles could be serviced in depots, and thus no extensive infrastructure would be necessary. Moreover, an increasing demand for low emission vehicles will arise in urban areas.

3. THE HYPASSE PROJECT

HYPASSE is an abbreviation of "hydrogen powered applications using seasonal and weekly surplus of electricity". This abbreviation illustrates that both the production of hydrogen and its intermediate storage and utilisation are studied in this project. The German part of this Swiss-German Eureka project comprises the infrastructure for the storage and refuelling of the hydrogen produced by electrolysis, whereby the electricity for this process is attained from surpluses of hydropower, as well as the construction and demonstrating operation of a hydrogen bus (figure 1). The bus will be designed for inner city operation and will be able to cover a range between 200 km and 300 km. Since electrolytically produced hydrogen will not be liquefied for reasons of energy, gaseous hydrogen will be fuelled and stored on board the bus. New, light-weight tanks which can sustain a pressure of 300 bar are envisaged, in order to increase the storage capacity. The engine is intended to have internal mixture formation, to achieve a high power to size ratio and to avoid the disadvantages of external mixture formation, as mentioned above. It is stipulated that the exhaust gas emissions of this new engine be better than the limit values for conventional diesel engines.

Implementation of the project is divided in two phases. Phase 1 lasts from 1990 to 1995. The German part of the project is sponsored by the Bundesministerium für Forschung und Technik (BMFT). During this phase all components which are necessary for the production of a bus (e. g. tanks, hydrogen supply system, engine, etc.) have been designed, tested and certified. Moreover, the concepts for the bus design, the fuelling station and the demonstrating operation are worked out and questions of safety and licensing are clarified. In phase 2 which is planned to last from 1995 to 1999, a bus will be built, tested and run in a demonstrating operation.

The results achieved in phase 1 in the fields of

- safety and licensing
- hydrogen storage tanks for a pressure of 300 bar
- hydrogen supply system for a pressure of 300 bar
- engine with internal mixture formation
- vehicle concept

will be presented in the following.

4. SAFETY AND LICENSING

The safety of hydrogen-powered vehicles is a subject that is discussed again and again (References 4, 5). For this reason we have secured the participation of TÜV Bayern Sachsen (Technical Inspection Authority, Bavaria) which has specialised in this field in the HYPASSE project. It is the task of TÜV Bayern Sachsen to carry out a safety analyses during the design phase and to assess the components and the overall system for registration in public road transport.

Both the safety study of the Büro für Technikfolgen-Abschätzung des Deutschen Bundestages "risks of a more comprehensive hydrogen usage", and the "safety study during the design phase" of TÜV Bayern Sachsen showed that "the risks for man and material can be controlled" and that "the dangers posed by hydrogen are different from those of conventional fuels, but not necessarily higher".

5. HYDROGEN STORAGE TANKS

The choice of the type of storage depends on the type of the vehicle, its field of application and the state of hydrogen (gaseous, liquid). Since hydrogen is available in gaseous form, and the engine needs a hydrogen injection pressure of at least 30 bar, a gaseous storage on board the bus is the best solution. The pressure of the tanks was increased from the usual 200 bar to 300 bar, to have a better storage density. The tanks should be made of fibre-reinforced plastics to arrive at the lowest possible weight.

Figure 2 shows the technical data of such a tank. It has a volume of 147 litters and a weight of approx. 100 kg. This tank can store 3,1 kg of hydrogen at 300 bar, which is equivalent to 11 litters of diesel. It has a 6 mm thick, gas-tight inner container made of aluminium, the so-called liner, which is wrapped with a composite of aramide fibre and epoxy resin for better stability. The aramide fibre was chosen because of its low weight combined with high tensile strength. Its high ductility results in a favourable burst behaviour: the tanks ripped up in the burst test (pressure greater than 1200 bar) instead of disintegrating into several parts. In addition, this fibre is relatively insusceptible to impact stress and it shows no contact corrosion with aluminium. The tanks were examined by TÜV Bayern Sachsen for an application in a vehicle. In addition to furthers burst tests (pressure higher than 1200 bar), burning tests, trap tests (height 1.2 m), fire tests and load cycle tests (more than 30000 cycles) were carried out. The requirements thereby defined in detail have all been fulfilled by the tanks.

6. HYDROGEN SUPPLY SYSTEM

The hydrogen supply system includes the whole gas management in the vehicle, from refuelling to the delivery of hydrogen to the engine. An electronic system which is included in a superordinate safety system monitors and controls the hydrogen flow.

The storage tanks are refuelled at 300 bar. A suitable fuelling system must be chosen, which ensures that the tanks are not fuelled at a pressure higher than 300 bar. The tanks are protected against excessive pressure increase in the case of fire by a temperature fuse which responds at approx. 100 °C and provides for pressure relief. The hydrogen engine needs an injection pressure of approx. 40 bar. A pressure reducer for all tanks together reduces the pressure of the tanks to this injection pressure. A safety device avoiding excess mass flow in the pipes allows only the maximum mass flow of 5 m^3/min which is required by the engine to flow in the supply system. In the case of larger flows the valve closes automatically. The temperature fuse with a rupture disc (45 bar) and the excess flow valve are contained in the cylinder valve. Shutoff valves, which are responsible for several tanks, are connected to the computer-controlled gas management system via pneumatically assisted solenoid valves. The safety system (gas sensors, crash sensor, etc.) acts directly on these shutoff valves.

7. ENGINE WITH INTERNAL MIXTURE FORMATION

The goal of engine development was to build a hydrogen engine with high power density and low emission on the basis of a standard-production bus engine. The OM 447 hLA, which is used in city busses, was to be taken as a basis. This is a horizontally installed 12 liter six-cylinder in-line engine with turbocharger and intercooler.

In spite of the greater technical effort the internal mixture formation was chosen to avoid the well known disadvantages of external mixture formation, such as backfiring into the intake manifold, and to achieve a higher specific power output. With internal mixture formation the hydrogen is injected directly into the combustion chamber at a pressure between 40 bar and 110 bar. A new injection system for hydrogen had to be developed in which the injection valve is hydraulically controlled. Engine-related parameters, such as power, torque and exhaust gas emissions were optimised with a single cylinder engine (Institut für Motorenbau Prof. Huber GmbH, München). With almost the same torque, the same power output and the same energy consumption the hydrogen engine shows essentially better exhaust gas emission characteristics (figure 4). The European 13-point test mode shows for the hydrogen engine only 20 % of the currently permitted limit values of nitrogen oxide, carbon monoxide and hydrocarbons. The emission of hydrocarbons and carbon monoxide were caused by the lubricating oil. The limits expected for the year 2000 can also be fulfilled without any problems (References 6, 7).

The results achieved up to now shall be transferred to the complete engine and then optimised. For this purpose a modern test bench for transient testing of hydrogen engines was built (Forschungsinstitut für Kraftfahrwesen und Fahrzeugmotoren Stuttgart) (Reference 8). With this test bench it is possible to make stationary developments, and to attain important parameters for the acceleration behaviour of a turbocharged hydrogen engine. The 'road load simulation system' allows an early determination of transmission parameters.

8. VEHICLE CONCEPT

The location of the hydrogen storage tanks on board the vehicle is a difficult and safety-critical task in the case of gaseous storage. Moreover, it is impossible in the HYPASSE project to design a new bus for reasons of costs, and therefore we have to modify a standard-production bus. This design should constitute an optimum between range (weight of the tanks), gross vehicle weight, maximum axle load, additional bus structure reinforcements and passenger capacity, at which the permissible gross vehicle weight and the maximum axle load have already been fully utilised.

For example, to cover a range of 300 km with a fully occupied bus, it is necessary to house 20 of the above-mentioned tanks in the bus. Without bus structure reinforcements this means an additional weight of more than 2000 kg. Different vehicle concepts, such as placing the tanks on the roof of the bus, in a trailer or in the interior of an articulated bus have been worked out and optimised. The concept we regard as best is shown in figure 5. It is based on an O 405 N low-floor bus, which is used for inner city regular-service transport. The storage tanks are placed on the roof of the bus. The optimum between range and gross vehicle weight is achieved at 200 km with 67 persons. For this purpose, 13 of the above-mentioned tanks have to be placed on the roof of the bus. Thereby it must be considered that the three emergency exits remain free. Including all additional measures there is in this case an additional weight of approx. 2500 kg. The range of 200 km corresponds to the average daily distance covered by an inner city regular-service bus.

9. SUMMARY

The results obtained in phase 1 of the HYPASSE project up to now, show that the expected goal can be reached. The 300 bar pressure tanks are available, the exhaust gas emissions of the engine fulfill the specifications. Our further objective is to build a ready-to-drive bus in 1995, and to test it in a subsequent demonstration cycle.

The question about a future use of hydrogen in road traffic is difficult to answer. The lack of a comprehensive infrastructure and the system-inherent disadvantages with reference to costs, pay load and range permit application only in niches, where the use of conventional fuels is no longer allowed for environmental reasons. Of course, such an application will take place first in urban areas which are highly burdened by emissions, as there the missing infrastructure can be replaced by central refuelling stations.

ACKNOWLEDGEMENT

This work is supported by Daimler-Benz AG (70 %) and the Bundeministerium für Forschung und Technik, no. 0326716A (30 %). I want to thank all my colleagues and other persons involved, who have contributed to the success of this work.

REFERENCES

1. K. Feucht, G. Hölzel, W. Hurich (Daimler-Benz)
   Perspectives of Mobile Hydrogen Application; Proceedings of the 7th World Hydrogen Conference, Moscow 1988

2. Alternative Energy Sources for Road Transport: Hydrogen Drive Test
   Verlag TÜV Rheinland 1990, ISBN 3-88585-775-8

3. H.-U. Huss (Daimler-Benz)
   Wasserstoffanwendung in Stadtbussen; Wasserstoff-Energietechnik III, VDI-Berichte 912; 1992, ISPN 3-18-090912-9

4. M. Socher, Th. Rieken (TAB)
   Risiken bei einem verstärkten Wasserstoffeinsatz; Büro für Technikfolgen-Abschätzung des Deutschen Bundestages, November 1992, TAB-Arbeitsbericht Nr. 13 - Kurzfassung

5. R. Szamer (TÜV Bayern Sachsen)
   Kein Sicherheitsrisiko; Energiespektrum, Oktober 1993

6. K. Binder (Daimler-Benz), K. Prescher, W. Bauer, R. Decker (IMH)
   Hydrogen Combustion Engine with Internal Mixture Formation, ISATA-Conference, Aachen 1993

7. R. Jorach, C. Enderle (Daimler-Benz), K. Prescher, R. Decker (IMH)
   Development of a Low-NOx-Truck-Hydrogen Engine with High Specific Power Output; Proceedings of the 10th World-Hydrogen Conference, Cocoa Beach 1994

8. B. Mahr, M. Briem, A. Sterner, U. Essers (FKFS)
   Test Bench for the Examination of High-Power Hydrogen Engines, Proceedings of the 10th World Hydrogen Conference, Cocoa Beach 1994

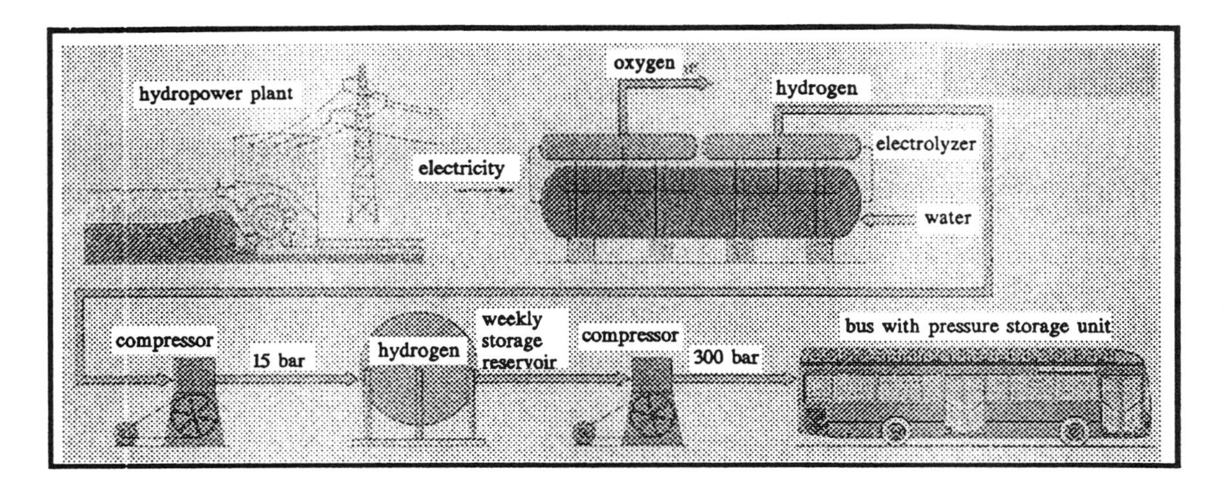

Fig. 1. Schematic Diagram of the HYPASSE Project

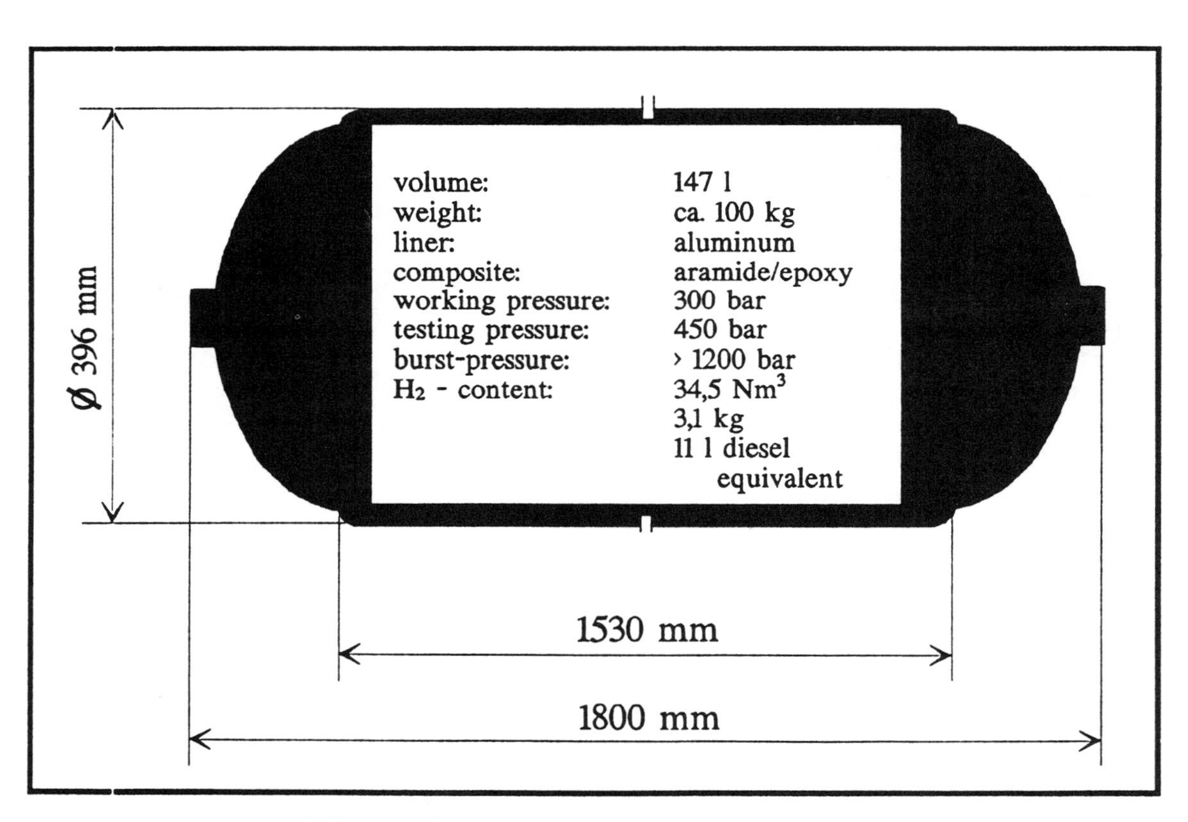

Fig. 2. 300 bar Hydrogen Storage Tank

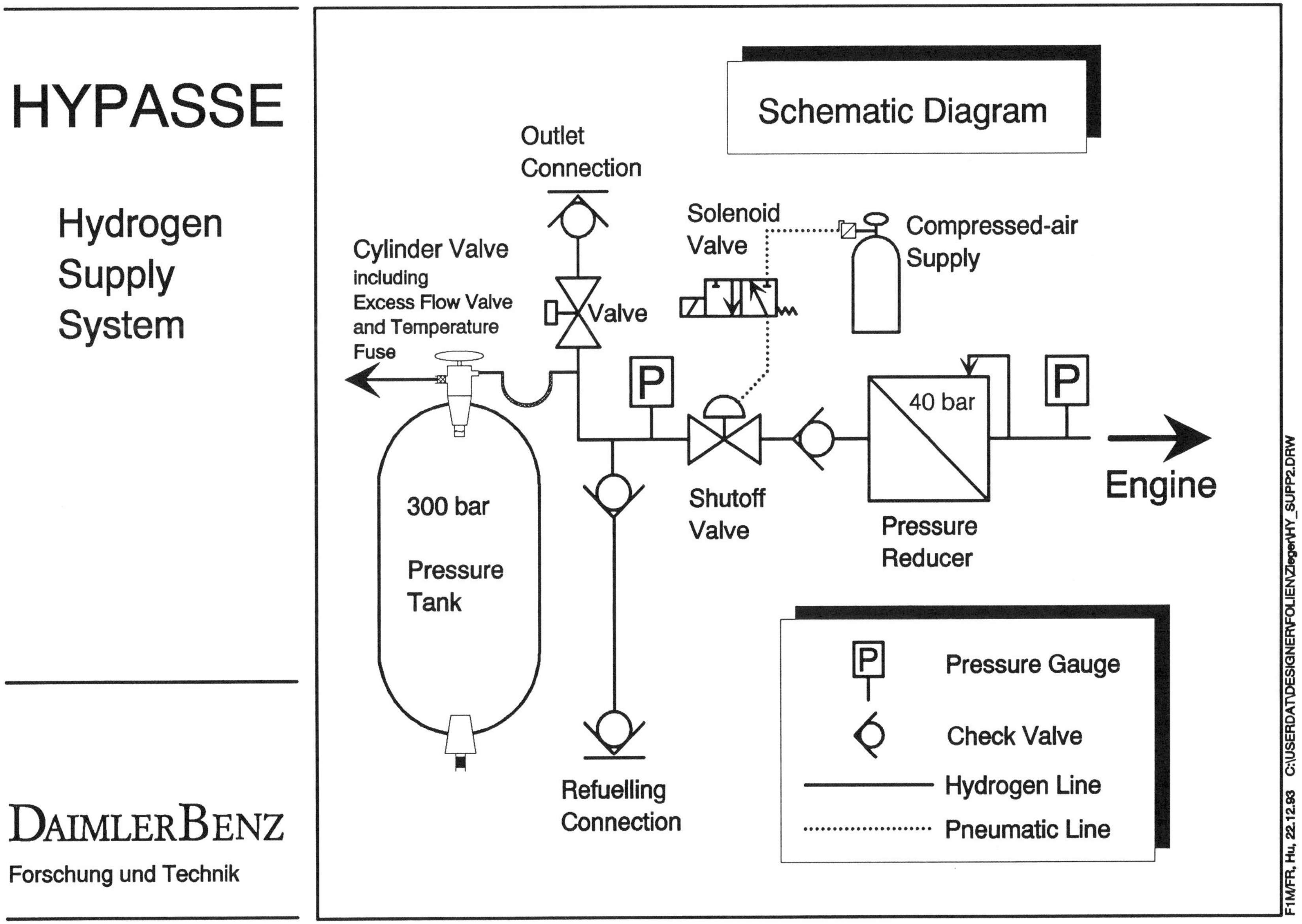

Fig. 3. Hydrogen Supply System

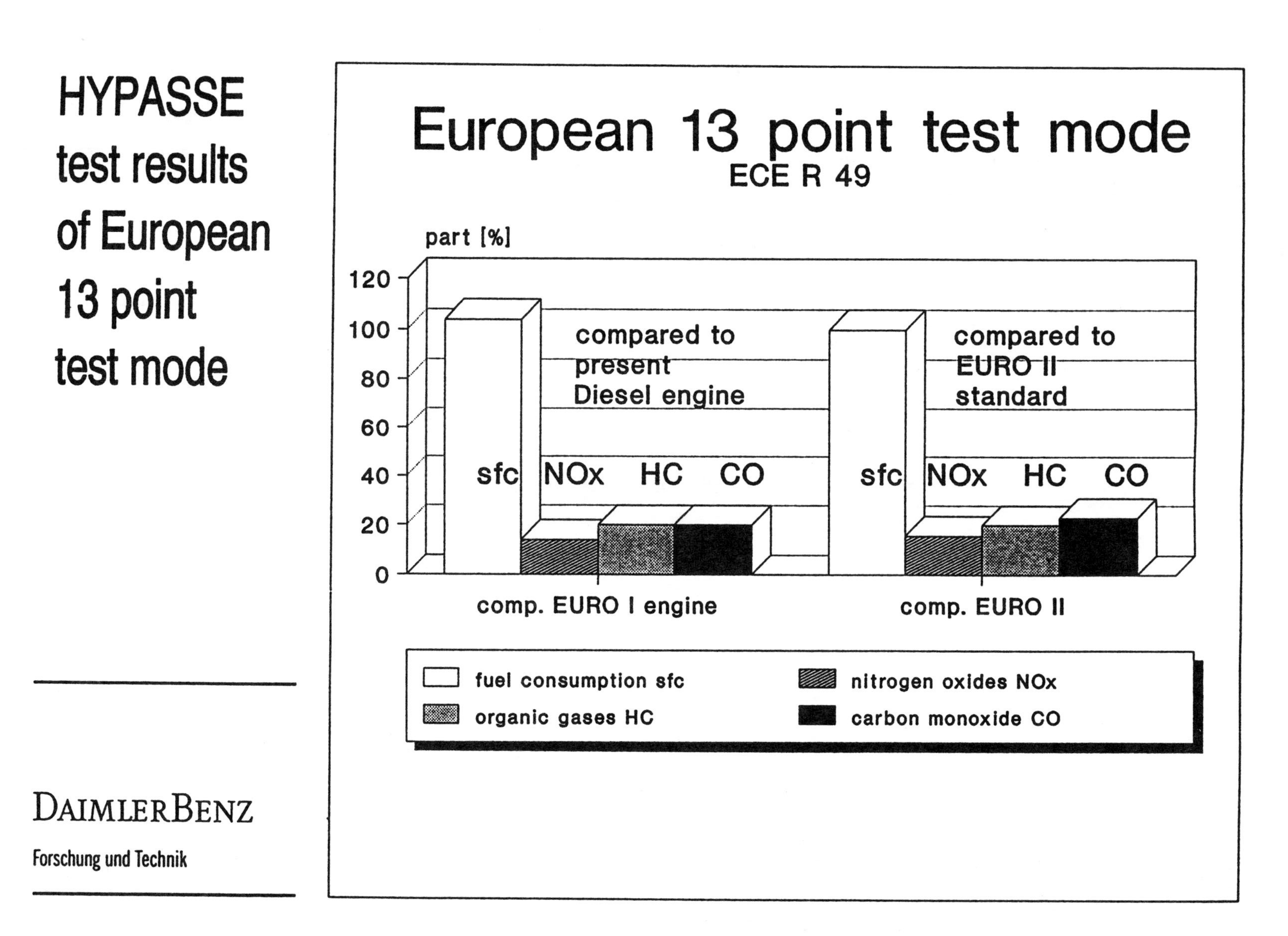

Fig. 4. Emission of the Hydrogen Engine

Hypasse
O405N
Vehicle Concept

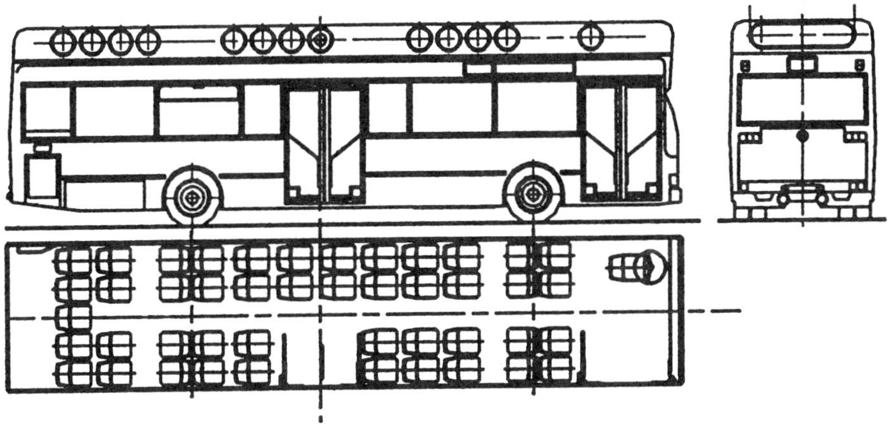

Storage:	13 bottles, dia. Ø400*1800, volume each 147l
Weight of bottles (incl. bus structure reinforcements):	2550 kg
Gross vehicle weight (fully occupied/empty):	17200 kg/13100 kg
Range (fully occupied/empty):	200 km/ 240 km
Passenger capacity:	67

DAIMLERBENZ
Forschung und Technik

Fig. 5. Vehicle Concept

Current Development and Outlook for the Hydrogen-Fuelled Car

Dr. D. Reister and W. Strobl
BMW AG

Synopsis

By its newest hydrogen-fuelled test vehicles, BMW has proven that this energy carrier is in principle a plausible alternative to fossil fuels in terms of vehicle technology. Test set-ups are already capable of achieving operating ranges of more than 300 km and of emulating the power output of an average car.

BMW will continue the development of this drive system technology in collaboration with expert partners, with particular emphasis on crucial questions such as safety, user acceptance and approval by inspection authorities. This approach will then permit field testing of small fleets. The importance of demonstration projects is that field results are obtained at an early stage and moreover that the public can be made aware of future energy systems.

A gradual market introduction is conceivable for vehicles designed to operate on gasoline and an optional alternative fuel. These could offer the advantage of minimal pollutant emissions in conurbations, for instance, where the need to avoid pollution is particularly pressing. The cost and effort of creating the required energy supply infrastructure in such areas would moreover be rendered transparent.

1. Introduction

Today, we can look back on over 100 years of automobile history. Throughout this period developments evolved according to user requirements, with the general good as an added consideration in recent years, in part steered by legal specifications.

The passenger car represents the ultimate consumer article in industrialized nations. Individual mobility is on the one hand an agreeable accomplishment of our affluent society, and on the other one of the fundamental conditions of an efficient and competitive economy. This explains why, for instance, passenger cars currently account for over 80 % of passenger transport in Germany. On the strength of the passenger car's almost unbeatable qualities, this statistic is unlikely to change much in the foreseeable future [1].

The future focus of automotive development will be on further improving traffic safety, the efficiency of road traffic and its environmental compatibility. The challenging aims of avoiding emissions and preserving natural resources will continue to exert a considerable influence on the development and production processes, the way in which a vehicle is used and the disposal of scrap vehicles in the future.

The amount of pollution generated when energy is converted by a vehicle and during the preceding stages of energy recovery and distribution is to be kept to the necessary minimum. It will probably become necessary to cut emissions of carbon dioxide, the main product of the combustion of fossil energy carriers, by a drastic degree in the medium term. These two requirements cannot be met simply by the increasingly cost-intensive measures of weightsaving vehicle design and efficient energy conversion. This is why scientists and the industry worldwide are seeking new, cleaner energy paths involving new forms of energy storage and energy conversion for propelling passenger cars.

2. Alternative energies for the automobile

Almost exclusively the mineral oil products gasoline and diesel oil are used as fuel for roadgoing vehicles, as they are easy to handle and exhibit a high energy density.

As a result of regionally varying supply, certain other fuels produced from fossil energy sources have been and are used: if methanol, liquid gas (LPG) or methane (CNG, LNG) are used, the levels of certain emission components are cut. Isolated market niches can therefore be exploited in order to alleviate certain regional problems caused by pollutant emissions. However, the above hydrocarbons are scarcely likely to gain global significance as vehicle fuels because natural resources are limited. In addition, when the energy conversion chain and cumulative emissions are considered as a whole, especially those emissions which are not yet governed by legal limits, no particular advantages over conventional fuels are to be expected [2].

The situation is the same for the production of fuels from biomass. These energy carriers on the one hand are hardly likely to achieve a very low level of pollutants and odoriferous components, and on the other total CO_2 emissions are not as favorable as is often assumed when the overall process is taken into account [3].

Whereas only a minute fraction of the total energy requirements of the world's population can be met by biologically utilized energy generated by sunlight, excluding cost considerations and in theory at least, it would be possible to cover the world's energy requirements several times over by exploiting solar energy by technical means.

There are numerous ways in which solar energy can be technically exploited (hydro power, wind energy, thermal collectors, photo-voltaic systems etc.), many of which still offer ample scope for development [4, 5]. However, these energy paths too involve intervening in nature, with industrial-scale plants in particular imposing major burdens. Although it provisionally appears to be much more environmentally compatible to utilize regenerative energy forms than to burn fossil energy resources, it is not yet possible to draw up a substantiated assessment of all the undesirable effects of the various techniques.

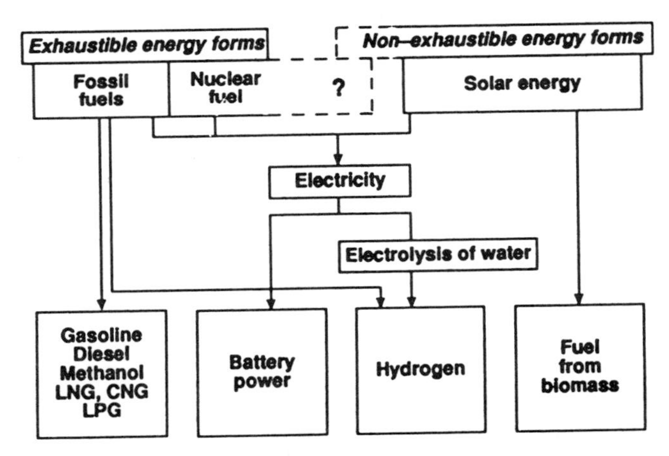

Fig. 1: Potential energy forms for automotive propulsion

Furthermore, it is not possible to produce easy-to-handle energy carriers for solar energy which are anything as convenient as hydrocarbons, which are available in liquid form at ambient temperatures. Electricity can be generated from solar energy. Batteries are required to store this solar energy; alternatively the secondary energy carrier hydrogen, which is easier to store and transport, can be produced by the electrolysis of water.

The water required in the electrolysis process is returned to the natural cycle when the hydrogen fuel oxidizes with oxygen. The hydrogen itself, a gas which is approx. 15 times lighter than air in ambient conditions, is colorless, odorless and non-toxic. In suitable combustion conditions with air, the only product of the combustion process is water. Water can safely be termed the cleanest fuel for internal-combustion engines if the problem of how to produce and distribute this energy carrier is solved in a satisfactory manner.

The use of hydrogen fuel, here liquid hydrogen, is hindered by the disadvantages that primary energy consumption is higher and that there is no appropriate infrastructure compared with electric power, whereas its potential lies in large operating ranges, high vehicle performance and rapid refuelling of the energy store. Even in the long term, electric cars are likely to remain a special option for emission-free short-distance driving. Hydrogen, on the other hand, will probably come into use in universal vehicles which are intended to achieve minimum emissions combined with a high operating range, provided it can be obtained economically from renewable energy sources.

In the short term, both drive principles (electric and hydrogen drive) can help open up clean energy paths which are independent of mineral oil resources, contributing towards economic and political global stability in the long term. BMW has therefore been conducting research into the potential of vehicles with electric and hydrogen drive, together with the operating context they require, for more than 10 years [6, 7]. The state of development now reached means that both the fundamental scope and the limitations of the above drive principles have been defined. There nevertheless remain a plethora of details concerning the vehicle, energy storage and energy conversion to be resolved. In addition, particular overriding questions relating to safety, approval, energy generation and energy distribution are far from answered.

3. Current state of development of hydrogen storage systems for passenger cars

Compared to the conventional fuels gasoline or diesel oil, storing hydrogen as the fuel for a passenger car is a very complicated matter. Following testing in research vehicles, metal hydride tanks and liquid tanks have proven fundamentally suitable. High-pressure tanks cannot be considered as an option for passenger cars in view of their unacceptably high hazard potential, weight and volume.

Metal hydride storage is based on the property of certain high-grade metal alloys of binding or releasing hydrogen at certain temperatures and pressures. Metal hydride storage, which has already reached a relatively advanced state of development [8], can be used for special applications where tank weight, vehicle range and cost are of only secondary importance.

Hydrogen stored in liquid form (LH_2) offers the highest energy density of all hydrogen storage systems in terms of mass and volume, and is therefore the only form of storage with scope for eventual universal use. Vehicles with a low range of up to 200 km are likely to have electric drive.

With the exception of liquid hydrogen technology, all other types of hydrogen storage for an operating range of more than 200 km would increase the vehicle's overall weight by more than 400 kg. Such an increase in passenger-car weights is untenable because of the adverse effects on road behavior and safety levels in a collision. The resulting drastic increase in fuel consumption would be far higher than the energy input for liquefying the hydrogen, which is slightly in excess of the energy input for other storage forms.

Ultra-efficient insulating technology is required for the tanks and lines in order to prevent evaporation losses as far as possible when liquid hydrogen is handled (at -253°C). In commencing its activities with liquid hydrogen (LH_2), BMW was able to build on the invaluable groundwork performed by DLR and profit from DLR's experience in the field of aerospace [9, 10, 11]. Messer Griesheim GmbH joined in at a very early stage, assuming an increasingly vital role in research work [12, 13]. Since 1989, the Process Engineering and Contracting Division of Linde AG has also been engaged in the development of tank systems for use in motor vehicles [14]. Now, several development phases later, tank systems representing a relatively high standard of technical maturity have been designed as a result of this collaboration between expert development partners:

The current cylindrical tanks have a capacity of 120 l LH_2 and are of a double-wall design, with an evacuated insulating jacket approx. 3 cm thick, containing up to 70 layers of aluminium foil and interlayered fiberglass matting. The weight of the tanks when full is about 60 kg, and the maximum operating overpressure 5 bar. Such tank systems permit operating ranges of up to approx. 400 km for average universal vehicles. As a result of the cylindrical shape and the required volume, which exceeds that of a conventional fuel tank by almost a factor of four, the area above the rear wheels is effectively the only possible location on a passenger car. For test purposes, it can just about be accommodated by a modern large-size saloon without drastic modifications to the vehicle package.

Fig. 2: Test vehicle with liquid-hydrogen tank system (120 l) being refuelled.

If the LH_2 tank layout is taken into account in the concept of a new vehicle type, very large operating ranges should be possible. For example, an LH_2 tank occupying 5 % of the overall volume of an average saloon car should be sufficient for a range of approx. 500 km.

In addition to the elementary requirements comprising maximum mass- and volume-related energy density, passenger-car tank systems must also satisfy a large number of technical, environmental and economic criteria.

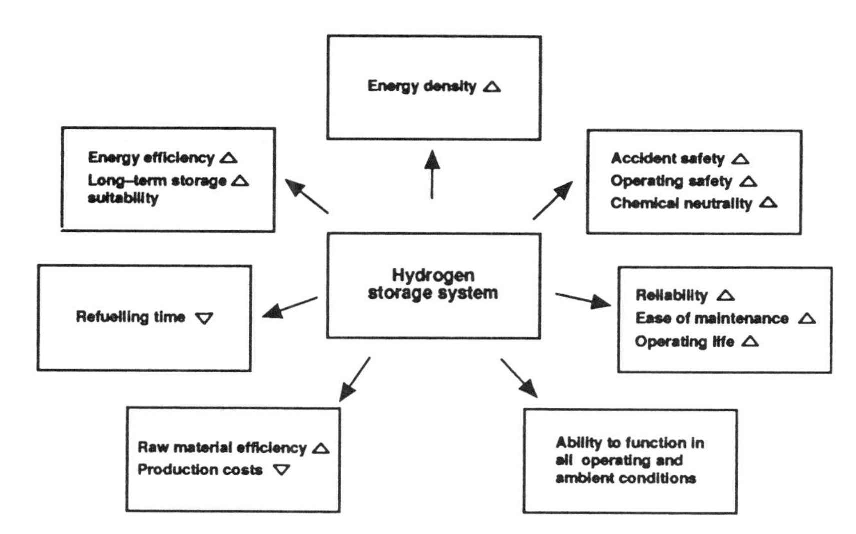

Fig. 3: A selection of important requirements of hydrogen storage systems for passenger cars.
(△ ▽: as high or low as possible)

Tanks for conventional fuels satisfy many of the requirements shown in Fig. 3 more effectively than hydrogen storage systems. However, analysis of the current development status leads us to conclude that liquid hydrogen technology in all probability offers the potential for satisfying all requirements at an acceptable cost or with defensible restrictions on use.

A factor of central importance in the design of LH_2 tank systems was and is the need to treat safety aspects as a priority. Valuable indicators of how liquid hydrogen can be handled safely in passenger cars have been available for some time now [13]. In the meantime, further progress has been achieved especially in the field of operating safety:

The durability of several system components has been confirmed in simulations of extremely tough operating conditions.
In order to protect the tanks against an impermissible buildup of pressure, more efficient and reliable valve combinations have been developed.
The incidence of residual heat through the LH_2 tank's insulation and the operating and maximum pressures have been optimized with the result that the vehicle can be left out of use for several days without energy losses. If the vehicle is not used for longer than this, evaporation losses of up to 2 % per

day must be anticipated. A no-risk blow-off process capable of handling this very low flow of hydrogen is being tested.

4. Essential development steps for liquid-hydrogen tank systems

The task of optimizing the function and economy of tank systems is clearly one of the responsibilities of car manufacturers and their component-manufacturer partners. This aspect of development work is progressing fruitfully.
Problems have been encountered in the organizational and financial processing of more general task areas. In the field of storage technology, this primarily means preliminary work for filing official approval and overall consideration of methods of transportation and retanking, in other words problem areas which extend beyond the tasks of individual vehicle manufacturers. These constitute predominantly essential processes whose significance is general and equally applicable to non-mobile uses. Even if the necessary level of consensus between the worlds of politics and industry for these tasks to be tackled is far from evident, BMW's many years of work are coming to fruition:

Via the Solar-Wasserstoff-Bayern GmbH pilot project, in which BMW is actively involved, the method of tanking liquid hydrogen is being further developed with the aim of bringing it to practical maturity [15]. The constellation of energy generating company and supplier, gases producer and distributor and vehicle manufacturer constitutes a highly promising interface between energy supply and energy utilization. Once suitable components and the process technology have been developed, the entire refuelling process will probably take less than 10 minutes (compared with the previous 1 hour) with virtually negligible energy losses.

Research into the behavior of LH_2 vehicle tank systems in very severe accidents again supports the approach adopted by BMW. In conjunction with MAN Nutzfahrzeuge AG, BMW is planning and conducting experimental research into the effect of fire on LH_2 vehicle tanks and into the destruction of tanks by excessive internal pressure or mechanical force. Suitable research departments, industrial companies and the bodies responsible for issuing official approval are involved in this project. The Commission of the European Community is to assist this work via the Euro-Québec Hydro-Hydrogen Pilot Project [16].

In assessing these activities, it should not however be overlooked that they will merely represent the start of a comprehensive safety research programme which is necessary in order to prepare society specifically for a new, unfamiliar and non-visible energy carrier in the form of hydrogen. The cost and effort of obtaining an initial operating permit for LH_2 tank systems in passenger cars are not yet foreseeable.

5. State of development of energy converters for passenger cars with LH_2 tank systems

At the start of the 20th century, the reciprocating internal-combustion engine supplanted all other versions as a result of the breakneck development of its benefits. Even when run on hydrogen, this energy converter will not see its dominant status dwindle in the foreseeable future. The internal-combustion engine's principle is the one which best satisfies the wide-ranging requirements of an energy converter for the utilization of hydrogen as a vehicle fuel.

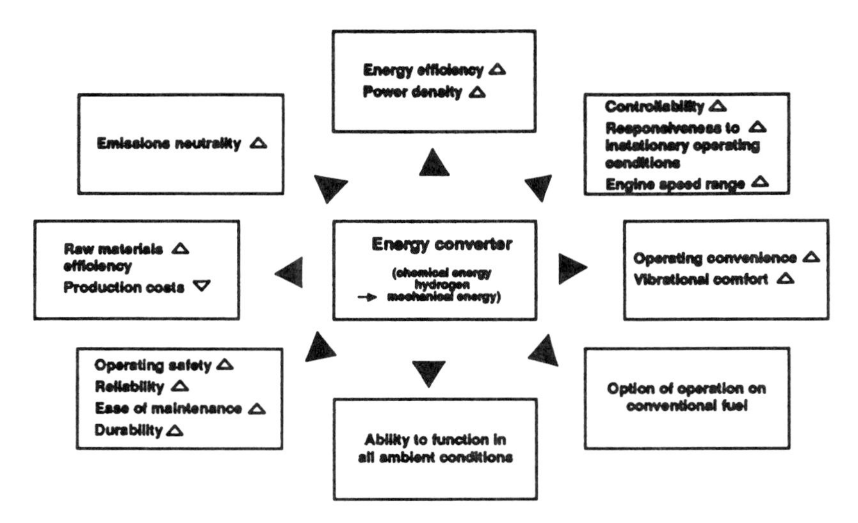

Fig. 4: Several important requirements for energy converters for hydrogen-powered passenger cars
$\Delta \nabla$: as high or low as possible)

Modern spark-ignition engines designed to run on unleaded gasoline can also run on hydrogen if a new mixture formation system is fitted and if modifications are made to the ignition system and the gas exchange organs. The operating data of future engines which are fully optimized for operation on hydrogen will, however, differ from current spark-ignition engines. The mixture formation system is nevertheless initially the most important development target. The way the hydrogen is prepared and metered exerts a greater influence on engine functions and system complexity than other assemblies:

High-energy, i.e. stoichiometric hydrogen/air mixtures can at present only be governed by technically very advanced internal mixture formation, or with the aid of additional measures such as water injection in the case of external mixture formation. Such engines can emulate spark-ignition engines in terms of power output. BMW has teamed up with DLR on the development of components for the direct injection of cryogenic hydrogen. The liquid-hydrogen delivery pump and injectors are particularly complex components in view of the low fuel temperature and the fuel's lack of lubricating properties. These components have to be driven hydraulically, for instance, necessitating auxiliary systems on the vehicle [17].

Less complex mixture formation concepts can be realized if hydrogen is combusted with excess air across the entire load range. The energy required to ignite the mixture increases, with the result that the hydrogen already in the intake passages can be added to the fuel mixture without the need for extra water injection. The inerting excess air keeps the combustion temperature relatively low, largely preventing nitrogen oxides from being generated.

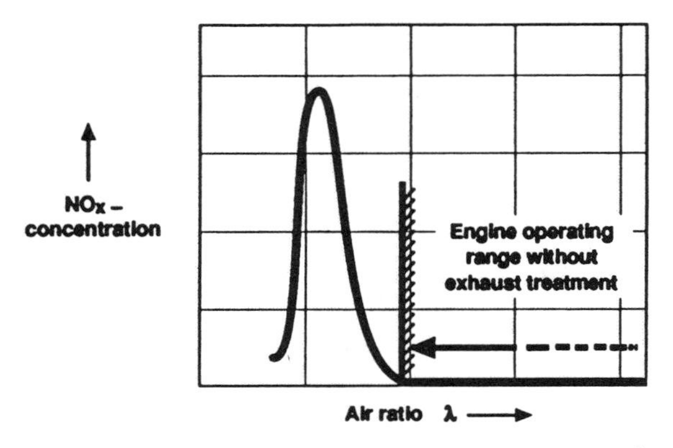

Fig. 5: Nitrogen oxide emissions from the combustion of homogeneous hydrogen/air mixtures in high-speed spark-ignition engines.

However, specific power output also falls in lean conditions (air ratio $\lambda > 1$). This power loss can be compensated for in part by mechanical supercharging [13].

BMW is conducting basic investigations into energy conversion with the aid of one-cylinder research engines and modified series engines with and without supercharging. The engines suitable for the current LH_2 research vehicles have capacities ranging from 2.5 to 5.0 liters and, as developments stand, cover an output range from 80 kW to 150 kW. Although smaller hydrogen-powered engines are of course also feasible in principle, in their current technical form they do not achieve sufficient road performance output if installed in the large test vehicles (which are used in order to accommodate the LH_2 tank).

Fig. 6: Hydrogen-powered 6-cylinder engine (displacement 2.5 l) on test rig.

Gasoline operation is generally retained as an option when hydrogen-powered vehicle engines are developed. Although this two-fuel concept necessitates additional input and restricts the parameters for hydrogen operation, it will be essential in the medium term until an adequate energy supply infrastructure has been established.

Development results obtained to date from hydrogen-powered engines reveal that reciprocating gasoline engines which can also run on liquid hydrogen exhibit similar operating characteristics to present-day car engines, and moreover achieve extremely low pollutant emission levels without exhaust-gas treatment. However, these results must not be allowed to obscure the fact that numerous engine details still need to be solved and fundamental matters concerning the approval procedure remain to be clarified before a hydrogen engine which is suitable for everyday use can be developed.

6. The need for further development work on energy converters for cars with LH_2 tank systems

Development work on hydrogen-powered spark-ignition engines must be pursued in order to confirm the approach which has been adopted for the engine design concept, with a view to achieving ultimate series maturity and suitability for everyday driving. Our main task is to resolve the conflict between high performance density and the need to prevent nitrogen oxide emissions even more effectively. The mixture formation system still offers considerable scope for development in this respect.

Alongside its own development activities and its joint work with DLR, BMW has teamed up with Robert Bosch GmbH on the development of sequential hydrogen metering into the intake air. Mixture formation using cryogenic hydrogen moreover offers crucial advantages in its scope for boosting efficiency and performance, but also harbors a large number of complications.

In addition to the mixture formation system, the gas exchange layout together with a suitable supercharging process, the ignition system and the tribological properties are in need of further development, always bearing in mind that optional gasoline operation should be assured.

Appropriate tools and methods first had to and have to be devised for development work to proceed efficiently. For example, BMW planned and built the first test rig for hydrogen engines in the world. This complex setup incorporating particularly advanced safety technology was started up early in 1989. A large number of new testing and measuring methods will need to be devised in future as flanking measures to development work proper. Improved exhaust emission measuring techniques capable of recording and assessing emissions by mass, which differ considerably from those on conventional engines, are particularly important.

Although current findings make it a reasonable assumption that undesirable emissions will be kept very low if the spark-ignition hydrogen combustion process can be suitably governed and that exhaust emissions treatment systems will not be necessary, hydrogen-powered internal-combustion engine concepts have not yet been unequivocally acknowledged as clean. It is therefore necessary to team up with environmental authorities and official bodies in order to make clear statements on the scope of hydrogen-powered engines for avoiding pollution. Only once trace concentrations of undesirable emissions have been assessed can hydrogen engines be definitively evaluated in terms of emissions and guidelines for further engine development work issued. It should be pointed out that optional gasoline operation will complicate clarification of the approval procedure.

Together with the scope for further development of hydrogen-powered internal-combustion engines, the prospects of electrochemical energy converters, i.e. fuel cell systems coupled with electric motors, should also be examined in future with a view to mobile use, as these concepts at least in theory offer potential for reducing vehicle fuel consumption.

Although steady progress has been made on fuel cell technology in recent years [18], no assessable complete systems which convert hydrogen self-sufficiently into mechanical work and are suitable for use in passenger cars are known as yet. It must currently be assumed that in view of the high level of auxiliary assemblies required, excessive weight and volume and the problems of governing engine power, air-operated fuel cell complete systems do not bring a sufficiently great improvement in the energy balance to justify the considerable extra cost compared to vehicles driven by internal-combustion engines. Via various fuel cell projects, BMW is nevertheless contributing the expert knowledge of a car manufacturer of this type of technical system with a view to defining the requirements and design criteria for the overall vehicle.

7. Comments on the economy of hydrogen

It is not yet possible to make a substantiated forecast of progress in the economy of the hydrogen car as one component in an overall energy generation, distribution and utilization system where all its effects are taken into account. An attempt will therefore be made here to assess in a plausibly pragmatic way the overall operating costs for a universal vehicle which can be run on hydrogen, compared with what is currently considered a conventional vehicle, on the basis of current costs and technologies which can be evaluated to some degree.

The following deliberations and assumptions are made:

a) Of the various possible energy paths, the basis is taken to be hydrogen production at solar-thermal power stations in high-sunlight areas involving relatively large transportation distances and the fuel supplied in liquid form. One feature of this structure is that its development is relatively advanced and harbors considerable potential for global availability, with the result that increasing demand would cause energy costs to fall rather than rise (in contrast to hydro power or wind energy). The energy generating costs - assuming industrial-scale plants are available - can be stated relatively accurately at approx. 0.50 DM/kWh LH_2, based on 1992 prices [4, 5]. This includes all investment costs for energy generation and distribution.

b) The assumed vehicle technology and form of utilization are characterized by the current preference for universal utility, comfort/convenience and safety, such as is offered by an attractive car in the deluxe midsize category (e.g. BMW 520i). Conventional operation is presupposed, with an average annual distance of 15,000 km covered and three-year period of ownership.

   The vehicle concept is based on the 2-fuel principle (hydrogen and gasoline) with a free choice of operating mode. The range on hydrogen should be in excess of 300 km and on gasoline in excess of 500 km. This concept ensures that it can on the one hand operate with even fewer emissions (virtually pollution-free) for special purposes, for instances in conurbations, and on the other be practicable even while the supply network for hydrogen fuel is still underdeveloped. Hydrogen operation is assumed to account for 50 % of total distance covered.

c) In estimating vehicle costs, it is assumed that the LH_2 tank system and the additional or modified components for the engine and safety technology have reached series maturity. Once the production totals reach the customary magnitude for the car industry, efficient production techniques can be implemented. Subject to this proviso, the manufacturing costs for a two-fuel vehicle will be around 30 - 40 % higher than those for a comparable conventional vehicle. Loss of value, based on acquisition cost, ought to be roughly of the same proportional order for both types of vehicle.

d) User-relevant gasoline costs are assumed to be 0.15 DM/kWh, reflecting the average market price for gasoline, inclusive of taxes, at the start of 1992.

e) Liquid hydrogen is taxed in the same way as gasoline (energy equivalent). A vehicle operating on two fuel types is taxed on the same basis as a low-pollution gasoline-engined vehicle of equivalent engine displacement.

A comparison of user-relevant costs of upkeep (Fig. 7) for both the aforementioned drive systems reveals that although a hydrogen car is appreciably more costly to run, the added expense is not out of all proportion even though the hydrogen itself is comparatively expensive.

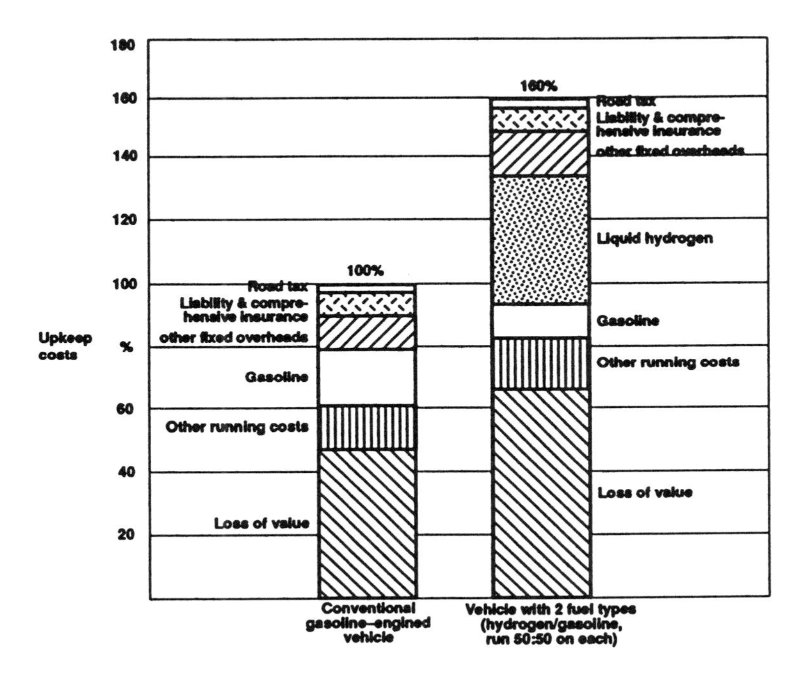

Fig. 7: Estimate of user-relevant upkeep costs for a two-fuel vehicle (hydrogen/gasoline) compared to the conventional gasoline vehicle. Prices based on 1992 levels.

The above estimates are conservative. The steady progress being made in technologies for utilizing renewable energy sources and in introducing environmental legislation is likely to reduce the difference in cost between conventional vehicles and hydrogen-powered vehicles.

8. Conclusion

The energy supply has always been one of man's major problems. The Gulf War at the start of 1991 once again drove home the fact that the question of resources holds the power to destabilize society, both now and in the future. The growing global population and the desire of an increasing number of people to share in the prosperity of industrialized nations are causing the world's energy requirements to spiral, with the result that it is becoming an increasingly pressing concern to establish an environmentally compatible energy supply.

The experts agree that new, pro-environmental energy paths which are immune to political crises must be opened up. However, our energy supply structures can only be changed to any significant degree with lengthy preparation and high expenditure.

It is therefore essential to push through the development of the necessary new technologies as soon and as concertedly as possible, so that mature systems suitable for everyday use will be available in good time. Bearing in mind the lengthy preparation and market penetration times and in order to generate impulses for other industries, at considerable expense to itself BMW has for more than a decade been working together with expert partners to devise drive systems which can run on electricity either directly in battery form or after its conversion into liquid hydrogen. In the course of these research and development activities, a number of impressive and occasionally very surprising facts which scarcely anyone would have thought possible only a few years ago have been discovered.

One such discovery is the confirmation that passenger cars which run on two different fuels (hydrogen/gasoline) are no longer a utopian notion. The vehicle concept of bivalent operation preserves the universality of modern-day passenger cars and offers realistic prospects of exploiting market niches in the medium term. All system components, from primary energy to final energy, are progressively being rendered more functional and economical according to the "learning by doing" principle.

Political organs are now called upon to provide assistance in dealing with overriding problems and in establishing the basic conditions for systematically developing these highly promising technologies.

Literature

[1] Shell 1991 forecast (No. 22): "Aufbruch zu neuen Dimensionen" (Heralding in New Dimensions). Hamburg, Deutsche Shell AG, 1991.

[2] Heitland, Herbert; Hiller, Heinz; Hoffmann, Hans-Jürgen: "Einfluss des zukünftigen Pkw-Verkehrs auf die CO_2-Emissionen" (Influence of Future Automobile Traffic on CO_2 Emissions). MTZ Motortechnische Zeitschrift 51 (1990) 2, pp. 66-72

[3] Förster, Hans-Joachim: "Entwicklungsreserven des Verbrennungsmotors zur Schonung von Energie und Umwelt - Teil 2" (Development Reserves of the Internal-Combustion Engine for the Protection of Energy and the Environment). ATZ Automobiltechnische Zeitschrift 93 (1991) 6, pp. 342-352

[4] An Interlaboratory White Paper. The Potential of Renewable Energy. SERI/TP-260-3674, UC Category: 233, DE 90000322. Golden, Colorado, 1990.

[5] Kronenberg, Friedrich et al.: "Bedingungen und Folgen von Aufbaustrategien für eine solare Wasserstoffwirtschaft: Bericht der Enquete-Kommission

'Gestaltung der technischen Entwicklung, Technikfolgen-Abschätzung und -Bewertung' des Deutschen Bundestages."
(Conditions and Consequences of Development Strategies for a Solar Hydrogen Industry: Report of the Federal German Government's Enquete Commission 'Shape of Technical Development, Estimate and Evaluation of Technical Consequences'.)
Bonn: Deutscher Bundestag, Public Relations Department, 1990

[6] Regar, Karl-Nikolaus; Braess, Hans-Hermann; Reister, Dietrich: "Die neuen Elektro-3er als Glied einer langen Entwicklungskette" (The New Electric 3 Series: One Link in a Long Chain of Developments). 3rd Aachen Colloquium on Vehicle and Engine Technology
Aachen: List Verlag 1991, pp. 551-597

[7] Braess, Hans-Hermann; Regar, Karl-Nikolaus; Strobl, Wolfgang: "Drive Systems to Protect the Environment". Automotive Technology International '90.
London: Sterling Publications International Limited, 1989, p. 249-255

[8] Quadflieg, Hansgert, et al.: "Alternative Energien für den Strassenverkehr - Wasserstoffantrieb in der Erprobung" (Alternative Energies for Road Traffic - Testing the Hydrogen Drive Principle).
Cologne: TÜV Rheinland GmbH, 1989.

[9] Peschka, Walter: "Liquid Hydrogen for Automotive Vehicles, Status and Development in Germany". Cryogenic Processes and Equipment.
New York: ASME 1984 Libr. of Congr. Catal. Card No. 84-72999, pp. 97-104.

[10] Strobl, Wolfgang; Peschka, Walter: "Liquid Hydrogen as a Fuel of the Future for Individual Transport". Hydrogen Energy Progress VI. Proceedings of the 6th World Hydrogen Energy Conference, Vienna.
New York: Pergamon Press, 1986, pp. 1161-1173.

[11] Peschka, Walter: "Flüssiger Wasserstoff als Energieträger" (Liquid Hydrogen as an Energy Carrier). Vienna/New York: Springer Verlag, 1984.

[12] Ewald, Rolf; Kesten, Martin: "Cryogenic Equipment of Liquid Hydrogen Powered Automobiles". Advances in Cryogenic Engineering (1989) No. 35.
New York: Plenum Press.

[13] Regar, Karl-Nikolaus; Fickel, H. Christian; Pehr, Klaus: "Der neue BMW 735i mit Wasserstoffantrieb" (The New Hydrogen Powered BMW 735i). VDI Reports No. 725, pp. 187-198. Düsseldorf: VDI Verlag 1989.

[14] Rüdiger, Horst: "Design Characteristics and Performance of a Liquid Hydrogen Tank System for Motor Cars". To be published in Cryogenics, 1992.

[15] Szyszka, Axel: "Demonstrationsanlage Neunburg vorm Wald zur Erprobung von Solar-Wasserstoff-Techniken" (The Demonstration Plant in Neunburg vorm Wald for Testing Solar Hydrogen Techniques).
Elektrizitätswirtschaft, periodical of the VDEW (Association of German Generating Plants), 90 (1991) 24, pp. 1438-1446.

[16] Kluyskens, Dominique; Ullmann, Oskar; Wurster, Reinhold: Executive Summary of the Euro-Québec Hydro-Hydrogen Pilot Project, Phase II.
Montreal and Ottobrunn, Hydro-Québec and Ludwig Bölkow Foundation, 1991.

[17] Strobl, Wolfgang; Peschka, Walter: "Forschungsfahrzeuge mit Flüssigwasserstofftechnik" (Research Vehicles with Liquid Hydrogen Technology).
VDI Reports No. 602, pp. 203-216. Düsseldorf: VDI Verlag, 1989.

[18] Strasser, Karl: "PEM Fuel Cells for Energy Storage Systems". IECEC '91, 26th Intersociety Energy Conversion Engineering Conference.
Boston, Massachusetts, 1991

Some Performance of Engine and Cooling System on LH_2 Refrigerator Van Musashi-9

Kimitaka Yamane, Masaru Hiruma, Takeshi Watanabe, Takashi Kondo, Kiyoharu Hikino,*
Tatsuhiko Hashimoto, and Shoichi Furuhama**
Hydrogen Energy Research Center, Musashi Institute of Technology
**Engine Research Div., Technical Research Lab., Hino Motors, Ltd.*
***Technical Development Dept., Technology and Safety Div.,*
Iwatani International Corporation

Abstract

As the practical utilization of the potential cold energy of liquid hydrogen(LH_2), the authors have been developing a LH_2 refrigerator van called Musashi-9 by modifying a 4-ton diesel truck. It has been found that LH_2 on board refrigerator van Musashi-9 can be used not only as the cold energy to keep the cooled goods at the temperature of 0 ± 5 °C but also as the fuel for the van. It has been also found that the NOx emission can be reduced to so a quite large extent as 53% of the future target value required in the 13-mode driving pattern for emission evaluation requirement in Japan.

1. INTRODUCTION

Hydrogen fueled vehicles have been developed at Musashi Institute of Technology for many years. In 1990, at the 8th World Hydrogen Energy Conference in Hawaii, Musashi-8 sport car was demonstrated with a 100-leter liquid hydrogen tank and a high pressure liquid pump, whose delivery pressure was 10MPa on board in the rear portion[1]. The high pressure hydrogen was warmed up by the engine coolant and was lastly injected into the engine cylinder near the top dead center from an injection valve actuated hydraulically.

It has been well understood[2, 3] that hydrogen in liquid state is most appropriate for the storage on vehicles in respect of the weight and the volume which are significant parameters for vehicles. However, it is also known that the specific production energy per liter of LH_2 is about 0.95 kWh[4], namely this amounts 40% of the low heat value of one liter LH_2. The amount is so great that the cold energy of LH_2 should be used effectively without releasing it to atmosphere as Musashi-8 did before.

Therefore, the objective of the development of Musashi-9 is to study whether LH_2 can be used not only as the cold energy but also as the fuel for the van aiming at keeping the temperature in the refrigerator box at 0 ± 5 °C in order to carry fresh vegetables, fish or flowers to their markets.

2. ENGINE AND COOLING SYSTEM

The LH_2 fuel supply system for Musashi-9 is basically the same as that for Musashi-8. A refrigerant reservoir, a refrigerant delivery pump, a heat exchanger with two fans to adjust the temperature in a refrigerator box and a refrigerator box have been added to the system. Figure 1 shows the schematic diagram and the layout of the system on the truck is shown in Fig. 2.

The 360-liter LH_2 in the tank, which is about the same as that of 100-liter diesel oil for the original truck in caloric basis is pumped out at the pressure of 10MPa by a LH_2 high pressure pump to the reservoir where the cold energy of LH_2 is transferred to the 150-liter refrigerant of 30% calcium chloride aqueous solution, which is fed to the heat exchanger by the refrigerant delivery pump to cool down the refrigerator box. In our preliminary study, it was found that an engine output power of 17PS was necessary to overcome the heat input of 870kcal/h into the refrigerator box. Therefore, it was necessary for the truck to run at more than 50km/h on a flat road or at more than 18km/h on a road with 3% slope. The truck is subject to the various conditions and sometimes parking. Therefore, the reservoir is required for this system. The capacity of the reservoir was determined so that the truck would park for 3 hours in the one charge of LH_2. The refrigerant is always circulated in the heat exchanger installed in the refrigerator box to cool down the temperature by a refrigerant delivery pump with a flow rate of $0.5m^3/h$. On the other hand, the warmed hydrogen gas at the pressure of 10MPa is fed to the high pressure hydrogen injector through a surge tank. The high pressure hydrogen injector is actuated hydraulically with working fluid of diesel oil by a diesel plunger pump.

No piping of hydrogen gas is permitted in the driver's cabin and the refrigerator box. Every valve necessary for the driver to handle can be operated remotely from the driver's cabin. The LH_2 tank locates at the place where it will not be hard to be damaged when a car crash occurs, namely the tank locates between the driver's cabin and the refrigerator box.

The base engine was a water-cooled, six-cylinder, four-stroke, direct injection(DI) engine made by Hino Motors. It had the bore of 104mm, the stroke of 113mm, the total piston displacement of 5.76 liters and the compression ratio of 17.9:1.

The modified engine for refrigerator van Musashi-9 has six high pressure hydrogen injectors with a 9-hole nozzle each in place of the diesel ones, a spark plug has been installed in each cylinder and the compression ratio has been reduced to 13:1. The piston has a cavity with the squish area ratio of 0.76. Figure 3 shows the relative location of the injection nozzle and spark plug employed for Musashi-9. These values were determined by the previous experiments carried out for the engine of Musashi-8. An electronic controlled injection pump has been employed for an accurate injection timing because it was difficult to control an accurate injection timing by the conventional mechanically controlled injection pump employed for Musashi-8.

3. RESULTS

Capability of Cooling System

The preliminary evaluation of the capability of the cooling system to keep the temperature in the refrigerator box at 0 ± 5 ℃ was made on the basis of the fact that a conventional refrigerator van with a refrigerator on board is cooled down in advance of loading goods which are also cooled down before hand.

Refrigerator vans run in various conditions so that a 13-mode driving pattern for the emission evaluation of trucks and buses with the gross vehicle weight over 2.5 tons as shown in Table I effective from the 1st of October, 1993 in Japan was adopted to figure out the average flow rate of LH_2. The average flow rate of LH_2 was found out at 3kg/h.

In the experiment, the temperatures at the outlet of the LH_2 pump and in the refrigerator box with the two fans in operation were measured. The revolution speed of the LH_2 pump was set at 170rpm to obtain the average flow rate of 3kg/h because it was thought that, if the temperature in the refrigerator box reached to the target temperature of 0 ± 5 ℃ at the average flow rate, the capability of the cooling system was verified. Figure 4 shows the result of the temperature history measured in the refrigerator box. It is found that the temperature reaches 0 ℃ in about 2.5h from the start of cooling. The temperature at the outlet of the LH_2 pump was measured at $-$ 220 ℃ through the experiment. Table II shows the heat transfer rates in the system estimated by the previous experiments and calculations done by Iwatani International Corporation, one of the cooperators for the development of Musashi-9 which were mainly in charge of the cooling system. By using the heat transfer rates and the temperature at the outlet of the LH_2 pump, a calculation was made to verify that the temperature gradient to time calculated was about the same as that measured, about $-$ 10 ℃ /h equivalent to $-$ 41kal/h at the temperature of 10 ℃ in the refrigerator box. As the result, the calculated one was found about $-$ 8 ℃ /h equivalent to $-$ 33kal/h. This means that the estimated heat transfer rates were reasonable and that the flow rate of 3kg/h was enough to cool down the refrigerator box.

Emission Performance

Hydrogen fueled engines have only one pollutant such as nitric oxides(NOx) in exhaust. It has been found that the system of a high pressure hydrogen injection near the top dead center with spark ignition allow us to eliminate completely the abnormal combustion such as pre-ignition and backfire with a high engine output power. However, in low load, the system is subject to larger amount of NOx than when external mixture formation is used in hydrogen fueled engines. Many experiments have been carried out to study the effective measures for reduction of NOx by using the system developed by Musashi Institute of technology. It has been found that a significant reduction of NOx in exhaust was obtained by using injection timing retardation, cold exhaust gas recirculation(EGR)[5] which can make use of the latent heat of water to decrease the maximum combuston temperature. When the two NOx reduction measures and intake choking to make the mixture as rich as possible at the same time, it was found that the concentration of NOx was obtained at 95ppm when the engine was operated in a 6-mode emission evaluation method, which was 37% of the value of 260ppm with only a 2% reduction in the effective thermal efficiency[6].

As the good result for reduction of NOx was obtained in the literature[6], those reduction measures were applied to the engine system of Musashi-9. The engine was operated in a test bench with no NOx reduction measure, injection timing retardation of 2.5 deg. crankangle from the injection point at the maximum torque, and injection timing retardation of 2.5 deg. crankangle, hot-EGR of 10~20% and intake choking making the mixture as rich as possilbe such as excess air ratio λ =1.3~1.5. The reason why hot-EGR was employed in this study was that the method is more practical and simpler than the cold-EGR one because a complicated EGR device is required to cool down the exhaust gas for the cold-EGR.

The engine was operated according to the 13-mode driving pattern to obtain the NOx emissions. Figure 5 shows the results obtained in the three operating conditions mentioned above. It is found that even the operation with no NOx reduction measure produced slightly larger amount of NOx than the value of 6g/kWh for a short-term requirement. The operation with only the injection timing retardation produced 5.25g/kWh and that the operation with injection timing retardation of 2.5 deg. crankangle, hot-EGR of 10~20% and intake choking in excess air ratio λ =1.3~1.5 produced 2.39g/kWh, namely 40% of the short-term requirement and 53% of even the future target value of 4.5g/kWh. It is found that this drastic reduction of NOx is attributed largely to the combination of the hot-EGR and the intake choking, namely the exhaust gas used for EGR was the product obtained when a fairly rich mixture of hydrogen and air at excess air ratio λ =1.3~1.5 was burnt.

4. CONCLUSION

(1) It has been found from the results obtained that LH_2 on board refrigerator van Musashi-9 can be used not only as the cold energy to keep the cooled goods at the temperature of 0 ± 5 ℃ but also as the fuel for the van.

(2) By applying the injection timing retardation of 2.5 deg. crankangle, the hot-EGR of 10~20% and the intake choking in the excess air ratio λ =1.3~1.5 at the same time, the NOx emission can be reduced to so a quite large extent as 53% of the future target value of the 13-mode driving pattern for the emission evaluation method newly established by the government in Japan.

Acknowledgements

This work was supported by Hino Motors Ltd. and Iwatani International Corporation. The authors would like to express their sincere appreciation for the companies' contributions. And the authors also would like to express many thanks to the students involved in this study.

REFERENCES

1. S. Furuhama, K. Koyanagi, N. Tomisawa and K. Yamaura, "The power system of a computer controlled hydrogen car — GH_2 injection and spark ignition engine with LH_2 tank and pump", the Proceedings of the Institute of Mechanical Engineers, 1991 — 11, pp.161 — 178

2. S. Furuhama, "Hydrogen engine technology "R & D at Musashi I. T."", Project Hydrogen '91 Conference Proceedings of I. A. H. E., 1991, pp.161 — 172

3. H. Böhnisch and et la, "Bedeutung Einstzbereiche und technische — ökonomische Entwicklungspotentiale von Wasserstoffnutzungstechniken", Band I , Energetisch Nutzungstechniken, Stuttgart, Ottobrum, März 1992, pp.161 — 178

4. M. Bracha, G. Lorenz, A. Palzelt, M. Wanner, "Large scale hydrogen liquefaction in Germany", the Proceedings of the 9th World Hydrogen Energy Conference, 1992 — 6, pp.1001 — 1009

5. Y. Ninomiya, Y. Hosono, H. Hashimoto, M. Hiruma and S. Furuhama, "NOx control in LH_2 -pump high pressure hydrogen injection engines", the Proceedings of the 9th World Hydrogen Energy Conference, 1992 — 6, pp.1295 — 1304

6. Y. Ninomiya, T. Koshiishi, M. Hiruma, T. Someya, "NOx control in high pressure hydrogen injection engines", the Proceedings of the 70th JSME Spring Annual Meeting, 1993 — 3,4, pp.731 — 733

TABLE I : 13-MODE DRIVING PATTERN

MODE NO.	DRIVING CONDITION		WEIGHT FACTOR
	ENGINE SPEED (Ratio to the engine speed at the max. power)	LOAD (Ratio to the full load)	
1	Idling speed	0%	0.205
2	40%	20%	0.037
3	40%	40%	0.027
4	Idling speed	0%	0.205
5	60%	20%	0.029
6	60%	40%	0.064
7	80%	40%	0.041
8	80%	60%	0.032
9	60%	60%	0.077
10	60%	80%	0.055
11	60%	95%	0.049
12	80%	80%	0.037
13	60%	5%	0.147

TABLE II : HEAT TRANFER RATES

SECTION		HEAT TRANSFER RATE(kcal/h)	REMARKS
LH_2 Pump → Reservoir		100	Adapters for super-insulated tube is considered.
Reservoir Cold hydrogen → Refrigarant		20△T	Overall heat transfer cofficient=37 (kcal/m²•h• °C)
Reservoir Heat input		43	Design value
Reservoir → Refrigerator box		80	Heat input at the delivery pump is considered.
Refrigerator box → Reservoir		20	
Heat exchanger Refrigerant → Air	Fan: ON	90△T	Overall heat transfer cofficient=8 (kcal/m²•h• °C)
	Fan: OFF	46△T	Overall heat transfer cofficient=4 (kcal/m²•h• °C)
Refrigerator box Heat input		870	Outer temp.=35 °C Inner temp.=0 °C

△T: Mean temperature

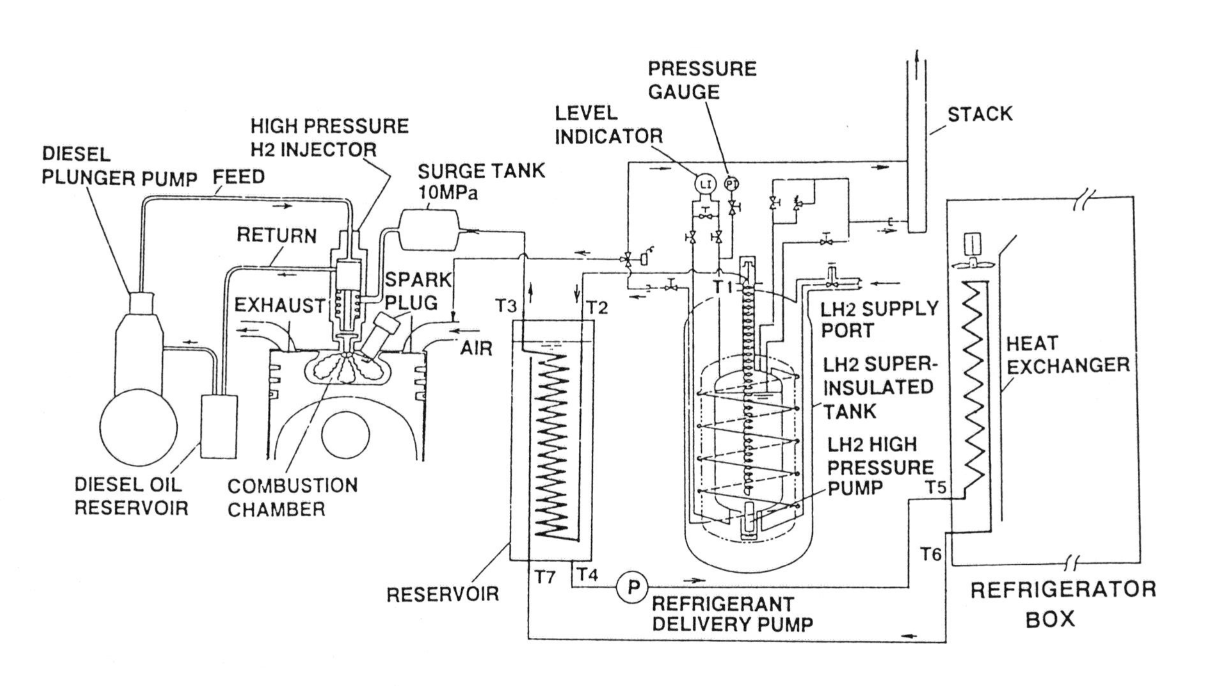

Fig. 1 Schematic Diagram of LH_2 Engine and Cooling System of Refrigerator Van Musashi-9

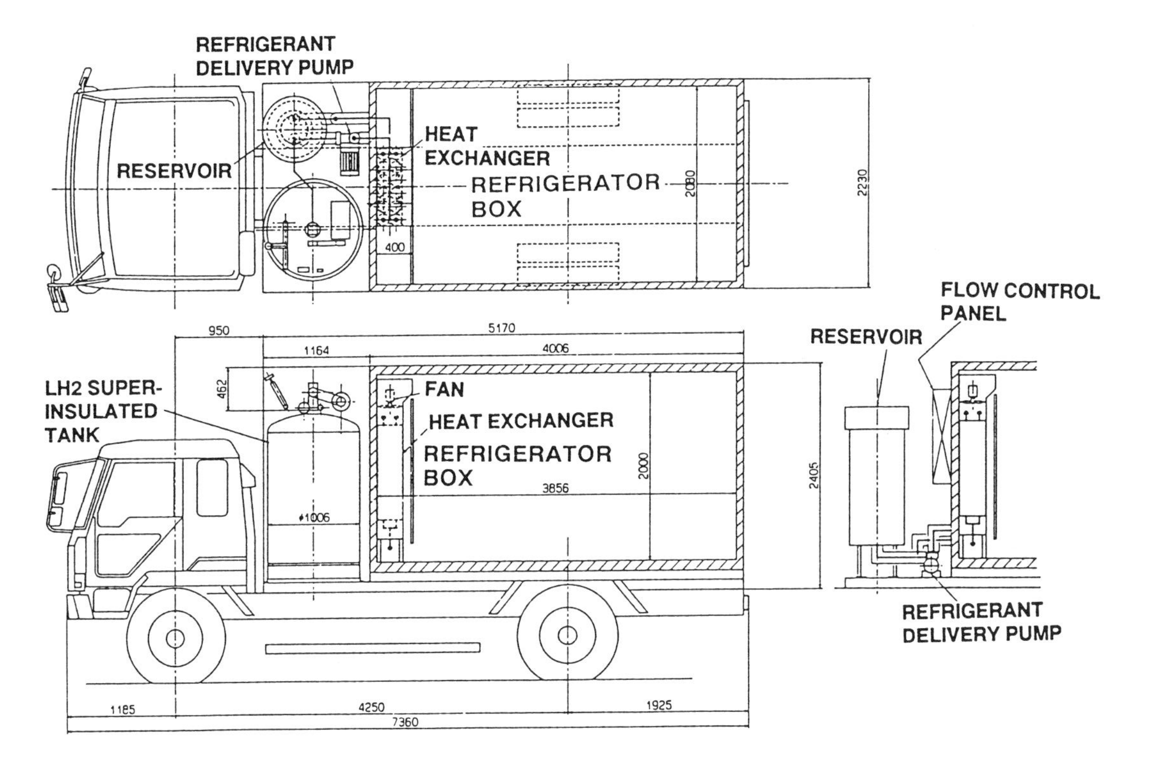

Fig. 2 Layout of LH_2 Engine and Cooling System of Refrigerator Van Musashi-9

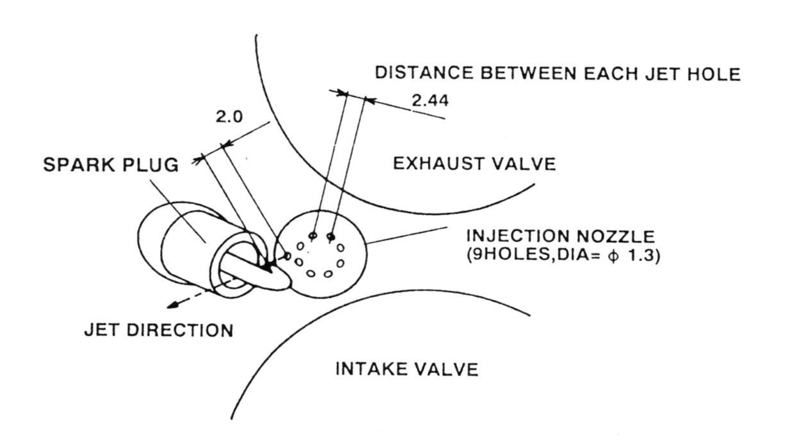

Fig. 3 Relative Location of Injector Nozzle and Spark Plug

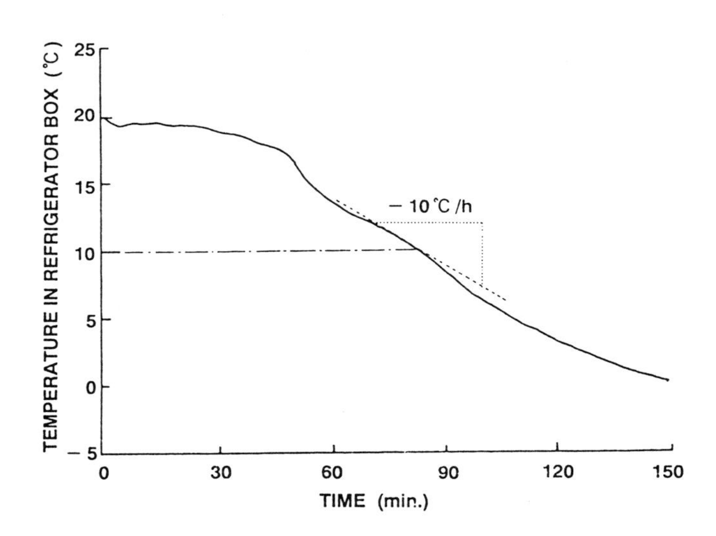

Fig. 4 Temperature History in Refrigerator Box

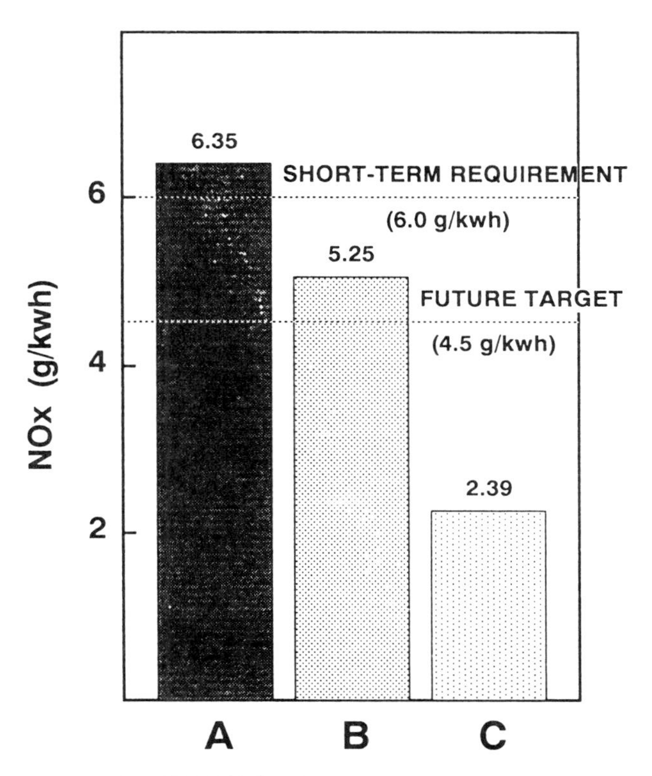

Fig. 5 NOx Emission Obtained in 13-mode Driving Pattern

Combustion Characteristics in Hydrogen Fueled Rotary Engine

Kenji Morimoto, Takafumi Teramoto, and Yuji Takamori
Mazda Motor Corporation

ABSTRACT

A hydrogen-fueled rotary engine was investigated with respect to the effects of the fuel supply method, spark plug rating and spark plug cavity volume on abnormal combustion. It was found that abnormal combustion was caused by pre-ignition from the spark plugs and gas leakage through the plug hole cavity. The hydrogen-fueled rotary engine could function through a wide operating range at a theoretical air-to-fuel ratio by optimising the above factors. Consequently, the hydrogen-fueled rotary engine achieved output power of up to 63%-75% of the gasoline specification, while the hydrogen-fueled reciprocating engine only reached 50%.

INTRODUCTION

Oil crises, global warming and other needs of the times show the necessity of practical alternative energy resources for the petroleum-based fuel engines. Hydrogen is being watched with keen interest as a clean, affordable and long-lasting fuel resource, which does not generate CO_2, CO, or HC. Although a great deal of research has been done on the practical use of hydrogen in conventional reciprocating engines, backfire and abnormal combustion are still reported to reduce engine power in the case of the conventional reciprocating engine (or CE)(1)*.

By contrast, the rotary engine (or RE) is deemed to have advantages over the CE in its characteristic structure of isolated induction and combustion chambers as well as not having a high temperature exhaust valve. These features may contribute to the solution of the backfire problem.

In this paper, the result of our investigations and experiments regarding the compatibility of the RE and hydrogen will be reported.

* Numbers in parentheses designate references at end of paper

COMPATIBILITY OF THE RE AND HYDROGEN

The RE is considered to have good compatibility with hydrogen in the following respects:

LESS OCCURRENCE OF BACKFIRE -High temperature heat sources such as spark plugs, exhaust valves, and residual heated gases are regarded as the probable causes of backfire(2)(3). The RE can prevent the fuel mixture in the induction cycle from being exposed to high temperature heat sources

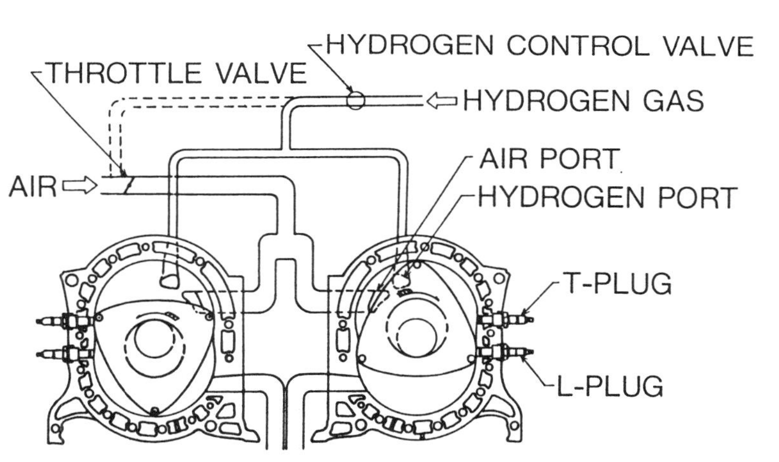

Fig. 1 - Schematic Diagram of Experimental Engine

because it is structurally freed from exhaust valves and its induction, compression, combustion and exhaust cycles are carried out in separate locations. Consequently, the RE is thought to have advantages over others in avoiding backfire.

EASE OF CONVERTING TO DIRECT INJECTION-It has been reported that limiting the supply of hydrogen fuel during the period the inlet valve is open would be an effective method to avoid backfire(1). In case of the RE, this method can easily be implemented by making a port on the side housing.

EXPERIMENTAL APPARATUS AND PROCESS

A MAZDA 13B RE, and a MAZDA FE CE, were used in this experiment. The principal technical data of the tested RE are shown in Table 1. As for testing the fuel system, we made a comparative study of the pre-mixed mixture method and the induction-cycle-injection method (the method of supplying hydrogen fuel within the duration of the induction cycle/stroke only), which is reportedly effective in avoiding backfire in the CE. In case of the pre-mixed mixture method, hydrogen was supplied upstream of the throttle valve. On the other hand, hydrogen was supplied from a hydrogen feeding port provided on the side-housing in the case of the induction-cycle-injection method. The timings of this hydrogen port and the induction port are shown in Table 3. The hydrogen supply starts with a delayed time slightly behind the induction, and ends 20° behind the eccentric angle after the induction cycle is completed. Fig.1 is an outline of the experimental apparatus for fuel system. The principal technical data of the tested CE are shown in Table 2. Regarding the fuel system, the pre-mixed mixture method was adopted with the hydrogen fuel supplied upstream of the throttle. Also, to prevent ignition by the intense high voltage induced from one set of spark leads in an adjacent one, in the case of multi-cylindered engine, the shielding of high tension cords and reduction of the plug-gaps (0.3mm) were performed to control the induced current(4). Output power, fuel consumption, and exhaust emissions were measured to evaluate the engine performance. At the same time, the profile of abnormal combustion was investigated by monitoring the pressure within the combustion chamber.

Table 1 - Hydrogen Rotary Engine Specification

ENGINE TYPE	2-ROTOR ROTARY ENGINE
SWEPT VOLUME	1308cc
COMPRESSION RATIO	9.4

Table 2 - Hydrogen Reciprocating Engine Specification

ENGINE TYPE	4-CYLINDER 4-STROKE CYCLE ENGINE
BORE × STROKE	86mm × 86mm
SWEPT VOLUME	1998cc
COMPRESSION RATIO	9.8

Table 3 - Inlet Port Timing

	INLET OPEN (°ATDC)	INLET CLOSE (°ABDC)
AIR	32	80
HYDROGEN	40	100

RESULT OF INVESTIGATIONS AND EXPERIMENTS

HYDROGEN RE AVOIDS ABNORMAL COMBUSTION -RELATED PERFORMANCE LIMITATION

The Effect of the Fuel Supply System -The influence of differences in the nature of fuel supply systems upon the power output limit caused by abnormal combustion has been studied. The result of a comparison made on the pre-mixed mixture system and the induction-cycle-injection is shown in Fig.2(in the succeeding figures, the pre-mixed mixture method, induction-cycle-injection method, and excess air ratio(λ) at normal combustion limit will be represented by "Pre-mixed", " Direct", and "Rich Limit λ", respectively). However, the evaluation was limited to regions of λ=1 or greater, since the power would be reduced at λ=1 or less. In case of the pre-mixed mixture method applied to the RE, the engine was operative up to 5000rpm at λ=1 with its throttle wide open under a full load condition. However, the power became unstable at 6500rpm at λ=1.4, and accordingly, it was impossible to decrease λ any more.

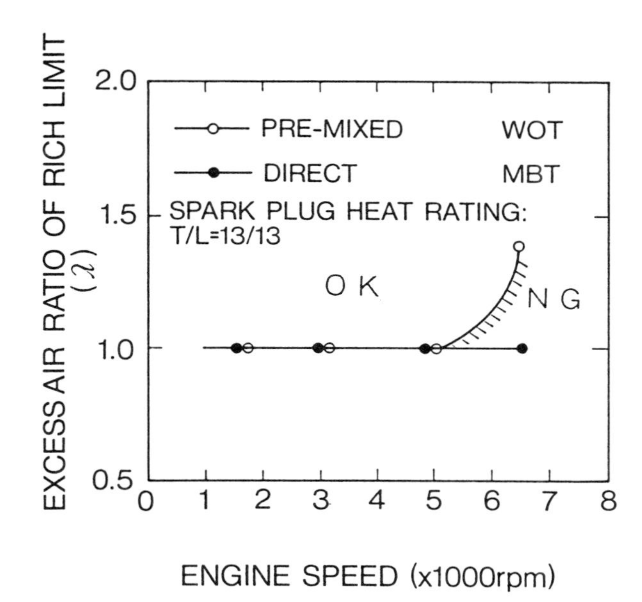

Fig. 2 - Comparison of Fuel Supply Methods on Rich Limit λ

On the other hand, 6500rpm at λ=1 was possible in case of the induction-cycle-injection method. The primary cause of the above situation seems to lie in the fact that at the location of the abnormal combustion initiation point ,the air-to-fuel ratio of the mixture created by the induction-cycle-injection could be leaner than that occurring with the pre-mixed mixture method, because stratified charge may occur in the former. It should be noted that the following investigation was based on the induction-cycle-injection method.

<u>The Effect of Spark Plug Heat Rating</u>- The optimum heat rating for the spark plugs to be used for hydrogen fuel was studied. Regarding the conventional plugs specified for regular gasoline specs, such as the leading-side spark plugs (hereafter called L plug) of heat rating 5 (rating as per item# by NGK Spark Plug) and the trailing-side plugs (hereafter called T plug) of heat rating 10 , the instability phenomenon of engine power occurred at λ=1.6. In order to investigate the cause of this phenomenon, we have evaluated its indicator diagram. Fig. 3 shows the transition of the indicator diagram during power instability. It reveals the transient developed from normal combustion to pre-ignition. Consequently, the instability of power proved to be attributable to pre-ignition. The effect of the heat rating of

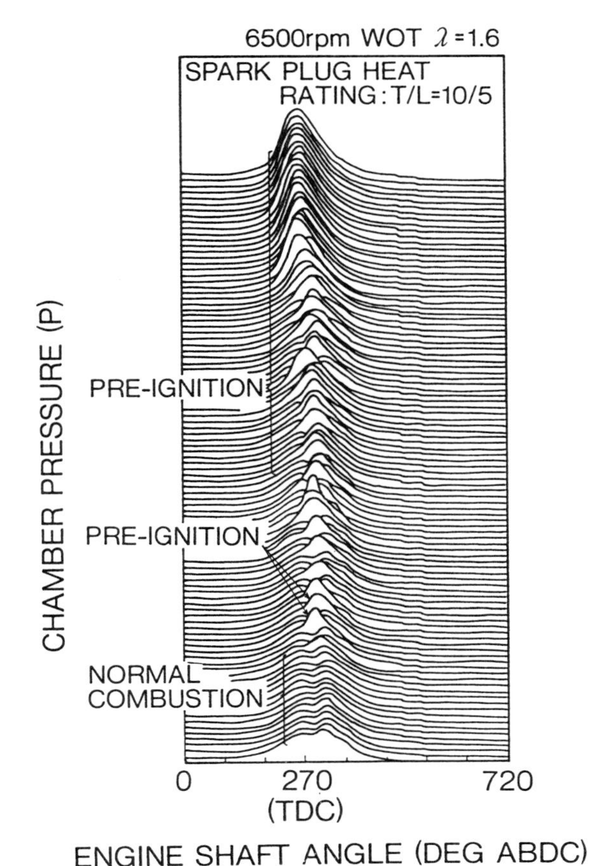

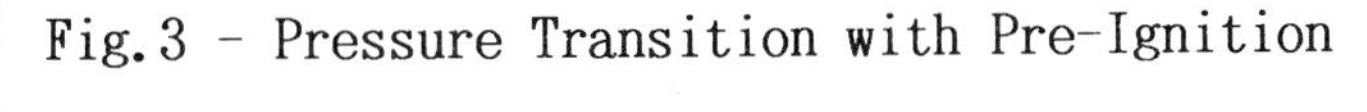

Fig. 3 - Pressure Transition with Pre-Ignition

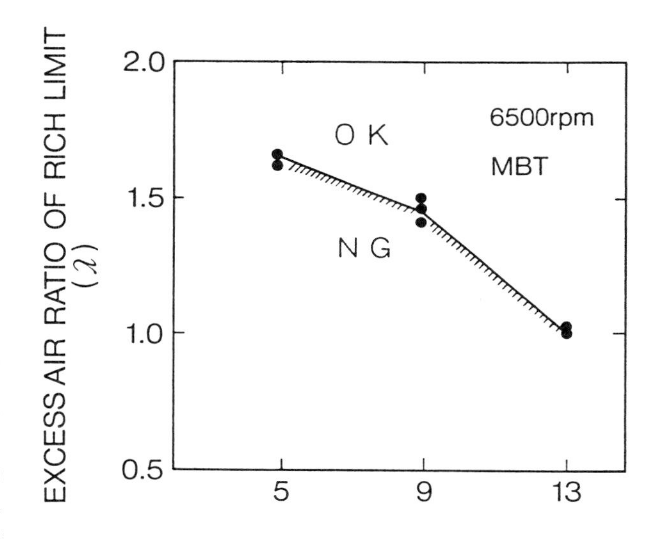

Fig. 4 - Effect of Spark Plug Heat Rating (T-Plug=L-Plug) on Rich Limit λ

spark plug was also evaluated, since the probable cause for the pre-ignition was predicted to be that pre-ignition originated on the spark plug's hot surface. The result is shown in Fig.4. The value of λ at the point of power instability occurrence could be improved in accordance with the increase in the heat rating, and at a heat rating of 13, the engine functioned satisfactorily through its operating range at around $\lambda=1$.

In addition, a supplementary experiment demonstrated the fact that the heat rating of the L plug was predominant in the occurrence of pre-ignition.

The Effect of L Plug Hole Volume - Although the engine using spark plugs of heat rating 13 could function through most of the full operating range at $\lambda=1$, it may occasionally be prone to becoming unstable at its max output of 6500rpm near $\lambda=1$. Therefore, a detailed analysis of the indicator diagram was made to determine the main cause. Fig.5 is the transition of the indicator diagram within the duration of this phenomenon. In contrast to Fig.3, it was noticed that a pre-ignition occurs suddenly, while the combustion proceeded normally.

This suggests that the unignited fuel mixture in the compression cycle had contacted the burning hot gas in the explosion cycle, via the L plug hole, and ignite it because there was no symptom of the pre-ignition by the spark plugs. In addition, the apex seal was passing through the L plug hole directly before the pre-ignition occurred. If such a phenomenon took place, the size and volume of the L plug cavity should have influenced it. To observe the effect, the volume of the L plug hole was made to vary in accordance with the projected volume of the L plug. As a result, it was made clear that the value of λ during the instability phenomenon occurrence could be improved by lengthening the L plugs; that is, reducing the volume of the L plug hole, as is shown in Fig.6.

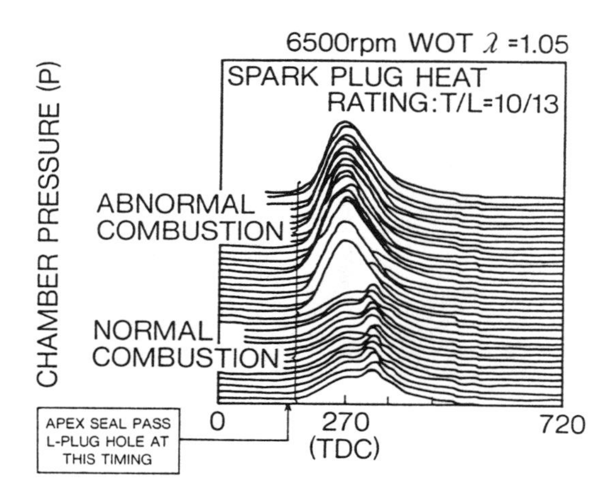

Fig.5 - Pressure Transition with Abnormal Combustion

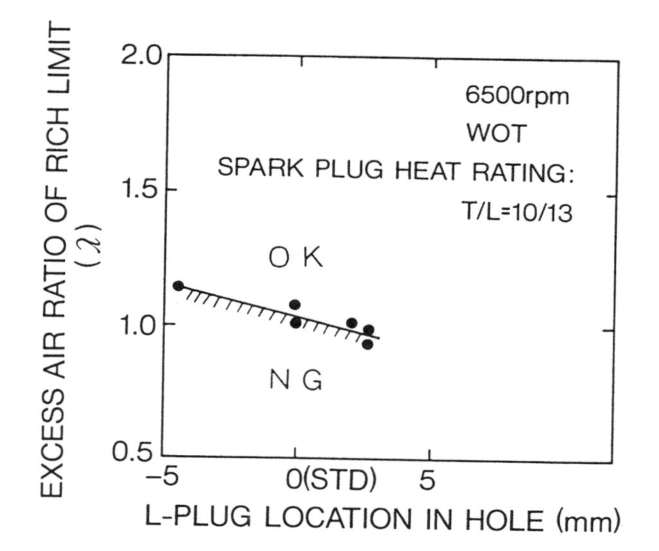

Fig.6 - Effect of L-Plug Location on Rich Limit λ

OUTPUT POWER- Fig.7 shows the correlation between excess air ratio and output power at 2500rpm. Power is enhanced as the value of λ goes down. In case of the CE, however, the excess air ratio could not be decreased over $\lambda=1.8$ because backfire occurred. By contrast, the RE functioned satisfactorily to the extent $\lambda=1$, and naturally achieved a high output power. Next, Fig.8 shows the maximum power ratio of the hydrogen engine to the gasoline engine. While the hydrogen driven CE(pre-mixed method) could yield only 50% of the power attainable by gasoline specs, the hydrogen driven RE achieved 63% by the pre-mixed method and 75% by the induction-cycle-injection. Incidentally, the hydrogen RE shows lower output when compared to the gasoline specs, even when driven at $\lambda=1$, because volumetric efficiency is low. In the present induction-cycle-injection system, the air and hydrogen which are simultaneously taken in result in the restriction of the volume of inhaled air by 30% due to the bulk of hydrogen.

In future developments, the method of direct injection of hydrogen on completion of the induction cycle is expected to be effective in enchancing the power.

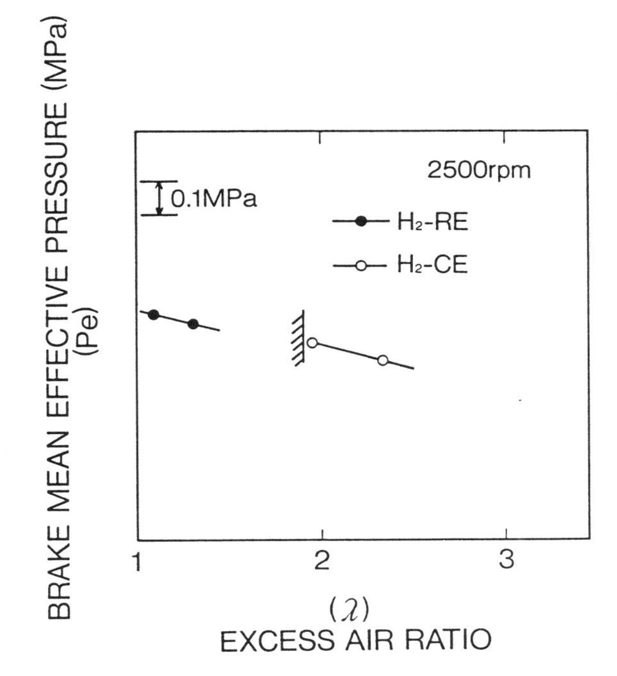

Fig.7 - Comparison between H2-RE and H2-CE on Rich Limit λ

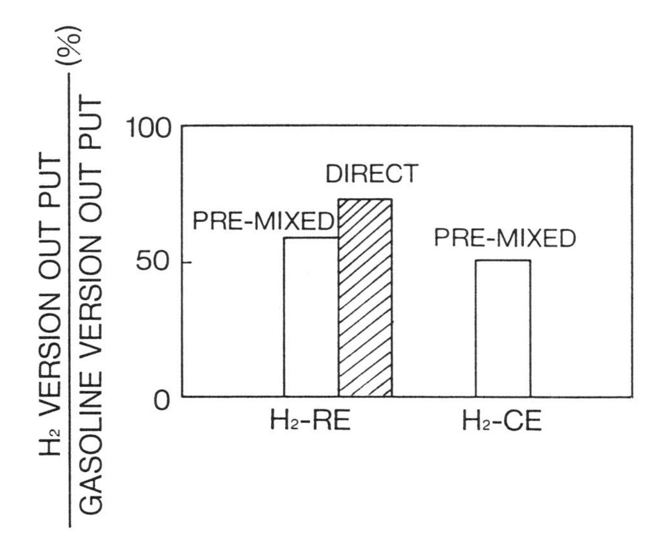

Fig.8 - Comparison between H_2-RE and H_2-CE for Out Put Power

COMBUSTION CHARACTERISTICS AT PART LOAD -The combustion characteristics in induction-cycle-injection method RE at part load was examined regarding indicator diagram, combustion fluctuation, thermal efficiency, and exhaust gas emission. The tests were conducted at N(engine speed)=1500rpm and Pe(brake mean effective pressure) =0.1MPa or 0.29MPa. Fig.9 shows indicator diagrams of various excess air ratio conditions. The indicator diagram indicates two peaks when excess air ratio is small. This is because combustion starts after top dead center due to a short combustion period when λ is small, as shown in Fig.10. The combustion fluctuation ratio was calculated to examine the combustion stability by EQ(1).

$$CFR=\sigma Pmax/PmaxM \quad (1)$$

where
CFR :combustion fluctuation ratio
σPmax:standard deviation of Pmax (350cycle)
PmaxM:mean of Pmax (350cycle)

Fig.11 indicates the combustion fluctuation ratio. This figure shows that hydrogen RE gets stable combustion even if the excess air ratio is over 4. Fig.12 shows excess air ratio vs. brake thermal efficiency. The maximum value at Pe=0.1MPa is around λ=3. From this it can be reasoned that while the pumping loss and specific heat ratio become favorable in

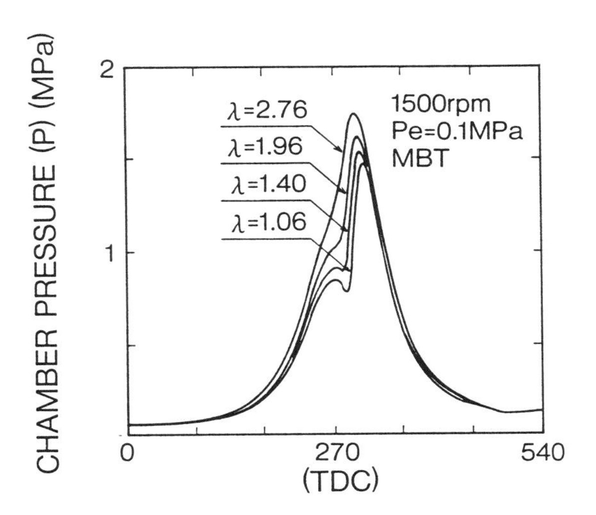

Fig.9 - Effect of Excess Air Ratio on Indicator Diagram

proportion to mixture leanness ,the value should be compensated with inferior burning, thereby settling the maximum value around their balancing point. Fig.13 shows NOx vs. excess air ratio . NOx becomes approximately 0 when λ=1.5 or more at Pe=0.1MPa and λ=2 or more at Pe=0.29MPa.

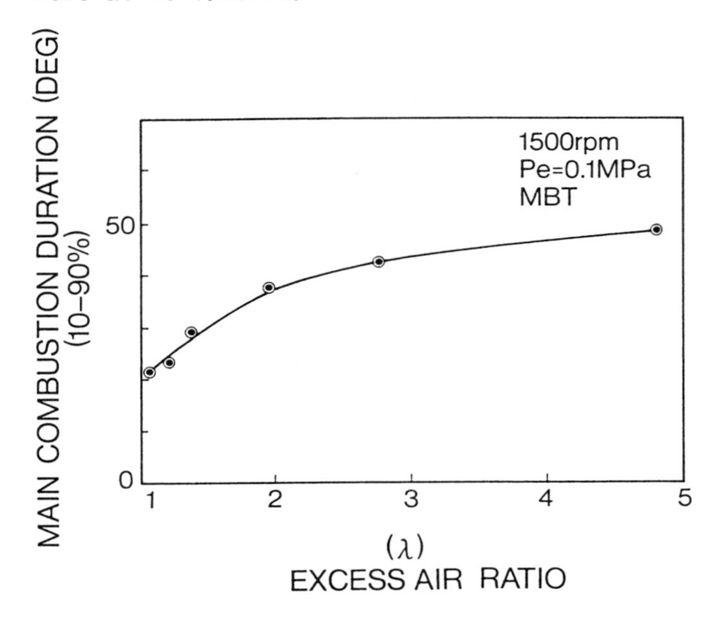

Fig.10 - Main Combustion Duration vs. Excess Air Ratio

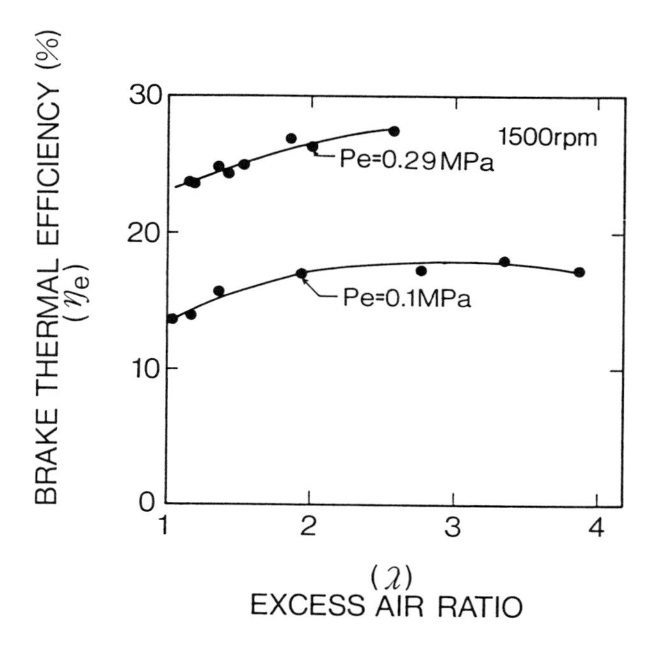

Fig.12 - Brake Thermal Efficiency vs. Excess Air Ratio

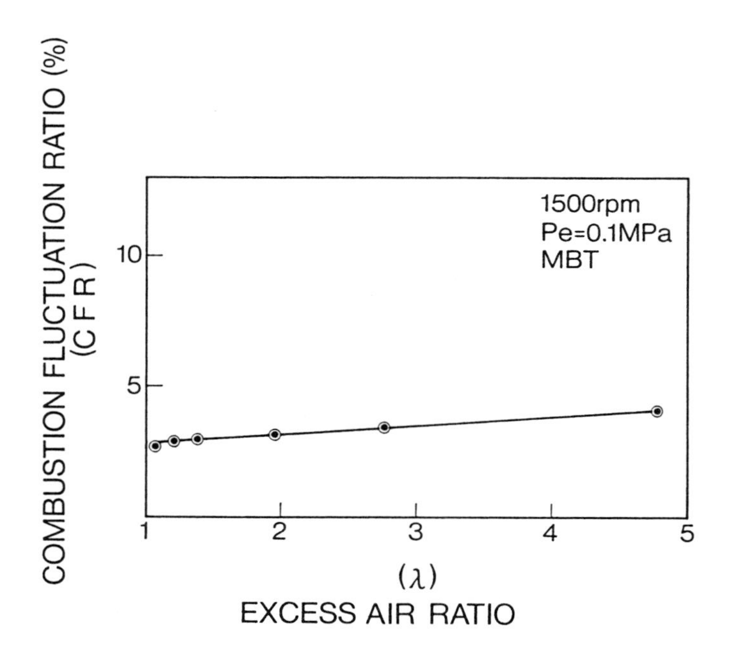

Fig.11 - Combustion Fluctuation Ratio vs. Excess Air Ratio

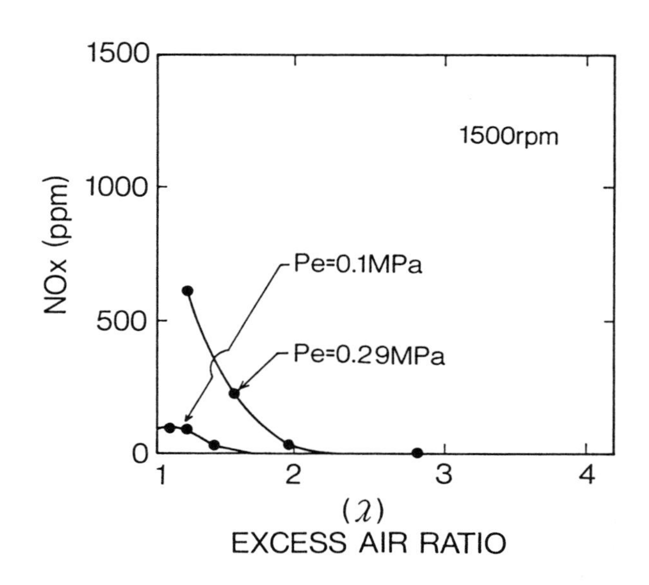

Fig.13 - NOx Emission vs. Excess Air Ratio

CONCLUSIONS

The compatibility of the RE with hydrogen has been demonstrated and the following were also clarified:

(1)The RE is inherently backfire preventive, and can function through a wide operating range at $\lambda=1$, if spark plugs of high heat rating tuned with hydrogen are employed. It was experimentally proved that the pre-mixed hydrogen RE achieved an output power up to 63% of the gasoline specs, while the hydrogen CE with pre-mixed mixture method marked only 50% of the gasoline specs. Furthermore, in case of the RE, the induction-cycle-injection could be implemented easily with an achievement of an output power of 75% of the gasoline specs.

(2)When operated around $\lambda=2\sim3$ in the partial load range, it was proved that a high thermal efficiency together with clean exhaust gas was attainable.

REFERENCES

1. S. Furuhama et al., Transaction of the Society of Automotive Engineers of Japan, No. 6, p. 12, 1973.
2. S. Furuhama, Journal of the Fuel Society of Japan, Vol. 56, No. 605, p. 731, 1977.
3. R. O. King et al., Canadian J. of Tech., Vol. 33, p. 455, 1955.
4. L. M. Das, Proc. 8th WHEC, Vol. 3, p. 1379, 1990.

Appendix A
Bibliography by Subject

Distribution and Transmission

1. Angus, H.C., "Storage, Distribution and Compression of Hydrogen," *Chemistry and Industry (London)*, 68, 1984.

2. Au, M., Chen, C.P., Ye, Z., Fang, T.S., Wu, J. and Wang, Q.D., "An Industrial Scale Experiment on the Recovery, Purification, Storage and Transport of Hydrogen Abstracted from the Purge Gas of a Synthetic Ammonia Plant and the Application of Hydrogen to the Tin Bath in a Float Glass Plant by Means of Mobile Hydride Hydrogen Containers," in *Hydrogen Energy Progress IX*, Edited by T.N. Veziroglu and C. Derive-J. Pottier, International Association for Hydrogen Energy, Coral Gables, FL, p. 1031, 1992.

3. Berry, G.D., Pasternak, A.D., Rambach, G.D., Smith, J.R., and Schock, R.N., "Hydrogen as a Future Transportation Fuel," Report No. UCRL-JC-117945, Lawrence Livermore National Laboratory, 1994.

4. Blazek, C.F., Daniels, E.J., Donakowski, T.D. and Novil, M., "Economics of Hydrogen in the '80s and Beyond," in *Recent Developments in Hydrogen Technology*, Edited by K. D. Williamson Jr. and F. J. Edeskuty, CRC Press, Inc., Boca Raton, FL, p. 1, 1986.

5. Bockris, J. O'M., *Energy Options*, Australia & New Zealand Book Company, Sydney, Australia, 1980.

6. Borel, P. and Mignard, B., "Hydrogen-powered Vehicles: Methodology for the Assessment of Production, Transportation and Distribution Processes," in *Hydrogen Energy Progress IX*, Edited by T.N. Veziroglu and C. Derive-J. Pottier, International Association for Hydrogen Energy, Coral Gables, FL, p. 1041, 1992.

7. Bulckaen, V., "Energy Losses in the Transport in 4000 km Pipelines of Liquid Hydrogen and Oxygen Derived from the Splitting of Water, and of Liquid Methane," *Int. J. Hydrogen Energy*, 17, p. 613, 1992.

8. Bunger, U.H., Andreassen, K., Henriksen, N. and Oyvann, A., "Hydrogen as an Energy Carrier, Production and Liquefacton of Hydrogen in Norway for Transportation to and Storage/Distribution," in *Hydrogen Energy Progress IX*, Edited by T.N. Veziroglu and C. Derive-J. Pottier, International Association for Hydrogen Energy, Coral Gables, FL, p. 1913, 1992.

9. Cacciola, G. and Giordano, N., "Hydrogen Transportation by a Chemical Closed-Loop Cycle, an Alternative Option for Long Distance Transport of Energy," in *Hydrogen Energy Progress V*, Edited by T.N. Veziroglu and J.B. Taylor, Pergamon Press, Elmsford, NY, p. 1237, 1984.

10. Chuveliov, A.V., *Hydrogen in Motor Vehicles: A Case Study of Hydrogen Utilization in Motor Vehicles as a Supplementary Fuel in Southern California Air Basin*, I.V. Kurchatov Institute of Atomic Energy, Report prepared for SCQAMD, 1989.

11. Cialone, H.J., Scott, P.M., Holbrook, J.H., et al, "Hydrogen Effects on Conventional Pipeline Steels," in *Hydrogen Energy Progress V*, Edited by T.N. Veziroglu and J.B. Taylor, Pergamon Press, Elmsford, N.Y., p. 1855, 1984.

12. D'Ajuz, A., Conti, A.M., Mattos, M.C., et al, "Electrical Energy Transmission from the Amazon Region: Hydrogen as a Promising Alternative in Brazil," *Int. J. Hydrogen Energy*, 14, p. 515, 1989.

13. Dalle Donne, M., "Hydrogen as an Energy Carrier in Substituting Petroleum. Demonstration Project:

Automobiles Driven by Nuclear Energy," *Int. J. Hydrogen Energy*, 8, p. 949, 1983.

14. Dutton, R., "Materials Degradation Problems in Hydrogen Energy Systems," *Int. J. Hydrogen Energy*, 9, p. 147, 1984.

15. Giacomazzi, G., "Prospects for Intercontinental Seaborne Transportation of Hydrogen," *Int. J. Hydrogen Energy*, 14, p. 603, 1989.

16. Gretz, J., Baselt, J.P., Ullmann, O. and Wendt, H., "The 100MW Euro-Quebec Hydro-Hydrogen Pilot Project," *Int. J. Hydrogen Energy*, 15, p. 419, 1990.

17. Grob, G.R., "Implementation of a Standardized World Hydrogen System," in *Hydrogen Energy Progress VIII*, Edited by T.N. Veziroglu and P.K. Takahashi, Pergamon Press, Elmsford, N.Y., p. 195, 1990.

18. Kaske, G., Schmidt, P. and Kanngießer, K.W., "Comparison Between High-Voltage Direct-Current Transmission and Hydrogen Transportation," *Int. J. Hydrogen Energy*, 16, p. 105, 1991.

19. Kelley, J.H. and Hagler, R., "Storage, Transmission and Distribution of Hydrogen," *Int. J. Hydrogen Energy*, 5, p. 35, 1980.

20. Kukkonen, C.A. and Shelef, M., "Hydrogen as an Alternative Automobile Fuel: 1993 Update," SAE Technical Paper No. 940766, Society of Automotive Engineers, Warrendale, PA, 1994.

21. Lewis, F.A., "Storage and Distribution of Hydrogen: Materials and Reliability," in *Hydrogen Energy Progress VI*, Edited by T.N. Veziroglu, N. Getoff, and P. Weinzierl, Pergamon Press, Elmsford, NY, p. 979, 1986.

22. Matringe, L. and Busquet, F., "High Flow Hydrogen Pipes Development and Operations," in *Hydrogen Energy Progress IX*, Edited by T.N. Veziroglu and C. Derive-J. Pottier, International Association for Hydrogen Energy, Coral Gables, FL, p. 1057, 1992.

23. Peschka, W., *Liquid Hydrogen—Fuel of the Future*, Springer-Verlag, New York, 1992.

24. Peterson, U., Wursig, G. and Krapp, R., "Design and Safety Considerations for Large Scale Sea-Borne Hydrogen Transport," in *Hydrogen Energy Progress IX*, Edited by T.N. Veziroglu and C. Derive-J. Pottier, International Association for Hydrogen Energy, Coral Gables, FL, p. 1021, 1992.

25. Russell, G., "Hawaii Hydrogen Energy Economy: Production and Distribution of Hydrogen and Oxygen in the District of North Kohala, the Big Island of Hawaii: a Global Prototype," in *Hydrogen Energy Progress IX*, Edited by T.N. Veziroglu and C. Derive-J. Pottier, International Association for Hydrogen Energy, Coral Gables, FL, p. 1863, 1992.

26. Russell, G.W. and Russell, A., "Hawaiian Hydrogen Mass Transit System: A Global Prototype," in *Hydrogen Energy Progress VIII*, Edited by T.N. Veziroglu and P.K. Takahashi, Pergamon Press, Elmsford, N.Y., p. 1587, 1990.

27. Sperling, D. and Kitamura, R., "Refueling and New Fuels: An Exploratory Analysis," *Int. J. Hydrogen Energy* 20A, p. 15, 1986.

28. Styrikovich, M.A. and Malyshenko, S.P., "Bulk Storage and Transmission of Hydrogen," in *Hydrogen Energy Progress VI,* Edited by T.N. Veziroglu, N. Getoff, and P. Weinzierl, Pergamon Press, Elmsford, NY, p. 765, 1986.

29. Veziroglu, T.N. and Barbir, F., "Hydrogen: The Wonder Fuel," *Int. J. Hydrogen Energy*, 17, p. 391, 1992.

30. Williams, L.O., *Hydrogen Power: An Introduction to Hydrogen Energy and its Applications*, Pergamon Press, Oxford, 1980.

Economic and Related Issues

1. Apostolescu, N. and Sfinteanu, D., "Efficiency, Pollution Control and Performances of Hydrogen Fueled Passenger Cars," in *Hydrogen Energy Progress VIII*, Edited by T.N. Veziroglu and P.K. Takahashi, Pergamon Press, Elmsford, N.Y., p. 1323, 1990.

2. Barbir, F. and Veziroglu, T.N., "Effective Costs of the Future Energy Systems," *Int. J. Hydrogen Energy*, 17, p. 299, 1992.

3. Barbir, F., Veziroglu, T.N. and Plass, Jr., H.J., "Environmental Damage due to Fossil Fuels Use," *Int. J. Hydrogen Energy*, 15, p. 739, 1990.

4. Baykara, S.Z. and Bilgen, E., "An Overall Assessment of Hydrogen Production by Solar Thermolysis," *Int. J. Hydrogen Energy*, 14, p. 881, 1989.

5. Berry, G.D., Pasternak, A.D., Rambach, G.D., Smith, J.R., and Schock, R.N., "Hydrogen as a Future Transportation Fuel," Report No. UCRL-JC-117945, Lawrence Livermore National Laboratory, 1994.

6. Bicelli, P., "Hydrogen: A Clean Energy Source," *Int. J. Hydrogen Energy*, 11, p. 555, 1986.

7. Blazek, C.F., Daniels, E.J., Donakowski, T.D. and Novil, M., "Economics of Hydrogen in the '80s and Beyond," in *Recent Developments in Hydrogen Technology*, Edited by K. D. Williamson Jr. and F. J. Edeskuty, CRC Press, Inc., Boca Raton, FL, p. 1, 1986.

8. Block, D.L. and Melody, I., "Efficiency and Cost Goals for Photoenhanced Hydrogen Production Processes," in *Hydrogen Energy Progress VIII*, Edited by T.N. Veziroglu and P.K. Takahashi, Pergamon Press, Elmsford, N.Y., p. 217, 1990.

9. Bockris, J. O'M., *Energy Options*, Australia & New Zealand Book Company, Sydney, Australia, 1980.

10. Bockris, J. O'M., "Hydrogen Economy," *Science*, 176, p. 1323, 1972.

11. Braun, H., *The Phoenix Project*, 1st Edition, Research Analysts, Phoenix, AZ, 1990.

12. Brinner, A., Bussmann, H., Hug, W. and Seeger, W., "Test Results of the Hysolar 10kW PV-Electrolysis Facility," *Int. J. Hydrogen Energy*, 17, p. 187, 1992.

13. Brun-Tsekhovoi, A.R., "On the Concept of Transition Period in Hydrogen Energy Development," *Int. J. Hydrogen Energy*, 17, p. 555, 1992.

14. Buchner, H., "The Question of the Hydrogen Infrastructure for Motor Vehicles," *Int. J. Hydrogen Energy*, 8, p. 373, 1983.

15. Bunger, U.H., Andreassen, K., Henriksen, N. and Oyvann, A., "Hydrogen as an Energy Carrier, Production and Liquefacton of Hydrogen in Norway for Transportation to and Storage/Distribution," in *Hydrogen Energy Progress IX*, Edited by T.N. Veziroglu and C. Derive-J. Pottier, International Association for Hydrogen Energy, Coral Gables, FL, p. 1913, 1992.

16. Chuveliov, A.V., *Hydrogen in Motor Vehicles: A Case Study of Hydrogen Utilization in Motor Vehicles as a Supplementary Fuel in Southern California Air Basin*, I.V. Kurchatov Institute of Atomic Energy, Report prepared for SCQAMD, 1989.

17. Curtis, D.L., "World Energy—An International Solution for a Hydrogen Economy," in *Hydrogen Energy Progress VIII*, Edited by T.N. Veziroglu and P.K. Takahashi, Pergamon Press, Elmsford, N.Y., p. 49, 1990.

18. Darrow, K., Biederman, N. and Konopka, A., "Commodity Hydrogen from Off-Peak Electricity," *Int. J. Hydrogen Energy*, 2, p. 175, 1977.

19. DeLuchi, M.A. and Ogden, J.M., "Solar-Hydrogen Fuel-Cell Vehicles," *Transpn. Res.-A*, 27A, p. 255, 1993.

20. DeLuchi, M.A., "Hydrogen Fuel-Cell Vehicles," Report No. UCD-ITS-RR-92-14, Institute of Transportation Studies, University of California, Davis, September, 1992.

21. DeLuchi, M.A. "Emissions of Greenhouse Gases from the Use of Transportation Fuels and Electricity," Report No. ANL/ESD/TM-22, Argonne National Laboratory, Argonne, IL, November, 1991.

22. DeLuchi, M.A., "Hydrogen Vehicles: an Evaluation of Fuel Storage, Performance, Safety, and Cost," *Int. J. Hydrogen Energy*, 14, p. 81, 1989.

23. Donakowski, T., Blazek, C., Novil, M., and Daniels, E., "Hydrogen Use in a Rural Alaskan Community," in *Hydrogen Energy Progress IV*, Edited by T.N. Veziroglu, W.D. Van Vorst, and J.H. Kelley, Pergamon Press, Elmsford, NY, p. 1453, 1982.

24. Drolet, B., Gretz, J., Kluyskens, D., Sandmann, F., Wurster, R., "The Euro-Quebec Hydro-Hydrogen

Pilot Project (EQHHPP)," in *Hydrogen Energy Progress X*, Edited by D.L. Block and T.N. Veziroglu, International Association of Hydrogen Energy, Coral Gables, 1994.

25. Dutta, S., Block, D.L. and Port, R.L., "Economic Assessment of Advanced Electrolytic Hydrogen Production," *Int. J. Hydrogen Energy*, 15, p. 387, 1990.

26. Fulkerson, W., Judkins, R.J. and Sanghvi, M.K., "Energy from Fossil Fuels," *Scientific American*, 263, p. 129, 1990.

27. Gaines, L.L. and Wolsky, A.M., "Economics of Hydrogen Production: The Next Twenty-Five Years," in *Hydrogen Energy Progress V*, Edited by T.N. Veziroglu and J.B. Taylor, Pergamon Press, Elmsford, NY, p. 259, 1984.

28. Goltsov, V.A., Goltsova, L.F. and Garkusheva, V.A., "Hydrogen Energy Development in the World," in *Hydrogen Energy Progress IX*, Edited by T.N. Veziroglu and C. Derive-J. Pottier, International Association for Hydrogen Energy, Coral Gables, FL, p. 1975, 1992.

29. Goltsova, L.F., Garkusheva, V.A., Alimova, R.F. and Goltsov, V.A., "Scientometric Studies of the Problem of 'Hydrogen Energy and Technology' in the World," *Int. J. Hydrogen Energy*, 15, p. 655, 1990.

30. Gregory, D.P., "The Hydrogen Economy," *Scientific American*, 228, p. 13, 1973.

31. Gretz, J., Baselt, J.P., Ullmann, O. and Wendt, H., "The 100MW Euro-Quebec Hydro-Hydrogen Pilot Project," *Int. J. Hydrogen Energy*, 15, p. 419, 1990.

32. Grob, G.R., "Implementation of a Standardized World Hydrogen System," in *Hydrogen Energy Progress VIII*, Edited by T.N. Veziroglu and P.K. Takahashi, Pergamon Press, Elmsford, N.Y., p. 195, 1990.

33. Hammerli, M., "When Will Electrolytic Hydrogen Become Competitive?" *Int. J. Hydrogen Energy*, 9, p. 25, 1984.

34. Hancock, O.G.J., "A Photovoltaic-Powered Water Electrolyzer: Its Performance and Economics," in *Hydrogen Energy Progress V*, Edited by T.N. Veziroglu and J.B. Taylor, Pergamon Press, Elmsford, NY, p. 335, 1984.

35. Harkin, T., "Proposal for a Sustainable Energy Future Based on Renewable Hydrogen," Washington, DC, June 1993.

36. Hinds, H.R., "IGT activities," *Int. J. Hydrogen Energy*, 7, p. 753, 1982.

37. Hurtak, J.J., "The Future of Hydrogen Technology for Commercial, Military and Environmental Science in Remote Regions," in *Hydrogen Energy Progress IX*, Edited by T.N. Veziroglu and C. Derive-J. Pottier, International Association for Hydrogen Energy, Coral Gables, FL, p. 1949, 1992.

38. Huston, E.L., "Liquid and Solid Storage of Hydrogen," in *Hydrogen Energy Progress V*, Edited by T.N. Veziroglu and J.B. Taylor, Pergamon Press, Elmsford, NY, p. 1171, 1984.

39. Kaske, G., Schmidt, P. and Kanngießer, K.W., "Comparison Between High-Voltage Direct-Current Transmission and Hydrogen Transportation," *Int. J. Hydrogen Energy*, 16, p. 105, 1991.

40. Kloeppel, J. and Rogerson, S., "The Hydrogen Economy," *Electronics World & Wireless World*, 97, p. 668, 1991.

41. Kukkonen, C.A., "Hydrogen as an Alternative Automotive Fuel," SAE Technical Paper No. 810349, Society of Automotive Engineers, Warrendale, PA, 1981.

42. Kukkonen, C.A. and Shelef, M., "Hydrogen as an Alternative Automotive Fuel: 1993 Update," SAE Technical Paper No. 940766, Society of Automotive Engineers, Warrendale, PA, 1994.

43. LeRoy, R.L. and Stuart, A.K., "Present and Future Costs of Hydrogen Production by Unipolar Water Electrolysis," in *Symposium on Industrial Water Electrolysis*, Electrochemical Society, p. 117, 1978.

44. Lin, F.N., Moore, W.I. and Walker, S.W., "Economics of Liquid Hydrogen from Water Electrolysis," in *Hydrogen Energy Progress V*, Edited by T.N. Veziro-

glu and J.B. Taylor, Pergamon Press, Elmsford, NY, p. 249, 1984.

45. Marchetti, C., "Central Place Theory and the Key to Hydrogen Dominance," in *Hydrogen Energy Progress VI,* Edited by T.N. Veziroglu, N. Getoff, and P. Weinzierl, Pergamon Press, Elmsford, N.Y., p. 3, 1986.

46. Németh, N., "Environment and Energy: Problems, Resolutions, Solutions," *Int. J. Hydrogen Energy*, 15, p. 457, 1990.

47. Ogden, J.M., "Cost and Performance Sensitivity Studies for Solar Photovoltaic/Electrolytic Hydrogen Systems," *Solar Cells*, 30, p. 515, 1991.

48. Ogden, J.M., "Renewable Hydrogen Transportation Fuels," in *Solar and Electric Vehicles '92*, 1992.

49. Ogden, J.M. and Nitsch, J., "Solar Hydrogen," in *Renewable Energy: Sources for Fuels and Electricity*, Edited by T.B. Johansson, H. Kelly, A.K.N. Reddy, R.H. Williams, and L. Burnham, Island Press, Washington, D.C., 1993.

50. Ogden, J.M. and Williams, R.H., *Solar Hydrogen: Moving Beyond Fossil Fuels*, World Resources Institute, Washington, D.C., 1989.

51. Okken, P.A., "Costs of Reducing CO_2 Emissions by Means of Hydrogen Energy," *Int. J. Hydrogen Energy*, 18, p. 319, 1993.

52. Petkov, T., Veziroglu, T.M., and Sheffield, J.W., "An Outlook of Hydrogen as an Automotive Fuel," *Int. J. Hydrogen Energy*, 14, p. 449, 1989.

53. Pimentel, D., Rodrigues, G., Wang, T., Abrams, R., Goldberg, K., Staecker, H., Ma, E., Brueckner, L., Trovato, L., Chow, C., Govindarajulu, U., and Boerke, S., "Renewable Energy: Economic and Environmental Issues," *BioScience*, 44, p. 536, 1994.

54. Plass, H.J., Barbir, F., Miller, H.P. and Veziroglu, T.N., "Economics of Hydrogen as a Fuel for Surface Transportation," *Int. J. Hydrogen Energy*, 15, p. 663, 1990.

55. Russell, G., "Hawaii Hydrogen Energy Economy: Production and Distribution of Hydrogen and Oxygen in the District of North Kohala, the Big Island of Hawaii: A Global Prototype," in *Hydrogen Energy Progress IX*, Edited by T.N. Veziroglu and C. Derive-J. Pottier, International Association for Hydrogen Energy, p. 1863, 1992.

56. Russell, G.W. and Russell, A., "Hawaiian Hydrogen Mass Transit System: A Global Prototype," in *Hydrogen Energy Progress VIII*, Edited by T.N. Veziroglu and P.K. Takahashi, Pergamon Press, Elmsford, N.Y., p. 1587, 1990.

57. Siegel, A. and Schott, T., "Optimization of Photovoltaic Hydrogen Production," *Int. J. Hydrogen Energy*, 13, p. 659, 1988.

58. Sosna, M.H., Baichtok, Y.K., Mordkovich, V.Z. and Korostyshevsky, N.N., "Energetics and Economics of the Process for Hydrogen Recovery by Means of the Membranes Made of Palladium Alloys," in *Hydrogen Energy Progress VIII*, Edited by T.N. Veziroglu and P.K. Takahashi, Pergamon Press, Elmsford, N.Y., p. 1201, 1990.

59. Sperling, D. and DeLuchi, M.A., *Alternative Transportation Fuels: An Environmental and Energy Solution*, Quorum Books, New York, 1989.

60. Sperling, D. and DeLuchi, M.A., "Transportation Energy Futures," *Annual Review of Energy*, 14, p. 375, 1989.

61. Stewart, W.F. and Edeskuty, F.J. *Logistics, Economics, and Safety of a Liquid Hydrogen System for Automotive Transportation*, The American Society of Mechanical Engineers, 73-ICT-78, 1973.

62. Stolyarevskii, A.Y. and Chuveliov, A.V., "Economical Damage Evaluation to Population and Environment in the Comparsion of Alternative Production and Application Technologies of Perspective Energy Carriers," in *Hydrogen Energy Progress VI,* Edited by T.N. Veziroglu, N. Getoff, and P. Weinzierl, Pergamon Press, Elmsford, N.Y., p. 94, 1986.

63. Stucki, S., "The Cost of Electrolytic Hydrogen from Off-Peak Power," *Int. J. Hydrogen Energy*, 16, p. 461, 1991.

64. Stuminsky, V., "On the Eve of the Hydrogen Era," in *Hydrogen Energy Progress IX*, Edited by T.N.

Veziroglu and C. Derive-J. Pottier, International Association for Hydrogen Energy, Coral Gables, FL, p. 1959, 1992.

65. Takahashi, P.K., McKinley, R., Antal, M.J., et al, "The Hawaii Hydrogen Plan," in *Hydrogen Energy Progress VIII*, Edited by T.N. Veziroglu and P.K. Takahashi, Pergamon Press, Elmsford, N.Y., p. 135, 1990.

66. Taylor, J.B., Alderson, J.E.A., Kalyanam, K.M., Lyle, A.B. and Phillips, L.A., "Technical and Economic Assessment of Methods for the Storage of Large Quantities of Hydrogen," *Int. J. Hydrogen Energy*, 11, p. 5, 1986.

67. Trimble, K. and Woods, R., "Fuel Cell Applications and Market Opportunities," *Journal of Power Sources*, 29, p. 1990.

68. Veziroglu, T.N. and Barbir, F., "Hydrogen: The Wonder Fuel," *Int. J. Hydrogen Energy*, 17, p. 391, 1992.

69. Winter, C., Klaib, H. and Nitsch, J., "Hydrogen as an Energy Carrier: What is Known? What Do We Need To Learn?" *Int. J. Hydrogen Energy*, 15, p. 79, 1990.

70. Winter, C.J. and Nitsch, J., "Hydrogen Energy—A 'Sustainable Development' Towards a World Energy Supply System for Future Decades," *Int. J. Hydrogen Energy*, 14, p. 785, 1989.

71. Zweig, R.M., "The Hydrogen Economy: Phase I," in *Hydrogen Energy Progress IX*, Edited by T.N. Veziroglu and C. Derive-J. Pottier, International Association for Hydrogen Energy, Coral Gables, FL, p. 1995, 1992.

72. Zweig, R.M., "Health Benefits of a Chinese Hydrogen Economy," *Int. J. Hydrogen Energy*, 12, p. 267, 1987.

73. Zweig, R.M., "Hydrogen: Prime Candidate Alternative Fuel for Solving Air-Pollution Problems," in *Hydrogen Energy Progress IV*, Edited by T.N. Veziroglu, W.D. Van Vorst, and J.H. Kelley, Pergamon Press, Elmsford, NY, p. 1789, 1982.

74. Zweig, R.M., "Hydrogen—Solution to Union of Soviet Socialist Republics Environmental Problems," in *Hydrogen Energy Progress VII*, Edited by T.N. Veziroglu and A.N. Protsenko, Pergamon Press, Elmsford, NY, p. 23, 1988.

75. Zweig, R.M., "Hydrogen—Solution for an Environmental Crisis," in *Hydrogen Energy Progress X*, Edited by D.L. Block and T.N. Veziroglu, International Association for Hydrogen Energy, Coral Gables, FL, p. 151, 1994.

Engines

1. Apostolescu, N. and Sfinteanu, D., "Efficiency, Pollution Control and Performances of Hydrogen Fueled Passenger Cars," in *Hydrogen Energy Progress VIII*, Edited by T.N. Veziroglu and P.K. Takahashi, Pergamon Press, Elmsford, N.Y., p. 1323, 1990.

2. Baker, N., Lynch, F., Mejia, L., and Olvason, L., "A Hydrogen Engine for Underground Mining Vehicles," in *Proceedings of the 18th Intersociety Energy Conversion Engineering Conference*, American Institute of Chemical Engineers, NY, p. 569, 1983.

3. Belogub, A., Kutsenko, A., Epifanov, S. and Kutznetsov, B., "Usage of Hydrogenous Gases as Additional Fuel for Gasoline Vehicle Engines," in *Hydrogen Energy Progress IX*, Edited by T.N. Veziroglu and C. Derive-J. Pottier, International Association for Hydrogen Energy, Coral Gables, FL, p. 1245, 1992.

4. Belogub, A.V. and Talda, G.B., "Fuel Supply System Constructions of Gasoline/Hydrogen Automobiles," *Int. J. Hydrogen Energy*, 16, p. 417, 1991.

5. Belogub, A.V. and Talda, G.B., "Petrol-Hydrogen Truck with Load-carrying Capacity 5 Tons," *Int. J. Hydrogen Energy*, 16, p. 423, 1991.

6. Billings, R.E., "Advances in hydrogen engine conversion technology," *Int. J. Hydrogen Energy*, 8, p. 939, 1983.

7. Billings, R.E., "A hydrogen-powered mass transit system," *Int. J. Hydrogen Energy*, 3, p. 49, 1978.

8. Billings, R.E. and Lynch, F.E., "Performance and Nitric Oxide Control Parameters of the Hydrogen

Engine," Publ. No. 73002, Energy Research, Provo, Utah, 1973.

9. Binder, K. and Withalm, G., "Mixture formation and combustion in a hydrogen engine using hydrogen storage technology," *Int. J. Hydrogen Energy*, 7, p. 651, 1982.

10. Bindon, J., Hind, J., Simmons, J., Mahlknecht, P. and Williams, C., "The Development of a Lean-Burning Carburetor for a Hydrogen-Powered Vehicle," *Int. J. Hydrogen Energy*, 10, p. 297, 1985.

11. Buchner, H., "Hydrogen Use—Transportation Fuel," in *Hydrogen Energy Progress IV*, Edited by T.N. Veziroglu, W.D. Van Vorst, and J.H. Kelley, Pergamon Press, Elmsford, NY, p. 3, 1982.

12. Buchner, H., "The hydrogen/hydride energy concept," *Int. J. Hydrogen Energy*, 3, p. 385, 1978.

13. Buchner, H. and Povel, R., "The Daimler-Benz Hydride Vehicle Project," *Int. J. Hydrogen Energy*, 7, p. 259, 1982.

14. Burstall, A.F., "Experiments on the Behavior of Various Fuels in a High Speed Internal Combustion Engine," *Proc. Inst. Automob. Eng. London*, 22, p. 358, 1927.

15. Cecil, W., "On the application of hydrogen gas to produce a moving power in machinery; with a description of an engine which is moved by the pressure of the atmosphere, upon a vacuum caused by explosions of hydrogen and atmospheric air," *Trans. Cambridge Philos. Soc.*, 1, p. 217, 1822.

16. Dahl, G.W. and Elsing, R., "Modification of the Fuel System of a Turboshaft Engine from Kerosine to Hydrogen," in *Hydrogen Energy Progress IX*, Edited by T.N. Veziroglu and C. Derive-J. Pottier, International Association for Hydrogen Energy, Coral Gables, FL, p. 1341, 1992.

17. Das, L.M., "Hydrogen Engines: A View of the Past and a Look into the Future," *Int. J. Hydrogen Energy*, 15, p. 425, 1990.

18. Das, L.M., "Fuel Induction Techniques for a Hydrogen Operated Engine," *Int. J. Hydrogen Energy*, 15, p. 833, 1990.

19. Das, L.M., "Exhaust Emission Characterization of Hydrogen-Operated Engine System: Nature of Pollutants and Their Control Techniques," *Int. J. Hydrogen Energy*, 16, p. 765, 1991.

20. Das, L.M., "Safety Aspects of a Hydrogen-Fuelled Engine System Development," *Int. J. Hydrogen Energy*, 16, p. 619, 1991.

21. Das, L.M., "Abnormal Combustion in Hydrogen Engines: Causes and Remedies," in *Hydrogen Energy Progress VIII*, Edited by T.N. Veziroglu and P.K. Takahashi, Pergamon Press, Elmsford, N.Y., p. 1379, 1990.

22. Das, L.M. "Studies on timed manifold injection in hydrogen operated spark ignition engine: performance, combustion and exhaust emission characteristics," Ph.D. Thesis, Institute of Indian Technology, New Delhi, India, 1987.

23. Davidson, D., Fairlie, M. and Stuart, A.E., "Development of a Hydrogen-Fuelled Farm Tractor," in *Hydrogen Energy Progress V*, Edited by T.N. Veziroglu and J.B. Taylor, Pergamon Press, Elmsford, N.Y., p. 1623, 1984.

24. Davidson, D., Fairlie, M. and Stuart, A.E., "Development of a Hydrogen-Fuelled Farm Tractor," *Int. J. Hydrogen Energy*, 11, p. 39, 1986.

25. De Boer, P.C.T. and Hulet, J.-F., "Performance of a hydrogen-oxygen-noble gas engine," *Int. J. Hydrogen Energy*, 5, p. 439, 1980.

26. De Boer, P.C.T., McLean, W.J. and Homan, H.S., "Performance and emissions of hydrogen fueled internal combustion engines," *Int. J. Hydrogen Energy*, 1, p. 153, 1976.

27. Desoky, A.A., Halaf, A.S.K. and El-Mahallawy, F.M., "Combustion Process in a Gas Turbine Combustor When Using H_2, NH_3, and LPG Fuels," *Int. J. Hydrogen Energy*, 15, p. 203, 1990.

28. Dini, D., "Energy/Environment Potential Impact of Hydrogen Fueled Engines Operating on Road Vehicles," in *Eighteenth FISITA Congress - The Promise of New Technology in the Automotive Industry*, Society of Automotive Engineers, Warrendale, PA, p. 129, 1990.

29. Dini, D., "Closed Cycle Water Steam (from $H_2/O_2/H_2O$ Combustion) Car," in *Hydrogen Energy Progress IX*, Edited by T.N. Veziroglu and C. Derive-J. Pottier, International Association for Hydrogen Energy, Coral Gables, FL, p. 1351, 1992.

30. Dini, D., "Nitrogen/Hydrogen Automotive Engine to Avoid Atmospheric Pollution," in *Hydrogen Energy Progress VIII*, Edited by T.N. Veziroglu and P.K. Takahashi, Pergamon Press, Elmsford, N.Y., p. 1363, 1990.

31. Dini, D., Botarelli, C., Nardi, G., Pagni, G., Sglavo, V., Sterzo, B.L. and Mazza, D., "Conversion to Hydrogen Operation of an Electronically Controlled Gasoline Supply System," in *Hydrogen Energy Progress X*, Edited by D.L. Block and T.N. Veziroglu, International Association for Hydrogen Energy, Coral Gables, FL, p. 1389, 1994.

32. Drexl, K.W., Holzt, H.P., and Gutmann, M., "Characteristics of a Single-Cylinder Hydrogen-fueled IC Engine using Various Mixture Formulation Methods," presented at the NATO/CCMS 4th Int. Symp. Automotive Propulsion Systems, Vol. 2, Session 6, Arlington, VA, April, 1977.

33. Du, T.-S., Li, J.-D. and Lu, Y.-Q., "Experimental study on spark engine burning methanol-hydrogen mixed fuel," in *Hydrogen Energy Progress VI,* Edited by T.N. Veziroglu, N. Getoff, and P. Weinzierl, Pergamon Press, Elmsford, N.Y., p. 1073, 1986.

34. Eichert, H. and Fischer, M., "Dynamics of Combustion of Hydrogen-Air and Hydrogen-Methane-Air Mixtures," in *Hydrogen Energy Progress VIII*, Edited by T.N. Veziroglu and P.K. Takahashi, Pergamon Press, Elmsford, N.Y., p. 1183, 1990.

35. Erren, R.A. and Hastings-Campbell, W., "Hydrogen: a commercial fuel for internal combustion engines and other purposes," *Journal of the Institute of Fuel*, 6, p. 277, 1933.

36. Escher, W.J.D., "Cooperative international liquid hydrogen automotive progress report," *Int. J. Hydrogen Energy*, 7, p. 519, 1982.

37. Escher, W.J.D., "Hydrogen-fueled Internal Combustion Engine, a Technical Survey of Contemporary U.S. Projects," Report No. TEC-75/005, U.S. Energy Research and Development Administration, Washington, D.C., 1975.

38. Escher, W.J.D., Adt Jr., R.R., Swain, M.R., Papas, J.M., et al, "Hydrogen Engine and Fuel Containment R&D: A Progress Report on the U. S. Department of Energy Program," in *International Symposium on Automotive Propulsion Systems, 5th*, DOE, p. 638, 1980.

39. Feucht, K., Hölzel, G. and Hurich, W., "Perspectives of Mobile Hydrogen Application," in *Hydrogen Energy Progress VII*, Edited by T.N. Veziroglu and A.N. Protsenko, Pergamon Press, Elmsford, N.Y., p. 1963, 1988.

40. Finegold, J.G., "Hydrogen: primary or supplementary fuel for automotive engines," *Int. J. Hydrogen Energy*, 3, p. 83, 1978.

41. Fulton, J., Lynch, F. and Marmaro, R., *Hydrogen for Reducing Emissions from Alternative Fuel Vehicles*, SAE Technical Paper No. 931813, Society of Automotive Engineers, Warrendale, PA, 1993.

42. Furuhama, S., "Two-stroke hydrogen injection engine," *Int. J. Hydrogen Energy*, 4, p. 571, 1979.

43. Furuhama, S., "A Liquid Hydrogen Car with a Two-Stroke Engine and LH_2 Pump," *Int. J. Hydrogen Energy*, 7, p. 809, 1981.

44. Furuhama, S., "State of the Art and Future Trend of Hydrogen-Fueled Engines," *JSAE Soc. Automot. Eng. Rev.*, 53, 1981.

45. Furuhama, S., "Trend of Social Requirements and Technological Development of Hydrogen-Fueled Automobiles," *JSAE review*, 13, p. 4, 1992.

46. Furuhama, S., "Hydrogen Engine Systems for Land Vehicles," *Int. J. Hydrogen Energy,* 14, p. 907, 1989.

47. Furuhama, S., "Hydrogen Engine Systems for Land Vehicles," in *Hydrogen Energy Progress VII*, Edited by T.N. Veziroglu and A.N. Protsenko, Pergamon Press, Elmsford, N.Y., p. 1841, 1988.

48. Furuhama, S. and Fukuma, T., "High output power hydrogen engine with high pressure fuel injection, hot surface ignition and turbocharging," *Int. J. Hydrogen Energy*, 11, p. 399, 1986.

49. Furuhama, S. and Fukuma, T., "High Output Power Hydrogen Engine with High Pressure Fuel Injection, Hot Surface Ignition and Turbo-Charging," in *Hydrogen Energy Progress V*, Edited by T.N. Veziroglu and J.B. Taylor, Pergamon Press, Elmsford, N.Y., p. 1493, 1984.

50. Furuhama, S., Hiruma, M. and Enomoto, Y., "Development of a liquid hydrogen car," *Int. J. Hydrogen Energy*, 3, p. 61, 1978.

51. Furuhama, S. and Kobayashi, Y., "Hydrogen Cars with LH_2-Tank, LH_2-Pump and Cold GH_2-Injection Two-Stroke Engine," SAE Technical Paper No. 820349, Society of Automotive Engineers, Warrendale, PA, 1983.

52. Furuhama, S. and Kobayashi, Y., "Development of a hot-surface-ignition hydrogen injection two-stroke engine," *Int. J. Hydrogen Energy*, 9, p. 205, 1983.

53. Furuhama, S., Matushita, T., Nakajima, T. and Yamaura, K., "Hydrogen Injection Spark Ignition Engine with LH_2 Pump Driven by High Pressure Hydrogen Expander," in *Hydrogen Energy Progress VII*, Edited by T.N. Veziroglu and A.N. Protsenko, Pergamon Press, Elmsford, N.Y., p. 1975, 1988.

54. Furuhama, S., Nakajima, T. and Honda, T., "Rankine Cycle Engines for Utilization of LH_2- Car Fuel as a Low Temperature Source," in *Hydrogen Energy Progress VIII*, Edited by T.N. Veziroglu and P.K. Takahashi, Pergamon Press, Elmsford, N.Y., p. 1399, 1990.

55. Furuhama, S., Ninomiya, Y. and Hruma, M., "NO_x Control in LH_2 -Pump High Pressure Hydrogen Injection Engines," in *Hydrogen Energy Progress IX*, Edited by T.N. Veziroglu and C. Derive-J. Pottier, International Association for Hydrogen Energy, Coral Gables, FL, p. 1295, 1992.

56. Gentili, R., "Lean air-fuel mixtures supplemented with hydrogen for S.I. engines: a possible way to reduce specific fuel consumption?" *Int. J. Hydrogen Energy*, 10, p. 491, 1985.

57. Glasson, N. and Green, R., "High Pressure Hydrogen Injection," in *Hydrogen Energy Progress IX*, Edited by T.N. Veziroglu and C. Derive-J. Pottier, International Association for Hydrogen Energy, Coral Gables, FL, p. 1285, 1992.

58. Gopal, G., Rao, P.S., Gopalakrishnan, K.V. and Murthy, B.S., "Use of hydrogen in dual-fuel engines," *Int. J. Hydrogen Energy*, 7, p. 267, 1982.

59. Gopalakrishnan, K.V., Prabhukumar, G.P. and Nagalingam, B., "Experimental Investigations on the Performance of a Hydrogen-Diesel Dual-Fuel Engine with the Addition of Auxiliary Fuel in the Intake Charge," in *Hydrogen Energy Progress VI,* Edited by T.N. Veziroglu, N. Getoff, and P. Weinzierl, Pergamon Press, Elmsford, NY, p. 1114, 1986.

60. Green, R.K. and Glasson, N.D., "Direct Injector Development for Hydrogen Fuelled Internal Combustion Engines," in *Hydrogen Energy Progress VIII*, Edited by T.N. Veziroglu and P.K. Takahashi, Pergamon Press, Elmsford, N.Y., p. 1285, 1990.

61. Hacohen, J., Pinhasi, G., Puterman, Y. and Sher, E., "Driving Cycle Simulation of a Vehicle Motored by a SI Engine Fueled with H_2-Enriched Gasoline," *Int. J. Hydrogen Energy*, 16, p. 695, 1991.

62. Hoeskstra, R.L., Collier, K. and Mulligan, N., "Demonstration of Hydrogen Mixed Gas Vehicles," in *Hydrogen Energy Progress X*, Edited by D.L. Block and T.N. Veziroglu, International Association for Hydrogen Energy, Coral Gables, FL, p. 1781, 1994.

63. Homan, H.S., De Boer, P.C.T., and McLean, W.J., "The effect of fuel injection on NO_x emissions and undesirable combustion for hydrogen-fuelled piston engines," *Int. J. Hydrogen Energy*, 8, p. 131, 1983.

64. Homan, H.S., Reynolds, R.K., De Boer, P.C.T., and McLean, W.J., "Hydrogen-fuelled diesel engine without timed ignition," *Int. J. Hydrogen Energy*, 4, p. 315, 1979.

65. Houseman, J. and Hoehn, F.W., "A Two-Charge Engine Concept: Hydrogen Enrichment," SAE Technical Paper No. 741169, Society of Automotive Engineers, Warrendale, PA, 1974.

66. Houseman, J. and Voecks, G.E., "Hydrogen engines based on liquid fuels," in *Hydrogen Energy Progress*, Edited by T.N. Veziroglu, K. Fueki, and T. Ohta, Pergamon Press, Elmsford, NY, p. 949, 1981.

67. Hydrogen Consultants Incorporated, "Gaseous Fuel System Development," 1993.

68. Ikegami, M., Miwa, K. and Shioji, M., "A study of hydrogen fuelled compression ignition engines," *Int. J. Hydrogen Energy*, 7, p. 341, 1982.

69. Jing-Ding, L., Ying-Qing, L. and Tian-Shen, D., "Improvement on the combustion of a hydrogen fueled engine," *Int. J. Hydrogen Energy*, 11, p. 661, 1986.

70. King, R.O. and Allan, A.B., "The oxidation, decomposition, ignition, and detonation of fuel vapours and gases. XXIX. The role of nuclei in the ignition by compressiom of gaseous heptane-air mixtures: first paper," *Canadian J. Technol.*, 34, p. 316, 1956.

71. King, R.O. and Rand, M., "The oxidation, decomposition, ignition, and detonation of fuel vapours and gases. XXVII. The hydrogen engine," *Canadian J. Technol.*, 33, p. 445, 1955.

72. King, R.O., Wallace, W.A. and Mahapatra, B., "The oxidation, ignition and detonation of fuel vapours and gases-V. The hydrogen engine and detonation of the end gas by the ignition effect of carbon nuclei formed by pyrolysis of lubricating oil vapor," *Canadian J. Technol.*, 34, 1957.

73. King, R.O., Hayes, S.V., Allan, A.B., Anderson, R.W.P., and Walker, E.J., "The Hydrogen Engine: Combustion Knock and Related Flame Velocity, *Trans. Engineering Istitute of Canada*, 2, p. 143, 1958.

74. Kludjian, V.Z., "Hydrogen for Vehicles—Mazda's Hydrogen Vehicle Development Program," presented at the *5th Annual U.S. Hydrogen Meeting*, Washington, D.C., 1994.

75. Koelsch, R.K. and Clark, S.J., "A comparison of hydrogen and propane fuelling of internal combustion engine," SAE Paper No. 790677, Society of Automotive Engineers, Warrendale, PA, 1979.

76. Kolbenev, I., "Hydrogen in Small Power Engineering," in *Hydrogen Energy Progress IX*, Edited by T.N. Veziroglu and C. Derive-J. Pottier, International Association for Hydrogen Energy, Coral Gables, FL, p. 751, 1992.

77. Kolbenev, I.L., Kolpakov, V.E. and Soldatkin, A.V., "Hydrogen Diesel Fuel Engine for Application on a Universal Tractor," in *Hydrogen Energy Progress VII*, Edited by T.N. Veziroglu and A.N. Protsenko, Pergamon Press, Elmsford, N.Y., p. 2095, 1988.

78. Kordzinski, C., Rudkowski, M., and Papuga, T., "An Experimental Investigation of the Optimum Air-to-Fuel Ratio of a Two-Stroke High-Speed Hydrogen Engine with Spark-Ignition, in *Hydrogen Energy Progress*, Edited by T.N. Veziroglu, K. Fueki, and T. Ohta, Pergamon Press, Elmsford, NY, p. 1231, 1981.

79. Krepec, T., Carrese, G. and Miele, D., "Further Investigations on Electronically Controlled Hydrogen Storage and Injection System for Automotive Applications," in *Hydrogen Energy Progress VIII*, Edited by T.N. Veziroglu and P.K. Takahashi, Pergamon Press, Elmsford, N.Y., p. 925, 1990.

80. Krepec, T., Giannacopoulos, T. and Miele, D., "New electronically controlled hydrogen-gas injector development and testing," *Int. J. Hydrogen Energy*, 12, p. 855, 1987.

81. Krepec, T., Giannacopoulos, T. and Miele, D., "New Electronically Controlled Hydrogen-Gas Injector Development and Testing," in *Hydrogen Energy Progress VI*, Edited by T.N. Veziroglu, N. Getoff, and P. Weinzierl, Pergamon Press, Elmsford, NY, p. 1087, 1986.

82. Krepec, T., Tebelis, T. and Kwok, C., "Fuel Control Systems for Hydrogen-Fueled Automotive Combus-

tion Engines—A Prognosis," *Int. J. Hydrogen Energy*, 9, p. 109, 1984.

83. Kudryash, A.P., Marakhovsky, V.P. and Kaidalov, A.A., "Pecularities of Hydrogen-Fueled Diesel Engine Performance," in *Hydrogen Energy Progress VII*, Edited by T.N. Veziroglu and A.N. Protsenko, Pergamon Press, Elmsford, N.Y., p. 2057, 1988.

84. Lambe, S.M., Watson, H.C. and Milkins, E.E., "Hydrogen Engine with Pilot Diesel Fuel Ignition," in *Hydrogen Energy Progress VIII*, Edited by T.N. Veziroglu and P.K. Takahashi, Pergamon Press, Elmsford, N.Y., p. 1333, 1990.

85. Li, J.-D., Lu, Y.-Q. and Du, T.-S., "Improvement on the Combustion of Hydrogen Fueled Engine," in *Hydrogen Energy Progress V*, Edited by T.N. Veziroglu and J.B. Taylor, Pergamon Press, Elmsford, NY, p. 1579, 1984.

86. Liu, Z. and Karim, G.A., "Knock Characteristics of Dual Fuel Engines Fuelled with Hydrogen," in *Hydrogen Energy Progress X*, Edited by D.L. Block and T.N. Veziroglu, International Association for Hydrogen Energy, Coral Gables, FL, p. 1807, 1994.

87. Lynch, F., "Hydrogen Engines: Emissions and Performance," in *SAE TOPTEC "Emissions from Alternative Fueled Engines*," p. 15, 1991.

88. Lynch, F.E., "Parallel induction: a simple fuel control method for hydrogen engines," *Int. J. Hydrogen Energy*, 8, p. 721, 1983.

89. MacCarley, C.A. "Electronic Fuel Injection Techniques for Hydrogen Fueled I.C. Engines," M.S. Thesis in Engineering, University of California, Los Angeles, CA, 1978.

90. MacCarley, C.A. and Van Vorst, W.D., "Electronic fuel injection techniques for hydrogen powered I.C. engines," *Int. J. Hydrogen Energy*, 5, p. 179, 1980.

91. MacDonald, J.S., "Evaluation of the Hydrogen Supplemented Fuel Concept with an Experimental Multicylinder," SAE Technical Paper No. 760101, Society of Automotive Engineers, Warrendale, PA, 1976.

92. Martorano, L. and Dini, D., "Hydrogen injection in two-stroke reciprocating gas engines," *Int. J. Hydrogen Energy*, 8, p. 935, 1983.

93. Mathur, H.B., "Hydrogen Fueled Internal Combustion Engines," in Proceedings of the National Workshop on Hydrogen Energy, New Delhi, India, July, 1985, p. 159.

94. Mathur, H.B. and Das, L.M., "Performance Characteristics of a Hydrogen Fuelled S.I. Engine Using Timed Manifold Injection," *Int. J. Hydrogen Energy*, 16, p. 115, 1991.

95. Mathur, H.B., Das, L.M. and Patro, T.N., "Effects of Dilutents on Combustion and Exhaust Emission Characteristics of a Hydrogen Operated Diesel Engine," in *Hydrogen Energy Progress IX*, Edited by T.N. Veziroglu and C. Derive-J. Pottier, International Association for Hydrogen Energy, Coral Gables, FL, p. 1333, 1992.

96. Mathur, H.B., Das, L.M. and Patro, T.N., "Hydrogen Fuel Utilization in CI Engine Powered End Utility Systems," *Int. J. Hydrogen Energy*, 17, p. 369, 1992.

97. Mathur, H.B. and Khajuria, P.R., "A computer simulation of hydrogen fueled spark ignition engine," *Int. J. Hydrogen Energy*, 11, p. 409, 1986.

98. Mathur, H.B. and Khajuria, P.R., "Performance and emission characteristics of hydrogen fueled spark ignition engine," *Int. J. Hydrogen Energy*, 9, p. 729, 1984.

99. May, H. and Gwinner, D., "Possibilities of improving exhaust emissions and energy consumption in mixed hydrogen-gasoline operation," *Int. J. Hydrogen Energy*, 8, p. 121, 1983.

100. Mishchenko, A.I., "The Ways of the Setting Up of Automobile-Type Hydrogen Engine Performance," in *Hydrogen Energy Progress V*, Edited by T.N. Veziroglu and J.B. Taylor, Pergamon Press, Elmsford, N.Y., p. 1529, 1984.

101. Morimoto, K., Teramoto, T. and Takamori, Y., "Combustion characteristics in hydrogen fueled rotary engine," SAE Paper No. 920302, Society of Automotive Engineers, Warrendale, PA, 1992.

102. Mulready, R.C., "Liquid Hydrogen Engines, in Technology and Uses of Liquid Hydrogen," Edited by R.B. Scott, W.H. Denton, and C.M. Nicholls, Pergamon Press, Oxford, 1964.

103. Murray, R.G., Schoeppel, R.J. and Gray, C.L. *The Hydrogen Engine in Perspective*, SAE Paper No. 729216, in Proc. *7th Intersociety Energy Conversion Engineering Conference*, American Chemical Society, Washington, D.C., 1971.

104. Nagalingam, B., Duebel, M. and Schmillen, K., "Performance of the Supercharged Spark Ignition Hydrogen Engine," in *Alternate Fuels for Spark Ignition Engines*, Society of Automotive Engineers, Warrendale, PA, p. 97, 1983.

105. Nagalingam, P., Prabhukumar, G.P. and Gopalakrishnan, K.V., "Exprimental Investigations on the Performance of a Hydrogen-Diesel Dual-Fuel Engine with the Addition of Auxilliary Fuels at the Intake Charge," in *Hydrogen Energy Progress VI,* Edited by T.N. Veziroglu, N. Getoff, and P. Weinzierl, Pergamon Press, Elmsford, NY, p. 1114, 1986.

106. Oehmichen, M., "Wasserstoff als Motortreib-mittel," in *Verein Deutsche Ingenieur, Deutsche Kraft-fahrtforschung*, Verlag GMBH, Berlin, 1942.

107. Olavson, L.G., Baker, N.R., Lynch, F.E., and Mejia, L.C., "Hydrogen Fuel for Underground Mining Machinery, SAE Technical Paper No. 840233, Society of Automotive Engineers, Warrendale, PA, 1984.

108. Parks, F.B., "A Single-Cylinder Engine Study of Hydrogen-Rich Fuels," SAE Technical Paper No. 760099, Society of Automotive Engineers, Warrendale, PA, 1976.

109. Peschka, W., "Liquid Hydrogen Pumps for Automotive Application," *Int. J. Hydrogen Energy*, 15, p. 817, 1990.

110. Peschka, W., "The status of handling and storage techniques for liquid hydrogen in motor vehicles," *Int. J. Hydrogen Energy*, 12, p. 753, 1987.

111. Peschka, W., "Liquid hydrogen fueled automotive vehicles in Germany—Status and Development," *Int. J. Hydrogen Energy*, 11, p. 721, 1986.

112. Peschka, W., "Operating characteristics of a LH_2-fuelled automotive vehicle and of a semi-automatic LH_2-refuelling station," *Int. J. Hydrogen Energy*, 7, p. 661, 1982.

113. Peschka, W., "Hydrogen Combustion in Tomorrow's Energy Technology," in *Hydrogen Energy Progress VI*, Edited by T.N. Veziroglu, N. Getoff, and P. Weinzierl, Pergamon Press, Elmsford, NY, p. 1019, 1986.

114. Peschka, W., "Hydrogen Combustion in Tomorrow's Energy Technology," *Int. J. Hydrogen Energy*, 12, p. 481, 1987.

115. Peschka, W. and Nieratschker, W., "Experience and Special Aspects on Mixture Formation of an Otto-Engine Converted for Hydrogen Operation," in *Hydrogen Energy Progress V*, Edited by T.N. Veziroglu and J.B. Taylor, Pergamon Press, Elmsford, N.Y., p. 1537, 1984.

116. Peschka, W. and Nieratschker, W., "Experience and special aspects on mixture formation of an Otto engine converted for hydrogen operation," *Int. J. Hydrogen Energy*, 11, p. 653, 1986.

117. Petkov, T., Veziroglu, T.N. and Sheffield, J.W., "An Outlook of Hydrogen as an Automotive Fuel," *Int. J. Hydrogen Energy*, 14, p. 449, 1989.

118. Pichainarong, P., Iwata, T. and Furuhama, S., "Study of Thermodynamic Analysis in Hydrogen Injection Engines," in *Hydrogen Energy Progress VIII*, Edited by T.N. Veziroglu and P.K. Takahashi, Pergamon Press, Elmsford, N.Y., p. 1275, 1990.

119. Prabhukumar, G.P., Swaminathan, S., Nagalingam, B. and Gopalakrishnan, K.V., "Water induction studies in a hydrogen-diesel dual-fuel engine," *Int. J. Hydrogen Energy*, 12, p. 177, 1987.

120. Qing, M. and Hou-Sheng, L., "A Study of Electronically Controlled Hydrogen Supplement System for a Gasoline-Hydrogen Engine and Improving its Combustion Characteristics," in *Hydrogen Energy Progress VII*, Edited by T.N. Veziroglu and A.N. Protsenko, Pergamon Press, Elmsford, N.Y., p. 1885, 1988.

121. Rainey, S.M. and Veziroglu, T.N., "Computer Modeling and Comparison of Hydrogen-Fueled and

Methane-Fueled Hypersonic Vehicles," *Int. J. Hydrogen Energy*, 17, p. 53, 1992.

122. Raman, V., Hansel, J., Fulton, J., Lynch, F. and Bruderly, D., "Hythane—An Ultraclean Transportation Fuel," in *Hydrogen Energy Progress X*, Edited by D.L. Block and T.N. Veziroglu, International Association for Hydrogen Energy, Coral Gables, FL, p. 1797, 1994.

123. Rao, B.H., Shrivastava, K.N. and Bhakta, H.N., "Hydrogen for dual fuel engine operation," *Int. J. Hydrogen Energy*, 8, p. 381, 1983.

124. Rao, K.S., Ganesan, V., Gopalakrishnan, K.V. and Murthy, B.S., "Modelling of combustion process in a spark ignited hydrogen engine," *Int. J. Hydrogen Energy*, 8, p. 931, 1983.

125. Rao, B.H., Bhakta, H.N., and Shrivastava, K.N., "Hydrogen Utilization in Diesel Engines," in *Alternative Energy Sources II*, Edited by T.M. Veziroglu, Hemisphere Publishing, Washington, D.C., p. 3589, 1981.

126. Ricardo, H.R., "Further Note on Fuel Research I," *Proc. Inst. of Automob. Eng. of London*, 5, p. 327, 1924.

127. Sampath, P. and Shum, F., "Combustion Performance of Hydrogen in a Small Gas Turbine Combustor," in *Hydrogen Energy Progress V*, Edited by T.N. Veziroglu and J.B. Taylor, Pergamon Press, Elmsford, N.Y., p. 1467, 1984.

128. Selvam, P., "Alcohol Fuels—The Question of Their Introduction: A Comparison with Conventional Vehicular Fuels and Hydrogen," *Int. J. Hydrogen Energy*, 17, p. 237, 1992.

129. Sheipak, A. and Isayev, E., "Effective Engine for Gasoline-Hydrogen Fuel," in *Hydrogen Energy Progress IX*, Edited by T.N. Veziroglu and C. Derive-J. Pottier, International Association for Hydrogen Energy, Coral Gables, FL, p. 1253, 1992.

130. Sheipak, A.A. and Kabalkin, V.N., "Adaptation of Truck to Gasoline-Hydrogen Fuel," in *Hydrogen Energy Progress VIII*, Edited by T.N. Veziroglu and P.K. Takahashi, Pergamon Press, Elmsford, N.Y., p. 1355, 1990.

131. Sher, E. and Hacohen, Y., "On the modeling of a SI 4-stroke cycle engine fueled with hydrogen-enriched gasoline," *Int. J. Hydrogen Energy*, 12, p. 773, 1987.

132. Sinclair, L.A. and Wallace, J.S., "Lean limit emissions of hydrogen-fueled engines," *Int. J. Hydrogen Energy*, 9, p. 123, 1984.

133. Smith, J.R., "The Hydrogen Hydrid Option," in *Proceedings of the Workshop on Advanced Components for Electric and Hybrid Electric Vehicles*, Gaithersburg, MD, NIST special publication 860, 1993.

134. Smith, J.R., "Optimized Hydrogen Piston Engines," Report No. UCRL-JC-116894, Lawrence Livermore National Laboratory, 1994.

135. Sorusbay, C. and Veziroglu, T.N., "Mixture Formation Techniques for Hydrogen-Fueled Internal Combustion Engines," in *Hydrogen Energy Progress VII*, Edited by T.N. Veziroglu and A.N. Protsenko, Pergamon Press, Elmsford, N.Y., p. 1909, 1988.

136. Stebar, R.F. and Parks, F.B., *Emission Control with Lean Operation using Hydrogen-Supplemented Fuel*, General Motors Research Publication No. 1537, 1974.

137. Stewart, W.F., "Hydrogen as a Vehicular Fuel," in *Recent Developments in Hydrogen Technology*, Edited by K. D. Williamson Jr. and F. J. Edeskuty, CRC Press, Inc., Boca Raton, p. 69, 1986.

138. Suzuki, K., Uchiyama, Y., and Hama, J., "Research of Hydrogen Fueled Spark Ignition Engine," in *Hydrogen Energy Progress*, Edited by T.N. Veziroglu, K. Fueki, and T. Ohta, Pergamon Press, Elmsford, NY, p. 1027, 1981.

139. Swain, M.R. and Adt, Jr., R.R., "The Hydrogen-Air Fueled Automobile," in *7th Intersociety Energy Conversion Engineering Conference*, American Chemical Society, Washington, D.C., p. 1382, 1972.

140. Swain, M.R., Adt, Jr., R.R., and Pappas, J.M., *Experimental Hydrogen-Fueled Automotive Engine Design Data-base Project*, Report DOE/CS/31212-

1, U.S. Department of Energy, Washington D. C., 1983.

141. Swain, M.R., Pappas, J.M., Adt Jr., R.R. and Escher, W.J.D., "Hydrogen-Fuelled Automotive Engine Experimental Testing to Provide an Initial Design Database," SAE Technical Paper No. 810350, Society of Automotive Engineers, Warrendale, PA, 1981.

142. Swain, M.R., Pappas, J.M., Adt Jr., R.R. and Escher, W.J.D., "Hydrogen engine design data-base summary," in *18th Intersociety Energy Conversion Engineering Conference*, American Institute of Chemical Engineers, New York, NY p. 536, 1983.

143. Swain, M.R., Swain, M.N., and Adt, R.R., "Considerations in the Design of an Inexpensive Hydrogen-Fueled Engine," SAE Technical Paper No. 881630, Society of Automotive Engineers, Warrendale, PA, 1988.

144. Swain, M.R., Swain, M.N., Leisz, A. and Adt Jr., R.R., "Hydrogen Peroxide Emissions from a Hydrogen-Fueled Engine," *Int. J. Hydrogen Energy*, 15, p. 263, 1990.

145. Swain, M.R., Yusuf, M.J., Dulger, Z. and Swain, M.N., "The Effects of Hydrogen Addition on Natural Gas Engine Operation," SAE Technical Paper No. 932775, Society of Automotive Engineers, Warrendale, PA, 1993.

146. Takano, E., Takamori, Y., Morimoto, K. and Teramoto, T., "Development of the Direct Injection Hydrogen-Fueled Rotary Engine," in *Hydrogen Energy Progress IX*, Edited by T.N. Veziroglu and C. Derive-J. Pottier, International Association for Hydrogen Energy, Coral Gables, FL, p. 1235, 1992.

147. Takiguchi, M. and Furuhama, S., "Combustion Improvement of Liquid Hydrogen Engine for Medium Duty Trucks," SAE Technical Paper No. 870535, Society of Automotive Engineers, Warrendale, PA, 1987.

148. Teramoto, T., Takamori, Y. and Morimoto, K., "Hydrogen Fuelled Rotary Engine," SAE Technical Paper No. 925011, Society of Automotive Engineers, Warrendale, PA, 1992.

149. Underwood, O.L. and Dieges, P.B., "Hydrogen and Oxygen Combustion for Pollution-Free Operation of Existing Standard Automotive Engines," in *Proceedings of the 6th Intersociety Energy Conversion Engineering Conference*, SAE Technical Paper No. 719046, 1971.

150. Van Vorst, W.D. and Woolley, R.L., "Hydrogen-fueled surface transportion," in *Hydrogen: Its Technology and Implications*, Edited by K. E. Cox and D. K. Williamson Jr., CRC Press, Boca Raton, FL, 1979.

151. Varde, K.S. and Frame, G.A., "Hydrogen aspiration in a direct injection diesel engine—its effects on smoke and other engine performance parameters," *Int. J. Hydrogen Energy*, 8, p. 549, 1983.

152. Varde, K.S. and Frame, G.A., "Development of a High Pressure Hydrogen Injection for SI Engine and Results of Engine Behavior," in *Hydrogen Energy Progress V*, Edited by T.N. Veziroglu and J.B. Taylor, Pergamon Press, Elmsford, N.Y., p. 1505, 1984.

153. Varde, K.S. and Frame, G.A., "Development of a high-pressure hydrogen injection for SI engine and results of engine behavior," *Int. J. Hydrogen Energy*, 10, p. 743, 1985.

154. Varde, K.S. and Frame, G.M., "A study of combustion and engine performance using electronic hydrogen fuel injection," *Int. J. Hydrogen Energy*, 9, p. 327, 1984.

155. Wallace, J.S. and Cattelan, A.I., "Hythane and CNG Fuelled Engine Exhaust Emission Comparison," in *Hydrogen Energy Progress X*, Edited by D.L. Block and T.N. Veziroglu, p. 1761, 1994.

156. Watson, H.C. and Lambe, S.M., "A study of ultra fast combustion in a hydrogen engine with pilot diesel fuel ignition," in *Hydrogen Energy Progress IX*, Edited by T.N. Veziroglu and C. Derive-J. Pottier, International Association for Hydrogen Energy, Coral Gables, FL, p. 1271, 1992.

157. Weigang, W. and Lainfang, Z., "The Research on Internal Combustion Engine with the Mixed Fuel of Diesel and Hydrogen," in *International Symposium*

on Hydrogen Systems, China Academic Publishers, p. 83, 1985.

158. Weil, K.H., "The hydrogen I.C. engine—its origin and future in the emerging-transportation-environment system," in *7th Intersociety Energy Conversion Engineering Conference*, American Chemical society, Washington, D.C., p. 1355, 1972.

159. Welch, A.B. and Wallace, J.S. *Performance characteristics of a hydrogen-fueled diesel engine with ignition assistance, Final Report*, NRCC, DSS Contract File No. 24 SU.31155-2-2664, Serial No. ISU 82-00340, 1986.

160. Winfield, B., "Hydrogen: It's Come a Long Way Since the Hindenburg," *Automotive Magazine*, p. 60, April, 1992.

161. Withalm, G. and Gelse, W., "The Mercedes-Benz Hydrogen Engine for Application in a Fleet Vehicle," in *Hydrogen Energy Progress VI,* Edited by T.N. Veziroglu, N. Getoff, and P. Weinzierl, Pergamon Press, Elmsford, NY, p. 1185, 1986.

162. Wolpers, F., Gelse, W. and Withalm, G., "Comparative Investigation of a Hydrogen Engine with External Mixture Formation Which Can Either be Operated with Cryogenic Hydrogen or Non-Cryogenic Hydrogen and Water Injection," in *Hydrogen Energy Progress VII*, Edited by T.N. Veziroglu and A.N. Protsenko, Pergamon Press, Elmsford, N.Y., p. 2119, 1988.

163. Wong, J.K.S., "Compression Ignition of Hydrogen in a Direct Injection Diesel Engine Modified to Operate as a Low-Heat-Rejection Engine," *Int. J. Hydrogen Energy*, 15, p. 507, 1990.

164. Woolley, R.L. and Anderson, V.R., "Hydrogen Engine NO_x Control by Water Induction," Billings Energy Corporation Publication No. 77001.

165. Woolley, R.L., "Hydrogen Engine NO_x Control by Water Induction," presented at the NATO/CCMS 4th Int. Symp. Automotive Propulsion Systems, Arlington, VA, April, 1977.

166. Woolley, R.L. and Hendrickson, D.L., "Water Induction in Hydrogen-Powered IC Engines," *Int. J. Hydrogen Energy*, 1, p. 401, 1977.

167. Zhaoxin, M., Tianqiang, H. and Fanru, M., "Study on Hydrogen as Subsidiary Fuel for Swirl Chamber Diesel Engine," in *Hydrogen Energy Progress VIII*, Edited by T.N. Veziroglu and P.K. Takahashi, Pergamon Press, Elmsford, N.Y., p. 1343, 1990.

Fuel Cells

1. Adcock, P.L., Barton, R.T., Dudfield, C.D., Mitchell, P.J. and Naylor, P., "Prospects for the Application of Fuel Cells in Electric Vehicles," *Journal of Power Sources*, 37, p. 201, 1990.

2. Adcock, P.L., Barton, R.T., Dudfield, C.D., et al, "Prospects for the Application of Fuel Cells in Electric Vehicles," *Journal of Power Sources*, 37, p. 201, 1992.

3. Altseimer, J.H., Nochumson, D.H., and Frank, J.A., "Fuel Cell Propulsion Systems for Large Trasportation Vehicles: Buses, Freight Locomotives, and Marinecraft," in *Proc. 18th Intersociety Energy Conversion Engineering Conference*, American Institute of Chemical Engineers, New York, p. 1435, 1983.

4. Amphlett, J.C., Mann, R.F., Peppley, B.A., and Stokes, P.M., "Some Design Considerations for a Catalytic Methanol Steam Reformer," in Proceedings of the 26th Intersociety Energy Conversion Engineering Conference, American Nuclear Society, La Grange, IL, p. 642, 1991.

5. Anahara, R., "Phosphoric Acid Fuel Cells (PAFCs) for Commercialization," *Int. J. Hydrogen Energy*, 17, p. 375, 1992.

6. Anand, N.K., Appleby, A.J., Dhar, H.P., Ferreiera, A.C., et al, "Recent Progress in Proton Exchange Membrane Fuel Cells at Texas A&M University," in *Hydrogen Energy Progress X*, Edited by D.L. Block and T.N. Veziroglu, International Hydrogen Energy Association, Coral Gables, FL, p. 1669, 1994.

7. Appleby, A.J. and Foulkes, F.R., *Fuel Cell Handbook*, Van Nostrand Reinhold, New York, 1989.

8. Appleby, A.J., Richter, G.J., Selman, J.R. and Winsel, A., "Conversion of Hydrogen in Fuel Cells," in *Electrochemical Hydrogen Technologies*, Edited by H. Wendt, Elsevier Science Publishing Company Inc., New York, p. 373, 1990.

9. Appleby, A.J., Richter, G.J., Selman, J.R. and Winsel, A., "Current Technology of PAFC, MCFC and SOFC Systems: Status of Present Fuel Cell Power Plants," in *Electrochemical Hydrogen Technologies*, Edited by H. Wendt, Elsevier Science Publishing Company Inc., New York, p. 425, 1990.

10. Appleby, A.J., "Fuel Cells and Hydrogen Fuel," in *Hydrogen Energy Progress IX*, Edited by T.N. Veziroglu and C. Derive-J. Pottier, International Association for Hydrogen Energy, Coral Gables, FL, p. 1375, 1992.

11. Appleby, A.J., "Fuel Cell Technology and Innovation," *Journal of Power Sources*, 37, p. 223, 1992.

12. Appleby, A.J., "Fuel Cell Electrolytes: Evolution, Properties and Future Prospects," *Journal of Power Sources*, 49, p. 15, 1994.

13. Appleby, A.J., "Fuel Cells and Hydrogen Fuel," *Int. J. Hydrogen Energy*, 19, p. 175, 1994.

14. Appleby, A.J., "Fuel Cells for Tactical Battlefield Power," *IEEE Aerospace and Electronics Systems Magazine*, 6, p. 49, 1991.

15. Appleby, A.J., "From Sir William Grove to Today: Fuel Cells and the Future," *Journal of Power Sources*, 29, p. 3, 1990.

16. Appleby, A.J., "Grove Anniversary Fuel Cell Symposium—Closing Remarks," *Journal of Power Sources*, 29, p. 267, 1990.

17. Appleby, A.J., Velev, O.A., et al, "Polymeric Perfluoro Bis-sulfonimides as Possible Fuel Cell Electrolytes," *Journal of the Electrochemical Society*, 140, p. 109, 1993.

18. Arshinov, A.N., Vaskov, N.I., Golin, Y.L., Kozin, V.G., Kornilov, V.F., Kuznetsov, L.M., Ovchinnikov, A.T. and Pospelov, B.S., "The Electrochemical Direct Current Generator for Space System 'Buran,'" in *Hydrogen Energy Progress IX*, Edited by T.N. Veziroglu and C. Derive-J. Pottier, International Association for Hydrogen Energy, Coral Gables, FL, p. 1485, 1992.

19. Baldwin, R., Pham, M., Leonida, A., et al, "Hydrogen-Oxygen Proton-Exchange Membrane Fuel Cells and Electrolyzers," *Journal of Power Sources*, 29, p. 399, 1990.

20. Bernardi, D.M., "Water-Balance Calculations for Solid-Polymer Electrolyte Fuel Cells," *Journal of Electrochemical Society*, 137, p. 3344, 1990.

21. Billings, R.E., Sanchez, M., Cherry, P. and Eyre, D.B., "LaserCell Prototype Vehicle," *Int. J. Hydrogen Energy*, 16, p. 829, 1991.

22. Blomen, L.J.M. and Mugerwa, M.N., *Fuel Cell Systems*, Plenum Press, New York, NY, 1993.

23. Broers, G.H.J. "High Temperature Galvanic Fuel Cells," Dissertation, Univ. Amsterdam, 1958.

24. Cameron, D.S., "World Developments of Fuel Cells," *Int. J. Hydrogen Energy*, 15, p. 669, 1990.

25. Chawla, S.K. and Ghosh, K.K., "Thermodynamic Analysis of Hydrogen Production from Biogas for Phosphoric Acid Fuel Cell," *Int. J. Hydrogen Energy*, 17, p. 405, 1992.

26. Chi, C.V., Glenn, D.R., and Abens, S.G., "Development of a Fuel-Cell Power Source for Bus," in *Proceedings of the 25th Intersociety Energy Conversion Engineering Conference*, American Institute of Chemical Engineers, NY, p. 308, 1990.

27. Crane, P. and Scott, D.S., "Efficiency and CO_2 Emission Analysis of Pathways by Which Methane Can Provide Transportation Services," *Int. J. Hydrogen Energy*, 17, p. 543, 1992.

28. Creveling, H.F., "Proton Exchange Membrane (PEM) Fuel Cell System R&D for Transportation Applications," in *The Annual Automotive Technology Development Contractors' Coordination Meeting*, Society of Automotive Engineers, Inc., Warrendale, PA, p. 485, 1992.

29. Crowe, B.J., "Fuel Cells—A Survey," NASA SP-5115, NASA, Washington, D.C., 1973.

30. Dees, D.W. and Kumar, R., "Status of Solid Oxide Fuel Cells for Transportation," in *Fuel Cell, 1990 Fuel Cell Seminar, Program and Abstracts*, Courtesy Associates, Inc., Washington, D.C., p. 71, 1990.

31. DeGeeter, E., Van den Broeck, H., Bout, P., Woortmann, M., Cornu, J., Peski, V., Dufour, A. and Marcenaro, B., "Eureka Fuel Cell Bus Demonstration Project," in *Hydrogen Energy Progress X*, Edited by D.L. Block and T.N. Veziroglu, International Association for Hydrogen Energy, Coral Gables, FL, p. 1457, 1994.

32. DeLuchi, M. *Hydrogen Fuel-Cell Vehicles*, Research Report UCD-ITS-RR-92-14, Institute of Transportation Studies, University of California, Davis, 1992.

33. DeLuchi, M.A. and Ogden, J.M., "Solar-Hydrogen Fuel-Cell Vehicles," *Transportation Research, Part A, Policy and Practice*, 27A, p. 255, 1993.

34. DeLuchi, M.A., Larson, E.D., and Williams, R.H., "Hydrogen and Methanol: Production from Biomass and Use in Fuel Cell and Internal Combustion Engine Vehicles," PU/CEES Report No. 263, Center for Energy and Environmental Studies, Princeton University, NJ, 1991.

35. Demin, A.K., Alderucci, V., Ielo, I., et al, "Thermodynamic Analysis of Methane Fueled Solid Oxide Fuel Cell System," *Int. J. Hydrogen Energy*, 17, p. 451, 1992.

36. Dimin, A.K., "The Method of Increasing of SOFC Efficiency," in *Hydrogen Energy Progress IX*, Edited by T.N. Veziroglu and C. Derive-J. Pottier, International Association for Hydrogen Energy, Coral Gables, FL, p. 1527, 1992.

37. Douglas, J., "Fuel Cells for Urban Power," *EPRI Journal*, September 1991.

38. Douglas, J., "Solid Futures in Fuel Cells," *EPRI Journal*, March, p. 6, 1994.

39. Frank, J.A., Altseimer, J.H., and Nochumson, D.H., "Fuel Cell Propulsion Systems for Small Transportation Vehicles, in *Proc. 18th Intersociety Energy Conversion Engineering Conference*, American Institute of Chemical Engineers, New York, p. 1425, 1983.

40. Freni, S., Cavallaro, S., Aquino, M., Ravida, D. and Giordano, N., "Life-time Limiting Factors for a MCFC," in *Hydrogen Energy Progress IX*, Edited by T.N. Veziroglu and C. Derive-J. Pottier, International Association for Hydrogen Energy, Coral Gables, FL, p. 1405, 1992.

41. Fuhrer, O., Rieke, S., Schmitz, C., Willer, B. and Wollny, N., "Alkaline Eloflux Fuel Cells and Electrolysis Cells," in *Hydrogen Energy Progress IX*, Edited by T.N. Veziroglu and C. Derive-J. Pottier, International Association for Hydrogen Energy, Coral Gables, FL, p. 1445, 1992.

42. Ganser, B. and Höhlein, B., "Hydrogen from Methanol: Fuel Cells in Mobile Systems," in *Hydrogen Energy Progress IX*, Edited by T.N. Veziroglu and C. Derive-J. Pottier, International Association for Hydrogen Energy, Coral Gables, FL, p. 1321, 1992.

43. Gao, Y., Noguchi, F., Mitamura, T., et al, "Thermally Prepared Air Electrode for Fuel Cell," *Electrochimica Acta*, 37, p. 1327, 1992.

44. Ghouse, M., Aba-oud, H., Bajunaid, M., Iftikhar, M. and Al-Garni, M., "Phosphoric Acid Fuel Cell R&D Activities at KACST," in *Hydrogen Energy Progress IX*, Edited by T.N. Veziroglu and C. Derive-J. Pottier, International Association for Hydrogen Energy, Coral Gables, FL, p. 1559, 1992.

45. Gillis, E., "Molten Carbonate Fuel Cell Technology," *EPRI Journal*, April/May 1994, p. 34, 1994.

46. Goldstein, R., "Solid Oxide Fuel Cell Development," *EPRI Journal*, October/November 1992, p. 32, 1992.

47. Gonzalez, E.R. and Srinivasan, S., "Electrochemistry of Fuel Cells for Transportation Applications," in *Hydrogen Energy Progress IV*, Edited by T.N. Veziroglu, W.D. Van Vorst, and J.H. Kelley, Pergamon Press, Elmsford, NY, p. 1189, 1982.

48. Gottesfield, S., Wilson, M.S., Zawodzinski, T. and Lemons, R.A., "Core Technology R&D for PEM

Fuel Cells," in proceedings of *The Annual Automotive Technology Development Contractors' Coordination Meeting*, Society of Automotive Engineers, Inc., Warrendale, PA, p. 511, 1992.

49. Harmon, R., "Alternative Vehicle-Propulsion Systems," *Mechanical Engineering*, 114, p. 58, 1992.

50. Hischenhofer, J.H., "Fuel Cell Technology Status," *IEEE AES Systems Magazine*, p. 21, November 1993.

51. Hischenhofer, J.H., "Fuel Cell Status, 1994," *IEEE AES Systems Magazine*, November 1994.

52. Hoogeveen, P., Marcenaro, B.G., Vermeeren, L. and Cornu, J.P., "Eureka Fuel Cell Bus," in *Hydrogen Energy Progress IX*, Edited by T.N. Veziroglu and C. Derive-J. Pottier, International Association for Hydrogen Energy, Coral Gables, FL, p. 1227, 1992.

53. Howard, P., "Ballard Fuel Cell Powered ZEV Bus," in *World Car 2001 Conference*, College of Engineering Center for Environmental Research and Technology, University of California, Riverside, 1993.

54. Hsu, M. and Tai, H., "Planar Solid Oxide Fuel Cell Technology Development, in *Fuel Cell, 1990 Fuel Cell Seminar, Program and Abstracts*, Courtesy Associates, Inc., Washington, D.C., p. 115, 1990.

55. Jones, L.E., Hayward, G.W., Kalyanam, K.M., et al, "Fuel Cell Alternative for Locomotive Propulsion," *Int. J. Hydrogen Energy*, 10, p. 505, 1985.

56. Karth, S. and Grimes, P., "Fuel Cells: Energy Conversion for the Next Century," *Physics Today*, Vol 46, p. 54, 1994.

57. Kaufman, A., "Phosphoric Acid Fuel Cell Bus Development," in proceedings of *The Annual Automotive Technology Development Contractors' Coordination Meeting*, Society of Automotive Engineers, Inc., p. 517, 1992.

58. Kelly, H. and Williams, R.H. *Fuel Cells and the Future of the US Automobile*, Report for the Office of Technology Assessment, U.S. Congress, 1992.

59. Kimble, M.C. and White, R.E., "A Mathematical Model of a Hydrogen/Oxygen Alkaline Fuel Cell," *Journal of Electrochemical Society*, 138, p. 3370, 1991.

60. Kimble, M.C. and White, R.E., "Parameter Sensitivity and Optimization Predictions of a Hydrogen/Oxygen Alkaline Fuel Cell Model," *Journal of the Electrochemical Society*, 139, p. 478, 1992.

61. Kordesch, K.V., "Hydrogen-Air/Lead Battery Hybrid System for Vehicle Propulsion," Abstr. No. 10, Electrochemical Society, 1970.

62. Kordesch, K.V., "City Car with H_2-Air Fuel Cell Lead Battery (One Year Operating Experience)," Intersociety Energy Conversion Engineering Conference, Boston, p. 103, August, 1971.

63. Kordesch, K., "Fuel Cells: The Present State of the Technology and Future Applications, with Special Consideration of the Alkaline Hydrogen/Oxygen (AIR) Systems," in *Hydrogen Energy Progress VI,* Edited by T.N. Veziroglu, N. Getoff, and P. Weinzierl, Pergamon Press, Elmsford, NY, p. 1201, 1986.

64. Kordesch, K., Holzleithner, K., Kalal, P., et al, "Hydrogen Fuel Cells as Direct Energy Converters," in *Hydrogen Energy Progress VI,* Edited by T.N. Veziroglu, N. Getoff, and P. Weinzierl, Pergamon Press, Elmsford, NY, p. 1233, 1986.

65. Kordesch, K., Gruber, C., Gsellmann, J., Kalal, P., Oliveira, J.C.T., et al, "Fuel Cell Research and Development Projects in Austria," *Int. J. Hydrogen Energy*, 14, p. 915, 1989.

66. Kordesch, K., Oliveira, J.C.T., Kalal, P. et al, "Fuel Cell R&D—Toward a Hydrogen Economy," *Electric Vehicle Developments*, 8, p. 25, 1989.

67. Korovin, N.V., "Fuel Cells and Hydrogen Fuel," in *Hydrogen Energy Progress IX*, Edited by T.N. Veziroglu and C. Derive-J. Pottier, International Association for Hydrogen Energy, Coral Gables, FL, p. 1385, 1992.

68. Korovin, N.V., "Electrochemical Power Plants Based on Fuel Cells: Their State and Prospects," *Thermal Engineering*, Vol. 41, p. 22, 1994.

69. Kusunoki, A., Matsubara, H., Kakuoka, Y., Yanagi, C., Kugimiya, K., Yoshino, M., Tokura, M.,

Watanabe, K., Ueda, S., Sumi, M., Miyamoto, H. and Tokunaga, S., "Development of Pressure Electrolyser and Fuel Cell with Polymer Electrolyte," in *Hydrogen Energy Progress IX*, Edited by T.N. Veziroglu and C. Derive-J. Pottier, International Association for Hydrogen Energy, Coral Gables, FL, p. 1419, 1992.

70. Lehman, P.A. and Chamberlin, C.E., "Design of a Photovoltaic-Hydrogen-Fuel Cell Energy System," *Int. J. Hydrogen Energy*, 16, p. 349, 1991.

71. Lehman, P.A. and Chamberlin, C.E., "A Photovoltaic-Hydrogen Fuel Cell Energy System: Control Strategy, Monitoring System Design, and Preliminary Results," in *Project Hydrogen '91: World Hydrogen Energy Conference*, p. 12, 1991.

72. Lemons, R.A., "Fuel Cells for Transportation," *Journal of Power Sources*, 29, p. 251, 1990.

73. Liebhafsky, H.A. and Cairns, E.J., *Fuel Cells and Fuel Batteries*, John Wiley & Sons, 1968.

74. Lindstrom, O., "Fuel Cell Markets: A Look at Economics and Commercial Installations," *Chemtech*, p. 44, January, 1989.

75. Lindstrom, O., "Scenarios for Fuel Cell Utilities," *Chemtech*, p. 122, February, 1989.

76. Lipkin, R., "Firing Up Fuel Cells," *Science News*, 144, p. 314, 1993.

77. Lloyd, A.C., "The California Plan," in *2nd Annual U.S. Hydrogen Meeting*, p. 8, 1991.

78. Lloyd, A.C., "Second Grove Fuel Cell Symposium," in *Second Grove Fuel Cell Symposium*, 1992.

79. Lovering, D.G., "Second Grove Fuel Cell Symposium," *Int. J. Hydrogen Energy*, 17, p. 559, 1992.

80. Lynn, D.K., McCormick, J.B., Bobbett, R.E., Srinivasan, S., and Huff, J.R., "Design Considerations for Vehicular Fuel Cell Power Plants, in *Proc. 16th Intersociety Energy Conversion Engineering Conference*, American Society of Mechanical Engineers, New York, p. 722, 1981.

81. Lynn, D.K., McCormick, J.B., Bobbett, R.E., Derouin, C. and Kerwin, W.J., "Fuel Cell Technologies for Vehicular Applications," SAE Technical Paper No. 800059, Society of Automotive Engineers, Warrendale, PA, 1980

82. Lynn, D.K., McCormick, J.B., Bobbett, R.E., Huff, J.R., and Srinivasan, S., "Acid Fuel Cell Technologies for Vehicular Power Plants," in *Proc. 17th Intersociety Energy Conversion Engineering Conference*, Institute of Electrical and Electronics Engineers, New York, p. 663, 1982.

83. Makansi, J., "Technology Shifts, Support Strengthens for Fuel-Cell Powerplants," *Power Magazine*, September 1991.

84. Marcenaro, B.G. and Andreoli, G.L., "Eureka Fuel Cell Bus," in *Hydrogen Energy Progress IX*, Edited by T.N. Veziroglu and C. Derive-J. Pottier, International Association for Hydrogen Energy, Coral Gables, FL, p. 1227, 1992.

85. Marcenaro, B., "EQHHPP FC BUS: Status of the Project and the Presentation of the First Experimental Results," in *Hydrogen Energy Progress X*, Edited by D.L. Block and T.N. Veziroglu, International Association for Hydrogen Energy, Coral Gables, FL, p. 1447, 1994.

86. Mark, J., Corbus, D. and Hudson, D.V., "Energy, Economic, and Environment Benefits of Light Duty Fuel Cell Vehicles," in proceedings of *The Annual Automotive Technology Development Contractors' Coordination Meeting*, Society of Automotive Engineers, Warrendale, PA, p. 543, 1992.

87. Marks, C., Rishavy, E.A., and Wycazalek, F.A., "Electrovan—A Fuel Cell Powered Vehicle," SAE Technical Paper No. 670176, Society of Automotive Engineers, Warrendale, PA, 1967.

88. McElroy, J.F. and Nuttall, L.J., "Status of Solid Polymer Electrolyte Fuel Cell Technology and Potential for Transportation Applications," in *Proc. 17th Intersociety Energy Conversion Engineering Conference*, Institute of Electrical and Electronics Engineers, New York, p. 667, 1982.

89. Metkemeyer, R. and Achard, P., "Comparison of Ammonia and Methanol Applied Indirectly in a Hydrogen Fuel Cell," in *Hydrogen Energy Progress IX*, Edited by T.N. Veziroglu and C. Derive-J. Pottier, International Association for Hydrogen Energy, Coral Gables, FL, p. 1517, 1992.

90. Minh, N.Q., "High-temperature Fuel Cells, Part 1: How the Molten Carbonate Cell Works and the Materials that Make it Possible," *Chemtech*, 21, p. 32, 1991.

91. Minh, N.Q., "High-Temperature Fuel Cells Part 2: The Solid Oxide Cell," *Chemtech*, 21, p. 120, 1991.

92. Minh, N.Q., "Ceramic Fuel Cells," *J. Am. Ceram. Soc.*, 76, p. 563, 1993.

93. Morehouse, J.H., "Parametric Study of a Hydrogen-Oxygen High Temperature Electrolyser-Fuel Cell Power Plant," *Int. J. Hydrogen Energy*, 15, p. 349, 1990.

94. Myles, K.M. and McPheeters, C.C., "Monolithic Solid Oxide Fuel Cell Development," *Journal of Power Sources*, 29, p. 311, 1990.

95. Nadal, M. and Barbir, F., "Development of a Hydride Fuel Cell/Battery Powered Electric Vehicle," in *Hydrogen Energy Progress X*, Edited by D.L. Block and T.N. Veziroglu, International Association for Hydrogen Energy, Coral Gables, FL, p. 1427, 1994.

96. Nakajima, H., "A Methanol Fuel Cell Having a Molybdenum-Modified Platinum-SPE Membrane Electrode," *Journal of Chemical Technology and Biotechnology*, 50, p. 555, 1991.

97. Newman, A., "Fuel Cells Come of Age," *Environ. Sci. Technol.*, 26, p. 2085, 1992.

98. Parkinson, G., "Need for Clean Energy Drives New Fuel Cell Development," *Chemical Engineering*, Vol. 98, p. 44, 1991.

99. Patil, P.G., "Fuel Cells for Transportation," presented at the 71st Annual Meeting of the Transportation Research Board, Washington, D.C., January, 1992.

100. Patil, P.G. and Huff, J., "Fuel-Cell/Battery Hybrid Power Source for Vehicles, in *Proceedings of the 22nd Intersociety Energy Conversion Engineering Conference*, American Institute of Aeronautics and Astronautics, NY, p. 993, 1987.

101. Petrov, K., Xiao, K., Gonzalez, E.R., Srinivasan, S., et al, "An Advanced Proton Exchange Membrane with an Improved Three-Dimensional Reaction Zone," *Int. J. Hydrogen Energy*, 18, p. 907, 1993.

102. Plowman, K.R., Cisar, A. and Carl, W.P., "Proton Exchange Membrane and Electrode Optimization," in *The Annual Automotive Technology Development Contractors' Coordination Meeting*, Society of Automotive Engineers, Warrendale, PA, p. 505, 1992.

103. Prater, K., "The Renaissance of the Solid Polymer Fuel Cell," *Journal of Power Sources*, 29, p. 239, 1990.

104. Prater, K., Testimony on S.1269, the Renewable Hydrogen Research and Development Act of 1991. Hearings before the United States Senate Committe on Energy and Natural Resources, Subcommittee on Energy Research and Development, June 25, 1991.

105. Prater, K.B., "Solid Polymer Developments at Ballard," *Journal of Power Sources*, 37, p. 181, 1992.

106. Prater, K.B., "Polymer Electrolyte Fuel Cells: A Review of Recent Developments," *Journal of Power Sources*, Vol. 51, p. 129, 1994.

107. Rawat, J.P. and Ansari, A.A., "Redox Ion Exchanger Fuel Cell," *Journal of Chemical Education*, 67, p. 808, 1990.

108. Riley, B., "Solid Oxide Fuel Cells—The Next Stage," *Journal of Power Sources*, 29, p. 223, 1990.

109. Romano, S. and Price, D., "Installing a Fuel Cell in a Transit Bus," SAE Technical Paper 900178, Society of Automotive Engineers, Warrendale, PA, 1990.

110. Romano, S., "The DOE/DOT Fuel Cell Cell Bus Program and Its Application to Transit Missions, in *Proceedings of the 25th Intersociety Energy Conver-*

sion Engineering Conference, American Institute of Chemical Engineers, NY, p. 293, 1990.

111. Rosen, M.A., "Comparison Based on Energy and Energy Analyses of the Potential Cogeneration Efficiencies for Fuel Cells and Other Electricity Generation Devices," *Int. J. Hydrogen Energy*, 15, p. 267, 1990.

112. Roy, P., Slamah, S.A. and Rodgers, D.N., "Hytec: A Thermally Regenerative Fuel Cell," in *Hydrogen Energy Progress IX*, Edited by T.N. Veziroglu and C. Derive-J. Pottier, International Association for Hydrogen Energy, Coral Gables, FL, p. 1623, 1992.

113. Salomon, R.E., "Hydrogen Concentration Cells as Energy Converters," in *Hydrogen Energy Progress V*, Edited by T.N. Veziroglu and J.B. Taylor, Pergamon Press, Elmsford, N.Y., p. 1693, 1984.

114. Schubak, G. and Scott, D.S., "H_2 - O_2 Fuel Cell and Advanced Battery Power Systems for Autonomous Underwater Vehicles: Performance Envelope Comparisons," in *Hydrogen Energy Progress IX*, Edited by T.N. Veziroglu and C. Derive-J. Pottier, International Association for Hydrogen Energy, Coral Gables, FL, p. 1475, 1992.

115. Scott, D.S. and Rogner, H., "Fuel Cell Locomotives in Canada," *Int. J. Hydrogen Energy*, 18, p. 253, 1993.

116. Serfass, J.A., "The Future of Fuel Cells for Power Production," *Journal of Power Sources*, 29, p. 119, 1990.

117. Silva, E.D., Gallo, W., Szajner, J., Amaral, E.D. and Bezerra, C., "A Solar/Hydrogen/Electricty Hybrid Vehicle," in *Hydrogen Energy Progress X*, Edited by D.L. Block and T.N. Veziroglu, International Association for Hydrogen Energy, Coral Gables, FL, p. 1441, 1994.

118. Springer, T.E., Zawodzinki, T.A. and Gottesfeld, S., "Polymer Electrolyte Fuel Cell Model," *Journal of Electrochemical Society*, 138, p. 2334, 1991.

119. Springer, T.E.; Zawodzinski, T.A.; Gottesfeld, S., "Modeling water content effects in polymer electrolyte fuel cells," in *Proceedings of the Symposium on Modeling of Batteries and Fuel Cells*, Edited by R.E. White and M.W. Verbrugge, J.F. Stockel, N.J. Pennington, Electrochem. Soc., USA, p. 209, 1991.

120. Springer, T.E. and Gottesfeld, S., "Pseudohomogeneous catalyst layer model for polymer electrolyte fuel cell," in *Proceedings of the Symposium on Modeling of Batteries and Fuel Cells*, Edited by R.E. White and M.W. Verbrugge, J.F. Stockel, N.J. Pennington, Electrochem. Soc, USA, p. 197, 1991.

121. Springer, T., Wilson, M., Zawodzinski, T., Derouin, C., Valerio, J. and Gottesfeld, S., "PEM and Direct Methanol Fuel Cell R&D," in *Preprints of the Annual Automotive Technology Development Contractors' Coordination Meeting*, Society of Automotive Engineers, Warrendale, PA, 1994.

122. Srinivasan, S., "Fuel Cells for Extraterrestrial and Terrestrial Applications," *Journal of the Electrochemical Society*, 136, p. 41, 1989.

123. Srinivasan, S., Manko, D.J., Koch, H., et al, "Recent Advances in Solid Polymer Electrolyte Fuel Cell Technology with Low Platinum Loading Electrodes," *Journal of Power Sources*, 29, p. 367, 1990.

124. Srinivasan, S., Velev, O.A., et al, "High Efficiency and High Power Density Proton Exchange Membrane Fuel Cells—Electrode Kinetics and Mass Transport," *Journal of Power Sources*, 36, p. 299, 1991.

125. Srivansan, et al, "Overview of Fuel Cell Technology," in *Fuel Cell Systems*, Edited by L.J.M. Blomen and M.N. Mugerwa, Plenum Press, New York, NY, p. 37, 1992

126. Staiti, P., Poltarzewski, Z., Alderucci, V., Maggio, G. and Giordano, N., "Solid Polymer Electrolyte Fuel Cell (SPEFC) Research and Development at the Institute CNR-TAE of Messina," in *Hydrogen Energy Progress IX*, Edited by T.N. Veziroglu and C. Derive-J. Pottier, International Association for Hydrogen Energy, Coral Gables, FL, p. 1425, 1992.

127. Staschewski, D., "Hydrogen-Air Fuel Cells of a New Alkaline Matrix Type Designed for Vehicular Applications," in *Hydrogen Energy Progress V*, Edited by

T.N. Veziroglu and J.B. Taylor, Pergamon Press, Elmsford, N.Y., p. 1677, 1984.

128. Staschewski, D., "Hydrogen-Air Fuel Cells of the Alkaline Matrix Type: Development of Components," in *Hydrogen Energy Progress VI,* Edited by T.N. Veziroglu, N. Getoff, and P. Weinzierl, Pergamon Press, Elmsford, NY, p. 1266, 1986.

129. Staschewski, D., "Hydrogen-Air Fuel Cells of the Alkaline Matrix Type: Manufacture and Impregnation of Electrodes," *Int. J. Hydrogen Energy*, 17, p. 643, 1992.

130. Strasser, K., "The Design of Alkaline Fuel Cells," *Journal of Power Sources*, 29, p. 149, 1990.

131. Strasser, K., "PEM-Fuel Cells: State of the Art and Development Possibilities," *Berichte der Bunsengesellschaft fur Physickalische Chemie*, 94, p. 1000, 1990.

132. Strasser, K., "Fuel Cells as a Power Source of Propulsion Systems," *Berichte der Bunsengesellschaft fur Physickalische Chemie*, 94, p. 922, 1990.

133. Strasser, K., "PEM Fuel Cells for Energy Systems," in Proceedings of the 26th Intersociety Energy Conversion Engineering Conference, p. 630, 1991.

134. Strasser, K., "Mobile Fuel Cell Development at Siemens," *Journal of Power Sources*, 37, p. 209, 1992.

135. Struthers, R., "Hydrogen Diffusion Fuel Cell," Patent No. 4,684,581, 1987.

136. Sutton, R.D. and Vanderborgh, N.E., "Electrochemical Engine System Modeling and Development," in *The Annual Automotive Technology Development Contractors' Coordination Meeting*, Society of Automotive Engineers, Inc., p. 493, 1992.

137. Swan, D.H. "Fuel-Cell Powered Electric Vehicles," SAE Technical Paper No. 891724, Society of Automotive Engineers, Warrendale, PA, 1989.

138. Swan, D.H., Velev, O.A., et al, "Proton Exchange Membrane Fuel Cell—A Strong Candidate as a Power Source for Electric Vehicles," Center for Electrochemical Systems and Hydrogen Research, Texas A&M, September 1991.

139. Teachman, M.E. and Scott, D.S., "A Performance Comparison of Urban Utility Vehicles with IC Engine and Solid Polymer Fuel Cell Technologies," in *Hydrogen Energy Progress IX*, Edited by T.N. Veziroglu and C. Derive-J. Pottier, International Association for Hydrogen Energy, Coral Gables, FL, p. 1465, 1992.

140. Trimble, K. and Woods, R., "Fuel Cell Applications and Market Opportunities," *Journal of Power Sources*, 29, p. 1990.

141. United States Congress Senate, *Renewable Hydrogen Energy Research and Development Act of 1991: Hearing Before the Subcommittee on Energy Research and Development of the Committee on Energy and Natural Resources, United States Senate, One Hundred and Second Congress*, Washington, D.C., 1991.

142. Van den Broek, H., Fuel Cells for Transportation, in *Energy Storage*, Edited by J. Silverman, Pergamon Press, Elmsford, NY, p. 230, 1980.

143. Van Bibber, L., Summers, W.A. and Feret, J.M., "An Industrial Application of Phosphoric Acid Fuel Cells Using By-Product Hydrogen," in *Hydrogen Energy Progress IX*, Edited by T.N. Veziroglu and C. Derive-J. Pottier, International Association for Hydrogen Energy, Coral Gables, FL, p. 1497, 1992.

144. Verbrugge, M.W. and Hill, R.F., "Analysis of Promising Perfluorosulfonic Acid Membranes for Fuel-Cell Electrolytes," *Journal of Electrochemical Society*, 137, p. 3770, 1990.

145. Vielstich, W., *Fuel Cells*, John Wiley & Sons Ltd., London, 1970.

146. Viledganin, G.A., Galyshev, U.V. and Jarov, V.F., "Development and Investigation of the Hydrogen Fueling Systems Used for Automobile Engines," in *Hydrogen Energy Progress VII*, Edited by T.N. Veziroglu and A.N. Protsenko, Pergamon Press, Elmsford, NY, p. 2105, 1988.

147. Walsh, M.P., "The Importance of Fuel Cells to Address the Global Warming Problem," *Journal of Power Sources*, 29, p. 13, 1990.

148. Watanabe, M. and Stoneheart, P., "International Fuel Cell Workshop," *Int. J. Hydrogen Energy*, 16, p. 63, 1991.

149. Wilkinson, D.P., Voss, H.H., and Prater, K., "Water Management and Stack Design for Solid Polymer Fuel Cells," *Journal of Power Sources*, 49, p. 117, 1994.

150. Wilson, M.S., Springer, T.E., Zawodzinski, T.A. and Gottesfeld, S., "Recent Achievements in Polymer Electrolyte Fuel Cell (PEFC) Research at Los Alamos National Laboratory," in *26th Intersociety Energy Conversion Engineering Conference*, 1991.

151. Yoshida, T., Hoshina, T., Mukaizawa, I., et al, "Properties of Partially Stabilized Zirconia Fuel Cell," *Journal of the Electrochemical Society*, 136, p. 2604, 1989.

152. Zawodzinski, T.A., Jr.; Springer, T.E.; Davey, J.; Valerio, J., et al, "Water transport properties of fuel cell ionomers," in *Proceedings of the Symposium on Modeling of Batteries and Fuel Cells*, Edited by R.E. White and M.W. Verbrugge, J.F. Stockel, N.J. Pennington, Electrochem. Soc., USA, p. 187, 1991.

153. Zawodzinski, T.A., Jr.; Springer, T.E.; Davey, J.; Jestel, R., et al, "A comparative study of water uptake by and transport through ionomeric fuel cell membranes, *Journal of the Electrochemical Society*, 140, p. 1981, 1993.

154. "Fuel Cell Demonstration Planned," *EPRI Journal*, September, p. 35, 1993.

155. "Fuel Cells: An Overview," Automotive Engineering, 100, p. 13, 1992.

156. "Cell of the Century," *Electric Review*, 28 October-10 November, 1994.

157. "Chrysler, Ford win fuel cell contracts," *Automotive News*, 6, July 18, 1994.

158. "Alternative Fuels—The Fuel Cells are Coming," H_2 *Digest*, 1994.

159. "Daimler Unveils Electric Vehicle Using Fuel Cells," H_2 *Digest*, 1994.

160. "Hydrogen Fuel Cell Vehicle: Lasercel 1™," article by the American Academy of Science, Independence, Mo.

Production

1. Abdel-Aal, H.K., "The Conversion of High-Pollutant Undesirable Oil Residues into Environmentally-Acceptable Hydrogen Fuel," in *Hydrogen Energy Progress VIII*, Edited by T.N. Veziroglu and P.K. Takahashi, Pergamon Press, Elmsford, N.Y., p. 357, 1990.

2. Abdel-Aal, H.K., "Water: Feed Stock for Hydrogen Production," in *Hydrogen Energy Progress VI,* Edited by T.N. Veziroglu, N. Getoff, and P. Weinzierl, Pergamon Press, Elmsford, N.Y., p. 30, 1986.

3. Abdel-Aal, H.K., Hassan, H.H., Mohamed, M.A. and Khairy, S.A., "Polymeric Materials for Photoelectrochemical Cells for Hydrogen Production," *Int. J. Hydrogen Energy*, 16, p. 745, 1991.

4. Abdel-Aal, H.K. and Hussein, I.A., "Electrolysis of Saline Water under Simulated Conditions of P.V. Solar Cell," in *Hydrogen Energy Progress IX*, Edited by T.N. Veziroglu and C. Derive-J. Pottier, International Association for Hydrogen Energy, Coral Gables, FL, p. 439, 1992.

5. Abdel-Aal, H.K. and Mohamed, M.A., "Potentials of Storing Solar Energy in the Form of Hydrogen for Egypt," *Energy Sources*, 11, p. 95, 1989.

6. Aihara, M., Sakurai, M. and Yoshida, K., "Reaction Improvement in the UT-3 Thermochemical Hydrogen Production Process," in *Hydrogen Energy Progress VIII*, Edited by T.N. Veziroglu and P.K. Takahashi, Pergamon Press, Elmsford, N.Y., p. 493, 1990.

7. Anderson, D. and K. Ahmed, "Where We Stand with Renewable Energy," *Finance & Development*, June, p. 40, 1993.

8. Andolfatto, F., Durand, R., Michas, A., et al, "Solid Polymer Electrolyte Water Electrolysis: Electrocatalysis and Long Term Stability," in *Hydrogen Energy Progress IX*, Edited by T.N. Veziroglu and C. Derive-J. Pottier, International Association for Hydrogen Energy, Coral Gables, FL, p. 429, 1992.

9. Aochi, A., Tadokoro, T., Yoshida, K., et al, "Economical and Technical Evaluation of UT-3 Thermochemical Hydrogen Production Process for an Industrial Scale Plant," *Int. J. Hydrogen Energy,* 14, p. 421, 1989.

10. Arashi, H., Naito, H. and Miura, H., "Hydrogen Production from High-Temperature Steam Electrolysis Using Solar Energy," *Int. J. Hydrogen Energy,* 16, p. 603, 1991.

11. Arnason, B., Sigfusson, T.I. and Jonsson, V.K., "New Concepts in Hydrogen Production in Iceland," in *Hydrogen Energy Progress IX*, Edited by T.N. Veziroglu and C. Derive-J. Pottier, International Association for Hydrogen Energy, Coral Gables, FL, p. 1863, 1992.

12. Baichtok, Y.K., Mordkovich, V.Z., Korostyshevsky, N.N., et al, "Thermosorption Compressor-Hydrogen Separator," in *Hydrogen Energy Progress VIII*, Edited by T.N. Veziroglu and P.K. Takahashi, Pergamon Press, Elmsford, N.Y., p. 1219, 1990.

13. Balajka, J., "The Analysis of Integrated Energy Systems," in *Hydrogen Energy Progress VIII*, Edited by T.N. Veziroglu and P.K. Takahashi, Pergamon Press, Elmsford, N.Y., p. 627, 1990.

14. Baykara, S.Z. and Bilgen, E., "An Overall Assessment of Hydrogen Production by Solar Thermolysis," *Int. J. Hydrogen Energy,* 14, p. 881, 1989.

15. Beenackers, A.A.C.M. and Van Swaaij, W.P.M., "Gasification, Synthesis Gas Production and Direct Liquefaction of Biomass," in *Energy from Biomass 1*, Elsevier Applied Science, p. 15, 1986.

16. Blank, M. and Szyszka, A., "Solar-hydrogen Demonstraton Plant in Neunburg vorm Wald," in *Hydrogen Energy Progress IX*, Edited by T.N. Veziroglu and C. Derive-J. Pottier, International Association for Hydrogen Energy, Coral Gables, FL, p. 677, 1992.

17. Blazek, C.F., Donakowski, T.D., Novil, M. and Rupinskas, R.L., "Users Guide for On-Site Electrolytic Hydrogen Production for Turbine Generator Cooling Applications," in *Hydrogen Energy Progress VI,* Edited by T.N. Veziroglu, N. Getoff, and P. Weinzierl, Pergamon Press, Elmsford, N.Y., p. 215, 1986.

18. Block, D.L. and Melody, I., "Efficiency and Cost Goals for Photoenhanced Hydrogen Production Processes," in *Hydrogen Energy Progress VIII*, Edited by T.N. Veziroglu and P.K. Takahashi, Pergamon Press, Elmsford, N.Y., p. 217, 1990.

19. Bockris, J. O'M., *Energy Options*, Australia & New Zealand Book Company, Sydney, Australia, 1980.

20. Bockris, J.O'M. and González, A., "The Photoproduction of Hydrogen," in *Hydrogen Energy Progress VIII*, Edited by T.N. Veziroglu and P.K. Takahashi, Pergamon Press, Elmsford, N.Y., p. 791, 1990.

21. Bockris, J.O. and Kainthla, R.C., "The Conversion of Light and Water to Hydrogen and Electric Power," in *Hydrogen Energy Progress VI,* Edited by T.N. Veziroglu, N. Getoff, and P. Weinzierl, Pergamon Press, Elmsford, NY, p. 449, 1986.

22. Borel, P. and Mignard, B., "Hydrogen-powered Vehicles: Methodology for the Assessment of Production, Transportation and Distribution Processes," in *Hydrogen Energy Progress IX*, Edited by T.N. Veziroglu and C. Derive-J. Pottier, International Association for Hydrogen Energy, Coral Gables, FL, p. 1041, 1992.

23. Bothe, H. and Kentemich, T., "Potentialities of H_2-Production by Cyanobacteria for Solar Energy Conversion Programs," in *Hydrogen Energy Progress VIII*, Edited by T.N. Veziroglu and P.K. Takahashi, Pergamon Press, Elmsford, N.Y., p. 729, 1990.

24. Bowman, M.G., "A Thermochemical Cycle for Splitting Carbon Dioxide or Water," in *Hydrogen Energy Progress VIII*, Edited by T.N. Veziroglu and P.K. Takahashi, Pergamon Press, Elmsford, N.Y., p. 615, 1990.

25. Bradke, M.V. and Schnurnberger, W., "Surface Analysis of Electrodes for Alkaline Water Electrolysis,"

in *Hydrogen Energy Progress VI,* Edited by T.N. Veziroglu, N. Getoff, and P. Weinzierl, Pergamon Press, Elmsford, N.Y., p. 438, 1986.

26. Brinner, A., Bussmann, H., Hug, W. and Seeger, W., "Test Results of the Hysolar 10kW PV-Electrolysis Facility," *Int. J. Hydrogen Energy*, 17, p. 187, 1992.

27. Bunger, U.H., Andreassen, K., Henriksen, N. and Oyvann, A., "Hydrogen as an Energy Carrier, Production and Liquefacton of Hydrogen in Norway for Transportation to and Storage/Distribution," in *Hydrogen Energy Progress IX*, Edited by T.N. Veziroglu and C. Derive-J. Pottier, International Association for Hydrogen Energy, Coral Gables, FL, p. 1913, 1992.

28. Casal, F.G., *Solar Thermal Power Plants: Achievements and Lessons Learned Exemplified by the SSPS Project in Almeria/Spain*, Springer-Verlag, Berlin, Germany, 1987.

29. Cervera-March, S. and Smotkin, E.S., "A Photoelectrode Array System for Hydrogen Production from Solar Water Splitting," *Int. J. Hydrogen Energy*, 16, p. 243, 1991.

30. Cevera-March, S., Borrell, L., Giménez, J. and Simarro, R., "Solar Hydrogen Photoproduction from Sulphide+Sulphite Substrate," in *Hydrogen Energy Progress VIII*, Edited by T.N. Veziroglu and P.K. Takahashi, Pergamon Press, Elmsford, N.Y., p. 853, 1990.

31. Chen, D. Z. and Huang, J.Y., "Prospects of a Hysolar Energy System Using Photovoltaic Electrolytic Conversion," in *Hydrogen Energy Progress IX*, Edited by T.N. Veziroglu and C. Derive-J. Pottier, International Association for Hydrogen Energy, Coral Gables, FL, p. 763, 1992.

32. Chen, D.Z. and Yu, C.P., "Thermodynamic Analysis of a Nuclear-Hydrogen Power System Using H_2/O_2 Direct Combustion Product as a Working Substance in the Bottom Cycle," in *Hydrogen Energy Progress VIII*, Edited by T.N. Veziroglu and P.K. Takahashi, Pergamon Press, Elmsford, New York, p. 1455, 1990.

33. Chen, J.-W. and Ti Tien, H., "Hydrogen Production from Water by Semiconductor Septum Electrochemical Photovoltaic Cell Using Visible Light," *Int. J. Hydrogen Energy*, 15, p. 563, 1990.

34. ChengLin, P., Yie, Y., Weimen, G., Long, Z. and Jun, P., "Thermochemical Generation of Hydrogen from Water," in *Hydrogen Energy Progress IX*, Edited by T.N. Veziroglu and C. Derive-J. Pottier, International Association for Hydrogen Energy, Coral Gables, FL, p. 271, 1992.

35. Christiansen, K., Andreassen, K. and Midjo, M., "Industrial Electrolytic Hydrogen Production," in *Hydrogen Energy Progress V*, Edited by T.N. Veziroglu and J.B. Taylor, Pergamon Press, Elmsford, NY, p. 715, 1984.

36. Chrysostome, G., Lemasle, J.M. and Ascab, G., "Syngas Production from Wood: Gasification Unit of Clamecy," in *Energy from Biomass 1*, Elsevier Applied Science, p. 407, 1986.

37. Chuveliov, A.V. *Hydrogen in Motor Vehicles: A Case Study of Hydrogen Utilization in Motor Vehicles as a Supplementary Fuel in Southern California Air Basin*. I.V. Kurchatov Institute of Atomic Energy, Report prepared for SCQAMD, 1989.

38. Clarke, A., "Wind Energy: Progress and Potential," *Energy Policy*, 19, p. 742, 1991.

39. Copeland, R.J., "Photochemical Hydrogen Production System," in *Hydrogen Energy Progress VIII*, Edited by T.N. Veziroglu and P.K. Takahashi, Pergamon Press, Elmsford, N.Y., p. 207, 1990.

40. Cromarty, B., "Modern Aspects of Steam Reforming for Hydrogen Plants," in *Hydrogen Energy Progress IX*, Edited by T.N. Veziroglu and C. Derive-J. Pottier, International Association for Hydrogen Energy, Coral Gables, FL, p. 13, 1992.

41. D'Ajuz, A., Conti, A.M., Mattos, M.C. et al, "Electrical Energy Transmission from the Amazon Region: Hydrogen as a Promising Alternative in Brazil," *Int. J. Hydrogen Energy*, 14, p. 515, 1989.

42. De Giz, M.J., Da Silva, J.P.C., Ferreira, M. et al, "Progress on the Development of Activated Cath-

odes for Water Electrolysis," in *Hydrogen Energy Progress VIII*, Edited by T.N. Veziroglu and P.K. Takahashi, Pergamon Press, Elmsford, N.Y., p. 405, 1990.

43. DeLaquil, P., Kearny, D., Geyer, M., and Diver, R., "Solar-Thermal Electric Technology," in *Renewable Energy: Sources for Fuels and Electricity*, Edited by T.B. Johansson, H. Kelly, A.K.N. Reddy, R.H. Williams, and L. Burnham, Island Press, Washington, D.C., 1993.

44. DeLuchi, M.A., "Nuclear-Electolytic Hydrogen as a Transportation Fuel," *American Nuclear Society*, 60, p. 490, 1989.

45. DeLuchi, M.A., "Hydrogen Vehicles: An Evaluation of Fuel Storage, Performance, Safety, Environmental Impacts, and Cost," *Int. J. Hydrogen Energy*, 14, p. 81, 1989.

46. Divisek, J., "Water Electrolysis is a Low- and Medium-Temperature Regime," in *Electrochemical Hydrogen Technologies*, Edited by H. Wendt, Elsevier Science Publishing Company Inc., New York, p. 137, 1990.

47. Divisek, J., Malinowski, P., Mergel, J. and Schmitz, H., "Improved Construction of an Electrolytic Cell for Advanced Alkaline Water Electrolysis," in *Hydrogen Energy Progress V*, Edited by T.N. Veziroglu and J.B. Taylor, Pergamon Press, Elmsford, NY, p. 655, 1984.

48. Divisek, J., Malinowski, P., Mergel, J. and Schmitz, H., "Advanced Techniques for Alkaline Water Electrolysis," in *Hydrogen Energy Progress VI*, Edited by T.N. Veziroglu, N. Getoff, and P. Weinzierl, Pergamon Press, Elmsford, N.Y., p. 258, 1986.

49. DOE, *Hydrogen Program Plan*, U.S. Department of Energy, Office of Conservation and Renewable Energy, FY1993-FY1997, DOE/CH10093-147.

50. Dönitz, W. and Erdle, E., "High Temperature Electrolysis of Water Vapour—Status of Development and Perspectives for Application," in *Hydrogen Energy Progress V*, Edited by T.N. Veziroglu and J.B. Taylor, Pergamon Press, Elmsford, NY, p. 767, 1984.

51. Dönitz, W., Erdle, E. and Streicher, R., "High Temperature Electrochemical Technology for Hydrogen Production and Power Generation," in *Electrochemical Hydrogen Technologies*, Edited by H. Wendt, Elsevier Science Publishing Company Inc., New York, p. 213, 1990.

52. Drolet, B., Gretz, J., Kluyskens, D., Sandmann, F., Wurster, R., "The Euro-Quebec Hydro-Hydrogen Pilot Project (EQHHPP)" in *Hydrogen Energy Progress X*, Edited by D.L. Block and T.N. Veziroglu, International Association of Hydrogen Energy, Coral Gables, 1994.

53. Dutta, S., "Technology Assessment of Advanced Electrolytic Hydrogen Production," *Int. J. Hydrogen Energy*, 15, p. 379, 1990.

54. Dutta, S., Block, D.L. and Port, R.L., "Economic Assessment of Advanced Electrolytic Hydrogen Production," *Int. J. Hydrogen Energy*, 15, p. 387, 1990.

55. Edye, L.A., Richards, G.N. and Zheng, G., "Transition Metals as Catalysts for Pyrolysis and Gasification of Biomass," in *Clean Energy from Waste & Coal*, Edited by R. M. Khan, American Chemical Society, Washington, D.C., p. 90, 1991.

56. El-Bassuoni, A.A., "Electrolysis of Sea Water Using a Packed Bed Reactor," in *Hydrogen Energy Progress VI*, Edited by T.N. Veziroglu, N. Getoff, and P. Weinzierl, Pergamon Press, Elmsford, N.Y., p. 282, 1986.

57. El-Osta, W.B. and Veziroglu, T.N., "Solar-Hydrogen Energy System for a Libyan Coastal County," *Int. J. Hydrogen Energy*, 15, p. 33, 1990.

58. Eljrushi, G. and Zubia, J., "Solar Hydrogen: The Great Sahara Project," in *Hydrogen Energy Progress IX*, Edited by T.N. Veziroglu and C. Derive-J. Pottier, International Association for Hydrogen Energy, Coral Gables, FL, p. 1877, 1992.

59. Eljrushi, G.S. and Veziroglu, T.N., "Solar-Hydrogen Energy System for Libya," *Int. J. Hydrogen Energy*, 15, p. 885, 1990.

60. Escher, W.D.J. and Ecklund, E.E., "The Hydrogen-via-Electricity Concept: Critique Report," prepared

for the U.S. Department of Energy, Washington, D.C., 1981.

61. Escher, W.J.D., "The Case for Solar/Hydrogen Energy," *Int. J. Hydrogen Energy*, 8, p. 479, 1983.

62. Escher, W.J.D., "Hydrogen-via-Electricity (HvE): A Transitional Transportation-Energy Concept," in *Proceedings of the American Power Conference*, Chicago, IL, p. 1270, 1991.

63. Fischer, M., "Utilization of Solar and Nuclear Energy for Hydrogen Production," in *Hydrogen Energy Progress VI,* Edited by T.N. Veziroglu, N. Getoff, and P. Weinzierl, Pergamon Press, Elmsford, NY, p. 460, 1986.

64. Funk, J.E., "Thermochemical Water Decomposition: Current Status," in *Recent Developments in Hydrogen Technology*, Edited by K. D. J. Williamson and F. J. Edeskuty, CRC Press, Inc., Boca Raton, p. 1, 1986.

65. Gaines, L.L. and Wolsky, A.M., "Economics of Hydrogen Production: The Next Twenty-Five Years," in *Hydrogen Energy Progress V*, Edited by T.N. Veziroglu and J.B. Taylor, Pergamon Press, Elmsford, NY, p. 259, 1984.

66. Garcia-Conde, A.G. and Rosa, F., "Solar Hydrogen Production: A Spanish Experience," in *Hydrogen Energy Progress IX*, Edited by T.N. Veziroglu and C. Derive-J. Pottier, International Association for Hydrogen Energy, Coral Gables, FL, p. 723, 1992.

67. Getoff, N., "Photoelectrochemical and Photocatalytic Methods of Hydrogen Production: A Short Review," *Int. J. Hydrogen Energy*, 15, p. 407, 1990.

68. Getoff, N., Stockenhuber, H., Kotchev, K. and Li, G., "Development of Multi-layer Semiconductor-Electrodes on Silicon Basis for H_2 Production," in *Hydrogen Energy Progress IX*, Edited by T.N. Veziroglu and C. Derive-J. Pottier, International Association for Hydrogen Energy, Coral Gables, FL, p. 537, 1992.

69. Ghosh, S., "Net Energy Production in Anaerobic Digestion," in *Energy from Biomass and Wastes V*, Edited by T.N. Veziroglu and J.B. Taylor, Institute of Gas Technology, p. 253, 1981.

70. Ghosh, S. and Klass, D.L., "Conversion of Urban Refuse to Substitute Natural Gas by the BIOGAS® Process," in *Fourth Mineral Waste Utilization Symposium*, 1974.

71. Ghosh, S. and Klass, D.L., "SNG From Refuse and Sewage Sludge by the BIOGAS® Process," in *Symposium on Clean Fuels From Biomass, Sewage, Urban Refuse, and Agricultural Wastes*, 1976.

72. Goldemberg, J., Johansson, T.B., Reddy, A.K.N., and Williams, R.H., *Energy for a Sustainable World*, World Resources Institute, Washington, D.C., September, 1987.

73. Goldemberg, J., Johansson, T.B., Reddy, A.K.N., and Williams, R.H., *Energy for a Sustainable World*, Wiley-Eastern, Delhi, 1988.

74. Gongxuan, L. and Shuben, L., "Hydrogen Production by H_2S Photocatalytic Decomposition," in *Hydrogen Energy Progress VIII*, Edited by T.N. Veziroglu and P.K. Takahashi, Pergamon Press, Elmsford, N.Y., p. 863, 1990.

75. Gonzalez, E.R., Ticianelli, E.A., Tanaka, A.A. and Avaca, L.A., "A Project for the Electrochemical Production and Utilization of Hydrogen in Brazil," *Energy Sources*, 11, p. 53, 1989.

76. Goyal, A. and Rehmat, A., "Fuel Evaluation for a Fluidized-Bed Gasification Process (U-Gas)," in *Clean Energy from Waste & Coal*, Edited by R. M. Khan, American Chemical Society, Washington, D.C., p. 58, 1991.

77. Graham, A.B. and Hynds, J.A., "On-Site Production of Hydrogen by Electrolysis," in *Hydrogen Energy Progress V*, Edited by T.N. Veziroglu and J.B. Taylor, Pergamon Press, Elmsford, NY, p. 665, 1984.

78. Grasse, W., Oster, F. and Aba-Oud, H., "Hysolar: The German-Saudi Arabian Program on Solar Hydrogen—5 Years of Experience," *Int. J. Hydrogen Energy*, 17, p. 1, 1992.

79. Greenbaum, E., "Hydrogen Production by Photosynthetic Water Splitting," in *Hydrogen Energy Progress VIII*, Edited by T.N. Veziroglu and P.K. Takahashi, Pergamon Press, Elmsford, N.Y., p. 743, 1990.

80. Gretz, J., Baselt, J.P., Ullmann, O. and Wendt, H., "The 100MW Euro-Quebec Hydro-Hydrogen Pilot Project," *Int. J. Hydrogen Energy*, 15, p. 419, 1990.

81. Grubb, M.J. and Meyer, N.I., "Wind Energy: Resources, Systems, and Regional Strategies," in *Renewable Energy: Sources for Fuels and Electricity*, Edited by T.B. Johansson, H. Kelly, A.K.N. Reddy, R.H. Williams, and L. Burnham, Island Press, Washington, D.C., 1993.

82. Hall, D.O., "Biomass Energy," *Energy Policy*, 19, p. 711, 1991.

83. Hammache, A. and Bilgen, E., "Energy Analysis of Solar Hydrogen Production Based on Sulphuric Acid Decomposition and Synthesis," in *Hydrogen Energy Progress VIII*, Edited by T.N. Veziroglu and P.K. Takahashi, Pergamon Press, Elmsford, N.Y., p. 327, 1990.

84. Hammache, A. and Bilgen, E., "Performance of Large Scale Photovoltaic Electrolyser Systems," in *Hydrogen Energy Progress VI*, Edited by T.N. Veziroglu, N. Getoff, and P. Weinzierl, Pergamon Press, Elmsford, N.Y., p. 287, 1986.

85. Hammerli, M., "When Will Electrolytic Hydrogen Become Competitive?" *Int. J. Hydrogen Energy*, 9, p. 25, 1984.

86. Hamrick, J.T., "Biomass-Fueled Gas Turbines," in *Clean Energy from Waste & Coal*, Edited by R. M. Khan, American Chemical Society, Washington, D.C., p. 78, 1991.

87. Hancock, O.G.J., "A Photovoltaic-Powered Water Electrolyzer: Its Performance and Economics," in *Hydrogen Energy Progress V*, Edited by T.N. Veziroglu and J.B. Taylor, Pergamon Press, Elmsford, NY, p. 335, 1984.

88. Hanson, J.A., Foster, P.W., Escher, W.J.D., Tison, R.R., "Solar/Hydrogen Systems for the 1985-2000 Time Frame: A Review and Assessment," *Int. J. Hydrogen Energy*, 7, p. 3, 1982.

89. Hao Shu-ren Mao, P.-S. and Wang, J.-S., "The Performance of Naphtha Steam Reforming Catalyst Z409/Z405 and its Applications in Production of Ammonia and Hydrogen," in *Hydrogen Energy Progress IX*, Edited by T.N. Veziroglu and C. Derive-J. Pottier, International Association for Hydrogen Energy, Coral Gables, FL, p. 23, 1992.

90. Hassmann, K. and Kühne, H., "Primary Energy Sources for Hydrogen Production," *Int. J. Hydrogen Energy*, 18, p. 635, 1993.

91. Hiraoka, K., Watanabe, K., Morishita, T. and et al, "Energy Analysis and CO_2 Emission Evaluation of a Solar Hydrogen Energy System for the Transportation System in Japan-II. Evaluation of the System," *Int. J. Hydrogen Energy*, 16, p. 755, 1991.

92. Hoffmann, P., *The Forever Fuel: The Story of Hydrogen*, Westview Press, Boulder, Colorado, 1981.

93. Houdart, A., "Liquid Hydrogen Producton Plant," in *Hydrogen Energy Progress IX*, Edited by T.N. Veziroglu and C. Derive-J. Pottier, International Association for Hydrogen Energy, Coral Gables, FL, p. 79, 1992.

94. Huang, S.D., Secor, C.K., Ascione, R., Zweig, R.M., "Hydrogen Production by Non-Photosynthetic Bacteria," *Int. J. Hydrogen Energy*,10, p. 227, 1985.

95. Hubbard, H.M., "Photovoltaics Today and Tomorrow," *Science*, 244, p. 297, 1989.

96. Hughes, L. and Scott, S., "Using Hydrogen from Hydro-Electricity to Alleviate Carbon Dioxide Emissions in Eastern Canada," *Energy Conversion Management*, 33, p. 151, 1992.

97. Huot, J.Y., Trudeau, M.L. and Schulz, R., "Low Hydrogen Overpotential Nanocrystalline Ni-Mo Cathodes for Alkaline Water Electrolysis," *Journal of the Electrochemical Society*, 138, p. 1316, 1991.

98. Jackson, T., "Renewable Energy: Great Hope or False promise," *Energy Policy*, 19, p. 2, 1991.

99. Jaksic, M.M. and Johansen, B., "Electrocatalytic in Situ Activation of Noble Metals for Hydrogen Evolution," in *Hydrogen Energy Progress VIII*, Edited by T.N. Veziroglu and P.K. Takahashi, Pergamon Press, Elmsford, N.Y., p. 461, 1990.

100. Johansson, T.B., Kelly, H., Reddy, A.K.N., Williams, R.H., and Burnham, L., "Renewable Fuels and Electricity for a Growing World Economy: Defining and Achieving the Potential," in *Renewable Energy: Sources for Fuels and Electricity*, Edited by Johansson, T.B., Kelly, H., Reddy, A.K.N., Williams, R.H., and Burnham, L., p. 1, Island Press, Washington, D.C., 1993.

101. Johansson, T.B., Kelly, H., Reddy, A.K.N., Williams, R.H., and Burnham, L., eds., *Renewable Energy: Sources for Fuels and Electricity*, Island Press, Washington, D.C., 1993.

102. Kauranen, P.S., Lund, P.D. and Vanhanen, J.P., "Development of a Self-Sufficient Solar-Hydrogen Energy System," in *Hydrogen Energy Progress IX*, Edited by T.N. Veziroglu and C. Derive-J. Pottier, International Association for Hydrogen Energy, Coral Gables, FL, p. 733, 1992.

103. Kauranen, P.S., Lund, P.D. and Vanhanen, J.P., "Control of Battery Backed Photovoltaic Hydrogen Production," *Int. J. Hydrogen Energy*, 18, p. 383, 1993.

104. Klass, D.L., "Energy from Biomass and Wastes: 1980 Update," in *Energy from Biomass and Wastes V*, Institute of Gas Technology, p. 27, 1981.

105. Knoch, P.H., "Energy Without Pollution: Solar-Wind-Hydrogen Systems: Some Consequences on Urban and Regional Structure and Planning," *Int. J. Hydrogen Energy*, 14, p. 903, 1989.

106. Kondrikov, N.B., "Problems of Sea Water Electrolysis for Ocean Energy Utilization & Hydrogen Production," in *Hydrogen Energy Progress VIII*, Edited by T.N. Veziroglu and P.K. Takahashi, Pergamon Press, Elmsford, N.Y., p. 649, 1990.

107. Kosanic, M.M. and Topalov, A.S., "Photochemical Hydrogen Production from $CdS/RhO_x/Na_2S$ Dispersions," *Int. J. Hydrogen Energy*, 15, p. 319, 1990.

108. Kreysa, G. and Sell, D., "Bioelectrochemical Hydrogen Production," in *Hydrogen Energy Progress IX*, Edited by T.N. Veziroglu and C. Derive-J. Pottier, International Association for Hydrogen Energy, Coral Gables, FL, p. 695, 1992.

109. Kronberger, H., Fabjan, C. and Frithum, G., "Development of High Performance Cathode for Hydrogen Production from Alkaline Solutions," *Int. J. Hydrogen Energy*, 16, p. 219, 1991.

110. Kuhne, H.M. and Hassmann, K., "Primary Energy Sources for Hydrogen Production," in *Hydrogen Energy Progress IX*, Edited by T.N. Veziroglu and C. Derive-J. Pottier, International Association for Hydrogen Energy, Coral Gables, FL, p. 1967, 1992.

111. Kukkonen, C.A., "Hydrogen as an Alternative Automobile Fuel," SAE Technical Paper No. 810349, Society of Automotive Engineers, Warrendale, PA, 1981.

112. Kukkonen, C.A. and Shelef, M., "Hydrogen as an Alternative Automobile Fuel: 1993 Update," SAE Technical Paper No. 940766, Society of Automotive Engineers, Warrendale, PA, 1994.

113. LeRoy, R.L. and Stuart, A.K., "Present and Future Costs of Hydrogen Production by Unipolar Water Electrolysis," in *Symposium on Industrial Water Electrolysis*, Electrochemical Society, p. 117, 1978.

114. LeRoy, R.L. and Stuart, A.K., "Unipolar Water Electrolysers: A Competitive Technology," in *Proceedings of the 2nd World Hydrogen Energy Conference*, Edited by T.N. Veziroglu and W. Seifritz, Pergamon Press, Elmsford, NY, p. 359, 1978.

115. Lin, F.N., Moore, W.I. and Walker, S.W., "Economics of Liquid Hydrogen from Water Electrolysis," in *Hydrogen Energy Progress V*, Edited by T.N. Veziroglu and J.B. Taylor, Pergamon Press, Elmsford, NY, p. 249, 1984.

116. Lin, G.H., Kapur, M., Kainthla, R.C. and Bockris, J.O'M., "One Step Method to Produce Hydrogen by a Triple Stack Amorphous Silicon Solar Cell," *Appl. Phys. Lett.*, 55, p. 386, 1989.

117. Lucchesi, A., Maschio, G., Rizzo, C. and Stoppato, G., "A Pilot Plant for the Study of the Production of Hydrogen-Rich Syngas by Gasification of Biomass," in *Research in Thermochemical Biomass Conversion*, Edited by A. V. Bridgewater and J. L. Kuester, Elsevier Applied Science, New York, p. 642, 1988.

118. Lutfi, N. and Verziroglu, T.N., "Solar-Hydrogen Demonstration Project for Pakistan," *Int. J. Hydrogen Energy*, 17, p. 339, 1992.

119. MacQueen, B.D. and Peterson, J.D., "Competitive Hydrogen Production and Emission Through the Photochemistry of Mixed-Metal Bimetallic Complexes," *Inorganic Chemistry*, 29, p. 2313, 1990.

120. Manarungson, S., Mok, W.S. and Antal, M.J.J., "Hydrogen Production by Gasification of Glucose and Wet Biomass in Supercritical Water," in *Hydrogen Energy Progress VIII*, Edited by T.N. Veziroglu and P.K. Takahashi, Pergamon Press, Elmsford, N.Y., p. 345, 1990.

121. Markov, S.A., Rao, K.K. and Hall, D.O., "A Hollow Fibre Photobioreactor for Continuous Production of Hydrogen by Immobilized Cyanobacteria Under Partial Vacuum," in *Hydrogen Energy Progress IX*, Edited by T.N. Veziroglu and C. Derive-J. Pottier, International Association for Hydrogen Energy, Coral Gables, FL, p. 1992.

122. Martin, J.H., "Operation of a Commercial Farm-Scale Plug-Flow Manure Digester Plant," in *Energy from Biomass and Wastes V*, Edited by T.N. Veziroglu and J.B. Taylor, Institute of Gas Technology, p. 439, 1981.

123. McAuliffe, C.A., *Hydrogen and Energy*, Gulf Book Division, Houston, 1980.

124. McKinley, K.R., Browne, S.H., Neill, R.D., et al, "Hydrogen Fuel from Renewable Resources," *Energy Sources*, 12, p. 105, 1990.

125. Meissner, D., "Key Problems in Solar Hydrogen Production through Photoelectrochemistry," in *Hydrogen Energy Progress IX*, Edited by T.N. Veziroglu and C. Derive-J. Pottier, International Association for Hydrogen Energy, Coral Gables, FL, p. 517, 1992.

126. Minet, R.G. and Desai, K., "Cost-effective Methods for Hydrogen Production," *Int. J. Hydrogen Energy*, 8, p. 285, 1983.

127. Mitchell, D.H. et al, "Methane/Methanol by Catalytic Gasification of Biomass," *CEP*, 53, 1980.

128. Miyake, J., "Application of Photosynthetic Systems for Energy Conversion," in *Hydrogen Energy Progress VIII*, Edited by T.N. Veziroglu and P.K. Takahashi, Pergamon Press, Elmsford, N.Y., p. 755, 1990.

129. Mizuta, S. and Kumagai, T., "Progress Report on the Mg-S-I Thermochemical Water-Splitting Cycle—Continuous Flow Demonstration," in *Hydrogen Energy Progress VI,* Edited by T.N. Veziroglu, N. Getoff, and P. Weinzierl, Pergamon Press, Elmsford, NY, p. 696, 1986.

130. Moreau, C., "Development Status of an Advanced Electrolyser," in *Hydrogen Energy Progress IX*, Edited by T.N. Veziroglu and C. Derive-J. Pottier, International Association for Hydrogen Energy, Coral Gables, FL, p. 345, 1992.

131. Morimoto, Y., Hayashi, T. and Maeda, Y., "Mobile Solar Energy Hydrogen Generating System," in *Hydrogen Energy Progress VI,* Edited by T.N. Veziroglu, N. Getoff, and P. Weinzierl, Pergamon Press, Elmsford, N.Y., p. 326, 1986.

132. Neill, D., Holst, B., Yu, C., Huang, N. and Wei, J., "Hnei Wind-Hydrogen Program," in *Hydrogen Energy Progress VIII*, Edited by T.N. Veziroglu and P.K. Takahashi, Pergamon Press, Elmsford, N.Y., p. 71, 1990.

133. Nuttall, L.J. and Russell, J.H., "Solid Polymer Electrolyte Water Electrolysis Development Status," in *Hydrogen Energy System*, Edited by T.N. Veziroglu and W. Seifritz, Pergamon Press, Elmsford, NY, p. 391, 1978.

134. Ogden, J.M., "Cost and Performance Sensitivity Studies for Solar Photovoltaic/Electrolytic Hydrogen Systems," *Solar Cells*, 30, p. 515, 1991.

135. Ogden, J.M. and Williams, R.H., "Electrolytic Hydrogen from Thin-Film Solar Cells," *Int. J. Hydrogen Energy*, 15, p. 155, 1990.

136. Ogden, J.M., "Renewable Hydrogen Transportation Fuels," in *Solar and Electric Vehicles '92*, 1992.

137. Ogden, J.M. and Nitsch, J., "Solar Hydrogen," in *Renewable Energy: Sources for Fuels and Electricity*, Edited by T.B. Johansson, H. Kelly, A.K.N. Reddy, R.H. Williams, and L. Burnham, Island Press, Washington, D.C., 1993.

138. Ogden, J.M. and Williams, R.H., *Solar Hydrogen: Moving Beyond Fossil Fuels*, World Resources Institute, Washington, D.C., 1989.

139. Ogden, J.M., "Studies of Solar Photovoltaic/Electrolytic Hydrogen Systems," presented at Photovoltaic Advanced Research and Development Project Conference, Denver, CO, 1992.

140. Ohta, T., "Photochemical and Photoelectrochemical Hydrogen Production from Water," in *Hydrogen Energy Progress VI,* Edited by T.N. Veziroglu, N. Getoff, and P. Weinzierl, Pergamon Press, Elmsford, NY, p. 484, 1986.

141. Ohta, T. and Sastri, M.V.C., "Hydrogen Energy Research Programs in Japan," *Int. J. Hydrogen Energy*, 4, p. 489, 1979.

142. Onstott, E.I., "Cerium Dioxide as a Recycle Reagent for Thermochemical Hydrogen Production by Modification of the Sulfur Dioxide-Iodine Cycle," in *Hydrogen Energy Progress VIII*, Edited by T.N. Veziroglu and P.K. Takahashi, Pergamon Press, Elmsford, N.Y., p. 531, 1990.

143. Onuki, K., Shimizu, S., Nakajima, H., Fujita, S., et al, "Studies on an Iodine-Sulfur Process for Thermochemical Hydrogen Production," in *Hydrogen Energy Progress VIII*, Edited by T.N. Veziroglu and P.K. Takahashi, Pergamon Press, Elmsford, N.Y., p. 547, 1990.

144. Otagawa, T., Kovach, A.J., Larkins, J.T. and Schubert, F.H., "Static Feed Water Electrolysis for Hydrogen Generation: Recent Advances," in *Hydrogen Energy Progress VI,* Edited by T.N. Veziroglu, N. Getoff, and P. Weinzierl, Pergamon Press, Elmsford, N.Y., p. 341, 1986.

145. Pacheco, F., "Auto Electrolytic Generation of Hydrogen from Seawater Cogeneration of Surplus Electric and Thermal Energy," in *Hydrogen Energy Progress VIII*, Edited by T.N. Veziroglu and P.K. Takahashi, Pergamon Press, Elmsford, N.Y., p. 641, 1990.

146. Panteleimonova, A.A. and Novikov, A.A., "The Present State of Sulphuric Acid Thermal Decomposition," in *Hydrogen Energy Progress VIII*, Edited by T.N. Veziroglu and P.K. Takahashi, Pergamon Press, Elmsford, N.Y., p. 557, 1990.

147. Parmon, V.N., "Photoproduction of Hydrogen (An Overview of Modern Trends)," in *Hydrogen Energy Progress VIII*, Edited by T.N. Veziroglu and P.K. Takahashi, Pergamon Press, Elmsford, N.Y., p. 801, 1990.

148. Peschka, W., *Liquid Hydrogen—Fuel of the Future*, Springer-Verlag, New York, 1992.

149. Petit, C., Libs, S., Roger, A.C., et al, "Hydrogen Production by Catalytic Oxidation of Methane," in *Hydrogen Energy Progress IX*, Edited by T.N. Veziroglu and C. Derive-J. Pottier, International Association for Hydrogen Energy, Coral Gables, FL, p. 53, 1992.

150. Pimentel, D., Rodrigues, G., Wang, T., Abrams, R., Goldberg, K., Staecker, H., Ma, E., Brueckner, L., Trovato, L., Chow, C., Govindarajulu, U., Boerke, S., "Renewable Energy: Economic and Environmental Issues," *BioScience*, 44, p. 536, 1994.

151. Prokopiev, S.I., Aristov, Y.I., Parmon, V.N. and Giordano, N., "Intensification of Hydrogen Production Via Methane Reforming and the Optimization of H_2 : CO Ratio in a Catalytic Reactor with a Hydrogen-Permeable Membrane Wall," *Int. J. Hydrogen Energy*, 17, p. 275, 1992.

152. Ptasinski, K.J. and Van Swaaij, W.P.M., "Development of a New Method for Hydrogen Recovery from Lean Gas Mixtures Using Metal Hydride Slurries," in *Energy from Biomass 1*, Elsevier Applied Science, p. 412, 1986.

153. Qiyuan, L., "A New Kind of Hydrogen-Producing Catalyst for Hydrocarbon Steam Reforming," *Int. J. Hydrogen Energy*, 17, p. 97, 1992.

154. Rosen, M.A., "Thermodynamic Analysis of Hydrogen Production by Thermochemical Water Decomposition Using the ISPRA MARK-10 Cycle," in *Hydrogen Energy Progress VIII*, Edited by T.N. Veziroglu and P.K. Takahashi, Pergamon Press, Elmsford, N.Y., p. 701, 1990.

155. Rosen, M.A. and Scott, D.S., "Energy Analysis of Hydrogen Production from Heat and Water by Electrolysis," *Int. J. Hydrogen Energy*, 17, p. 199, 1992.

156. Rosen, M.A. and Scott, D.S., "Energy Analysis of a Current-Technology Process for Hydrogen Production by Water Electrolysis," in *Hydrogen Energy Progress VIII*, Edited by T.N. Veziroglu and P.K. Takahashi, Pergamon Press, Elmsford, N.Y., p. 583, 1990.

157. Ruckman, J.H. *Progress Report on Hydrogen Production and Utilization for Community and Automotive Power*, Society of Automotive Engineers, Research Report 789214, 1978.

158. Rusli, A., Sato, T., Nagai, H., et al, "Design of Solid Reactants and Reation Kinetics Concerning the Fe-Compounds in the UT-3 Thermochemical Cycle," in *Hydrogen Energy Progress VIII*, Edited by T.N. Veziroglu and P.K. Takahashi, Pergamon Press, Elmsford, N.Y., p. 503, 1990.

159. Russell, G., "Hawaii Hydrogen Energy Economy: Production and Distribution of Hydrogen and Oxygen in the District of North Kohala, the Big Island of Hawaii: A Global Prototype," in *Hydrogen Energy Progress IX*, Edited by T.N. Veziroglu and C. Derive-J. Pottier, International Association for Hydrogen Energy, Coral Gables, FL, p. 1863, 1992.

160. Salamov, O.M., Rzayev, P.F., Efendieva, N.G. and Gasanova, S.G., "Usage of Solar Energy to get Hydrogen and Oxygen by Electrolysis of Water," in *Hydrogen Energy Progress IX*, Edited by T.N. Veziroglu and C. Derive-J. Pottier, International Association for Hydrogen Energy, Coral Gables, FL, p. 687, 1992.

161. Salman, O.A., Bishara, A., Marafi, A., Abu-Khalifeh, H. and Al-Sayegh, A., "Hydrogen from Hydrogen Sulphide by Solar Energy: Pilot Plant Studies," in *Hydrogen Energy Progress VIII*, Edited by T.N. Veziroglu and P.K. Takahashi, Pergamon Press, Elmsford, N.Y., p. 539, 1990.

162. Salzano, F.J., Skaperdas, G. and Mezzina, A., "Water Vapor Electrolysis: Systems Considerations and Cost/Performance Benefits," in *Hydrogen Energy Progress V*, Edited by T.N. Veziroglu and J.B. Taylor, Pergamon Press, Elmsford, NY, p. 787, 1984.

163. Sasikala, K., Ramana, C.V. and Raghuveer Rao, P., "Environmental Regulation for Optimal Biomass Yield and Photoproduction of Hydrogen by *Rhodobacter Sphaeroides* O. U. 001," *Int. J. Hydrogen Energy*, 16, p. 597, 1991.

164. Sasikala, K., Ramana, C.V. and Raghuveer Rao, P., "Photoproduction of Hydrogen from the Waste Water of Distillery by *Rhodobacter Sphaeroides* O. U. 001," *Int. J. Hydrogen Energy*, 17, p. 23, 1992.

165. Sasikala, K., Ramana, C.V., Rao, P.R. and Subrahmanyam, M., "Effect of Gas Phase on the Photoproduction of Hydrogen and Substrate Conversion Efficiency in the Photosynthetic Bacterium *Rhodobactersphaeroides* O.U. 001," *Int. J. Hydrogen Energy*, 15, p. 795, 1990.

166. Schmittinger, P., "Hydrogen by Chlor-Alkali Electrolysis," in *Electrochemical Hydrogen Technologies*, Edited by H. Wendt, Elsevier Science Publishing Company Inc., New York, p. 261, 1990.

167. Scott, D.S., "Hydrogen and the Environment," in *Hydrogen Energy Progress VI,* Edited by T.N. Veziroglu, N. Getoff, and P. Weinzierl, Pergamon Press, Elmsford, N.Y., p. 159, 1986.

168. Shakhbazov, S.D. and Yusupov, I.M., "Non-traditional sources of energy for hydrogen," in *Hydrogen Energy Progress IX*, Edited by T.N. Veziroglu and C. Derive-J. Pottier, International Association for Hydrogen Energy, Coral Gables, FL, p. 285, 1992.

169. Siegel, A. and Schott, T., "Optimization of Photovoltaic Hydrogen Production," *Int. J. Hydrogen Energy*, 13, p. 659, 1988.

170. Sims, G.P., "Hydroelectric Energy," *Energy Policy*, 19, p. 776, 1991.

171. Singh, S.P., Srivastava, S.C. and Pandey, K.D., "Photoproduction of Hydrogen by a Non-Sulfur Bacterium Isolated from Root Zones of Water Fern *Azolla Pinnata*," *Int. J. Hydrogen Energy*, 15, p. 403, 1990.

172. Singh, S.P., Srivastava, S.C. and Pandey, K.D., "Hydrogen Production by Rhodoseudomonas at the Expense of Vegetable Starch, Sugarcane Juice and Whey," in *Hydrogen Energy Progress IX*, Edited by T.N. Veziroglu and C. Derive-J. Pottier, International Association for Hydrogen Energy, Coral Gables, FL, 1992.

173. Sircar, S., "Production of Hydrogen and Ammonia Synthesis Gas by Pressure Swing Adsorption," *Separation Science and Technology*, 25, p. 1087, 1990.

174. Smith, G.D., Ewart, G.D. and Tucker, W., "Hydrogen Production by Cyanobacteria," in *Hydrogen Energy Progress VIII*, Edited by T.N. Veziroglu and P.K. Takahashi, Pergamon Press, Elmsford, N.Y., p. 735, 1990.

175. Solar Energy Research Institute, "Central Receiver Technology: Status and Assessment," Report No. SERI/SP-220-3314, Department of Energy, September, 1989.

176. Sosna, M.H., Baichtok, Y.K., Mordkovich, V.Z. and Korostyshevsky, N.N., "Energetics and Economics of the Process for Hydrogen Recovery by Means of the Membranes Made of Palladium Alloys," in *Hydrogen Energy Progress VIII*, Edited by T.N. Veziroglu and P.K. Takahashi, Pergamon Press, Elmsford, N.Y., p. 1201, 1990.

177. Specht, M., Bandi, A., Maier, C.U., et al, "Energetics of Solar Methanol Synthesis from Atmospheric Carbon Dioxide Compared to Solar Liquid Hydrogen Generation," *Energy Conversion Management*, 33, p. 537, 1992.

178. Specht, M., Kühne, H. and Schefold, J., "Hydrogen Production with Silicon Photoelectrodes," in *Hydrogen Energy Progress VIII*, Edited by T.N. Veziroglu and P.K. Takahashi, Pergamon Press, Elmsford, N.Y., p. 815, 1990.

179. Srinivasan, S. and Salzano, F.J., "Prospects for Hydrogen Production by Water Electrolysis to be Competitive with Conventional Methods," *Int. J. Hydrogen Energy*, 2, p. 53, 1977.

180. Stannard, J.H., Fitzpatrick, N.P., Rao, B.M.L. and Anderson, W.M., "A Portable Hydrogen Generator for Vehicles," in *Hydrogen Energy Progress VIII*, Edited by T.N. Veziroglu and P.K. Takahashi, Pergamon Press, Elmsford, N.Y., p. 935, 1990.

181. Steeb, H., Seeger, W. and Oud, H.A., "Hysolar: An Overview on the German-Saudi Arabian Program on Solar Hydrogen," in *Hydrogen Energy Progress IX*, Edited by T.N. Veziroglu and C. Derive-J. Pottier, International Association for Hydrogen Energy, Coral Gables, FL, p. 1845, 1992.

182. Steinberg, M. and Cheng, H.C., "Modern and Prospective Technologies for Hydrogen Production from Fossil Fuels," *Int. J. Hydrogen Energy*, 14, p. 797, 1989.

183. Stewart, D.J., "Methane from Crop-Grown Biomass," in *Fuel Gas Systems*, Edited by D. L. Wise, CRC Press, Inc., Boca Raton, p. 85, 1983.

184. Stojic, D.L., Miljanic, S.S., Grozdic, T.D. and Jaksic, M.M., "Improvement of Hydrogen Isotope Separation in Water Electrolysis," in *Hydrogen Energy Progress IX*, Edited by T.N. Veziroglu and C. Derive-J. Pottier, International Association for Hydrogen Energy, Coral Gables, FL, p. 393, 1992.

185. Struck, B.D., Schütz, G.H. and Van Velzen, D., "Cathodic Hydrogen Evolution in Thermochemical-Electrochemical Hybrid Cycles," in *Electrochemical Hydrogen Technologies*, Edited by H. Wendt, Elsevier Science Publishing Company Inc., New York, p. 301, 1990.

186. Stucki, S., "The Cost of Electrolytic Hydrogen from Off-Peak Power," *Int. J. Hydrogen Energy*, 16, p. 461, 1991.

187. Szyszka, A., "Technical Communication: Realization of the Solar-Hydrogen Project at Neunburg Vorm Wald, F.R.G.," *Int. J. Hydrogen Energy*, 15, p. 597, 1990.

188. Tadokoro, Y., Kajiyama, T., Yamaguchi, T., et al, "Cycle Simulation of the UT-3 Thermochemical Hydrogen Production Process," in *Hydrogen Energy Progress VIII*, Edited by T.N. Veziroglu and P.K. Takahashi, Pergamon Press, Elmsford, N.Y., p. 513, 1990.

189. Tagawa, H. and Mitzusaki, J., "Thermal Decomposition of Metal Sulfates and Sulfuric Acid and Its Applicability to Thermochemical Hydrogen Production," in *Hydrogen Energy Progress VIII*, Edited by T.N. Veziroglu and P.K. Takahashi, Pergamon Press, Elmsford, N.Y., p. 593, 1990.

190. Tamkins, T., "Tilting at Wind Power," *Audubon*, Sept.-Oct., p. 24, 1993.

191. Taqui Khan, M.M., Adiga, M.R. and Bhatt, J.P., "Photogeneration of Hydrogen by *Halobacterium Halobium* MMT_{22} Coupled with Silicon PN Junction Semiconductor without External Bias Potential-II," *Int. J. Hydrogen Energy*, 17, p. 89, 1992.

192. Tarnay, D.S., "Hydrogen Production at Hydro Power Plants," in *Hydrogen Energy Progress V*, Edited by T.N. Veziroglu and J.B. Taylor, Pergamon Press, Elmsford, NY, p. 323, 1984.

193. Tennakone, K., "Hydrogen from Brine Electrolysis: A New Approach," *Int. J. Hydrogen Energy*, 14, p. 681, 1989.

194. Tennakoon, C.L.K., Bhardwaj, R.C. and Bockris, J.O., "Space-Based Bacterial Production fo Hydrogen," in *Hydrogen Energy Progress IX*, Edited by T.N. Veziroglu and C. Derive-J. Pottier, International Association for Hydrogen Energy, Coral Gables, FL, p. 1641, 1992.

195. Tien, H.T. and Chen, J., "Semiconductor Septum Electrochemical Photovoltaic Cells: A New Approach to Solar Hydrogen Production from Sea Water," in *Hydrogen Energy Progress VIII*, Edited by T.N. Veziroglu and P.K. Takahashi, Pergamon Press, Elmsford, N.Y., p. 833, 1990.

196. Timpe, R.C., Sears, R.E. and Malterer, T.J., "Pine and Willow as Carbon Sources in the Reaction Between Carbon and Steam to Produce Hydrogen Gas," in *Energy from Biomass and Wastes XII*, Edited by D. L. Klass, Institute of Gas Technology, Chicago, IL, p. 763, 1989.

197. Uemiya, S., Sato, N., Ando, H., Matsuda, T. and Kikuchi, E., "Steam Reforming of Methane in a Hydrogen-Permeable Membrane Reactor," *Applied Catalysis*, 67, p. 223, 1991.

198. Upadhyay, R.K., "On the Formulation of Parameters for Evaluation of Thermochemical Cyclic Processes for the Production of Hydrogen," in *Hydrogen Energy Progress VI,* Edited by T.N. Veziroglu, N. Getoff, and P. Weinzierl, Pergamon Press, Elmsford, NY, p. 744, 1986.

199. Vainer, A.J., Igoshin, A.I., Kazaryan, V.A. and Smirnov, V.I., "The Utilization of Underground Caverns in Rock Salt Formations for Hydrogen Production and Storage," in *Hydrogen Energy Progress IX*, Edited by T.N. Veziroglu and C. Derive-J. Pottier, International Association for Hydrogen Energy, Coral Gables, FL, p. 159, 1992.

200. Vandenborre, H., Leysen, R., Nackaerts, H., et al, "Advanced Alkaline Water Electrolysis Using Inorganic-Membrane-Electrolyte (I.M.E.) Technology," in *Hydrogen Energy Progress V*, Edited by T.N. Veziroglu and J.B. Taylor, Pergamon Press, Elmsford, NY, p. 703, 1984.

201. Vandenborre, H., Leysen, R., Tollenboom, J.P. and Baetslé, L.H., "On the Inorganic-Membrane-Electrolyte (IME) Water Electrolysis," in *Hydrogen Energy System*, Pergamon Press, p. 2379, 1978.

202. Vatsalia, T.M., "Hydrogen Production from (Cane-Molasses) Stillage by *C. Freundi* and Its Use in Improving Methanogenesis," in *Hydrogen Energy Progress VIII*, Edited by T.N. Veziroglu and P.K. Takahashi, Pergamon Press, Elmsford, N.Y., p. 775, 1990.

203. Venkatarman, C. and Vatsala, T.M., "Hydrogen Production from Whey by Phototrophic Bacteria," in *Hydrogen Energy Progress VIII*, Edited by T.N. Veziroglu and P.K. Takahashi, Pergamon Press, Elmsford, N.Y., p. 781, 1990.

204. Veziroglu, T.N. and Barbir, F., "Hydrogen: The Wonder Fuel," *Int. J. Hydrogen Energy*, 17, p. 391, 1992.

205. Ward, R.F., "Potential of Biomass," in *Fuel Gas Systems*, Edited by D. L. Wise, CRC Press, Inc., Boca Raton, FL, p. 1, 1983.

206. Whug, W., Bussmann, H. and Brinner, A., "Intermittent Operation and Operation Modelling of an Alkaline Electrolyser," in *Hydrogen Energy Progress IX*, Edited by T.N. Veziroglu and C. Derive-J. Pottier, International Association for Hydrogen Energy, Coral Gables, FL, p. 783, 1992.

207. Williams, L.O., *Hydrogen Power: An Introduction to Hydrogen Energy and its Applications*, Pergamon Press, Oxford, 1980.

208. Winter, C. and Fuchs, M., "Hysolar and Solar-Wasserstoff-Bayern," *Int. J. Hydrogen Energy*, 16, p. 723, 1991.

209. Wurster, R. and Stiftung, L.-B., "The Euro-Quebec Hydro-Hydrogen Pilot Project EQHHPP," in *Hydrogen Energy Progress VIII*, Edited by T.N. Veziroglu and P.K. Takahashi, Pergamon Press, Elmsford, N.Y., p. 59, 1990.

210. Yalçin, S., "A Review of Nuclear Hydrogen Production," *Int. J. Hydrogen Energy*, 14, p. 551, 1989.

211. Zahed, A.H., Bashir, M.D., Alp, T.Y. and Najjar, Y.S.H., "A Perspective of Solar Hydrogen and its Utilization in Saudi Arabia," *Int. J. Hydrogen Energy*, 16, p. 277, 1991.

212. Zweibel, K., *Harnessing Solar Power: The Photovoltaics Challenge*, Plenum Press, New York, NY, 1990.

213. "A Growth Market in Wind Power," EPRI Journal, December, p. 4, 1992.

Safety

1. Balthasar, W. and Schödel, J.P., *Hydrogen Safety Manual*, Commision of the European Communities, Luxembourg, 1983.

2. Beavais, R., Mayinger, F. and Strube, G., "Turbulent Flame Acceleration: Mechanisims and Significance for Safety Considerations," in *Hydrogen Energy Progress IX*, Edited by T.N. Veziroglu and C. Derive-J. Pottier, International Association for Hydrogen Energy, Coral Gables, FL, p. 1093, 1992.

3. Beer, F.P. and Johnston, E.R., *Mechanics of Materials*, McGraw Hill, New York, 1981.

4. Chemical Propulsion Information Agency, "Fuel to the Future—Safety in Ground-Based LH_2 Logistics," Johns Hopkins University, Applied Physics Laboratory, 1988.

5. Chuveliov, A.V. *Hydrogen in Motor Vehicles: A Case Study of Hydrogen Utilization in Motor Vehicles as a Supplementary Fuel in Southern California Air Basin*, I.V. Kurchatov Institute of Atomic Energy, Report prepared for SCQAMD, 1989.

6. Daous, M.A., Bashir, M.D. and El-Naggar, M.A., "Experiences with the Safe Operation of a Two KW Solar Hydrogen Plant," in *Hydrogen Energy Progress IX*, Edited by T.N. Veziroglu and C. Derive-J. Pottier, International Association for Hydrogen Energy, Coral Gables, FL, p. 1155, 1992.

7. Das, L.M., "Safety Aspects of a Hydrogen-Fuelled Engine System Development," *Int. J. Hydrogen Energy*, 16, p. 619, 1991.

8. DeLuchi, M.A., "Hydrogen Vehicles: An Evaluation of Fuel Storage, Performance, Safety, and Cost," *Int. J. Hydrogen Energy*, 14, p. 81, 1989.

9. Durand, H., "Hydrogen Safety Aspects on Test Facilities," in *Hydrogen Energy Progress IX*, Edited by T.N. Veziroglu and C. Derive-J. Pottier, International Association for Hydrogen Energy, Coral Gables, FL, p. 1141, 1992.

10. Edeskuty, F.J., "Safety," in *Hydrogen: Its Technology and Implications*, Edited by K. E. Cox and K. D. J. Williamson, CRC Press, Inc., Boca Raton, FL, p. 208, 1979.

11. Edeskuty, F.J., Haugh, J.J. and Thompson, R.T., "Safety Aspects of Large-Scale Combustion of Hydrogen," in *Hydrogen Energy Progress VI,* Edited by T.N. Veziroglu, N. Getoff, and P. Weinzierl, Pergamon Press, Elmsford, N.Y., p. 147, 1986.

12. Eichert, H. and Fischer, M., "Hydrogen Safety in Energy Application Compared with Natural Gas," in *Hydrogen Energy Progress V*, Edited by T.N. Veziroglu and J.B. Taylor, Pergamon Press, Elmsford, N.Y., p. 1869, 1984.

13. Eichert, H., Kratzel, T. and Wetzel, F., "Instationary High-Turbulent Hydrogen Combustion Processes," in *Hydrogen Energy Progress IX*, Edited by T.N. Veziroglu and C. Derive-J. Pottier, International Association for Hydrogen Energy, Coral Gables, FL, p. 1181, 1992.

14. Finegold, J.G. and Van Vorst, W.D., "Crash test of liquid hydrogen automobile," in *1st World Hydrogen Energy Conference*, University of Miami, Coral Gables, FL, p. 6, 1976.

15. Fischer, M., "Safety Aspects of Hydrogen Combustion in Energy Systems," *Int. J. Hydrogen Energy*, 11, p. 593, 1986.

16. Fromageau, R., Droniou, C. and Rubinstein, M., "Hydrogen Data: A Numerical and Factual Data Bank on Hydrogen-Material Interactions," in *Hydrogen Energy Progress IX*, Edited by T.N. Veziroglu and C. Derive-J. Pottier, International Association for Hydrogen Energy, Coral Gables, FL, p. 1175, 1992.

17. Hansel, J.G., Mattern, G.W. and Miller, R.N. *Safety Considerations in the Design of Hydrogen-Powered Vehicles*, Air Products and Chemicals, Inc., 1993, W864A1.

18. Harkin, T., "Proposal for a Sustainable Energy Future Based on Renewable Hydrogen," 1993.

19. Hoffmann, P., *The Forever Fuel: The Story of Hydrogen*, Westview Press, Boulder, CO, 1981.

20. Hord, J., "Is Hydrogen a Safe Fuel?" *Int. J. Hydrogen Energy*, 3, p. 157, 1978.

21. Karim, G.A. and Panlilio, V., "Flame Propagation and Extinction within Stratified Mixtures Involving Hydrogen and Dilutent Inert Gases in Air," in *Hydrogen Energy Progress IX*, Edited by T.N. Veziroglu and C. Derive-J. Pottier, International Association for Hydrogen Energy, Coral Gables, FL, p. 1191, 1992.

22. Khristenko, Y. and Tomilim, K., "Safe Operating Conditions Research for Hydrogen Systems with Periodical and Continuous Gas Venting," in *Hydrogen Energy Progress IX*, Edited by T.N. Veziroglu and C. Derive-J. Pottier, International Association for Hydrogen Energy, Coral Gables, FL, p. 1165, 1992.

23. Kirillov, I.A., Rusanov, V.D. and Fridman, A.A., "The Weak Shock Wave Evolution in the Zone of Spontaneous Propagation of Exothermal Chemical Reaction," in *Hydrogen Energy Progress IX*, Edited by T.N. Veziroglu and C. Derive-J. Pottier, International Association for Hydrogen Energy, Coral Gables, FL, p. 1103, 1992.

24. Knowlton, R.E., "An Investigation of the Safety Aspects in the Use of Hydrogen as a Ground Transportation Fuel," *Int. J. Hydrogen Energy*, 9, p. 129, 1984.

25. Knowlton, R.E., "An Investigation of the Safety Aspects in the Use of Hydrogen as a Ground Transportation Fuel," in *Hydrogen Energy Progress V*, Edited by T.N. Veziroglu and J.B. Taylor, Pergamon Press, Elmsford, NY, p. 1881, 1984.

26. Knowlton, R.E., "Safety in New Uses of Hydrogen Energy," in *Advances in Cryogenic Engineering*, Plenum Press, New York, NY, p. 1057, 1985.

27. Nelson, H.G., "A Review—Materials and Safety Problems Associated with Hydrogen Containment," in *Hydrogen Energy Progress V*, Edited by T.N. Veziroglu and J.B. Taylor, Pergamon Press, Elmsford, N.Y., p. 1841, 1984.

28. National Fire Protection Association *Gaseous Hydrogen Systems*, NFPA 50A, 1989.

29. Ordin, P.M., "Review of Hydrogen Accidents and Incidents in NASA Operations," NASA-X71565, 1974.

30. Pehr, K., "Aspects of Safety and Acceptance of LH_2 Tank Systems in Passenger Cars," in *Hydrogen Energy Progress X*, Edited by D.L. Block and T.N.

Veziroglu, International Association for Hydrogen Energy, Coral Gables, FL, p. 1399, 1994.

31. Peschka, W., *Liquid Hydrogen—Fuel of the Future*, Springer-Verlag, New York, 1992.

32. Peterson, U., Wursig, G. and Krapp, R., "Design and Safety Considerations for Large Scale Sea-Borne Hydrogen Transport," in *Hydrogen Energy Progress IX*, Edited by T.N. Veziroglu and C. Derive-J. Pottier, International Association for Hydrogen Energy, Coral Gables, FL, p. 1021, 1992.

33. Pfriem, H.-J., "Overview of the Cooperative Program on Hydrogen Storage, Conversion and Safety of the International Energy Agency," *Int. J. Hydrogen Energy*, 16, p. 329, 1991.

34. Quadflieg, H., "From Research to Market Application? Experience with the German Hydrogen Fuel Project," in *Hydrogen Energy Progress VI,* Edited by T.N. Veziroglu, N. Getoff, and P. Weinzierl, Pergamon Press, Elmsford, New York, 1986.

35. Swain, M.R. and Swain, M.N., "A Comparison of H_2, CH_4 and C_3H_8 Fuel Leakage in Residential Settings," in *Hydrogen Energy Progress IX*, Edited by T.N. Veziroglu and C. Derive-J. Pottier, International Association for Hydrogen Energy, Coral Gables, FL, p. 1121, 1992.

36. Union Carbide—LINDE, *Material Safety Data Sheet*, Union Carbide-LINDE Division, 1986, report L-4604-A.

37. Wierzba, I., Oladpio, A.B. and Karim, G.A., "The Limits for Flame Flashback Within Streams of Lean Homogeneous Fuel-Dilutent-Air Mixtures Involving Hydrogen," in *Hydrogen Energy Progress IX*, Edited by T.N. Veziroglu and C. Derive-J. Pottier, International Association for Hydrogen Energy, Coral Gables, FL, p. 1111, 1992.

38. Youngblood, W.W., "Safety Criteria for the Operation of Gaseous Hydrogen Pipelines (Final Report)," W. W. Youngblood Wyle Labs, Huntsville, AL, 1984.

39. Zweig, R.M., "Safety Video, 1992," 1992.

Storage

1. Akiba, E., Ishido, Y.Y., Hayakawa, H., et al, "The Cyclic Life Tests of Mg-Ni Hydrogen Absorbing Alloys," *Zeitschrift für Physikalische Chemie Neue Folge*, 164, p. 1319, 1989.

2. Alcock, C.B., Hewitt, J.S., Khatamian, D., et al, "Research on Hydrogen Storage Alloys and Their Uses," in *Hydrogen Energy Progress V,* Edited by T.N. Veziroglu and J.B. Taylor, Pergamon Press, Elmsford, NY, p. 1309, 1984.

3. Amankwah, K.A.G., Noh, J.S., and Schwarz, J.A., "Hydrogen Storage on Superactivated Carbon at Refigeration Temperatures," *Int. J. Hydrogen Energy*, 14, p. 437, 1989.

4. Angus, H.C., "Storage, Distribution and Compression of Hydrogen," *Chemistry and Industry (London)*, 68, 1984.

5. Anzulovic, I., "Optimization of Gaseous Hydrogen Storage System," *Int. J. Hydrogen Energy*, 17, p. 129, 1992.

6. Au, M., Chen, C.P., Ye, Z., Fang, T.S., Wu, J., and Wang, Q.D., "An Industrial Scale Experiment on the Recovery, Purification, Storage and Transport of Hydrogen Abstracted from the Purge Gas of a Synthetic Ammonia Plant and the Application of Hydrogen to the Tin Bath in a Float Glass Plant by Means of Mobile Hydride Hydrogen Containers," in *Hydrogen Energy Progress IX*, Edited by T.N. Veziroglu and C. Derive-J. Pottier, International Association for Hydrogen Energy, Coral Gables, FL, p. 1031, 1992.

7. Bernauer, O. "Use of Hydrides in Motor Vehicles," in *International Symposium on Automotive Propulsion Systems, 5th*, DOE, p. 668, 1980.

8. Bernauer, O., "Metal Hydride Storages," *Zeitschrift für Physikalische Chemie Neue Folge*, 164, p. 1381, 1989.

9. Binder, K. and Withalm, G., "Mixture formation and combustion in a hydrogen engine using hydrogen storage technology," *Int. J. Hydrogen Energy*, 7, p. 651, 1982.

10. Blazek, C.F., Daniels, E.J., Donakowski, T.D., and Novil, M., "Economics of Hydrogen in the '80s and Beyond," in *Recent Developments in Hydrogen Technology*, Edited by K. D. Williamson Jr. and F. J. Edeskuty, CRC Press, Inc., Boca Raton, p. 1, 1986.

11. Bockris, J. O'M., *Energy Options*, Australia & New Zealand Book Company, Sydney, Australia, 1980.

12. Bowman, R.C.J., "Preparation and Properties of Amorphous Hydrides," in *Materials Science Forum*, Edited by R. G. Barnes, Trans Tech Publications Ltd., Brookfield, p. 197, 1988.

13. Bracha, M., Lorenz, G., Patzelt, A., and Wanner, M., "Large-Scale Hydrogen Liquefaction in Germany," in *Hydrogen Energy Progress IX*, Edited by T.N. Veziroglu and C. Derive-J. Pottier, International Association for Hydrogen Energy, Coral Gables, FL, p. 1001, 1992.

14. Bradhurst, D.H. and Heuer, P.M., "The Properties of a Metal Hydride Fuel for use in an Urban Vehicle," *J. Less Common Met.*, 89, p. 575, 1983.

15. Buchner, H., "The hydrogen/hydride energy concept," *Int. J. Hydrogen Energy*, 3, p. 385, 1978.

16. Buchner, H. and Povel, R., "The Daimler-Benz Hydride Vehicle Project," *Int. J. Hydrogen Energy*, 7, p. 259, 1982.

17. Buchner, H., "Hydrogen Use—Transportation Fuel," in *Hydrogen Energy Progress IV*, Edited by T.N. Veziroglu, W.D. Van Vorst, and J.H. Kelley, Pergamon Press, Elmsford, NY, p. 3, 1982.

18. Bunger, U.H., Andreassen, K., Henriksen, N. and Oyvann, A., "Hydrogen as an Energy Carrier, Production and Liquefacton of Hydrogen in Norway for Transportation to and Storage/Distribution," in *Hydrogen Energy Progress IX*, Edited by T.N. Veziroglu and C. Derive-J. Pottier, International Association for Hydrogen Energy, Coral Gables, FL, p. 1913, 1992.

19. Carpetis, C., "Comparison of the Expenses Required for the On-Board Fuel Storage Systems of Hydrogen Powered Vehicles," *Int. J. Hydrogen Energy*, 7, p. 61, 1982.

20. Carpetis, C. and Peschka, W., "On the Storage of Hydrogen by use of Cryo-Adsorbants," in *Proc. 1st World Hydrogen Energy Conference*, University of Miami, Coral Gables, FL, 9C-45, 1976.

21. Carpetis, C. and Peschka, W., "A Study on Hydrogen Storage by use of Cryoadsorbants," *Int. J. Hydrogen Energy*, 5, p. 539, 1980.

22. Cheng, J.S., Durand, R., Faure, R. and Jorge, G., "Study of Hydrogen in Metals, Alloys and Hydrides by Electrochemical Impedance and Transient Measurements," in *Hydrogen Energy Progress IX*, Edited by T.N. Veziroglu and C. Derive-J. Pottier, International Association for Hydrogen Energy, Coral Gables, FL, p. 795, 1992.

23. Chuveliov, A.V. *Hydrogen in Motor Vehicles: A Case Study of Hydrogen Utilization in Motor Vehicles as a Supplementary Fuel in Southern California Air Basin*, I.V. Kurchatov Institute of Atomic Energy, Report prepared for SCQAMD, 1989.

24. Cialone, H.J., Scott, P.M., Holbrook, J.H., et al, "Hydrogen Effects on Conventional Pipeline Steels," in *Hydrogen Energy Progress V*, Edited by T.N. Veziroglu and J.B. Taylor, Pergamon Press, Elmsford, N.Y., p. 1855, 1984.

25. Cicconardi, S.P., Jammelli, E., and Spazzafumo, G., "Hydrogen Energy Storage: Preliminary Analysis," in *Hydrogen Energy Progress IX*, Edited by T.N. Veziroglu and C. Derive-J. Pottier, International Association for Hydrogen Energy, Coral Gables, FL, p. 2011, 1992.

26. Dalle Donne, M., "Hydrogen as an Energy Carrier in Substituting Petroleum. Demonstration Project: Automobiles Driven by Nuclear Energy," *Int. J. Hydrogen Energy*, 8, p. 949, 1983.

27. DeLuchi, M.A., "Hydrogen Vehicles: An Evaluation of Fuel Storage, Performance, Safety, Environmental Impacts, and Cost," *Int. J. Hydrogen Energy*, 14, p. 81, 1989.

28. Deyou, B., "Hydrogen and Hydride Technology Activity in China," in *Hydrogen Energy Progress VIII*, Edited by T.N. Veziroglu and P.K. Takahashi, Pergamon Press, Elmsford, N.Y., p. 1081, 1990.

29. Dienhart, H. and Siegel, A., "Hydrogen Storage in Isolated Electrical Energy Systems with Photovoltaic and Wind-Energy," in *Hydrogen Energy Progress IX*, Edited by T.N. Veziroglu and C. Derive-J. Pottier, International Association for Hydrogen Energy, Coral Gables, FL, p. 713, 1992.

30. Dini, D. and Dini, G., "Thermochemical Computation of Hydrogen Liquefaction Plants," in *Hydrogen Energy Progress VIII*, Edited by T.N. Veziroglu and P.K. Takahashi, Pergamon Press, Elmsford, N.Y., p. 873, 1990.

31. Dini, D. and Martorano, L., "Cryogenic Fuel Storage in Cars using Hydrogen as Power Source in Road Transport," in *Proc. 15th Int. Congr. Refrigeration*, Venice, September 23 to 29, 1979.

32. Duret, B. and Saudin, A., "Microspheres for onboard hydrogen storage," in *Hydrogen Energy Progress IX*, Edited by T.N. Veziroglu and C. Derive-J. Pottier, International Association for Hydrogen Energy, Coral Gables, FL, p. 137, 1992.

33. Dutta, K. and Srivastava, O.N., "Investigations on Synthesis, Characterization and Hydrogenation Behaviour of La_2Mg_{17} and Related Intermetallics," in *Hydrogen Energy Progress VIII*, Edited by T.N. Veziroglu and P.K. Takahashi, Pergamon Press, Elmsford, N.Y., p. 1027, 1990.

34. Eklund, G. and von Krusenstierna, O., "Storage and Transportation of Merchant Hydrogen," *Int. J. Hydrogen Energy*, 8, p. 949, 1983.

35. Ertl, G., "Interaction of Hydrogen with Metal Surfaces," *Zeitschrift für Physikalische Chemie Neue Folge*, 164, p. 1115, 1989.

36. Escher, W.J.D., Adt, Jr., R.R., Swain, M.R., Papas, J.M., et al, "Hydrogen Engine and Fuel Containment R&D: A Progress Report on the U. S. Department of Energy Program," in *International Symposium on Automotive Propulsion Systems, 5th*, DOE, p. 638, 1980.

37. Escher, W.J.D., "Survey of Liquid Hydrogen Container Techniques for Highway Fuel System Applications," Report No. HCP/M2753-01, U.S. Department of Energy, Washington, D.C., 1979.

38. Ewald, R., "Liquid Hydrogen Fuelled Automobiles: On-Board and Stationary Cryogenic Installations," *Cryogenics*, 30, p. 38, 1990.

39. Feucht, K., Hölzel, G. and Hurich, W., "Perspectives of Mobile Hydrogen Application," in *Hydrogen Energy Progress VII*, Edited by T.N. Veziroglu and A.N. Protsenko, Pergamon Press, Elmsford, N.Y., p. 1963, 1988.

40. Finegold, J.G., McKinnon, J.T., and Karpuk, M.E., "Dissociated Methanol as a Consumable Hydride for Automobiles and Gas Turbines," in *Hydrogen Energy Progress IV*, Edited by T.N. Veziroglu, W.D. Van Vorst, and J.H. Kelley, Pergamon Press, Elmsford, NY, p. 1359, 1982.

41. Fraenkel, D., Lazar, R., and Shabtai, J., "The Potential of Zeolite Molecular Sieves as Hydrogen Storage Media," in *Alternative Energy Sources*, Edited by T.N. Veziroglu, Hemisphere Publishing, Washington, D.C., p. 3771, 1978.

42. Furuhama, S., Hiruma, M. and Enomoto, Y., "Development of a liquid hydrogen car," *Int. J. Hydrogen Energy*, 3, p. 61, 1978.

43. Furuhama, S., "A Liquid Hydrogen Car with a Two-Stroke Engine and LH_2 Pump," *Int. J. Hydrogen Energy*, 7, p. 809, 1981.

44. Furuhama, S. and Kobayashi, Y., "Hydrogen Cars with LH_2-Tank, LH_2-Pump and Cold GH_2-Injection Two-Stroke Engine," SAE Technical Paper No. 820349, Society of Automotive Engineers, Warrendale, PA, 1982.

45. Furuhama, S., "Hydrogen Engine Systems for Land Vehicles," in *Hydrogen Energy Progress VII*, Edited by T.N. Veziroglu and A.N. Protsenko, Pergamon Press, Elmsford, N.Y., p. 1841, 1988.

46. Furuhama, S., Matushita, T., Nakajima, T. and Yamaura, K., "Hydrogen Injection Spark Ignition Engine with LH_2 Pump Driven by High-Pressure Hydrogen Expander," in *Hydrogen Energy Progress VII*, Edited by T.N. Veziroglu and A.N. Protsenko, Pergamon Press, Elmsford, N.Y., p. 1975, 1988.

47. Furuhama, S., "Hydrogen Engine Systems for Land Vehicles," *Int. J. Hydrogen Energy*, 14, p. 907, 1989.

48. Furuhama, S., Sakurai, T. and Shindo, M., "Study of Evaporation Loss of Liquid Hydrogen Storage Tank with LH_2-Pump," in *Hydrogen Energy Progress VIII*, Edited by T.N. Veziroglu and P.K. Takahashi, Pergamon Press, Elmsford, N.Y., p. 1087, 1990.

49. Gordan, R., "Composite Pressure Vessels for Gaseous Hydrogen Powered Vehicles," in *Hydrogen Energy Progress V*, Edited by T.N. Veziroglu and J.B. Taylor, Pergamon Press, Elmsford, NY, p. 1225, 1984.

50. Groll, M., Isselhorst, A. and Wierse, M., "Metal Hydride Devices for Environmentally Clean Energy Technology," in *Hydrogen Energy Progress IX*, Edited by T.N. Veziroglu and C. Derive-J. Pottier, International Association for Hydrogen Energy, Coral Gables, FL, p. 773, 1992.

51. Grünenfelder, N.F. and Schucan, T.H., "Seasonal Storage of Hydrogen in Liquid Organic Hydrides: Description of the Second Prototype Vehicle," *Int. J. Hydrogen Energy*, 14, p. 579, 1989.

52. Gupta, M., "Electronic Structure of Intermetallic Hydrides for Hydrogen Storage," in *Materials Science Forum*, Edited by R. G. Barnes, Trans Tech Publications Ltd., Brookfield, p. 77, 1988.

53. Hettinger, W., Tocha, K. and Kesten, M., "Concept, Design and Performance of an Automatically Controlled Refuelling System for Liquid Hydrogen Powered Vehicles," in *Hydrogen Energy Progress IX*, Edited by T.N. Veziroglu and C. Derive-J. Pottier, International Association for Hydrogen Energy, Coral Gables, FL, p. 1315, 1992.

54. Holstvoogd, R.D., Van Swaaij, W.P.M., Versteeg, G.F. and Snijder, E.D., "Continuous Absorption of Hydrogen in Metal Slurries," *Zeitschrift für Physikalische Chemie Neue Folge*, 164, p. 1429, 1989.

55. Houdart, A., "Liquid Hydrogen Producton Plant," in *Hydrogen Energy Progress IX*, Edited by T.N. Veziroglu and C. Derive-J. Pottier, International Association for Hydrogen Energy, Coral Gables, FL, p. 79, 1992.

56. Huang, Y.C., Goto, H., Sato, A., et al, "Solar Energy Storage by Metal Hydride," *Zeitschrift für Physikalische Chemie Neue Folge*, 164, p. 1391, 1989.

57. Huiyou, Z., Jianyin, C., Xinyi, P., et al, "Studies of Effect of Alloyed Elements and the Second Phases on the Properties of TiFe-Based Alloys."

58. Huston, E.L., "Liquid and Solid Storage of Hydrogen," in *Hydrogen Energy Progress V*, Edited by T.N. Veziroglu and J.B. Taylor, Pergamon Press, Elmsford, NY, p. 1171, 1984.

59. Hynek, S., Fuller, W., Bentley, J., and McCullough, J., "Hydrogen Storage by Carbon Sorption," in *Hydrogen Energy Progress X*, Edited by D.L. Block and T.N. Veziroglu, International Association for Hydrogen Energy, Coral Gables, FL, p. 985, 1994.

60. Ibrasheva, R.K., Solomina, T.A., Leonova, G.I., et al, "Role of Active Surface in Processes of Hydrogen Sorption-Desorption by Intermetallic Compounds," in *Hydrogen Energy Progress VIII*, Edited by T.N. Veziroglu and P.K. Takahashi, Pergamon Press, Elmsford, N.Y., p. 1097, 1990.

61. Imamura, H., Miura, H., Futsuhara, M. and Tsuchiya, S., "Efficient Photoassisted Hydrogen Storage in Rare Earth Intermetallic Compounds by the Use of Chemical Hydrogen Carriers," *Zeitschrift für Physikalische Chemie Neue Folge*, 164, p. 1403, 1989.

62. Imamura, H., Kasahara, S., Takada, T. and Tsuchiya, S., "Hydrogen Storage in Rare Earth Intermetallic Compounds by the Use of Chemical Hydrogen Carriers," *Zeitschrift für Physikalische Chemie Neue Folge*, 164, p. 1397, 1989.

63. Ishikawa, H., Oguro, K., Kato, A., et al, "Microencapsulation and Compaction of Hydrogen Storage Alloy," *Zeitschrift für Physikalische Chemie Neue Folge*, 164, p. 1409, 1989.

64. Jones, L.W., "Liquid Hydrogen as a Fuel for the Future," *Science*, 174, p. 367, 1971.

65. Kelley, J.H. and Hagler, R., "Storage, Transmission and Distribution of Hydrogen," *Int. J. Hydrogen Energy*, 5, p. 35, 1980.

66. Kharkats, Y.I. and Pleskov, Y.V., "A Plant for Solar Energy Conversion and Storage: "Solar Array + Electrolyser + Storage Battery," Computation of the Non-Steady-State Operating Conditions and Design

Optimization," *Int. J. Hydrogen Energy*, 16, p. 653, 1991.

67. Khrussanova, M. and Peshev, P., "The Effect of the d-Electron Concentration on the Absorption of Some Systems of Hydrogen Storage," *Materials Research Bulletin*, 26, p. 1291, 1991.

68. Kolachev, B.A., "Properties of Hydrogen Storage Alloys," *Soviet Materials Science*, 26, p. 642, 1990.

69. Krepec, T., Miele, D. and Lisio, C., "Improved Concept of Hydrogen On-Board Storage and Supply for Automotive Applications," *Int. J. Hydrogen Energy*, 15, p. 27, 1990.

70. Krepec, T., Carrese, G. and Miele, D., "Further Investigations on Electronically Controlled Hydrogen Storage and Injection System for Automotive Applications," in *Hydrogen Energy Progress VIII*, Edited by T.N. Veziroglu and P.K. Takahashi, Pergamon Press, Elmsford, N.Y., p. 925, 1990.

71. Kukkonen, C.A. and Shelef, M., "Hydrogen as an Alternative Automobile Fuel: 1993 Update," SAE Technical Paper No. 940766, Society of Automotive Engineers, Warrendale, PA, 1994.

72. Los Alamos National Lab *Refueling Considerations for Liquid-Hydrogen Fueled Vehicles*, LA-UR-84-1490, 1984.

73. Lanyin, S., Fangjie, L. and Deyou, B., "An Advanced TiFe Series Hydrogen Storage Material with High Hydrogen Capacity and Easily Activated Properties," *Int. J. Hydrogen Energy*, 15, p. 259, 1990.

74. Ledjeff, K., "Comparison of Storage Options for Photovoltaic Systems," *Int. J. Hydrogen Energy*, 15, p. 629, 1990.

75. Lewis, F.A., "Storage and Distribution of Hydrogen: Materials and Reliability," in *Hydrogen Energy Progress VI*, Edited by T.N. Veziroglu, N. Getoff, and P. Weinzierl, Pergamon Press, Elmsford, NY, p. 979, 1986.

76. Libowitz, G.G., "An Introduction to Metallic Hydrides and Their Applications," in *Symposium on Hydrogen Storage Materials, Batteries, and Electrochemistry*, The Electrochemical Society, Inc., p. 3, 1991.

77. Lin, F.N., Moore, W.I. and Walker, S.W., "Economics of Liquid Hydrogen from Water Electrolysis," in *Hydrogen Energy Progress V*, Edited by T.N. Veziroglu and J.B. Taylor, Pergamon Press, Elmsford, NY, p. 249, 1984.

78. Liventsov, V.M. and Kuznetsov, A.V., "Simplified Method of Analysis of Metal Hydride Accumulators of Hydrogen," in *Hydrogen Energy Progress IX*, Edited by T.N. Veziroglu and C. Derive-J. Pottier, International Association for Hydrogen Energy, p. 843, 1992.

79. Lund, P.D., "Optimization of Stand-Alone Photovoltaic Systems with Hydrogen Storage for Total Energy Self-Sufficiency," *Int. J. Hydrogen Energy*, 16, p. 735, 1991.

80. Marinescu-Pasoi, L., Behrens, U., Langer, G., Gramatte, W., et al, "Hydrogen Metal Hydride Storage with Integrated Catalytic Recombiner for Mobile Application," *Int. J. Hydrogen Energy*, 16, p. 407, 1991.

81. Matsushita, A. and Matsumoto, T., "Compressibility and Hydrogen Storage Properties in Haucke Compounds," *Zeitschrift für Physikalische Chemie Neue Folge*, 163, p. 491, 1989.

82. Mattsoff, S. and Noréus, D., "Hydriding Kinetics of Mg_2Ni at Low Temperatures Where Intrinsic Processes Dominate the Reaction Rates," *Int. J. Hydrogen Energy*, 12, p. 33, 1987.

83. Matysina, Z.A., Zaginaichenko, S.Y. and Pogorelova, O.S., "The Solubility and Distribution of Hydrogen Atoms in Ordering Alloys," in *Hydrogen Energy Progress IX*, Edited by T.N. Veziroglu and C. Derive-J. Pottier, International Association for Hydrogen Energy, p. 979, 1992.

84. Mayersohn, N.S. 'The Outlook for Hydrogen," *Popular Science*, 1993, 67.

85. McAuliffe, C.A., *Hydrogen and Energy*, Gulf Book Division, Houston, 1980.

86. Nelson, H.G., "A Review—Materials and Safety Problems Associated with Hydrogen Containment," in *Hydrogen Energy Progress V*, Edited by T.N. Veziroglu and J.B. Taylor, Pergamon Press, Elmsford, N.Y., p. 1841, 1984.

87. O'Connel, L.G., "A Comparison of Energy-Storage Devices for use in Future Automobiles," Report UCRL-85654, Lawrence Livermore Laboratory, Livermore, CA, 1981.

88. Ohta, T. and Sastri, M.V.C., "Hydrogen Energy Research Programs in Japan," *Int. J. Hydrogen Energy*, 4, p. 489, 1979.

89. Peschka, W. and Carpetis, C., "Cryogenic Hydrogen Storage and Refueling for Automobiles," *Int. J. Hydrogen Energy*, 5, p. 619, 1980.

90. Peschka, W., "Liquid Hydrogen as a Vehicular Fuel—A Challenge for Cryogenic Engineering," in *Hydrogen Energy Progress IV*, Edited by T.N. Veziroglu, W.D. Van Vorst, and J.H. Kelley, Pergamon Press, Elmsford, NY, p. 1053, 1982.

91. Peschka, W. and Winter, C.J., "The Secondary Energy Carrier Hydrogen Review of DFVLR-Activities," *Int. J. Hydrogen Energy*, 9, p. 319, 1984.

92. Peschka, W., "The status of handling and storage techniques for liquid hydrogen in motor vehicles," *Int. J. Hydrogen Energy*, 12, p. 753, 1987.

93. Peschka, W., "Liquid Hydrogen Pumps for Automotive Application," *Int. J. Hydrogen Energy*, 15, p. 817, 1990.

94. Peschka, W., *Liquid Hydrogen—Fuel of the Future*, Springer-Verlag, New York, 1992.

95. Peschka, W. and Escher, W.J.D., "Germany's Contribution to the Demonstrated Technical Feasibility of the Liquid-Hydrogen Fueled Passenger Automobile," SAE Technical Paper No. 931812, Society of Automotive Engineers, Warrendale, PA, 1993.

96. Peterson, U., Wursig, G. and Krapp, R., "Design and Safety Considerations for Large-Scale Sea-Borne Hydrogen Transport," in *Hydrogen Energy Progress IX*, Edited by T.N. Veziroglu and C. Derive-J. Pottier, International Association for Hydrogen Energy, Coral Gables, FL, p. 1021, 1992.

97. Petkov, T., Veziroglu, T.N. and Sheffield, J.W., "An Outlook of Hydrogen as an Automotive Fuel," *Int. J. Hydrogen Energy*, 14, p. 449, 1989.

98. Pfriem, H.-J., "Overview of the Cooperative Program on Hydrogen Storage, Conversion and Safety of the International Energy Agency," *Int. J. Hydrogen Energy*, 16, p. 329, 1991.

99. Podgorny, A.N., "Hydride Systems in Power Engineering and Motor Transport," *Int. J. Hydrogen Energy*, 14, p. 599, 1989.

100. Post, M., Murray, J.J. and Taylor, J.B., "Metal Hydride Studies at the National Research Council of Canada," *Int. J. Hydrogen Energy*, 9, p. 137, 1984.

101. Reilly, J. J., "Chemistry of Intermetallic Hydrides," in *Symposium on Hydrogen Storage Materials, Batteries, and Electrochemistry*, The Electrochemical Society, Inc., p. 24, 1991.

102. Robinson, S.L. and Handrock, J.L., "Hydrogen Storage for Vehicular Applications: Technology Status and Key Development Areas," Sandia National Laboratories, White Paper, Livermore, CA, February, 1994.

103. Sarangi, S., "Cryogenic Storage of Hydrogen," in *Progress in Hydrogen Energy*, D. Reidel Publishing Company, p. 123, 1985.

104. Schucan, T.H. and Taube, M., "Seasonal Storage Hydrogen as a Fuel for Heavy Vehicles," in *Hydrogen Energy Progress VI,* Edited by T.N. Veziroglu, N. Getoff, and P. Weinzierl, Pergamon Press, Elmsford, NY, p. 826, 1986.

105. Selvam, P., Viswanathan, B., Swamy, C.S. and Srinivasan, V., "Surface Studies of Some Hydrogen Storage Materials," *Zeitschrift für Physikalische Chemie Neue Folge*, 164, p. 1199, 1989.

106. Shemet, V.Z. and Pomytkin, A.P., "Thermal Decomposition and High-Temperature Oxidation of Metal Hydrides," in *Hydrogen Energy Progress VIII*, Edited by T.N. Veziroglu and P.K. Takahashi, Pergamon Press, Elmsford, N.Y., p. 1071, 1990.

107. Specht, M., Bandi, A. and Maier, C.U., "Methanol: A Solar Hydrogen Storage Fuel," in *Hydrogen Energy Progress IX*, Edited by T.N. Veziroglu and C. Derive-J. Pottier, International Association for Hydrogen Energy, Coral Gables, FL, p. 599, 1992.

108. Sperling, D. and DeLuchi, M.A., *Alternative Transportation Fuels: An Environmental and Energy Solution*, Quorum Books, New York, 1989.

109. Stewart, W.F. and Edeskuty, F.J. *Logistics, Economics, and Safety of a Liquid Hydrogen System for Automotive Transportation*, The American Society of Mechanical Engineers,73-ICT-78, 1973.

110. Stewart, W.F., "Hydrogen as a Vehicular Fuel," in *Recent Developments in Hydrogen Technology*, Edited by K. D. Williamson Jr. and F. J. Edeskuty, CRC Press, Inc., Boca Raton, FL, p. 69, 1986.

111. Stewart, W.F., "Experimental Investigation of Onboard Storage and Refueling Systems for Liquid-Hydrogen-Fueled Vehicles," Report DOE/CE-0039, U.S. Department of Energy, Washington, D.C., 1982.

112. Strobl, W. and Peschka, W., "Liquid Hydrogen as a Fuel of the Future for Individual Transport," in *Hydrogen Energy Progress VI,* Edited by T.N. Veziroglu, N. Getoff, and P. Weinzierl, Pergamon Press, Elmsford, NY, p. 1161, 1986.

113. Styrikovich, M.A. and Malyshenko, S.P., "Bulk Storage and Transmission of Hydrogen," in *Hydrogen Energy Progress VI,* Edited by T.N. Veziroglu, N. Getoff, and P. Weinzierl, Pergamon Press, Elmsford, NY, p. 765, 1986.

114. Suda, S., "Metal Hydrides," *Int. J. Hydrogen Energy*, 12, p. 323, 1987.

115. Suda, S., "Energy Conversion Systems Using Metal Hydrides," *Zeitschrift für Physikalische Chemie Neue Folge*, 164, p. 1463, 1989.

116. Taube, M., Rippin, D.W.T., Cresswell, D.L., and Knecht, W., "A System of Hydrogen-Powered Vehicles with Liquid Organic Hydrides," *Int. J. Hydrogen Energy*, 8, p. 213, 1983.

117. Taylor, J.B., Alderson, J.E.A., Kalyanam, K.M., Lyle, A.B. and Phillips, L.A., "Technical and Economic Assessment of Methods for the Storage of Large Quantities of Hydrogen," *Int. J. Hydrogen Energy*, 11, p. 5, 1986.

118. Teitel, R.J., "Microcavity Hydrogen Storage—Final Progress Report," Report BNL-51439, Brookhaven National Laboratory, Upton, NY, 1981.

119. Tong, X.O. and Lewis, F.A., "Hydrogen Permeation in Stressed Membranes of Palladium Alloys," in *Hydrogen Energy Progress VIII*, Edited by T.N. Veziroglu and P.K. Takahashi, Pergamon Press, Elmsford, N.Y., p. 1175, 1990.

120. Töpler, J. and Feucht, K., "Results of a Test Fleet with Metal Hydride Motor Cars," *Zeitschrift für Physikalische Chemie Neue Folge*, 164, p. 1451, 1989.

121. Turillon, P.P., "Design of Hydride Containers for Hydrogen Storage," in *Hydrogen Energy Progress IV*, Edited by T.N. Veziroglu, W.D. Van Vorst, and J.H. Kelley, Pergamon Press, Elmsford, NY, p. 1289, 1982.

122. Uchida, H., Terao, K. and Huang, Y.C., "Current Problems in the Development and Application of Hydrogen Storage Materials," *Zeitschrift für Physikalische Chemie Neue Folge*, 164, p. 1275, 1989.

123. Ullman, A.Z. and Van Vorst, W.D., "Methods of On-Board Generation of Hydrogen for Vehicular Use," in *Proc. 1st World Hydrogen Energy Conference*, Miami, FL, 1976.

124. Vainer, A.J., Igoshin, A.I., Kazaryan, V.A. and Smirnov, V.I., "The Utilization of Underground Caverns in Rock Salt Formations for Hydrogen Production and Storage," in *Hydrogen Energy Progress IX*, Edited by T.N. Veziroglu and C. Derive-J. Pottier, International Association for Hydrogen Energy, Coral Gables, p. 159, 1992.

125. Varghese, A.P. and Herring, R.H., "Transient Shielded Liquid Hydrogen Containers," in *Hydrogen Energy Progress VIII*, Edited by T.N. Veziroglu and P.K. Takahashi, Pergamon Press, Elmsford, N.Y., p. 1145, 1990.

126. Veziroglu, T.N. and Barbir, F., "Hydrogen: The Wonder Fuel," *Int. J. Hydrogen Energy*, 17, p. 391, 1992.

127. Walker, G., Weiss, M., Reader, G. and Fauvel, O.R., "Small-Scale Liquefaction of Hydrogen," in *Hydrogen Energy Progress VIII*, Edited by T.N. Veziroglu and P.K. Takahashi, Pergamon Press, Elmsford, N.Y., p. 891, 1990.

128. Wallace, J.S. and Ward, C.A., "Hydrogen as a fuel," *Int. J. Hydrogen Energy*, 8, p. 255, 1983.

129. Wallace, J.S., "A Comparision of Compressed Hydrogen and CNG Storage," *Int. J. Hydrogen Energy*, 9, p. 609, 1984.

130. Wang, Q.-D., Wu, J. and Chen, C.-P., "Development of New Mischmetal-Nickel Hydrogen Storage Alloys According to the Specific Requirements of Different Applications," *Zeitschrift für Physikalische Chemie Neue Folge*, 164, p. 1293, 1989.

131. Wang, Q.-d., Wu, J. and Gao, H., "Vacuum Sintered Porous Metal Hydride Compacts," *Zeitschrift für Physikalische Chemie Neue Folge*, 164, p. 1367, 1989.

132. Wang, X.L. and Suda, S., "Reaction Kinetics of Hydrogen-Metal Hydride Systems," *Int. J. Hydrogen Energy*, 15, p. 569, 1990.

133. Williams, L.O., *Hydrogen Power: An Introduction to Hydrogen Energy and its Applications*, Pergamon Press, Oxford, England, 1980.

134. Williams, L.O. and Spond, D.E., "A Storage Tank for Vehicular Storage of Liquid Hydrogen," *Appl. Energ.*, 6, p. 99, 1980.

135. Wolpers, F., Gelse, W. and Withalm, G., "Comparative Investigation of a Hydrogen Engine with External Mixture Formation Which Can Either be Operated with Cryogenic Hydrogen or Non-Cryogenic Hydrogen and Water Injection," in *Hydrogen Energy Progress VII*, Edited by T.N. Veziroglu and A.N. Protsenko, Pergamon Press, Elmsford, N.Y., p. 2119, 1988.

136. Woolley, R.L., "The Hydrogen Homestead Project," in *Hydrogen for Energy Distribution*, The Institute, p. 589, 1978.

137. Young, K.S., "Advanced Composites Storage Containment for Hydrogen," *Int. J. Hydrogen Energy*, 17, p. 505, 1992.

138. Young, K.S., "Advanced Composites Storage Containment for Hydrogen & Methane," in *Hydrogen Energy Progress VIII*, Edited by T.N. Veziroglu and P.K. Takahashi, Pergamon Press, Elmsford, N.Y., p. 967, 1990.

139. "New Method for Storing Hydrogen Fuel Developed," *New Technology Week*, February 4, 1991.

Vehicles

1. Altseimer, J.H., Nochumson, D.H., and Frank, J.A., "Fuel Cell Propulsion Systems for Large Trasportation Vehicles: Buses, Freight Locomotives, and Marinecraft, in *Proc. 18th Intersociety Energy Conversion Engineering Conference*, American Institute of Chemical Engineers, New York, p. 1435, 1983.

2. American Academy of Science, "Hydrogen Fuel Cell Vehicle: Lasercel 1™," American Academy of Science.

3. Baker, N., Lynch, F., Mejia, L., and Olvason, L., "A Hydrogen Engine for Underground Mining Vehicles," in *Proceedings of the 18th Intersociety Energy Conversion Engineering Conference*, American Institute of Chemical Engineers, NY, p. 569, 1983.

4. Belogub, A.V. and Talda, G.B., "Petrol-Hydrogen Truck with Load-carrying Capacity 5 Tons," *Int. J. Hydrogen Energy*, 16, p. 423, 1991.

5. Billings, R.E., "A hydrogen-powered mass transit system," *Int. J. Hydrogen Energy*, 3, p. 49, 1978.

6. Billings, R.E., "Hydrogen Homestead," in *2nd World Hydrogen Energy Conference*, Edited by T.N.

Veziroglu and W. Seifritz, Pergamon Press, Elmsford, NY, p. 1709, 1978.

7. Billings, R.E., "Advances in hydrogen engine conversion technology," *Int. J. Hydrogen Energy*, 8, p. 939, 1983.

8. Billings, R.E., Hatch, S.M. and DiVacky, R.J., "Conversion and Testing of Hydrogen-powered Post Office Vehicle," *Int. J. Hydrogen Energy*, 8, p. 943, 1983.

9. Billings, R.E., Sanchez, M., Cherry, P. and Eyre, D.B., "LaserCell Prototype Vehicle," *Int. J. Hydrogen Energy*, 16, p. 829, 1991.

10. Buchner, H., "Hydrogen Use—Transportation Fuel," in *Hydrogen Energy Progress IV*, Edited by T.N. Veziroglu, W.D. Van Vorst, and J.H. Kelley, Pergamon Press, Elmsford, NY, p. 3, 1982.

11. Cannon, J.S., "Alternative Transportation Fuels in the USA: Government Hydrogen Vehicle Programs," in *9th World Hydrogen Energy Conference*, International Association for Hydrogen Energy, Coral Gables, FL, p. 1811, 1992.

12. Chuveliov, A.V. *Hydrogen in Motor Vehicles: A Case Study of Hydrogen Utilization in Motor Vehicles as a Supplementary Fuel in Southern California Air Basin*, I.V. Kurchatov Institute of Atomic Energy, Report prepared for SCQAMD, 1989.

13. Ciancia, A., Pede, G., Brighigna, M. and Perrone, V., "A Compressed Hydrogen Fuelled Vehicle at ENEA: Status and Development," in *Hydrogen Energy Progress X*, Edited by D.L. Block and T.N. Veziroglu, p. 1415, 1994.

14. Davidson, D., Fairlie, M. and Stuart, A.E., "Development of a Hydrogen-Fuelled Farm Tractor," *Int. J. Hydrogen Energy*, 11, p. 39, 1986.

15. DeGeeter, E., Van den Broeck, H., Bout, P., Woortmann, M., Cornu, J., Peski, V., Dufour, A. and Marcenaro, B., "Eureka Fuel Cell Bus Demonstration Project," in *Hydrogen Energy Progress X*, Edited by D.L. Block and T.N. Veziroglu, International Association for Hydrogen Energy, Coral Gables, FL, p. 1457, 1994.

16. DeLuchi, M. *Hydrogen Fuel-Cell Vehicles*, Research Report UCD-ITS-RR-92-14, Institute of Transportation Studies, University of California, Davis, 1992.

17. DeLuchi, M.A., "Hydrogen Vehicles: An Evaluation of Fuel Storage, Performance, Safety, Environmental Impacts, and Cost.," *Int. J. Hydrogen Energy*, 14, p. 81, 1989.

18. DeLuchi, M.A., "Hydrogen Vehicles," in *Alternative Transportation Fuels*, Edited by D. Sperling, Quorum Books, New York, 1989.

19. DeLuchi, M.A. and Ogden, J.M., "Solar-Hydrogen Fuel-Cell Vehicles," *Transportation Research, Part A, Policy and Practice*, 27A, p. 255, 1993.

20. Donnelly, Jr., J.J., Escher, W.J.D., Greayer, W.C., and Nichols, R.J., "Study of Hydrogen-Powered Automobiles," Report ATR-79 (7759)-1, U.S. Department of Energy, Washington, D.C., 1979.

21. Dozier, K. "Aussies Tout Hydrogen Alternative Watson: Tests Show Significant CO, CO_2 and HC Reductions," *Environment Week*, November 22, 1990.

22. Erren, R.A. and Hastings-Campbell, W., "Hydrogen: a commercial fuel for internal combustion engines and other purposes," *Journal of the Institute of Fuel*, 6, p. 277, 1933.

23. Escher, W.J.D., "Cooperative international liquid hydrogen automotive progress report," *Int. J. Hydrogen Energy*, 7, p. 519, 1982.

24. Escher, W.J.D., *Hydrogen-Fueled Internal Combustion Engine, A Technical Survey of Contemporary U.S. Projects*, Escher Technology Associates, Survey report TEC-75/005, 1975.

25. Escher, W.J.D., "Hydrogen as a Vehicular Fuel: Review of Progress Made by U.S. Department of Energy-Supported Projects," presented at the Inst. Gas Technol. Symp. Nonpetroleum Vehicular Fuels," Washington, D.C., February 11 to 13, 1980.

26. Feucht, K., Hölzel, G. and Hurich, W., "Perspectives of Mobile Hydrogen Application," in *Hydrogen Energy Progress VII*, Edited by T.N. Veziroglu and

A.N. Protsenko, Pergamon Press, Elmsford, N.Y., p. 1963, 1988.

27. Feucht, K., Hurich, W., Komoschinski, N. and Povel, R., "Hydrogen Drive for Road Vehicles—Results from the Fleet Test Run in Berlin," in *Hydrogen Energy Progress VI,* Edited by T.N. Veziroglu, N. Getoff, and P. Weinzierl, Pergamon Press, Elmsford, NY, p. 1079, 1986.

28. Feucht, K., Hurich, W., Komoschinski, N. and Povel, R., "Hydrogen Drive for Road Vehicles—Results from the Fleet Test Run in Berlin," *Int. J. Hydrogen Energy*, 13, p. 243, 1988.

29. Finegold, J.G., Lunch, F.E. and Bush, A.F., "The UCLA Hydrogen Car: Design, Construction, and Performance," SAE Technical Paper No. 730507, Society of Automotive Engineers, Warrendale, PA, 1973.

30. Finegold, J.G. and Van Vorst, W.D., "Crash test of liquid hydrogen automobile," in *1st World Hydrogen Energy Conference*, University of Miami, Coral Gables, FL, p. 6, 1976.

31. Frank, J.A., Altseimer, J.H., and Nochumson, D.H., "Fuel Cell Propulsion Systems for Small Transportation Vehicles, in *Proc. 18th Intersociety Energy Conversion Engineering Conference*, American Institute of Chemical Engineers, New York, p. 1425, 1983.

32. Freiwald, D.A. and Barattino, W.J., "Alternative Transportation Vehicles for Military-Base Operations," *Int. J. Hydrogen Energy*, 6, p. 631, 1981.

33. Furuhama, S., "A Liquid Hydrogen Car with a Two-Stroke Engine and LH_2 Pump," *Int. J. Hydrogen Energy*, 7, p. 809, 1981.

34. Furuhama, S., "Hydrogen Engine Systems for Land Vehicles," *Int. J. Hydrogen Energy*, 14, p. 907, 1989.

35. Furuhama, S., "Trend of Social Requirements and Technological Development of Hydrogen-Fueled Automobiles," *JSAE review*, 13, p. 4, 1992.

36. Furuhama, S., "Hydrogen Engine Systems for Land Vehicles," in *Hydrogen Energy Progress VII*, Edited by T.N. Veziroglu and A.N. Protsenko, Pergamon Press, Elmsford, N.Y., p. 1841, 1988.

37. Furuhama, S., Hiruma, M. and Enomoto, Y., "Development of a liquid hydrogen car," *Int. J. Hydrogen Energy*, 3, p. 61, 1978.

38. Furuhama, S. and Kobayashi, Y., "Hydrogen Cars with LH_2-Tank, LH_2-Pump and Cold GH_2-Injection Two-Stroke Engine," SAE Technical Paper No. 820349, Society of Automotive Engineers, Warrendale, PA, 1983.

39. Heffel, J.W. and Norbeck, J.M., "Preliminary Evaluation of UC Riverside's Hydrogen Truck," presented at the Sixth Annual U.S. Hydrogen Meeting, Alexandria, VA, 1995.

40. Hoffmann, P., "Europe unveils proof of concept hydrogen bus," *The Hydrogen Letter*, 9, p. 1994.

41. Hoffmann, P., *The Forever Fuel: The Story of Hydrogen*, Westview Press, Boulder, Colorado, 1981.

42. Hoffmann, P., "HYPASSE-Bus to operate next year, maybe with new H-engine," *The Hydrogen Letter*, 1994.

43. Hoogeveen, P., Marcenaro, B.G., Vermeeren, L. and Cornu, J.P., "Eureka Fuel Cell Bus," in *Hydrogen Energy Progress IX*, Edited by T.N. Veziroglu and C. Derive-J. Pottier, International Association for Hydrogen Energy, Coral Gables, FL, p. 1227, 1992.

44. Howard, P., "Ballard Fuel Cell Powered ZEV Bus," in *World Car 2001 Conference*, College of Engineering Center for Environmental Research and Technology, University of California, Riverside, 1993.

45. Just, J.S., "Hydrogen as a Substitute Fuel," *Gas Oil Power Am. Tech. Rev.*, p. 326, 1944.

46. Kludjian, V.Z., "Hydrogen for Vehicles—Mazda's Hydrogen Vehicle Development Program," in presented at the 5th Annual U.S. Hydrogen Meeting, Washington, D.C., 1994.

47. Kordesch, K.V., "Hydrogen-Air/Lead Battery Hybrid System for Vehicle Propulsion," Abstr. No. 10, Electrochemical Society, 1970.

48. Kordesch, K.V., "City Car with H_2-Air Fuel Cell Lead Battery (One Year Operating Experience)," Intersociety Energy Conversion Engineering Conference, Boston, p. 103, August, 1971.

49. Kukkonen, C.A., "Hydrogen as an Alternative Automobile Fuel," SAE Technical Paper No. 810349, Society of Automotive Engineers, Warrendale, PA, 1981.

50. Kukkonen, C.A. and Shelef, M., "Hydrogen as an Alternative Automobile Fuel: 1993 Update," SAE Technical Paper No. 940766, Society of Automotive Engineers, Warrendale, PA, 1994.

51. Lemons, R.A., "Fuel Cells for Transportation," *Journal of Power Sources*, 29, p. 251, 1990.

52. Lloyd, A.C., "The California Plan," in *2nd Annual U.S. Hydrogen Meeting*, p. 8, 1991.

53. Lynch, F.E., "Parallel induction: a simple fuel control method for hydrogen engines," *Int. J. Hydrogen Energy*, 8, p. 721, 1983.

54. Lynn, D.K., McCormick, J.B., Bobbett, R.E., Srinivasan, S., and Huff, J.R., "Design Considerations for Vehicular Fuel Cell Power Plants, in *Proc. 16th Intersociety Energy Conversion Engineering Conference*, American Society of Mechanical Engineers, New York, p. 722, 1981.

55. Lynn, D.K., McCormick, J.B., Bobbett, R.E., Derouin, C. and Kerwin, W.J., "Fuel Cell Technologies for Vehicular Applications," SAE Technical Paper No. 800059, Society of Automotive Engineers, Warrendale, PA, 1980.

56. Lynn, D.K., McCormick, J.B., Bobbett, R.E., Huff, J.R., and Srinivasan, S., "Acid Fuel Cell Technologies for Vehicular Power Plants," in *Proc. 17th Intersociety Energy Conversion Engineering Conference*, Institute of Electrical and Electronics Engineers, New York, p. 663, 1982.

57. Marcenaro, B., "EQHHPP FC BUS: Status of the Project and the Presentation of the First Experimental Results," in *Hydrogen Energy Progress X*, Edited by D.L. Block and T.N. Veziroglu, International Association for Hydrogen Energy, Coral Gables, FL, p. 1447, 1994.

58. Marcenaro, B.G. and Andreoli, G.L., "Eureka Fuel Cell Bus," in *Hydrogen Energy Progress IX*, Edited by T.N. Veziroglu and C. Derive-J. Pottier, International Association for Hydrogen Energy, Coral Gables, FL, p. 1227, 1992.

59. Marks, C., Rishavy, E.A., and Wyczalek, F.A., "Electrovan—A Fuel Cell-Powered Vehicle," SAE Technical Paper No. 670176, Society of Automotive Engineers, Warrendale, PA, 1967.

60. McElroy, J.F. and Nuttall, L.J., "Status of Solid Polymer Electrolyte Fuel Cell Technology and Potential for Transportation Applications," in *Proc. 17th Intersociety Energy Conversion Engineering Conference*, Institute of Electrical and Electronics Engineers, New York, p. 667, 1982.

61. Mishchenko, A.I., Belogub, A.V., Talda, G.B., Savistsky, V.D. and Baikov, V.A., "Hydrogen as a Fuel for Road Vehicles," in *Hydrogen Energy Progress VII*, Edited by T.N. Veziroglu and A.N. Protsenko, Pergamon Press, Elmsford, NY, p. 2037, 1988.

62. Nadal, M. and Barbir, F., "Development of a Hydride Fuel Cell/Battery Powered Electric Vehicle," in *Hydrogen Energy Progress X*, Edited by D.L. Block and T.N. Veziroglu, International Association for Hydrogen Energy, Coral Gables, FL, p. 1427, 1994.

63. Newkirk, M.S. and Abel, J.L. *The Boston Reformed Fuel Car*, SAE Technical Paper No. 720670, Society of Automotive Engineers, Warrendale, PA, 1972.

64. Normile, D. "Mazda's Hydrogen Miata," *Popular Science*, 1993, 40.

65. Ohta, T. and Sastri, M.V.C., "Hydrogen Energy Research Programs in Japan," *Int. J. Hydrogen Energy*, 4, p. 489, 1979.

66. Olavson, L.G., Baker, N.R., Mejia, L.C. and Lynch, F.E., "Hydrogen Fuel for Underground Mining Machinery," in *International Congress & Exposition*, Society of Automotive Engineers, Warrendale, PA, p. 1, 1984.

67. Olavson, L.G., Baker, N.R., Lynch, F.E., and Mejia, L.C., "Hydrogen Fuel for Underground Mining Machinery," SAE Technical Paper No. 840233, Society of Automotive Engineers, Warrendale, PA, 1984.

68. Perris Smogless Automobile Association, "An Answer to the Automotive Air Pollution Problem," First Annual Report, Perris, CA, 1971.

69. Peschka, W., "Operating characteristics of a LH_2-fuelled automotive vehicle and of a semi-automatic LH_2-refuelling station," *Int. J. Hydrogen Energy*, 7, p. 661, 1982.

70. Peschka, W., "Liquid Hydrogen as a Vehicular Fuel—A Challenge for Cryogenic Engineering," in *Hydrogen Energy Progress IV*, Edited by T.N. Veziroglu, W.D. Van Vorst, and J.H. Kelley, Pergamon Press, Elmsford, NY, p. 1053, 1982.

71. Peschka, W., "Hydrogen Combustion in Tomorrow's Energy Technology," in *Hydrogen Energy Progress VI*, Edited by T.N. Veziroglu, N. Getoff, and P. Weinzierl, Pergamon Press, Elmsford, NY, p. 1019, 1986.

72. Peschka, W., "Liquid hydrogen fueled automotive vehicles in Germany—Status and development," *Int. J. Hydrogen Energy*, 11, p. 721, 1986.

73. Peschka, W., *Liquid Hydrogen—Fuel of the Future*, Springer-Verlag, New York, 1992.

74. Peschka, W. and Escher, W.J.D., "Germany's Contribution to the Demonstrated Technical Feasibility of the Liquid-Hydrogen Fueled Passenger Automobile," SAE Technical Paper No. 931812, Society of Automotive Engineers, Warrendale, PA, 1993.

75. Peschka, W., "The Status of Handling and Storage Techniques for Liquid Hydrogen in Motor Vehicles," *Int. J. Hydrogen Energy*, 12, p. 753, 1987.

76. Peschka, W., "Hydrogen Combustion in Tomorrow's Energy Technology," *Int. J. Hydrogen Energy*, 12, p. 481, 1987.

77. Peschka, W., "Liquid Hydrogen Pumps for Automotive Application," in Adv. Cryo. Eng., Plenum Press, New York, Vol. 35B, p. 1783, 1990.

78. Peschka, W., "Liquid Hydrogen Pumps for Automotive Application," *Int. J. Hydrogen Energy*, 15, p. 817, 1990.

79. Peschka, W., "Cryogenic Fuel Technology and Elements of Automotive Propulsion Systems," Adv. in Cryogenic Engineering 37, Plenum Press, New York, 1992.

80. Povel, R., Topler, J., Withalm, G. and Halene, C., "Hydrogen drive in field testing," in *Hydrogen Energy Progress V*, Edited by T.N. Veziroglu and J.B. Taylor, Peramon Press, p. 1563, 1984.

81. Prater, K., "Solid Polymer Developments at Ballard," *Journal of Power Sources*, 37, p. 181, 1992.

82. Reister, D. and Strobl, W., "Current Development and Outlook for the Hydrogen-Fuelled Car," in *Hydrogen Energy Progress IX*, Edited by T.N. Veziroglu and C. Derive-J. Pottier, International Association for Hydrogen Energy, Coral Gables, FL, p. 1201, 1992.

83. Silva, E.D., Gallo, W., Szajner, J., Amaral, E.D. and Bezerra, C., "A Solar/Hydrogen/Electricty Hybrid Vehicle," in *Hydrogen Energy Progress X*, Edited by D.L. Block and T.N. Veziroglu, International Association for Hydrogen Energy, Coral Gables, FL, p. 1441, 1994.

84. Sorenson, H., "The Boston Reformed Fuel Car—A Low Polluting Gasoline Fuel System for Internal Combustion Engines," in *Proceedings of the 7th Intersociety Energy Conversion Engineering Conference*, San Diego, CA, 1972.

85. Stewart, C.F. *The Riverside Hydrogen Powered Bus*, Final Report DMT-048, California Department of Transportation, Division of Mass Transportation, 1979.

86. Stewart, W.F. *A Liquid-Hydrogen Fueled Buick*, Los Alamos Scientific Laboratory, Research Report LA-8605-MS, 1980.

87. Stewart, W.F., "Hydrogen as a Vehicular Fuel," in *Recent Developments in Hydrogen Technology*, Edited by K. D. J. Williamson and F. J. Edeskuty, CRC Press, Inc., Boca Raton, p. 69, 1986.

88. Stewart, W.F., "Operating Experience with a Liquid-Hydrogen Fueled Buick and Refueling System, in *Hydrogen Energy Progress IV*, Edited by T.N. Veziroglu, W.D. Van Vorst, and J.H. Kelley, Pergamon Press, Elmsford, NY, p. 1071, 1982.

89. Stewart, W.F., Edeskuty, F.J., Williamson, K.D., Jr., and Lutgen, H.M., "Operating Experiences with a Liquid Hydrogen Fueled Vehicle," in *Advances in Cryogenic Engineering*, Vol. 20.

90. Stewart, W.F., "Operating Fueled Buick and Refueling System," *Int. J. Hydrogen Energy*, 9, p. 525, 1984.

91. Strobl, W. and Peschka, W., "Liquid Hydrogen as a Fuel of the Future for Individual Transport," in *Hydrogen Energy Progress VI,* Edited by T.N. Veziroglu, N. Getoff, and P. Weinzierl, Pergamon Press, Elmsford, NY, p. 1161, 1986.

92. Takano, E., Takamori, Y., Morimoto, K. and Teramoto, T., "Development of the Direct Injection Hydrogen-Fueled Rotary Engine," in *Hydrogen Energy Progress IX*, Edited by T.N. Veziroglu and C. Derive-J. Pottier, International Association for Hydrogen Energy, Coral Gables, FL, p. 1235, 1992.

93. Taube, M., Rippin, D., Knecht, W., Hakimifard, D., Milisavlejevic, B. and Grunenfelder, N., "A Prototype Truck Powered by Hydrogen from Organic Liquid Hydrides," *Int. J. Hydrogen Energy*, 10, p. 595, 1985.

94. Van Vorst, W.D. and Woolley, R.L., "Hydrogen-fueled surface transportion," in *Hydrogen: Its Technology and Implications*, Edited by K. E. Cox and D. K. Williamson Jr., CRC Press, Boca Raton, FL, 1979.

95. Vandenborre, H. and Sierens, R., "Greenbus: A Hydrogen Fuelled City Bus," in *Hydrogen Energy Progress X*, Edited by D.L. Block and T.N. Veziroglu, International Association of Hydrogen Energy, Coral Gables, FL, p. 1959, 1994.

96. Viledganin, G.A., Galyshev, U.V. and Jarov, V.F., "Development and Investigation of the Hydrogen Fueling Systems Used for Automobile Engines," in *Hydrogen Energy Progress VII*, Edited by T.N. Veziroglu and A.N. Protsenko, Pergamon Press, Elmsford, NY, p. 2105, 1988.

97. Watson, H.C., Milkins, E.E., Martin, W.R.B. and Edsell, J., "An Australian Hydrogen Car," in *Hydrogen Energy Progress V*, Edited by T.N. Veziroglu and J.B. Taylor, Pergamon Press, Elmsford, NY, p. 1549, 1984.

98. Weil, K.H., "The hydrogen I.C. engine—its origin and future in the emerging-transportation-environment system," in *7th Intersociety Energy Conversion Engineering Conference*, American Chemical Society, Washington, D.C., p. 1355, 1972.

99. Williams, L.O., "Hydrogen Powered Automobiles Must use Liquid Hydrogen," *Cryogenics*, 13, p. 693, 1973.

100. Winfield, B. "Hydrogen: It's come a long way since the Hindenburg," *Automobile Magazine*, 1992, 60.

101. Withalm, G. and Gelse, W., "The Mercedes-Benz Hydrogen Engine for Application in a Fleet Vehicle," in *Hydrogen Energy Progress VI,* Edited by T.N. Veziroglu, N. Getoff, and P. Weinzierl, Pergamon Press, Elmsford, NY, p. 1185, 1986.

102. Woolley, R.L., "Design Considerations for the Riverside Hydrogen Bus," in *2nd World Hydrogen Energy Conf.*, Edited by T.N. Veziroglu and W. Seifritz, Pergamon Press, Elmsford, NY, 1978.

103. Woolley, R.L., "Performance of a Hydrogen-Powered Transit Vehicle," in *Proc. 11th Intersociety Energy Conversion Conference*, Lake Tahoe, CA, 1976.

104. Wurster, R., Bracha, M., Braedt, J., Knorr, H. and Strobl, W., "Application of LH_2 cars and Urban

Buses in Munich," in *Hydrogen Energy Progress IX*, Edited by T.N. Veziroglu and C. Derive-J. Pottier, International Association for Hydrogen Energy, Coral Gables, FL, p. 1215, 1992.

105. Yamane, K., Hiruma, M., Watanabe, T., Kondo, T., Hikino, K., Hashimoto, T. and Furuhama, S., "Some performance of engine and cooling system on LH_2 refrigerator van Musashi-9," in *Hydrogen Energy Progress X*, Edited by D.L. Block and T.N. Veziroglu, FL, p. 1825, 1994.

106. Zieger, J., "HYPASSE—Hydrogen powered automobiles using seasonal and weekly surplus of electricity," in *10 World Hydrogen Energy Conference*, International Association for Hydrogen Energy, Coral Gables, FL, p. 1367, 1994.

107. Zweig, R.M. and Lynch, F.E., "Hydrogen Vehicle Progress in Riverside, California," in *World Car Conference 2001*, CE-CERT, University of California, Riverside, p. 1993.

108. Zweig, R.M., "Environmental Impact of One Year Experience with a Hydrogen Bus," in *Hydrogen Energy Progress*, Pergamon Press, Elmsford, NY, p. 2273, 1980.

109. Zweig, R.M. and Lynch, F.E., "Hydrogen Vehicle Progress in Reverside, California," in *Hydrogen Energy Progress VII*, Edited by T.N. Veziroglu and A.N. Protsenko, Pergamon Press, Elmsford, NY, p. 1923, 1988.

110. "Daimler Unveils Electric Vehicle Using Fuel Cells," *H_2 Digest*, 1994.52.

111. "The Green Car," Energy Partners, West Palm Beach, Florida, 1992.

Appendix B

Unit Conversion Factors

The following tables provide conversion factors between units in the U.S. Customary system, the metric system, and the International System (SI). To convert a quantity expressed in a unit in the left-hand column to the equivalent in a unit in the top row of a table, multiply the quantity by the factor listed as common to both units. Numbers followed by an asterisk are definitions of the relation between the two units.

Units of Pressure

Units	Pa ($N \cdot m^{-2}$)	dyn • cm^{-2}	bar	atm	mmHg (torr)	in. Hg	lbf • $in.^{-2}$
1 Pa ($N \cdot m^{-2}$)	= 1	10	10^{-5}	9.869×10^{-6}	7.501×10^{-3}	2.953×10^{-4}	1.450×10^{-4}
1 dyn • cm^{-2}	= 0.1	1	10^{-6}	9.869×10^{-7}	7.501×10^{-4}	2.953×10^{-5}	1.450×10^{-5}
1 bar	= 10^5*	10^6	1	0.9869	750.0617	29.530	14.504
1 atm	= 101325.0*	1013250	1.013250	1	760	29.9213	14.6959
1 mmHg (torr)	= 133.3224	1333.224	1.333×10^{-3}	1.316×10^{-3}	1	0.0394	0.0193
1 in. Hg	= 3386.388	33863.88	0.03386388	0.03342105	25.4	1	0.4911541
1 lbf • $in.^{-2}$	= 6894.757	68947.57	0.06894757	0.06804596	51.71493	2.036021	1

Units of Length

Units	μm (micron)	cm	m	mil	in.	mile
1 μm (micron)	= 1	10^{-4}	10^{-6}	0.03937	3.937×10^{-5}	6.2137×10^{-10}
1 cm	= 10^4	1	0.01*	3.937×10^2	0.3937	6.2137×10^{-6}
1 m	= 10^6	100	1	3.937×10^4	39.3701	6.2137×10^{-4}
1 mil	= 25.4	2.54×10^{-3}	2.54×10^{-5}	1	0.001	1.5783×10^{-8}
1 in.	= 2.54×10^4	2.54*	0.0254	1000	1	1.5783×10^{-5}
1 mile	= 1.6093×10^9	1.6093×10^5	1.6093×10^3	6.336×10^7	6.336×10^4	1

Units of Area

Units	μm^2	cm^2	m^2	mil^2	$in.^2$	$mile^2$
1 μm^2	= 1	10^{-8}	10^{-12}	1.550×10^{-3}	1.550×10^{-9}	3.861×10^{-19}
1 cm^2	= 10^8	1	10^{-4}*	1.550×10^5	0.1550	3.861×10^{-11}
1 m^2	= 10^{12}	10^4	1	1.550×10^9	1550	3.861×10^{-7}
1 mil^2	= 645.16	6.452×10^{-6}	6.452×10^{-10}	1	10^{-6}	2.491×10^{-16}
1 $in.^2$	= 6.452×10^8	6.452*	6.452×10^{-4}	10^6	1	2.491×10^{-10}
1 $mile^2$	= 2.590×10^{18}	2.590×10^{10}	2.590×10^6	4.014×10^{15}	4.014×10^9	1

Units of Volume

Units	m^3	cm^3	liter	$in.^3$	ft^3	qt
1 m^3	= 1	10^6	10^3	6.103×10^4	35.3147	1.0567×10^3
1 cm^3	= 10^{-6}	1	10^{-3}	0.06103	3.532×10^{-5}	1.0567×10^{-3}
1 liter	= 10^{-3}	1000*	1	61.0237	0.0353	1.0567
1 $in.^3$	= 1.639×10^{-5}	16.3871*	0.0164	1	5.787×10^{-4}	0.0173
1 ft^3	= 2.832×10^{-2}	28316.85	28.31685	1728*	1	2.9922
1 qt	= 9.464×10^{-4}	946.353	0.9464	57.75	0.0342	1

Units of Mass

Units	g	kg	oz	lb	metric ton	ton
1 g	= 1	10^{-3}	0.0353	2.2046×10^{-3}	10^{-6}	1.1023×10^{-6}
1 kg	= 1000	1	35.2740	2.2046	10^{-3}	1.1023×10^{-3}
1 oz	= 28.3495	0.0283	1	0.0625	2.8350×10^{-5}	3.125×10^{-5}
1 lb	= 453.5924	0.4536	16*	1	4.5359×10^{-4}	0.0005
1 metric ton	= 10^6	1000*	35273.96	2204.623	1	1.1023
1 ton	= 907184.7	907.1847	32000	2000*	0.9072	1

Units of Density

Units	$g \cdot cm^{-3}$	$g \cdot L^{-1}$ ($kg \cdot m^{-3}$)	$oz \cdot in.^{-3}$	$lb \cdot in.^{-3}$	$lb \cdot ft.^{-3}$	$lb \cdot gal^{-1}$
$g \cdot cm^{-3}$	= 1	1000	0.5780	0.0361	62.4280	8.3454
$g \cdot L^{-1}$ ($kg \cdot m^{-3}$)	= 10^{-3}	1	5.7804×10^{-4}	3.6127×10^{-5}	0.0624	8.3454×10^{-3}
$oz \cdot in.^{-3}$	= 1.7300	1729.994	1	0.0625	108	14.4375
$lb \cdot in.^{-3}$	= 27.6799	27679.91	16	1	1728	231
$lb \cdot ft.^{-3}$	= 0.0160	16.0185	9.2592×10^{-3}	5.7870×10^{-4}	1	0.1337
$lb \cdot gal^{-1}$	= 0.1198	119.8264	4.7495×10^{-3}	4.3290×10^{-3}	7.4805	1

Reprint Acknowledgments

CE-CERT wishes to thank the following publishers for their permission to reprint the following articles in *Hydrogen as a Fuel for Surface Transportation*:

International Association for Hydrogen Energy
P.O. Box 248266
Coral Gables, FL 33124

Das, L.M., "Fuel Induction Techniques for a Hydrogen Operated Engine," *Int. J. Hydrogen Energy*, Vol. 15, p. 833, 1990.

Raman, V., Hansel, J., Fulton, J., Lynch, F., Bruderly, D., "Hythane—An Ultraclean Transportation Fuel," in *Hydrogen Energy Progress X*, International Association for Hydrogen Energy, Coral Gables, FL, p. 1797, 1994.

Straschewski, D., "Hydrogen-Air Fuel Cells of the Alkaline Matrix Type: Manufacture and Impregnation of Electrodes," *Int. J. Hydrogen Energy*, Vol. 17, p. 643, 1992.

Ganser, B. and Hohlein, B., "Hydrogen from Methanol: Fuel Cells in Mobile Systems," in *Hydrogen Energy Progress IX*, International Association for Hydrogen Energy, Coral Gables, FL, p. 1321, 1992.

Anand, N.K., Appleby, A.J., Dhar, H.P., Ferreira, A.C., Kim, J., Mukerjee, S., Nandi, A., Parthasarathy, A., Rho, Y.W., Somasandaran, S., Srinivasan, S., Velev, O.A., and Wakizoe, M., "Recent Progress in Proton Exchange Membrane Fuel Cells at Texas A&M University," in *Hydrogen Energy Progress X*, International Association for Hydrogen Energy, Coral Gables, FL, p. 1669.

Steinberg, M. and Cheng, H.C., "Modern and Prospective Technologies for Hydrogen Production from Fossil Fuels," *Int. J. Hydrogen Energy*, Vol. 14, p. 797, 1989.

Dutta, S., "Technology Assessment of Advanced Electrolytic Hydrogen Production," *Int. J. Hydrogen Energy*, Vol. 15, p. 379, 1990.

Aochi, A., Tadokoro, T., Yoshida, K., Kameyama, H., Nobue, M., and Yamaguchi, T., "Economical and Technical Evaluation of UT-3 Thermochemical Hydrogen Production Process for an Industrial Scale Plant," *Int. J. Hydrogen Energy*, Vol. 14, p. 421, 1989.

Getoff, N., "Photoelectrochemical and Photocatalytic Methods of Hydrogen Production: A Short Review," *Int. J. Hydrogen Energy*, Vol. 15, p. 407, 1992.

Smith, G.D., Edward, G.D., and Tucker, W., "Hydrogen Production by Cyanobacteria," in *Hydrogen Energy Progress VIII*, Pergamon Press, Elmsford, NY, p. 735, 1990.

Pehr, K. "Aspects of Safety and Acceptance of LH_2 Tank systems in Passenger Cars," in *Hydrogen Energy Progress X*, International Association for Hydrogen Energy, Coral Gables, FL, p. 1399, 1994.

Hord, J., "Is Hydrogen a Safe Fuel?" *Int. J. Hydrogen Energy*, Vol. 3, p. 157, 1978.

Das, L.M., "Safety Aspects of a Hydrogen-Fueled Engine System Development," *Int. J. Hydrogen Energy*, Vol. 16, p. 619, 1991.

Fischer, "Safety Aspects of Hydrogen Combustion in Hydrogen Energy Systems," *Int. J. Hydrogen Energy*, Vol. 11, p. 593, 1986.

Ogden, J.M. and DeLuchi, M.A., "Renewable Hydrogen Transportation Fuels," in *Hydrogen Energy Progress IX*, International Association for Hydrogen Energy, FL, p. 1363, 1992.

Plass Jr., H.J., Barbir, F., Miller, H.P., and Veziroglu, T.N., "Economics of Hydrogen as a Fuel for Surface Transportation," *Int. J. Hydrogen Energy*, Vol. 15, p. 663, 1990.

Taylor, J.B., Alderson, J.E.A., Kalyanam, K.M., Lyle, A.B., Phillips, L.A., "Technical and Economic Assessment of Methods for the Storage of Large Quantities of Hydrogen," *Int. J. Hydrogen Energy*, Vol. 11, p. 5, 1986.

Nadal, M. and Barbir, F., "Development of a Hydride Fuel Cell/Battery Powered Electric Vehicle," in *Hydrogen Energy Progress X*, International Association for Hydrogen Energy, Coral Gables, FL, Vol. 3, p. 1427, 1994.

Zieger, J., "HYPASSE—Hydrogen Powered Automobiles using Seasonal and Weekly Surplus of Electricity," in *Hydrogen Energy Progress X*, International Association for Hydrogen Energy, Coral Gables, FL, p. 1367, 1994.

Reister, D. and Strobl, W., "Current Development and Outlook for the Hydrogen-Fuelled Car," in *Hydrogen Energy Progress IX*, International Association for Hydrogen Energy, Coral Gables, FL, p. 1202, 1992.

Yamane, K., Hiruma, M., et al., "Some Performance of Engine and Cooling Systems on LH_2 Refrigerator Van Musashi-9," in *Hydrogen Energy Progress X*, International Association for Hydrogen Energy, Coral Gables, FL, p. 1825, 1994.

The Society of Automotive Engineers
400 Commonwealth Dr.
Warrendale, PA 15096-0001

Peschka, W. and Escher, W.J.D., "Germany's Contribution to the Demonstrated Technical Feasibility of the Liquid Hydrogen-Fueled Passenger Automobile," SAE Technical Paper No. 931812, Society of Automotive Engineers, Warrendale, PA, 1993.

Kaufman, A., "Phosphoric Acid Fuel Cell Bus Development," in Proceedings of *The Annual Automotive Technology Development Contractor's Coordination Meeting*, Society of Automotive Engineers, Warrendale, PA, 1992.

Creveling, "Proton Exchange Membrane (PEM) Fuel Cell System R&D for Transportation Application," in the Proceedings of *The Annual Automotive Technology Development Contractor's Coordination Meeting*, Society of Automotive Engineers, Warrendale, PA, 1992.

Gottesfeld, S., Wilson, M.S., Zawodzinski, T., and Lemons, R.A., "Core Technology R&D for PEM Fuel Cell," in the Proceedings of *The Annual Automotive Technology Development Contractor's Coordination Meeting*, Society of Automotive Engineers, Warrendale, PA, 1992.

Morimoto, K. and Teramoto, T., "Combustion Characteristics in Hydrogen-Fueled Rotary Engine," SAE Technical Paper No. 920302, Society of Automotive Engineers, Warrendale, PA, 1992.

SAE of Japan
10-2 Goban-cho
Chiyoda-ku
Tokyo 102
Japan

Furuhama, S., "Trend of Social Requirements and Technological Development of Hydrogen-Fueled Automobiles," *JSAE Review*, Vol 13, p. 4, 1992.

American Chemical Society
1155 Sixteenth Street, N.W.
Washington, D.C. 20036

Minh, N.Q., "High-Temperature Fuel Cells, Part 1: How the Molten Carbonate Cell Works and the Materials that Make it Possible," *ChemTech*, Vol. 21, p. 32, 1991.

Minh, N.Q., "High-Temperature Fuel Cells, Part 2: Solid Oxide Cell," *ChemTech*, Vol. 21, p. 120, 1991.

Institute of Gas Technology
Headquarters
1700 S. Mt. Prospect Road
Des Plaines, IL 60018

Timpe, R.C., Sears, R.E., Malterer, T.J., "Pine and Willow as Carbon Sources in the Reaction between Carbon and Steam to Produce Hydrogen Gas," in *Energy from Biomass and Wastes XII*, Institute of Gas Technology, Chicago, IL, p. 763, 1989.

Elsevier Science S.A.
P.O. Box 564
1001 Lausanne
Switzerland

Prater, K.B., "Solid Polymer Fuel Cell Developments at Ballard," *Journal of Power Sources*, Vol. 37, p. 181, 1992.

Index

H

I

J